Climate Impacts on Agricultural and Natural Resource Sustainability in Africa

Bal Ram Singh • Andy Safalaoh
Nyambilila A. Amuri • Lars Olav Eik
Bishal K. Sitaula • Rattan Lal
Editors

Climate Impacts on Agricultural and Natural Resource Sustainability in Africa

Springer

Editors
Bal Ram Singh
Faculty of Environmental Sciences
and Natural Resource Management
Norwegian University of Life Sciences
Ås, Norway

Nyambilila A. Amuri
Department of Soil and Geological Sciences
Sokoine University of Agriculture
Morogoro, Tanzania

Bishal K. Sitaula
Department of International Environment
and Development Studies
Norwegian University of Life Sciences
Ås, Norway

Andy Safalaoh
Animal Science Department
Lilongwe University of Agriculture
and Natural Resources (LUANAR)
Lilongwe, Malawi

Lars Olav Eik
Department of International Environment
and Development Studies
Norwegian University of Life Sciences
Ås, Norway

Rattan Lal
School of Environment and Natural
Resources
Carbon Management and Sequestration
Center
The Ohio State University
Columbus, OH, USA

ISBN 978-3-030-37536-2 ISBN 978-3-030-37537-9 (eBook)
https://doi.org/10.1007/978-3-030-37537-9

This Springer imprint is published by the registered company Springer Nature Switzerland AG
The registered company address is: Gewerbestrasse 11, 6330 Cham, Switzerland

Foreword

Climate change and sustainable management of natural resources remain the major issues for all interventions to improve agricultural production, food access, and agriculture-based livelihoods in sub-Saharan Africa (SSA). Agriculture in SSA is predominated by small farms and subsistence farming by hand tools and limited use of other inputs including fertilizers, pesticides, and irrigation. There is also an increasing concern on declining land resources due to rapid soil degradation, harsh and uncertain climate, and the rapidly increasing population. Population of 800 million in 2010 is projected to increase to 1.1 billion in 2020 and to 1.8 billion in 2050 in Africa.

Despite these limitations, signs of agronomic yield increases and noticeable promise with impressive annual growth rates have been observed. However, sustaining the growth rate will become harder in the future due to increasing population, warmer climate, limited water resources, soil erosion and contamination, and more pervasive pests and pathogens. Furthermore, the IPCC Special Report on Global Warming indicates an increase of 1.5 °C change in temperature in SSA, creating a threat to ecosystems, biodiversity, and human health. These threats are more challenging and apparent in SSA than elsewhere. These challenges create a need of generating new knowledge on natural resource management and climate change to provide an enabling environment for smallholder farmers for engaging in sustainable agricultural practices.

Recognizing the value of agricultural production, the problems of natural resource degradation, and the challenge of climate change in SSA, a project entitled "Capacity Building for Managing Climate Change in Malawi" (CABMACC) was supported by the Royal Kingdom of Norway and implemented during the period 2013–2018. The Lilongwe University of Agriculture and Natural Resources (LUANAR) and the Norwegian University of Life Sciences (NMBU) jointly implemented the program. CABMACC was aimed to strengthen the teaching, training, research, technology development, and outreach for climate change adaptation and mitigation planning. A long-term and outstanding collaboration of LUANAR, NMBU, and Sokoine University of Agriculture (SUA) is further extended in this knowledge and experience-sharing platform to enhance dissemination of research findings from CABMACC project and beyond. The research under the abovementioned project focused primarily in Malawi, and therefore, contributors beyond the project were invited to cover wider geographical regions and their physical and social heterogeneities.

This book, *Climate Impacts on Agricultural and Natural Resource Sustainability in Africa*, deals with both the natural science and social science aspects, under dwindling natural resources, changing climate, and increasing climate uncertainties in SSA.

We convey our thanks to the successful authors, editors, and reviewers of the chapters in this book. We believe that the knowledge presented here is a crucial piece in the ingredients required for sustainable resource management under changing and uncertain climate in SSA.

Sjur Baardsen
Rector, Norwegian University of Life Sciences (NMBU)
Ås, Norway

George Kanyama-Phiri
Vice Chancellor, Lilongwe University of Agriculture
and Natural Resources (LUANAR)
Lilongwe, Malawi

Raphael Tihelwa Chibunda
Vice Chancellor, Sokoine University
of Agriculture (SUA)
Morogoro, Tanzania

Preface

Most countries in sub-Saharan Africa (SSA) are dependent primarily on agriculture for economic growth and livelihoods. Majority of the households, especially rural smallholder farmers, are perpetually food insecure due to unsustainable practices in agriculture, degrading agroecology, poor natural resource management, and political and institutional challenges. Agriculture in SSA countries is dominated by small farms, often less than 2 ha, and is primarily based on hand tools and manual operations with limited use of farm machinery and soil amendments, insufficient supplemental irrigation, and inadequate measures for soil and water conservation.

The harsh and changing climate has further aggravated the situation, adversely affected the natural resources, jeopardized agricultural production, and marginalized the livelihood opportunities. Adverse effects of climate change on agricultural production and the environment have made the SSA region as one of the hot spots leading to severe degradation of soil, drastic depletion of nutrients and soil organic matter stocks, water scarcity and contamination, and reduction of the above- and below-ground biodiversity.

The overall goal of the project "Capacity Building for Managing Climate Change in Malawi" (CABMACC) was to improve livelihoods and food security through innovative responses and enhance the capacity of adaptation to climate change. It was conducted at the Lilongwe University of Agriculture and Natural Resources (LUANAR) in Malawi in cooperation with the Norwegian University of Life Sciences (NMBU). The project was implemented in several districts of Malawi, which are considered the hot spots for climate change-related vulnerability.

To deliberate some of the challenging issues stated above, an international conference on Sustainable Agriculture and Natural Resource Management under Changing Climate in sub-Saharan Africa was organized at LUANAR, Lilongwe, Malawi, from 16 to 18 October 2018. The conference was an avenue to bring in researchers who conducted research in SSA and share findings that can be documented to provide scientific evidences to form policies to attain sustainable agriculture and natural resource management under changing climate. The major objectives of the conference were to bring new knowledge on sustainable use of natural resources to enhance agricultural productivity under changing climate and explore new avenues of policies, value added chain, and adoption of innovative technologies on smallholder's farms.

The 34-chapter book represents the oral presentations made during the conference. The book includes, in addition to introductory and concluding chapters, five thematic parts, namely, (i) Conservation Agriculture, Carbon Sequestration, and Soil and Water Management, (ii) Sustainable Crop/Livestock/Aquaculture/Fish Production, (iii) Policy and Institutions for Sustainable Agriculture and Natural Resource Management, (iv) Value Added Options for Smallholder Market Access and Integration, and (v) Upscaling Innovative Technologies on Smallholder Farms.

Nearly 150 participants attended the conference from Malawi, Rwanda, Ethiopia, Tanzania, Kenya, Norway, and the USA. The steering committee involved in the organization of the conference included representatives from LUANAR, Malawi; NMBU, Norway; Ohio State University (OSU), USA; and Sokoine University of Agriculture (SUA), Tanzania. The conference was a concluding activity of the project "CABMACC" in Malawi funded by the Royal Kingdom of Norway.

We, the editors, wish to thank all the authors for their outstanding contributions for the book. We also thank the staff at Springer for following the proposed publication schedule and bringing out the publication on time. Our special thanks are due to PCO staff at LUANAR for their help in the organization of the conference and managing the flow of manuscripts between the authors and the editors.

Ås, Norway — Bal Ram Singh
Lilongwe, Malawi — Andy Safalaoh
Morogoro, Tanzania — Nyambilila A. Amuri
Ås, Norway — Lars Olav Eik
Ås, Norway — Bishal K. Sitaula
Columbus, OH, USA — Rattan Lal

Contents

Editors Biographies

Bal Ram Singh, PhD is a professor emeritus at the Norwegian University of Life Sciences. He earned his PhD degree from G.B. Pant University of Agriculture and Technology, India. His program focuses on bioavailability and mobility of heavy metals in the soil and plant system, fertility management and agricultural sustainability in soils of the tropics, and carbon sequestration in soils. He has served as chairman of the program board "Soils and Plants" of the Research Council of Norway and as deputy head of the Department, in addition to many national and international committees. He chaired the Cost Action FA0905 (EU) "Mineral Improved Crop Production for Healthy Food and Feed," in which more than 200 scientists from 31 countries participated. He has supervised 76 graduate students and 16 visiting fellows/scientists from 20 countries and published 456 articles, of which 260 are in peer-reviewed journals and books. He is a fellow of ASA (2004) and SSSA (2005) and recipient of the International Soil Science Award (SSSA) in 2011. He is currently chair of Division 3 of the International Union of Soil Science, president of the Norwegian Society of Soil Science, and member of the Geomedicine Committee – Food, Environment, and Health of the Norwegian Academy of Science and Letters.

Andy Safalaoh, PhD is an associate professor of Animal Nutrition at the Lilongwe University of Agriculture and Natural Resources (LUANAR), Lilongwe, Malawi. He earned his PhD degree in Science and Technology Studies from the University of Nottingham, UK, and his Master of Science in Animal Science from Oklahoma State University, USA. His research interests focus on nutrient evaluation of unconventional feedstuffs such as sorghum and millet and recently on the use of insects as feeds. In addition, he has developed interest in the development and exploration of climate-resilient agricultural technologies and innovations as instruments for climate change adaptation and mitigation with a focus on the food-feed nexus. He has previously served as deputy head, Animal Science Department, and postgraduate and seminar coordinator and chairperson, Research and Publications Committee, LUANAR. He is currently the university program coordinator at LUANAR and has been leading the implementation of the 5-year (2013–2018) Norwegian Government-funded Capacity Building for Managing Climate Change in Malawi (CABMACC) Program and other projects. He is also a Leadership in Environment and Development (LEAD) Cohort 12 fellow (2004), Imperial College London. At regional level, he has facilitated several training sessions on agriculture, science, technology, and innovation (ASTI) in collaboration with Technical Centre for Agricultural and Rural Cooperation (CTA), Wageningen, Netherlands, and under the RAEIN-Africa Innovation Systems Approach Competency Building Training Program in eight countries. Before joining the university, he worked with Save the Children USA-Malawi Country Office in various portfolios as training and development coordinator, food production coordinator, and program manager.

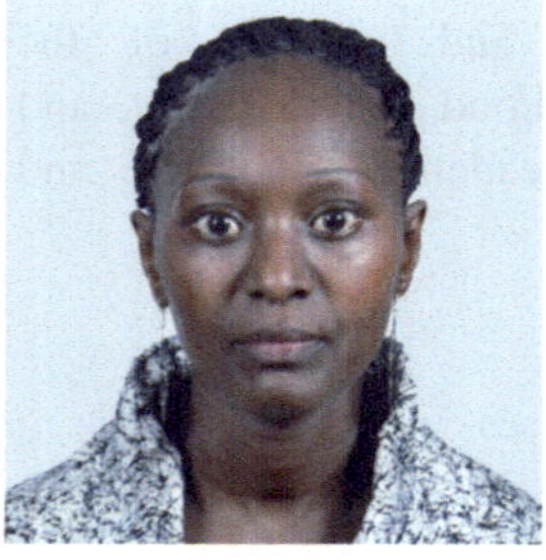

Nyambilila A. Amuri, PhD is a senior lecturer at Sokoine University of Agriculture (SUA), Morogoro, Tanzania. Currently, she serves as a coordinator for research and publication and head of Department of Soil and Geological Sciences at SUA. She earned her PhD in Soil Science from the University of Arkansas, Fayetteville, USA, and her Bachelor of Science in Horticulture and MSc in Soil Science and Land Management from SUA, Morogoro, Tanzania. Her research, university teaching, and outreach experience and interest are on C and N dynamics and residue management in agricultural soils, integrated soil fertility and agroecological management, appropriate fertilizer uses in agriculture, agronomic micronutrient fortification and mineral nutritive quality of food crops, site-specific fertilizer recommendations and innovative and cost-effective soil testing methods, and soil chemistry. Currently, she serves as a secretary general for the Soil Science Society of East Africa (SSSEA) and also chaired East Africa Fertilizers and Soil Conditioners of the Agriculture and Agrochemicals Technical Committee. She has supervised 17 graduate students and published 50 papers in peer-reviewed journals, conference proceedings, chapters in books, and extension manuals. She is a recipient of the Fulbright Scholar Award, Margaret MacNamara Memorial Fund Award, and NORAD Sponsorship Award.

Lars Olav Eik is a professor at the Norwegian University of Life Sciences (NMBU) specialized in animal nutrition and small ruminant production systems. After graduating from NMBU, he joined Sokoine University of Agriculture (SUA), Tanzania, working on dry season feeding of ruminants and introduction of dairy goat keeping in Tanzania. After this assignment, he returned to NMBU and completed his PhD based on work with dairy goats in Norway.

Since 2005, he has coordinated three major research programs in collaboration with SUA. He has also participated in research projects in Ethiopia, Malawi, and South Africa. Often working together with farmers and private sector, his main interest is developing multifunctional production systems and value chains for small ruminants, both in tropical and temperate regions. He has supervised a number of PhD students, particularly from East Africa. His teaching covers small ruminant nutrition and production systems and tropical animal husbandry and aquaculture.

Bishal K. Sitaula, PhD is a professor at the Norwegian University of Life Sciences, where he earned his PhD degree. His program focuses on various institutional collaboration programs in higher education and research in South Asia, Africa, and Western Balkan. His international collaboration experiences in diverse environmental and development issues, in inter- and multidisciplinary framework, mainly focus on ecological and socioeconomic issues influencing the environment and global changes. The specific topics covered are anthropogenic influences in soil water and air, land use and changes, agricultural intensification, GHG fluxes from land uses, carbon dynamics, land degradation, system analyses, environmental education, conflict, peace and development studies including wisdom and personal transformation relevant for ecosystem management, and global change and development. He has field research experiences from Europe, Asia, Africa, and North America through institutional collaboration, educational program, and networking projects with national and international organization. He has various program leadership experiences. He has contributed in developing various educational program and course curricula and created several educational documentary films. Despite more than 190 scientific publications, he has wider public engagements/social work with extensive delivery of public talks with media coverage.

Rattan Lal, PhD is a distinguished university professor of Soil Science and director of the Carbon Management and Sequestration Center, Ohio State University, and an adjunct professor of the University of Iceland. He was president of the World Association of Soil and Water Conservation (1987–1990), International Soil and Tillage Research Organization (1988–1991), Soil Science Society of America (2006–2008), and International Union of Soil Sciences (2017–2018). His professional research interests include soil carbon sequestration for food and climate security, conservation agriculture, principles and practices of soil erosion control, eco-intensification of agroecosystems, soil restoration, and sustainable management of soils. He authored 950 journal articles, authored/edited 100 books, mentored 350 researchers, has 144 h index and total citations of 95,000, and is editor of the *Advances in Soil Science* and *Encyclopedia of Soil Science*. He is laureate of the GCHERA World Agriculture Prize 2018, Glinka World Soil Prize 2018, and Japan Prize 2019.

Part I
Introduction

Agricultural and Natural Resource Sustainability Under Changing Climate in Africa

Bal Ram Singh, Andy Safalaoh, Nyambilila A. Amuri, Lars Olav Eik, Bishal K. Sitaula, and Rattan Lal

Abstract Sustaining soil and forest resources, ensuring food security, and reducing poverty under changing climate in Africa are major challenges. The ever-increasing population, which may reach 1.4 billion by 2030, demands that food production must increase by 20% as compared to the present production. The population growth rate of 2.7% in 2017 in SSA region is also the highest in the world. This has intensified the problem of food insecurity as nearly 34% of population in SSA appears to be food insecure. Soils of SSA are prone to a range of soil degradation processes, and this is further aggravated by the increase in frequency and intensity of extreme events. The biophysical process of soil degradation is strongly influenced by the socio-economic, political, and cultural factors including the land tenure and gender-related issues. Indeed, soil-climate-human factors are intricately interlinked, and humanity's impact on soil is increasing with the increase in population, affluence with which the population lives, and changes in technologies. Use of innovative technologies, financing value-added investments, and promotion of value addition to agricultural products are the ways of increasing and ensuring sustainable agricultural and livestock production. The adoption of new technologies, such as integrated

B. R. Singh (✉)
Faculty of Environmental Sciences and Natural Resource Management,
Norwegian University of Life Sciences, Ås, Norway
e-mail: balram.singh@nmbu.no

A. Safalaoh
Animal Science Department, Lilongwe University of Agriculture and Natural Resources (LUANAR), Lilongwe, Malawi

N. A. Amuri
Department of Soil and Geological Sciences, Sokoine University of Agriculture, Morogoro, Tanzania

L. O. Eik · B. K. Sitaula
Department of International Environment and Development Studies, Norwegian University of Life Sciences, Ås, Norway

R. Lal
School of Environment and Natural Resources, Carbon Management and Sequestration Center, The Ohio State University, Columbus, OH, USA

B. R. Singh et al. (eds.), *Climate Impacts on Agricultural and Natural Resource Sustainability in Africa*, https://doi.org/10.1007/978-3-030-37537-9_1

dairy and cropping systems, leading to increased farm productivity and income, is spreading in central Malawi and the southern highlands of Tanzania. For example, integrated dairy and cropping systems are spreading and farmers are keeping few but more high-yielding cows in both regions. These cows produce milk and meat for household consumption and sale, but the dung provided is used for biogas production for cooking and light, as well as fertilizer for the crops. Currently, dairy goats provide milk and meat for approximately 100,000 Tanzanian smallholder farmers. The idea of personal transformation for changing our attitude for greater good and for sustainable agriculture is also a part of this process.

1 Introduction

Providing food security, reducing soil degradation, maintaining forests and improving soil and ecosystem pools, adapting climate smart agriculture, and reducing poverty are the major challenges in Africa. These challenges are further aggravated by the ever-increasing population, which is expected to rise from 0.87 billion in 2010 to 1.4 billion in 2030 and 2.1 billion in 2050 in sub-Saharan Africa (SSA) (United Nations 2015). This increased population also leads to increased food demands, and FAO (2015) has forecasted that the food demands will increase by 20% by 2030 in SSA. Nearly 95% of agriculture in Africa is rainfed, and it is highly susceptible to drought and the projected climate change; and the increasing temperature has further aggravated the situation, adversely affecting the natural resources, agricultural production, and the livelihood opportunities for small farmers. Adverse effects of climate change on agricultural production and the environment have made the SSA region one of the global hot spots, leading to severe soil degradation, depletion of nutrients and soil organic matter (SOM) stocks, scarcity and contamination of water resources, and reduction of the above- and below-ground biodiversity.

Soil resources in sub-Saharan Africa (SSA) are diverse but suitable for agricultural land use and can support the growing of a wide range of food crops, horticultural crops, plantations, and livestock-based systems. However, these resources have experienced the severity of degradation by diverse processes (i.e., erosion, salinization, soil organic carbon or SOC, and nutrient depletion); and this has resulted into stagnated crop productivity, insecurity in food supply (one in four Africans), and mal- and undernourishment. The problem of food and nutritional insecurity is further aggravated by drought and heat waves caused by changing climate.

One of the ways of increasing and ensuring sustainable agricultural production is financing value-added investments and promotion of value addition to agricultural products (World Bank 2018). For example, in Malawi, the agricultural sector is limited by market access constraints and limited integration/coordination of most agricultural value chains. Therefore, the Malawi government has identified investments in agribusiness, value addition, and investments into the domestic markets as priority intervention areas, with emphasis on crops (Malawi Government 2017). Recent developments are, however, encouraging as livestock is also gaining prominence. The adoption of new technologies, such as integrated dairy and cropping systems, leading to increased farm productivity and income, is spreading in central Malawi and the southern highlands of Tanzania. For example, farmers keep few but more high-yielding cows for producing milk and meat for household consumption and sale. These cows also provide dung which provides biogas for cooking and light, as well as fertilizer for the crops. Combined with appropriate use of compound fertilizer and lime, crop yields have doubled or even tripled.

The objective of this chapter is to introduce briefly the issues pertaining to natural resource management, crop and livestock production, value addition, and innovative technologies for small farmers as they are affected by the changing climate in Africa.

2 Agricultural Productivity and Population Growth

Agriculture is a strategic sector that provides great potential for economic transformation to achieve inclusive development through poverty reduction and ensuring food security in Africa. This is affirmed by the African Union (AU) Comprehensive Africa Agriculture Development Programme (CAARDP), which highlights the importance of agriculture as the employer of more than half of the population of Africa and hence re-emphasizes agriculture in development programs and investment choices (NEPAD 2013). The role of agriculture as an employer of the majority of population, including youths, in Africa cannot be overemphasized. A study by Kafle et al. (2019) demonstrated that farming is still the major employment opportunity for a large section of the population in Tanzania and Malawi, with 59% and 56% of youth, respectively, consistently engaging in farming. Agriculture being the main employer in SSA is rated highly as a potential contributor to poverty reduction, especially in small-scale farming, if well developed (Dorosh and Thurlow 2018). The population growth of SSA has been steadily increasing, reaching 1.061 billion people in 2017 (Fig. 1; World Bank Data 2019); this makes SSA the region with the highest average world population growth rate of 2.7% in 2017 (NEPAD 2013; 1.06). The FAO forecasted an increase in food demand by 20% in 2030 in SSA due to increase in population (FAO 2015; FinMark Trust 2016). The increase in production must go hand in hand with increase in agricultural productivity, to ensure sufficient food and fiber and improve livelihood of the majority who depend on agriculture.

The population growth in SSA can both have positive and negative impacts to agriculture. SSA is a region with a strategic potential for rapid economic growth due to its increasing population; it is endowed with natural resources, including water and land (Deininger et al. 2014), and it has a high market potential. The increased population corresponds to increased food demand (Jayne et al. 2017; NEPAD 2013), making this an opportunity for agro-food system development to drive other non-farming sectors, since agriculture would be more productive and profitable. Considering SSA land coverage and diversity in agro-ecological zone (eco-region and soil types) (Lal 2015), SSA is strategically both the supply and the market for

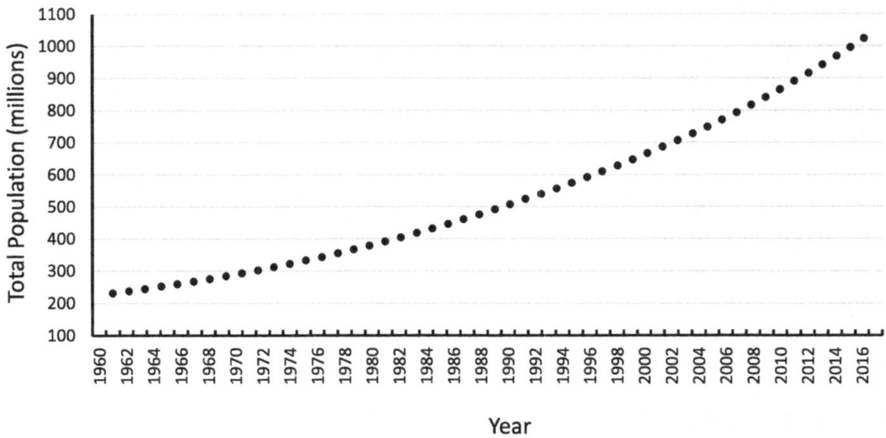

Fig. 1 Increase of total population in sub-Saharan Africa (excluding high-income countries) over 56-year period. (Data source: https://data.worldbank.org/indicator/SP.POP.TOTL?display=graph &locations=ZF)

agricultural produce, including food, fiber, and fuel. There is a great potential for SSA to trade within the region, considering the comparative advantages in natural resources and climate. Areas with high potential for crop production (humid and sub-humid areas) can trade with areas of high potential for extensive livestock keeping (semi-arid areas) of SSA. Developing agriculture begins with increasing its productivity and profitability, followed by distribution and agro-processing for value addition and preservation. This has been evidenced in a study by Dorosh and Thurlow (2018), who showed that non-agriculture sector, especially trade, transportation, and manufacturing (especially agro-processing), can effectively contribute to poverty reduction. However, Jayne et al. (2017) urge that agriculture is the major driver of development for the non-farming economy because if made profitable, it will provide higher income to a large population, resulting into high demand of goods and services. Therefore, agriculture and population growth are potentially major drivers of other sectors through provision of market, ensuring the availability of agro-produce as raw materials for agro-processing, and access and affordability of food to the growing population. Thus, the agricultural and non-agricultural sectors in SSA depend on each other for development.

The major challenge that remains in SSA is to increase agricultural productivity to meet the requirement of the growing population, in the face of climate change. Food insecurity is still prevalent in SSA with about 33.8% of population appearing to have severe prevalence of food insecurity in 2017, which represents an increase of 18% from 2016 (FAO and NECA 2018). Food supply from domestic agricultural production is not sufficient to meet the domestic demand in most of the SSA countries due to low yields and declining productivity, caused by low input use (Jayne et al. 2010). Thus, intensification of agriculture to increase yield per unit area is inevitable. Many SSA countries have attempted to increase productivity through ensuring supply/accessibility of agricultural inputs (improved seeds and fertilizers) using different approaches. Common approaches have included provision of subsidies to resource poor small-scale farmers in Tanzania, Malawi, Zambia, and Ghana (FinMark Trust 2016), with varied success, including improved productivity and ensuring the countries are food secure (FinMark Trust 2016). Further, performance of these programs is documented, discussed, and presented in this book as a way to provide evidence to improve similar future endeavors. Alternative sustainable intensification performances are also presented.

Another major threat caused by population growth is the increase in pressure on land, which may constrain farming system. One way of ensuring increased productivity is agricultural intensification, which can be done by increasing yield per unit area through increased use of agricultural inputs (improved seeds, fertilizers, pesticides), labor, and technology (irrigation, mechanization, frequency of cultivation) (Carswell 1997), as opposed to increasing production by expanding the area under cultivation or rangeland. A study in Ethiopia by Josephson et al. (2014) showed an increase in agriculture intensification and a decrease in farm size with increase in population. Similarly, a community approach to the study of increased population and climate change impact in SSA revealed that these two stresses cause shift in livestock-keeping practice from purely migratory pastoralism to agro-pastoralist coupled with adoption of conservation agriculture (CA) and have a positive impact by improving food security in SSA dryland areas (Burian et al. 2019). However, improved food security due to change of agro-pastoral system has not been universally achieved (Tache and Oba 2010; Josephson et al. 2014), especially in areas with less than 500 mm annual rainfall and without CA. Further analysis by Burian et al. (2019) cautioned that with continued increase in population, it would reach a stage when the household land area will be too small to sustain household food security, leading to food shortage, to the extent that increased productivity might not solve the food shortage problem. Another study in Kenya reported that 500 people/km^2 is the threshold for intensification in farming; beyond this population density, even agricultural intensification will also decrease (Muyanga and Jayne 2014), and the agricultural productivity may not sustain the population demand. These findings suggest that the current increase in population may cause serious problems in food supply and increase in hunger and poverty. The warning is that increase in productivity per unit area alone may not solve the problem. In such a scenario, sustainable intensification that ensures increase in productivity without compromising environmental and social benefits is needed (Dicks et al. 2019). Therefore, a more integrative approach that includes other sectors such as health and environment is necessary.

Although agricultural production increased during the 30 -year period from 1983 to 2013, the anticipated productivity is yet to be realized (NEPAD 2013). The increased production has been due to expansion of agricultural land (Amuri 2015; Charles et al. 2010), leaving less or no improvement in intensification of land and labor (NEPAD 2013). The trend of expansion of agricultural land can be exemplified by the increase in land area under cereal production (major staple food for most of SSA population) at the rate of 1 million ha (Mha) per year (Fig. 2). The trend of increase in cropland is similar to that of increase in population in SSA, indicating

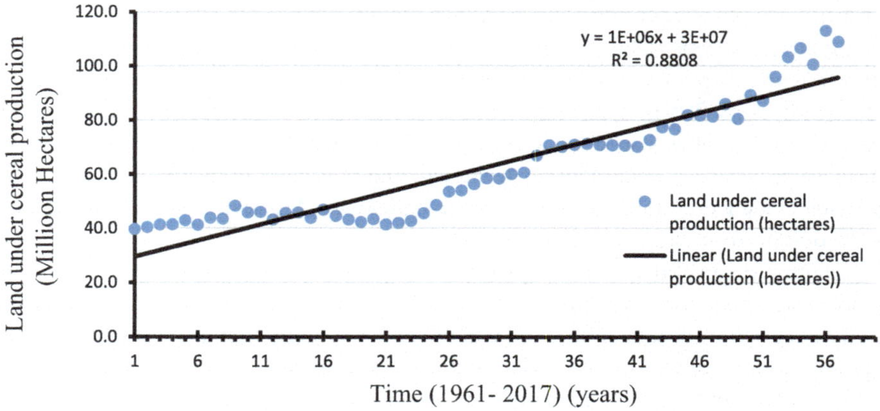

Fig. 2 Increase of land area under cereal production in sub-Saharan Africa (excluding high-income countries) over 56-year period. (Data source: https://data.worldbank.org/indicator/AG.LND.CREL.HA?view=map)

that the major coping strategy to satisfy food demand has been through expansion of agricultural land. The productivity per unit area is low, ranging from 1.1 to 1.7 Mg/ha for cereals for 2009–2017 (FAOStat 2019a, b) and 0.2–0.7 Mg/ha for cotton under rainfed conditions (Mathews and Tunstall 2006; VIB Fact Series 2017). The dominant agricultural production system in most of SSA is still subsistence, with small farm size and limited use of fertilizers and/or other agrochemicals. These farming systems are usually uncoordinated, with weak farmer's institutions and cooperatives, which render farmers less competitive in the market and usually fetch lower prices than investment in agriculture. As a result, the farming system in SSA has less promise to sustain employment, especially for youth. Therefore, a first step toward agricultural development in SSA should be increased productivity and empowerment of farmers to implement modern agricultural technology effectively to improve productivity.

The farming systems and population pressure in SSA tend to increase the extent of soil degradation, causing depletion of SOM (Lal 2016a), nutrient mining (Amuri 2015; Lal 2015), overgrazing, and accelerated erosion. Soil degradation is further exacerbated by poor land use planning, with little or no scientific evidence to back up land use and fertilizer use. It further reduces the ability to utilize adequately the land resources to support demands from the growing population. The land degradation and expansion of agricultural land has also been linked to land conflicts among multiple users. Some studies have suggested that land tenure may manage conflicts, but Kalabamu (2019) reported persistence of land conflict despite some success in land tenure system in Botswana. Thus, land tenure system may not be effective in solving land conflicts without soil conservation and restoration of degraded land.

The steady increase of population in SSA creates market for produce and commodities and increases demand for food and fiber. To sustain the steadily growing population, agricultural productivity must be increased, and the latter must take into account sustainable use of resources to ensure neither soil degradation nor expansion of agricultural land encroaches on areas such as water sources, wetlands, natural forests, and wildlife. Thus, ecological advocacies must be increased to ensure sustainable intensification. Further, a more comprehensive analysis is needed to come up with an innovative and sustainable intensification package that is sound in terms of policy, technical, and environmental considerations. The papers in this book provide some potential technologies and practices toward sustainable crop and livestock production.

Climate change has become an additional challenge to rainfed agriculture among smallholder farmers. Weather variability has been experienced, with some seasons having adequate rainfall, others having it in insufficient amounts, while in other seasons the onset of rain has been changing. Under such climate variability, efforts to increase productivity by intensifying sustainable use of input have not yielded results. In some cases, use of fertilizers and pesticides has rendered no yield due to moisture deficit. Therefore, sustaining increase in population calls for innovative intensifications that are resilient to climate change impact and weather variability. Burian et al. (2019) suggested more changes in the farming system to bring on board irrigation agriculture, mechanization, and increase in off-farm jobs. These innovative technologies must be adaptive to the farming system in SSA, if the positive effect of increased population as market for agricultural produce is to be realized.

3 Soil and Water Resources

Soil resources in general and those of SSA in particular are diverse and suitable for agricultural land use and growing a wide range of food crops, horticultural crops, and plantations and for livestock-based systems. Further, soil resources of Africa have been mapped (EC 2017). The extent and severity of degradation by diverse processes (i.e., erosion, salinization, soil organic carbon, and nutrient depletion) is assessed, and their causal factors are identified (ITPS 2018). Yet, the agronomic productivity of food staples has stagnated since the 1960s, soil degradation is rampant, food insecurity affects one in four Africans, and malnourishment and undernourishment are serious issues. Further, the problems of food and nutritional insecurity may be exacerbated by the uncertain and changing climate characterized by extreme events including the pedological drought and heat waves. Further, the risks of the already serious problem of soil erosion by water and wind may be drastically aggravated because of the change in climate and the agronomic yields of food staples adversely affected because of the increase in frequency and intensity of drought, the prevalence of heat wave, and the increase in frequency of other extreme events. Soil degradation is caused and aggravated by land misuse and soil mismanagement and perpetuated by the prevalence of extractive farming practices.

Soil and water resources in SSA, as anywhere else, are intricately interlinked through climate. Soil is the largest reservoir of fresh water. Therefore, the soil hydrological cycle and its partitioning into different components are determined by the climatic parameters at micro-, meso-, and macroscales. Retention, denaturing, filtration, and purification are done during percolation of water through the soil solum. Land use and management is a strong control of the hydrological cycle and its partitioning into different components.

The low agronomic yield and perpetual hunger have persisted in SSA despite the widespread adoption of improved varieties developed by the national and international programs including those released by the CGIAR centers (e.g., CIMMYT, ICRISAT, ICARDA, IITA, WARDA, IRRI, ILCA, ICRAF). Indeed, the Green Revolution (GR) of the 1960s by-passed SSA despite the availability of improved varieties because other components (i.e., fertilizers, irrigation, appropriate farm machinery, and energy-based inputs) were either not available to the resource-poor farmers and small landholders (<5 Ha) or farmers were not sure of their usefulness because of the fragile soils and harsh environments.

Soils of SSA are prone to a range of soil degradation processes – physical, chemical, biological, and ecological. The extent and severity of degradation (especially the physical degradation by accelerated erosion) will exacerbate due to the ever-growing pressure of the rapidly increasing population in conjunction with the increase in frequency and intensity of extreme events (e.g., drought stress, heat wave, high incidence of pests and pathogens including infestation by weeds). Further, the extent and severity of soil degradation is related to the interaction between the factors and causes of soil degradation. Among the causes, the human dimensions are the critical determinants of soil degradation. Indeed, the biophysical process of soil degradation is strongly influenced by the socio-economic, political, and cultural factors including the land tenure and the gender-related issues (Lal 2007a). A high percentage of farms in SSA are operated by women, and often the technological advances are neither available because of being prohibitively expensive nor are these often appropriate for their specific needs.

Rather than the seed-centric approach of the conventional GR package of the 1960s, the constraints to narrowing the yield gaps must be addressed through a soil-centric approach (Lal 2019). The latter must be based on restoration and sustainable management of soil health (physical, chemical, biological, and ecological). In this context, key determinants of soil health must be identified and judiciously managed (Lal 2016c). As is evident, the yield potential of elite varieties can only be realized, if grown under optimal conditions of soil environments especially those of soil physical conditions (e.g., structure, tilth, plant available water capacity, and water infiltration rate and soil temperature regime).

Most soils of agroecosystems of SSA are severely depleted of their soil organic carbon (SOC) concentration and stock. The SOC concentration in the root zone of upland soils may be as low as 1 g/kg or lower in comparison with the threshold level of >11 g/kg (Aune and Lal 1997) and preferably 20 g/kg (Loveland and Webb 2003). Severe depletion of the SOC concentration may be attributed to a widespread adoption of extractive farming practices such as residue removal for competing uses (e.g., feed, fuel, fencing and house construction, and uncontrolled grazing), in-field burning, and none or low rate of inputs of organic and inorganic amendments.

Susceptibility to a range of soil degradation processes is another cause of the depletion of SOC concentration and stock in soils of SSA. The SOC-related degradation is set in motion by decline in soil structure, reduction in proportion and strength of soil aggregates, crusting, compaction, decline in water infiltration, high runoff, erosion and non-point source pollution, and the attendant drought-flood syndrome. Indeed, drought and flood are the two sides of the same coin. Thus, restoration and judicious management of SOC concentration and stock is the most critical strategy to reverse the degradation trends.

4 Impact of Climate Change

Climate is one of the five active factors of soil formation (Jenny 1941), and "human" is now the sixth factor, hence the name "Anthrosols" (Bidwell and Hole 1965; Dudal 2004). Leopold (1949) stated that "Human beings are all interlocked with plants, animals, soils and waters in one humming community of cooperation and competition: one biota. They are related and bound into a seamless fabric." Indeed, soil-climate-human factors are intricately interlinked, and humanity's impact on soil is increasing with increase in population, affluence with which the population lives, and changes in technologies. Thus far, from the onset of agriculture about 10–12 millennia ago till now, the impact of humanity on climate and soil has been negative because both of these resources have been taken for granted. Indeed, climate and soil, considered and used as global commons, are prone to the tragedy of commons and are a phenomenon of the Anthropocene (Lal 2007b).

Global abundance of greenhouse gases (GHGs) in 2017 was 405.5 ppm for carbon dioxide (CO_2), 1859 ppb for methane (CH_4), and 329.9 ppb for nitrous oxide (N_2O) (Lal 2018a; WMO 2019).The atmospheric concentration of CO_2 attained the value of 415.6 ppm during May 2019 (NOAA-ESRL 2019). Carbon dioxide levels at present are more than those at any time during the past 800,000 years (Lindsey 2018). Three principal sources of emission of GHGs are fossil fuel combustion, tropical deforestation, and accelerated soil erosion. For the decade of 2008–2017, annual average emissions were 9.4 ± 0.5 Pg C/year from fossil fuel combustion and 1.5 ± 0.7 Pg C/year from tropical deforestation. Of this, 4.7 ± 0.02 Pg C/year was absorbed by the atmosphere, 2.4 ± 0.5 Pg C/year by the ocean, and 3.2 ± 0.8 Pg C/year by the land-based sinks with a budget imbalance of 0.5 Pg C/year (Le Quéré et al. 2018). For the year 2017, emission was 9.9 ± 0.5 Pg C/ year from fossil fuel combustion and 1.4 ± 0.7 Pg C/ year from tropical deforestation. In comparison, uptake by different sinks was 4.6 ± 0.2 Pg C/year by the atmosphere, 2.5 ± 0.5 Pg C /year by the ocean, and 3.8 ± 0.8 Pg C / year by the land-based sinks, with a *budget imbalance* of 0.3 Pg C (Le Quéré et al. 2018). Thus, the anthropogenic activities have approximately caused global warming of 1.0 °C above the pre-industrial level (with a range of 0.8–1.2 °C). Further, with the business-as-usual attitude, global warming is likely to reach 1.5 °C between 2030 and 2052 (IPCC 2018).

The importance of land-based solutions, such as sequestration of atmospheric CO_2 in the terrestrial biosphere (Lal et al. 2018a), cannot be overemphasized. The soil C sink capacity of 135 Pg (Lal 2018a), equivalent to the historic C loss from soils of the managed ecosystems, is equivalent to CO_2 drawdown of 63 ppm. Re-carbonization of the soil and the terrestrial biosphere has numerous co-benefits, especially those related to advancing specific Sustainable Development Goals (SDGs) of the United Nations (Lal 2018b; Lal et al. 2018b).

5 Value Addition for Smallholders

Malawi is a landlocked agro-based economy in sub-Saharan Africa with a per capita GNP of just US \$320 in 2017 (World Bank 2017) and is vulnerable to several external and internal shocks. About 84% of the population live in rural areas, and around 70% of the population live below the international poverty line of US \$1.90 per day (World Bank 2017).

Over the past decade, Malawi's development progress has been negatively affected by climate-induced shocks leaving the country in a cycle of vulnerability. Malawi has experienced extreme droughts and floods, which have negatively impacted agricultural production. This makes adaptation a priority area, as climate change impacts disproportionately affect those least able to bear them. Climate-related shocks, market failure, domestic political, and governance shocks have collectively contributed to economic stagnation and a slow pace of poverty reduction (IMF 2018). The consequences are being felt through declining crop yields, low animal productivity, increasing costs of production, and difficulties for the economy to keep pace with population growth.

A significant proportion of the Malawi population faces serious nutritional deficiencies. Only 8% of children aged between 6 and 23 months meet their minimum meal frequency, and 37% under the age of 5 are stunted due to malnutrition. Malnutrition affects children's health condition and learning possibilities and is a severe challenge for Malawi's overall development. This calls for an integrated food systems approach that would address both food and nutrition security, including dietary diversification and nutrition education.

Smallholder farmers face significant production and market constraints, poor access to information, and inadequate public services, which adversely affect their welfare and employment opportunities. Current land use practices are eroding the capacity of the natural resource base to sustain productivity for current and future generations, with extractive land use practices causing alarming rates of land degradation: loss of top soil averages 29 tons per ha per annum, and tree cover is declining at over 1% per annum.

According to Malawi's Issues Report submitted to the UNFCCC, agriculture is the key sector to achieve food security, economic growth, and wealth creation. More than 80% of the country's population is directly or indirectly employed in the sector. This also accounts for nearly 90% of foreign exchange earnings and 39% of gross domestic product (GDP). Despite its importance, the agriculture sector faces a number of challenges including (i) overdependence on rainfed farming, (ii) limited absorption of improved technologies, (iii) weak private sector engagement, (iv) limited farmer organization, and (v) lack of investment capacity in mechanization. About 99% of Malawi agricultural land is under rainfed cultivation, and smallholder farmers cultivate 69% of that land.

One way of increasing and ensuring sustainable agricultural production is financing value-added investments and promotion of value addition to agricultural products (World Bank 2018). For example, according to the Malawi National Agricultural Investment Plan (NAIP), commercialization of the agricultural sector is limited by market access constraints and limited integration/coordination of most agricultural value chains. As such, the Malawi government has identified investments in agribusiness, value addition, and investments into the domestic markets as priority intervention areas, with emphasis on crops (Malawi Government 2017). Recent developments are however encouraging as livestock is also gaining prominence.

In both Malawi and neighboring Tanzania, pioneer farmers have adopted new technologies leading to increased farm productivity and income. For example, in central Malawi and the southern highlands of Tanzania, integrated dairy and cropping systems are spreading. Farmers are keeping more high-yielding cows for producing milk and meat for household consumption and sale. Combined with appropriate use of compound fertilizer and lime, crop yields have doubled or even tripled.

Most farmers cannot afford buying a dairy cow. Therefore, in Tanzania, dairy goats are becoming increasingly popular. Since the introduction four decades ago, Norwegian dairy goats and crosses have reached a number of about half a million, which is more than 10 times the numbers of goats in Norway.

For farmers, attaining stable and fair commodity prices is a key success factor. *Njombe Milk* Factory Company Limited in Tanzania and Lilongwe Dairy in Malawi are examples of successful enterprises contributing to increased farmer's income where milk is bought at an agreed price and farmers are assured of ready market. Smallholder farmers may also increase their income through sale of meat as exemplified by the Nyama World with its newly established slaughter facilities in Mzuzu, Malawi. The company founded in 2010 is a growing agribusiness with a known national brand of high-quality meat. It has since developed into a fully integrated value chain with a feedlot, a fattening farm, a halal-certified processing plant, and a wholesale distribution system combined with six retail outlets. The company supports farmers with improved Bonsmara bulls to crossbreed with the Malawi Zebu and then buys 6- to 8-month-old bulls from farmers for fattening, thereby providing farmers with a market outlet. The strategy is to expand the domestic business and shortly to build exports into neighboring countries and the Middle East markets. Norfund has made a 23.3 million NOK loan commitment (2.75 million USD) to Afrisphere Worldwide Limited, who trades as Nyama World with a fully integrated beef production company in Malawi. All livestock are sourced, processed, and sold predominantly in Malawi. Once the business is fully developed, the aim is to engage as much as up to 30,000 smallholder farmers in an integrated value chain.

This example demonstrates that if the successful value chains are established for different commodities, farmers are capable of responding to market demand.

6 Innovative Technologies on Smallholder Farms

In countries like Malawi and Tanzania, outside the national parks and forest reserves, small-scale farmers are keepers of the land, engaged in small-scale subsistence agriculture, producing food mainly for household consumption and an increasing amount of cash crops for domestic and export markets. In most cases, low agricultural productivity remains a challenge, and it is very important that Tanzanian farmers improve productivity in order to provide adequate amounts of food for a growing population. At the same time, conservation and sustainable management of natural resources are vital in order to mitigate climate change, maintain biodiversity, and ensure sustained supply of water. There is a vast and virtually untapped potential for improved agricultural productivity and sustainable environmental management, which can be attained through improved agricultural and livestock practices, as well as commercial smallholders' tree planting (including agroforestry) and natural resource conservation activities. The fundamental rationale envisaged is that if farming communities increase productivity, household income will subsequently increase, resulting in reduced poverty and hunger. Hence, the continuation of up-scaling and developing more efficient production and natural resource management systems can enable the small-scale farmers to escape the poverty trap.

Since soils in Malawi and Tanzania are mostly depleted and crop yields are low, continued research on use of compost, manure, lime, and appropriate compound fertilizer with micronutrients needs to be emphasized. Incorrect application of fertilizers and pesticides is a major concern in the region. Therefore, we recommend the "One Health, One World Approach" with a priority on food safety throughout the food value chain. On the best farms, appropriate fertilizer and herbicide use (*best practices*) increased maize yield from 1.3 tons/ha (national average) to 5.4 tons.

Introducing new technologies takes time. In southern highlands, Njombe, Tanzania, the work started 20 years ago with a focus on dry-season feeding of cows for improved annual distribution of milk production. Today farmers from many areas in Tanzania as well as other countries are coming to our demonstration farms in order to learn about integrated dairy farming practices. The use of "farmer-to-farmer learning," by means of lead farmers educating others based on own experiences, has become an efficient tool for up-scaling new farming practices.

The Njombe Milk Factory, which is partly owned by the dairy farmers, currently distributes milk, yoghurt, and an assortment of cheese types locally as well as to food stores as far as Dar es Salaam and Zanzibar. In Malawi, pilot biogas plants were also recently introduced in Bolero under the Capacity Building for Managing Climate Change (CABMACC). However, it is manure and not milk, which is the most important product from this integrated livestock system. Manure used as biogas initially provides energy and subsequently bio-slurry, which is used as a soil improver. In the proposed new phase, we will continue system research both on the technical and social aspects of this technology. This work is important, as studies on the value of bio-slurry, both as a soil improver and for plant protection, are presently lacking. Furthermore, in "controversial" new biogas plants, a possibility exists for including human waste. Research is required on this technology to establish whether such an option will be culturally acceptable. The sale of hay by farmers in the Njombe region as a cash crop is also an innovation.

Before starting, extension staff from Njombe participating in the EPINAV-CA project visited and learned from practicing farmers in Zambia. Upon returning to Tanzania, unlike in Zambia, farmers decided to keep dairy cows as a core part of the system. To improve fertility of the poor soils in Njombe, farmers add bio-slurry, lime, and specific soil type-adapted mineral fertilizer (Yara). This integrated system has provided households with clean energy and income from milk. Furthermore, the unique combination of inputs has increased crop yields by two- to threefold. As vegetable production is no longer seasonal in Njombe, positive outcomes such as improved health status and increased household incomes are evident. Our studies have demonstrated that Njombe farmers now produce more food per area of land. Because of this, farmers are increasingly establishing own woodlots, which in turn are diminishing the pressure on existing forest reserves.

Professional dairy manufacturing is a vital link in securing a functioning market for milk products. Granarolo and Cefa Dairy Companies (both based in Italy) jointly with NGOs established the Njombe Milk Factory in 2004. Today the collaborating dairy farmers are shareholders in the company, which distributes milk, yoghurt, and cheeses both locally and nationally. Through our collaboration with the Njombe Milk Factory, we have focused on farmer training as a means for improved management, particularly during the dry season. Improving feeding and milk quality is an important element in securing economic sustainability, by allowing the factory to operate at full capacity for longer periods of the year. According to research conducted by Roma Tre University, the income levels of the 800 livestock farmers delivering milk to the dairy have increased by on average 140% over a 3-year period. The use of "farmer-to-farmer" training has proven highly successful in introducing new technologies to other farmers in Njombe as well as in other parts of Tanzania. Yara has already made use of Njombe project information in international marketing campaigns. In Malawi, a REDCAP project, funded under the CABMACC program in Linthipe (Dedza), introduced and promoted new technologies among dairy farmers under the Dzaonewekha Milk Bulking Group (soon to be a registered cooperative). This dairy cooperative included the development and use of balanced total mixed rations to meet nutrient requirements and reduce carbon emissions, growing of improved pasture, use of artificial insemination, and community animal health workers/volunteers. These technologies have helped farmers to increase milk yield from 4 to 12 liters per day. Despite this improvement, more needs to be done to attain the full potential of milk yields. The farmers were also assisted with a homogenizer so that they can process, package, and sell their milk themselves for increased profitability instead of going through an intermediary.

Currently, dairy goats provide milk and meat for approximately 100,000 Tanzanian smallholder farmers. The goats are mainly descendants of the 80 Norwegian dairy goats imported to Mgeta district, Morogoro, Tanzania, in the mid-1980s. Thirty-five years on, the number of milking goats in Tanzania is more than 500,000, distributed all over the country.

How have the vegetable farmers of Mgeta in collaboration with the SUA-NMBU alliance been able to pioneer this success? The answer is rather simple. We introduced a superior and at the same time easily adaptable production system. Tanzanian farmers keep around 16 million local East African goats for meat only, and they provide an annual meat yield of 8–12 kg giving an income of 30,000 TZS per female goat (doe) with offspring. A Norwegian doe, on the other hand, will provide the same amount of meat plus 240 kg milk, raising the traditional annual income by almost tenfold, to 270,000 TZS per doe. This total may even rise to 340,000 TZS per doe for breeders selling 3–4-month-old live goat kids, which come as additional income to milk. Based on these scenarios, it is no wonder that so many farmers testify about dairy-goat keeping transforming their lives.

In terms of biological efficiency, milking goats are far superior to meat producers. We have calculated using farm data from Mgeta that one dairy goat may provide the same amount of edible animal protein as 33 local East African goats combined.

Aquaculture is an important component of the EPINAV program. Training of farmers has proven to be successful, and fish farming is increasing in popularity across the country, where fish farmers mostly produce tilapia. The productivity of the native Tanzanian tilapia that is based on wild stocks is low compared to the genetically improved fast-growing strains. SUA can play a critical role using its research expertise to play a leading role in selective breeding to develop a better-growing tilapia strain.

Integrated aquaculture has great potential in improving food security and ensuring diversification of diets, as the systems generate both animal proteins and vegetables. However, there is a need to reach out to medium- and large-scale farmers in order to achieve meaningful gains in aquaculture production. Private companies can play a critical role in disseminating developed technologies by SUA. For instance, they can serve as multiplication centers for quality fingerlings from broodstock developed at SUA.

7 Transformative Thinking in Agriculture

We may have learned how humans contribute to climate change at proximal level, but we rarely ask why we leave a larger-than-necessary carbon footprint. The reality is that, due to the urgency of emerging climate crises, incremental change in technology may not address the problems at their source. To address these problems at their source, we introduce the idea of personal transformation for changing our attitude for greater good and for sustainable agriculture. We presented few educational perspectives, where a transformation refers to the change or shift of meaning on dominant ideas, beliefs, and assumptions. We argue that the transformation takes place within the self after the individual views and the ideas from different perspectives. In other words, we cannot transform ourselves viewing the beliefs and assumptions in a taken-for-granted manner.

8 Conclusions

The biophysical process of soil degradation is strongly influenced by the socio-economic, political, and cultural factors, including land tenure and the gender-related issues. Indeed, soil-climate-human factors are intricately interlinked, and humanity's impact on soil is increasing with increase in population, affluence with which the population lives, and changes in technologies. Therefore, re-carbonization of the soil and the terrestrial biosphere has numerous co-benefits, especially those related to advancing specific Sustainable Development Goals (SDGs) of the United Nations.

Degradation of soil and forest resources and reduction in poverty of ever-increasing population of SSA call for innovative intensifications that are resilient to climate change impact and weather variability. Thus, there is a need for adoption and intensification of farming systems such as irrigation agriculture, mechanization, and increase in off-farm jobs. However, these innovative technologies must be adaptive to the farming system in SSA if the positive effect of increased population as market for agricultural produce has to be realized. It has been cautioned that the continued increase in population would reach a stage when the household land area will be too small to sustain household food security, leading to food shortage, to the extent that increased productivity might not solve the food insecurity problem. Adoption of new technologies such as integrated dairy and cropping systems leading to increased farm productivity and income is spreading in central Malawi and the southern highlands of Tanzania. These developments are encouraging as an integrated system of crop-livestock-aquaculture is gaining importance.

Integrated aquaculture has great potential in improving food security and ensuring diversification of diets, as the systems generate both animal proteins and vegetables. However, there is a need to reach out to medium- and large-scale farmers in order to achieve meaningful gains in aquaculture production. Private companies and NGOs can play a crucial role in disseminating developed technologies by SUA to different regions in Tanzania and other SSA countries, and the idea of personal transformation for changing small farmers' attitude for greater good should also be probed further.

References

Amuri NA (2015) Enhancing resilience of food production systems under changing climate and soil degradation in semi-arid and highlands of Tanzania. In: Lal R et al (eds) Sustainable intensification to advance food security and enhance climate resilience in Africa. Springer International Publishing Switzerland, Cham, pp 385–405. https://doi.org/10.1007/978-3-319-09360-4

Aune J, Lal R (1997) Agricultural productivity in the tropics and critical limits of properties of Oxisols, Ultisols and Alfisols. Trop Agric (Trinidad) 74:96–103

Bidwell OW, Hole FD (1965) Man as a factor of soil formation. Soil Sci 99:65–72

Burian A, Karaya R, Wernerssond JEV, Egberthf M, Lokorwah B, Nyberg G (2019) A community-based evaluation of population growth and agro-pastoralist resilience in sub-Saharan drylands. Environ Sci Pol 92:323–330

Carswell G (1997) Agricultural intensification and rural sustainable livelihoods: A think piece. IDS working paper 64. Sustainable Livelihood Programme, IDS, University of Sussex, Brighton

Charles H, Godfray J, Beddington JR, Crute IR, Haddad L, Lawrence D, Muir JF, Pretty J, Robinson S, Thomas SM, Toulmin C (2010) Food security: the challenge of feeding 9 billion people. Science 327:812–818

Deininger K, Hilhorst T, Songwe V (2014) Identifying and addressing land governance constraints to support intensification and land market operation: evidence from 10 African countries. Food Policy 38:76–83

Dicks LV, Rose DC, Ang F, Aston S, Birch ANE, Boatman N, Bowles EL, Chadwick D, Dinsdale A, Durham S, Elliott J, Firbank L, Humphreys S, Jarvis P, Jones D, Kindred D, Knight SM, Lee MRF, Leifert C, Lobley M, Matthews K, Midmer A, Moore M, Morris C, Mortimer S, Murray TC, Norman K, Ramsden S, Roberts D, Smith LG, Soffe R, Stoate C, Taylor B, Tinker D, Topliff M, Wallace J, Williams P, Wilson P, Winter M, Sutherland WJ (2019) What agricultural practices are most likely to deliver "sustainable intensification" in the UK? Food Energy Secur 8:e00148

Dorosh P, Thurlow J (2018) Beyond agriculture versus non-agriculture: decomposing sectoral growth–poverty linkages in five African countries. World Dev 109:440–451

Dudal R (2004) The sixth factor of soil formation. Paper presented at the international conference on soil classification 2004, Petrozavodsk, Russia, 3–5 August 2004

E.C. (2017) Evaluation Study of the Payment of Agricultural Practices Beneficial for the Climate and the Environment. Alliance Environment and Thunen Institute

FAO (2015) The State of Food Insecurity in the World 2015. Meeting the 2015 international hunger targets: taking stock of uneven progress. Rome, FAO

FAOstat (2019a) Crops. June 2019. http://www.fao.org/faostat/en/#data/QC

FAOstat (2019b) Land use. Available at http://www.fao.org/faostat/en/#data/RL

FinMark Trust (2016) Report agricultural subsidies in SADC countries – current status and impact. Agricultural Subsidies in SADC: Final Report, p 187

Food and Agriculture Organization of the United Nations (FAO) and United Nations Economic Commission for Africa (ECA) (2018) Regional Overview of Food Security and Nutrition. Addressing the threat from climate variability and extremes for food security and nutrition. Accra. 116 pp

IMF (International Monetary Fund) (2018) Malawi, IMF country report no. 18/336. https://www.imf.org, site visited 30.8.2019

IPCC (2018) Global warming of 1.5 °C. Summary for Policy Makers. https://report.ipcc.ch/sr15/pdf/sr15_spm_final.pdf

Jayne TS, Mather D, Mghenyi E (2010) Principal challenges confronting smallholder agriculture in Sub-Saharan Africa. World Dev 38:1384–1398

Jayne TS, Yeboah FK, Henry C (2017) The future of work in African agriculture: trends and drivers of change. International Labour Office. Research Department working paper no. 25, p 32

Jenny H (1941) Factors of soil formation: a system of quantitative pedology. McGraw-Hill, New York

Josephson AL, Ricker-Gilbert J, Florax RJGM (2014) How does population density influence agricultural intensification and productivity? Evidence from Ethiopia. Food Policy 48:142–152

Kafle AK, Garcia X, Wang PE, Pfeffer GD, Strahan HB (2019) Nutrient demand and fungal access to resources control the carbon allocation to the symbiotic partners in tripartite interactions of Medicago truncatula. Plant Cell Environ 42:270–284

Kalabamu FT (2019) Land tenure reforms and persistence of land conflicts in Sub-Saharan Africa – the case of Botswana. Land Use Policy 81:337–345

Lal R (2007a) Promoting technology in Sub-Saharan Africa, South Asia. CSA News 52(7):10–13

Lal R (2007b) Tragedy of the global commons: soil, water and air. CSA News 52:10–11

Lal R (2015) Sustainable intensification for adaptation and mitigation of climate change and advancement of food security in Africa. In: Lal R et al (eds) Sustainable intensification to advance food security and enhance climate resilience in Africa. Springer International Publishing Switzerland, Cham, pp 1–17. https://doi.org/10.1007/978-3-319-09360-4

Lal R (2016a) Beyond COP 21: potential and challenges of the "4 per thousand" initiative. J Soil Water Conserv 71:20A–25A

Lal R (2016c) Soil health and carbon management. Food Energy Secur 5(4):212–222

Lal R (2018a) Digging deeper. A wholistic perspective of factors affecting SOC sequestration. Glob Chang Biol 24(8). https://doi.org/10.1111/gcb.14054

Lal R (2018b) Sustainable development goals and the IUSS. In: Lal R, Horn R, Kosaki T (eds) Soil and sustainable development goals. Catena Schweizerbart, Stuttgart, pp 189–196

Lal R (2019) Ecointensification through soil carbon sequestration: harnessing ecosystem services and advancing sustainable development goals. J Soil Water Conserv 74(3):55A–61A

Lal R, Smith P, Jungkunst HF, Mitsch W et al (2018a) The carbon sequestration potential of terrestrial ecosystems. J Soil Water Conserv 73:145A–152A

Lal R, Horn R, Kosaki T (eds) (2018b) Soil and sustainable development goals. Catena Schweizerbart, Stuttgart, p 196

Le Quéré C, Andrew RM, Friedlingstein O, Sitch S, Hauck J, Pongratza J et al (2018) Global carbon budget. Earth Syst Sci Data 10:2141–2194

Leopold A (1949) The land ethic. In: A sand county Almanac. University of Colorado, Boulder

Lindsey R (2018) Climate change: atmospheric carbon dioxide. NOAA. Climate.gov

Loveland P, Webb J (2003) Is there critical level of organic matter in the agricultural soils of temperate regions: review. Soil Tillage Res 70:1–18

Malawi Government (2017) Malawi national agricultural investment plan (NAIP)

Mathews GA, Tunstall JP (2006) Smallholder cotton production in sub-Saharan Africa: an assessment of the way forward. Int J Pest Manag 52:149–153

Muyanga M, Jayne TS (2014) Effects of rising rural population density on smallholder agriculture. Food Policy 48:98–113

New Partnership for Africa's Development (NEPAD) (2013) Agriculture in Africa. Transformation and outlook. NEPAD, Midrand

NOAA-ESRL (2019) All time records high of for CO_2 daily averages. Daily CO_2. https://www.co2.earth/daily-co2

Tache B and Oba G (2010) Is poverty driving borana herders in southern ethiopia to crop cultivation? Human Ecology Vol. 38, No. 5 (OCTOBER 2010), pp. 639–649

United Nations, Department of Economic and Social Affairs, Population Division (2015) World Population Prospects: The 2015 Revision, Key Findings and Advance Tables. Working Paper No. ESA/P/WP.241

VIB Fact Series (2017) IBPO's fact series -Maize in Africa. http://www.vib.be/en/about-vib/Documents/VIB_MaizeInAfrica_EN_2017.pdf

World Bank (2017) World Development Indicators 2017. Washington, DC. © World Bank. https://openknowledge.worldbank.org/handle/10986/26447 License: CC BY 3.0 IGO., accessed June 5, 2019.

World Bank (2018) Malawi systematic country diagnostic: breaking the cycle of low growth and slow poverty reduction. The World Bank, Washington, DC

ITPS and FAO 2018. Global soil organic carbon map. Technical Report. Rome. 162 pp.

World Bank data (2019) https://data.worldbank.org/indicator/SP.POP.GROW?locations=ZG-ZF

WMO (2019) Greenhouse Gas Bulletin #15. The State of Greenhouse Gases in the Atmosphere Based on Global Observations Through 2018. World Meteorological Organization, Geneva, Switzerland.

Part II
Conservation Agriculture, Carbon Sequestration, and Soil and Water Management

The Prospects for Conservation Agriculture in Ethiopia

Jens B. Aune

Abstract About 90% of the farmland in Ethiopia is tilled by the ard plough (the *maresha*) pulled by a pair of oxen. Animal traction is therefore the major output from the livestock system in Ethiopia, whilst the output in terms of milk and meat production is significantly less. This study summarises the effects of conservation agriculture (CA) on yield, soil properties, soil erosion, labour use and economic return in Ethiopia, as well as factors determining the adoption of CA in the country. An approach for the upscaling of CA in Ethiopia is proposed. By summarising results from different field experiments that compare conventional tillage to reduced tillage, the change in yield (%) was −4.0, +0.8 and −0.1 for maize (*Zea mays*), teff (*Eragrostis tef*) and wheat (*Triticum aestivum*), respectively. The yield difference between conventional tillage and zero tillage was −24.8% in maize. Factors that have been found to promote the adoption of CA include access to inputs and credit, the competence of CA amongst extension agents, membership of farm organisations, farmers' education and farmers' assets. Studies on technology adoption in Ethiopia show that it is preferable to bundle technologies together in the form of a package. CA should therefore not be promoted as a stand alone approach, but rather as part of an integrated farming system focusing on crop and livestock intensification through the use of improved germplasm, integrated soil fertility management, weed and pest management and fodder production as well as partial stall feeding and changes in the composition of livestock. Conservation agriculture has implications in farming systems beyond crop production, and this should be reflected in the approach for the upscaling of CA.

1 Introduction to Conservation Agriculture (CA)

Conservation agriculture (CA) is based on the principles of minimum soil disturbance, permanent soil cover and diversification through rotation/associations (Thierfelder et al. 2018). CA can be considered as a form of climate-smart agriculture

---

J. B. Aune (✉)
Department of International Environment and Development Studies (Noragric), Norwegian University of Life Sciences (NMBU), Ås, Norway
e-mail: jens.aune@nmbu.no

B. R. Singh et al. (eds.), *Climate Impacts on Agricultural and Natural Resource Sustainability in Africa*, https://doi.org/10.1007/978-3-030-37537-9_2

and as an approach to reverse soil degradation (Lal 2015). Crop establishment methods under CA include direct sowing using planters (mechanised), planting with the use of a dibble stick (manual), ripping of the soil followed by sowing (mechanised) and the use of permanent planting basins (manual). The use of permanent raised beds (mechanised) is a method employed on heavy soil prone to waterlogging. CA has been proposed as an approach to increase yields, for carbon sequestration, to enhance biodiversity and to save on labour and costs (FAO 2018). Access to appropriate farm machinery is a key factor for upscaling of CA across the continents (Lal 2015).

2 Conservation Agriculture in Africa

CA is progressing in Africa at a pace of around 150,000 hectares (ha) per year, and the area under CA has tripled during the last 10 years reaching around 1.5 million ha (Kassam et al. 2018). This represents about 1% of the farmland in Africa. South Africa is the African country with the most land farmed using CA methods (mechanised CA). Despite this progress, many at the *2nd African Congress on Conservation Agriculture* in 2018 were concerned about the slow uptake of CA in Africa. Only limited research has been conducted on CA by the national research institutions, with most of the research being undertaken by international research centres such as CIMMYT.[1]

The knowledge base on CA in Africa has increased rapidly during the last 10 years with regard to yield response, environmental impact and factors determining adoption. A meta-study including 42 studies from Africa showed that zero tillage (ZT) has a slightly lower yield (−24 kg ha^{-1}) than conventional tillage (CT), but zero tillage (ZT) with mulch compared to conventional tillage gave an average yield increase of 294 kg ha^{-1} in the first 3 years, whilst this yield increase was 487 kg ha^{-1} when ZT with mulch was practised for more than 3 years (Corbeels et al. 2014). A meta-study across the Earth's semi-arid and sub-humid conditions also showed that ZT without mulching gave a lower yield than CT (Rusinamhodzi et al. 2011), and mulching improved yield over time. This shows that mulching is a key factor in the success of CA. There are contradictory results regarding the most suitable rainfall conditions for CA. Corbeels et al. (2014) showed that CA gave higher yields than conventional tillage when rainfall was above 1000 mm compared to below 600 mm, whilst Rusinamhodzi et al. (2011) showed the opposite effect. CA has been found to perform better on loamy soil compared to sandy or clay soil (Corbeels et al. 2014). The reason is probably that CA will not work well where there are problems with waterlogging or in dry areas where it is difficult to produce sufficient quantities of mulch. Lal (2015) has proposed to develop a soil suitability guide to

[1] International Maize and Wheat Improvement Center.

identify the soil types most suited for CA. Another important result from the meta-study by Corbeels et al. (2014) is that there is a stronger (+391 kg ha^{-1}) effect of CA compared to CT when more than 100 kg N/ha was applied, whilst the corresponding effect was low (+85 kg ha^{-1}) when less than 100 kg N/ha was applied. These meta-studies clearly show the importance of combining mulching with nitrogen application. Nitrogen application is important in order to reduce nitrogen immobilisation in the soil, as a result of the organic inputs, and to ensure sufficient mulch production every year.

A major challenge for the development of CA in Africa is the retention of crop residues as mulch, as livestock graze freely in the dry season. A comparative study of the use of crop residues in Africa and Asia showed that the use of crop residues as a soil amendment is most likely to succeed if alternative, good-quality fodder is available, or in areas where biomass production is high and the demand for stover is low (Valbuena et al. 2012). Farmers will not practise mulching if the perceived benefit of using straw as mulch is higher than the perceived benefit of using stover for fodder or other purposes. Hence, the agro-ecological and socio-economic conditions must be taken into consideration when promoting CA (Lal 2015). Options for increasing the use of crop residues as mulch include reducing the use of crop residues as livestock fodder by developing alternative fodder resources and reducing the number of animals by giving emphasis to more milk- and meat-producing animals. In such an intensification pathway, more of the fodder resources can be used for promoting growth of the animal and not just body maintenance. There is also a growing market for livestock products as a result of population growth and increasing wealth. The intensification of crop and livestock production should therefore occur in parallel (Duncan et al. 2016).

The major reasons for the adoption of CA across the world are reduced fuel costs, saved labour, timely sowing, improved profitability, carbon sequestration, improved water-use efficiency and reduced erosion (Baudron et al. 2015a; Lal 2016). CA has rarely been adopted to increase yield. For this reason, CA has been promoted in areas where:

- Tillage is costly.
- Power for ploughing is limited.
- Planting is delayed as a result of power shortage.
- Water deficiency limits yield.
- Erosion is a threat to the productivity of farmland.

There are different views regarding the potential for CA in Africa. Giller et al. (2009) claim that the potential for CA is very limited in Africa whilst Baudron et al. (2015a) argue that CA has potential in Africa due to increasing tillage costs as a result of higher energy prices combined with feed shortages for traction animals. The areas affected by erosion will also increase due to increased cultivation of steep slopes and reduced soil cover. Furthermore, climate change will also necessitate the development of cropping systems that can ensure timely sowing and more drought-resistant farming systems.

There are multiple constraints to the adoption of CA in Africa (Thierfelder et al. 2018) including:

- Limited knowledge and capacity amongst farmers
- Lack of sufficient mulch material
- Lack of profitable rotation systems
- Lack of inputs (fertiliser, seeds, herbicides and machinery)
- High costs of inputs
- Lack of credit
- Lack of output market for rotational crops (legumes)
- Farmers' attitudes

Some of these constraints are not specific to CA, but are general constraints to the uptake of agricultural technologies. All of these factors must be taken into consideration when promoting CA. For successful adoption of CA, it is therefore not only necessary to address the three core principles of CA but to also include complementary practices and enabling factors (Thierfelder et al. 2018). CA can best be promoted by combining technological, socio-economic and institutional approaches. In order for farmers to adopt CA, they need to be convinced about the agronomic effect, stover trade-off, labour requirements, financial requirements and the market for grain and livestock products.

3 The Ox-Plough System in Ethiopia

In Ethiopia, more than 90% of the land is tilled using the *maresha,* which is pulled by a pair of oxen (Astatke et al. 2003). The maresha does not turn the soil as the mouldboard plough does, but rather breaks the soil. Fields are usually ploughed several times and each passing of the maresha should be perpendicular to the previous pass. The first pass with the maresha typically reaches a depth of 8 cm, whilst the last pass can go down to 20 cm (Astatke et al. 2003). The first pass is normally performed following the first rains to reduce the formation of clods on the soil surface and because it is more energy-demanding to plough dry soil. When tillage follows a rainfall, it will also serve as a form of weed control. According to farmers, the reasons for tillage are weed control, moisture conservation, soil warm-up and manure incorporation (Tamesgen et al. 2008). The purpose of the last pass with the maresha is also to bury the seeds. The intensity of ploughing depends on several factors; teff fields are usually ploughed 3–5 times whilst maize fields are ploughed 2–4 times. Teff fields are ploughed intensively because the crop requires a smooth seedbed and because the crop is a poor competitor against weeds. Heavier soil is also usually ploughed more often than lighter soil. If rainfall starts early and is distributed over a longer period, more tillage is performed. More intensive tillage is also performed on large farms and on farms with more male labour (Tamesgen et al. 2008). A pair of oxen can plough around 0.25 ha per day (Nyssen et al. 2011).

The ox-plough system is deeply rooted in Ethiopia and is strongly associated with the cultivation of cereal crops such as teff. This system was well adapted to former conditions in Ethiopia, when grazing areas were abundant, farms were larger and population density was lower (Aune et al. 2001). Large grazing areas are required to uphold the system, as in addition to the two oxen, two young 'apprentice' bulls/calves are required in addition to a stud bull and four or five cows (McCann 1995). This livestock system therefore consists of at least 11 animals in order to maintain it over time.

The ox-plough system puts female-headed households and the poorest households without access to a pair of oxen at a disadvantage. Farmers without oxen need to rent them, and surveys show that the farmer who ploughs usually claims 50% of the harvest and may, in addition, take the straw (Aune et al. 2001). This is a constraint for many farmers in Ethiopia as 48% of households own zero or one ox. Female-headed households are required to hire a man for ploughing, as it is not culturally acceptable for women to plough.

Without important changes in the livestock system in Ethiopia, it will be very demanding to introduce CA at a large scale. A survey of 643 households in the Rift Valley in Ethiopia and Western Kenya showed that the major output from the livestock system in Ethiopia is traction power, whilst milk production is the major output in Western Kenya (Baudron et al. 2014). In the central Rift Valley, there is an average of 5.69 TLU ha^{-1} cultivated land with oxen constituting 42% of the livestock, whilst cows represent 36% of the livestock and the average milk production is 160 litres cow^{-1} $year^{-1}$ (Baudron et al. 2014). In Western Kenya, where more intensive livestock production is practised, there are on average 4.04 TLU ha^{-1} cultivated land and oxen represent 12% of the livestock population, whilst cows represent 62%. It is therefore clear that milk production per farm is much higher in Western Kenya compared to the Ethiopian Rift Valley. Manure application per ha is also four times higher in Kenya compared to Ethiopia, despite higher livestock number per ha in Ethiopia. These figures show that livestock production could become more productive in Ethiopia if the livestock production system changes from an emphasis on producing traction power to more emphasis on milk and meat production. Intensive milk production requires fodder of good quality, and this may allow more crop residues to be used as mulch, as straw is not a good fodder resource for lactating animals. A reduction in the number of animals would also make it possible for farmers to retain more straw as mulch.

4 Effect of CA on Soil Properties, Soil Erosion and Yields in Ethiopia

CA can improve soil properties, both as a result of mulching and by introducing nitrogen fixation crops in the rotation system or in intercropping. Mulching improves soil properties by increasing water infiltration, by improving the soil's water-holding capacity, and by reducing evaporation through lowering the soil's surface temperature. A long-term experiment on Vertisol in Tigray showed that soil chemical properties such as organic carbon, nitrogen and soil microbial life were higher in permanent raised bed compared to plots under conventional tillage (Areya et al. 2016). The soil's physical properties such as aggregate stability, infiltration rate (time to pond), penetration resistance and bulk density were also more favourable under permanent raised beds compared to conventional tillage. This improvement in soil's physical properties will also make the soil less prone to erosion.

In the Rift Valley of Ethiopia, it was shown that soil volumetric content was higher at the grain filling stage in mulched plots compared to non-mulched plots (Sime et al. 2015). Mulching has been found critical in order to improve soil quality, as the amount of organic matter in the topsoil was twice as high in plots with residue retention, compared to those with no retention of straw as mulch (Baudron et al. 2015b).

Soil erosion is a great threat to the long-term productivity of Ethiopian soil. CA is known to reduce erosion by improving water infiltration and by increasing resistance to erosion. Soil erosion was reduced from 30 t/ha in conventional tillage plots without mulch to 16 Mg ha^{-1} in zero tillage plots combined with 3 mg mulch ha^{-1} (Adimassu et al. 2019) (average erosion across 3 years). However, zero tillage without mulch did not reduce soil erosion compared to conventional tillage. The soil loss in CA plots with mulching is still quite high especially in erosion-prone areas. It is therefore necessary to combine CA with other measures to control soil erosion.

Table 1 shows yield effect of convention tillage compared to different forms of minimum tillage, with and without mulch. The tillage systems included were reduced tillage, zero tillage with and without mulch, planting basins and permanent raised beds. It appears that that there is no clear yield difference between conventional tillage and reduced tillage. Comparing conventional tillage to reduced tillage gave an average yield change in percent of −4.0, +0.8 and −13.6% for maize, teff and wheat, respectively. Zero tillage compared to conventional tillage in maize reduced yield by 24.8%, and raised permanent beds in teff on Vertisol reduced yield by 13.6% compared to conventional tillage. These results show that reduced tillage could be an option for Ethiopian farmers, as this method maintains yield whilst reducing production costs. Zero tillage seems to reduce yields compared to conventional tillage. There was a tendency for yields under zero tillage and reduced tillage to improve over time compared to conventional tillage (Areya et al. 2016).

The effect of mulching on yield has been recorded of up to 3 Mg ha^{-1} under dryland conditions, whilst under more humid conditions, there is less impact of mulching on yield (Baudron et al. 2014).

These results may underestimate the yield advantage of CA at the farm level. Results from other African countries show that sowing is often delayed when conventional tillage is practised due to repeated tillage operations (Baudron et al. 2015b). This will inevitably penalise yield under conventional tillage.

Vertisols (heavy clay soils) are particularly prone to waterlogging. Farmers in Ethiopia have traditionally used ridges or raised beds to reduce problems with waterlogging on Vertisols, and research promoted further modification in traditional

Table 1 Effect of tillage method on yields (kg ha^{-1}) of maize, teff and wheat in Ethiopia

Crop	Site	Soil	Conventional tillage (CT)	Minimum tillage	Minimum tillage with mulch	Zero tillage	Raised bed	Source
Maize	Ziway	Loamy sand/ sandy loam	5787	4828		3722		Sime et al. (2015)
Maize	Melkassa	Loamy sand/ sandy loam	5888	5202		4087		Sime et al. (2015)
Maize	Bako Tibe	Nitisol, sandy loam/loam	3250			3425		Zerihun et al. (2014).
Maize		Different soil types	4525	4990				Ito et al. (2007)
Maize	Western Ethiopia, Bako Tibe	Nitisol, silty clay	5768	6073	6473			Tolessa et al. (2007)
Yield change % compared to CT – Maize				**−4.0**		**−24.8**		
Teff	Gare Arera	Nitisol	871	829		725		Tulema et al. (2008)
Teff	Gare Arera	Vertisol	1494	1583		1427	1644	Tulema et al. (2008)
Teff		Different soil types	1100	1187				Ito et al. (2007)
Teff	Gumselasa	Vertisol	1173				678	Oicha et al. (2010)
Teff	Mekelle	Cambisol (clay loam)	760	723				Habtegebrial et al. (2007)
Teff	Adigudem, Tigray		1290	1240			1100	Areya et al. (2016)

(continued)

Table 1 (continued)

Crop	Site	Soil	Conventional tillage (CT)	Minimum tillage	Minimum tillage with mulch	Zero tillage	Raised bed	Source
Yield change% compared to control – Teff				**+0.8**			**−13.6**	
Wheat		Different soil types	2255	2460				Ito et al. (2007)
Wheat	Holeta	Clay, Nitisol	1843	1406		1066		Adimassu et al. (2019)
Wheat	May Zegzeg, Tigray	Vertisol	3200				3400	Araya et al. (2012)
Wheat		Vertisol	1610	2020			2460	Areya et al. (2016)
Wheat	Kulumsa	Clay soil	3360			3390		Taa et al. (2004)
Wheat	Asasa	Clay loam	3216	3033				Taa et al. (2004)
Yield change% compared to control – Wheat				**−0.1**				

Average percent change in yield is presented where there are at least data from three sites farming practices. One such traditional tillage system is the Derdero method, which consists of broadcasting the seed after repeated tillage followed by one pass with the maresha to make ridges or narrow raised beds (Nyssen et al. 2011). Most of the seeds will therefore germinate on the ridge or raised bed, making the plants less exposed to waterlogging. The distance between the furrows is about 60 cm. In the following years, only one pass with the maresha is made to bury the seeds and rebuild the ridge, thus representing a minimum tillage system. An improvement of the system has been developed using the application of glyphosate to control weeds and retaining 50% of the standing straw for use as mulch. The retained straw should not be grazed. A major advantage is a reduced need for traction power, as only one pass with the maresha is practised in the years following the establishment of the ridge/raised bed. Results from Tigray show that it took at least 3 years before there was a yield benefit using the Derdero tillage method compared to conventional tillage (Areya et al. 2016). The soil properties such as organic carbon, total N, available P and soil microbial biomass, infiltration rate and aggregate stability improved as a result of the raised bed.

5 Effect of CA on Labour Use and Economic Return

Reduced tillage can decrease labour demand if minimum tillage is combined with the use of herbicides. Total labour use in maize production in the South Achefer region (northwestern Ethiopia) was reduced from 79 hours (h) ha^{-1} in conventional tillage to 57 h ha^{-1}using minimum tillage with herbicides, and oxen hours were reduced from 17 h ha^{-1} with traditional tillage to 5 h ha^{-1} with minimum tillage (Jaleta et al. 2016). However, in an experiment in maize (in the Rift Valley) where herbicides were not used, labour use for land preparation and weeding was 37, 41 and 51 h ha^{-1} for conventional tillage, reduced tillage and zero tillage, respectively (Sime et al. 2015). Labour use for tillage was less in zero tillage and reduced tillage but higher weeding compared to conventional tillage. This higher labour use combined with lower yield resulted in a lower gross margin for reduced tillage and zero tillage compared to conventional tillage (Sime et al. 2015). The gross margin in zero tillage was 52% less than in conventional tillage and 22% less in reduced tillage compared to conventional tillage. Mulching in zero tillage plots compared to non-mulch ZT plots increased gross margin with 74% as a result of higher yields and less labour use for weeding (Sime et al. 2015).

Field experiments with teff show that reduced tillage gives a satisfactory economic return on Vertisol (Tulema et al. 2008). This study showed that reduced tillage gave an equal gross margin to conventional tillage on a Vertisol, whilst on a Nitisol, conventional tillage gave a 47% higher gross margin than reduced tillage. Experiments conducted on Vertisols in Tigray also confirm that reduced tillage gives yields as high as conventional tillage (Habtegebrial et al. 2007), with the only difference being that there is more weed infestation in reduced tillage plots. Experiences of *Sasakawa Global 2000* also showed that reduced tillage with mulching and herbicides gave an average gross margin in teff of 144 US$ ha −1 across 4 years, whilst conventional tillage with herbicides gave an average gross margin of 125 US$ha−1 (Ito et al. 2007). Reduced tillage appears to be a particularly interesting option if herbicides are used to control weeds in teff grown on Vertisol. Reduced tillage can also be an interesting option for farmers without access to oxen, as these farmers may have to forgo 50% of their harvest to get their land ploughed.

6 Adoption of CA in Ethiopia

The adoption of CA in Ethiopia depends on socio-economic, social, institutional and biophysical factors (Dabi et al. 2017; Teklewold et al. 2013). Socio-economic factors have a strong bearing on the uptake of new tillage methods and other agricultural technologies. Studies from Ethiopia, Kenya, Tanzania and Malawi show that credit-constrained farmers were less likely to adopt minimum tillage with mulching, whilst farmers with more assets were more likely to adopt (Marenya et al. 2017). About 57% of the farmers are credit-constrained in Ethiopia, and availability of assets is important, making it easier for farmers to purchase inputs. Furthermore, experiences from the *SIMLESA*[2] project in Ethiopia show that the more distant the farmers are from the market and the poorer the condition of the roads, the less likely the farmers are to adopt minimum soil tillage and soil and water conservation methods (Jaleta et al. 2016). Access to herbicides was another important factor for the uptake of minimum tillage. Policy simulation shows that the adoption of CA can be stimulated by increasing the number of extension agents per farmer and by providing subsidies on farm inputs such as fertiliser and seed (Marenya et al. 2017). Subsidies have a particularly strong impact on adoption (Marenya et al. 2017).

Social and institutional factors influencing the adoption of CA include the education of household heads, the social network of the head of household and spouse, membership of farm organisations, number of relatives, access to male labour, land tenure and plot size (Tamesgen et al. 2008; Tessema et al. 2016; Jaleta et al. 2016; Marenya et al. 2017; Kassie et al. 2015). Membership of farm organisations builds trust in CA as it gave farmers an opportunity to consult each other. Livestock ownership, particularly oxen ownership, reduced the adoption of minimum tillage with mulching, probably because easy access to traction animal makes it less attractive for the farmers to switch to minimum tillage (Marenya et al. 2017).

Biophysical conditions also affect the uptake of tillage methods. In Ethiopia, it has been found that CA was adopted more on fields with medium to good soil fertility (Marenya et al. 2017) and on plots distant from the homestead (Tessema et al. 2016). Other studies have shown that the uptake of reduced tillage was not dependent on soil fertility (e.g. Jaleta et al. 2016). It appears that more residues are used

[2] Sustainable Intensification of Maize and Legume Systems for Food Security in Eastern and Southern Africa.

as mulch on smaller farms and on lands more intensively cultivated combined with a higher intensity in livestock production (Brown et al. 2018). Crop residues are also increasingly privatised and sold as fodder in Ethiopia (Valbuena et al. 2012).

A study assessing the adoption potential of CA across Ethiopia used low livestock density or controlled grazing, good market access, high population density and moderate rainfall (above 500 mm) as criteria for high potential for CA (Tesfaye et al. 2015). Based on these criteria, it was assessed that 9% of the cultivated area had a high socio-economic and biophysical potential for CA, whilst 24% of the cultivated area was considered unsuitable for CA. In Ethiopia, it is high livestock density in particular that limits the opportunities for CA (Tesfaye et al. 2015).

7 Practical Experiences of CA

Experiences on CA in Ethiopia have accumulated over the last 20 years through research and development projects. CA has been promoted in Ethiopia through *Sasakawa Global 2000*, the Sustainable Land Management Programme of the Ministry of Agriculture, the *Canadian Foodgrains Bank*, the Drylands Coordination Group and the *SIMLESA* project. The introduction of CA with mulching is a challenge in Ethiopia as farmers prefer to use crop residue as fodder because it gives an immediate benefit, whilst the benefits from using straw for mulching may not necessarily be seen in the first years. The free grazing system also makes the use of residues as mulch difficult in Ethiopia. Only 3% of farmers retain as much as 1 Mg mulch ha^{-1}in Ethiopia (Baudron et al. 2014).

CA has been introduced in the Wolaita and Sidamo regions through the NGO Terepeza Development Association, with funding from the Canadian Foodgrains Bank. This NGO has been promoting CA since 2013. CA was initially part of an integrated development programme, but since 2015, there has been a specific project on CA promotion. CA has been adopted by around 7000 households in the Wolaita area. The Wolaita area has favourable rainfall conditions as it receives about 1300 mm rainfall $year^{-1}$. The adoption seems to be genuine as there is no input or other forms of subsidies following the training on CA. In the first years, CA was demonstrated in training centres set up by the Ministry of Agriculture and on the cultivated land of schools, but as confidence in CA has grown, the approach is now demonstrated in the farmers' own fields. In Wolaita, the project has been able to reach 10 of the 40 villages. Project objectives were erosion control, increased soil fertility and to address climate variability. CA is practised in annual crops such as maize, lab-lab (*Lablab purpureus*), pigeon pea (*Cajanus cajan*) and taro (*Colocasia esculenta*) and in perennial crops such as enset (*Ensete ventricosum*) and papaya (*Carica papaya*). Bylaws have been introduced in the area to restrict free grazing of crop residues and thereby ensure that a significant amount of crop residues are used as mulch. The mulching materials used include crop residues, household waste and other types of organic material. Farmers use crop residues for both mulching and fodder. The upper part of the maize straw is used for fodder, whilst the lower part of less fodder quality is used as mulch. Cover crops are also cultivated to produce fodder and mulch. The major cover crops are lab-lab, cowpea (*Vigna unguiculata*) and pigeon pea. When farmers start CA, they first mulch the field. There are different ways of preparing the soil for planting. One way is to open a small pit for planting followed by planting. Alternatively, farms split the mulch by hand according to distance between lines, followed by one pass of the maresha along this line to open a furrow to facilitate sowing. The farmers summarised the benefits of CA as follows:

- Able to grow a wider variety of crops because of better water conditions in the soils, such as taro, papaya and kale.
- Less predation from the stalk borer and fall armyworm as a result of CA.
- Able to cultivate the land without oxen, of particular benefit to female farmers.
- Farmers invest more in fattening animals.
- Reduced erosion.
- Less labour use.
- Able to manage weeds without herbicides when land is 0.5 ha, as in Wolaita.
- Benefits from cover crops for fodder, fuel and mulch.
- The yield benefits increased over time (Fig. 1).

The initial experiences of the Sustainable Land Management Programme and the Development Fund in the Gimbi area of the Oromia region are also positive, as farmers expressed an interest to continue the CA practices after the initial phase of the project (Gjengedal 2016). The CA practices introduced were ripping, mulching with crop residues, herbicides to control weeds, fertilisers and intercropping. Farmers observed that CA reduced labour, prevented erosion and increased yield. The project also benefited the poorest segments of the population, as poor farmers without oxen were able to cultivate their land. However, for the long-term sustainability of the project, the farmers need secured supplies of fertiliser, seeds and herbicides.

CIMMYT's *SIMLESA* project, which tested CA in various locations in Ethiopia, found that CA increased yield compared to farmers' practices by 40% and 28% for common beans (*Phaseolus vulgaris*) and maize, respectively (CIMMYT 2018).

Fig. 1 Mulching in papaya (left) and a field mulched and prepared for sowing (right) in Wolaita, Ethiopia. (Photo: Jens B. Aune)

Results from other projects on CA in Ethiopia have not always been positive. A Drylands Coordination Group project in the Ziway (Rift Valley) in the Oromia area failed because it was impossible for farmers to retain the crop residues left in the field due to free grazing (Sime et al. 2015). This shows the importance of the establishment of bylaws for the protection of mulch.

8 Upscaling of CA in Ethiopia

CA should not be promoted as a stand alone approach, but rather as part of an integrated farming system that focuses on sustainable intensification of livestock combined with improved crop husbandry methods. One key finding from studies on technology adoption in Ethiopia is that increased yield and farm income can be achieved if technologies are bundled or presented as a package, as there are synergies between the technologies (Kassie et al. 2015; Marenya et al. 2017). A study focused on a combination of different technologies showed that the adoption of only conservation tillage increased labour and pesticide use to compensate for the effect of increased weed and pest infestation. The adoption of all the technologies did not increase pesticide use and labour because system diversification reduced the need for pesticide application and weeding (Marenya et al. 2017). Yield increased as the number of technologies adopted increased from one to six. Different technologies should therefore not be presented in isolation, but rather as part of a package. This study also showed that smaller farms had a higher tendency to adopt the full package compared to larger farms, and the skill of the extension agents increased the adoption of improved practices (Marenya et al. 2017).

Most CA projects in Africa have been promoted without considering the implications for livestock. This may be one of the reasons for the moderate uptake of CA. As discussed above, CA without mulching will not give the desired benefit in terms of yield increase and improved soil fertility. In order to ensure that sufficient quantities of mulch are available, it is critical to ensure sufficient straw production through the use of good agronomic practices such as use of appropriate varieties, good seed quality, seed priming, use of organic and mineral fertiliser of sufficient quality and use of integrated pest management. By ensuring increased straw production, the 'cake' will increase, thereby reducing the competition for crop residues. However, once this residue is produced, it is equally important that part of the crop residues are retained as mulch. Across Ethiopia, livestock graze freely on crop residues left in fields, making the retention of crop residues impossible. Bylaws should be established, as in Wolaita, to ensure that parts of the crop residue are retained in the fields. CA should not be promoted unless such agreements with the farmers are in place. In parallel to promoting CA, it is essential that new fodder resources are developed that can reduce the need for using straw as fodder. These new fodder resources can be cover crops such as lab-lab or *Crotalaria juncea*, fodder trees and the planting of fast-growing grass such as Napier grass (*Pennisetum purpureum)* along farm borders. Crop and livestock intensification should go hand in hand.

The need for traction animals will be reduced with the introduction of CA, and farmers can replace oxen with milk- and meat-producing animals. An approach for mechanisation is to open a planting line with the use of rippers followed by the use of simple seeders pulled by donkey or ox to deliver seeds and fertiliser. There are very positive experiences with the use of low-cost seeders in Mali (Aune et al. 2019). Such intensification of the livestock system is not likely to occur unless farmers change to CA. The benefit of CA is therefore not only reduced erosion and more crop production but, equally important, an increase in livestock production. Another approach for ensuring less competition for crop residues is the introduction of motorised tillage, which would reduce the need for farmers to keep oxen.

The introduction of CA is not likely to succeed unless backed by appropriate policy measures. Key elements would be a strengthening of the competence of extension agents on CA in Ethiopia and ensuring that the extension agents have sufficient resources to reach out to farmers. CA should become part of a wider extension programme focusing on crop and livestock intensification. Another critical policy measure is to improve farmers' access to key inputs such as seeds, fertiliser and equipment necessary for practising CA and to ensure functioning markets for crop and livestock products. Credit or subsidy programmes can make these production factors much more easily accessible to the farmers. An option for improving the uptake of resource conservation technologies such as CA could be to make access to credit contingent upon practising CA (Kassie et al. 2015).

In order to succeed with CA, the following factors must be in place:

- Increased crop yield to improve availability of crop residues
- Development of alternative fodder resources such as improved pastures and fodder trees
- Development of bylaws to prevent free grazing of crop land
- Improved management of communal land
- Introduction of partial stall feeding of animals
- Increased emphasis on milk- and meat-producing animals
- Reduced need for oxen ploughing by promoting reduced tillage or zero tillage
- Easy access to seeds, fertiliser, herbicides and farm equipment
- Policy intervention such as credit and extension

For CA to be adopted on a large scale in Ethiopia, it is important to target areas with conducive conditions for rapid CA uptake. From the discussion above, it appears that such conditions exist in areas with good market access for crop and livestock products. CA may be less suited to low-rainfall areas where it may be challenging to produce sufficient mulch and to diversify the production system.

9 Conclusions

The field experiments on CA show that reduced tillage gives equal yield to conventional tillage whilst zero tillage does not function well under some conditions. However, despite these neutral or negative yield effects, CA may still be interesting for farmers because of lower production costs and less labour input, particularly if combined with herbicides. Reduced tillage appears to be an option on Vertisols in teff. The development of tillage methods with less need for traction animals will facilitate livestock intensification based on milk or meat production and promote pro-poor development pathways, as the tillage cost for farmers without oxen and female-headed households is particularly high. Factors that have been found to promote the adoption of CA include access to inputs and credit, competence on CA amongst extension agents, membership of farm organisations, farmers' education and farmers' assets.

CA should not be promoted as a stand alone approach but rather as part of an integrated farming system, focusing on crop and livestock intensification including improved germplasm, integrated soil fertility management, fodder production, change in composition of livestock and stall feeding. Such an integrated approach can make it possible for farmers to retain part of the crop residue for mulch, as mulching is essential for maintaining long-term productivity of the soil. Experience shows that bundling agricultural technologies into a package gives far better results than promoting the technologies individually. However, such a change in the production system will not take place unless backed by appropriate policies that strengthen the extension systems for CA and improve access to subsidies/credit for purchasing inputs necessary for the uptake of CA.

References

Adimassu Z, Alemu G, Tamene L (2019) Effects of tillage and crop residue management on runoff, soil loss and crop yield in the Humid highlands of Ethiopia. Agr Sys 168:11–18

Areya T, Nyssen J, Govaerts B, Deckers J, Sommer R, Bauer H, Gebrehiwot K, Cornelis WM (2016) Seven years resource-conserving agriculture effect on soil quality and crop productivity in the Ethiopia drylands. Soil Tillage Res 163:99–109

Araya T, Cornelis WM, Nyssen J, Govaerts B, Getnet F, Bauer H, Amare K, Raes D, Haile M, Deckers J (2012) Medium-term effects of conservation agriculture based cropping systems for sustainable soil and water management and crop productivity in the Ethiopian highlands. Field Crops Res 132:53–62

Astatke A, Jabbar A, Tanner D (2003) Participatory conservation tillage research: an experience with minimum tillage on an Ethiopian highland vertisol. Agr Ecosyst Environ 95:401–415

Aune JB, Bussa MT, Asfaw FG, Ayele AA (2001) The ox ploughing system in Ethiopia: can it be sustained? Outlook Agr 30:275–280

Aune JB, Coulibaly A, Woumou K (2019) Intensification of dryland farming in Mali through mechanisation of sowing, fertiliser application and weeding. Arch Agron Soil Sci 65:400–410

Baudron F, Jaleta M, Okitoi O, Tegegn A (2014) Conservation agriculture in Africa crop-livestock systems: expanding the niche. Agric Ecosyst Environ 187:171–182

Baudron F, Thierfelder C, Nyagumbo I, Gérard B (2015a) Where to target conservation agriculture for African smallholders. How to overcome challenges associated with its implementation? Experiences from Eastern and Southern Africa. Environment 2:338–357

Baudron F, Mamo A, Tirfessa D, Argaw M (2015b) Impact of farmland exclosure on the productivity and sustainability of a mixed crop-livestock system in the Central Rift Valley of Ethiopia. Agr Ecosyst Environ 207:109–118

Brown B, Nuberg I, Llewellyn R (2018) Research capacity for local innovations: the case of conservation agriculture in Ethiopia, Malawi and Mozambique. J Agr Res Extens 24:249–262

CIMMYT (2018) SIMLESA end of project external review and stakeholders' meeting, 6–9 March 2018. https://bit.ly/2Kj53IM
Corbeels M, Sakyi RK, Kühne RF, Whitbread A (2014) Meta-analysis of crop responses to conservation agriculture in sub-Saharan Africa. CCAFS report No. 12. CGIAR Research Program on Climate Change, Agriculture and Food Security, Copenhagen
Dabi N, Fikirie K, Mulualem T (2017) Soil and water conservation practices on crop productivity and its economic implications in Ethiopia: A review. Asian J Agr Res.11:128–136
Duncan AJ, Bachewe F, Mekonnen K, Valbuena D, Rachier G, Lule D, Bahta M, Erenstein O (2016) Crop residue allocation to livestock feed, soil improvement and other uses along a productivity gradient in Eastern Africa. Agr Ecosyst Environ 228:101–116
FAO (2018) Conservation agriculture. Why we do it. http://www.fao.org/conservation-agriculture/overview/why-we-do-it/en/
Giller KE, Witter E, Corbeels M, Tittonell P (2009) Conservation agriculture and smallholder farming in Africa: The heretics' view. Field Crops Res 114:23–34
Gjengedal M (2016) Conservation agriculture in Gimbi, Ethiopia. Benefits and barriers to adoption. M.Sc. thesis, University of Oslo, p 111
Habtegebrial K, Singh BR and Haile M (2007) Impact of tillage and fertilization on yield, nitrogen use efficiency of tef (Eragrostis tef (Zucc.) Trotter) and soil properties. Soil Tillage Res 94:55–63
Ito M, Matsumoto T, Quinones MA (2007) Conservation tillage practices in sub-Saharan Africa: The experience of Sasakawa Global 2000. Crop Protection 26:417–423
Jaleta M, Kassie M, Tesfaye K, Teklewold T, Jena PR, Marenya P, Erenstein O (2016) Resource saving and productivity enhancing impacts of crop management innovation packages in Ethiopia. Agric Econ 47:513–522
Kassam A, Friederich T, Derpch R (2018) Global spread of conservation agriculture. Int J Environ Stud 76:29. https://doi.org/10.1080/00207233.2018.1494927
Kassie M, Teklewold H, Jaleta M, Marenya P, Erenstein O (2015) Understanding the adoption of a portfolio of sustainable intensification practices in eastern and southern Africa. Land Use Policy 32:400–411.0
Lal R (2015) A system approach to conservation agriculture. J Soil Water Conserv 70:82A–88A
Lal R (2016) Potential and challenges of conservation agriculture in sequestration of atmospheric CO_2 for enhancing climate-resilience and improving productivity of soil of small landholders farms. CAB Rev 11:1–16
Marenya PP, Kassie M, Jaleta M, Rahut DB, Erenstein O (2017) Predicting minimum tillage adoption among smallholder farmers using micro-level and policy variables. Agr Food Econ 5:12
McCann JC (1995) People of the plough. An agricultural history of Ethiopia, 1800–1900. University of Wisconsin Press, Madison, p 298
Nyssen J, Govaerts B, Araya T, Cornelis WM, Bauer H, Haile M, Sayre K, Deckers J (2011) The use of the maresha ard plough for conservation agriculture in Northern Ethiopia. Agr Sust Dev 31:287–297
Oicha T, Cornelis W, Verplancke H, Nyssen J, Deckers J, Behailu, M., Haile M, Govaerts B (2010) Short-term effects of conservation agriculture on vertisols under tef (Eragrostis tef (Zucc.) Trotter) in the Northern Ethiopian highlands. Soil Tillage Res 106:294–302
Rusinamhodzi L, Corbeels M, van Wijk MT, Rufino MC, Nyamangara J, Giller KE (2011) A meta-analysis of long-term effects of conservation agriculture on maize yield under rain-fed conditions. Agr Sust Dev 31:657–673
Sime G, Aune JB, Mohammed H (2015) Agronomic and economic response of tillage and water conservation management in maize, central rift valley Ethiopia. Soil Tillage Res 148:20–30
Taa A, Tanner D, Bennie ATP (2004) Effects of stubble management, tillage and cropping sequence on wheat production in the south-eastern highlands of Ethiopia. Soil Till Res 76:69–82
Tamesgen M, Rockstrom J, Savenije HHG, Hoogmoed WB, Alemu D (2008) Determinants of tillage frequency among smallholder farmers in two semi-arid areas in Ethiopia. Phys Chem Earth 33:183–191

Teklewold H, Kassie M, Shiferaw B, Köhlin G (2013) Cropping system diversification, conservation tillage and modern seed adoption in Ethiopia: impacts on household income, agrochemical use and demand for labour. Ecol Econ 93:85–93

Tesfaye K, Jaleta M, Jena P, Mutenje M (2015) Identifying potential recommendation domains for conservation agriculture in Ethiopia, Kenya and Malawi. Environ Manag 55:330–346

Tessema YM, Asafu-Adjaye J, Kassie M, Mallawaarachchi T (2016) Do neighbours matter in technology adoption? The case of conservation tillage in northwest Ethiopia. Afr J Agr Res Econ 11:211–225

Thierfelder C, Baudron F, Setimela P, Nyagumbo I, Mupangwa W, Mhlanga B, Lee N, Gerard B (2018) Complementary practices supporting conservation agriculture in Southern Africa. A Rev Agr Sust Dev 38:16

Tulema B, Aune JB, Johnsen FH, Vanlauwe B (2008) The prospects of reduced tillage in tef (Eragrostis tef Zucca) in Gare Arera, West Shawa Zone of Oromiya, Ethiopia. Soils Tillage Res 99:58–65

Tolessa D, du Preez CC, Ceronio GM (2007) Effect of tillage system and nitrogen fertilization on yield and yield components of maize in Western Ethiopia. South Afr J Plant Soil 24:63–69

Valbuena D, Erenstein O, Tui S H-K, Abdoulaye T, Claessens L, Duncan AJ, Gerard B, Rufino MC, Teufel N, Rooyen A, Wijk MT van (2012) Conservation agriculture in mixed crop-livestock systems: Scoping trade-offs in Sub-Saharan Africa and South Asia. Field Crops Res 132:175–184

Zerihun A, Tadesse B, Shiferaw T, Kiftle D (2014) Conservation Agriculture: Maize-legume intensification for yield, profitability and soil fertility improvement in maize belt areas of Western Ethiopia. Inter J Plant Soil Sci 3:969–985

Land Use Changes and Sustainable Land Management Practices for Soil Carbon Sequestration in Sub-Saharan African Agro-ecosystems

Kennedy Were, Bal Ram Singh, and George Ayaga

Abstract Land use plays an important role in the global system. It alters in response to changing human needs driven by an array of socio-economic, technological, political, and environmental factors. However, the rate of land use change in sub-Saharan Africa (SSA) remains alarming due to a population that is rapidly growing at the rate of 2.6% annually and technological advances leading to the transformation of the land surface. For example, between 2010 and 2015, the rate of forest decline was about 2.8 million ha year^{-1}. The prominent mode of transformation has been the conversion of fragile native ecosystems (i.e., forests, woodlands, savannahs, grasslands, and steppe) into agro-ecosystems in order to fulfill the escalating demand for food, fiber, fuel, and shelter. This has had ramifications on the principal carbon (C) pools, especially the soil organic C (SOC) pool. Studies have shown that a decline in SOC stocks following deforestation in SSA can at times exceed 50%. However, the exact magnitude of SOC depletion in SSA is still uncertain owing to scarcity of reliable long-term data needed for accounting. Replenishing SOC stocks in SSA agro-ecosystems calls for adoption of a combination of appropriate sustainable land management (SLM) practices (e.g., conservation agriculture (CA) and integrated nutrient management), which can enhance the capture and storage of C in plants and soils, as well as mitigate GHG emissions and climate change. For example, soil C sequestration under CA and fertilizer use in Africa has, on average, been 0.37 Mg C ha^{-1} year^{-1}and 0.63 Mg C ha^{-1} year^{-1}, respectively. In this chapter, we

K. Were (✉)
Kenya Agricultural and Livestock Research Organisation, Food Crops Research Centre - Kabete, Nairobi, Kenya

B. R. Singh
Faculty of Environmental Sciences and Natural Resource Management, Norwegian University of Life Sciences, Ås, Norway

G. Ayaga
Kenya Agricultural and Livestock Research Organization, Food Crops Research Centre - Alupe, Busia, Kenya

B. R. Singh et al. (eds.), *Climate Impacts on Agricultural and Natural Resource Sustainability in Africa*, https://doi.org/10.1007/978-3-030-37537-9_3

have provided a synthesis of the impact of land use changes (i.e., conversion of forests to croplands) on the SOC pool, as well as the SLM strategies with a potential for sequestering C in SSA agricultural soils.

Abbreviations and Acronyms

Al^{3+}	Aluminum
ASALs	Arid and semi-arid lands
BF	Bamboo forest
C	Carbon
CA	Conservation agriculture
Ca^{2+}	Calcium
CH_4	Methane
CO_2	Carbon dioxide
DOM	Dissolved organic matter
DRC	Democratic Republic of Congo
FAO	Food and Agriculture Organization of the United Nations
Fe^{2+}	Iron
GHG	Greenhouse gases
INM	Integrated nutrient management
ITPS	Intergovernmental Technical Panel on Soils
K	Potassium
Kg	Kilogram
Mg	Megagram
Mg^{2+}	Magnesium
MRTs	Mean residence times
N	Nitrogen
N_2O	Nitrous oxide
NF	Natural forest
P	Phosphorus
PF	Plantation forest
Pg	Petagram
POM	Particulate organic matter
SIC	Soil inorganic carbon
SLM	Sustainable land management
SOC	Soil organic carbon
SOM	Soil organic matter
SSA	Sub-Saharan Africa
UN	United Nations
UNFCCC	United Nations Framework Convention on Climate Change
USDA	United States Department of Agriculture

1 Introduction

Land use is an important determinant of the physical and human environment. It refers to the functional role of the biophysical land attributes (i.e., land cover) in the realization of different human purposes or economic activities (Bibby 2009). Land use is dynamic and alters in response to changing human needs driven by an array of socio-economic, technological, political, and environmental factors. However, the rate of land use change in sub-Saharan Africa (SSA) – the part of Africa below the Sahara Desert (Fig. 1) – has been unprecedented due to the rapidly growing population and technological advances. SSA is grappling with a high population growth rate (i.e., 2.6% per year) and the attendant pressure on services provided by the principal ecosystems. The official United Nations (UN) population estimates and projections show that Africa's population reached 1.3 billion as of mid-2017 and is projected to be 1.7 billion in 2030, 2.5 billion in 2050, and 4.5 billion in 2100 (UN 2017). The predominant type of land use change has been the conversion of fragile natural ecosystems (i.e., forests, woodlands, savannahs, grasslands, and steppe) into agro-ecosystems in order to meet the escalating demands for food, fiber, and fuel (Were et al. 2016; Bombelli et al. 2009).

According to the literature, the world's forests shrunk by 16.1 million ha year^{-1} on average in the 1990s with the greatest losses occurring in the tropics where about 15.2 million ha were lost annually (Rahman 2004). At this time, Africa had about 680 million ha of natural forests (i.e., about 30% of the world's tropical forests), which reduced to about 600 million ha in 2015 owing to land use changes, biomass burning, over-harvesting, and shifting cultivation (FAO 2016). This represents a net loss of about 3.2 million ha of forest each year between 1990 and 2015. These land use dynamics have had significant effects on the major carbon (C) pools, especially the soil organic C (SOC) pool. For example, Ciais et al. (2011) estimated the net release of C from forest degradation and deforestation in SSA at 0.24 Pg C year^{-1}

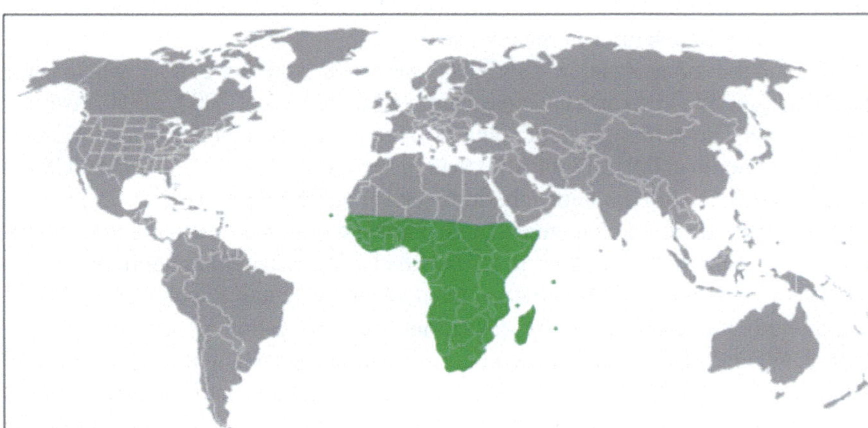

Fig. 1 Geographical location of sub-Saharan Africa (in green). (Source: USDA)

(1 Pg is equal to 10^{15} g), which is equivalent to 20% of global land use CO_2 emissions to the atmosphere. Vågen et al. (2005) also noted from their review of soil C sequestration literature that SOC stocks can diminish by 0.90 Mg C ha^{-1} $year^{-1}$ (1 Mg is equal to 10^6 g) following the clearance and cultivation of forest areas in SSA. These figures highlight the critical role of land use change and forestry in the SSA and global C balance; however, this role is still loaded with large uncertainties attributable to scarcity of studies and reliable long-term data needed for complete accounting.

The objective of this chapter is to provide a synthesis of the impact of land use changes (i.e., conversion of forests to croplands) on the SOC pool, as well as the sustainable land management (SLM) strategies with potential for sequestering C in SSA agricultural soils. The goal is to increase awareness and inform agricultural development programs that aim at mitigating climate change, advancing food and nutritional security, and sustaining life in SSA. We precede the synthesis with an overview of SOC (i.e., its components, dynamics, and functions) for better understanding of the subject matter.

2 An Overview of SOC

2.1 Components of SOC

Walcott et al. (2009) defined soil as the thin mantle comprising organic materials, inorganic materials, and living organisms that cover the Earth's surface to a depth of about two (2) meters. The constituent organic materials are derived from leaf litter, branches, plant roots, soil organisms, and manure, which together form SOC. The plants, animals, and microbes constitute the *living organic matter*, which upon death and decomposition become the *non-living organic matter*. The latter is measured either as particulate organic matter (POM), dissolved organic matter (DOM), humus, or inert (recalcitrant) organic matter.

SOC and the soil inorganic C (SIC) form the soil C (pedologic) pool, which is the largest C reservoir in terrestrial ecosystems (FAO 2004; Lal 2002, 2004; Post and Kwon 2000). The pedologic pool contains three (3) times more C (i.e., 1550 Pg of SOC to 1 m depth and 950 Pg of SIC) than in the biotic C pool (i.e., 560 Pg C) and twice as much C as in the atmospheric C pool (i.e., 760 Pg C) (Lal 2008). Unlike SOC, SIC is made up of elemental C and carbonate minerals, such as calcite, dolomite, and gypsum, derived either from the weathering of parent material or the dissolution of atmospheric CO_2 into carbonic acid and its reaction with Ca^{2+} and Mg^{2+} brought into the soil system by calcareous dust, irrigation water, fertilizer, or manure. Whereas SIC is an important constituent of the soils in arid and semi-arid lands (ASALs), SOC is high in the soils of temperate regions and extremely high in the organic (peat) soils. SOC also varies widely among eco-regions, being higher in the cool and moist areas than in the warm and dry ones.

2.2 Dynamics of SOC

The magnitude of organic C in the soil system is a dynamic balance between the rate of C gains (i.e., inputs of dead plant, animal, and microbial residues) and the rate of C losses from decomposition, mineralization, leaching, and erosion processes (FAO 2001, 2004; Walcott et al. 2009). Under aerobic conditions, most of the C entering the soil is released to the atmosphere through heterotrophic respiration. According to FAO (2001, 2004) reports, only about 1% of C entering the soil (55 Pg C $year^{-1}$) accumulates into more stable fractions (i.e., 0.4 Pg C $year^{-1}$) with long mean residence times (MRTs). The turnover rate of the different SOC components also depends on their composition and complex interactions between the biological, chemical, and physical processes in the soil (Post and Kwon 2000; Walcott et al. 2009). Some components, such as lignin and charcoal, are difficult to digest biochemically, while others, such as carbohydrates and proteins, break down rapidly. Therefore, there is a continuum of soil organic compounds in terms of their decomposability and turnover times (MRTs). Walcott et al. (2009) presented a simple approach that separates SOC pools into fractions depending on how fast the components are broken down and replaced (i.e., recalcitrance). These SOC pools are:

(a) *Fast, labile, or active SOC pool*, which has a short turnover time (i.e., days to years), with fast decomposition
(b) *Slow, stable, or humus SOC pool*, which has a longer turnover time (i.e., years to decades and centuries), with slower decomposition
(c) *Passive, refractory, or recalcitrant SOC pool*, which has a much longer turnover time (e.g., centuries to millennia)

The amount of SOC in the different pools is controlled by complex interaction of climatic, pedologic, agronomic, and biotic factors. Soil type, depth, and texture are crucial in terms of soil properties. It has been established that the content of SOC is generally greater at the surface and diminishes exponentially with depth (Walcott et al. 2009). This is because organic materials that input to forest and agricultural soils (i.e., litter fall, exudates, leachates, dead roots, crop residues, manures, and fertilizers) mostly reside in the upper layers, with only small amounts penetrating much deeper. However, in some Vertisols where the shrink-swell nature of the soils encourages downward movement of organic matter, high SOC levels can be found at depths greater than 50 cm. Moreover, soil texture plays a role in the stabilization of organic compounds; hence, textural variations can have significant effects on SOC content in the different pools. In fine-textured soils, about 30% of SOC tends to be found in the passive pool (in the form of charcoal and physically protected C), while in the coarse-textured soils it is only about 4%. It has also been reported that, in some instances, decomposition of humus in the slow C pool is slower in clays and silts than in coarse sandy soils and that the presence of Fe^{2+}, Al^{3+}, and Ca^{2+} in clays can help to protect soil humus from further decomposition (Walcott et al. 2009).

Climate is another important determinant of SOC content in the different pools. Biological processes, such as the amount of organic matter inputs, as well as the transfer and transformation of SOC are affected by soil temperature, oxygen, and soil moisture (Walcott et al. 2009; Post and Kwon 2000; FAO 2001). Higher temperatures coupled with adequate water supply result in faster decomposition of soil organic matter (SOM), less storage of C in the slow and passive pools, and greater loss of C through respiration. This provides the rationale behind the occurrence of thick surface accumulation of light fraction organic C in boreal and tundra ecosystems where temperatures are low. Since rainfall stimulates plant growth, soils in the humid regions contain more SOC than soils in the dry regions. Microbial activity is also enhanced in wet soils under aerobic conditions, which causes more breakdown of organic matter than in dry soils. However, in continually saturated soils, decomposition rates are reduced and highly organic soils, such as peats, develop.

Lastly, management practices and microbial activities also influence the amount, decomposability, and placement of organic matter inputs in SOC pools. For instance, cultivation impacts on SOC by causing soil disturbance leading to the release of C into the atmosphere. However, this depends on the initial SOC content, the intensity of cultivation, and the level and type of plant residue inputs. For example, conservation tillage loses less C, and adding plant residues with higher C:N and lower N:lignin ratios reduces decomposition rates and increases SOC. Similarly, earthworms, ants, and termites increase the amount of stable organic C in some soils and enhance the decomposition of plant residues. Overall, altering land management practices can create a system where SOC levels either become unstable or stable in the long term.

2.3 *Functions of SOC*

Ecosystem functions and services delivered by SOC are manifold and integral to securing food and fiber production. The main functions of SOC, which have found their lucid expression in the work of Lal (2004), Bationo et al. (2007), and FAO (2001), are as follows:

(a) Maintaining soil quality for agricultural and environmental purposes (e.g., sustaining agronomic productivity, C sequestration, and creating climate-smart soils and agro-ecosystems).
(b) Improving the dynamics and bio-availability of main plant nutrient elements (e.g., N, P, and K) and enhancing cation-exchange capacity.
(c) Supporting biological activity in the soil (i.e., the amount, diversity, and activity of soil biota).
(d) Determining the physical, chemical, and biological soil properties. For instance, aggregation and stability of the soil structure increases with SOM content. This in turn improves the infiltration capacity, plant available water capacity, resistance of soils against erosion, and soil tilth.
(e) Enhancing efficiency in fertilizer and water use due to reduction in losses by drainage, evaporation, and volatilization.

(f) Regulating various processes underlying the supply of nutrients, hence creating a favorable environment for plant growth.
(g) Regulating various processes governing the creation of soil-based ecosystem services.
(h) Buffering against sudden fluctuations in soil reaction (pH) due to the application of agricultural chemicals.
(i) Moderating soil temperature through its effect on soil color and albedo.
(j) Reducing sediment load in streams and rivers.
(k) Filtering pollutants of agricultural chemicals, as well as complexing and immobilizing metals.
(l) Buffering the emissions of greenhouse gases (GHGs) from soil to the atmosphere.

It is obvious from the above functions that SOC is a key indicator of ecosystem health and productivity; therefore, conserving, restoring, and enhancing SOC concentration to above the threshold level of 1.5–2% is a critical determinant of food and nutritional security worldwide.

2.4 *Soil C Sequestration*

When land is converted from native ecosystems to agro-ecosystems, the SOC pool gets depleted if it is not managed sustainably. This is because the bulk of the crop biomass is usually removed from the fields after harvest for use as food or fuel. Only a small amount of readily decomposable residues remain on the fields to accumulate SOM. Removal of crop biomass also aggravates the degradation processes (i.e., erosion) initiated by the land use changes. In addition, frequent tillage and other perturbations disintegrate soil aggregates, redistribute crop residues, and alter soil aeration, moisture, and temperature. This accelerates microbial decomposition and the oxidation of SOM to CO_2, which is ultimately emitted to the atmosphere (Were et al. 2015).

Restoration of the SOC pool and concomitant ecosystem services can only be achieved when more SOM is gained than lost through the process of C sequestration. Soil C sequestration per se is the uptake and conversion of atmospheric CO_2 to organic matter through photosynthesis by plants and the subsequent transfer of the organic matter into the soil reservoir for storage in a way that prolongs its mean residence time (Lal et al. 2015) (Fig. 2). SOC sequestration affords double wins because it restores ecosystem services, which stimulate productive and stable agro-ecosystems, and curbs the emission of GHGs, which contribute to climate change.

Lal (2014) argued that an effective soil C sequestration strategy should create a positive ecosystem C budget through (i) increasing the soil application of biomass C, (ii) decreasing losses of SOC by soil erosion, (iii) moderating soil temperature and reducing the rates of mineralization, and (iv) enhancing the MRTs of SOC by increasing soil aggregation and stability (Fig. 3).

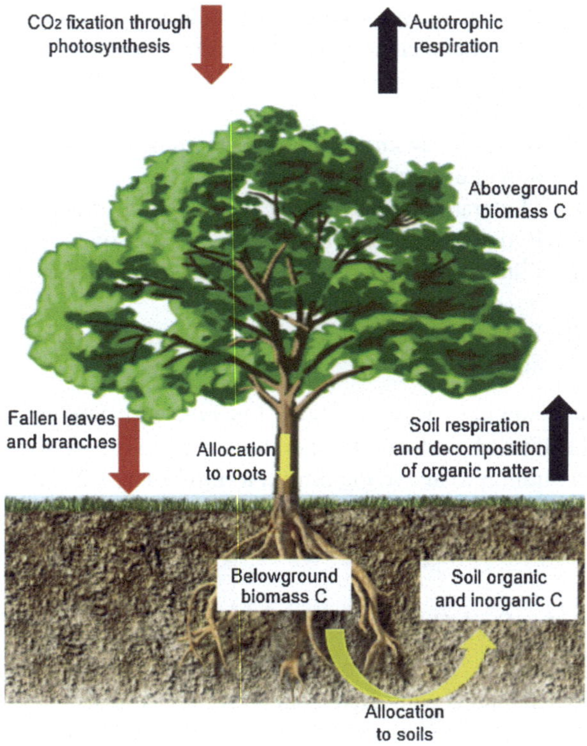

Fig. 2 Losses and gains of C in plants and soils

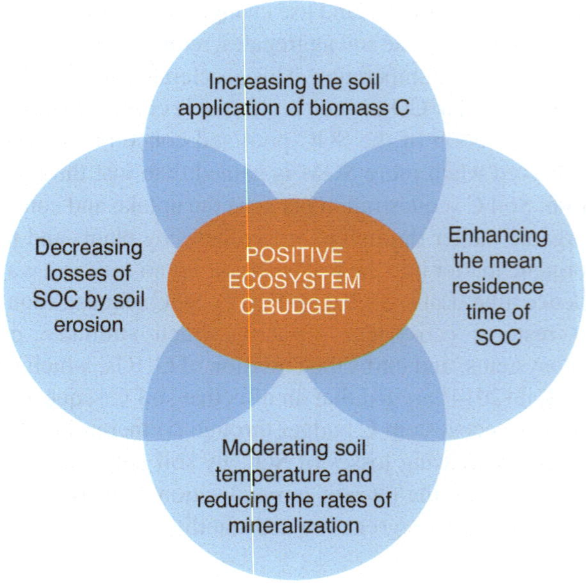

Fig. 3 Strategies for soil C sequestration

3 SOC Stocks in SSA

There is no consensus on the exact estimates of SOC stocks in SSA as a whole. This can mainly be ascribed to a shortage of (reliable) soil information and methodological discrepancies. Nevertheless, the existing estimates can still offer insights into the potential of SSA to mitigate climate change through sequestration of C and reduction of CO_2 emissions. The digital SOC map (Fig. 4) developed by FAO and ITPS (2018) indicates that shrub lands and barren and sparsely vegetated areas in the Sahel region (i.e., Mauritania, Mali, Niger, Chad, Eritrea, and North Sudan) and southern Africa (i.e., Namibia, Botswana, and parts of South Africa) (Fig. 5) contain between 0 and 20 Mg Cha^{-1} in their soils at 30 cm depth, which is the lowest in

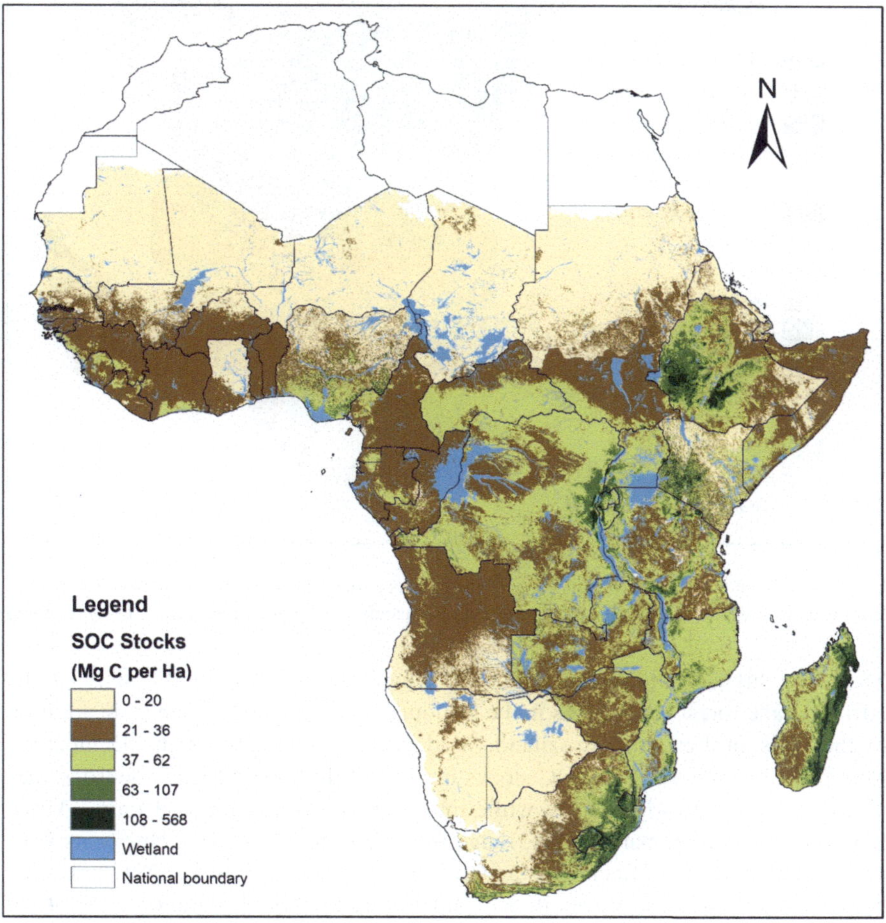

Fig. 4 SOC stocks of SSA's topsoil (0–0.3 m depth) in Mg C ha^{-1}. The map was generated using digital data of SOC stocks from FAO and ITPS (http://54.229.242.119/GSOCmap/). (Data source: Food and Agriculture Organization of the United Nations)

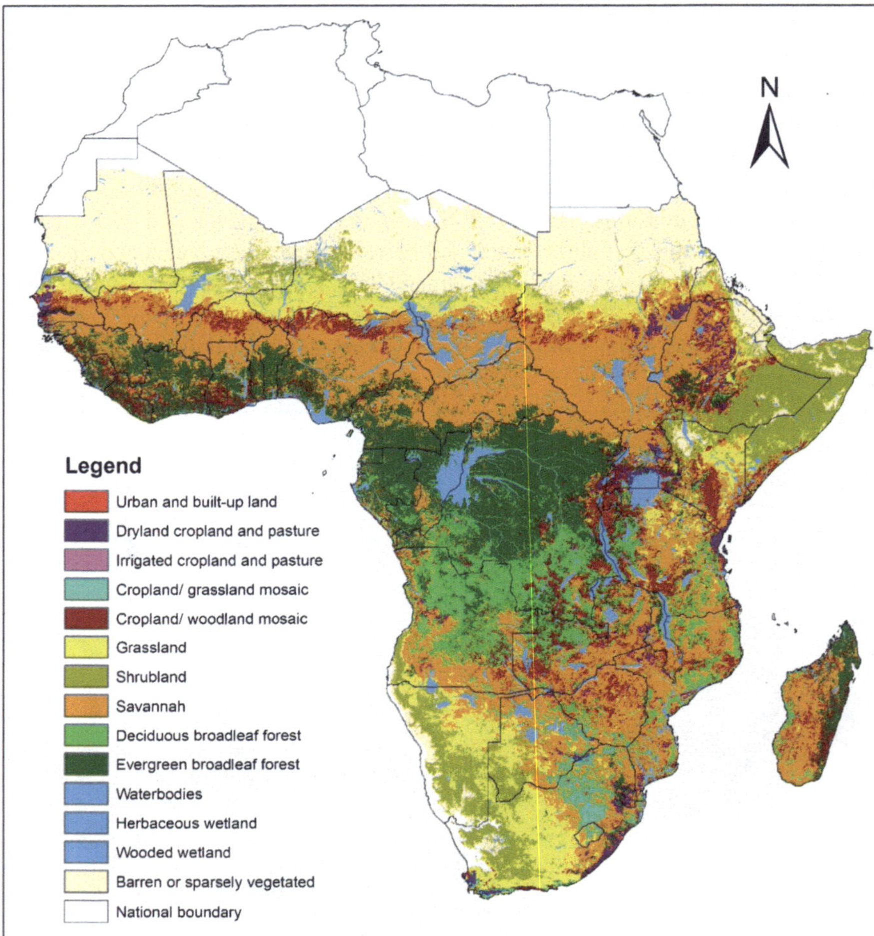

Fig. 5 Land use and land cover in SSA. The map was generated using digital data of land cover from FAO (www.fao.org). (Data source: Food and Agriculture Organization of the United Nations)

SSA. The highest amounts of C (>100 Mg C ha^{-1}) occur in the soils of the Afromontane forest ecosystems in the Ethiopian and Kenyan highlands, in addition to the soils of Lesotho, Swaziland, Madagascar, and parts of the Democratic Republic of Congo (DRC). SOC stocks in most of the west African countries vary between 21 and 36 Mg C ha^{-1}, while the stocks in Mozambique, Central Africa Republic, and a large part of DRC range from 37 to 62 Mg C ha^{-1}. Due to its vastness and abundance of deciduous and evergreen broadleaf forests, DRC holds the highest amount of SOC stocks in SSA amounting to about 9.4 Pg, or 1.4% of the global SOC stock. Though not exact, these estimates depict the fundamental role of SOC pool in controlling CO_2 emissions and annual C budgets at the regional and global scales.

4 Impact of Land Use Changes on SOC Stocks in SSA

The IPCC (2007) attributed the dramatic increases in global surface temperatures and other observed climatic changes to anthropogenic emissions of GHGs through land use changes, fossil fuel combustion, and cement production. POM is, especially, sensitive to land use changes (FAO 2004), which greatly influence its amount, quality, and turnover (Post and Kwon 2000; Birch-Thomsen et al. 2007). The attendant climatic changes are also strongly linked to soil degradation (i.e., loss of SOM and soil quality) because of the feedback between SOM and climatic elements. Climate change aggravates depletion of SOM by altering the patterns of temperature, rainfall, solar radiation, and winds.

A few studies have been conducted in SSA with a view to understanding the anthropogenic effects, particularly land use changes on the storage and fluxes of C in the agro-ecosystems. The focus has mostly been on the impact of converting natural forests to cultivated lands. For example, Demessie et al. (2013) evaluated the changes in SOC stocks and concentrations under the chronosequences of 12, 20, 30, 40, and 50 years after conversion of natural forests to agro-forestry and agricultural lands in southern Ethiopia. The results showed that depletion of SOC stock after conversions from natural forests varied from 28.2 to 98.9 Mg C ha^{-1} under the chronosequences of 12–50 years of agro-forestry and agricultural lands. The rate of SOC loss after 12 years of agro-forestry was 6.2 Mg C ha^{-1} year^{-1}, which declined to 0.9 Mg C ha^{-1} year^{-1} after 50 years. The corresponding losses for agricultural lands were slightly higher (i.e., 6.6 and 1.3 Mg C ha^{-1} year^{-1}). A similar trend was observed by Awiti et al. (2008) along a forest-cropland chronosequence in Kakamega Forest and its environs in Kenya, where the topsoil SOC content declined from 7.27 kg C m^{-2} in the forests to 2.67 kg C m^{-2} in the croplands over a period of 60 years.

In addition, Were et al. (2015) analyzed the variations of SOC stocks under natural forests (NF), plantation forests (PF), bamboo forests (BF), and croplands that had been converted from such forests (i.e., NF2C, PF2C, and BF2C) in a Kenyan Afromontane ecosystem. The results indicated significant differences in SOC stocks between NF and NF2C ($p < 0.0001$) and between PF and PF2C ($p < 0.0001$). Specifically, the surface soils (0–15 cm) of NF had the highest SOC stocks (71.6 Mg ha^{-1}), while NF2C had the lowest (35.4 Mg ha^{-1}) signifying a decline in SOC stocks by about 51% after NF conversion. These results compare with others from cognate studies; for example, Amanuel et al. (2018), Lemma et al. (2006), and Kimigo et al. (2008) also examined SOC variations in relation to land use changes in the upper Blue Nile River basin in Ethiopia, southwestern highlands of Ethiopia, and Sasumua catchment in Kenya and found that SOC stocks reduced by 36%, 43%, and 8%, respectively, after NF conversion. All these results confirm the observation by Vågen et al. (2005) that SOC contents decrease by 063% following deforestation in SSA.

Besides natural forest conversions, a few other studies have also attempted to shed light on the impact of converting natural savannahs to cultivated lands in

SSA. Birch-Thomsen et al. (2007) examined the trends in the spatial distribution of SOC stocks following the conversion of natural and semi-natural savannahs in the semi-arid parts of Tanzania over a period of 50 years. Predictably, the results showed that soils that had been cultivated for 50 years had on average less than 50% SOC compared to the uncultivated ones. The documented rate of SOC losses from such conversions is, on average, 2.77 and 0.82 Mg C ha^{-1} $year^{-1}$ in the East Sudanian savannah and southern Africa regions, respectively (Vågen et al. 2005). Detwiler (1986) attributed the shrinkage of the SOC pool after conversions of natural forests and savannahs to the changes in quantity and quality of organic material inputs to the soil, soil erosion, leaching, and soil disturbances.

In contrast, Majaliwa et al. (2010) reported an increase in SOC after conversion of forests to tea (*Camellia sinensis*) plantations around Kibale National Park in Uganda. The observed differences in SOM were ascribed to the high level of organic material input (i.e., residues from regular pruning and leaf fall) in tea plantations than in natural forests coupled with better soil management practices, such as application of inorganic fertilizer, which stimulates decomposition and mineralization of plant residues compared to natural conditions. In other words, the rate of SOC loss after conversion depends on multiple factors, including the method of clearing, terrain, climate, soil types, and land use (type and duration) after clearing (Vågen et al. 2005; Girmay et al. 2008).

5 Sustainable Land Management Options for Soil C Sequestration in SSA

From Sect. 4, it is evident that the conversion of native ecosystems to agro-ecosystems in SSA depletes SOC, which in turn influences the concentrations of C in the atmosphere. However, the SSA agricultural landscapes can be sustainably managed with a view to mitigating CO_2 emissions, enhancing C sinks by removing CO_2 from the atmosphere for later storage as SOM, and creating resilient systems among other socio-economic and environmental benefits. There are several SLM technological solutions for conserving, replenishing, and enhancing C in the SSA agricultural soils (Fig. 6).

To begin with, a fundamental SLM practice for soil C sequestration in SSA is conservation agriculture (CA), which aims at accumulating SOC and creating a healthy soil ecosystem by not tilling the soil prior to planting. That is, crops are established on the residues left on the ground after harvesting. By minimizing soil disturbance, retaining crop residues as surface mulch, managing nutrients, and diversifying the cropping system through intercropping and incorporating cover crops in the rotation cycle (Aune and Coulibaly 2015; Lal 2015; FAO 2013), CA increases SOC, which in turn improves soil tilth, fertility, biological activity, and infiltration capacity while curtailing soil erosion, soil compaction, and the release of C to the atmosphere. CA also ameliorates soil moisture capacity, hence increasing

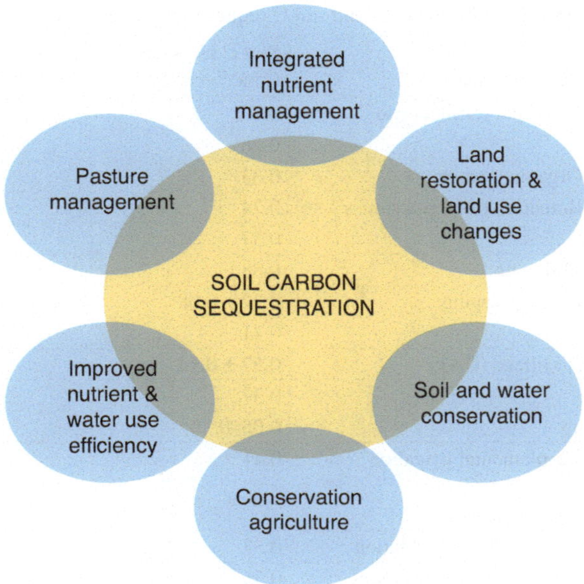

Fig. 6 Summary of the SLM options for soil C sequestration in SSA

water use efficiency. Other co-benefits of CA include minimal fossil fuel consumption and labor requirements by avoiding plowing and increased flexibility in planting and harvesting. A meta-analysis of soil C sequestration rates by the World Bank (2012) revealed that soils under CA in Africa sequestered about 0.37 Mg C ha^{-1} $year^{-1}$on average (Table 1), while other authors estimated sequestration in the range of 0.05–0.36 Mg C ha^{-1} $year^{-1}$ (Vågen et al. 2005) and 0.57 ± 0.14 Mg C ha^{-1} $year^{-1}$ (West and Post 2002). Variations in SOC sequestration rates under reduced tillage can be ascribed to site characteristics, such as soil type, climate, crops grown, previous intensity of tillage, and degree of degradation.

Besides CA, integrated nutrient management (INM) is also a crucial practice for soil C sequestration in SSA. INM is not characterized by unique field practices, but is rather an approach that combines the best of organic and conventional technologies in a way that is environmentally appropriate, productive and sustainable (Agriculture for Impact 2014). That is, it maximizes the use of organic resources (e.g., manure and compost) and improved germplasm, minimizes nutrient losses, allows timely and judicious use of inorganic fertilizers based on need and economic viability, and maintains and enhances beneficial soil organisms and biological processes (Oladele and Braimoh 2014; Sanginga and Woomer 2009). A practical example of INM practice includes intercropping and rotating improved maize with a legume (e.g., beans), and applying both organic and inorganic fertilizers in conjunction with water harvesting, or soil conservation technologies, such as tied ridges and retention of crop residues (Zingore et al. 2015; Lal 2009). By boosting the soil's physical, chemical, and biological characteristics (e.g., enhanced supply of N and other essential nutrients in the soil, reduced soil water and nutrient losses, as well as improved soil porosity and water infiltration), INM strategies stimulate biomass

Table 1 SLM practices and soil C sequestration rates in Africa

SLM practice	Soil C sequestration rate (Mg C ha^{-1} $year^{-1}$)	Source
Inorganic fertilizer use	0.63	Oladele and Braimoh (2014)
	0.26	The World Bank (2012)
Organic manure use	0.33	The World Bank (2012)
Retention of crop residues	0.24	Raji and Ogunwole (2006)
	0.37	The World Bank (2012)
Mulching	0.38	The World Bank (2012)
Cover cropping	0.32 ± 0.08	Poeplau and Don (2015)
	0.41	The World Bank (2012)
No tillage (CA)	0.57 ± 0.14	West and Post (2002)
	0.37	The World Bank (2012)
	0.05–0.36	Vågen et al. (2005)
Supplemental irrigation	0.41	Gebeyehu and Soromessa (2018)
Crop rotation:		
Diversify crop rotation	0.38	The World Bank (2012)
Intensify crop rotation	0.34	The World Bank (2012)
Terracing	0.42	The World Bank (2012)
Rainwater harvesting	0.84	The World Bank (2012)
Cross-slope barriers	1.19	The World Bank (2012)
Agro-forestry:		
Alley cropping	1.46	The World Bank (2012)
Improved fallow	2.41	The World Bank (2012)
	0.07–1.37	Vågen et al. (2005)
Cacao and salmwood	0.06	Lorenz and Lal (2014)
Parklands, live fences, and home gardens	0.20–0.80	Luedeling et al. (2011); Gelaw et al. (2014)
Rotational woodlots	2.20–5.80	Luedeling et al. (2011)
Improved grazing and pasture management	0.47	Conant et al. (2017)
	0.80	The World Bank (2012)

production, which determines the amount of C stored in the plant biomass and the amount returned to the soil. The World Bank (2012) reported that the average effect size of fertilizer use on soil C was an additional 0.26 Mg C ha^{-1} $year^{-1}$, while manure application yielded 0.06 Mg C ha^{-1} $year^{-1}$ more in Africa. Oladele and Braimoh (2014) found a slightly higher C sequestration rate of 0.63 Mg C ha^{-1} $year^{-1}$ after reviewing different fertilizer experiments in Africa.

In addition, general improvements in nutrient and water use efficiency through technologies, such as drip irrigation, micro-dosing of mineral fertilizer, and precision farming, can be effective in providing sinks of C in the soil and reducing GHG emissions in SSA. These technologies ensure that the right amounts of water and fertilizer are placed close to the growing plant at the right time and that less drops of water are used per crop, hence sustaining net primary (biomass) production. This can boost the amount of organic materials returned to the soil, especially in the ASALs, and build up SOC pools both in the short and long term. For instance, a study by Gebeyehu and Soromessa (2018) in northwest Ethiopia showed that the mean SOC stock was higher by 2.85 Mg C ha^{-1} (3.44%) in the surface soils (0–30 cm depth) of irrigated compared to rain-fed farming systems. The irrigated farming system sequestered C at the rate of 0.41 Mg C ha^{-1} $year^{-1}$.

Grazing and pasture management is another SLM practice that can contribute significantly to soil C sequestration in the SSA agricultural systems. By introducing improved grass species with higher productivity and C allocation to deeper roots; reducing the frequency and extent of fires through effective fire suppression means; burning at a time of the year when less amounts of CO_2, CH_4, and N_2O are emitted; reducing the fuel load through appropriate vegetation management; restoring the degraded grazing lands (through exclosures); improving soil *fauna* (e.g., earthworms and termites), irrigation, fertilization, and legume integration; managing the stocking rates; and adjusting the timing of grazing, optimal soil C sequestration and re-carbonization can be realized (UNFCCC 2008; Smith et al. 2007). The World Bank (2012) indicated that attainable rate of soil C sequestration through improved management was about 0.8 Mg C ha^{-1} $year^{-1}$ on average in Africa. Conant et al. (2017) also synthesized data from various studies and confirmed that C sequestration rates were positive for most improved grazing and pasture management practices, with an average of 0.47 Mg C ha^{-1} $year^{-1}$ across all studies. Derner and Schuman (2007), Rimhanen et al. (2016), and Gebregergs et al. (2019) are among the other authors who have reported similar results.

Furthermore, restorative land uses that favor conservation and improvement of soil, water, and air quality have potential for soil C sequestration in SSA agro-ecosystems. For instance, introducing agro-forestry and reverting marginal croplands to native systems can culminate in an accrual of SOC owing to higher aboveground biomass C, reduced soil disturbance, and minimal removal of residue and harvested products. The C sequestration rates in agro-forestry soils can significantly differ among agro-forestry systems (e.g., type, age, tree species, and management), climates, and soils (e.g., type, depth, and severity of degradation). Vågen et al. (2005) reported high potentials for C sequestration after establishment of improved (tree) fallow systems with the attainable rates ranging from 0.07 to 1.37 Mg C ha^{-1} $year^{-1}$. Lorenz and Lal (2014) also established that about 0.06 Mg C ha^{-1} $year^{-1}$ was sequestered in the topsoil (0–15 cm) of a 25-year-old agro-forestry system consisting of cacao (*Theobroma cacao*) and salmwood (*Cordia alliodora*) in Ghana. Moreover, Luedeling et al. (2011) noted that agro-forestry systems, such as parklands, live fences, and home gardens, in Africa accumulated C at the rate of 0.2–0.8 Mg C ha^{-1} $year^{-1}$, while rotational woodlots sequestered at the rate of 2.2–5.8 Mg C ha^{-1} $year^{-1}$. Apart from agro-forestry, restoration of drained wetlands (or peat lands) and degraded soils can also enhance soil C sequestration and mitigate CO_2 emissions. Wetlands are characterized by peaty soils, which accumulate C because of the slow rate of decomposition under anaerobic conditions even though such conditions can also instigate CH_4 emissions if not properly managed. For the degraded soils, re-carbonization can be achieved through practices such as retention of crop residues, judicious use of nutrient amendments and organic substrates (e.g. manuresand composts), minimal tillage, re-vegetation, and soil and water conservation measures (e.g., terracing, mulching, and contour plowing). For example, Raji and Ogunwole (2006) observed sequestration at the rate of 0.24 Mg C ha^{-1} $year^{-1}$ in the 0–15 cm depth of soils after incorporation of crop residues in a study conducted in Nigeria.

Finally, inclusion of green manure and leguminous cover crops (e.g., cow peas, groundnuts, and velvet beans) in rotation cycle is also a promising SLM option for soil C sequestration in SSA agricultural landscapes. Poeplau and Don (2015) conducted a meta-analysis to assess C sequestration in agricultural soils through cover cropping and found that the time since introduction of cover crops in crop rotations was linearly correlated with changes in SOC stocks with an annual change rate of 0.32 ± 0.08 Mg C ha^{-1} $year^{-1}$ in a mean soil depth of 22 cm and during the observed period of up to 54 years. Legume-based cropping systems reduce C and N losses from the soil by improving soil aggregate stability and reducing erosion, improve soil quality by increasing SOC through their biomass, and enhance the quality of residue input and soil biodiversity (Lal 2004). Agro-ecosystems with high biodiversity are also known to absorb and sequester more C.

6 Concluding Remarks

To conclude, soils store substantial amounts of ecosystem C; hence, even slight changes in SOC pool through land use changes can impact significantly on the C cycle, climate, and soil quality. In SSA, the conversion of forests to cultivated lands owing to population expansion has resulted in a reduction of the SOC stocks. Restoring, conserving, and enhancing SOC stocks in SSA agro-ecosystems calls for the adoption of a combination of appropriate SLM practices (e.g., CA), which can capture and store C in plants and soils, as well as mitigate GHG emissions and climate change. SLM practices for soil C sequestration are well-documented and ample information is available. This chapter has highlighted a number of such practices with potential to replenish SOC stocks in SSA; however, there is no one-size-fits-all or a magic bullet practice considering the diversity and heterogeneity of SSA environments. Agricultural development and C sequestration projects in the region should objectively select and scale out the best-fit SLM practices for specific contexts taking into account not only their potential for C sequestration (climate benefit) but also their technical feasibility, social acceptability, economic viability, environmental benefits, and biodiversity benefits. Climate benefits of the selected SLM technologies should be gauged both in terms of C sequestration rates and N_2O and CH_4 emissions associated with the technologies (i.e., abatement rates). The agricultural development and C sequestration projects can also gain immensely from policy frameworks, which foster multi-sectoral and multi-stakeholder partnerships, including the government, private sector, non-governmental organizations, and farmers, for the effective, systematic, and coordinated scaling of SLM practices.

References

Agriculture for Impact (2014) No ordinary matter: conserving, restoring and enhancing Africa's soils. A Montpellier Panel report

Amanuel W, Yimer F, Karltun E (2018) Soil organic carbon variation in relation to land use changes: the case of Birr watershed, upper Blue Nile River Basin, Ethiopia. J Ecol Environ 42:16

Aune JB, Coulibaly A (2015) Micro-dosing of mineral fertilizer and conservation of agriculture in sustainable agricultural intensification in sub-Saharan Africa. In: Lal R, Singh BR, Mwaseba DL, Kraybill D, Hansen DO, Eik LO (eds) Sustainable intensification to advance food security and enhance climate resilience in Africa. Springer, Berlin, pp 223–234

Awiti AO, Walsh MG, Kinyamario J (2008) Dynamics of topsoil carbon and nitrogen along a tropical forest–cropland chronosequence: evidence from stable isotope analysis and spectroscopy. Agric Ecosyst Environ 127:265–272

Bationo A, Kihara J, Vanlauwe B, Waswa B, Kimetu J (2007) Soil organic carbon dynamics, functions and management in west African agro-ecosystems. Agric Syst 94(1):13–25

Bibby P (2009) Land use change in Britain. Land Use Policy 26S:S2–S13

Birch-Thomsen T, Elberling B, Fog B, Magid J (2007) Temporal and spatial trends in soil organic carbon stocks following maize cultivation in semi-arid Tanzania, East Africa. Nutr Cycl Agroecosyst 79:291–302

Bombelli A, Henry M, Castaldi S, Adu-Bredu S, Arneth A, de Grandcourt A, Grieco E, Kutsch WL, Lehsten V, Rasile A, Reichstein M, Tansey K, Weber U, Valentini R (2009) An outlook on the Sub-Saharan Africa carbon balance. Biogeosciences 6:2193–2205

Ciais P, Bombelli A, Williams M, Piao SL, Chave J, Ryan CM, Henry M, Brender P, Valentini R (2011) The carbon balance of Africa: synthesis of recent research studies. Phil Trans R Soc A 369:2038–2057

Conant RT, Cerri CEP, Osborne BB, Paustian K (2017) Grassland management impacts on soil carbon stocks: a new synthesis. Ecol Appl 27(2):662–668

Derner JD, Schuman GE (2007) Carbon sequestration and rangelands: A synthesis of land management and precipitation effects Journal of Soil and Water Conservation, 62(2):77–85

Demessie A, Singh BR, Lal R (2013) Soil carbon and nitrogen stocks under chronosequence of farm and traditional agroforestry land uses in Gambo District, Southern Ethiopia. Nutr Cycl Agroecosyst 95:365–375

Detwiler RP (1986) Land use change and the global carbon cycle: the role of tropical soils. Biogeochemistry 2:67–93

FAO (2001) Soil carbon sequestration for improved land management. World Soil Resources Reports 102, Rome

FAO (2004) Carbon sequestration in dry land soils. World Soil Resources Reports 102, Rome

FAO (2013) Climate-smart agriculture sourcebook. Food and Agriculture Organization of the United Nations, Rome. http://www.fao.org/docrep/018/i3325e/i3325e.pdf. Accessed 13 Mar 2016

FAO (2016) Global forest resources assessment 2015. How are the world's forests changing? 2nd edn. Food and Agriculture Organization of the United Nations, Rome, p 54

FAO and ITPS (2018) Global soil organic carbon map. Technical Report. Rome. 162 pp

Gebeyehu G, Soromessa T (2018) Status of soil organic carbon and nitrogen stocks in Koga Watershed Area, Northwest Ethiopia. Agric Food Sec 7:9

Gebregergs T, Tessema ZK, Solomon N, Birhane E (2019) Carbon sequestration and soil restoration potential of grazing lands under exclosure management in a semi-arid environment of Northern Ethiopia. Ecol Evol 9:6468–6479. https://doi.org/10.20944/preprints201803.0207.v1

Gelaw AM, Singh BR, Lal R (2014) Soil organic carbon and total nitrogen stocks under different land uses in a semi-arid watershed in Tigray, Northern Ethiopia. Agric Ecosyst Environ 188:256–263

Girmay G, Singh BR, Mitiku H, Borresen T, Lal R (2008) Carbon stocks in Ethiopian soils in relation to land use and soil management. Land Degrad Dev 19:351–367

IPCC (2007) Summary for Policymakers. In: Solomon S, Qin D, Manning M, Chen Z, Marquis M, Averyt KB, Tignor M, Miller HL (eds) Climate change 2007: The physical science basis. Contribution of Working Group I to the fourth assessment report of the Intergovernmental Panel on Climate Change. Cambridge University Press, Cambridge, NY

Kimigo J, Mbuvi JP, Kironchi G, Gicheru PT (2008) Effects of land use change on soil qualities in the Sasumua Catchment of Tana River drainage system in Nyandarua District, Kenya. East Afr Agric For J 74(2):85–94

Lal R (2002) Soil carbon dynamics in cropland and rangeland. Environ Pollut 116:353–362

Lal R (2004) Soil carbon sequestration to mitigate climate change. Geoderma 123:1–22

Lal R (2008) Carbon sequestration. Phil Trans R Soc B 363:815–830

Lal R (2009) Soil degradation as a reason for inadequate human nutrition. Food Sec 1:45–57

Lal R (2014) Climate strategic soil management. Challenges 5:43–74

Lal R (2015) Restoring soil quality to mitigate soil degradation. Sustainability 7:5875–5895

Lal R, Negassa W, Lorenz K, (2015) Carbon sequestration in soil. Current Opinion in Environmental Sustainability 15:79–86

Lemma B, Kleja DB, Nilsson I, Olsson M (2006) Soil carbon sequestration under different exotic tree species in the southwestern highlands of Ethiopia. Geoderma 136(3–4):886–898

Lorenz K, Lal R (2014) Soil organic carbon sequestration in agroforestry systems. A review. Agron Sustain Dev. Springer Verlag/EDP Sciences/INRA 34(2):443–454

Luedeling E, Sileshi G, Beedy T, Diez J (2011) Carbon sequestration potential of agro-forestry in Africa. In: Kumar BM, Nair PH (eds) Carbon sequestration potential of agro-forestry systems: opportunities and challenges. Springer, Dordrecht, pp 61–84

Majaliwa JGM, Twongyirwe R, Nyenje R, Oluka M, Ongom B, Sirike J, Mfitumukiza D, Azanga E, Natumanya R, Mwerera R, Barasa B (2010) The effect of land cover change on soil properties around Kibale National Park in south-western Uganda. Appl Environ Soil Sci 2010:1

Oladele O, Braimoh A (2014) Potential of agricultural land management activities for increased soil carbon sequestration in Africa – a review. Appl Ecol Environ Res 12(3):741–751

Poeplau C, Don A (2015) Carbon sequestration in agricultural soils via cultivation of cover crops – a meta-analysis. Agric Ecosyst Environ 200:33–41

Post WM, Kwon KC (2000) Soil carbon sequestration and land use change: processes and potentials. Glob Chang Biol 6:317–327

Rahman MM (2004) Estimating carbon pool and carbon release due to tropical deforestation using high-resolution satellite data. PhD dissertation, Dresden University of Technology, Germany, pp 191

Raji BA, Ogunwole JO (2006) Potential of soil carbon sequestration unver various land use in the sub-humid and semi-arid savannah of Nigeria: lessons from long-term experiments. Int J Soil Sci 1(1):33–43

Rimhanen K, Ketoja E, Yli-Halla M, Kahiluoto H (2016) Ethiopian agriculture has greater potential for carbon sequestration than previously estimated. Global Change Biology 22 (11):3739–3749

Sanginga N, Woomer PL (2009) Integrated Soil Fertility Management in Africa: Principles, Practices and Development Process. Tropical Soil Biology and Fertility Institute of the Centre for Tropical Agriculture. pp. 263

Smith P, Martino D, Cai Z, Gwary D, Janzen H, Kumar P, McCarl B, Ogle S, O'Mara F, Rice C, Scholes B, Sirotenko O (2007) Agriculture. In: Metz B, Davidson OR, Bosch PR, Dave R, Meyer LA (eds) Climate change 2007: mitigation. Contribution of Working Group III to the fourth assessment report of the Intergovernmental Panel on Climate Change. Cambridge University Press, Cambridge, NY

The World Bank (2012) Carbon sequestration in agricultural soils. Report No. 67395-GLB

UN, Department of Economic and Social Affairs, Population Division (2017) World population prospects: the 2017 revision, key findings and advance tables. Working paper no. ESA/P/WP/248. https://esa.un.org/unpd/wpp/publications/Files/WPP2017_KeyFindings.pdf. Accessed 27 May 2018

UNFCCC (2008) Challenges and opportunities for mitigation in the agricultural sector. Technical Paper GE-08-64351

Vågen TG, Lal R, Singh BR (2005) Soil carbon sequestration in sub-Saharan Africa: a review. Land Degrad Dev 16:53–71

Walcott J, Bruce S, Sims J (2009) Soil carbon for carbon sequestration and trading: a review of issues for agriculture and forestry. Bureau of Rural Sciences, Department of Agriculture, Fisheries and Forestry, Canberra

Were KO, Singh BR, Dick ØB (2015) Effects of land cover changes on soil organic carbon and nitrogen stocks in the Eastern Mau Forest Reserve, Kenya. In: Lal R, Singh BR, Mwaseba DL, Kraybill D, Hansen DO, Eik LO (eds) Sustainable intensification to advance food security and enhance climate resilience in Africa. Springer, Berlin, pp 113–133

Were KO, Gelaw AM, Singh BR (2016) Smart strategies for enhanced agricultural resilience and food security under a changing climate in sub-Saharan Africa. In: Lal R, Kraybill D, Hansen DO, Singh BR, Mosogoya T, Eik LO (eds) Climate change and multi-dimensional sustainability in African agriculture. Springer, Cham, pp 431–453

West TO, Post WM (2002) SOC sequestration rates by tillage and crop rotation. A global data analysis. Soil Sci Soc Am J 66:1930–1946

Zingore S, Mutegi J, Agesa B, Tamene L, Kihara J (2015) Soil degradation in sub-Saharan Africa and crop production options for soil rehabilitation. Better Crops 99(1):24–26

Gendered Adaptation and Coping Mechanisms to Climate Variability in Eastern Uganda Rice Farming Systems

Thelma Akongo and Charity Chonde

Abstract This chapter addresses the dynamics and realities of gender-differentiated effects of climate variability on men and women in rice-growing systems in Uganda, based on their capacity to adapt and cope. Findings show that climate variability has reduced yields, cultivable area, and cropping sequence of major crops. Sixty-five percent of the respondents perceived that yields for all the major crops reduced while 25% perceived yield increase as a result of climate variability ($p = 0.023 < 0.05$). Production of lowland rice variety Super reduced from 2100 kg/hectare in a normal year to 200 kg/hectare in a drought year ($p = 0.001 < 0.01$) while K5 variety decreased from 2625 kg/hectare to 1750 kg/hectare ($p = 0.006 < 0.01$). Upland rice variety *Kaiso* declined from 2375 kg/hectare to 52.5 kg/hectare ($p = 0.009 < 0.01$). More female respondents reported a decrease in cultivable area for almost all crops, with the exception of cassava, compared to their male counterparts. There were significant differences between the proportions of men and women who perceived decreases in cultivable areas for rice ($p = 0.015 < 0.05$), maize ($p = 0.03 < 0.05$), and groundnuts ($p = 0.009 < 0.05$). The study determines that both men and women are affected by climate variability, becoming poorer with very limited economic, human, and social resources to build resilience to climate change. It further determines that both men and women rely more on coping mechanisms to respond to effects of climate variability, which are more short-term survival strategies compared to long-term adaptation strategies, given the nature of resources at their disposal. The study concludes by proposing appropriate institutional intervention strategies to be integrated into rice commodity development.

T. Akongo (✉)
Environment and Social Safeguards Unit, National Agricultural Research Organisation, Entebbe, Uganda

C. Chonde
Extension Department, Lilongwe University of Agriculture and Natural Resources, Lilongwe, Malawi
e-mail: cchonde@luanar.ac.mw

B. R. Singh et al. (eds.), *Climate Impacts on Agricultural and Natural Resource Sustainability in Africa*, https://doi.org/10.1007/978-3-030-37537-9_4

1 Introduction

1.1 Gender Dynamics, Climate Change, and Food Security

Gender and food insecurity challenges are increasingly emerging as priority concerns, particularly in developing countries, where efforts to improve agricultural production are being undermined by the effects of climate change (FAO 2007). Extreme weather conditions associated with the prevailing climatic conditions have had diverse effects on the poor and women who derive their livelihood from the agricultural sector. The unexpected incidences of floods, submergence, and drought and the increasing incidences of pest and diseases pose threats to food security (Porter et al. 2014). Notably frequent flash floods have left people homeless and displaced and have destroyed crops, public property, and infrastructure. These climatic events are more visible in severe stress-prone areas that entirely depend on rains for agriculture (Nwafor 2007; Jagtap 2007), and the effects are more varied between women and men, with women being more affected. The differential effects are attributed to the existing gender inequalities that are in existence. They manifest in various forms including limited access to resources, gender divisions of labor, limited decision-making ability, and power relations. These inequalities lower women's social and economic status, reinforce their vulnerabilities, and make it difficult for them to effectively respond to or cope with the adverse effects of climate change (Olsson et al. 2014).

Despite the numerous efforts geared toward this cause, however, the belief in women as change agents still remains strong on the basis that they have long been promoters of change. They have been prominent in natural resource management, innovations, farming, and household welfare. Hence, women may not only have valuable knowledge that can enable them identify appropriate adaptation techniques but also knowledge that could help design effective climate change mitigation strategies (UNDP 2010). Given the argument above therefore, development interventions promoting women's engagement are still emphasized not only to promote their status but also enable them benefit from their disposition as change agents.

Currently, however, the differential information on the effect of climate change on women and men is not sufficient to enable policy and decision-makers to address the existing climate change, gender, and food insecurity challenges effectively for the benefit of rural communities. This information gap calls for more studies on the existing challenges and vulnerabilities of women and men in relation to the adverse effects of climate change and their capacity to cope and adapt. In this regard, a study on gender, food security, and climate change variations with respect to rice (*Oryza sativa*) production in Uganda was undertaken.

1.2 Rice Production in Uganda

Although rice production was introduced in Uganda way back in 1904 (Bigirwa et al. 2005), its role in the country's economy only became noticed in the late 1940s as part of the then government's efforts to incorporate rice-based rations in the feeding of soldiers during and after the Second World War. With the establishment of the Kibimba Rice Scheme in 1966 and Doho Rice Scheme in 1976, smallholder rice production, mainly in the eastern and northern parts of the country, was also spontaneously twigged but with emphasis on lowland rice varieties. It was only in the late 1980s that production rapidly increased to the current figure of nearly 95,000 ha. The country's total annual rice production now stands at 140,000 metric tons of milled rice, representing about 70% of the current national rice demand estimated at 190,000–200,000 metric tons. Oryokot et al. (2004) report that by 2004, Uganda's rice imports stood at about 45,000 metric tons. The very rapid increase in rice production in the country is mainly attributed to the release of improved rice varieties (especially the NERICA rice), conducive government policy, increase in demand and consumption of rice particularly among the urban and peri-urban populations, and current higher rate of returns on investment in rice production of 1.8 compared to such cereals as maize with a rate of return on investment of only 1.2 (NAADS 2004).

Although Uganda has tremendous potentials in terms of its good soils, favorable climate, two growing seasons, political support, and farmers' enthusiasm for increasing its rice production to self-sufficiency, the crop is still relatively new in the country's farming systems. Most of the rice in Uganda today is grown in the western and eastern parts of the country, with the eastern part producing the bulk of the rice (NAADS 2004). It has become an important food security and income-generating crop among the rural poor for alleviating poverty. This has further been supported by the rice-milling sub-sector, an important agro-processing development that is contributing largely to the efforts to address unemployment challenges. Efforts to increase rice production in Uganda to cater for the increasing demand are being emphasized.

1.3 Problem Statement and Rationale for the Study

Despite the current efforts to promote rice production in the country, there are concerns that the efforts may be worthless due to various challenges such as increasing population pressure, the diversion of agricultural land for bio-fuels and other human needs, unreliable traditional weather patterns as a result of climate change, and the desire by many rice-growing nations to build up buffer stocks (Ahmad 2012). Besides, the unforeseen changes associated with global warming, high temperatures, and unpredictable rainfall are expected to impact greatly on rice production (Ramírez and Kallarackal 2015). It is also worth noting that many scholars have already stated that climate change will lead to increasing levels of drought if temperatures continue to rise; and this will lead to diseases, starvation, landslides, rising sea levels, and floods, among others.

Amidst these foreseen challenges, there is still a belief that traditional coping and adaptation mechanisms against natural disasters can still be crucial for survival and development of marginalized communities, especially where the existing disaster management and development programs are not accessible or have failed to protect them against recurring floods and increasing poverty.

It is therefore important to analyze and understand how the poor, especially women, have been living with the adverse effects of climate change to date. There is also need to examine whether the traditional coping mechanisms and adaptation strategies are still useful in dealing with the vast effects of climate change in the social, economic, cultural, and political context of the societies they live in. The information generated will go a long way to enable policy makers to make informed decisions on gender, climate change, and food insecurity concerns and will also generate information that will provide researchers and service providers in the various institutions with a basis from which to design other studies and intervention programs.

1.4 The Overall Objective

The overall objective of the study was to examine climate variability risks faced by women and men and the adaptive strategies employed in relation to food insecurity in rice production systems in Uganda.

1.5 Specific Objectives of the Study

(a) To assess the existing gender differences with respect to accessing resources for adaptation and the livelihood strategies in the two rice production systems
(b) To assess the effects of climate variability on men and women in rice production systems
(c) To identify the coping and adaptation strategies used in response to extreme climate variations

1.6 Research Questions Addressed

(a) What effects of climate change or risks do men and women face?
(b) What strategies do men and women use to adapt to climate variations?
(c) Are there gender differences in access to resources that serve as constraints to improving their adaptation strategies?

2 Methodology

This methodology section presents the study area and the rationale for its selection; it goes further to discuss the sampling procedure and how the final respondents for the study were selected. It highlights the data collection process and methods used and the analysis and ethical considerations while conducting the study.

2.1 Rationale for the Selection of the Study Area

The Ministry of Agriculture, Animal Industry and Fisheries (MAAIF) demarcated the country into agricultural zones, each with specific production features. The purpose for the demarcation is for each zone to undertake a set of agricultural enterprises with the best comparative advantage. In this regard, the eastern agro-ecological zone was selected on the basis of rice production. The zone comprises of two rice production systems, namely, the upland rice production system found in Teso and Bukedea region and lowland rice production system found in Doho and Butelega.

2.2 Description of the Study Area

This region was selected for the survey because it is the largest rice production area and also because it is one of the earliest rice-growing regions in the country. Zoning of the area was based on rainfall, soil, vegetation, main crop, and population. Rainfall in this region is higher in the southern part, although drought-prone pockets exist and soils become lighter as one moves further north. The population density is highest in the south, falling toward the north, which borders the pastoral area of Karimojong. The Karimojong are often dependent on the region for food when they migrate in with cattle to search for water and pasture during the dry season. This dependence is partly what makes parts of the region prone to seasonal famine and civil strife. The region lies on a flat plateau with gentle slopes punctuated by isolated rocky granite outcrops. The altitude ranges from 1000 to 1880 meters above sea level. The area is divided by numerous wide grass swamps that fill with water in the rainy season and dry up in the dry season. These swamplands, which drain into Lake Kyoga, represent a valuable grazing area to local and neighboring livestock farmers, particularly in the dry season of November to February.

The area receives a total of 1000–1100 mm of rainfall per annum, with a bimodal distribution. This region is unique in Uganda in its use of draft animals for cultivation, and it is this that enabled it to have the highest cultivable acreage in the country, prior to 1986. As such, it is described as an integrated crop-livestock system in which livestock provided farm power and manure for crop production. In return, crop residues provided an important source of fodder for animals, especially during the dry season.

2.3 Map of the Study Area

Figure 1 shows the position of the study area on the map of Uganda indicated by the red area.

2.4 Sampling Procedure

To conduct the study, districts with support from extension and research personnel, and with the potential of producing up to 408,000 tons of rice per season, were purposively sampled from the two rice production systems. The aim was to obtain information for both rice production systems, in relation to production interventions by the supporting service providers. The districts selected included Kumi, Bukedea, Soroti, Kaberamaido, Iganga, and Mayuge. These districts are located in the shaded area indicated on the map of Uganda in Fig. 1.

Identification of Respondents

To identify the most appropriate respondents for the study, the research team consulted district agricultural personnel in the respective selected districts and sub-counties together with research officers working in the area. The district personnel provided lists of documented rice farming households who had grown rice for more than 10 years. This formed the basis for selection of households in the selected area. Given that the study sought to understand gender-related challenges in the rice farming systems, deliberate emphasis was put on selecting female-headed households as well as male-headed households to make comparisons and identify the prevailing gender differences. From each of the selected districts, at least two sub-counties and seven villages were selected. Selection was based on the ease of accessibility and the presence of organized rice-based farmer groups. In summary, a total of 5 districts, 10 sub-counties, 30 villages, and 60 households, comprising of 21

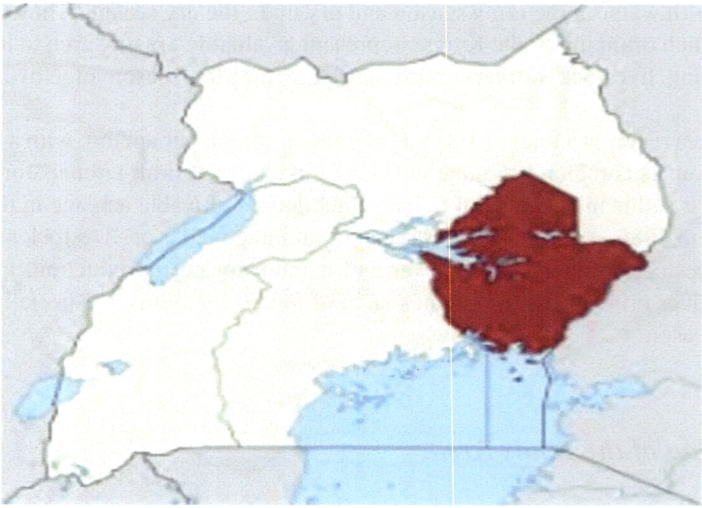

Fig. 1 Map of Uganda with the red area showing the eastern agro-ecological zone

female-headed and 39 male-headed households, were selected. In addition, key informants such as village chiefs, women councilors, elders, local council leaders, agricultural officers, and researchers were also identified for the study.

2.5 Data Collection Methods and Tools

The data collection methods used included household interviews, focus group discussions (FGDs), key informant interviews (KIIs), observations, and review of secondary data from similar studies conducted. A total of 60 household interviews, 15 focus group discussions (i.e., 7 within the upland and 8 within the lowland rice production system), and 21 key informant interviews were conducted. For quantitative data collection, household interviews were conducted using a semi-structured questionnaire with several open-ended questions to capture respondents' insights and perceptions. The questionnaire was reviewed and tailored to the Ugandan perspective since it was used in Asia to collect similar information. The questionnaire was designed in such a way that it contained different sections for both male and female rice growers, in order to make comparisons of their submissions. The questionnaire also sought gender-disaggregated information on access to land, land size, income sources, access to labor for rice production, access to resources, and decision- making power. In addition, information on extreme weather events for a 10-year period, and how the events affected rice and livestock production, was solicited. A considerable amount of time was spent revising, editing, and preparing the questionnaire with the help of several district and research officers in the zone.

To gather more qualitative data, 21 key informants and 15 focused group discussions were conducted. The FGDs were done with the youth, as well as male and female respondents. However, male groups were separated from female groups to avoid dominance of one group over the other and to allow voices from each group to be independently captured. Similar topics were discussed in both male and female groups with female FGDs held in afternoons, to give them time to attend to domestic duties. The issues shared were related to food security in the community, challenges, and coping mechanisms used to avert threats like crop diseases, land pressure, soil infertility, and lack of markets. Seasonal calendars were also used to elicit information on crop and livestock production activities for different seasons and climate variations. In addition, the gender roles in rice production as well as other prominent livelihood activities were also captured. Using the problem tree, respondents were asked to identify the different climate risk factors, their root causes, and adaptation strategies.

2.6 Data Analysis

The study employed both qualitative and quantitative technique of data analysis. Content analysis was used for qualitative data to analyze texts from the transcribed interviews. This involved searching for underlying themes contained in the transcribed data, containing information relating to particular themes identified in the research information. The basic steps involved concept development, coding of data, analysis, and interpretation. However, to aid the process, the QSR NVivo 8 computer software was used to simplify the data analysis. For the quantitative data, the SPSS software data program was used. Descriptive statistics were used to analyze quantitative data.

3 Results and Discussions

This section looks at the research findings based on the specific objectives of the study. Firstly, this section explores the existing gender inequalities in land ownership as well as those related to the social demographic characteristics of the people living in the different rice production systems. Next it provides the community profile and livelihoods strategies employed. Lastly, this section presents findings which facilitate understanding of the existing gender inequalities and livelihood strategies which form the basis for further understanding climate variability, as well as the risks and effects on women and men in the area and how they differently respond and adapt to these effects.

3.1 Existing Gender Differences and Livelihood Strategies Employed Within the Upland and Lowland Rice Farming Systems

Land Ownership Under Upland and Lowland Rice Production Systems

Table 1 shows land ownership status of the various household members under upland and lowland rice production systems.

Table 1 shows that the land area under lowland rice production is higher than that under upland rice production indicated by 93% and 81% of the total land area, respectively. However, 91% and 75% of the land under upland and lowland rice production are owned by male spouses compared to 2% and 1% of the land under upland and lowland rice production which are owned by the female spouses, significant at $p = 0.001 < 0.01$ and $p = 0.004 < 0.01$, respectively. Despite their significant role in the agricultural sector (Ugandan women provide 70–80% of agricultural and food production labor), only about 16% of Ugandan women own land in their own right (Rugadya 2010). Their ownership of registered land is even lower at 7% (Rugadya 2010; Bikaako and Ssenkumba 2003). There are no laws explicitly prohibiting women from owning land, but customary and inheritance laws make it difficult for Ugandan women to own land (Adoko and Levine 2005). Land area co-owned by spouses was more under upland rice production than it was for lowland rice production. The findings indicate that sons and daughters do not own land under the two rice production systems while the elderly household members under the upland rice production system owned 10% of the land.

Table 1 Household land ownership under upland and lowland rice production systems

Land ownership	Upland rice	Lowland/ swamp rice
Total area cultivated (ha)	10.5	15.2
% owned	81	93
% borrowed	3	30
% belonging to husband	91	75
% belonging to wife	2	1
% belonging to husband and wife	11	2
% belonging to sons	0	0
% belonging to daughters	0	1
% belonging to elderly persons	10	0

Socio-economic Characteristics of Respondents

Table 2 indicates that the total number of respondents sampled for the survey was 179, with 105 respondents (58.7%) and 74 respondents (41.3%) from upland and lowland rice production systems, respectively. The proportion of male-headed households was 66.5% while that of the female-headed households was 33.5%. The result reveals that about one-third of the sampled respondents were women which suggests that there were less female farmers involved in rain-fed lowland rice production compared to their male counterparts. This may be attributed to the fact that rice production is a labor-intensive activity, and considering the multiple roles of females, more males are engaged in rice production than females. This finding agrees with what Adekunle (2013) reported that there was low number of female rice farmers involved in upland rice production in sub-Saharan Africa. He attributed their low involvement to socio-economic constraints, including resource endowment, capital, and land. There were no marked differences in the socio-economic characteristics of the respondents for both women and men within the two rice production systems. For instance, the average age for males in upland and lowland rice systems was 41.5 years and 44.6 years, respectively, while that of women was 38.6 years and 40.3 years, respectively. This shows that rice farming is mainly practiced by the middle-aged group. The average years of schooling for respondents in the two rice systems were generally low (i.e., 6.2 and 6.5 for men and 3.4 and 3.0 for women), with women having lower educational levels than men in both farming systems. The low levels of formal education for many rural communities are mostly linked to the vicious cycle of poverty among the rural folks. The findings also clearly indicate that most of the respondents own the land that they cultivate and that renting out and renting in of land under the two production systems is done to a very limited extent.

The household size of the respondents varied from a minimum of 2 to a maximum of 20 members, with an average of 6 members per household. The number of dependents per household was relatively higher with the majority aged 20 and below, while the elderly were above 60 years. The Uganda Bureau of Statistics

Table 2 Socio-economic characteristics of respondents

Characteristic	Upland rice	Lowland rice	Total
Frequency	105(58.7)	74(41.3)	179
Average age (years)			
Principal male	41.5	44.6	119
Principal female	38.6	40.3	60
Average years in school			
Principal male	6.25	6.53	119
Principal female	3.4	3.0	60
Average years in farming			
Principal male	20.67	22.31	119
Principal female	20.1	19.2	60
Household dependency ratio	66.4	70.3	563
Tenure status (%)			
Owned	81	93	174
Rent-in	3	17	20
Rent-out	1	4.5	5.5
Type of ecosystem			
Rain fed	63	56	119
Irrigated	19	25	44
Partially irrigated	26	29	55
Average rice farm size (acres)	3.0	2.5	5.5

(UBOS) projects Uganda's total population in 2017 at 37.6 million, with youth (18–30 years) constituting roughly 22% (8.2 million) of that population. Children (below 18 years of age) constitute 55% of the population and youth numbers continue to bulge. This may explain the high dependency ratios. Therefore, there must be concerted efforts to develop this group of the population, given their numbers and potential to boost growth and development. Results also indicate that the majority of repondents were educated up to the primary level, followed by a few with secondary education. As noted, those with hardly any education were more than those with university and tertiary education put together. Figure 2 illustrates the disparities in educational levels of the respondents.

Community Profile and Livelihoods

Both upland and lowland rice farming systems are dependent on rain-fed agriculture. Crops, such as groundnuts (*Arachis hypogaea*) and cassava (*Manihot esculenta*), are among the main crops grown for home consumption, while maize (*Zea mays*), sweet potatoes (*Ipomoea batatas*), and rice are grown for sale. Rains in the zone come biannually from March to June and from August to December. In addition to crop production, households raise livestock such as cattle (*Bos taurus*), goats (*Capra aegagrus hircus*), and chickens (*Gallus gallus domesticus*). Currently, the numbers of livestock,

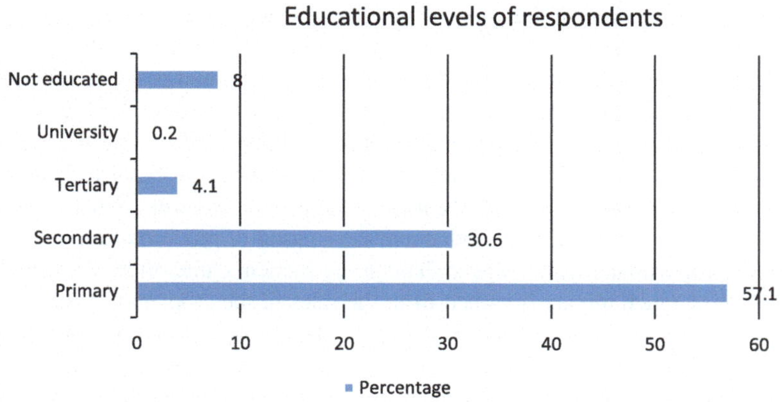

Fig. 2 Educational levels of the respondents

especially cattle, are low due to cattle rustling by the Karimojong. Those living near smaller lakes often engage in fishing for food and sale. With smaller land and livestock holdings, poorer households are unable to produce and raise enough to cover their annual food needs. Therefore, they rely on selling their labor to better-off households to supplement their income. Oftentimes, poor households engage in selling smaller livestock such as chickens, and those residing closer to lakes sell fish. Better-off households in this zone derive their food and income from crop production. They are able to cultivate more land by employing labor from the poor and oxen. They also purchase agricultural inputs to increase their production.

The better-off households supplement their annual food needs with food purchased from the market, in addition to milk and meat from their animals. Though the bulk of their income is earned from the sale of crops, the better-off also sell livestock and livestock products. The zone has good market access, with rural and larger road networks which allow a steady stream of commodity flow into and out of the zone. Crops are sold locally to traders who take them to larger trading centers outside the zone and into South Sudan. Traders from as far as Kampala purchase livestock from the local markets and transport them to Kampala, Lira, and/or South Sudan. The main hazards affecting livelihoods and food security in the zone are flooding, prolonged dry spells, birds, crops, and livestock diseases.

Rice Production and Gender Roles in the Zone

Rice is one of the major cash crops in the zone. The major rice varieties grown are Super, Super China, Super America, K5, Benenego, Kaiso, Kibuyu, Kigaire, Sena, Kibimba, Vietnam, Kigada, and Upland. About 30% of the respondents sampled in this zone indicated that they grow upland rice varieties, and the amount of land planted to rice per small-scale farmer is 0.4 hectares while the large-scale farmers grow 0.81–1.2 hectares per season. Rice in this zone is predominantly a man's crop, and women only provide labor to support production of the crop. Gender roles, however, vary with respect to the tasks performed in rice farming operations. Men are mainly responsible for land preparation, plowing, raising nursery beds, fertilizer and pesticide application, milling, and marketing. Women are mainly responsible for weeding, bird scaring, harvesting, and transporting the crop home for drying. Tilling the land and storage of the dried crop is normally done by both men and women. Men almost exclusively take over the responsibilities of milling and eventual marketing of the crop. Whereas women may have a greater input in rice production, the sharing of proceeds from the crop is usually disproportional, with a bigger share going to men. Quite often, the women do not know how much money is earned from the sales of the crop, and neither do they know how the proceeds are utilized. Studying gender constraints and rice varietal characteristic preferences in lowland rice ecosystem in Ghana, Addison et al. (2014) reported that rice production is labor intensive; hence, fewer females are involved. In terms of gender roles, Addison et al. further reported that women are mostly involved in transplanting, weeding, bird scaring, and post-harvest activities such as threshing, winnowing, drying, and to a lesser extent marketing.

Current Status of Rice Production in the Zone

Prior to the effects of extreme climate variability, both upland and lowland rice varieties were grown twice a year. Currently, as indicated by the respondents, trends have changed. Based on the fact that rice broadcasting and transplanting are dependent on the rainfall pattern, respondents indicated that it was no longer easy for them to determine the onset of rains to plant early and get good yields. They indicated that prior to the current variations in climate, farmers would prepare their seed beds in February and transplant rice seedlings between April and May or June and July. Broadcasting was done between February and March. Several rice varieties with different maturity periods were planted, depending on the rainfall intensity. For instance, the Superica variety, as mentioned by the respondents, took about 6 months to mature (longer than the others) and would be planted at the onset of the long rains. K5 rice variety took 3 months and would be planted in the shorter rainy season. In this way respondents indicated that they were able to tap farm gate prices that range between 1800 Ugandan shillings per kg of rice during peak harvest time and 3200 Ugandan shillings per kg when supply is decreasing.

3.2 Effects of Climate Variability (ECV) and Their Consequences on Women and Men Rice Farmers

The global influencers of agricultural production and their variability include technology, genetics, climate, soil, field management practices, and associated decisions such as fertilizer applications, tillage and crop hybrid selection, irrigation management, row spacing, planting date and depth, and population density (Kucharik and Ramankutty 2005; Andresen et al. 2002; Duvick and Cassman 1999; Duvick 1992, 1977). Among these factors, weather and climate are prominent drivers of agricultural production systems. It has been shown that recent trends in change of climate variables may be responsible for substantially affecting crop yield trends despite advances in technology and other fronts (IPCC 2013).

This study also investigated effects of climate variability on livelihood strategies of the men and women farmers in the study area. The main livelihood strategies discussed by the respondents in the zone in both production systems were crop and livestock production. Maize, groundnuts, cassava, and rice were the major crops produced while cows, bullocks, goats, pigs (*Sus domesticus*), and chickens were the main livestock reared. The crops and livestock are presented according to their order of importance to their livelihoods, as agreed upon by the respondents.

Effects of Climate Variability on Crop Production Sequence and Yields

In sub-Saharan Africa, sequential cropping in rain-fed areas has been one of the ways of increasing crop yields in addition to intercropping. This involves cultivation of two or more crops on the same field after each other or with overlapping growing periods (relay cropping) (Francis 1986). Prior to in-depth discussions, respondents were asked how their current crop production practices and challenges compare to the practices they had over 20 years before. The majority of both female and male respondents observed that due to the current ECV, it has become quite risky to depend on their cropping calendars as a guide in determining when to prepare their land parcels for crop and livestock production. They indicated that the current dry conditions have been prolonged and the rains are characterized by floods and hailstorms that prolong planting time and create favorable conditions for pest and disease infestations. Comparing the situation with earlier times, that is, about 25 years ago, respondents reported that the rains used to begin as early as February, which prompted early land preparation and planting. Crops would be harvested prior to peak pest infestation, hence higher yields. They indicated that in the present times, the rains delay as far as the month of March and beyond, which affects crop planting schedules. This delays planting time and reduces the growing season. Respondents cited past instances when farmers used to plant banana (*Musa acuminata*), coffee (*Coffea arabica*), and rice in April and maize, groundnuts, and cassava in March, and yet currently they cannot decide on what crops to rotate, given the variations in rainfall. This confirms that ECV events have indeed distorted their crop rotational sequence to the extent that they now find it very difficult to decide on what crops to rotate, if rotation happens at all. Additionally, they have to plant pest-, disease-, and drought-resistant crop varieties in order to overcome ECV. Findings here suggest that climate variability has resulted in reduced crop rotation and growing season and late planting with implications on soil fertility and pest and disease infestation. Waha et al. (2013) observed that sequential cropping in rain-fed systems of the large parts of sub-Saharan Africa is constrained by the length of growing period, high labor intensity, lack of knowledge, and lack of market access. Duku et al. (2018) studied the impacts of climate change on cropping patterns in tropical sub-humid watershed areas in central Benin, which are either used for rain-fed sequential cropping or can support rain-fed sequential cropping. They reported that about 41% of cultivated areas in the Upper Ouémé watershed will decrease to between 2% and 16% by 2050, depending on the climate change scenario. Duku et al. thus concluded that farmers will have to shift to single cropping systems or adopt improved agronomic practices, including drought-resistant and short-cycle cultivars.

Effects of Climate Variability on Main Crop Yields

Respondents were also asked to give their perceptions on how ECV affected yields of the main crops grown. Figure 3 shows the variations in yields of rice, maize, groundnuts, and cassava under drought, flood, and normal conditions.

With reference to Fig. 3, respondents indicated that compared to the normal year where most crops yields were relatively high, crop yields during drought and flood conditions were quite low due to ECV events. This result is consistent with the findings of Mariara and Karanja (2006) in Kenya and that of Eid et al. (2006) in Egypt. Maize yields, however, were slightly higher during the floods compared to the rest of the crops. Contrary to other crops that yield highly in the normal year, cassava crop yielded slightly higher in drought conditions than the rest of the crops. This demonstrates that cassava is more drought resistant than the rest of the crops. Shumetie and Yismaw (2018) reported in their study that households, on average, lost about 8.06 quintals of crop because of insufficient rainfall. Shumetie and Yismaw also found that an increase in summer temperature had a negative effect on productivity, and it caused an average loss of 8 quintals for 86.88% of the sample respondents. Further discussions to determine perceptions of respondents on whether individual crop yields had increased, decreased, or not changed due to ECV events indicated that there is a general decline in the yields of all crops. The perceptions were similar for both male and female respondents interviewed, further confirming the decline in crop yields due to ECV events. The perceptions are shown in Fig. 4. For instance, female farmers (68%, 60%, 60%, and 58%) perceived that yield decreased for groundnuts, cassava, maize, and rice, respectively, whereas male farmers (74%, 66% 66%, and 65%) perceived reduced yield for cassava, maize, groundnuts, and rice, respectively. Most women perceived that yields reduced for groundnut production while most men perceived reduced yield for cassava production. Further analysis shows that in both farming systems, an average of 65% of the respondents perceived that yields for all the major crops had reduced while an average of 25% perceived that the yield had increased as a result of climate variability significantly different at $p = 0.023 < 0.05$. About 10% of the respondents perceived that yield remained the same despite climate variability. Greg et al. (2011) reported that most studies revealed that climate change is likely to reduce agricultural productivity, production stability, and household income in some areas that already have high levels of food insecurity. A study on the impacts of climate change on rice production and causes of simulated

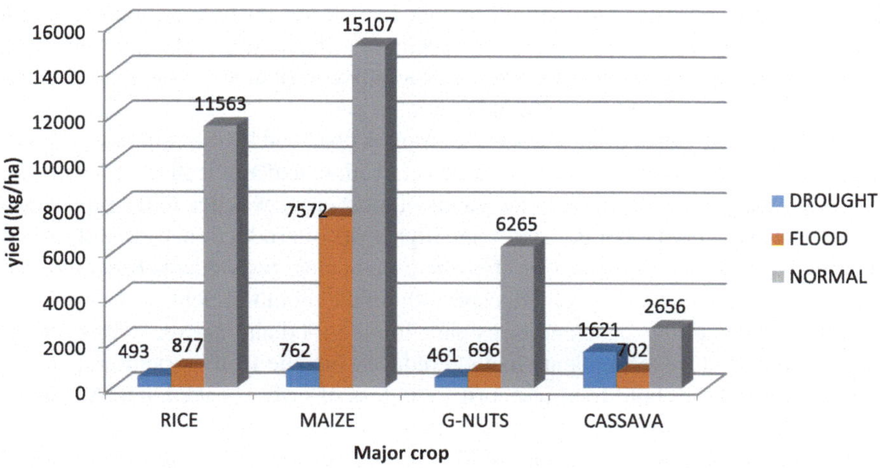

Fig. 3 Crop yields in extreme climatic conditions

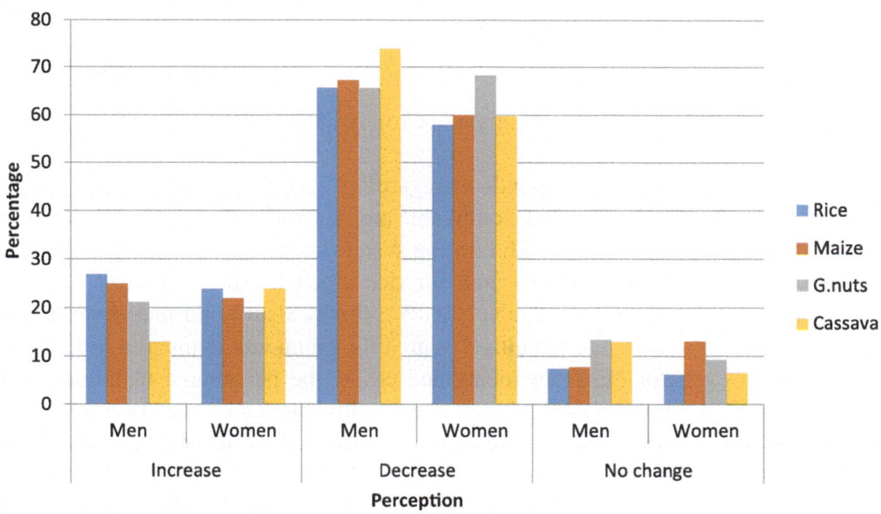

Fig. 4 Effects of ECV events on individual crop yields as perceived by both men and women

yield changes in Africa concluded that shortening of the growing period is the main cause of projected yield decline and that this effect was present in all sites and all growing environments in Africa (van Oort and Zwart 2017). On the contrary, Koudahe et al. (2018) in their study of the impact of climate variability on crop yields in southern Togo reported a slight increase in growing season precipitation with non-significant effect on crop yields. This is in contradiction with Sogbedji (1919), who revealed that in Togo a decrease in seasonal rainfall amount represented a serious threat to maize growth. It has also been reported by Adamgbe and Ujoh (2013) in

Nigeria that rainy days and rainfall amount had strong positive relationship with maize yield ($r = 0.747$ and $r = 0.599$, respectively). In addition, Adamgbe and Ujoh (2013) observed that rainfall characteristics jointly contributed 67.4% to explaining the variations in the yield of maize in Nigeria.

The increased yields of the respective crops as observed by the respondents were attributed to use of improved crop varieties, irrigation, and application of fertilizers. However, the general decrease in the yields of most crops was due to the prevalence of new diseases, high labor demands, and high costs of production, especially when planting is done late in the season. For the case of rice, respondents observed that the cost of "bird scaring" was high, especially in the month of September when the crop is at its tasselling stage and birds have limited food alternatives. Other effects of ECV reported by respondents include reduced income from agricultural activities, increased household food insecurity, and loss of investment capital. A respondent gave the example that before the 1990s, the yield of rice per hectare was 4250 kg compared to the current yields that are between 2250 and 2500 kg per hectare. Declining yields result in reduced commodity quantities available for the market, hence low income from agricultural activities, which in turn leads to low investment capital. Ultimately, declining yields, low income, and loss of investment capital result in household food insecurity.

Effect of Climate Variability on Cultivable Crop Area

During the discussions with respondents regarding ECV on cultivable crop area, it was observed that ECV affected the cultivable area for various crops, and the effects varied with the type of crop. In both farming systems, male and female respondents reported an increase in cultivable area for rice albeit the proportion of females (42.8%) was higher than that of men (30.8%). It was also found that more males (22.8%) reported an increase in cultivable area for groundnut compared to females (10.2%). However, for the other mentioned crops, the percentage of female and male respondents who observed an increase in cultivable land were lower and relatively similar as shown in Fig. 5. The increase in rice cultivable area was attributed to the introduction of new upland rice varieties, which came as a result of increased demand for food and the gains that were emerging from the highly demanded upland rice in the market.

Comparatively, the percentage of male and female respondents who indicated increases in crop cultivable areas were lower than those who reported decreases in the same (see Fig. 6). This shows that majority of both males and females reported a decline in cultivable areas for the mentioned crops. Findings further show that more female respondents reported a decrease in cultivable area for almost all crops, with the exception of the cassava crop, compared to their male counterparts. The percentage of male and female respondents who observed a decrease in fallow land were more, compared to the proportion of male and female respondents who reported a decrease in cultivable area for the crops. A two-sample t-test between proportions was performed to determine whether there was a significant difference

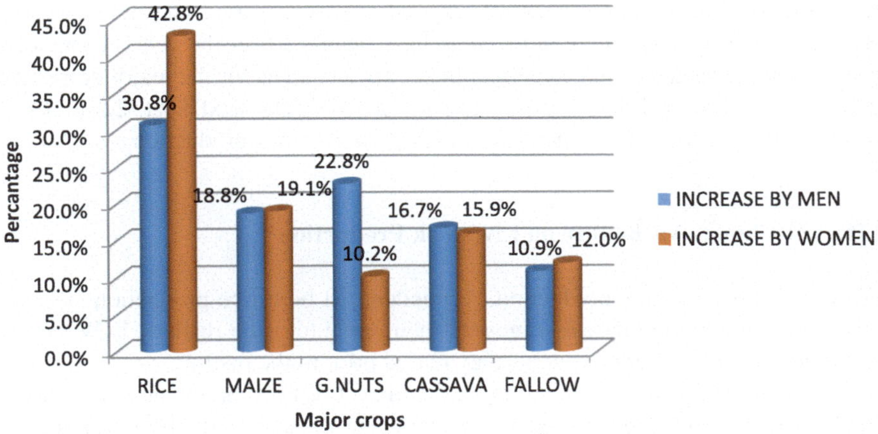

Fig. 5 Male and female respondents who perceived increased cultivable crop areas

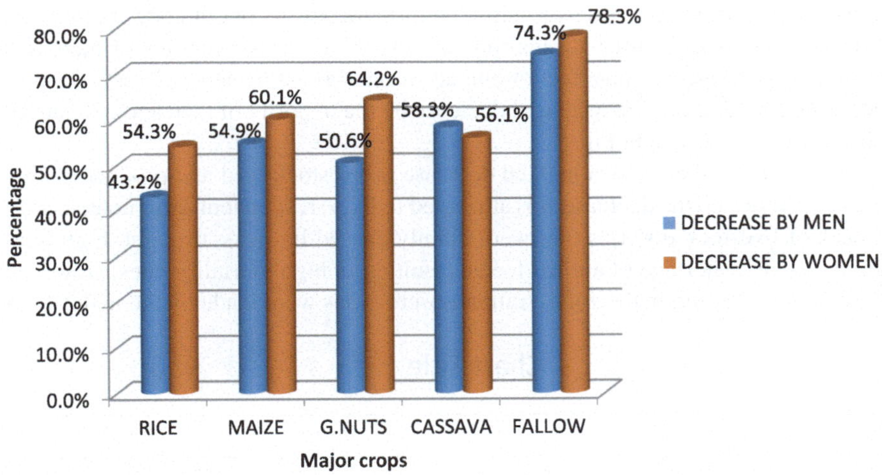

Fig. 6 Male and female respondents who perceived decreased cultivable crop areas

between men and women who reported a decrease in cultivable area for the major crops and fallow. The t-statistic was not significant for cassava and fallow at $p = 0.111 > 0.05$ and $p = 0.240 > 0.05$. On the other hand, there were significant differences between the proportions of men and women who perceived decreases in cultivable areas for rice ($p = 0.015 < 0.05$), maize ($p = 0.03 < 0.05$), and groundnuts ($p = 0.009 < 0.05$).

The decrease in the area cultivated for all the crops mentioned above, including rice, was generally attributed to prolonged drought, infection by pests and diseases, limited arable land, and infertile soils. Overall, the general perception given by respondents was that cultivable crop areas had generally decreased due to

ECV. Figure 7 gives an overview of respondents' perceptions on how cultivable areas for all crops and fallow in the zone have changed over the years. Generally, most of the respondents perceived that the cultivable areas for the major crops and practice of fallow had decreased as a result of ECV. The next Sect. 2.2 looks at respondents' perceptions on the effect of EVC on livestock production.

Effect of Climate Variability on Livestock Production

With regard to livestock production, respondents in both rice production systems observed that although they lost most of their cattle to cattle rustlers, ECV events had reduced the number of animals they reared prior to the rustling.

Based on the results shown in Fig. 8, the impact of ECV on the number of livestock owned by respondents does not come out clearly, with the exception of chicken. Chicken numbers were quite high in the normal years, but they reduced in the drought and flood conditions. This lack of clarity in the results owing to the different livestock could be attributed to the general lack of livestock in the area. Hence one cannot rely on these observations to confirm the effects of climate variability on livestock numbers. Further discussions to determine the perceptions of male and female respondents on the effect of climate variability on livestock, livestock products, and fodder indicated a general decrease by the majority of men and women for all products, as shown in Fig. 9.

As observed, there was a marked decrease in livestock feed and products owing to ECV. Although the decrease was attributed to ECV, respondents also noted occurrences of livestock diseases that were mainly caused by ticks, drought, high temperatures, and shortage of animal feeds, resulting in high mortality rates. However, in relation to fishing in the zone, similar observations were made by the majority of

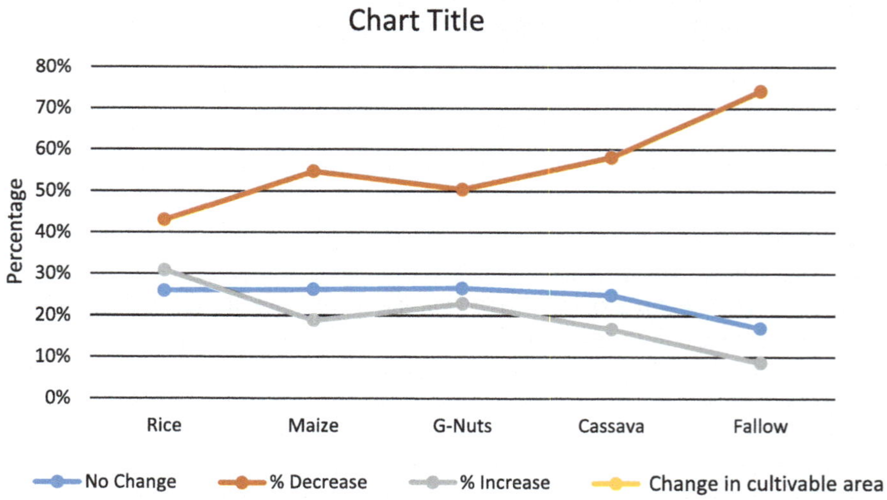

Fig. 7 Perceptions on the effects of ECV on cultivable area for major crops

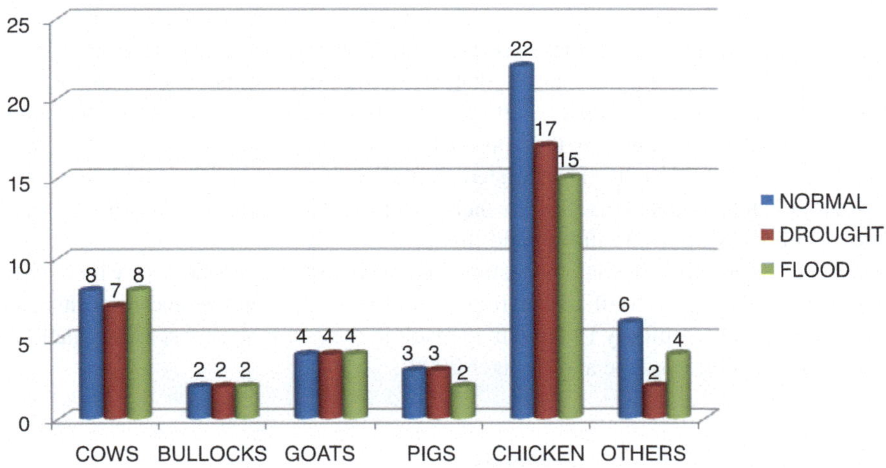

Fig. 8 Effect of climate variability on the average number of livestock owned by respondents

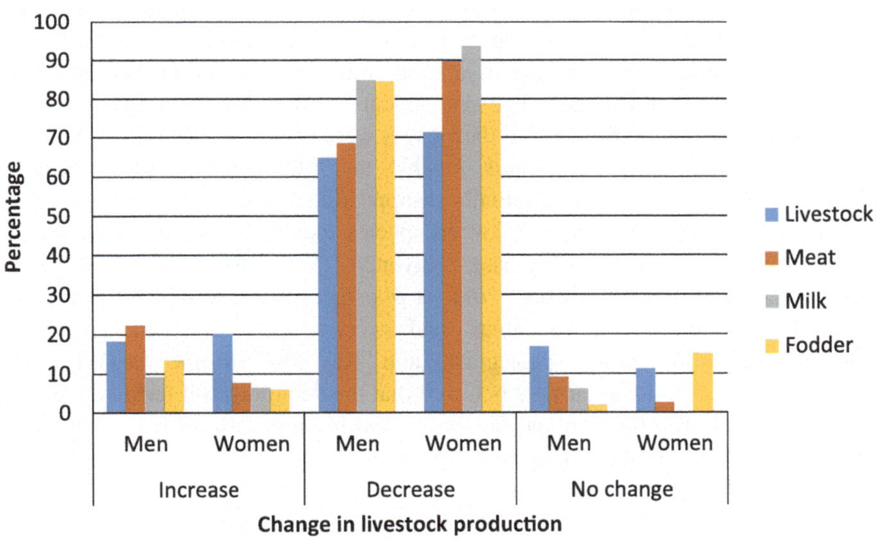

Fig. 9 Effect of EVC on the number of livestock and livestock products

the respondents regarding decreased numbers due to the effects of ECV events mainly attributed to high temperatures claiming the lives of many fish in the water bodies.

Literature shows that the direct effect of climate change as a result of increased ambient temperature and concurrent changes in heat exchanges causes heat stress which influences growth, reproduction performance, milk production, wool production, and animal health and welfare (Walter et al. 2010; Reilly 1996). Heat stress reduces feed intake and results in poor growth performance in animals, even though indigenous cattle are known to be tolerant to high temperatures (Walter et al. 2010). Extremely high temperature caused by extreme weather events experienced at this era may still affect *B. indicus*, resulting in reduced milk and meat production and reduced time for foraging, as they prefer to remain in the shade (Robertshaw and Finch 1976). Additionally, heat-related mortality and morbidity would increase. Climate change is also expected to increase the risks of drought and floods that occur with El Niño in the future, and this could result in serious high mortality of livestock. Drought will lead to pasture shortage and water scarcity, which will aggravate the existing conflicts rotating around natural resources and food insecurity in the region. Similarly El Niño may result in the outbreak of diseases related to flooding (Van den Bossche and Coetzer 2008).

Other Effects of ECV on Livelihoods

Other effects of ECV, as perceived by respondents across the zone, were deforestation and swamp reclamation as a result of population pressure and limited sources of income. The respondents reported that they are forced to clear natural forests for charcoal burning and reclaim swamps to increase their acreage for the production of rice and other crops. The changes in vegetation cover observed in the zone were attributed to weather changes. The respondents further observed that they had cleared off most natural forests and their rivers were drying while some weeds like "Olukozi" (couch grass – *Elymus repens*) and "Olubembe" (spear grass – *Heteropogon contortus*) were gradually disappearing. Most of the farm land was now covered by different types of weed species locally known as "Kafadanga" (milk grass), "Olukozi" (couch grass), "Kayonga" (*Striga* weed – *Striga hermonthica*), and "Obukala" (blackjack – *Bidens pilosa*) in the case of the upland rice production system. Swampy areas had weeds such as "Enkengeenene," "Olukalabwe," "Olubembe" (spear grass), and "Kayonga" (*Striga* weed). In addition to weather variations, farmers believed that the declining soil fertility, coupled with the lack of crop rotation and fallowing, and improper use of herbicides were also factors contributing to changes in vegetation cover.

3.3 Coping Strategies Employed Under Different Consequences of ECV

This section looks at the various coping strategies employed by the respondents to adapt to the effects of climate variability and hence improve their livelihoods. As observed, ECV greatly affected agricultural activities in the zone, resulting in low crop and livestock production. The ability of the respondents to respond to, cope with, or adapt to ECV events and stresses placed on their livelihoods and well-being depends on resources available to them. Results generated from focus group discussions and individual interviews (including key informants) revealed the following findings regarding men and women's capacity to cope with or adapt to climate change including climate variability. Both men and women in the zone from the two rice production systems indicated that although it had become difficult to access information and other natural resources such as water, medicinal plants, and different types of grass plants used for thatching houses and making baskets and mats, they could still rely on some of these resources to cope. The majority of the women indicated that they resorted to using grass plants to make mats and baskets which they sold to generate some income.

Men however indicated that the time spent collecting grass was too long and the activity was quite tedious. They added that in addition to earning very little income from the mats, one had to craft it first, an activity which needed patience and more time. However male respondents indicated that they were more interested in quicker income-generating strategies and wage employment. Consequently, the majority of them were migrating to other towns in search of employment opportunities. Other coping strategies included selling of labor. This was reported by the majority of women, who indicated that it was a faster means of obtaining money for daily food and health expenses, given that in most cases their spouses were away looking for jobs or in bars socializing and taking alcohol. The respondents however found it easier to discuss their coping strategies based on the conditions they were adjusting to. The conditions they mainly addressed were extreme drought, severe floods, and increased pest and disease incidences. Below are observations made in relation to the coping strategies and the conditions in which they best applied for women and men.

Based on the various coping strategies employed by the respondents during drought conditions, as indicated in Fig. 10, it is evident that majority of men turn to making bricks, digging latrines, selling assets and labor, and growing resistant varieties. On the other hand, majority of women resort to changing the types of crops grown, cultivating small areas, early planting, poultry production, and selling of labor. During the flooding conditions, the majority of men resort to digging and building water channels and bands, fishing, and selling assets. Shifting to rice production and draining farms of excess water was highly reported by both women and men, whereas migrating to relatives was reported by more women (see Fig. 11). These findings show that both during drought and flooding conditions, men generally resort to off-farm and non-farm activities, while women still stick to farming activities as means of coping. This may imply that men unlike women have resources to diversify away from agriculture when climatic conditions become too harsh for agricultural production, hence better able to sustain livelihoods than women.

Respondents indicated that during pest and disease outbreaks, they employed various strategies, as observed in Fig. 12. Among these strategies, the majority of men planted resistant varieties, performed pest and disease management, sought trainings in crop management, and made bricks. Female respondents reported more crop nutrition management, among other strategies that were similar to those employed by the majority of male respondents. The other coping strategies employed to address the negative effects of ECV were going to the neighboring villages to sell

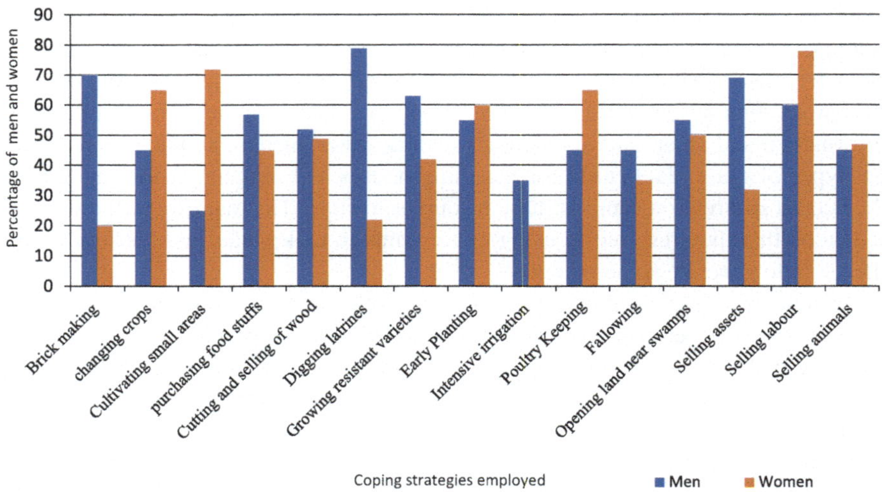

Fig. 10 Coping strategies for men and women in extreme drought conditions

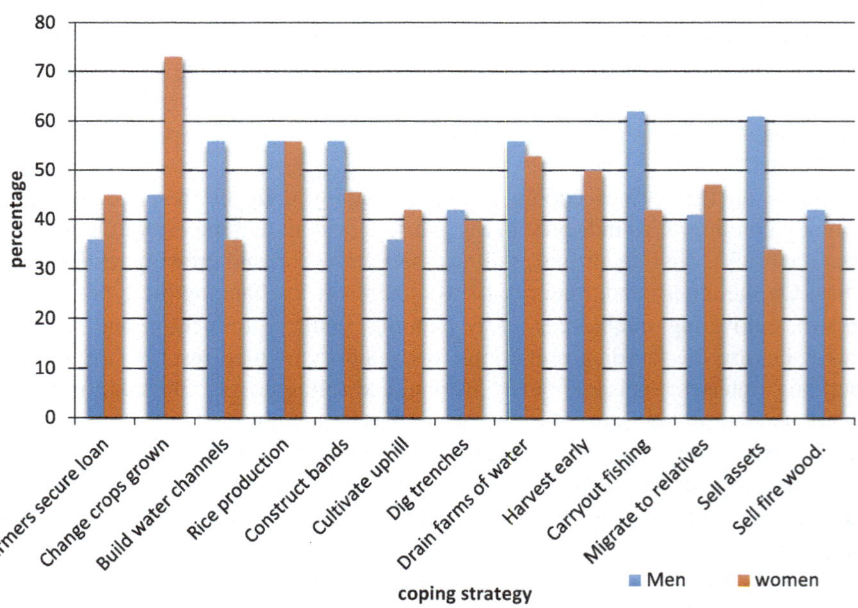

Fig. 11 Coping strategies for men and women in extreme flood conditions

labor, shop keeping, getting support from friends and relatives, storing food and other necessities, selling assets, and spending less.

Observations showed that the youth were not keen on agricultural activities due to the difficulties being experienced in their communities. The youth also noted that

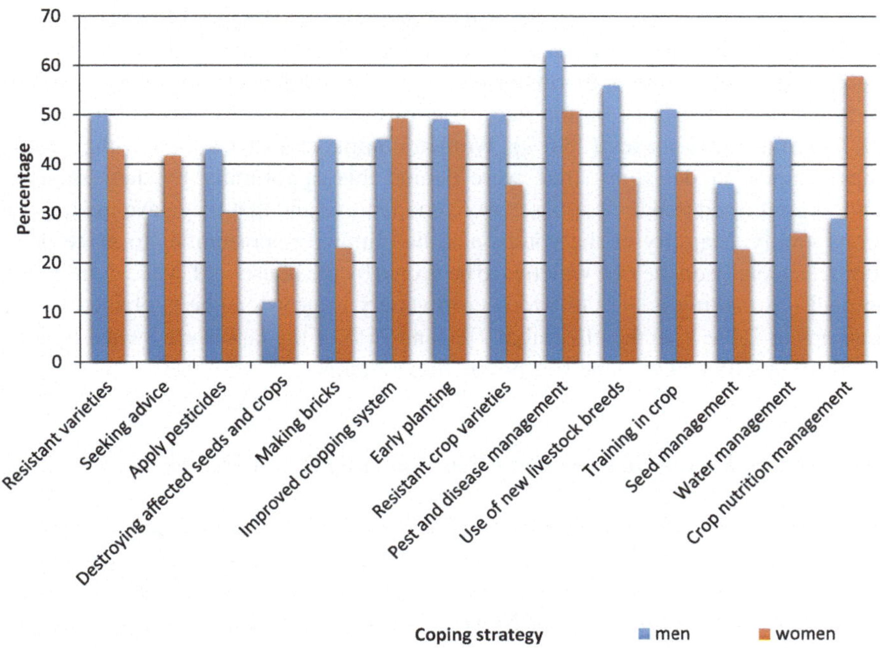

Fig. 12 Coping strategies for men and women in extreme drought conditions

the difficulties were not just about ECV but also about lack of knowledge and skills in entrepreneurship and acquiring capital for investment. These problems are attested to by both male and female youths and farmers. They indicated that they were keen to develop projects and enterprises to generate incomes; however, they did not have funds. The men and women pointed out that starting a craft business as an income-generating activity needs one to have access to grass that is used for mat and basket weaving. Both men and women talked about producing fruits and vegetables for commercial markets. They had many ideas about what to do but lacked the means. This is an indication that both men and women have the potential to benefit from natural resources, provided they had the capital. Both men and women require more than just having access to land and natural resources. It was also clear that they lack the technology and economic resources to reduce ECV. Climate change could therefore increase their poverty levels, reinforcing existing vulnerabilities between men and women.

3.4 *Decisions on Choice of Coping Strategies Employed by Households*

Despite the fact that most farming activities did not change, most of the coping strategies employed to address the effects of ECV such as extreme drought, severe floods, pests, and diseases are based on household deciding on what strategy to adopt and the resources available at their disposal. In extreme ECV, majority of male respondents indicated that they were responsible for making decisions in their households. For example, the majority of male respondents (60.8%) reported that they made the decisions on the rice varieties to be grown, as opposed to women (7.7%) and both spouses (26.6%). Women complained that although they could make important decisions, they were denied this opportunity by their spouses. Others confirmed that their spouses were cooperative and that they gave made decisions jointly. Majority of the youths also had limited opportunities to make decisions, especially on the rice varieties to be grown by the household. Male domination in decision-making related to which crop variety to grow can be explained by the findings in Table 1, to the effect that 91% and 75% of the land were owned by husbands in upland and lowland rice production systems.

Decisions on Factors to Consider when Changing Rice Varieties

Decisions to determine the choice of rice varieties to be grown as reported by both men and women were based on good yields, tolerance to stress, market demand for rice, recommendations from agriculture departments, and recommendations from non-governmental organizations (NGOs). In this regard, although the majority of men indicated that they were more informed and also had more access to these services compared to women, findings indicate that majority of women were also knowledgeable and informed, and they made similar observations regarding the criteria they used in determining change in rice varieties. Figure 13 indicates the percentage ratings by men and women of the factors that determined the decision to change a rice variety.

Figure 13 shows that majority of the respondents reported that good yields and tolerance of variety to stress were not only important but very important when making decisions on whether to change rice varieties for planting. The majority of respondents (73.9%) however indicated that they did not change their rice varieties during ECV event conditions. This was mainly because they lacked the new rice varieties with the preferred traits that were mentioned as a basis for decision-making. In comparison, 39.5% shifted from crop to livestock and 21.1% increased the number of crops they were growing to spread the risks of losing everything. Other factors also considered very important in determining the decisions made when choosing rice varieties include advice from NGOs, availability of capital, availability of water, market demand, suitability of soil and land, and farmers' access to seeds and fertilizers.

Factors to Consider when Changing Cropping Patterns

In relation to changes in cropping patterns, decisions to change were based on good yields, tolerance to stress, market demand, recommendations made by the National Agricultural Research Organisation (NARO), and availability of water. Both male

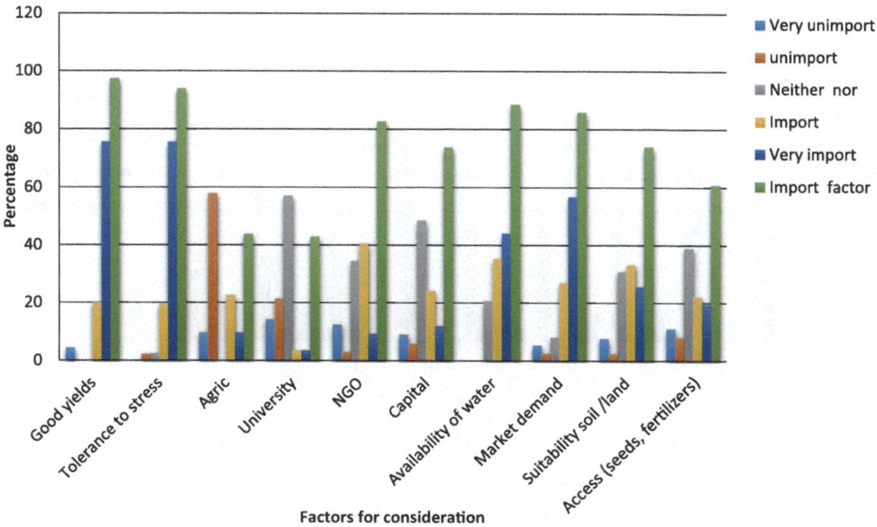

Fig. 13 Important factors to consider when changing rice varieties

and female respondents agreed on the criteria used. Based on the results in Fig. 14, the majority of respondents still noted that good yields and tolerance of variety to stress and advice from the universities were important considerations when making decisions on whether to change cropping patterns as a coping mechanism to ECV. Other factors considered in making decisions related to cropping patterns included availability of water (65.4%), market demand (70%), and suitability of soil and land (60%) (see Fig. 14).

Factors to Consider when Adopting Technologies

As far as the respondents were concerned, being a group member increased the chances for one to access new technologies. Group membership increased one's chances to network with other farmers. According to Mignouna et al. (2011), belonging to a social group improves social capital, which involves permitting trust, as well as idea and information exchange. Farmers within a social group benefit from peer learning and use of a new technology. Uaiene et al. (2009) argued that social network effects are key for individual decisions, especially in the context of agricultural innovations, as these enable farmers to share information and learn from each other. Examining the effect of community-based organization in uptake of corm-paired banana technology in Uganda, Katungi and Akankwasa (2010) observed that farmers who were members of community-based organizations were more likely to engage in social learning about the technology, hence increasing their likelihood to adopt the technologies. However, even though many authors have reported a positive association between social group and technology adoption,

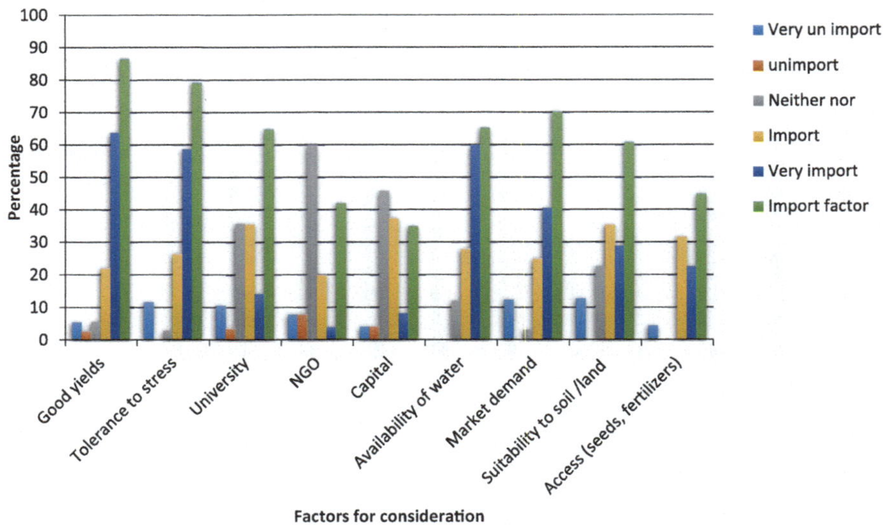

Fig. 14 Factors for determining change in cropping pattern

social groups may also have a negative influence on technology adoption, particularly where free-riding behavior exists. A study on adoption of green revolution technologies in India reported that learning externalities in social networks increased the gains of adoption; however, farmers appeared to social-loaf on their neighbors' costly experimentation with the new technology (Foster and Rosenzweig 1995).

Human capital of the farmer is understood to have an important effect on farmers' decision to adopt new technologies. Most studies on adoption have attempted to measure human capital through the farmer's education, gender, age, and household size (Fernandez-Cornejo et al. 1994, 2007; Mignouna et al. 2011; Keelan et al. 2014). In this study, findings showed that being of age with responsibilities to take care of also influenced one's ability to seek for rice technologies that promote increased production and hence income generation. Additionally, it has been documented that older farmers are assumed to have gained knowledge and experience over time and are better able to assess technology information than younger farmers (Mignouna et al. 2011; Kariyasa and Dewi 2011). On the other hand, age was found to have a negative association with technology adoption in this study. The elderly members of the communities were not keen to go in search of high-yielding crop varieties. As farmers advance in age, they tend to be risk-averse and reduce interest in long-term investment in the farm. On the contrary, younger farmers are generally risk-taking and more inclined to experiment with new technologies (Mauceri et al. 2005; Adesina and Zinnah 1993). Respondents also agreed that those who were more educated had better chances of understanding the importance and attributes of a particular technology, let alone wanting to get the skill in utilizing it, as compared to those who were less educated, the majority of whom were women. The respondents agreed that those who were more educated had higher chances of accessing credit facilities; hence, they could invest in high-yielding crop and animal technologies for improved production. The respondents also reported that if the technology being introduced did not interfere with their existing local practices, the chances that they would buy and utilize it were high. Another factor relating to adoption, as reported by respondents, was the size of the household. Respondents indicated that when one had many household members, the available labor would encourage them to adopt a new crop technology that was high yielding. The size and quality of cultivable land was also one of the factors reported to influence adoption of rice technologies, that is to say, if one had the land available, then the chances of adopting the rice technology were high. This is because when land is available, farmers are flexible to allocate more land to rice. Apparently, all the respondents interviewed indicated that rice field management practices are very laborious. The most critical needs of labor are in land preparation (since the land needs to be plowed 2–3 times before planting), planting (since making furrows with local materials such as sticks is very tedious), weeding (about 2–3 times are required for weeding), and bird scaring (if birds are not scared, total loss is experienced, and so far no effective alternative control methods are available).

Varietal Characteristics That Influence Adoption of a Technology

Other important factors to consider for increased adoption were varietal characteristics in terms of what farmers want with respect to yield and tillering capacity. Demand for the variety on the market was also important, in addition to resistance to rice pests and diseases. Also, concerns about the ability for rice to survive under lower rainfall conditions were important, in addition to the extent of demand attached to the growing cycle criterion, i.e., a shorter growing cycle was viewed as desirable. Other factors reported to influence the adoption of technologies by respondents were high-yielding attribute of a variety, tillers being viewed as able to survive the stress and damage by birds, early-maturing varieties, resistance to drought and disease/insects, tolerance to salinity, higher volume after cooking, ease of harvesting, and a good aroma and taste. Figure 15 indicates some of the factors listed by respondents perceived to influence the rate of technology adoption.

Based on the results in Fig. 15, majority of female respondents noted that the use of stress-tolerant crop varieties and planting of early-maturing varieties were highly likely to affect the adoption rate of technology. Other factors such as new land management techniques, pest and disease management techniques, development and use of crop varieties resistant to pests and diseases, new livestock breeds, and improved animal health management were also reported by a higher percentage of respondents as likely to influence the rate of technology adoption.

4 Conclusions and Recommendations

This section presents the conclusions and recommendations, based on the findings of the study. The study aimed at examining gender-differentiated effects of climate change among men and women. The study made use of existing experience and

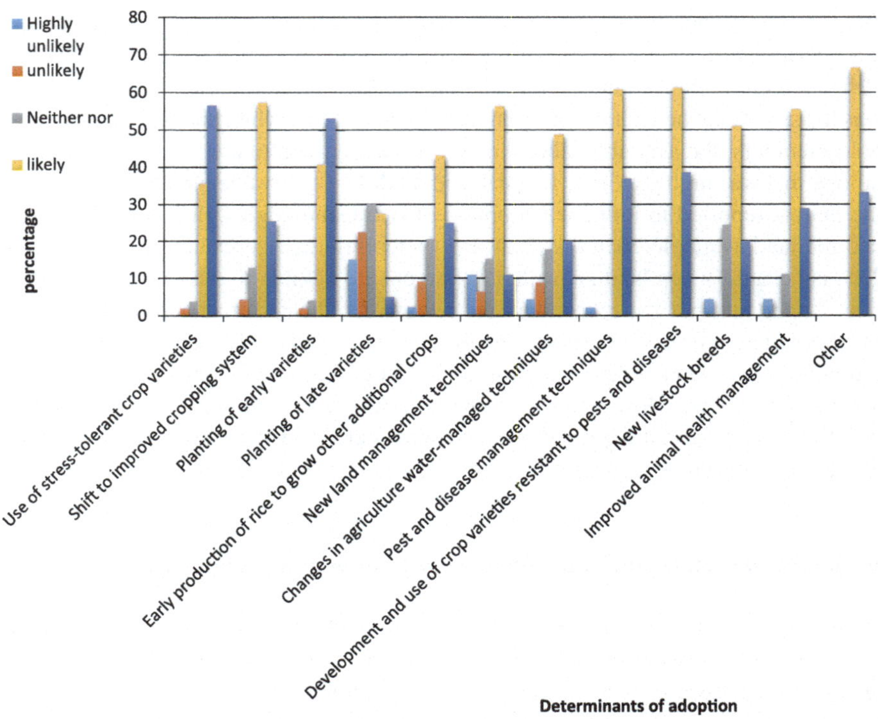

Fig. 15 Factors likely to influence adoption rate of technologies

exposure to climate variability effects to explore the current gender-differentiated effects in relation to rice growing in eastern Uganda. As already indicated in the literature, it is well established that while climate change is viewed to be global in nature, its impacts are not globally homogeneous but rather differentiated across regions, generations, social and cultural groups, age classes, income groups, and gender.

The results generated from this study are not representative of eastern Uganda as a whole, given that communities or social group are unique and differ culturally, socially, and economically and hence have different norms and values that influence gender-differentiated effects of climate change. The results of the study are an important step in unveiling the dynamics and realities of gender-differentiated effects of climate. It is evident that climate change and its effects are known by the respondents; however, they lack the capacity to adapt and build resilience to climate change. Both men and women are affected grossly by EVC effects, becoming more poverty stricken with very limited economic, human, and social resources to build resilience to climate change. It is also evident from the observations that both men and women rely more on coping mechanisms to respond to the ECV effects, which are more short-term survival strategies as opposed to long-term adaptation strategies, given the nature of resources at their disposal. To maintain their crop and livestock production for food and income security, coping strategies such as social grants, government relief, and seasonal employment that are critical for improvement are viewed as fallback alternatives in case their strategies failed. This could be a result of the mechanism used to provide services, which end up being skewed to the benefit of a few.

The ability of respondents to build resilience to climate change is further constrained by a combination of both physiological and physical factors. These include helplessness, stresses, and strains defined as elements of poverty, HIV/AIDS, governance, and the inability of service providers to effectively implement programs and policies designed to enhance the lives of men and women. These factors further affect adaptation as they hinder men and women's ability to recover from the impacts of climatic events. Gender-differentiated impacts of climate variability were manifested in the unequal distribution of roles and responsibilities of men and women in rice production systems in the zone, through socially constructed roles and responsibilities. Women seemed to bear the heavier burdens from climate variability effects, although they were not able to clearly appraise these effects in relation to the effects on men. Women were found to have extra workloads when faced with climatic stressors as they made efforts to cope by working longer hours than men to be able to cater for the children and the other household members who depend on them, which affected them both physically and emotionally. Married women also had the additional restriction of having to account to their husbands for their movements, even if it was in the cause of trying to cater for the family, in the face of climate change variability.

The effects on men were more psychological than physical. When climate change led to unemployment, men coped very poorly. The pressure of unemployment often resulted in alcohol, drug abuse, and violence. It was also evident that climate change impacted differently on rural men and women. Changes in rainfall patterns, for instance, instantly interrupted the working routine of rural men and women, while those in urban areas felt the impact much later.

Results from this study confirm that women play an important role in supporting households and communities to cope and adapt to climate variability to ensure food security. They have the opportunity to diversify their livelihoods to cope better with climate variability, but this would not be possible if women's mobility is restricted. Lessons can therefore be drawn from their knowledge on how women can be better assisted to adapt to climate change. With enough commitment and further support, they are in position to take advantage of new opportunities to address issues of gender and climate change.

Rice production in Uganda has high potential to improve farmers' livelihoods and contribute to economic growth and development. Promoting food security through rice production and addressing climate change issues is very critical and requires a holistic approach while addressing issues relating to gender inequality and climate-related impacts in rice-growing communities. There is a need to understand the historical and social disadvantages of both men and women and the multiple concerns emerging from historical and social disadvantages of both men and women that affect their ability to cope with current climate variability. In addition, it is important to understand the differential effects of extreme climate events on men and women as well as their adaptation strategies. Gender-specific barriers that pose constraints to adaptation strategies of men and women should be understood and addressed to enable them to better adapt to the effects of weather shocks. It is also important to build the skills of women in order to empower them to access crop (rice) technologies that are better adapted to ECV conditions and technologies, which will facilitate them in regard to labor shortages and quality production. Lastly, women should be engaged in other off-farm interventions that are income generating.

References

Adamgbe EM, Ujoh F (2013) Effect of variability in rainfall characteristics on maize yields in Gboko, Nigeria. J Environ Prot 4:881–887

Addison M, Edusah SE, Sarfo-Mensah P (2014) Gender constraints and rice varietal characteristics preferences in lowland rice ecosystem in Ghana. J Dev Country Stud 4(15):95

Adekunle W (2013) Improving smallholder incomes through intensification of upland rice. A case study produced by WREN media, funded by the Swiss Agency for Development and Cooperation (SDC) and implemented by the European Initiative on Agricultural Research for Development (EIARD)

Adesina A, Zinnah M (1993) Technology characteristics, farmers' perceptions and adoption decisions: a Tobit model analysis in Sierra Leone. Agric Econ 9:297–311

Adoko J, Levine S (2005) A land market for poverty eradication. A case of the impact of Uganda's land acts on policy hopes for development and poverty eradication. Land and Equity Movement in Uganda (LEMU), Kampala

Ahmed M (2012) Analysis of incentives and disincentives for rice in Uganda. Technical notes series, MAFAP, FAO, Rome

Andresen JA, Alagarswamy G, Rotz CA, Ritchie JT, LeBaron AW (2002) Weather impacts on maize, soybean, and alfalfa production in the Great Lakes region, 1895–1996. J Agron 93:1059–1070

Bigirwa G, Kikafunda J, Lamo J, Ochen J, Opio F, Tsuboi T (2005) Upland rice growing in Uganda: reasons for the rapid spread. Afr Crop Sci Conf Proc 7:151–154

Bikaako W, Ssenkumba J (2003) Gender, land and rights: contemporary contestations in law, policy and practice in Uganda. In: Wanyeki LM (ed) Women and land in Africa. Zed Books, New York, pp 232–277

Duku C, Zwart SJ, Hein L (2018) Impacts of climate change on cropping patterns in a tropical, sub-humid watershed. PLoS One 13(3):e0192642

Duvick D (1977) Genetic rates of gain in hybrid maize yields during the past 40 years. Maydica 22:187–196

Duvick DN (1992) Genetic contributions to advances in yield of US maize. Maydica (Italy) 37:69–79

Duvick DN, Cassman KG (1999) Post–green revolution trends in yield potential of temperate maize in the North-Central United States. Crop Sci 39:1622–1630

Eid M, El-Marsafawy SM, Ouda SA (2006) Assessing the economic impacts of climate change on agriculture in Egypt, a Ricardian approach. Centre for Environmental Economics and Policy in Africa, University of Pretoria, CEEPA Discussion Paper No. 16

FAO (2007) Climate change and food security. FAO, Rome

Fernandez-Cornejo J, Daberkow S, Huang H (1994) The adoption of IPM techniques by vegetable growers in Florida, Michigan and Texas. J Agric Appl Econ 26(1):158–172

Fernandez-Cornejo J, Mishra A, Nehring R, Hendricks C, Southern M, Gregory A (2007) Off-farm income, technology adoption, and farm economic performance. Agricultural Economics Report No. 36. USDA ERS, Washington, DC

Foster D, Rosenzweig M (1995) Learning by doing and learning from others: human capital and technical change in agriculture. J Polit Econ 103(6):1176–1209

Francis CA (1986) Multiple cropping systems. Macmillan Publishing Company, New York, pp 34–39

Greg EE, Anam BE, William MF, Duru EJC (2011) Climate change, food security and agricultural productivity in Africa: Issues and policy directions. International Journal of Humanities and Social Science 1(21) 205–223.

IPCC (2013) In: Stocker TF, Qin D, Plattner GK, Tignor M, Allen SK, Boschung J, Nauels A, Xia Y, Bex V, Midgley PM (eds) Climate change 2013: the physical science basis. Contribution of working group I to the fifth assessment report of the intergovernmental panel on climate change. Cambridge University Press, Cambridge, UK/New York, p 1535

Jagtap S (2007) Managing vulnerability to extreme weather and climate events: Implications for agriculture and food security in Africa. Proceedings of the international conference on climate change and economic sustainability held at Nnamdi Azikiwe University, Enugu, Nigeria. 12–14 June 2007

Kariyasa K, Dewi A (2011) Analysis of factors affecting adoption of integrated crop management farmer field school (icm-ffs) in swampy areas. Int J Food Agric Econ 1(2):29–38

Katungi E, Akankwasa K (2010) Community-based organizations and their effect on the adoption of agricultural technologies in Uganda: a study of banana (Musa spp.) pest management technology. Acta Hortic 879:719–726

Keelan C, Thorne F, Flanagan P, Newman C (2014) Predicted willingness of Irish farmers to adopt GM technology. J Agrobiotechnol Manag Econ 12(3):394–403

Koudahe K, Koffi D, Kayode JA, Awokola SO, Adebola AA (2018) Impact of climate variability on crop yields in southern Togo. Environ Pollut Clim Change 2:148

Kucharik CJ, Ramankutty N (2005) Trends and variability in US corn yields over the twentieth century. J Earth Interact 9:1–29

Mariara KJ, Karanja FK (2006) The economic impact of climate change on Kenyan crop agriculture, a Ricardian approach. Centre for Environmental Economics and Policy in Africa, University of Pretoria, CEEPA Discussion Paper No. 12

Mauceri M, Alwan J, Norton G, Barrera V (2005) Adoption of integrated pest management technologies: acase study of potato farmers in Carchi, Ecuador; selected paper prepared for presentation at the American Agricultural Economics Association annual meeting, Providence, Rhode Island, 24–27 July 2005

Mignouna B, Manyong M, Rusike J, Mutabazi S, Senkondo M (2011) Determinants of adopting imazapyr-resistant maize technology and its impact on household income in western Kenya. AgBioforum 14(3):158–163

NAADS (2004) National Agricultural Advisory Services, Annual Report 2003/04; Ministry of Agriculture Animal Industry and Fisheries, Entebbe

Nwafor JC (2007) Global climate change: the driver of multiple causes of flood intensity in Sub-Saharan Africa. Paper presented at the international conference on climate change and economic sustainability held at Nnamdi Azikiwe University, Enugu, Nigeria, 12–14 June 2007

Olsson L, Opondo M, Tschakert P, Agrawal A, Eriksen S, Ma S, Perch L, Zakeldeen S (2014) Livelihoods and poverty: climate change 2014: impacts, adaptation, and vulnerability. Part A: global and sectoral aspects. Contribution of working group II to the fifth assessment report of the intergovernmental panel on climate change. Cambridge University Press, Cambridge, UK, pp 793–832

Oryokot J, Forster M, Kayayo B, Pookat V (2004) Public/Private sector partnerships in sustainable development and promotion of rice for improved household incomes. Paper presented at the year of rice conference, Uganda

Porter JR, Xie L, Challinor AJ, Cochrane K, Howden SM, Iqbal MM, Lobell DB, Travasso MI (2014) Food security and food production systems. In: Field CB, Barros VR, Dokken DJ, Mach KJ, Mastrandrea MD, Bilir TE, Chatterjee M, Ebi KL, Estrada YO, Genova RC, Girma B, Kissel ES, Levy AN, MacCracken S, Mastrandrea PR, White LL (eds) Climate change 2014: impacts, adaptation, and vulnerability. Part a: global and sectoral aspects. Contribution of working group II to the fifth assessment report of the intergovernmental panel on climate change. Cambridge University Press, Cambridge, UK/New York, pp 485–533

Ramírez F, Kallarackal J (2015) The effect of increasing temperature on phenology. In: Responses of fruit trees to global climate change. Springer Briefs in Plant Science. Springer, Cham

Reilly J (1996) Agriculture in a changing climate: Impacts and adaptation. In Climate change 1995: Impacts, adaptation and mitigation of climate change:scientific-technical analysis. Contribution of a working group II to the 2nd assessment report of the intergovernmental panel on climate change (R. T. Watson, M. C. Zinyowera and R. H. Moss eds.). Cambridge University Press, Cambridge, 429–467.

Rugadya M (2010) Women's land rights in Uganda: status of implementation of policy and law on women's land rights for ECA, ACGS Addis Ababa. Maastricht University

Robertshaw D and Finch V (1976). The effects of climate change on productivity of beef cattle. In Cattle production in developing countries (A. J. Smith ed.). University of Edinburgh, Edinburgh, 132–137.

Shumetie A, Yismaw MA (2018) Effect of climate variability on crop income and indigenous adaptation strategies of households. Int J Clim Change Strateg Manag 10(4):580–595

Sogbedji JM (1919) Maize nitrogen utilization and nitrate leaching modeling in Togo and New York. Cornell University, New York

Uaiene R, Arndt C, Masters W (2009) Determinants of agricultural technology adoption in Mozambique. Discussion Papers No. 67E

UNDP (2010) Gender, climate change and community-based adaptation: a guidebook for designing and implementing gender-sensitive community-based adaptation programmes and projects. UNDP, New York

Van den Bossche P, Coetzer JAW (2008). Climate change and animal health in Africa. Review Science Technology Off. Int. Epiz, 27(2) 551–562.

van Oort P, Zwart SJ (2017) Impacts of climate change on rice production in Africa and causes of simulated yield changes. Glob Chang Biol 24(3):1029–1045

Waha K, Muller C, Bondeau A, Dietrich JP, Kurukulasuriya P, Heinke J et al (2013) Adaptation to climate change through the choice of cropping system and sowing date in sub-Saharan Africa. Glob Environ Change-Hum Policy Dimens 23(1):130–143

Walter O, Edgardo V, Patricia L (2010) Climate change and links to animal diseases and animal production. Conf. OIE (179-186). Retrieved on 21/6/ 2019 from http://www.oie.int./doc/gen/D11834.PDF

Integrated Soil Fertility Management Based on Pigeon Pea and Cowpea Cropping Systems Influences Nitrogen Use Efficiency, Yields and Quality of Subsequent Maize on Alfisols in Central Malawi

Keston O. W. Njira, Ernest Semu, Jerome P. Mrema, and Patson C. Nalivata

Abstract An integrated soil fertility management (ISFM) involving food grain legumes in predominantly maize systems is becoming popular in southern Africa. A study on diversified pigeon pea and cowpea cropping systems was conducted on Alfisols in Lilongwe and Dowa districts of Central Malawi. The objective of the study was to evaluate the residual effects of legume-legume and legume-cereal intercropping systems involving pigeon pea (PP) and cowpea (CP) when integrated with inorganic nitroen (N) fertilizer on N mineralization patterns, rotational maize (MZ) N uptake, maize yield, maize grain crude protein content (grain quality), and nitrogen use efficiency (NUE) in the forfm of partial factor productivity of applied N (PFP_N). The field experiment was arranged in a split plot design with the previous legume-based systems (with residues retained) as main plots and the 0, 45, 90 and 120 kg ha^{-1} inorganic N rates as sub-plots. Results showed significantly higher ($P < 0.05$) NUE in plots that were previously legume-based of 76.5 kg grain kg^{-1} N applied in a previously Sole CP, compared to 39.4 kg grain kg^{-1} N applied in a previously Sole MZ, for Lilongwe site and 78.4 kg grain kg^{-1} N applied in a preciously PP + CP intercrop, compared to 40.5 kg grain kg^{-1} N applied in a previously Sole MZ, for Dowa site. Similarly, results from treatments that were previously legume-based consistently showed significantly higher ($P < 0.05$) maize grain yield increases by ranges of 30–59% at 0 kg N ha^{-1} and 28–42% at 120 kg N ha^{-1}fertilizer application rate at the Lilongwe site. Similar trends were observed for the Dowa site, with the highest grain yield increase of 45% under treatment that was previously PP + CP intercrops. Maize grain protein significantly ($P < 0.05$) increased with additions of

K. O. W. Njira (✉) · P. C. Nalivata
Crop and Soil Sciences Department, Lilongwe University of Agriculture and Natural Resources, Lilongwe, Malawi
e-mail: knjira@luanar.ac.mw

E. Semu · J. P. Mrema
Department of Soil and Geological Sciences, Sokoine University of Agriculture, Morogoro, Tanzania

B. R. Singh et al. (eds.), *Climate Impacts on Agricultural and Natural Resource Sustainability in Africa*, https://doi.org/10.1007/978-3-030-37537-9_5

inorganic N by ranges of 9.0–11.4% and 10.2–13.1% for Lilongwe and Dowa sites, respectively. From this study, it is noted that the novel legume-legume cropping system involving pigeon pea and cowpea is amongst technologies that improve NUE, yields and quality of the subsequent maize crop and play a great role in sustainable agricultural intensification.

1 Introduction

Declining soil fertility continues to be one of the main challenges for sub-Saharan Africa crop production systems (Zingore et al. 2015). Continuous soil erosion and nutrient depletion are some of the most pressing challenges in managing agricultural soils in most parts of the region (Tamene and Le 2015). Earlier studies listed Burundi, Malawi, Kenya, Ethiopia, Lesotho and Rwanda as countries with very high nutrient depletion rates (Stoorvogel and Smailing 1990). The rising costs of inorganic fertilizers and growing human populations are exacerbating the challenges in managing soil fertility and increasing crop production (AGRA 2014).

For the case of Malawi, where agricultural production is dominated by smallholder maize production, affordability of inorganic fertilizers is one of the crucial challenges (African Development Bank 2011), with nitrogen being the most limiting element in maize production (Makumba 2003; Tamene et al. 2015). Over the years, the Malawi government has shown a political will to improve the situation by implementing a farm input subsidy policy, which reasonably improved yields in some years but with unproven sustainability (Dorward and Chirwa 2011; Ricker-Gilbert et al. 2014). Similar to many other countries, research and implementation of various soil management interventions, including use of various types of manures, agroforestry technologies and inclusions of legumes in the cropping systems, are promoted. Inclusion of food legumes seems to be the most popular in Malawi due to a number of factors including food diversity and security, increased income and soil fertility improvement (Kerr et al. 2007; Kamanga et al. 2010; Snapp et al. 2014).

Legumes such as common beans (*Phaseolus vulgaris*), groundnuts (*Arachis hypogaea*), pigeon pea (*Cajanus cajan*), soybean (*Glycine max*) and cowpea (*Vigna unguiculata*) are grown by many farmers in Malawi, both as food and cash crops (Kamanga et al. 2010). They are grown in various cropping systems, including sole or monocropping, legume-cereal intercrops and legume-legume intercrops, commonly referred to as "doubled-up" legume technology (ICRISAT/MAI 2000).

Soil fertility benefits from legume interventions evaluated through their effects on maize in short rotations have been reported from studies done in Malawi and elsewhere (Sakala et al. 2003; Mhango 2011). However, very few studies have reported soil fertility benefits from legumes that are grown as legume-legume intercrops. Soil fertility benefits on maize yields from doubled-ups of pigeon pea-groundnuts or pigeon pea-soybean were reported to be comparable to those of sole crops (Mhango 2011; Njira et al. 2013; Phiri et al. 2014).

However, information on effects of the combinations of crops, specifically pigeon pea-cowpea doubled-up on maize yields, and yield components such as quality is scanty. Furthermore, information on the mineralization patterns of legume residues in various combinations of either legume-legume or legume-cereal is scanty. Therefore, the objective of this study was to evaluate and compare the effects of the pigeon pea-cowpea "doubled-up", in comparison with sole and legume-cereal intercrops, on N mineralization patterns and their implications on N uptake, nitrogen use efficiency, yields and grain protein content of maize in short rotation when integrated with inorganic fertilizers.

2 Materials and Methods

2.1 Site Description and Characterization

The study was conducted as a continuation of the 2013/2014 legume-based study at Lilongwe and Dowa districts of central region of Malawi (Sect. 2.4). Assessment of the effects of legume-based treatments on subsequent maize was conducted in the 2014/2015 growing season. For the Lilongwe site, the experiment was conducted at the Lilongwe University of Agriculture and Natural Resources (LUANAR) research farm (14° 11′ S, 33° 46′ E) within the Mkwinda Extension Planning Area (EPA), specifically at Bunda Campus; and in Dowa district, the experiment was conducted at Nachisaka EPA (13° 37′ S, 33° 56′ E). Both Lilongwe and Dowa districts fall within the medium-altitude agro-ecological zone of Malawi. Lilongwe (Bunda Campus) is at 1157 m above sea level, whereas Dowa (Nachisaka EPA) is at 1190 m above sea level. Overall the medium-altitude areas receive an average rainfall of about 875 per annum. Soils of the two sites are deep well-drained Chromic Luvisols (WRB classification) or Alfisols (Typic Hapludalfs in the USDA Soil Taxonomy) (Chilimba et al. 2011; Mutegi et al. 2015).

2.2 Initial Soil Fertility Status at the Start of the Experiment

Soil analysis at the beginning of the 2013/2014 cropping season at the 0–20 cm depth indicated that the soils have sand-clay-loam texture, moderate acidity (pH of 6.0), low soil organic matter (1.1%), low total N (0.05%), high available Mehlich-3 P (57 mg kg^{-1}) and high amounts of basic cations showing 1.24, 4.24 and 0.35 $cmol_c$ kg^{-1}soil for Mg, Ca and K, respectively, for the Dowa site. On the other hand, the Lilongwe site soil have sand-clay-loam texture, slight acidity (pH of 6.2), medium soil organic matter (2.8), medium total N (0.14%), high available Mehlich-3 P (41 mg kg^{-1}) and adequate amounts of basic cations showing 0.99. 4.78 and 0.23 $cmol_c$ kg^{-1} soil for Mg, Ca and K, respectively.

2.3 Soil Sampling and Analysis

Soil sampling and analysis were also conducted in the 2014/2015 growing season. Soil samples were collected from each of the plots in the previous legume-based treatments in the depth ranges of 0–15 cm and 15–30 cm. This was after harvesting the legume-based treatments of 2013/2014 growing season, in preparation for the 2014/2015 growing season. Soils were analysed for total N, available P, soil organic carbon (SOC) and pH using the micro-Kjeldahl method, Mehlich-3 extraction procedures, wet oxidation and potentiometric method, respectively, as outlined by Anderson and Ingram (1989). Soil analysis results for the 2014/2015 growing season are presented in Table 1. Rainfall was recorded from each of the sites during the cropping season.

2.4 Treatments and Residue Quality Determination

The 2013/2014 legume-based treatments included sole pigeon pea (PP), sole cowpea (CP), pigeon pea-cowpea (PP + CP) intercrop, pigeon pea-maize (PP + MZ) intercrop, cowpea-maize CP + MZ) intercrop, sole maize (unfertilized – Sole MZ) and sole maize fertilized at 92 kg N ha^{-1} (Sole MZ + 92 N). The 92 kg N ha^{-1} is the current recommended N rate by Malawi's Ministry of Agriculture and Food Security (MoAFS) (2012). The plot size was 15 m × 7 m. The crop residues were incorporated in the soil soon after harvesting. The legume-based cropping systems' treatments were arranged in randomized complete block design (RCB).

In the 2014/2015 study year, treatments were superimposed on the 2013/2014 legume-based treatments. Each legume-based treatment plot was split into four sub-plots that received different levels of inorganic N as follows: 0, 45, 90 or 120 kg N ha^{-1} (in the 7 m × 3.75 m sub-plots). Therefore, the split plot arrangement in RCB was obtained. The sources of N applied were the basal NPK fertilizer (23:21:0 + 4S) and the top-dressing N fertilizer was urea (46% N). One maize seed was planted per each planting station at a spacing of 25 cm between planting stations along the ridge/row and 75 cm between ridges. The maize variety used was the DKC 8053. All conventional crop management practices such as land preparation and weeding were followed. Nitrogen concentration and C/N ratios were determined in the crop residues before incorporation, to assess their quality according to assess their quality according to Palm et al.(2001) and Brady and Weil (2008).

2.5 Soil Mineral N Assessment During the Growing Season

Soil samples were collected from the 0–20 cm depth by augering a minimum of three randomly selected points from each sub-plot. Composite samples were made from soils of each sub-plot. Then, the composite samples were kept in cooler boxes and

Table 1 Selected post-harvest soil properties from the previous legume-based cropping systems at Lilongwe and Dowa sites, Central Malawi

	Previous cropping system	pH		Total N (%)		Available Mehlich-3 P (mg/kg)		SOC	
Site	Depth (cm) →	0–15	15–30	0–15	15–30	0–15	15–30	0–15	15–30
Lilongwe	Sole MZ	6.1	6.1	0.057	0.054	40.9	40.2	1.14	1.07
	Sole MZ + 92 N	5.7	5.9	0.063	0.059	44.3	38.4	1.25	1.20
	Sole CP	6.1	6.0	0.062	0.058	41.8	39.3	1.24	1.16
	CP + MZ intercrop	6.0	6.1	0.065	0.054	45.5	38.6	1.31	1.08
	Sole PP	6.1	6.1	0.059	0.054	45.0	37.4	1.17	1.09
	PP + CP intercrop	5.9	6.0	0.057	0.055	41.2	39.2	1.14	1.12
	PP + MZ intercrop	6.1	6.0	0.058	0.059	37.9	41.2	1.17	1.19
Dowa	Sole MZ	6.2	6.1	0.12	0.09	28.3	27.5	2.45	1.80
	Sole MZ + 92 N	6.1	6.3	0.14	0.10	27.5	27.3	2.86	2.00
	Sole CP	6.3	6.3	0.15	0.10	32.1	27.5	2.96	2.09
	CP + MZ intercrop	6.2	6.4	0.14	0.09	34.2	29.3	2.88	1.90
	Sole PP	6.3	6.2	0.13	0.10	32.1	24.7	2.65	2.04
	PP + CP intercrop	6.1	6.0	0.14	0.10	28.3	25.4	2.85	1.95
	PP + MZ intercrop	6.2	6.1	0.13	0.09	25.4	26.8	2.63	1.87

SOC soil organic carbon, *MZ* maize, *CP* cowpea, *PP* pigeon pea, 92 N = 92 Kg N Ha^{-1}. Critical values: $N = 1.13$ (Landon 1991), $P = 19$ Mg/Kg (Chilimba 2007), SOC = 0.88% (Chilimba 2007), pH = 5.5 (Chilimba 2007)

taken to the laboratory and kept in a refrigerator to avoid further mineralization. The soil mineral N, both nitrate (NO_3^-) and ammonium (NH_4^+), were then extracted using 2 M KCl and then analysed colourimetrically according to the procedure as outlined by Anderson and Ingram (1989). The N mineralization assessments were done as a series of activities from a week of planting (wk 0) to the 11th week after planting (wk 11).

2.6 Harvesting and Determination of Yields, Harvest Indices and Plant Analysis to Determine Grain Protein Content

Harvesting was done after the maize had reached physiological maturity and dried. A net plot of 5.5 m x 1.5 m was demarcated from each sub-plot leaving guard rows, to avoid outside interference. The maize stalks were cut at ground level with cobs not removed. These were weighed as the total aboveground biomass. Then the cobs were detached and the grains were threshed. The net plot grain fresh weight was recorded and a sample of 100 seeds was taken using a quartering procedure and brought to the laboratory. Grain yield, total dry matter yield and harvest indices were determined according to Mloza-Banda (1994). Furthermore, grain N content was determined using the procedure described by Anderson and Ingram (1989). Maize grain crude protein content was determined by multiplying grain N concentration by a Jones factor of 6.25, according to FAO (2003).

2.7 Determination of N Uptake and Nitrogen Use Efficiency

Maize plant samples were collected at tasselling, and five plants from each sub-plot were sampled for determination of N concentration in maize plant, using the procedure described by Anderson and Ingram (1989). Nitrogen uptake by the maize plant was determined by multiplying the total aboveground dry matter (TDM) with plant N concentration. On the other hand, nitrogen use efficiency was determined for applied nitrogen, using the partial factor productivity of applied N (PFP_N), whereby the amount (kg) of grain yield produced is divided by amount (kg) of N applied (Dobermann 2005).

2.8 Rainfall Data Collection

Rainfall amounts in millimetres (mm) were collected from both Lilongwe and Dowa study sites. For the Lilongwe site, the experiment was conducted at a university research farm, where a rain gauge is installed and rainfall is systematically recorded every year. On the other hand, the experiment in Dowa district was implemented at a farmer's field; and therefore, a rain gauge was installed on the field and data were systematically recorded by an extension officer who was residing in the same village. Daily rainfall recordings of both sites were computed to be presented as monthly rainfall totals.

2.9 *Data Analysis*

Data was subjected to analysis of variance (ANOVA) using GenStat 15 statistical package. Separation of means was achieved using Tukey's honestly significant difference (HSD) test, where the cropping system and fertilizer N rate factors were analysed. Additionally, one-way ANOVA was used to analyse a specific factor, where means were separated by the least significant differences (LSD) at 5% significant level. Data for each site were analysed separately.

3 Results

3.1 *Monthly Rainfall at the Lilongwe and Dowa Sites in the 2014/2015 Cropping Season*

Monthly rainfall distribution for the Lilongwe and Dowa sites in the 2014/2015 cropping season is presented in Fig. 1. The Lilongwe site received rainfall from November to May, with the highest amount of 243 mm in the month of January and an annual total amount of 639 mm. On the other hand, the Dowa site received rainfall from the month of December to April, with the highest amount of 189 mm in the month of February and an annual total of 577 mm. The total amount rainfall was marginally adequate.

3.2 *Selected Post-harvest Soil Properties from the Previous Legume-Based Cropping Systems for Lilongwe and Dowa Sites*

Post-harvest (2013/2014 growing season) soil analysis for the Lilongwe site showed medium to slightly acidic soil reaction, with pH ranging from 5.7 in the previous sole maize plus 92 kg N ha^{-1} plot (Sole MZ + 92 N) to 6.1 in a number of previous

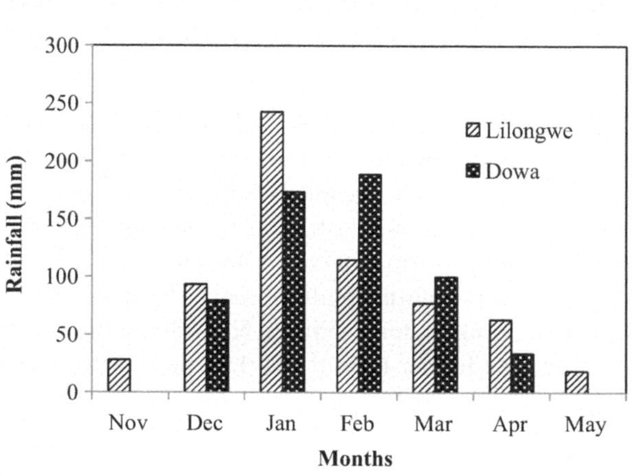

Fig. 1 Monthly rainfall (mm) during the 2014/2015 growing season for Lilongwe and Dowa sites, Central Malawi

legume-based treatments in the topsoil (Table 1). The Lilongwe topsoil pH ranged from 5.9 (Sole MZ + 92 N) to 6.1 in a number of previous legume-based treatments. Furthermore, the overall mean total soil N was 0.06 and 0.056 for the topsoil and subsoil, respectively. Soil organic carbon (C) was also low, with overall means of 1.2 and 1.1 for the topsoil and subsoil, respectively. On the other hand, the available (Mehlich-3) P was high with overall means of 42.4 and 39.2 mg kg^{-1} for the topsoil and subsoil, respectively (Table 1).

For the Dowa site, Table 1 shows a slightly acidic soil reaction with a pH range of 6.1–6.3 for the topsoil and a pH of 6.0–6.4 for the subsoil. Total soil N was low to medium, with overall mean values of 0.14 and 0.10 in the topsoil and subsoil, respectively. Soil organic C was medium with overall mean values of 2.75 and 1.95 in the topsoil and subsoil, respectively. On the other hand, available (Mehlich-3) P was medium with overall mean values of 29.7 and 26.9 mg kg^{-1} for the topsoil and subsoil, respectively.

3.3 Stover Quality from Previous Cropping Systems for Lilongwe and Dowa Sites

Results showed different characteristics of the maize stover, ranging from high quality (N % > 2.5 and C/N ratios <25) to low quality (N % < 2.5 and C/N ratios >25). For the Lilongwe site, sole cowpea (CP) stover was of the highest quality, with N concentration of 2.7% and C/N ratio of 16.4, whilst maize stover from the maize-cowpea intercrop showed the lowest quality with N concentration of 0.7% and C/N ratio of 59.4. For the Dowa site, high-quality stover was depicted in Sole CP and Sole PP, both with N concentration of 2.8 with C/N ratios of 15.9 and 16.4, respectively. Maize stover in the intercrops with cowpea and pigeon pea showed the lowest quality characteristics, with N concentrations of 0.8 and 0.7% and C/N ratios of 63.5 and 58.1, respectively.

3.4 Field Soil Mineral N Patterns as Influenced by the Previous Season at Lilongwe and Dowa Sites

Figure 2 shows mineral N (NO_3^-, NH_4^+) levels in soils sampled from planting week to week 11 after planting. Results are presented only for the legume-based sub-plots that did not receive inorganic N fertilizer application to elucidate the clear effect of the previous cropping systems. The general trend showed decrease in mineral N levels from the planting week to the later weeks.

There were significant differences ($P < 0.05$) in nitrate N levels due to the previous cropping systems. Nitrate N levels for previous legume-based systems were significantly higher ($P < 0.05$) than that in the previous Sole MZ by a mean of

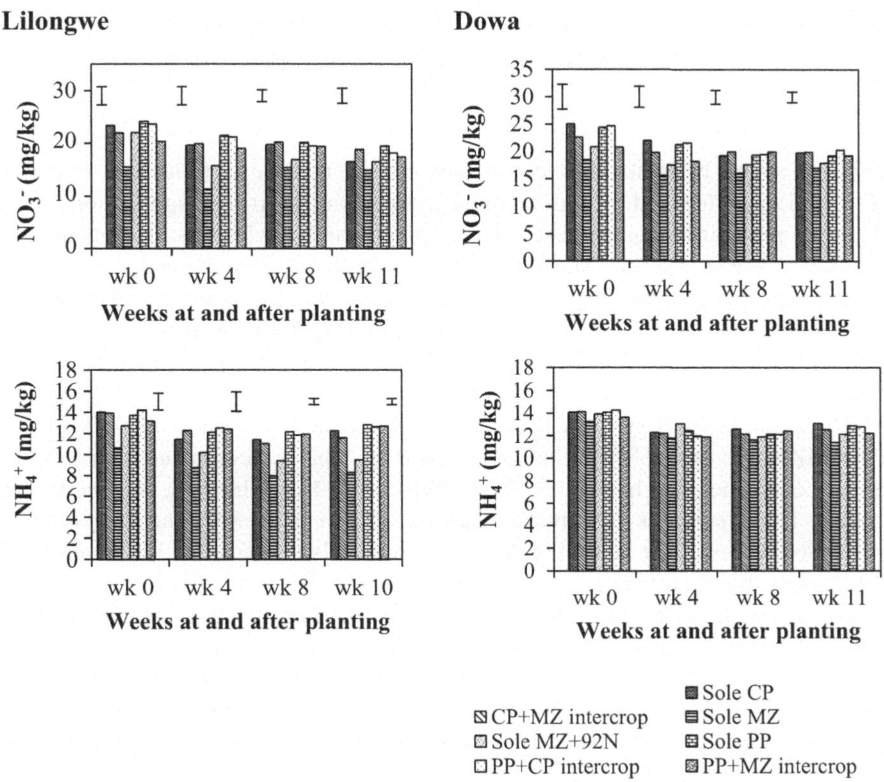

Fig. 2 Soil mineral N in the maize plots in the 2014/2015 cropping season as influenced by the previous season cropping systems at Lilongwe and Dowa sites, Central Malawi. The above analysis shows mineral N patterns in the sub-plots that did not receive fertilizer N application in the 2014/2015 cropping season (vertical bars within the charts represent LSD values for separation of means for each specific week. MZ Maize, CP Cowpea, PP Pigeon Pea, 92 N = 92 Kg N Ha^{-1})

29% in wk. 0–17% in wk. 11 for the Lilongwe site (Fig. 2). Furthermore, the plot that was previously Sole MZ + 92 N showed equally higher NO_3-N levels during the planting week but drastically dropped in the subsequent weeks. There were no significant differences in NO_3-N amongst plots previously planted to legumes as sole cropping, legume-legume intercrop or legume-cereal intercrop. The trend for NH_4-N was somewhat similar to that of NO_3-N. Similar to the Lilongwe site, in the Dowa site, the trend of mineral N in the soils of the study plots showed a decrease from the planting to week 11 after planting. The previous legume-based plots showed higher ($P < 0.05$) NO_3-N than that in the previous Sole MZ plots by 19% in wk. 0 to and 14% in wk. 11. The trend was similar for NH_4-N. However, no significant differences were observed in NH_4-N as influenced by the previous cropping systems.

3.5 Nitrogen Uptake by Maize Plants as Influenced by the Previous Cropping Systems at Lilongwe and Dowa Sites

Figure 3 shows N uptake by the maize plants. There were significant differences ($P < 0.05$) as influenced by both cropping system and N fertilizer application. Plots that were previously legume-based showed significantly higher ($P < 0.05$) N uptake than the Sole MZ and Sole MZ + 92 N. Increasing N application rate led to significant increases in N uptake by the maize plants. However, from plots that were previously legume-based, N uptake rates at 45, 90 and 120 kg N ha^{-1} were not significantly different; but all these were significantly higher ($P < 0.05$) than the treatment with 0 kg N ha^{-1}.

The legume-based systems led to higher N uptake by a range of 38% in the PP + MZ intercrop to 68% in the Sole CP at 0 kg N ha^{-1} as compared to that in the Sole MZ. On the other hand, at the highest rate of 120 kg N ha^{-1}, the legume-based systems led to increases in N uptake by a range of 28% in PP + MZ intercrop to 48% in the Sole CP over the Sole MZ. Plots that previously contained Sole CP and Sole PP showed slightly higher N uptakes than the rest of the treatment plots.

Similar to the Lilongwe site, in the Dowa site, plant N uptake by subsequent maize was significantly influenced by both the previous cropping systems and N fertilizer application. Increasing fertilizer N application led to significant increases in N uptake (Fig. 3). Nitrogen uptake rates at 45, 90 and 120 kg N ha^{-1} were all significantly higher ($P < 0.05$) than that at 0 kg N ha^{-1} but were not significantly higher amongst themselves. The legume-based systems led to higher N uptake by a range of 23% in the CP + MZ intercrop to 41% in the PP + MZ at 0 kg N ha^{-1} than that in the Sole MZ. On the other hand, at the highest rate of N (120 kg N ha^{-1}) the legume-based systems led to increases in N uptake by a range of 23% in PP + MZ intercrop to 36% in the PP + CP intercrop over the Sole MZ.

3.6 Maize Grain and Total Dry Matter Yields and Harvest Indices as Influenced by the Previous Cropping Systems at Lilongwe and Dowa Sites

Maize grain and total dry matter or aboveground biomass were significantly ($P < 0.05$) influenced by both cropping system and N application (Fig. 4). Increasing the rate of N application led to increase in yields. Treatments that were previously legume-based showed significantly higher ($P < 0.05$) maize grain yields at all the four rates of N fertilizers than that by the treatments under the previous Sole MZ. Amongst the previous legume-based treatments, the 0 kg N ha^{-1} showed significantly lower yields than those by the rest of the higher

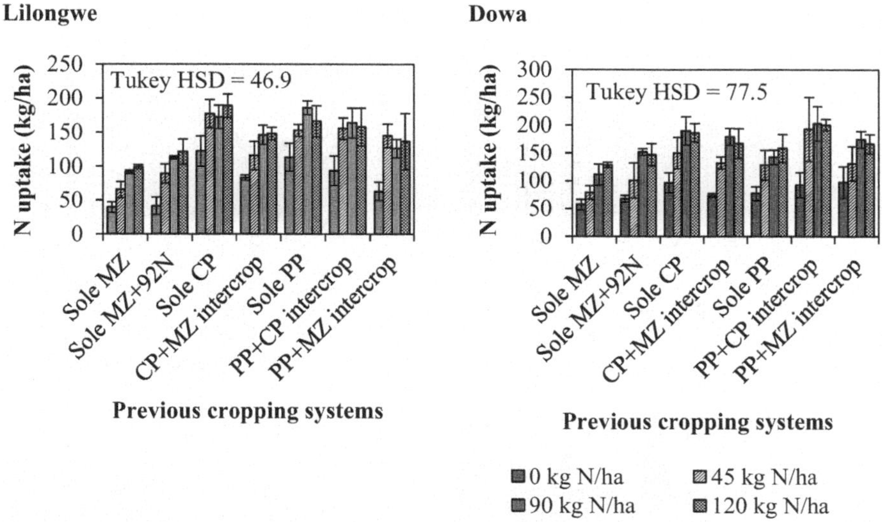

Fig. 3 (**a**, **b**) N uptake in maize crop as influenced by the previous cropping systems and N fertilizer rate at Lilongwe and Dowa sites. Means were separated using Tukey's HSD test at 5% significant level, and its value is shown on the chart; error bars represent standard errors of the means. MZ maize, CP cowpea, PP pigeon pea, 92 N = 92 kg N ha^{-1}

N rates of 45, 90 and 120 kg N ha^{-1}, which, amongst themselves, were not significantly different. The previous legume-based systems produced significantly higher ($P < 0.05$) grain yields than the previous Sole MZ by a range of 30% under previous CP + MZ intercrop to 59% under previous Sole CP at 0 kg N ha^{-1} fertilizer application. Similarly, at the highest rate of N application, 120 kg N ha^{-1}, the previous legume-based treatments were significantly higher ($P < 0.05$) than the previous Sole MZ by a range of 28% under previous CP + MZ intercrop to 42% under previous Sole CP. The TDM yields followed a similar trend. Furthermore, HI% by the treatments on the previous Sole MZ was significantly lower than the rest of the cropping systems (Fig. 5). Additionally, significant differences ($P < 0.05$) in HI% were also observed as influenced by the rate of N applied for the Lilongwe site.

Maize grain yields for the Dowa site were significantly different ($P < 0.05$) as influenced by both cropping system and fertilizer N rate applied. Treatments that were previously legume-based showed significantly higher ($P < 0.01$) grain yields than those that were previously Sole MZ at all rates of N fertilizer applied. At the 0 kg N ha^{-1}, they were significantly higher ($P < 0.05$) by a range of 24% under treatment that was previously CP + MZ intercrop to 48% under the treatment that was previously Sole PP, whereas at the 120 kg N ha^{-1}, they were significantly higher ($P < 0.05$) by a range of 35% under previous PP + MZ intercrop to 45% under treatment that was previously PP + CP. On the other hand, the previously Sole MZ + 92 N treatment yields were not significantly different from the legume-based plots' yields at 120 kg N ha^{-1}. Furthermore, the treat-

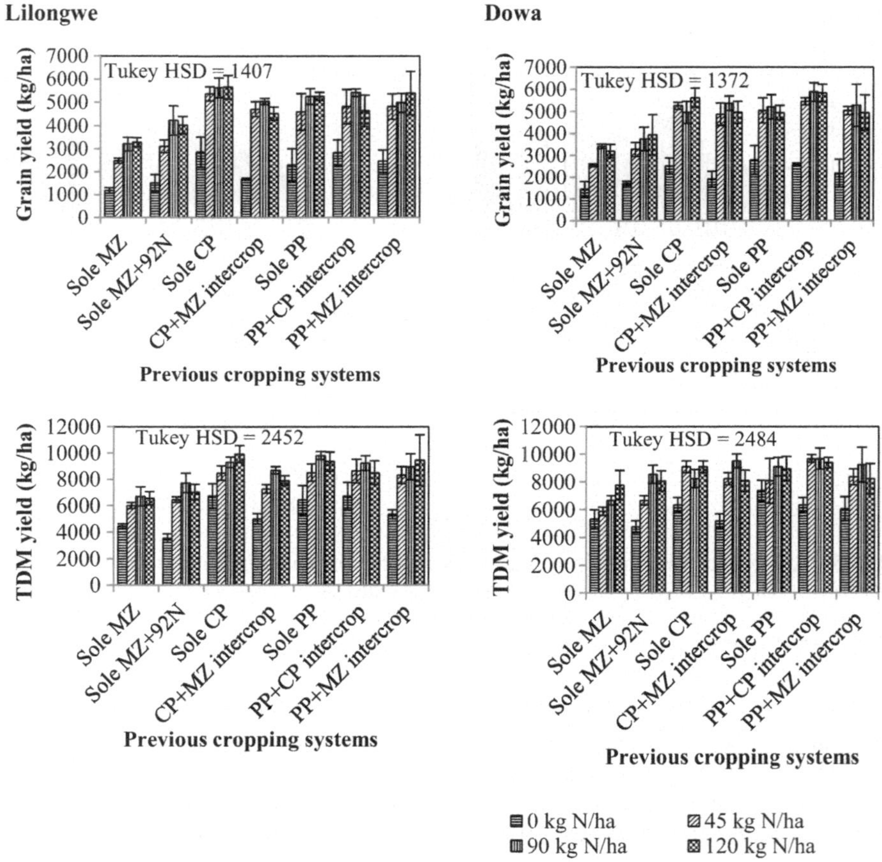

Fig. 4 Grain and total dry matter (TDM) yields of maize as influenced by the previous cropping systems and N fertilizer rate at Lilongwe and Dowa study sites. Means were separated using Tukey's HSD test at 5% significant level, and its value is shown on the bar charts; error bars represent standard errors of the means. MZ maize, CP cowpea, PP pigeon pea, 92 N = 92 kg N ha^{-1}, TDM total dry matter yield

ment that was previously Sole MZ + 92 N yield at 45 kg N ha^{-1} was not significantly different from the previous Sole MZ grain yield at the same rate of fertilizer. The harvest indices were also significantly affected by both cropping system and the amount of N applied (Fig. 5). Treatments that previously contained legumes produced significantly higher ($P < 0.05$) HI% than the treatments that were previously Sole MZ. However, the HI% by the previously PP + MZ, CP + MZ and MZ + 92 N were not significantly different from that under treatment that was previously Sole MZ. Increasing amount of N fertilizer led to significant increases in HI%.

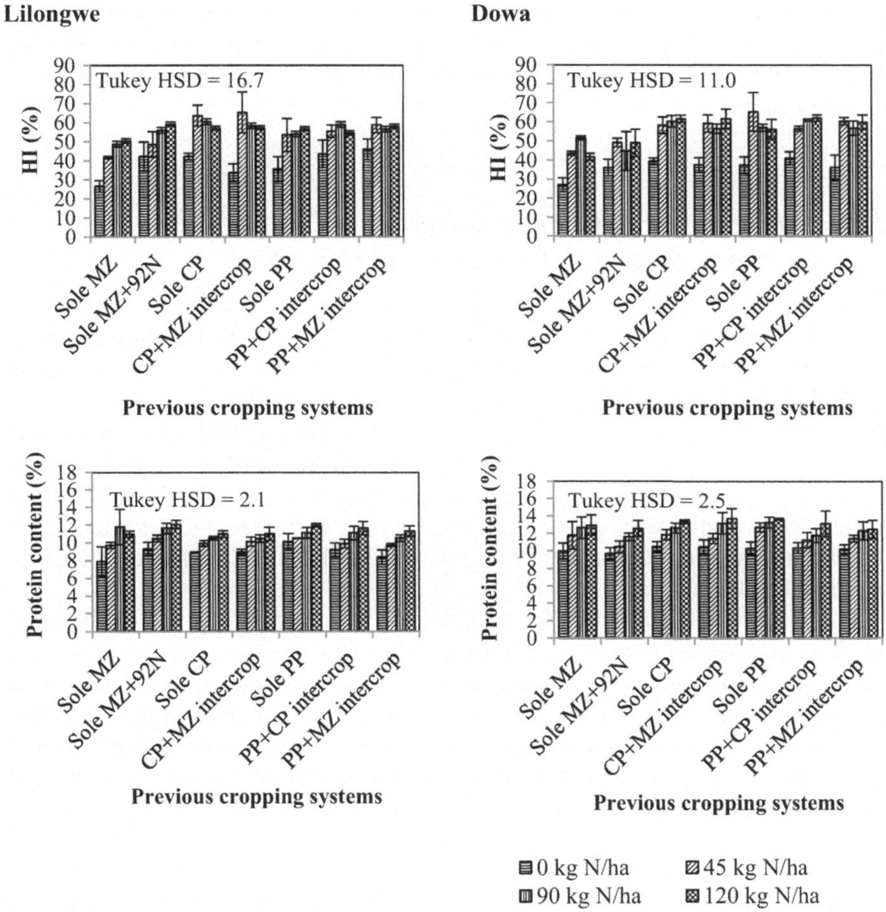

Fig. 5 Harvest index percentages (HI%) and grain protein content (%) of maize as influenced by the previous cropping systems and N fertilizer rate at Lilongwe and Dowa study sites. Means were separated using Tukey's HSD test at 5% significant level and its value is shown on the bar charts; error bars represent standard errors of the means. MZ = maize, CP = cowpea, PP = pigeon pea, 92 N = 92 kg N ha^{-1}

3.7 *Maize Grain Protein Content as Influenced by the Previous Cropping Systems at Lilongwe and Dowa Sites*

Maize grain protein content (Fig. 5) was significantly different ($P < 0.01$) due to N fertilizer rate applied. The trend of protein percentages increased in maize grain was from 9.0% in 0 kg N ha^{-1} to 11.4% protein in 120 kg N ha^{-1}inorganic fertilizer application (Fig. 5). In general, increasing the amount of N applied led to increase in maize grain protein content. For the Dowa site, the trend in maize grain protein percentage was similar to that of the Lilongwe site. Increasing amount of N led to significant increase in grain protein content (Fig. 5) with the lowest amount of 10.2% at 0 kg N $ha^{-1}$and highest amount of 13.1% at 120 kg N ha^{-1}inorganic fertilizer application.

3.8 Nitrogen Use Efficiency as Influenced by the Previous Cropping System at Lilongwe and Dowa Sites

The nitrogen use efficiency as demonstrated by the PFP_N showed significant differences ($P < 0.05$) as influenced by cropping systems and amount of fertilizer N applied (Table 2). Treatments that were previously legume-based systems were significantly higher in terms of their N use efficiency in both applied and indigenous N (PFP_N) than those that were previously Sole MZ and Sole MZ + 92 N. The overall mean values for treatments that were previously Sole MZ and Sole MZ + 92 N were not significantly different. The highest efficiency for PFP_N was at 45 kg N ha^{-1} in all the treatments, and the treatment that was previously Sole CP showed a slightly higher value than the rest.

For the Dowa site, the PFP_N was significantly ($P < 0.05$) influenced by the cropping systems and fertilizer N amount applied (Table 2). Similar to the Lilongwe site, the highest NUE was at 45 kg N ha^{-1} fertilizer application rate in all legume-based treatments. All the former legume-based treatments showed mean PFP_N values that were higher than the common ranges of 40–70 and were all significantly higher ($P < 0.05$) than that under the Sole MZ. The treatment that was previously PP + CP intercrop showed a slightly higher value than the rest.

4 Discussion

In this study an integrated nutrient management approach was implemented; crop residues were retained to the system and different levels of inorganic fertilizer were added. Levels of N concentrations and C/N ratios of the different crop residues used under the present study were similar to those of other past studies (Palm et al. 2001; Mtambanengwe et al. 2007; Abera et al. 2013). For both sites, sole cropped legume residues showed the highest quality (high N and lower C/N ratios) which can be attributed to the nutrient accumulations achieved during the growth of those crops. On the other hand, legume crop residues from intercropped species (Table 3) still satisfied the high-quality (i.e. high N concentration and low C/N ratios) criteria according to Palm et al. (2001) and Brady and Weil (2008).

For the Lilongwe and Dowa sites, soil mineral N (NO_3^- and NH_4^+) patterns showed ranges (Fig. 2) similar to those that have been reported by a number of authors in similar environments (Mhango 2011; Matusso et al. 2014). The signifi-

Table 2 Nitrogen use efficiency in the form of PFP_N as influenced by the previous cropping system and fertilizer rate on maize at Lilongwe and Dowa sites, Central Malawi

Site	Fertilizer rate	Previous cropping systems: Sole MZ	Sole MZ + 92 N	Sole CP	CP + MZ intercrop	Sole PP	PP + CP intercrop	PP + MZ intercrop	Mean (FR effect)
PFP_N (kg grain/kg N applied)									
Lilongwe	0	–	–	–	–	–	–	–	–
	45	55.2	69.1	119.5	104.9	102.3	107.3	107.3	95.1a
	90	35.6	46.9	62.7	56.1	58.4	60.3	55.4	53.6b
	120	27.3	33.6	47.2	37.7	43.8	38.7	45.1	39.1c
	Mean (CS effect)	39.4b	49.9b	76.5a	66.2a	68.1a	68.8a	69.3a	
	F.pr (CS)	0.001							
	F.pr (FR)	0.001							
	F.pr (CS*FR)	0.048							
	CV (%)	18.4							
Dowa	0	–	–	–	–	–	–	–	–
	45	56.7	72.8	116.9	108.4	112.1	121.3	112.0	100.0a
	90	38.1	41.7	55.1	59.7	57.9	65.3	58.7	53.8b
	120	26.7	32.8	46.8	41.3	41.2	48.6	41.2	39.8c
	Mean (CS effect)	40.5b	49.1b	72.9a	69.8a	70.4a	78.4a	70.6a	
	F.pr (CS)	0.001							
	F.pr (FR)	0.001							
	F.pr (CS*FR)	0.01							
	CV (%)	14.3							

Means were separated using the Tukey's HSD test at 5% significant level, – = not applicable, means with different letters are significantly different; *CS* cropping system, *FR* fertilizer rate, symbol * = interaction, $PFP_N > 70$ kg grain/kg N applied may imply very efficiently managed systems (Dobermann 2005), *MZ* maize, *CP* cowpea, *PP* pigeon pea, 92 N = 92 kg N ha^{-1}

cantly higher levels of mineral N in the plots that had legume residues (CP and PP), whether sole cropped or intercropped, can be attributed to high quality of the legume residues. The maize residues showed very low N concentrations and very high C/N ratios, whereas the legumes, both CP and PP, showed high quality (Table 3). Zeng et al. (2010) and Abbasi et al. (2015) studied various crop residues, including those of maize, groundnuts and soybean, and reported a positive correlation between N mineralization and initial N contents of the residues. This study confirms that addition of high-quality residues to the poor-quality ones improves the rate of nutrient mineralization. This can be attributed to the activation of microbial activities in the soil that need reasonable amounts of nitrogen to facilitate the mineralization process and reduce the period of N immobilization.

Table 3 Concentration of nitrogen and C/N ratios of stover from previous legume-based cropping systems for Lilongwe and Dowa sites, Central Malawi

Site	Crop in the previous cropping system	Stover N concentration (%)	C/N ratio
Lilongwe	Sole MZ	0.8	55.7
	Sole MZ + 92 N	1.0	46.4
	Sole CP	2.7	16.4
	Sole PP	2.3	19.5
	CP in CP + MZ intercrop	2.1	21.7
	CP in PP + CP intercrop	2.3	20.2
	PP in PP + CP intercrop	2.1	21.7
	PP in PP + MZ intercrop	2.2	20.8
	MZ in PP + MZ intercrop	0.8	61.3
	MZ in CP + MZ intercrop	0.7	59.4
Dowa	Sole MZ	0.8	55.9
	Sole MZ + 92 N	1.0	44.8
	Sole CP	2.8	15.9
	Sole PP	2.8	16.4
	CP in CP + MZ intercrop	2.5	18.4
	CP in PP + CP intercrop	2.7	16.7
	PP in PP + CP intercrop	2.5	18.1
	PP in PP + MZ intercrop	2.5	18.3
	MZ in PP + MZ intercrop	0.7	58.1
	MZ in CP + MZ intercrop	0.8	63.5

N (%) < 2.5 implies low quality (Palm et al. 2001); C/N > 25 implies low quality and more likely to induce immobilization of N (Brady and Weil 2008). *MZ* maize, *CP* cowpea, *PP* pigeon pea, 92 N = 92 kg N ha^{-1}, CP in PP + CP = cowpea stover from the pigeon pea + cowpea intercrop

Similarly, Partey et al. (2014) reported an increase in mineralization rate of 58% and 55% by mixing maize residues with *Vicia faba* and *Tithonia diversifolia*, respectively. Furthermore, significant increases in microbial biomass were also reported with *V. faba* and *T. diversifolia* alone or in combination with maize residues (Partey et al. 2014). On the other hand, Sakala et al. (2000) reported increased N mineralization with additional amounts of high-quality PP residues and N fertilizer to maize residues. It should be noted that the previously Sole MZ + 92 N plots showed relatively higher amounts of mineral N than that of previously Sole MZ plots. This can be attributed to the slightly higher quality of the residues as well as residual N from the previously applied urea; however, the highest levels were only more conspicuous at the start of the rains, and thereafter they dropped. This implies that the already readily available N from the previous season was leached quickly by the rains. Beauchamp (1987) reported residual benefits of urea on maize in a season immediately after the season in which urea was applied.

The maize grain yields determined in this study were in similar ranges as maize yields reported by Zingore (2011) for medium to highly fertile soils of Malawi and by Tamene et al. (2015) for farmers practising good management practices such as integration of NPK fertilizers and animal manure, which are relatively higher than the Malawi farmers' average. The significantly higher maize plant N uptake and grain and TDM yields for plots that were previously Sole CP, Sole PP and PP + CP intercrop than those that were previously Sole MZ and Sole MZ + 92 N at all fertilizer levels can be attributed to the combined effect of high-quality residues and inorganic N fertilizer, which increased N recovery and yields more than the low-quality MZ residues combined with inorganic N fertilizer. Mtambanengwe et al. (2007), in a similar study involving *Crotalaria juncea* green manure and *Calliandra calothyrsus* and *Pinus patula* residues as organic amendments in Zimbabwe, observed improvements in maize yields by between 24% and 104% from combined mineral fertilizer and leguminous resources over those from sole fertilizer. On the other hand, Mahama et al. (2016), in a similar study, reported greater benefits in maize N uptake due to cowpea and pigeon pea previous cropping systems in the USA than by that in the soybean system. The trends at the Dowa site were similar to those at Lilongwe, with slight differences. For instance, the amounts of N uptake in Dowa were above 50 kg N ha^{-1} in all the zero fertilizer plots and in all the previous legume-based cropping system plots, compared to Lilongwe levels that were lower in the treatment that was previously Sole MZ. This can be attributed to the initial soil total N and soil organic matter levels, which were higher in Dowa than in Lilongwe.

The significantly higher harvest indices due to the previous legume-based cropping systems can be attributed to increased synchrony in the release of N from the high-quality residues and the N requirements of the maize crop. This was complemented by the effect of applied N. However, apart from other treatments being significantly higher than the zero N rate, no differences were observed amongst the higher fertilizer rates. Maobe et al. (2010) noted the increase in HI with N application from both *Mucuna pruriens* residues and inorganic fertilizer, but at only medium rates of N from 30 to 60 kg N ha^{-1} and no further increase in HI at the fertilizer rates above 100 kg N ha^{-1}.

The crude protein percentages of maize grain in the present study are similar to those reported in other studies (Silva et al. 2005). Results in the present study show that increasing the rate of N application led to increase in crude protein content of the maize grain. Blumenthal et al. (2008) in a review of importance and effect of N on crop quality summarize a number of studies where increasing the N rate leads to increase in maize grain protein. However, in the present study, grain protein in maize differed significantly due to applied inorganic N fertilizer, but significant differences were not observed due to previous cropping systems in both study sites. This can be attributed to the timing of N application in that the protein content might have been influenced by top dressing N as fertilizer, which was applied closer to the tasselling stage and is more likely to have been channelled to kernel development and filling, as opposed to the N from the cropping system residuals which acted more as a basal fertilizer for vegetative development. In a similar study, increase in crude protein content of maize was reported to be influenced by late N application towards silking (Silva et al. 2005).

The previous legume-based systems resulted in significantly higher NUE in terms of PFP_N than by those that were previously Sole MZ or Sole MZ + 92 N in both study sites, and this can be attributed to interactive effects of the high-quality residues and the applied inorganic N. For maize residues with low quality, NUE can be reduced by the longer periods of N immobilization and, in turn, not meeting the synchrony with plant growth stages when the N is greatly needed. This implies that in the maize monoculture systems, much of the plant growth will only be dependent of the applied N, which is more susceptible to losses through leaching. According to Dobermann (2005), common ranges of NUE are 40–70 kg grain kg^{-1} N applied. The PP + CP intercrop and Sole CP NUE values were the highest, but the rest of the legume-based systems treatments in this study resulted in far much higher NUE values than the stated ranges. According to Dobermann (2005), NUE values above the stated ranges are obtained when there is good crop performance at lower rates of N and in well-managed systems. Therefore, the pigeon pea- and cowpea-based systems planted either in sole cropping, legume-legume intercrops or legume-cereal intercrops can all be considered good systems for increased NUE and increased yields for maize in rotation.

5 Summary and Conclusions

Integrated soil fertility management (ISFM) aims at achieving sustainable yields whilst minimizing costs on the part of a farmer. From this study it can be noted that integration of pigeon pea and cowpea either as sole crops, legume-legume intercrops or legume-cereal intercrops led to similarly higher N uptake, grain and TDM yields and harvest indices in the rotational maize than the maize-maize monoculture, with or without inorganic fertilizer. The interactive effect of the legume residues and inorganic fertilizer led to higher maize grain yields, with increases ranging from 30% under treatment that was previously CP + MZ intercrop (1689 kg ha^{-1}) to 59% under treatment that was previously Sole CP (2864 kg ha^{-1}) at 0 kg N ha^{-1} fertilizer application, than that under treatment that was previously Sole MZ (1178 kg ha^{-1}). Similarly, at the highest rate of N application, 120 kg N ha^{-1}, the previous legume-based treatments produced higher maize grain yields than the treatment that was previously Sole MZ (3277 kg ha^{-1}), by increases ranging from 28% under treatment that was previously CP + MZ intercrop (4525 kg ha^{-1}) to 42% under treatment that was previously Sole CP (5665 kg N ha^{-1}), in the Lilongwe site.

For the Dowa site, at 0 kg N ha^{-1}, maize grain yield increase over the treatment that was previously Sole MZ (1454 kg ha^{-1}) ranged from 24% under the previously CP + MZ intercrop (1908 kg ha^{-1}) to 48% under treatment that was previously Sole PP (2792 kg ha^{-1}), whereas at 120 kg N ha^{-1}, increase over the treatment that was previously Sole MZ (3204 kg N ha^{-1}) ranged from 35% under treatment that was previously PP + MZ intercrop (4945 kg ha^{-1}) to 45% under treatment that was previously PP + CP intercrop (5836 kg ha^{-1}).

From this study it is noted that both legume sole cropping and legume-cereal intercrops, as well as legume-legume intercropping systems involving pigeon pea and cowpea with residues retained increased N mineralization rates, N uptake and nitrogen use efficiency by the subsequent maize. Additionally, increasing inorganic N application increased maize grain yield, total dry matter yield and grain crude protein content.

Therefore, for the smallholder farmers on the Alfisols of Lilongwe and Dowa districts, Central Malawi, and in similar agro-ecological zones elsewhere, a rotation of pigeon pea or cowpea in legume-legume, legume-cereal and sole cropping systems can substantially increase maize yields with the implication on lowering the investment costs on inorganic fertilizers. Furthermore, an integrated approach can improve both quantitative yields and quality of the maize grain.

Acknowledgements The study was fully funded by the Alliance for a Green Revolution in Africa (AGRA). The authors also thank Yohane Ng'omacheza (a farmer from Dowa district) and the Lilongwe University of Agriculture and Natural Resources (LUANAR), Malawi, for providing land for the field trials. Benard Msukwa, a technician at LUANAR, is also appreciated for the assistance he provided during soil and plant chemical analysis in the laboratory.

References

Abbasi MK, Tahir MM, Sabir N, Khurshid M (2015) Impact of the addition of different plant residues on nitrogen mineralization–immobilization turnover and carbon content of a soil incubated under laboratory conditions. Solid Earth 6:197–205

Abera G, Wolde-Meskel E, Bakken LR (2013) Effect of organic residue amendments and soil moisture on N mineralization, maize (*Zea mays* L.) dry biomass and nutrient concentration. Arch Agron Soil Sci 59(9):1263–1277

African Development Bank (2011) Interim country strategy paper for Malawi. African Development Bank. Abidjan, Ivory Coast

Alliance for Green Revolution in Africa (2014) Africa Agriculture Status Report: climate change and smallholder agriculture in sub-Saharan Africa. Nairobi, Kenya

Anderson JM, Ingram JSI (1989) Tropical soil biology and fertility: a handbook of methods of analysis. CAB International, Wallingford

Beauchamp EG (1987) Corn response to residual N from urea and manures applied in the previous years. Can J Soil Sci 67:931–942

Blumenthal J, Battenspenrger D, Cassman KG, Mason KG, Pavlista A (2008) Importance of nitrogen on crop quality and health. In: Hatfield JL, Folett RF (eds) Nitrogen in the environment: sources, problems and management, 2nd edn. Elsevier, Amsterdam. https://digitalcommons.unl.edu/agronomyfacpub/200/

Brady NC, Weil RR (2008) The nature and properties of soils. Revised, 14th edn. Pearson Prentice Hall, New Jersey

Chilimba ADC (2007) Methods of soils, plants, fertilizers and miscellaneous analyses: a working manual for chitedze and bvumbwe research stations soils laboratories. Ministry of Agriculture. Lilongwe, Malawi

Chilimba ADC, Young SD, Black CR, Rogerson KB, Ander EL, Watts MJ, Lammel J, Broadly MR (2011) Maize grain and soil surveys reveal suboptimal dietary selenium intake is widespread in Malawi. Sci Rep 1:72. https://doi.org/10.1038/srep00072

Dobermann AR (2005) Nitrogen use efficiency-the state of the art. Agronomy –Faculty Publications Paper 316

Dorward A, Chirwa E (2011) The Malawi agriculture input subsidy programme: 2005/06 to 2008/09. Int J Agr Sustain 9(1):232–247

Food and Agricultural Organization of the United Nations (2003) Food energy–methods of analysis and conversion factors. FAO Food and Nutrition Paper 77. Rome

International Crops Research Institute for Semi-Arid Tropics (ICRISAT)/Ministry of Agriculture and Irrigation (MAI) (2000) Cost-effective soil fertility options for smallholder farmers in Malawi. ICRISAT, Bulawayo and MAI, Lilongwe

Kamanga BCG, Whitbread A, Wall P, Waddington SR, Almekinders C, Giller KE (2010) Farmer evaluation of phosphorus fertilizer application to annual legumes in Chisepo, Central Malawi. Afr J Agric Res 5(8):668–680

Kerr RB, Snapp S, Chirwa M, Shumba L, Msachi R (2007) Participatory research on legume diversification with Malawian smallholder farmers for improved human nutrition and soil fertility. Exp Agric 43:437–453

Landon JR (ed) (1991) Booker Tropical Soil Manual. A handbook for soil survey and agricultural land evaluation in the tropics and subtropics. Longman Scientific & Technical co-published with John Wiley and Sons, Inc, New York

Mahama GY, Prasad PVV, Roozeboom KL, Nippert JB, Rice CW (2016) Response of maize to cover crops, fertilizer nitrogen rates, and economic return. Agron J 108(1):17–31

Makumba W (2003) Nitrogen use efficiency and carbon sequestration in legume tree- based agroforestry systems: a case study of Malawi. PhD Thesis. Wageningen University and Research Center, Wageningen, The Netherlands

Maobe SN, Akundabweni LSM, Mburu MWK, Ndufa JK, Mureith JG, Gachene CKK, Makini FW, Okello JJ (2010) Effects of Mucuna green manure and inorganic urea nitrogen sources and application rates on harvest indices of maize (*Zea mays* L.). World J Agric Sci 6(5):532–539

Matusso JMM, Mungwe JN, Mucheru-Muna M (2014) Effects of different maize (*Zea mays* L.) soybean (*Glycine max* (L.) Merrill) intercropping patterns on soil mineral-N, N-uptake and soil properties. Afr J Agric Res 9(1):42–55

Mhango WG (2011) Nitrogen budgets in legume based cropping systems in Northern Malawi. PhD Dissertation. Michigan State University. East Lancing, USA

Ministry of Agriculture and Food Security (2012) Guide to agricultural production and natural resources management in Malawi. Agricultural Communications Branch. Lilongwe, Malawi

Mloza-Banda HR (1994) Principles and practices of crop management: field study guide. Crop Science Department, Bunda College of Agriculture, Lilongwe

Mtambanengwe F, Mapfumo P, Vanlauwe B (2007) Comparative short term effects of different quality organic resources on maize productivity under two different environments in Zimbabwe. In: Bationo A, Waswa B, Kihara J, Kimetu J (eds) Advances in integrated soil fertility management in Sub-Saharan Africa: challenges and opportunities. Springer. The Netherlands

Mutegi J, Kabambe V, Zingore S, Harawa R,Wairegi L (2015) The fertilizer recommendation issues in Malawi: gaps, challenges, opportunities and guidelines. Soil Health Consortium of Malawi, Lilongwe, Malawi

Njira KOW, Nalivata PC, Kanyama-Phiri GY, Lowole MW (2013) Effects of sole-cropped, doubled-up legume residues and inorganic nitrogen fertilizer on maize yields in Kasungu, Central Malawi. Agric Sci Res J 3(3):97–106

Palm CA, Gachengo CN, Delve RJ, Cadish G, Giller KE (2001) Organic inputs for soil fertility management: some rules and tools. Agric Ecosyst Environ 83:27–42

Partey ST, Preziosi RF, Robsosn GD (2014) Improving maize residue use in soil fertility restoration by mixing with residues of low C-to-N ratio: effects on C and N mineralization and soil microbial biomass. J Soil Sci Plant Nutr 14(3):518–531

Phiri AT, Msaky JJ, Mrema JP, Kanyama-Phiri GY, Msanya BM (2014) Effect of pigeon pea-groundnut system on selected soil properties. Int J Plant Soil Sci 3(4):397–407

Ricker-Gilbert J, Jumbe C, Chamberlin J (2014) How does population density influence agricultural intensification? Evidence from Malawi. Food Policy 48:114–128

Sakala WD, Cadisch G, Giller KE (2000) Interactions between residues of maize and pigeonpea and mineral N fertilizers during decomposition and N mineralization. Soil Biol Biochem 32:679–688

Sakala WD, Kumwenda JTD, Saka AR (2003) The potential of green manures to increase soil fertility and maize yields in Malawi. Biol Agric Hortic 21(2):121–130

Silva PRF, Strieder ML, RPS C, Rambo L, Sangoi L, Argenta G, Forsthofer EL, Silva AA (2005) Grain yield and kernel crude protein content increases of maize hybrids with late nitrogen side-dressing. Sci Agric (Piracicaba, Braz) 62(5):487–492

Snapp SS, Jayne TS, Mhango W, Benson T, Ricker-Gilbert J (2014) Maize yield response to nitrogen in Malawi's smallholder production systems. International Food Policy Research Institute. Washington DC, USA

Stoorvogel JJ,Smailing EMA (1990) Assessment of soil nutrient depletion in Sub-Saharan Africa: 1983–2000. Report 28. Wageningeni, The Netherlands

Tamene L, Le QB (2015) Estimating soil erosion in sub-Saharan Africa based on landscape similarity mapping and using the revised universal soil loss equation (RUSLE). Nutr Cycl Agroecosyst 102:17–31. The Netherlands

Tamene L, Mponela P, Ndengu G, Kihara J (2015) Assessment of maize yield gap and major determinant factors between smallholder farmers in the Dedza district of Malawi. Nutr Cycl Agroecosyst:1–18. The Netherlands

Zeng D, Mao R, Chang SX, Li L, Yang D (2010) Carbon mineralization of tree leaf litter and crop residues from poplar-based agroforestry systems in Northeast China: a laboratory study. Appl Soil Ecol 44:133–137

Zingore S (2011) Maize productivity and response to fertilizer use as affected by soil fertility variability, manure application, and cropping system. Better Crops 95(1):4–6

Zingore S, Mutegi J, Agesa B, Tamene L, Kihara J (2015) Soil degradation in sub-Saharan Africa and crop production options. Better Crops 99(1):24–26

A Hydrological Assessment of Wetlands in Lilongwe Peri-urban Areas: A Case of Njewa Catchment, Lilongwe, Malawi

Abel K. Mkulama, Austin Tibu, and Kenneth Wiyo

Abstract Assessment of wetlands and evaluation of related anthropic factors are crucial to monitoring impacts of land use changes on hydrological processes. The objective of this research was to assess the hydrological changes of Njewa wetland as it transforms from virgin wetland to industrial settlement. ArcGIS 10.6 was used to analyze the changes in land cover. Global positioning system (GPS) was used to collect geospatial data. Geographic information system (GIS) technologies such as digital elevation model (DEM) were used to delineate the wetland catchment area and produce maps. Velocity-area method was used to monitor streamflow discharge for 10 months, resulting in a seasonal hydrograph. Velocity was estimated using flotation method. The hydrograph was used to estimate baseflow (BF) and surface runoff volume, depth, and coefficient.

The land cover changes indicate that Njewa wetland has transformed from a virgin vegetative cover and farm land into a peri-urban center with lots of industrial buildings. Consequently, surface runoff coefficient has increased from 0.24 to 0.65, and the river flow was characterized with a low baseflow (close to zero). These results underscore that land use changes lead to changes in hydrology of rivers in Njewa wetland, leading to flooding in urban areas.

A. K. Mkulama (✉) · A. Tibu
Environmental Science and Management Department, Lilongwe University of Agriculture and Natural Resources (LUANAR), Lilongwe, Malawi
e-mail: abel.mkulama@bunda.luanar.mw; atibu@bunda.luanar.mw

K. Wiyo
Agricultural Engineering Department, Lilongwe University of Agriculture and Natural Resources (LUANAR), Lilongwe, Malawi
e-mail: kwiyo@bunda.luanar.mw

B. R. Singh et al. (eds.), *Climate Impacts on Agricultural and Natural Resource Sustainability in Africa*, https://doi.org/10.1007/978-3-030-37537-9_6

1 Introduction

Land use changes continue to pose severe threat to natural hydrological process of wetlands in urban Lilongwe (FAO 2013). In peri-urban wetland catchment areas, unprecedented changes in land use practices can alter hydrological processes; wetland development should therefore be monitored. Research has demonstrated that land use and cover change and its impacts on the existing environment should be investigated and monitored in a systematic manner in order to prevent flooding (Hadjimitsis 2010). Land use and land cover changes affect hydrological processes in general and increase flood risks (Khan 2005; Brath et al. 2006). Furthermore, because hydrological behavior of wetlands is inter-related with the catchment's topologic, meteorological, climatic, and biological factors, such unprecedented land use change through deforestation and construction of buildings can trigger off sequences of floods due to changes in the hydrological regime of the catchment (Hadjimitsis 2010).

It is also understood that human pressures on natural resources tend to exceed regenerative capacities of the natural processes, leading to their degradation (Munthali and Murayama 2011). For accurate monitoring of such pressures, natural resources managers recognize the use of high-resolution images to quantitate the rates at which changes occur. In most cases, satellite-acquired image data is recommended for land use and cover analysis. According to Hadjimitsis (2010), satellite imagery provides a synoptic overview of large regions, recorded with a standardized monitoring system.

Measurements used for predicting channel characteristics are needed to decide wetland management approaches in the face of land use changes. In particular, discharge is needed for planning flood control and designing engineering structures, including bridges and road culverts (Brooks et al. 2013). Flooding occurs when streamflow discharge exceeds the capacity of the channel or stream (Brooks et al. 2013). Field records of streamflow are either monitored using installed weirs or using velocity-area method (Brassington 1998). Since weirs are expensive to install and maintain, hydrologists prefer the latter, which uses flotation method to calculate water velocities along a control section and a steel tape to measure length and width of the channel at right angles to the direction of the flow to determine cross-sectional area. The product of cross-sectional area and velocity gives streamflow discharge. In theory, this method tends to overestimate streamflow discharge since surface velocity is significantly greater than average velocity. Therefore, estimated values are corrected with a factor of 0.75 (Brassington 1998).

Another critical aspect of gauging is selecting a control section, part of the stream for which a *rating curve*[1] is developed. The criteria for selecting a control section are that it must be stable, must have a sufficient depth for obtaining velocity measurements even at the lowest of streamflow, and should be located in a straight reach free from turbulent flow (Brooks et al. 2013). Such conditions are based on the wetland's natural configuration. Therefore, a reconnaissance field survey is required before gauging can begin.

[1] Graph of stage against discharge

Other researchers have also linked the floods in the wetland catchments of Lilongwe to urbanization. Wiyo et al. (2015) argue that anthropogenic activities on land use have increased surface runoff, triggering a sequence of floods in Lilongwe urban areas. Indeed, similar patterns have emerged in Likuni and Lingadzi catchments in the same district (Chikankheni 2009). Kambalame (2017) also establishes that due to land use changes between 1999 and 2004, surface runoff coefficient increased from 0.03 to 0.20 for Likuni and 0.08 to 0.24 in Lingadzi. This chapter attempts to establish information on hydrological changes of Njewa wetland, in order to provide basis for its management in the advent of rapid land use changes, estimate land cover change, determine streamflow discharge, and estimate baseflow and surface runoff coefficient.

2 Methodology and Study Area

2.1 Study Area

Malawi is located in Southern Africa and is part of the Southern Africa Development Community (SADC). The study was conducted in Njewa catchment, located at latitude 13°58′28.46’S and longitude 33°42′23.77E in Lilongwe District. The catchment area is at an elevation ranging from 1103 m in the wetland to 1134 m at the highest point. The area receives unimodal rainfall (mean annual rainfall of 980 mm), with most rains falling between November and March. The land use is characterized with a wetland now turned into warehouses and buildings constructed on agricultural land. The soil type is loam in the uplands and clay in the wetland (Fig. 1).

2.2 The Method

To establish the relationship between land use change and hydrological processes of the wetland, geographical information system technologies were applied during land use-related assessments while hydrological and meteorological analyses were based on hydrographs analyzed in Microsoft Excel.

The first step in land cover assessment involved delineating the wetland catchment area using topographic maps and digital elevation modeling (DEM). GPS was used to collect coordinates of the watershed line of the catchment, which were then used in Quantum GIS version 2.18.20 (https://qgis.org/en/site/forusers/download.html) to produce maps. The second step of land use assessment involved analyzing Landsat images (http://www.earthexploerer.com) in ArcGIS version 10.6.

Hydrological parameters such as stage and streamflow discharge were monitored for a period of 10 months (from July 1, 2017, to April 30, 2018) to cover both rainy and dry seasons. A control section of 11 meters of the reach to culvert boxes located

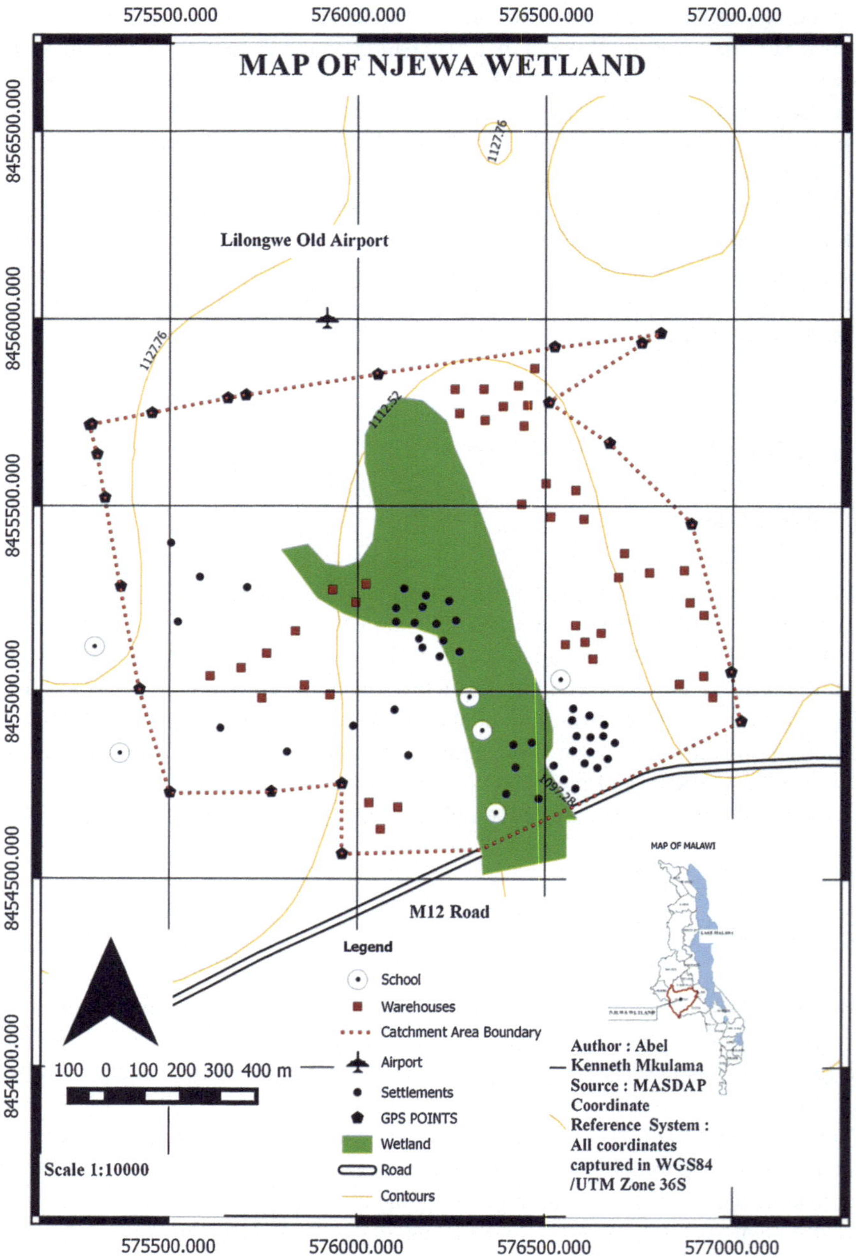

Fig. 1 Map showing Njewa wetland catchment in Lilongwe District, Malawi

Table 1 A synoptic view of the methodology used for collecting hydrological data (gauging)

Date (Julian calendar)	1	2	…
Distance (m)	1.3	1.3	…
Time (s)	40	20	…
Approaching velocity (m/s)	1.3/40 = 0.03	1.3/20 = 0.65	…

at the wetland outlet was selected based on the stream configuration and scientific recommendations,[2] along which fluctuating average water levels (stage) were measured using a steel tape in centimeters.

Water velocities were estimated by timing a float over a distance of 1.3 m as per recommendation.[3] Measurements were repeated five times, with the float placed at different distances along the section to correct potential human errors such as reading of the digital time. The average time was then calculated in seconds (s) and divided into the measured distance to give average velocity in meters per second (m/s).

To determine cross-sectional area of the stream, a steel tape was used to measure length and width of the two rectangular culverts located at the wetland outlet. The product of these values gave the constant cross-sectional area of the stream in square meters (m^2). Furthermore, for each culvert and on each day, the cross-sectional area was calculated by multiplying the average stage or depth of the water in meters with the culvert width measured earlier. In fact, wetland flows were the product of the new area and the approaching velocity (Table 1).

Since this was a surface water investigation, this method was appropriate to cope with a range of flows over the entire period of the investigation, including low flows in dry weather.

Daily rainfall amount for the catchment area during the period of study was collected from the Department of Climate Change and Meteorological Services (Table 2).

2.3 *Remote Sensing Data*

ArcGIS 10.6 was used to classify land use and cover changes occurring in the catchment between 1990, 2000, and 2010. A map was produced to depict land uses for 2010. Algorithms to delineate the catchment were applied in Quantum GIS to delineate and assess land cover changes within the catchment in 2009 and 2018.

[2] Control section must be stable, must have a sufficient depth for obtaining velocity measurements even at the lowest of streamflow, and should be located in a straight reach without turbulent flow.

[3] Hydrologists recommend 1–10 m distance for measuring water velocity, using the flotation method depending on speed of water.

Table 2 Shows how cross-sectional area and streamflow discharge were calculated

Calculating cross-sectional area and streamflow discharge				
Constant length	Culvert 1	0.092	…	…
Constant width(m)	Culvert 2	0.96	…	…
Depth of water (cm)	Culvert 1	8.0	…	…
	Culvert 2	6.0	…	…
Cross-sectional area (m^2)	Culvert 1	0.092 × (8.0/100 m) = 0.007	…	…
	Culvert 2	0.96 × (6.0/100 m) = 0.0058	…	…
Discharge (Q) (m^3/s)	Culvert 1	0.03 × 0.007 = 0.00021	…	…
	Culvert 2	0.03 × 0.0058 = 0.00017	…	…
Total Q (m^3/s)	Culvert 1 + Culvert 2	0.00021 + 0.00017 = 0.00038	…	…

2.4 *Meteorological Data and Hydrological Analysis*

Total rainfall volume for the catchment area obtained from the Department of Climate Change and Meteorological Services was used to calculate SRO coefficient.

Using calculated total discharge, an annual hydrograph was produced in Microsoft Excel to show the changes in streamflow discharge over a period of 10 months (Fig. 4). From this, baseflow was estimated using the "baseflow separation" technique (Brooks and Flolliott 2013). The estimated baseflow was then subtracted from total streamflow discharge (total Q) to give Qnet (m^3/s). Since flow of surface water tends to be higher than average flow, Qnet was corrected with a factor of 0.75 (Brassington 1998).

Since the area under hydrograph after baseflow separation equals the total surface runoff volume in m^3, the area under hydrograph was calculated using the trapezoidal method (Eq. 1).

$$A = \frac{a+b}{2} * h \tag{1}$$

where a and b are opposite sides of a trapezoid and h is height.

To depict total surface runoff volume for 24 hours, Eq. 2 was applied in Microsoft Excel:

$$A = \left\{ \left[\frac{(a+b)}{2} h \right] \times 24 \times 60 \times 60 \right\} \tag{2}$$

Total surface runoff volume was divided by the total catchment area in m^2 to obtain surface runoff depth in meters. Finally, the surface runoff depth was converted to millimeters and divided by the total annual rainfall volume that fell in the catchment area to obtain the SRO coefficient.

3 Results and Discussion

The land cover assessment showed insignificant change between 1990 and 2010; this means that a fraction of area used for rainfed herbaceous crops in 1990 was also used for rainfed herbaceous crops in 2000 and in 2010 (Table 3). The same was true for built-up urban areas and cultivated wetland (*dambo*[4]), as indicated in the output of Table 3. This somewhat contradicts the expected high land use change, resulting in high discharge. This would seem to suggest that the high discharge rates (Fig. 4) were caused by the increased urbanization that took place between 2011 and 2018. These findings are consistent with other findings (FAO 2013; Kambalame 2017) that show increased urbanization results in increased surface runoff (Fig. 2).

Even though the scope of the land cover analysis was limited to the year 2010, satellite images taken between 2011 and 2018 indicate a significant shift in land use in the wetland catchment. Land that was covered by rainfed herbaceous crops and the wetland is significantly replaced with built-up areas as shown in Fig. 3. Indeed, with reference to FAO (2013), the same pattern emerges for Njewa catchment of Lilongwe.

Discharge continued to decrease with stage in July, following the end of the 2016/2017 rainy season. Between August and toward the end of September (Julian dates 32–89), the baseflow reached its lowest (zero). This occurred in the middle of the dry season when streamflow is at its lowest and evaporation highest (Figs. 4 and 5).

A stage-streamflow discharge graph shows a direct linear relationship. This means that the depth of water in the wetland determines the rate at which water is discharged. These results correspond with findings from similar studies conducted elsewhere such as Kambalame (2017) in Likuni and Lingadzi catchments of Lilongwe District, Malawi, and Hadjimitsis (2010) in Cyprus. This finding implies that impacts of the land use changes are likely to cause shifts in hydrological regimes during both rainy season (high surface runoff leading to flooding) and dry season (low baseflows leading to drying up of rivers).

The surface runoff coefficient was 0.65 (Table 4). This was a significant rise compared to the ones recorded from neighboring streams in the same district such as Lingadzi (0.2) and Likuni (0.24) (Kambalame 2017). The rise in the surface runoff coefficient can be attributed to the unprecedented conversion of land cover from wetlands to warehouses and other buildings in the Njewa catchment. The research demonstrates that land cover changes in Njewa catchment, Lilongwe, are leading to reduced infiltration rates and high surface runoff coefficient, and this poses an increased risk of flooding in the catchment (Hadjimitsis 2010; Wiyo et al. 2015). Further, land cover changes reduce infiltration due to pavements and roofs (Wiyo et al. 2015). The low baseflow can be related to the low infiltration rate due to high speed of runoff and no vegetation (SADC 2016).

Therefore, Njewa catchment requires proper watershed management to handle increased surface runoff volumes such as storm drains, large culverts, and wetland reserve area as required.

---

[4] In FAO documents on land use change, the term *dambo* is used instead of *wetland* particularly in Southern Africa.

Table 3 Summary of land cover change analysis for Njewa catchment area, Lilongwe, Malawi, in years 1990, 2000, and 2011

AREA(Hectare)	1990	2000	AGGREGATED CLASS_2011	DISTRICT	CHANGE_1990	CHANGE00_2010	CLASS_ELEMENT
57.93128	1Hcs	1Hcs	AG[a]	Lilongwe	1Hcs-1Hcs	1Hcs-1Hcs	RAINFED HERBACEOUS CROP(s) Small (<2 ha)
14.8112	5Bu	5Bu	URB[b]	Lilongwe	5Bu-5Bu	5Bu-5Bu	Built-Up, Urban/Built-Up, Non-Urban
27.96768	1HcMspf	1HcMspf	AG	Lilongwe	1HcMspf-1HcMspf	1HcMspf-1HcMspf	CULTIVATED WETLAND (CULTIVATED *DAMBO*)

[a]AG is a generalized class for agriculture in terrestrial and aquatic/regularly flooded land
[b]URB is a generalized class for urban and rural areas (including not built-up area)

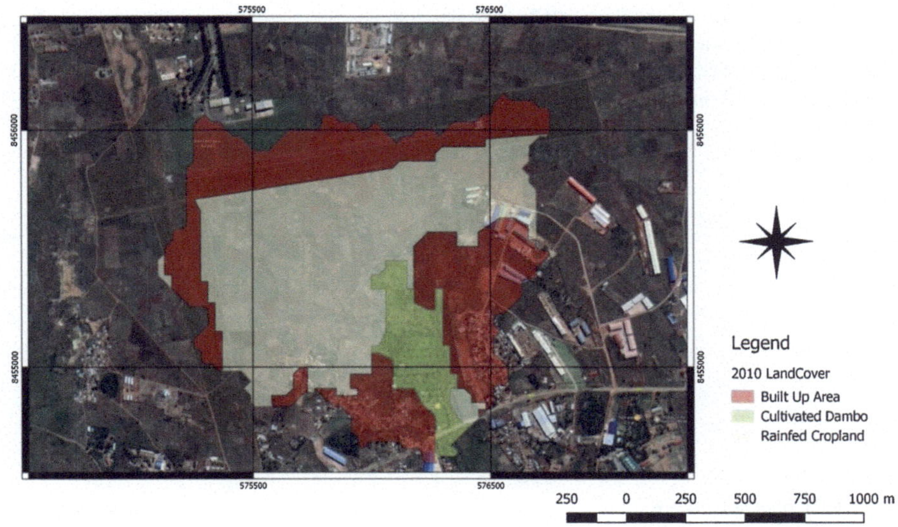

Fig. 2 Land cover classification of Njewa catchment for 2010, Lilongwe, Malawi

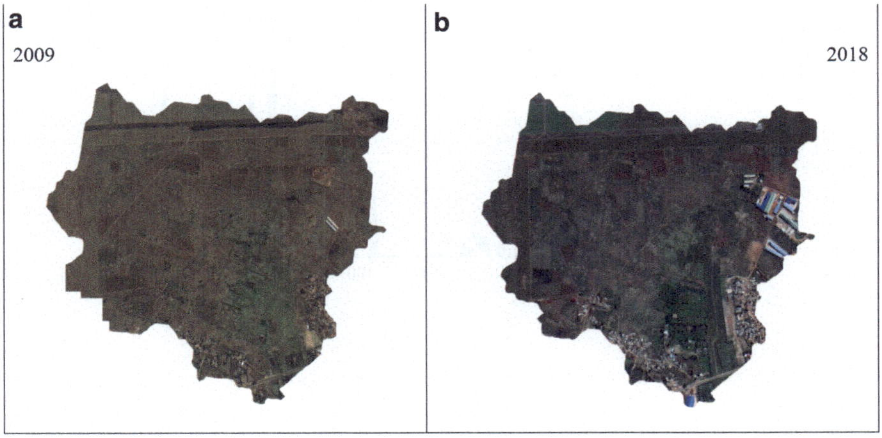

Fig. 3 (**a**) Google Earth image of Njewa catchment in Lilongwe, Malawi (2009); (**b**) Google Earth image of Njewa catchment in Lilongwe, Malawi (2018). Both images are linearly scaled at 304.8 m

4 Conclusions and Recommendation

The results illustrate that unprecedented land cover change has occurred from 2011 to 2018 for the Njewa catchment area in Lilongwe, Malawi. These results indicate that intense anthropogenic activities on wetland catchments translate into low infiltration and high surface runoff and corresponding streamflow response of wetlands posing flood risk. Future studies shall consist of assessing wetland's flood response using GIS and groundwater mapping for the wetland to determine groundwater-

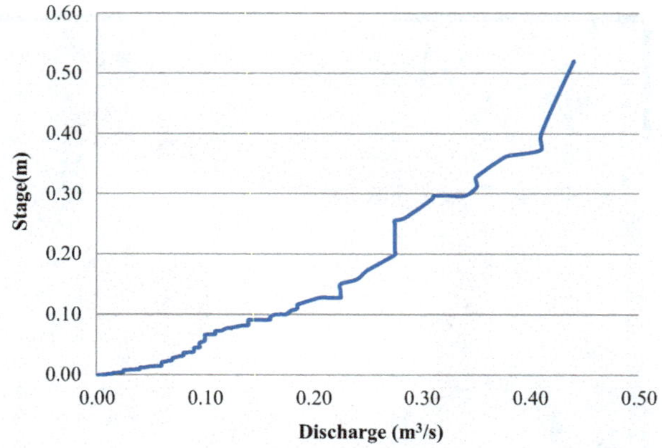

Fig. 5 The stage-discharge relationship of Njewa wetland in Lilongwe, Malawi, over 304 Julian days

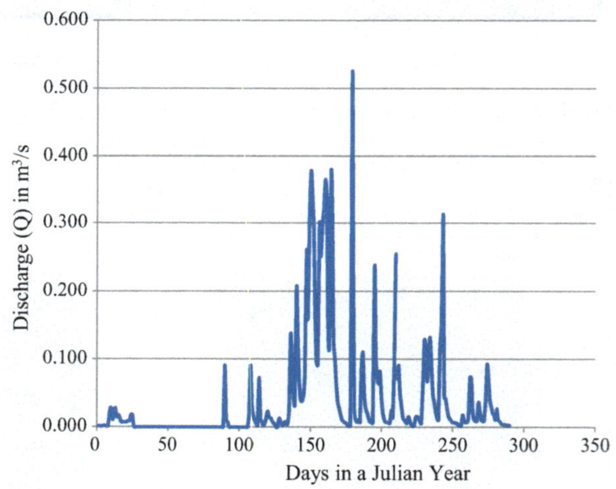

Fig. 4 A hydrograph showing annual streamflow discharge at Njewa wetland in Lilongwe, Malawi (2017/2018)

Table 4 Summary of results of the hydrological assessment of Njewa wetland in Lilongwe, Malawi

Serial number	Parameter	Unit	Value
1.	Surface runoff volume (SRO volume)	m^3	1,034,225.93
2.	Total area of the delineated catchment area	ha	171.00
3.	Surface runoff depth	m	0.60
4.	Rainfall	mm	932
5.	Surface runoff coefficient		0.65
6.	Estimated baseflow (BF)	m^3/s	0

surface water exchange. The application of computer simulations for modeling wetland flows could also be an interesting area to consider. Land use planners should monitor developments in urban wetlands to minimize the risk of urban flooding.

Acknowledgments Grateful acknowledgment is made to the Environmental Science and Management Department at LUANAR for the partial funding of this research.

References

Brassington R (1998) Field hydrogeology, 2nd edn. John Wiley & Sons Ltd, London

Brooks K, Flolliott M (2013) Hydrology and the management of watersheds, 4th edn. John Wiley & Sons, Inc., New Delhi

Brath A, Montanari A, Moretti G (2006) Assessing the effect on flood frequency of land use change via hydrological simulation (with uncertainty). Journal of Hydrology, 324(1–4), 141–153. https://doi.org/10.1016/j.jhydrol.2005.10.001

Brooks N Kenneth, Fflolliott F Peter, Magner A Joseph (2013) Hydrology and The Management of Watersheds. Fourth edi. New Delhi, India: John Wiley & Sons, Inc., Publication.

Chikankheni A (2009) Assessment of a groundwater gallery: a case study at Njewa village in Likuni. Lilongwe University of Agriculture and Natural Resources, Lilongwe

FAO (2013) Atlas for land cover and land cover change for Malawi from 1990, 2000 and 2010. Food and Agriculture Organization of the United Nations, Rome

Hadjimitsis DG (2010) Brief communication: "Determination of urban growth in catchment areas in Cyprus using multi-temporal remotely sensed data: risk assessment study". Nat Hazards Earth Syst Sci 10(11):2235–2240. https://doi.org/10.5194/nhess-10-2235-2010

Kambalame E (2017) Assessment of hydrological impacts of urban land use and environmental changes, on the hydrology of Lingadzi and Likuni Rivers, Lilongwe, BSc thesis. Lilongwe University of Agriculture and Natural Resources, Lilongwe

Khan SD (2005) Urban development and flooding in Houston Texas, inferences from remote sensing data using neural network technique. Environ Geol 47(8):1120–1127. https://doi.org/10.1007/s00254-005-1246-x

Munthali KG, Murayama Y (2011) Land use/cover change detection and analysis for Dzalanyama forest reserve, Lilongwe, Malawi. Procedia Soc Behav Sci 21:203–211. https://doi.org/10.1016/j.sbspro.2011.07.035

SADC (2016) Phase IIa final report, Gaborone, Botswana

Wiyo KA, Fiwa L, Mwase W (2015) Solving deforestation, protecting and managing key water catchments in Malawi using smart public and private partnership. J Sustain Dev 8(8); ISSN 1913-9063 E-ISSN 1913-9071 Published by Canadian Center of Science and Education. https://doi.org/10.5539/jsd.v8n8p251

Part III
Sustainable Crop/Livestock/Aquaculture/Fish Production

Productivity and Chemical Composition of Maize Stover and Rice Straw Under Smallholder Farming Systems Intensification in Tanzania

Ephraim J. Mtengeti, Eva Mtengeti, and Lars Olav Eik

Abstract Cereal crop-dairy production is one of the crop-livestock production systems with a high potential for increasing food security of smallholder farmers in Africa. The question is how to meet the challenge of increasing the productivity and production efficiency of crops and milk per unit land area in highly populated areas of Tanzania. In these areas, smallholder farmers own an average of 0.5–2 ha of land and use the bulk of crop residues to feed livestock. This 4-year study aimed at demonstrating the impact of intensification, through improved plant nutrition, on productivity and chemical composition of maize and rice crop residues. The demonstrated treatments were farmers' practice (FP) and improved plant nutrition (IPN). We recorded the yield of maize and rice crop residues and the concentration of macro- and micro-minerals in oven-dried samples of crop residue. The concentrations of N, P, K, Mg and Ca, which were not significantly ($P > 0.05$) different between the farmers' and improved plant nutrition practices, ranged from 0.48% to 0.83%, 0.03% to 0.08%, 1.11% to 2.23%, 0.07% to 0.083% and 0.10% to 0.26% for maize stover and 0.48% to 0.84%, 0.08% to 0.23%, 2.31% to 2.83%, 0.20% to 0.33% and 0.10% to 0.22% for rice straw, respectively. The concentrations of Cu and Zn ranged from 2.3 to 14.0 and 6.0 to 20.5 mg/kg for maize stover and 1.7 to 12.4 and 27.0 to 55.7 mg/kg for rice straw, respectively. Comparable nutrients in the two agronomic practices can be related to nutrient dilution under improved plant nutrition due to high biomass production that was significantly ($P < 0.05$) higher (on average 8.9 Mg DM/ha) than under farmers' practice (3.7 Mg DM/ha for maize stover). Since cereal crop residues from improved plant nutrition did not differ significantly ($P > 0.0.5$) from those of the farmers' practice in terms of nutrient concentration, we concluded that farmers' practice crop residues obtained their

E. J. Mtengeti (✉) · E. Mtengeti
Sokoine University of Agriculture, Morogoro, Tanzania

L. O. Eik
Department of International Environment and Development Studies, Norwegian University of Life Sciences, Ås, Norway

B. R. Singh et al. (eds.), *Climate Impacts on Agricultural and Natural Resource Sustainability in Africa*, https://doi.org/10.1007/978-3-030-37537-9_7

nutrients mainly from soil reserve. Therefore, soil fertility mining happens when farmers feed all crop residues to livestock without the return of manure to the farm. The smallholder farmers, therefore, should be trained on the improvement of plant nutrition and feeding of crop residues to reduce depletion of soil fertility. Except for Zn concentrations, all other mineral concentrations and nitrogen content were low to very low for dairy cattle fodder, and supplementations of nitrogen and mineral premix are important in optimising milk production in cereal crop residue-based diets.

1 Introduction

1.1 Importance of Maize and Rice on Food Security in Tanzania

Maize and rice account for 31% and 13%, respectively, of the total food production in Tanzania (WEMA 2010). Nearly 75% and 60% of the country's population consume maize and rice, respectively. The per capita consumption of maize and rice averages to approximately 74.5 and 16.5 kg/person/year (FAOSTAT 2012; PASS 2012), respectively. Maize provides 60% of dietary calories and more than 35% of utilisable protein to the Tanzanian population (USDA 2017).

Maize and rice crops are vital food and cash crops grown in almost all regions in the country (Mdoe et al. 2015; RATES 2003; USDA 2012). Maize crop comprises 45% of the cultivated area, and its production accounts for more than 70% of the cereal produced in Tanzania (USDA 2017). Since farmers sell about 40% of their maize locally and to neighbouring countries, it is a significant source of income for the majority of smallholder farmers in the southern highlands of Tanzania.

Nearly 90% of the two cereal crops are produced by smallholder farmers, with an average farm size ranging from 0.5 to 2 ha and keeping one to three cattle for milk production, draught power and manure for their farms. Farmers grow maize in various agro-ecologies ranging from near sea level to 2000 m above sea level depending on the variety. It is widely grown from infertile sandy soils towards the eastern and northern to loam-clay soils in the southern part of the country (Mdoe et al. 2015; USDA 2017). Despite its importance to the food security, maize productivity in the country has been persistently low (about 1.6 Mg/ha) because of lack of skill and capital for sustainable intensification of the crop (Mtengeti et al. 2015). Most rice is grown by smallholder farmers under rain-fed conditions, but some farmers grow 2–2.5 ha under irrigation schemes along the river valleys and alluvial plains with permanent water supplies. Mean yield in smallholder farms is generally low, ranging from 1.6 to 2.4 Mg/ha. With irrigation, production of rice can be more than 5 Mg/ha (USDA 2012).

1.2 Importance of Maize and Rice on Livestock Feed in Tanzania

In Tanzania, as in other countries in East and Southern Africa, 50% of the smallholder farmers also keep animals (Thornton et al. 2002). Livestock provides milk, meat, income from sales of highly demanded livestock products to the increasing urban populations (NEPAD 2005), manure for fertilisation of crops and draught animal power to ease the drudgery and burden of hand cultivation. Increase in human population has led to an expansion of cropping land especially in high agricultural potential areas along the high- and lowlands of the country. Available land for forage production and grazing has become less. Instead, more crop residues, mainly maize stover and rice straw, are contributing enormously to livestock feed in the larger part of the year. Due to population growth and land scarcity in Kenya highlands, for example, smallholder dairy farmers obtain up to 34% of total feed from crop residues (Romney et al. 2003).

The principal constraint of cereal crop residues as feed for livestock is nutrient deficiencies for growth, production and reproduction of the animals. Cereal crop residues have deficient crude protein and minerals and high fibre content. The digestibility of these residues is therefore low. Efforts to improve the quality of cereal crop residues include breeding, agronomical practices and utilisation strategies. Vaswani et al. (2016) reported that different varieties of maize fodder differed substantially in their chemical composition. They found higher concentrations of minerals and nitrogen content in improved as compared with regular types. In an assessment of the nutritive value of whole maize stover and its morphological fractions, Li et al. (2004) found that the leaf had the highest nutrients and the stem rind had the lowest. These workers supported the development of a device that can separate different stover parts so that they can be used more efficiently for different purposes. Amongst the strategies for utilisation of low-quality cereal crop residues are urea treatment and supplementation with leguminous multipurpose tree leaves. The benefit of improving stover and straw quality through improved plant nutrition is somewhat less reported. Since increased biomass of stover and straw can feed more animals and leave some portion for soil cover, it is an essential advantage of improved plant nutrition.

1.3 Addressing Sustainable Intensification of Maize and Rice Crops as Food and Feed in Tanzania

Intensification in this context means improved efficiency of crop and livestock production, with reduced greenhouse gas production per unit of crop and livestock products and more crop and livestock products per unit of water. Neglecting crop or livestock improvement in crop-livestock mixed systems may end up degrading the environment. Policy initiatives that tend to favour intensification of staple food crops, such as massive subsidies for fertiliser and irrigation without regard to livestock improvement, have grave implications for the environment. Keeping a large number of low productive animals will encourage free-range grazing and consumption of crop residues in the field, resulting in poor management of both crop residues and manure. Continual removal of crop residue biomass, either through burning or free-range grazing, will deplete soil organic matter and is environmentally unsustainable in the long term (Valbuena et al. 2012). Crop residue allocation to livestock feeding will enhance manure availability for crop production and thus lead to improved income from sales of milk and meat. The added revenue may be used to buy fertiliser to improve crop yield and farm income. Additional future food and feed production on smallholder farms can consequently take place through increased productivity per unit area. For this to happen, sustainable intensification of both crop and livestock is inevitable.

The question is, therefore, how can crop productivity be increased to meet increased food demand for human and feed for livestock and still reduce mining of soil nutrients?

We believe it can happen through intensification of crop production by adopting improved agronomic practices (IAPs) for strategic soil fertility management, water conservation, weed and pest control and use of improved crop varieties. There is a need, therefore, to breed and promote crop varieties with high grain yields and high yields of good-quality residues and resistance to the significant biotic and abiotic stresses such as pests, diseases, parasitic weeds, drought and low fertility. In Kenya highlands maize-dairy systems, feeding maize stover of improved varieties with high grain yield and resistant to maize streak virus disease resulted in higher milk yields and better animal health; and as a consequence, there was improved household income generation (Lenne and Thomas 2006).

Another question is why do smallholder farmers tend to be late in adopting the new improved agronomic practices (IAPs) for sustainable intensification of crop production?

The reasons could be high prices and lack of knowledge about the use of fertilisers, pesticides, herbicides, improved crop variety seeds and other inputs. As a result, most smallholder farmers in the country recycle crop seeds and apply a meagre amount of fertiliser (Westengen and Berg 2016; Mtengeti et al. 2016). Another constraint for the low adoption of IAPs by the smallholder farmers is that they come in various packages, each lacking the profit margin needed to motivate the poor farmers to adopt (Foster and Rosenweig 2010). We believe IAPs for sustainable intensification of crop production should come in a single package, promoted by a strong partnership between public technical advisory institutions/NGOs and agro-input industries. Consequently, from 2011 to 2014, we initiated a public-private partnership (PPP) between the two public universities, the Sokoine University of Agriculture and Norwegian University of Life Sciences, and two international agro-input companies, namely, Yara and Syngenta. The PPP aimed at conducting research and demonstrating how to achieve sustainable agricultural intensification under smallholder maize and rice farmers through effective plant nutrient management and protection for increasing crop productivity whilst preserving the environment. The paper presents part of the results of the PPP implementation, dealing with investigating the effects of improved plant nutrition on the productivity and chemical composition of whole maize stover and rice straw in the Njombe, Mvomero and Kilombero districts. For another aspect of the study, we refer to Mtengeti et al. (2015, 2016).

2 Materials and Methods

2.1 Study Sites and Crops

We established maize crop trials at one smallholder farm in each of four villages in Njombe district, southern highlands of Tanzania (Table 1). The area is under unimodal rainfall type starting from November to April. The annual amount of rainfall is 1000–2000 mm, and temperature ranges from 22 °C to 30 °C maximum and 15 °C to 20 °C minimum. Soils in Njombe are generally highly weathered, infertile and acidic with high phosphorus fixation and low organic matter content (Mtengeti et al. 2015).

We carried out the rice trials in Mvomero and Kilombero districts in Morogoro region. Average annual rainfall is 1000 mm and temperature ranges from 24 °C to 32 °C. Mvomero and Kilombero district rice-growing areas are floodplains of Wami and Ruaha rivers, respectively, and have vertisols, fluvisols and complex vertisols and fluvisols (Msanya et al. 2003).

2.2 Treatments

We conducted the study for 4 consecutive years to investigate the influence of improved plant nutrition on the productivity and chemical composition of maize stover and rice straw. One farmer in each village participated in having two treatments, namely, the farmers' practice (FP) and Yara/SUA/Syngenta (YSS), as improved agricultural practices (Tables 2 and 3). Maize trials were planted at the beginning of the long rainy season in December at a spacing of 90 cm by 30 cm and harvested in July.

Table 1 Location of the study sites

Region	District	Villages	Crop established	Altitude (masl)
MOROGORO In eastern Agricultural zone	Mvomero	Dihombo	Rice	370
		Dakawa	Rice	566
	Kilombero	Mkula	Rice	290
NJOMBE In southern Highlands Agricultural zone	Njombe	Ibumila	Maize	1820
		Matiganjola	Maize	1791
		Welela	Maize	1793
		Kichiwa	Maize	1798

Table 2 Inputs application for maize crop (treatments)

Activity	Input		When
	Yara/SUA/Syngenta (YSS)	Farmers' practice (FP)	
Seed treatment	Apron star at 10 g/4 kg seed	No seed treatment	Seed preparation during planting
1st fertiliser application	YaraMila CEREAL @350 kg/ha	(DAP) @ 62 kg/ha	During planting
Pre-emergence herbicide application	Primagram gold 3 lts/ha	No application. But first weeding third week after planting	Just after planting
1st insecticide application	Karate 5 EC for control of stalk borers @395mls/acre	Karate 5 EC for control of stalk borers @395mls/ha	3rd week after planting If the symptom of attack occurs
2nd fertiliser application. Sprayed on the leaves	YaraVita TRACEL™ BZ @ 2 kg/ha	No application	3rd week after planting
3rd & 4th fertiliser application	200 kg/ha YaraMila CEREAL	Application of urea fertiliser @ 123 kg/ha	5th week after planting (knee height)
4th fertiliser application	52 kg/ha YaraMila JAVA	No application	7th week after planting (tasselling)
2nd insecticide application	Karate 5 EC @ 395mls/ha	No application	8th week after planting
Herbicide application	Gramoxone @ 1240mls/ha	Weeding by use of draft animals	10th week after planting

Fertiliser composition:

1. YaraMila CEREAL = 23 N – 10P – 5 K+ 2MgO + 3S + 0.3Zn
2. TRACEL BZ = 5 N – 7.5 P_2O_5–5 K_2O + 5 mg + 5S + 5Zn + 5Bo + 0.1Cu + 0.1Fe + 0.1Mn + 0.1Mo
3. YaraMila JAVA = 22 N – 6P –12 K + 2CaO + 1MgO + 3S + 0.2B + 0.2Zn
4. DAP = 18% N – 46% P_2O_5
5. Urea = 46% N

Rice trials were planted twice per year at Dihombo and Mkula in August and March and harvested in December and June, respectively. At Dakawa, rice was planted only once in March and harvested in July. Rice in all trials was planted at 20 cm by 20 cm.

2.3 Crop Harvesting and Crop Residues Chemical Analysis

We demarcated every treatment plot on each farm into three sub-plots for each crop harvesting period. Two sampling units were then located at the middle of each sub-plot, hence making a total of six sampling units per treatment. Maize sampling plot was a 4-m-long line, and rice was 1 m^2. The farmers continued to harvest their crops

Table 3 Inputs application for rice crop (treatments)

Activity	Input		When
	Yara/SUA/Syngenta practice (YSS)	Farmers' practice (FP)	
Seed treatment	Apron star at 2.5 g/kg seed	No seed treatment	Seed preparation during planting
Fertiliser application in rice nursery	YaraMila CEREAL 3 kg/100 m^2	Fertiliser application (urea @ 1 kg/m^2)	For the nursery seedbed
Herbicide application	Touchdown forte @ 2.47lt/ha	No treatment	Clear weeds before paddling
1st fertiliser application	YaraMila CEREAL @ 200 kg/ha	No application	During transplanting/planting
Herbicide application	Solito 320 EC @ 1482mls/ha	Hand weeding	2–3 wks after transplanting
2nd fertiliser application on the leaves	YaraVita TRACEL™ BZ @ 2 kg/ha	Urea @ 123 kg/ha	4 wks after transplanting (4–6 leaves)
3rd fertiliser application	YaraMila JAVA at 247 kg/ha	No fertiliser	5th week after transplanting at the tillering stage of growth
4th fertiliser application	YaraLiva NITRABOR at 61.7 kg/ha	No fertiliser	Booting stage of growth
Insecticide application	Karate 5 EC @ 395mls/ha	Application of karate 5 EC @ 395mls/ha	4–5 weeks after transplanting If symptom of attack is noted
Fungicide application	Artea 330 EC @ 494mls/ha	No application of fungicide	4th–fifth week after transplanting If the symptom of attack is noted

Fertiliser composition:

1. YaraMila CEREAL = 23 N – 10P – 5 K+ 2MgO + 3S + 0.3Z
2. TRACEL BZ = 5 N – 7.5 P_2O_5–5 K_2O + 5 mg + 5S + 5Zn + 5Bo + 0.1Cu + 0.1Fe + 0.1Mn + 0.1Mo
3. YaraMila JAVA = 22 N – 6P – 12 K + 2CaO + 1MgO + 3S + 0.2B + 0.2Zn
4. YaraLiva NITRABOR = 15.4 N – 25.5 CaO – 0.3 Bo

after sampling. After every harvest, we collected two soil samples at 0–20 cm and 20–40 cm depth for physical and chemical property analysis. The Research Centre Hanninghof, Yara International, Dulmen, Germany, analysed contents of macro- and micronutrients in samples from the soils, maize stover, rice straw and their grain. Various agronomic data were also recorded including plant height, cob length, grain yield (Mg /ha at 14% MC) and whole stover biomass (Mg DM/ha) for maize and tillers/m^2, tiller height, number of panicles/m^2, grain yield (Mg/ha at 14% MC) and whole straw biomass (Mg DM/ha) for rice. In this chapter, however, only whole maize stover and rice straw chemical nutrients in terms of macro- and micronutrients and biomass yield are presented and discussed.

2.4 Data Analysis

The data was handled and analysed using Excel, and a t-test was used to check if the difference between farmers' (FP) and improved (YSS) practices was significant. The data for each site was analysed individually because of the environmental heterogeneity of the site.

3 Results and Discussion

3.1 Crop Residues Productivity

Generally, farmers' practice (FP) produced lower maize stover yield than improved agronomical practices (YSS) (Table 4). The difference between YSS and FP in rice straw was, however, rather small (Table 5). The reason could be the rather fertile floodplain soils in rice fields that receive frequent floods with plant nutrients from surrounding hills. Improved agronomical practices increased maize stover productivity in some cases more than three times the farmers' practice. Maize stover productivity ranged from 3.2 to 18.3 Mg DM/ha in improved agronomical practices and from 1.1 to 7.7 Mg DM/ha in farmers' practice. By enhanced agronomical practices, smallholder farmers can produce a high amount of crop residues that can feed more animals, and also some can be left in the field for soil cover to reduce land degradation and soil erosion.

The smallholder farmers require bulk feed for their 1–2 dairy cows in their homestead, as well as enough manure to fertilise about 0.4 ha of the maize crop. In a way, the dairy cow benefits from maize crop stover, and the maize plant gets fertiliser from the dairy cow. The smallholder farmer gets both milk and maize grain, thus sustaining household food and income security. The farmer can also preserve the environment by not expanding cultivated land towards the conserved forest. Therefore, mixed farming systems integrating crops and livestock can contribute to reduced poverty and enhanced livelihoods and at the same time contribute to environmental sustainability (Lenne and Thomas 2006). This conclusion is in agreement

Table 4 Maize stover dry matter yield (MgDM/ha)

	2011		2012		2013		2014	
Villages	YSS	FP	YSS	FP	YSS	FP	YSS	FP
Ibumila	9.33^a	2.72^b	7.29	5.97	7.0^a	3.7^b	5.27^a	2.18^b
Matiganjola	NR	NR	11.42^a	4.94^b	9.82^a	3.26^b	3.15^a	1.10^b
Welela	NR	NR	11.38^a	3.01^b	18.29^a	5.58^b	9.1^a	4.86^b
Kichiwa	NR	NR	10.9^a	7.7^b	14.4^a	1.07^b	5.34^a	2.70^b

FP farmers' practice, *YSS* Yara/SUA/Syngenta improved practice, *NR* not recorded. Values in the same row and year followed by different superscripts are significantly different at $P < 0.05$

Table 5 Rice straw dry matter yield (MgDM/ha)

	Long rain								Short rain	
	2011		2012		2013		2014		2013	
Villages	YSS	FP	YSS	FP	YSS	FP	YSS	FP	YSS	FP
Dihombo	8.74	7.03	8.74	6.03	7.54	7.33	8.31	7.86	11.14	10.48
Dakawa	NR	NR	8.82	7.84	8.92[a]	5.49[b]	6.75	6.20	NR	NR
Mkula	NR	NR	8.98[a]	5.98[b]	8.76	6.15	5.44	5.32	11.20	11.01

FP farmers' practice, *YSS* Yara/SUA/Syngenta improved practice, *NR* not recorded. Values in the same row and year followed by different superscripts are significantly different at $P < 0.05$

with Duncan et al. (2013) who noted that mixed crop-livestock systems produce 50% of global cereals, 34% beef and 30% milk.

It is important to realise that when cereal crop residues are left on the field, it may take them a season to mineralise and release nutrients to the plants. The process may be enhanced by incorporation of inorganic fertiliser or through decomposition in composting processes (Okalebo et al. 2006). However, this process may also take place in the rumen of a cow at a lower cost. Some smallholder farmers who do not keep livestock, especially those growing rice in Dihombo and Mkula villages, burn most of the cereal crop residues to pave the way for the cultivation of the next crop or to avoid free-range grazing livestock that graze on the residues and destroy their farm soil structure. Introduction of appropriate rice straw feeding technologies could enhance crop-livestock integration and sustain the environment better under smallholder farmers than burning the straw.

3.2 *The Chemical Composition of the Maize Stover from Different Demonstration Sites*

The chemical composition of the maize stover from different demonstration sites is shown in Table 6. Generally at Ibumila village, there was no significant difference ($P > 0.5$) between FP and YSS. At Matiganjola village, however, significant ($P < 0.05$) improvement of N content in maize stover due to YSS treatment was observed in 2012 and 2013, whilst that of Ca was observed in 2013 and 2014 and Cu and Zn contents were significantly depressed by YSS in 2014. In 2012 at Welela village, YSS improved P and depressed Zn and in 2013 depressed N, Cu and Zn contents in maize stover. Nitrogen content in the maize stover was improved in 2014 at Kichiwa village, whilst at the same village P content was improved in 2012 and 2013 and Cu and Zn contents were depressed in 2014. Regardless of years, K, Mg and S contents were improved and Fe, Mn and Mo contents were depressed by YSS in Welela village (Figs. 1 and 2). The depression of *N* content in Welela village and generally micronutrient contents in YSS treatment could be due to the soils of Welela that had some complications with fertiliser applications. Soil samples of the demonstration sites analysed before the study showed the lowest soil pH levels (4.1)

Table 6 Mean chemical composition of maize stover in different villages from 2012 to 2014

		Ibumila village			Matiganjola village			Welela village			Kichiwa village		
Nutrients	Practices	2012	2013	2014	2012	2013	2014	2012	2013	2014	2012	2013	2014
N %	FP	0.48	0.50	0.6	0.54^{b}	0.47^{b}	0.60	0.64	0.83^{a}	0.64	0.61	0.50	0.60^{b}
	YSS	0.55	0.54	0.6	0.64^{a}	0.65^{a}	0.62	0.57	0.63^{b}	0.57	0.64	0.56	0.72^{a}
P-value		0.104	0.378	0.882	0.012	0.002	0.620	0.248	0.018	0.665	0.758	0.06	0.091
P %	FP	0.05	0.07	0.09	0.043	0.044	0.078	0.032^{b}	0.035	0.044	0.04^{b}	0.03^{b}	0.05
	YSS	0.07	0.06	0.07	0.04	0.042	0.08	0.04^{a}	0.038	0.038	0.05^{a}	0.04^{a}	0.05
P-value		0.231	0.527	0.137	0.614	0.723	0.147	0.012	0.580	0.290	0.042	0.037	0.611
Ca%	FP	0.12	0.21	0.22	0.17	0.13^{b}	0.14^{b}	0.16^{a}	0.16	0.157	0.20^{a}	0.26	0.18
	YSS	0.15	0.26	0.17	0.19	0.16^{a}	0.19^{a}	0.13^{b}	0.14	0.153	0.15^{b}	0.21	0.22
P-value		0.149	0.278	0.180	0.651	0.004	0.074	0.024	0.230	0.823	0.046	0.034	0.216
Cu mg/kg	FP	2.84	13.02	2.85	2.97	14.04	3.03^{a}	2.75	12.67^{a}	3.13	3.22	2.58	3.18^{a}
	YSS	2.72	13.49	2.62	2.66	13.99	2.26^{b}	2.99	11.65^{b}	2.66	3.82	2.62	2.59^{b}
P-value		0.69	0.036	0.601	0.136	0.173	0.023	0.53	0.024	0.14	0.13	0.91	0.05
Zn mg/kg	FP	9.4	15.3	8.89	7.2^{b}	11.7	13.6^{a}	14.3^{a}	9.0^{a}	11.8	7.85	6.9	11.9^{a}
	YSS	11.9	21.2	8.52	12.4^{a}	11.7	7.3^{b}	7.2^{b}	6.0^{b}	6.33	12.37	8.4	6.8^{b}
P-value		0.083	0.24	0.106	0.66	0.047	0.99	0.001	0.0004	0.027	0.153	0.19	0.024

FP farmers' practice, *YSS* Yara/SUA/Syngenta improved practice. Values in the same column and nutrient followed by different superscripts are significantly different at $P < 0.05$

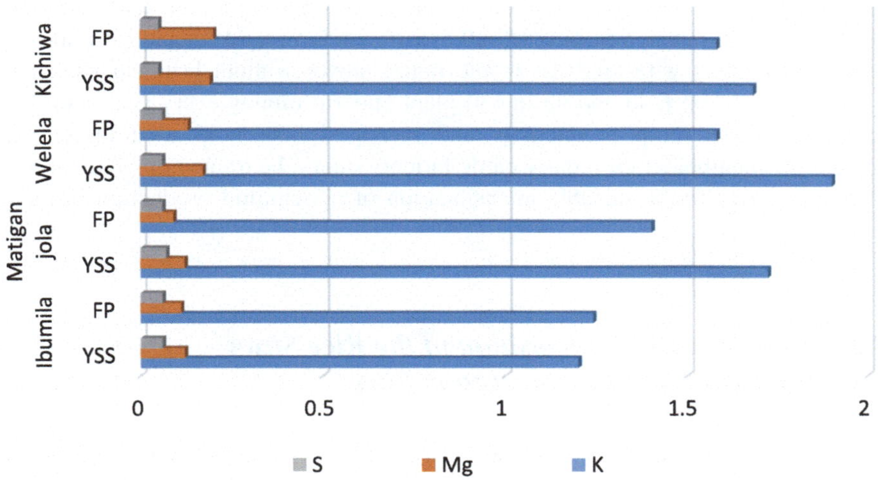

Fig. 1 Some macronutrients' concentration in maize stover (% DM)

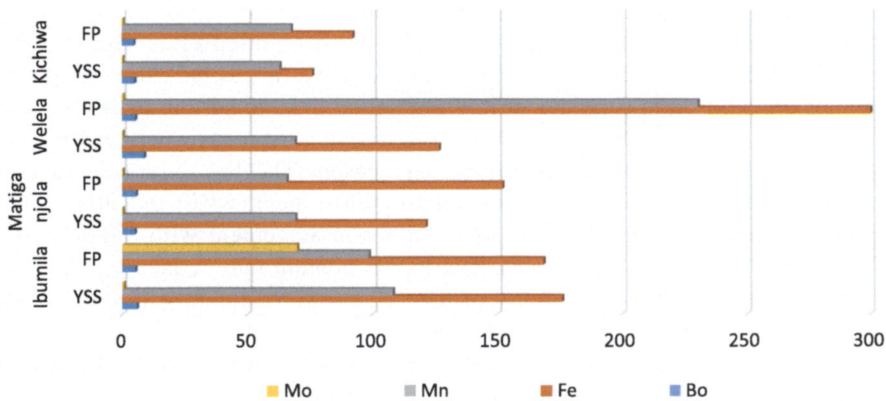

Fig. 2 Some micronutrients' concentration in maize stover (mg/kg)

and signs of Mn toxicity in Welela and Ibumila villages (Mtengeti et al. 2015). All assessed macro- and micronutrients except Fe and Mn were rather low in supporting growth, production and reproduction of a dairy cattle (NRC 2001) but are within the reported range (Mtengeti et al. 2008; Li et al. 2004; Vaswani et al. 2016). The farmers, however, use maize stover in bulk, leaving the animals to select the more nutritious parts of the stover especially the leaves. Li et al. (2004) reported twice as much N, P and Ca concentration (1.59%, 0.11% and 1.05%) in maize stover leaf as compared to whole maize stover (0.65%, 0.05% and 0.40%). Also farmers will supplement the dairy animals with maize bran, multipurpose tree legume leaves and grasses planted in contour bands along the homestead farm plot. This means maize stover can just be used as a bulk basal diet and not as the main feed for dairy cattle.

Since most nutrient concentrations in maize stover in different demonstration sites were rather not significantly ($P < 0.5$) different between YSS and FP, it means that most farmers who feed the whole maize stover without bringing back the manure to the crop field end up in soil plant nutrient mining every year, and this leads to reduced crop productivity and increased household food insecurity. Therefore, smallholder crop-dairy cattle farmers should be trained on efficient use of maize stover for sustainable intensification of agricultural production in their respective areas.

3.3 The Chemical Composition of the Rice Straw from Different Demonstration Sites

The mean chemical compositions of the rice straw from different demonstration sites and years are shown in Table 7. The improved agronomical practices had no significant ($P > 0.05$) effect on rice straw macro- and micronutrient concentrations. Except for K (Fig. 3), all other macronutrients in rice straw were too low to support growth, production and reproduction of cattle. However, such low macronutrient values of rice straw have also been reported elsewhere (Singh et al. 1995; Sath et al. 2013). Farmers who use rice straw as bulk feed for their animals are advised to supplement it with mineral licks to sustain their productivity. Rice straw is chaffed and fed as basal diet in most areas where green forage is rather scarce, especially in the dry season. The low N concentration ranging from 0.48% to 0.84% and the high lignin content (>8%) and silica content in the leaf (ranging from 8% to 14%) are in agreement with Singh et al. (1995) and Abd-Talib et al. (2018). The low N content may reduce the digestibility to improve their digestibility of the rice straw, treatment with urea, urea with lime or alkaline hydrogen peroxide is important. Also, supplementation with protein as an energy source to encourage and increase utilisation efficiency can be effective (Mo et al. 2008). High K content in rice straw signifies the need to train the smallholder rice growers not to remove all the rice straw from the fields (Mtengeti et al. 2016).

Generally, YSS has an insignificant impact for most of the micronutrient concentration of rice straw in several demonstration sites (Table 7, Figs. 3 and 4). An exception was for Cu, which was depressed by YSS treatments in Mkula village in 2012 and 2014, but was improved in Dihombo in 2013 and 2014. Another micronutrient that was lowered by YSS treatments was Zn in Mkula in 2013 and Dihombo in both short and long rainy seasons of 2013 and long rains of 2014. The two minerals, Cu and Zn, which are essential for cattle reproduction, were also reported deficient in Dihombo. Interestingly, dairy cattle had long calving interval in this village (Mtengeti et al. 2008; Phiri et al. 2008). Except for Fe and Mn, all other assessed micronutrients were rather low to support growth, production and reproduction of cattle. There is a need for strategic effort to improve rice straw

Table 7 Mean chemical composition of rice straw in different villages from 2012 to 2014

		Mkula village			Dihombo village					Dakawa agricultural research institute		
Nutrients	Practices	2012 LR	2013 SR	2014 LR	2012 LR	2013 SR	2013 LR	2014 SR	2014 LR	2012 LR	2013 LR	2014 LR
N %	FP	0.84	0.59	0.63	0.73	0.51	0.51	0.59	0.69	0.81	0.54	0.60
	YSS	0.78	0.57	0.62	0.71	0.48	0.49	0.56	0.69	0.75	0.63	0.56
P-value		0.242	0.34	0.885	0.867	0.108	0.303	0.154	0.97	0.47	0.008	0.469
P %	FP	0.23	0.14	0.13	0.17	0.14	0.12	0.17	0.16	0.17	0.11	0.09
	YSS	0.20	0.12	0.12	0.15	0.15	0.12	0.14	0.15	0.14	0.12	0.08
P-value		0.085	0.078	0.113	0.226	0.436	0.924	0.005	0.317	0.020	0.702	0.536
Ca %	FP	0.23	0.3	0.24	0.28	0.26	0.23	0.24	0.29	0.28	0.26	0.23
	YSS	0.25	0.20	0.24	0.28	0.25	0.25	0.24	0.21	0.33	0.32	0.28
Cu mg/kg	FP	3.47^a	2.85	3.12^a	3.20	2.28	11.79^b	2.21	1.7	2.04	9.61	5.36^a
	YSS	2.32^b	3.17	2.55^b	2.77	2.49	12.42^a	3.02	2.11	2.49	10.17	3.49^b
P-value		0.072	0.344	0.083	0.611	0.694	0.011	0.043	0.084	0.153	0.678	0.006
Zn mg/kg	FP	51.5	59.3^a	56.7	51.5	55.7^a	51.2^a	45.8^a	36.2	34.5	34.7	27
	YSS	53.54	49.7^b	48.4	40.6	41.5^b	41.3^b	36.4^b	43.4	34.2	30.9	29
P-value		0.63	0.041	0.143	0.102	0.029	0.008	0.039	0.256	0.926	0.275	0.539

FP farmers' practice, *YSS* Yara/SUA/Syngenta improved practice, *SR* short rain, *LR* long rain. Values in the same column and nutrient followed by different superscripts are significantly different at $P < 0.05$

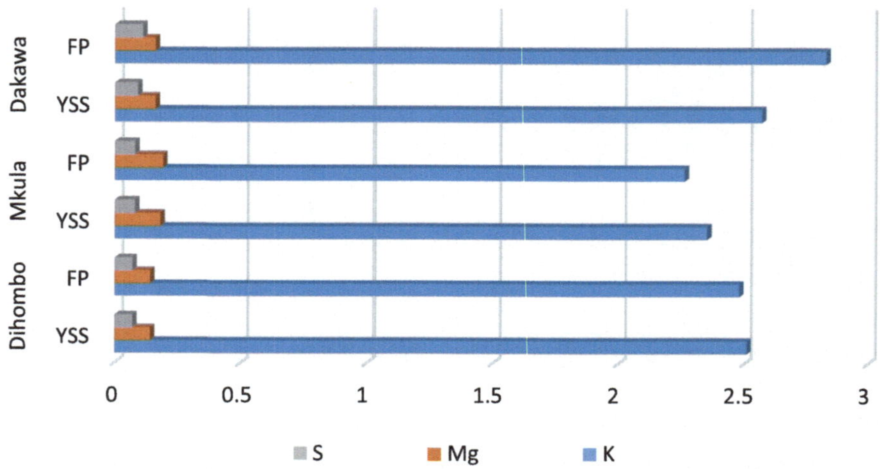

Fig. 3 Some macronutrients' concentration in rice straw (% DM)

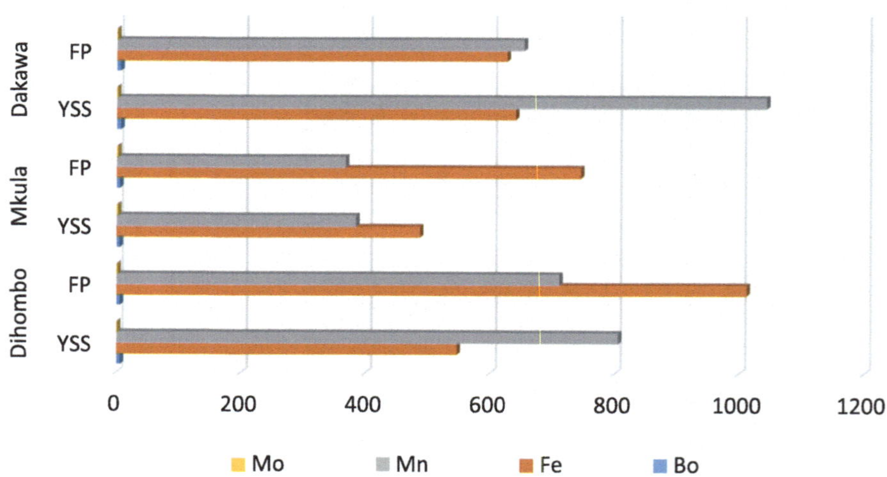

Fig. 4 Some micronutrients' concentration in rice straw (mg/kg)

utilisation efficiency as livestock feed, so as to strengthen sustainable intensification of crop-livestock amongst smallholder farmers. This is achievable because rice ranks second in production and consumption in Tanzania (Kashenge-Killenga et al. 2016), and current paddy production is about 2.7 million tonnes (USDA 2017), which could mean production of 2.7 million tonnes of rice straw. If the farmers are trained on how to treat the crop residues to improve digestibility, the crop residues could subsist dry cows and draught animals with minimum concentrate supplementation.

4 Conclusions and Recommendations

The results from this study have demonstrated that (1) smallholder maize farmers' practice produced less maize stover biomass than improved agronomic management practices; (2) rice straw productivity between the two methods did not differ appreciably; (3) both maize stover and rice straw macro- and micronutrient concentrations did not differ markedly between the improved agronomic management and farmer practices; (4) except for K in rice straw and Fe and Mn in both crop residues, all other assessed macro- and micronutrients were very low to support livestock growth, production and reproduction; and (5) even with the optimal and appropriate application of required plant nutrients and protection against pests, it is high biomass production that is realised rather than increased macro- and micronutrient concentration in the crop residues. Therefore the benefit of improved agronomical management practices is to increase crop residue biomass to feed more animals and leave some residues for soil protection against erosion. There is a need, therefore, to strategically train the smallholder farmers in treating the increased cereal crop residues from improved agronomical practices for efficient utilisation to enhance sustainable smallholder crop-livestock farming systems intensification in Tanzania.

Acknowledgements The authors of this chapter would like to thank the Norwegian government, Yara International and Syngenta International for the financial support and Sokoine University of Agriculture for the logistical support during the period of this study. Special thanks to Dr. Frank Brentrup of Research Centre Hanninghof, Yara International, Dulmen, Germany, for the sample handling and analysis. The authors are also grateful to the district council's extension workers for their assistance in data collection and smallholder maize and rice farmers in Njombe, Kilombero and Mvomero districts for allowing part of their farms to be used in this study.

References

Abd-Talib N, Ahamad A, Mohd-Setapar S, Khamis AK, Lokhat D, Rufatullah M (2018) Removal of silica from rice straw by using alkaline hydrogen peroxide solution in fixed bed column. J Mater Environ Sci 9:864–872

Duncan AJ, Tarawali SA, Thorne PJ, Valubuena D, Descheemaeker K, Homann-Keetut S (2013) Integrated crop-livestock systems – a key to sustainable intensification in Africa. Trop Grassl 1:202–206

FAOSTAT (2012) The food and agricultural organisation's statistical database. Land Use Area Database. Available Online at http://faostat.fao.org/site/377/DesktopDefault.aspx?PageID=377#ancor

Foster AD, Rosenweig MR (2010) Microeconomics of technology adaption. Centre discussion paper no. Economics Growth Centre, Yale University, New Haven, p 984

Kashenge-Killenga S, Meliyo J, Urassa G, Kongo V (2016) Extent of salt-affected soils and their effects in irrigated and lowland rain-fed rice growing areas of southwestern Tanzania. In: Lal R, Kraybill D, Hansen DO, Singh BR, Mosogoya T, Eik LO (eds) Climate change and multi-dimensional sustainability in African agriculture. Springer, Basel, pp 97–126. ISBN 978-3-319-41238-2

Lenne JM, Thomas D (2006) Integrating crop-livestock research and development in SuSaharan Africa. Option, Imperative and impossible? Outlook Agr 35:167–175

Li HY, Xu L, Liu J, Wang N (2004) Assessment of nutritive value of whole corn stover and its morphological fractions. Asian-Australas J Anim Sci 27(2):194–200

Mdoe NSY, Mlay GI, Kadigi M (2015) Farming systems in Tanzania. Empirical evidence of changes in livelihood patterns and poverty among smallholder farmers. In: Lal R, Singh BR, Mwaseba DL, Kraybill D, Hansen DO, Eik LO (eds) Sustainable intensification to advance food security and enhance climate resilience in Africa. Springer Basel, pp 555–571. ISBN 978-3-319-09359-8

Mo M, Weiby KV, Urio NA, Mtengeti EJ, Eik LO, Sundstøl F, Safari J (2008) Recent trends in upgrading crop residues for ruminant feeding: a review. Tanzan J Agric Sci 8(2):171–179

Mtengeti EJ, Phiri ECJH, Urio NA, Mhando DG, Mvena Z, Ryoba R, Mdegela R, Singh BR, Mo M, Wetlesen A, Lørken T, Reksen O (2008) Forage availability and its quality in the dry season on smallholder dairy farms in Tanzania. Acta Agric Scand Sect A Anim Sci 58(4):196–204

Mtengeti EJ, Brentrup F, Mtengeti E, Eik LO, Chambuya R (2015) Sustainable intensification of maize and rice in smallholder farming systems under climate change in Tanzania. In: Lal R, Singh BR, Mwaseba DL, Kraybill D, Hansen DO, Eik LO (eds) Sustainable intensification to advance food security and enhance climate resilience in Africa. Springer Basel, pp 441–465. ISBN 978-3-319-09359-8

Mtengeti EJ, Brentrup F, Mtengeti E, Eik LO, Chambuya R (2016) Effect of improved plant nutrition on maize (*Zea mays*) and rice (*Oryza sativa*) grain chemical nutrient content under smallholder farming systems in Tanzania. In: Lal R, Kraybill D, Hansen DO, Singh BR, Mosogoya T, Eik LO (eds) Climate change and multi-dimensional sustainability in African agriculture. Springer Basel, pp 485–504. ISBN 978-3-319-41238-2

Msanya BM, Kaaya AK, Araki S, Otsuka H, Nyadzi GI (2003) Pedological characteristics, general fertility and classification of some benchmark soils of Morogoro District, Tanzania. Afric J of Sci Techn 4(2):101–112

National Research Council (2001) Nutrient requirements for dairy cattle. 7th rev. ed. Natl. Acad. Sci., Washington, DC

NEPAD (2005) Integrating livestock, forestry and fisheries subsector into the CAADP, Companion document to the Comprehensive Africa Agriculture Development Programme, NEPAD, Nairobi

Okalebo JR, Othieno CO, Woomer PL, Karanja NK, Semoka JRM, Bekunda MA, Mugendi DN, Musaya RM, Bationo A, Mukhwana EJ (2006) Available technologies to replenish soil fertility in East Africa. Nutr Cycl Agroecosyst 76:153–170

Pass (2012) Private agricultural sector support: investment potential in Grain. http://www.pass.ac.tz/investmentpotential.html

Phiri ECJH, Chibunda RT, Pereka AE, Mtengeti EJ, Urio NA, Katakweba AA (2008) Haematological and metabolic profiles of smallholder dairy cows in relation to weather patterns in southern highlands of Tanzania. Tanzan Vet J 25(2):111–120

RATES (2003) Maize market assessment and baseline study for Tanzania. Regional Agricultural Trade Expansion Support Program (RATES). Nairobi, Kenya

Romney DL, Thorne P, Lukuyu B, Thornton PK (2003) Maize as food and feed in intensive small-scale systems: management options for improved integration in mixed farming systems of east and southern Africa. Field Crop Res 84:159–168

Sath K, Pauty, Holtenius K (2013) Mineral status in cattle fed rice straw and Para grass combined with different levels of protein derived from cassava foliage. Asian-Australas J Anim Sci 26:59–64

Singh RB, Saha RC, Singh M, Chandra D, Shukla SG, Walli TK, Pradhani PK, Kessels HPP (1995) Rice straw-its production and utilization in India. In: Sngh K, Schiere JB (eds) Handbook for straw feeding systems. ICAR, New Delhi pp 325–337

Thornton PK, Krushka RRL, Heminger N, Kristjanson P, Reid RS, Atieno AN, Ndegwa T (2002) Mapping poverty and livestock in the developing world. ILRI, Nairobi

USDA (2017) United States Agency for International Development Tanzania corn, wheat and rice report. http://gain.fas.usda.gov. Visited 13 Feb 2019

USDA (2012) United States Agency for International Development review of policy constraints on competitive East Africa Community rice production. http://agritrade.cta.int/en/layout/set/print/Agriculture/Commodities. Visited 12 Oct 2018

Valbuena D, Erenstein O, Homann-Kee Tui S, Abdoulaye T, Claessens L, Duncan AJ, Gerard B, Rufino MC, Teufel N, van Rooyen A, van Wijk MT (2012) Conservation agriculture in mixed crop-livestock systems: scoping crop residue trade-offs in sub-Saharan Africa and South Asia. Field Crop Res 132:175–184

Vaswani S, Kumar R, Kumar V, Roy R, Kumar M (2016) Nutritional and mineral composition of different varieties of normal and high quality protein maize fodder at post-cob stage. Int J Sci Environ Technol 5:2719–2727

WEMA (2010) Mitigating the impact of drought in Tanzania: The WEMA intervention. Policy brief Water Efficient Maize for Africa (WEMA). http://www.aatf-africa.org/userfiles/WEMA-Tz-policy. Visited 20 Oct 2018

Westengen OT, Berg T (2016) Crop adaptation to climate in SSA: the role of genetic resources and seed systems. In: Lal R, Kraybill D, Hansen DO, Singh BR, Mosogoya T, Eik LO (eds) Climate change and multi-dimensional sustainability in African agriculture. Springer Basel, pp 327–343. ISBN 978-3-319-41238-2

Intensification of Sorghum and Pearl Millet Production in the Sahel-Sudanian Climatic Zones of Mali

Adama Coulibaly and Jens B. Aune

Abstract The objective of this study was to explore options for intensification of pearl millet and sorghum production in the Sahelian and Sahel-Sudanian climatic zones of Mali. A multilocation experiment was conducted over 3 years, and the treatments were ordered according to increasing levels of intensification (increasing costs) by adding one new element on the top of the others for each new level of intensification. The treatments included a control, seed priming, and microdosing of 0.5 g of NPK fertilizer per pocket and 2 g of NPK fertilizer per pocket.

Sorghum yields increased by 55% in the first year, whereas the response for the following 3 years went up to 19%, but the response in pearl millet did not exceed 13% in these years. For sorghum, the treatment combining seed priming and 0.5 g of NPK fertilizer per pocket gave a yield increase compared to the control of 131% in the first year, while in the following two seasons, the corresponding yield increase was 26% and 41%, respectively. For pearl millet, the corresponding yield increase was 17% and 35%. There was generally only a small difference in gross margin between the treatments with 0.5 and 2 g of fertilizer per pocket.

By ordering the treatments according to increasing gross margin, it was shown that seed priming and microdosing do not increase farmers' risk of a reduced gross margin, but these methods increase the probability of a high gross margin.

This experiment shows that agricultural intensification in sorghum and pearl millet can be achieved at a low cost and that the economic return for this form of intensification is very good.

A. Coulibaly
Laboratoire Sol-Eau-Plante, Institut d'Economie Rurale, Bamako, Mali

J. B. Aune (✉)
Department of International Environment and Development Studies, Noragric, Norwegian University of Life Sciences, Ås, Norway
e-mail: jens.aune@nmbu.no

B. R. Singh et al. (eds.), *Climate Impacts on Agricultural and Natural Resource Sustainability in Africa*, https://doi.org/10.1007/978-3-030-37537-9_8

1 Introduction

Crop production is challenging in the Sahel-Sudanian zone of Mali owing to frequent droughts occurring throughout the growing season, which affect crop establishment, tillering, flowering, and grain filling of the cereals. The rainfall is also often unevenly distributed in time and across the landscape. Crop development is further constrained by the limited supply of plant nutrients, particularly nitrogen and phosphorus (Penning de Vries 1982). These factors make crop production a risky undertaking.

Several technologies have been proposed to improve crop production in drylands. Promising methods include seed priming (Harris 2006; Aune and Ousman 2011), microdosing (placing small amounts of mineral fertilizer adjacent to the plants) (Buerkert et al. 2001; Aune and Ousman 2011), ridging (Subbarao et al. 2000), and stone bunds and zais for water retention (Sawadogo 2011). These methods can ensure a uniform crop establishment, vigorous crop growth, and improved grain filling.

Intensification of crop production in the Sahel has previously been described as climbing a ladder (Aune and Bationo 2008). We build on this theoretical approach and test the approach under field conditions in Mali. The experiment (on-farm) included seed priming and increasing levels of fertilizer microdosing. The study explored the effect of these technologies on crop establishment, yield, gross margin, and value cost ratio. The objective of this study was to assess under which conditions intensification is likely to be beneficial and to identify appropriate levels of intensification in southern Mali (Koulikoro and Segou regions). The methods proposed here do not intend to maximize yields, but rather to intensify cereal production within the means that are available to the resource-poor farmers in the Sahel.

2 Materials and Methods

The experiments were located in Nossombougou and Didieni in the region of Koulikoro and Konobougou in the region of Segou. Nine villages were involved in the study. In each village, the project identified 15 voluntary farmers who agreed to test the treatments. The soils in the project areas were leached sandy loam and loamy sand (PIRT 1983) (Table 1).

Table 1 Agro-climatic characteristics of the experiment sites (Konate 1984)

Site	Rainfall (mm)	Agro-climatic zones	Crop	Villages
Didieni	500	Sahelian	Millet, groundnut	2
Nossombougou	600	Sahelian	Millet, sorghum, groundnut	4
Konobougou	700	Sahel-Sudanian	Millet, sorghum, groundnut	3

The treatments were as follows:

T1. Control
T2. Seed priming
T3. Seed priming + fertilizer microdosing at 0.5 g per pocket of NPK fertilizer 15-15-15 (12.5 kg/ha), a cereal blend. The fertilizer was placed adjacent to the seed planting pocket at sowing.
T4. Seed priming + fertilizer microdosing at 2 g per pocket of NPK fertilizer 15-15-15 (50 kg/ha) at 15 days after sowing (DAS)

The seeds in the seed priming treatments were soaked in water for 8 hours and thereafter dried in the open air for 2 hours prior to sowing. Seeds in all treatments were treated with the fungicide Caiman Rouge (endosulfan and thiram) in order to reduce attack of fungus and pests. In treatments T2, T3, and T4, the seeds were treated after priming and surface drying.

The treatments tested in 2006 were T1, T2, and T4, whereas in 2007 and 2008 all four treatments were tested. One farmer was considered as a replication. The size of each plot was 1000 m^2. Yield was measured by harvesting three randomly selected quadrats, each with an area of 10 m^2.

For pearl millet, the local variety Toroniou, which arrives at maturity in 90 days, was used, whereas for sorghum, the variety Djacoumbé was used. This variety takes about 100 days from sowing to maturity. The plots were sowed after a rain shower of about 15 and 20 mm for sandy soil and silty soil, respectively. Weeding was undertaken according to the needs in the plots, and the stand was thinned to two plants per sowing hill. The spacing of the pockets was 80 cm between rows and 50 cm within the row, giving 25,000 pockets/ha.

2.1 *Effect of Drying Time on Primed Seed Germination*

The farmers are not always able to sow all the seeds immediately following seed priming. In order to test for the effect of drying time, the seeds were dried for 2 hours, 24 hours, 48 hours, and 72 hours. The experimental design was block Fisher with four treatments in 2010 and three treatments in 2011. The size of each plot was 25 m^2. Planting distances were the same as detailed above.

Measurements and Data Analysis (Common to the Two Experimental Series)

Observations were taken on number of established plants, number of harvested heads, and grain and stover yield. A statistical analysis was undertaken using the SAS statistical program, and the Duncan test was used to rank differences between the treatments.

Fertilizer Use Efficiency

Fertilizer use efficiency (FUE) was calculated by dividing the increased grain yield due to fertilizer by the amount of fertilizer applied (Kelly et al. 2006). According to Kelly et al. (2006), the threshold value is 10 kg of grain (millet or sorghum) for each kg of fertilizer. The FUE was calculated for the combined effect of seed priming and microdosing and for microdosing alone. The yield difference between "priming + microdosing" minus "control" was used to calculate the combined effect of priming and microdosing, whereas the difference between "priming + microdosing" minus "priming" was used to calculate the effect of microdosing alone.

Analysis of Environmental Conditions on the Response to Seed Priming and Microdosing

The analysis assessed under which environmental condition seed priming and microdosing are performing the best. The relative yields between treatments and control were calculated and plotted against the yield of the control in the plot.

Economic Analysis

A partial gross margin was calculated for each treatment by calculating the difference between the revenue from the grains and the costs. The cost items were inputs and labor related to seed priming and microdosing. The other labor costs were not included. Local prices for the grains and input were used.

Cumulative Probability for Gross Margin

A probability analysis was undertaken by utilizing the results from all the sites in 2007 and 2008. For each treatment, the gross margins were arranged in ascending order to display the probability of reaching a certain gross margin for each treatment.

3 Results and Discussion

3.1 Yield Increase in Sorghum and Millet

Overall, there was a good response to seed priming and microdosing in pearl millet and sorghum, but sorghum was generally more responsive to priming and microdosing than millet.

Priming was able to increase grain yield in both sorghum and millet (Tables 2 and 3). In sorghum, priming increased yield by 55% in 2006, whereas the response in sorghum for 2007 and 2008 was up to 19%. For millet, the response to priming was a 49% yield increase in 2006, whereas the increase for 2007 and 2008 was up to 12%. These results show that the effect of seed priming is highly variable between years. Seed priming kick-starts plant growth and can ensure a better crop establishment (Harris et al. 2001; Aune and Ousman 2011). The crop can thereby better utilize available water early in the growing season and benefit from the nitrogen flush early in the growing season (Birch 1958). Use of seed priming can be considered as a "long-hanging fruit" for agricultural intensification as the method entails no additional cost and the additional labor is very low.

Yields were further increased when seed priming was combined with microdosing. As for priming, the effects of microdosing were more pronounced in sorghum than in millet in terms of percentage yield increase in kg/ha. In 2006 seed priming combined with 2 g of fertilizer per pocket increased yield by 92% and 131% for millet and sorghum, respectively (Table 2). In 2007 and 2008 there were two microdosing treatments: one used an application of 0.5 g at sowing (T3) and the other treatment used an application of 2 g of fertilizer per pocket about 15 days after sow-

Table 2 Effects of seed priming and microdosing of fertilizer on millet and sorghum yields (kg/ha), Konobougou-Nossombougou 2006

	Grain	
Treatment	Millet	Sorghum
T1 control	1107c	1087c
T2 seed priming	1644b	1686b
T3 seed priming + fertilizer microdosing at 2 g per pocket of cereal blend (50 kg/ha) 15 DAS	2128a	2506a
Treatment	**	**
Treatment × site	ns	**
CV (%)	41	32

Cereal blend contains 15-15-15 NPK; DAS days after sowing
** Significant less than $p<0.01$

Table 3 Effects of seed priming and fertilizer microdosing on millet and sorghum yields (kg/ha), 2007–2008 in Koulikoro and Segou regions

	Millet		Sorghum	
Treatment	Didi.	Kon.	Nosso	Kono.
T1 control	1135b	1537b	1646d	1549c
T2 seed priming	1240b	1736b	1965c	1732bc
T3 seed priming +0.5 g fertilizer per pocket at planting but in different hole by the seed	1325a	2074a	2331b	1978ab
T4 seed priming +2 g fertilizer per hole of cereal blend, 15 DAS	1567a	2034a	2816a	2011a
Treatment	*	**	**	**
Treatment × year	ns	ns	**	ns
CV (%)	34	30	25	20

DAS days after sowing
* Significant less than $p<0.05$,
**Significant less than $p<0.01$

Table 4 Effects of seed priming and microdosing on fertilizer use efficiency in millet and sorghum

	Pearl millet		Sorghum	
	2006	2007/2008	2006	2007/2008
0.5 g fertilizer + priming		29.1		44.6
2 g fertilizer + priming	20.4	9.2	28.3	16.3

ing (T4). For millet, there was no significant yield difference between the two fertilizer treatments. However, for sorghum in Nossombougou, yields were significantly higher with the 2 g treatment compared to the 0.5 g treatment (Table 4). The 2 g fertilizer treatment increased yield by 73% at this site, while the 0.5 g fertilizer treatment increased yield by 41%. At Konobougou there was no significant difference between the microdosing treatments, but the yield increase compared to the control for the two microdosing treatments was in the order of 30%. The lower response to microdosing in Konobougou compared to Nossombougou could be the result of a residual fertilizer effect of previous cotton field planting, as Konobougou is located in the cotton belt of Mali.

This response to seed priming and microdosing is in line with previous experiments with these methods (Aune et al. 2007; Aune et al. 2012; Aune and Ousman 2011). The yields are however higher in this study, and the percentage yield increase is therefore less than in the previous studies.

Fertilizer use efficiency The fertilizer use efficiency (FUE) ranged from 9.2 to 44.6 kg grains per kg of applied fertilizer (Table 4). FUE was higher in the 0.5 g treatment as compared to the 2 g fertilizer treatment. Sorghum also had a higher FUE than the millet. A ratio above 10 is considered as high FUE (Kelly et al. 2006). A seen in Table 4, the application of 0.5 g of fertilizer combined with priming gives an FUE of 29.1 and 44.6 for sorghum and millet, respectively. However, the application of 2 g of fertilizer per pocket combined with priming also gave a satisfactory result, particularly in sorghum. The FUE is typically reduced when the fertilizer rate is increased (Beyaert and Roy 2005).

Similar FUEs have been reported from dryland areas in Sudan. Priming and application of 0.3 g of mineral fertilizer per pocket gave an FUE of 16 and 35 in millet and sorghum, respectively (Aune and Ousman 2011). Results from Niger have shown that FUE was greatly increased when fertilizer application was combined with residue retention (Yamoah et al. 2002).

Economic analysis The economic analysis is based on the results from 2007 and 2008 seasons. As Table 5 shows, there was an increase in the gross margin in sorghum with increasing use of inputs. Seed priming alone increased net return by 18,325 and 10,900 CFA/ha in sorghum and pearl millet, respectively.

Compared with the control, the net benefit in sorghum increased by 15.4%, 32.1%, and 41.5% for T2, T3, and T4, respectively. In millet, the corresponding increase was 11.0%, 23.9%, and 23.3% for T2, T3, and T4, respectively. This shows that in millet, going beyond 0.5 g of NPK fertilizer/ha had no effect.

Table 5 Partial budget of seed priming and microdosing of fertilizer on farmer income with sorghum and millet (FCFA per hectare), 2007–2008

	Sorghum				Pearl millet			
	T1	T2	T3	T4	T1	T2	T3	T4
Average yield (kg/ha)	1598	1849	2155	2414	1336	1488	1700	1801
Price millet/sorghum (FCFA/kg)	75	75	75	75	75	75	75	75
Total income (FCFA/ha)	119,850	138,675	161,625	181,050	100,200	111,600	127,500	135,075
Seed costs (FCFA/ha)	600	600	600	600	600	600	600	600
Fertilizer cost (FCFA/ha)	0	0	3125	12,500	0	0	3125	12,500
Labor (FCFA/ha)	0	500	1000	1500	0	500	1000	1500
Variable costs (FCFA/ha)	600	1100	4725	14,600	600	1100	4725	14,600
Gross margin (FCFA/ha)	119,250	137,575	156,900	166,450	99,600	110,500	122,775	120,475
Value cost ratio		36.7	9.1	3.4		21.8	5.6	1.5

T1 = control, T2 = seed priming, T3 = seed priming +12.5 kg NPK/ha, T4 = seed priming +50 kg NPK/ha

The calculation of the value-cost-ration (VCR) in sorghum showed that it was 36.7, 9.1, and 3.4 for T2, T3, and T4, respectively. In millet the corresponding values were 21.8, 5.6, and 1.5. As the threshold value for VCR is 2 for introducing new technologies (Bationo et al. 2012), all treatments can be considered to give a satisfactory response except the T4 in millet. Seed priming is clearly the treatment with the highest VCR. However, seed priming plus 0.5 g of fertilizer per pocket (T3) will be the most interesting option in millet production because it has a high gross margin and a satisfactory VCR both in millet and sorghum. The T4 treatment gives a 6% higher gross margin than the T3 in sorghum, but the VCR is 9.1 for T3 as compared to 3.4 for T4.

Research in Niger showed a low economic return (based on VCR) related to the application of 2 g of DAP per pocket in pearl millet (Bielders and Gerard 2015). These researchers report that the risk of using microdosing is less when the yield in the control plot is low. There are therefore good reasons to question the recommendation by ICRISAT to use microdosing in the order of 2 to 6 grams of fertilizer per pocket in pearl millet (ICRISAT 2012). The application of 6 g of fertilizer per pocket results in an application of 150 kg of fertilizer/ha if there are 25,000 pockets/ha. This cannot be considered a low application rate under Sahelian conditions. Major development actors like the Alliance for Green Revolution in Africa (AGRA) have also been promoting the rates proposed by the International Crop Research Institute for Semi-Arid Tropics (ICRISAT) (FARA 2009).

3.2 Effects of Storage Time of Primed Seeds on Sorghum and Millet Plant Establishment and Yield

It may be difficult for the farmers to know what quantity of seeds to prime when the rain starts, because that depends on how much land they will be able to sow in the following day. If too many seeds have been primed, the question is what to do with the seed surplus. A study was therefore initiated to assess the length of time it is possible to store the primed seeds without losing the effect of priming on seed germination, emergence, and crop growth. Sorghum seeds in this study were dried in shade for 2, 24, 48, and 72 hours following seed priming, whereas millet seeds were dried for 2, 24, and 48 hours. Neither sorghum nor millet showed any effect of drying time on emerged plants, number of heads, and grain yield. This result demonstrates that it is possible for farmers to store the seeds for up to 3 days if they are not able to utilize all the seeds on the day of priming. This makes seed priming a more flexible approach than if all the seeds need to be sown immediately following seed priming.

3.3 Environmental Conditions on the Response to Seed Priming and Microdosing

It was important to determine the environmental conditions under which seed priming and microdosing would perform best. The relative yield between treatments (T2, T3, and T4) and control (T1) was calculated; if it was above one, this would indicate a positive effect of the treatment (priming and microdosing) as compared to the control. This relative yield for each treatment and site was plotted against yield of the control for the site. The yield of the control can be considered as an indicator for the environment at the site as it is determined mainly by rainfall, soil conditions, disease, pest, and weed pressure. The regression analysis showed that the relative benefit of seed priming and microdosing was reduced when environmental conditions become more favorable (higher yields in the control).

In sorghum, there was hardly any effect of seed priming and microdosing when the yield in the control was above 2000 kg/ha. Below a sorghum yield of 900 kg/ha in the control, there was a good response to seed priming and microdosing. Between 900 and 2000 kg/ha, seed priming and microdosing had a positive effect, but the effect was more variable. These results show that seed priming and microdosing are technologies that are mostly suited to marginal environments (Figs. 1, 2, and 3).

The result for pearl millet showed that, as with sorghum, there is a decline in relative yield between the treatment and the control when the yield of the control is increased. As Figs. 4, 5, and 6 indicate, there is a consistent response to these treatments when the yield in the control is below 500 kg/ha. Above 500 kg/ha in the control, the response in comparison to the control is still evident, but the effect is more variable.

However, when plotting the yield in the control against the absolute yield increase, no clear relationship was apparent (figure not shown). The same increase in yield (kg/ha), as a result of the treatment, can therefore be expected across different yield levels in the control.

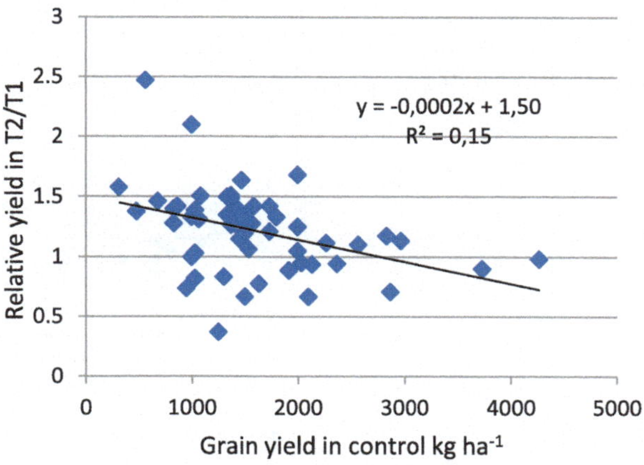

Fig. 1 Relative sorghum yield T2/T1 for the sites in 2007 and 2008

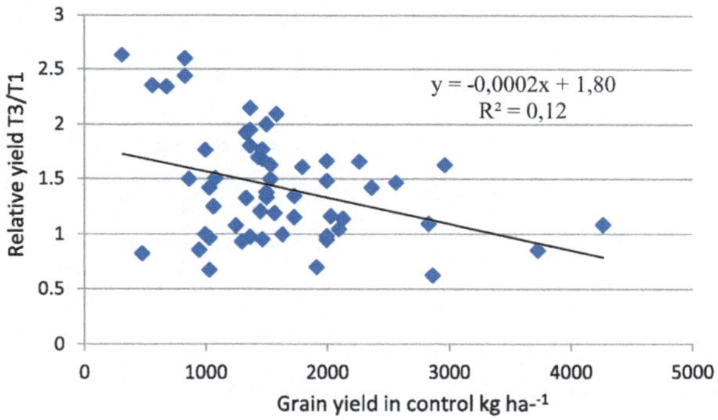

Fig. 2 Relative sorghum yield T3/T1 for the sites in 2007 and 2008

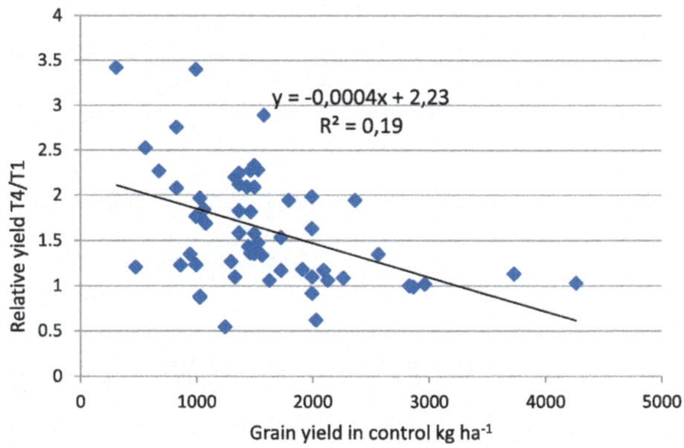

Fig. 3 Relative sorghum yield T4/T1 for the sites in 2007 and 2008

It would make more sense to introduce seed priming and microdosing in areas with low yield, even though the absolute yield increase is the same. The reason for this is that the yield increase associated with seed priming and microdosing is less visible at high yield levels, whereas the farmers can easily be convinced of the effect in areas where the yields are low. The additional grain and corresponding straw yield obtained under such marginal conditions may also be very valuable.

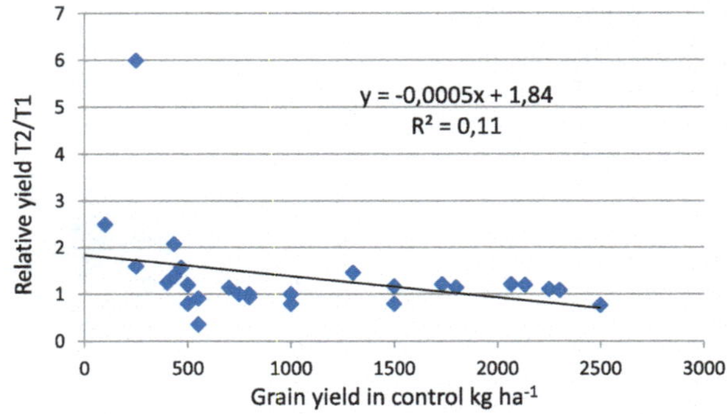

Fig. 4 Relative pearl millet yield T2/T1 for the sites in 2007 and 2008

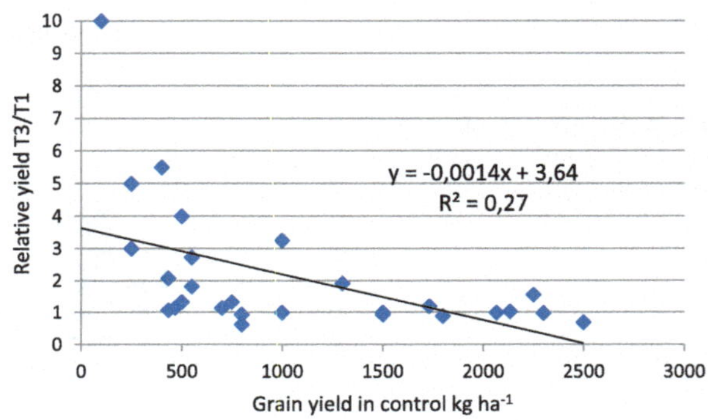

Fig. 5 Relative pearl millet yield T3/T1 for the sites in 2007 and 2008

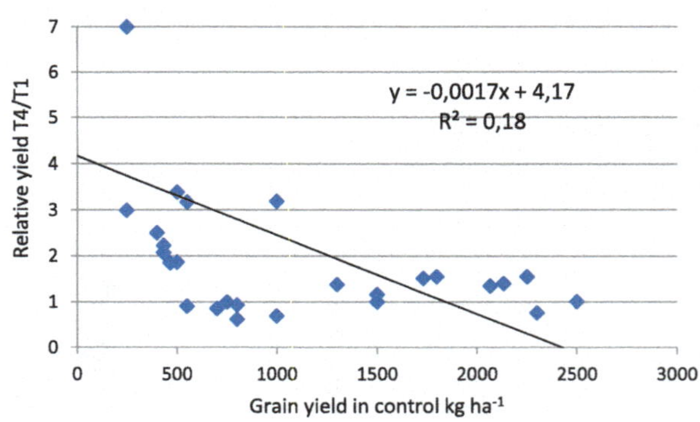

Fig. 6 Relative pearl millet yield T4/T1 for the sites in 2007 and 2008

3.4 *Risk of Seed Priming and Microdosing*

Figure 7 shows the cumulative probability for gross margin of different treatments in sorghum for the years 2007 and 2008. It appears that the probability of a low net benefit (below 75,000 CFA/ha) is the same in the different treatments. Use of microdosing and seed priming therefore did not increase farmers' risk. We observe that, for the control treatment (T1), about 20% of the farmers obtained a net benefit above 150,000 CFA/ha, whereas for treatments T3 and T4, about 50% of the farm-

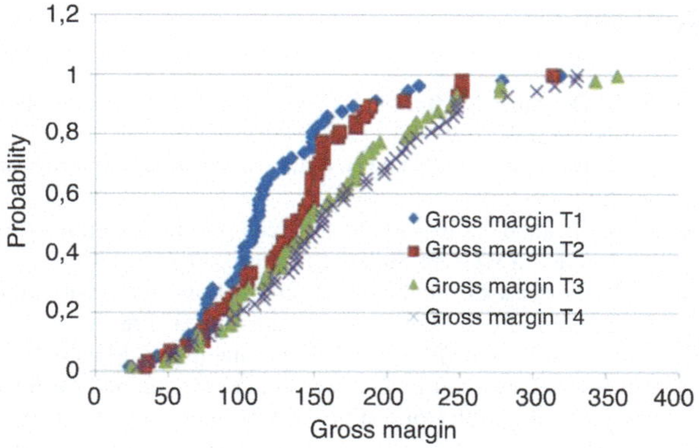

Fig. 7 Cumulative probability for gross margin/ha (CFA in thousands) of sorghum for the treatments T1, T2, T3, and T4 in the seasons 2007 and 2008

ers obtained a gross margin above 150,000. There is therefore a considerably higher probability that farmers will get a high benefit when combining seed priming and microdosing. There is only a minor difference in the distribution of income between the T3 and T4 treatments, showing that farmers may well choose an application of 0.3 g per pocket instead of 2 g per pocket. About 35% of the farmers practicing seed priming obtained an income above 150,000 CFA/ha.

The same distribution of probabilities of gross margin was observed with pearl millet as with sorghum, but the response to the treatments was less than that observed for sorghum.

4 Conclusion

The results show that there is a good opportunity of agricultural intensification in rain-fed agriculture in Mali. In both series of experiments, sorghum responded better than pearl millet to increasing levels of intensification. The experiment showed that the application of 0.5 g of fertilizer per pocket is a more appropriate rate than the application of 2–6 g of fertilizer per pocket as recommended by ICRISAT. This is particularly the case for fertilizer application in pearl millet. The relative yield increase as a result of the treatments was higher when the general yield levels at the sites were low. This makes seed priming and microdosing suitable technologies, particularly under marginal conditions. However, the effect of the treatments in terms of yield in kg/ha is the same at higher yield levels, indicating that intensification is also feasible in better environmental conditions. A risk assessment of the technologies showed that there is no difference between the treatments with regard to the probability of a low gross margin, whereas it appears that the probability of a high gross margin is significantly higher when seed priming and microdosing are practiced. Seed priming and microdosing therefore do not increase farmers' risk, but enhance the probabilities for a high gross margin.

References

Aune JB, Doumbia M, Berthe A (2007) Microfertilizing sorghum and pearl millet in Mali 2007. Outlook Agr 36:199–203

Aune JB, Bationo A (2008) Agricultural intensification in the Sahel – the ladder approach. Agric Syst 98:119–125

Aune JB, Ousman A (2011) Effect of seed priming and microdosing of fertilizer on sorghum and pearl millet in Western Sudan. Exp Agric 47:419–430

Aune JB, Traoré CO, Mamadou S (2012) Low-cost technologies for improved productivity of dryland farming in Mali. Outlook Agr 41:103–108

Bationo A, Fairhust T, Giller K, Kelly V, Lunduka R, Mando A, Mapfumo P, Odour G, Romney D, Vanlauwe B, Wairegi L, Zingore S (2012) Handbook for integrated soil fertility management. African Soil Health Consortium, CABI International, 149 pp

Beyaert RP, Roy RC (2005) Influence of nitrogen fertilization on multi-cut forage sorghum–sudan-grass. Yield and nitrogen use. Agron J 97:1493–1501

Birch HF (1958) The effect of soil drying on humus decomposition and nitrogen availability. Plant Soil 10:9–31

Bielders C, Gerard B (2015) Millet response to microdose fertilization in South-Western Niger: effect of antecedent fertility management and environmental factors. Field Crop Res 171:165–175

Buerkert A, Bationo A, Piepho H-P (2001) Efficient phosphorus application strategies for increased crop production in sub-Saharan West Africa. Field Crop Res 72:1–15

FARA (2009) AGRA launches micro dosing fertiliser project. http://farastaff.blogspot.no/2009/12/agra-launches-micro-dosing-fertiliser.html. Assessed 12 Sept 2016

Harris K, et al. (2001) Role of scaffolds in MAP kinase pathway specificity revealed by custom design of pathway-dedicated signaling proteins. Curr Biol 11(23):1815–24

Harris D (2006) Development and testing of "on-farm" seed priming. Adv Agron 90:129–178

ICRISAT (2012) Fertilizer microdosing. http://www.icrisat.org/jewels/Posters/Poster%20Fertilizer%20Microdosing.pdf. Assessed 12 Sept 2016

Kelly V, Reardon T, Yangen D, Naseem A (2006) Fertilizer in sub-Saharan Africa: Breaking the vicious circle of high prices and low demand. Policy Synthesis no. 32. MSU Agricultural Economics, USA

Konate M (1984) Climate of the sorghum and millet cultivation zones of the Semi-arid tropical regions of West Africa. In: Agrometeorology of sorghum and millet in the semi-arid tropics: Proceedings International Symposium, Nov. 15–20, 1982. ICRISAT Center, Patanacheru, India, pp 101–114

Penning de Vries FWT (1982) La fertilisation et l'irrigation. In: de Vries FWTP, Djitèye MA (eds) La productivity des pâturage sahéliens. Une étude des sols, des végétations et de l'exploitation de cette ressource naturelle. PUDOC, Wageningen, pp 425–434

PIRT (1983) Projet Inventaire des Ressources Terrestres du Mali. Les ressources terrestres du Mali. Government of Mali/USAID/TAMS, New York

Sawadogo H (2011) Using soil and water conservation techniques to rehabilitate degraded lands in northwestern Burkina Faso. Int J Agric Sustain 9:120–128

Subbarao GV, Renard C, Paynes WA, Bationo A (2000) Long-term effects of tillage, phosphorus fertilization and crop rotation on pearl millet-cowpea productivity in the west-African Sahel. Exp Agric 36:243–264

Yamoah CF, Bationo A, Shapiro B, Koala S (2002) Trend and stability of millet yields treated with fertilizer and crop residues in the Sahel. Field Crop Res 75:53–62

Impact of Climate Variability on the Use and Exposure of Pesticides in Sugarcane Production in Malawi

Trust Kasambala Donga, Richard Meadow, Bishal K. Sitaula, and Ole M. Eklo

Abstract It is widely accepted that climate change will affect sugarcane production and its associated pests. The aim of this chapter is to review the impact of climate variability on factors and processes affecting environmental exposure of pesticides used in sugarcane production in Malawi. We indicate that changes in temperature and rainfall will have a dual effect on pesticide risk. Temperatures higher than 30–35 °C affect pesticide toxicity, though effects will vary with pesticide-pest combination. Rapid degradation of pesticides such as acetamiprid and atrazine is expected at temperatures above 30 °C. Higher temperature may increase the incidence and severity of pests such as red spider mites, prompting farmers to use more pesticides. On the other hand, the amount and timing of rainfall in relation to pesticide application are important determinants on the amount of pesticide residue remaining in the environment. There is a higher likelihood of pesticide transport to surface (through runoff) and percolating to groundwater at higher rainfall intensity. A higher soil water content will result in increased pesticide degradation. There is a need to determine the occurrence of pesticide residue in sugarcane cropping and aquatic systems surrounding sugarcane plantations. We highly recommend building capacity in this sector, particularly in biological control of pest species using microbial agents such as insect pathogenic fungi pathogens.

T. K. Donga (✉)
Lilongwe University of Agriculture and Natural Resources, Lilongwe, Malawi

Norwegian University of Life Sciences, Ås, Norway
e-mail: tdonga@luanar.ac.mw

R. Meadow · O. M. Eklo
Norwegian University of Life Sciences, Ås, Norway

Norwegian Institute of Bioeconomy Research, Ås, Norway

B. K. Sitaula
Department of International Environment and Development Studies, Norwegian University of Life Sciences, Ås, Norway

B. R. Singh et al. (eds.), *Climate Impacts on Agricultural and Natural Resource Sustainability in Africa*, https://doi.org/10.1007/978-3-030-37537-9_9

1 Introduction

Africa is the world's most vulnerable region to climate change (Challinor et al. 2007; Dasgupta et al. 2014). However, spatiotemporal variation in terms of vulnerability and susceptibility exists among and within African countries (Adhikari et al. 2015). Vulnerability to climate change is defined as 'the degree to which geophysical, biological and socio-economic systems are susceptible to and unable to cope with, adverse impacts of climate change' (IPCC 2007). Brooks et al. (2005) outlined socioeconomic factors that determine a nation's vulnerability and adaptive capacity to climate change. These factors include economy, health and nutrition, literacy rate, infrastructure, geography and demography, and dependence on agriculture (Brooks et al. 2005). Malawi is one of the world's poorest countries with a gross domestic product (GDP) per capita of about USD480 (The World Bank Group 2017). The country's economy is highly dependent on rainfed agriculture. The majority of the population live in rural areas. About 55% of females are literate compared to 73% of males. The HIV/AIDS prevalence rate is 9.2% (The World Fact Book 2016). Since 2013/2014, food insecurity has been increasing (SADC/VAC 2016). Poverty rates are highest in southern Malawi where there is a higher risk of flooding and waterborne diseases (The World Fact Book 2016; Mwale et al. 2015). Malawi is thus highly vulnerable to climate change impacts.

Any change in climate over time, whether due to natural variability or because of human activity, will significantly affect agriculture (Delcour et al. 2015; Aktar et al. 2009; Noyes et al. 2009; USAID 2007). It is difficult to isolate climate change from normal climate variability, i.e. the way climate elements such as temperature and rainfall change from the mean value in a given month/season/year (McSweeney et al. 2010; Wood and Moreniere 2013). Three external atmospheric drivers bring about Malawi's climate: (1) the El Niño Southern Oscillation, an Indo-Pacific phenomenon that modulates circulation; (2) the Indian Ocean Dipole, an equatorial pattern that affects rainfall; and (3) the Subtropical Indian Ocean Dipole (McSweeney et al. 2010; Wood and Moreniere 2013). Climate impacts that Malawi is experiencing include changing onset and shortening of the rainfall season, increased frequency of riverine and flash floods, droughts and heat waves (Zulu et al. 2012). Understanding how climate change/weather variability affects specific components of the agricultural sector is important for the development and effective implementation of mitigation and coping strategies.

1.1 Importance of Sugarcane

Sugarcane is a source of livelihood to millions of people and is integral to the economic development program of sugar producing countries (Hess et al. 2016; FAOSTAT 2018). It grows well in areas with long periods of sunlight (12–14 hrs.) where the temperature range is between 20 °C and 35 °C. The crop requires high humidity (80–85%) and a minimum of 1100 mm of rain per year or equivalent water from irrigation during the main growth phase, while ripening requires a dry period (DAFF 2014). Well-drained, fertile sandy to clay soils with a pH between 6.0 and 7.7 are ideal for growing sugarcane (DAFF 2014).

1.2 Impact of Climate Change on Sugarcane Production

Many studies have focused on the impact climate change will have on various aspects of sugarcane production (Jones et al. 2015; Zhao and Li 2015; Marin et al. 2014; Chandiposha 2013; Knox et al. 2010; Gawander 2007; Deressa et al. 2005). Overall, these studies indicate that projected future temperatures will have no significant effect on sugarcane growth since the projected temperature increases are within the crop's optimum range (30–32 °C). High temperature scenarios will enhance sugarcane growth and yield (Gawander 2007). However, temperatures higher than 35 °C will negatively affect sugarcane germination and internode development (Rasheed et al. 2011; Bonnett et al. 2006). Higher temperature will also lead to high evapotranspiration resulting in increased irrigation demands to minimize crop losses. In addition, temperature under current climate change scenarios will favour insect pests, weeds and certain fungal diseases (Matthieson 2007). However, there is little attention on implications of climate variability on pesticide exposure in sugarcane production. Chandiposha (2013) provided an account of how climate change would influence pest occurrence and distribution. Responses to climate change are usually long term. However, farmers have to respond to climate variability every season. Responses may include choice of crop and decisions on which soil and water management practices to adopt, all of which have a bearing on pest occurrence and severity. In addition, there is little focus on how an increased frequency of floods or rising temperatures will affect the efficacy and hazards posed by pesticides and how farmers will respond to such changes. The main objective of this review is therefore to bring into focus how climate variability affects the sugarcane industry and the amount and exposure to pesticides used in sugarcane production in Malawi.

We obtained information on pesticides, climate change and/or variability and its effects on agriculture and pesticides in general from the wide variety of available published literature as well as official and private documents.

2 Properties of Pesticides Used in Sugarcane Production in Malawi

Several insects, weeds and pathogens, commonly referred to as pests, significantly affect the production of sugarcane. Furthermore, climate scientists predict that climate change and/or variability will affect sugarcane production and the associated pests. Pesticides are chemicals used to minimize yield losses caused by pests. Pesticides for killing weeds (herbicides) and insect pests (insecticides) are the main

Table 1 Pesticide use in Malawi for the major crops

Crops	Rank (1 = mostly used; 6 = least used)
Tobacco	1
Tea	2
Sugarcane	3
Coffee	4
Cotton	5
Maize	6

Malawi Pesticide Control Board (2017)

types of pesticides used in the sugarcane industry in Malawi (Kasambala Donga and Eklo 2018). Malawi's sugarcane industry is the third largest consumer of pesticides in the country, after tobacco and tea (Table 1).

The effects of pesticides after their application are determined by the inherent properties of the pesticides as well as several environmental factors (Delcour et al. 2015). Water solubility, persistence in soil (measured as soil half-life), potential for adsorption to soil particles and mobility (K_{oc}) and dissociation (pKa) are considered key properties when determining how a pesticide or its residue(s) behaves in the environment (Kerle et al. 2007). Table 2 provides an overview of the various properties of pesticides used by sugarcane farmers in Malawi. The solubility values listed in the table show the pesticides agromectin, chlorpyrifos and cypermethrin to be less soluble in water, while acetamiprid, dimethoate, monosodium methanearsonate (MSMA) and 2-methyl-4-chlorophenoxyacetic acid (MCPA) are highly soluble. Plants easily absorb pesticides that are highly water soluble (Kerle et al. 2007). Pesticides with a soil half-life of less than 30 days are nonpersistent. Moderately persistent pesticides such as glyphosate and cypermethrin have a soil half-life of between 31 and 100 days. MSMA is the most persistent pesticide listed in Table 2. The pesticides abamectin, chlorpyrifos, cypermethrin, fluazifop-P, glyphosate and profenofos have high K_{oc} values; this implies that they adhere strongly to soil particles and remain concentrated on the application site. Soil half-life values range from 1 to 7 days for acetochlor, to 200 days for MSMA. Some of the pesticides such as atrazine, ametryn and diuron have a high potential for contaminating groundwater through leaching. Glyphosate, MCPA and MSMA readily dissociate in solution (high solubility values) but differ in their degradation and organic carbon sorption constant. Profenofos, diuron, cypermethrin and chlorpyrifos do not readily ionize but have a high propensity for adsorption onto soil particles. There is a high probability that runoff will contain these chemicals. There is high risk of surface and groundwater contamination from pesticides with low sorption coefficients, such as acetamiprid, acetochlor, metolachlor, ametryn and atrazine.

Environmental factors, especially precipitation, temperature, wind and concentration of carbon dioxide gas (CO2), influence the amount of pesticide remaining after application, i.e. the pesticide residue (Fig. 1; Delcour et al. 2015). For example, precipitation may result in a reduction in pesticide residue through wash off (Fig. 1). The same factor greatly influences the contamination of soils by pesticides through deposition (Fig. 1).

Table 2 Overview of the active ingredients in commonly used pesticides, target pests, application rates and key environmental factors

Substance group Active ingredient (a.i.)	Target pests	Typical application rates [grams (g) of active ingredient (a.i.) per hectare (ha)] (g a.i. ha^{-1})	Mode of action	Factors determining pesticides fate in the environment			
				Solubility in water (mg L^{-1})[a]	Half-life in soil (DT50: days)[b]	Organic carbon sorption constant (K_{oc})[c]	Dissociation constant at 25 °C (pKa)
Avermectins (abamectin)	Arthropod pests: aphids (*Sipha flava*), thrips and red spider mites	21.6	Stimulate the chloride channels that are regulated by the neurotransmitter glutamate	Insoluble	1–7	4000	–[d]
Organophosphate (dimethoate)	Aphids	2.2	Acetylcholinesterase (AChE) inhibitor	39,800	2.6		No dissociation
Organophosphate (chlorpyrifos)	Soil and foliage arthropod pests	750	Acetylcholinesterase (AChE) inhibitor	1.05	21	8151	–
Chloroacetamide (S-metolachlor)	Grasses and some broad-leaved weeds	1536	Inhibition of VLCFA (inhibition of cell division)	480	21	110–369	–
Organometal (organic arsenical)	Sedges, grasses and broad-leaved weeds	2160	Inhibition of VLCFAs (inhibition of cell division)	580,000	200	–	9.02 weak acid
Phenoxyacetic acid (MCPA)	Annual and perennial weeds	1080	Synthetic auxin	29,390	25	–	3.73 weak acid
Triazines (ametryn) (atrazine)	Most annual and broad-leaved weeds	900–1200	Inhibits photosynthesis (photosystem II)	200	37	316	10.07 very week acid
		1125–1350		35	29	100	1.7 very weak base
Phenoxyaliphatic acids (fluazifop-p-butyl)	Ripener	55.5	Inhibits acetyl-CoA carboxylase	0.93	8.2	3394	–[5]

(continued)

Table 2 (continued)

Substance group Active ingredient (a.i.)	Target pests	Typical application rates [grams (g) of active ingredient (a.i.) per hectare (ha)] (g a.i. ha^{-1})	Mode of action	Factors determining pesticides fate in the environment			
				Solubility in water (mg L^{-1})[a]	Half-life in soil (DT50: days)[b]	Organic carbon sorption constant (K_{oc})[c]	Dissociation constant at 25 °C (pKa)
Ethylene generator (ethephon)	Flower suppressant	480	Plant growth regulator with systemic properties	1,000,000	13.4	–	2.82
Phenylpyrazole (fipronil)	Various insect pests and mites		Broad spectrum with contact and stomach action. GABA-gated chloride channel antagonist	3.78	142	–	No dissociation
Phenylureas (diuron)	Weeds and mosses	1600	Inhibits photosynthesis	35.6	89	813	–
Organophosphate (profenofos)	Lepidopteran pests and mites	440	Acetylcholinesterase (AChE) inhibitor	28	7	2016	–

[a]Highly soluble pesticides have large solubility values
[b]Soil half-life: <30 days implies nonpersistent, 30–100 days means moderately persistent and >100 days shows pesticide is highly persistent
[c]The higher the K_{oc} value, the more strongly the pesticide is sorbed
[d]"- "indicates data not available
Sources: PPDB (2017); Kerle et al. (2007); EU (2004); Kasambala Donga and Eklo (2018); EXTOXNET (1994)

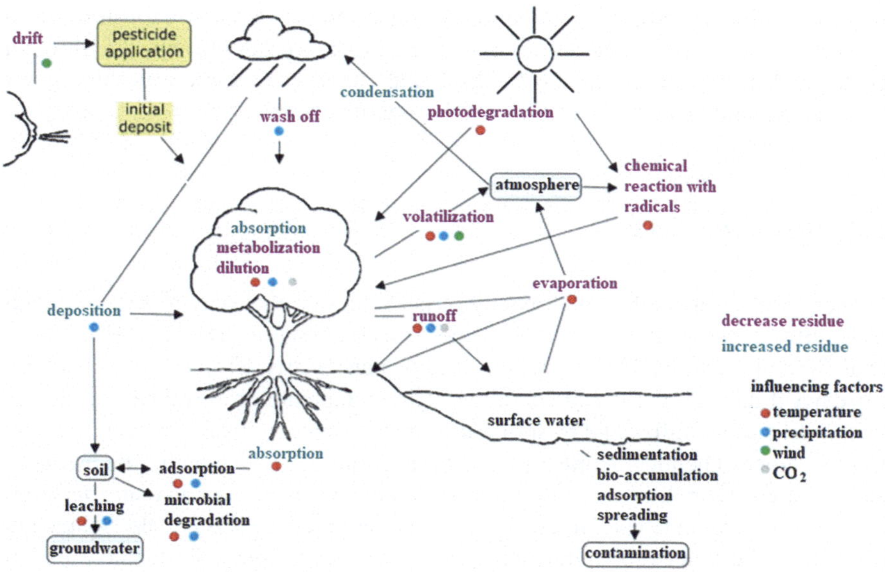

Fig. 1 Illustration of the environmental factors that influence pesticide fate after application. (Delcour et al. 2015)

3 Ways Through Which Climate Influences Pesticide Use and Exposure

3.1 Pest Occurrence

Climate-induced changes alter both pest development and levels of infestation. Wet and humid conditions favour the proliferation of fungal and bacterial diseases. Climate-induced aridity may increase the incidence of sugarcane diseases such as ratoon stunting disease, which causes shortening of sugarcane stems, and smut disease, which results in the blackening of shoot apices (Matthieson 2007). Although these are important diseases of sugarcane in Malawi, increase in their incidence will not affect pesticide exposure since these diseases are controlled using cultural methods. Higher temperatures may also increase the incidence and severity of insect pests. The severity of infestation of sugarcane by plant sap-sucking red spider mites in Chikwawa is closely linked to periods of dry, hot weather with low humidity and high evapotranspiration (Koloko 2016). Fipronil, a highly toxic pesticide, has been used in Malawi to manage an outbreak of African migratory locusts in the Lower Shire River Valley. During the 2014/2015 cropping season, additional amounts of acetamiprid and cypermethrin were sprayed to manage an outbreak of yellow sugarcane aphids.

Fipronil is highly toxic to terrestrial and aquatic life, does not dissociate and has high potential for bioaccumulation (PPDB 2017). These few examples illustrate the impact of climate-induced pest outbreaks on pesticide use and exposure. Farm workers and local communities are at increased risk of pesticide exposure through pesticide drift irrigation channels (Wilson et al. 2004) as they use water that flows through irrigation canals for bathing and household chores.

3.2 Pesticide Toxicity

Higher temperatures affect the toxicity of pesticides on their target pests although these effects vary with different pesticide-pest combinations (Fishel 2015; Noyes et al. 2009; Donahoe 2014). Pesticides may be particularly affected by temperature extremes if they are not stored correctly; higher temperatures may cause pesticides to expand and volatilize and to spill out of their containers when opened. Farmers lacking proper chemical storage systems may store pesticides within their homes and are at greater risk of pesticide exposure. Sadly, this is the case in many developing countries (Mengistie et al. 2015; Stadlinger et al. 2010; Kasambala Donga and Eklo 2018). Pesticides containing phosphoric acid (a.k.a. organophosphates) tend to be more toxic to insect and mite pests at 26–28 °C than at 20 °C, while pesticides derived from naturally occurring pyrethrin extracted from *Chrysanthemum* plants are more toxic at lower temperatures (Jegede et al. 2017; Noyes et al. 2009). Maximum temperatures in the sugarcane-growing areas of Malawi range between 27 °C and 37 °C (Phiri and Saka 2009), which is higher than the temperatures used in pesticide toxicity studies (Jegede et al. 2017; Noyes et al. 2009). Since cypermethrin is widely used in Malawi to control a range of insect pests infesting sugarcane, a reduction in efficacy is likely to result in either increased frequency or amount of pesticide application.

3.3 Pesticide Degradation

As shown in Fig. 1, temperature strongly influences the degradation of pesticides, and several reports exist on the effects of temperature on some of the pesticides examined in this study (de Beeck et al. 2017; Jegede et al. 2017). The rate of degradation of atrazine increased with increasing temperature (Dong and Sun 2016). Higher temperature also enhances the activities of microorganisms that degrade pesticides. At 30 °C and pH 7, bacteria degraded 90% of chlorpyrifos and profenofos within 8 days (John et al. 2016). Acetamiprid degradation was rapid in soils with higher temperatures (Vela et al. 2017). The sugarcane growing districts of Malawi experience high temperatures (above 30 °C) during most of the year. Hence, we expect the estimate of the risk of pesticide exposure to be significantly lower under rising temperatures, assuming all other degradation factors remain constant.

Soil moisture is also an important factor in pesticide degradation (Chai et al. 2013; Sebaï et al. 2010). With the exception of rainfed sugarcane (less than 20%), irrigation is essential to meet the crop's water demands. Under current climate scenarios, the demand for irrigation will rise. Irrigation may cancel temperature-induced drought effects on pesticide degradation (Gonczi 2016).

3.4 Pesticide Transport

The pesticides currently used in sugarcane production in Malawi utilize water as a solvent. High temperatures result in increased volatilization of highly and semi-volatile pesticides through evapotranspiration of the pesticides and their metabolites to the atmosphere (Bloomfield et al. 2006). Most of the pesticides in use in the sugarcane industry in Malawi are, however, less volatile (Kasambala Donga and Eklo 2018). Water-based pesticides such as monosodium methylarsenate (MSMA) and its metabolites show some persistence in soil and sediments because they tend to move slower than water and remain concentrated in shallow soil depths (Mahoney et al. 2015; Bloomfield et al. 2006), increasing the possibility of pesticide contamination of the environment after initial application. A study in Australian forests found residues of atrazine and its metabolite desethylatrazine in 1.8-m-deep groundwater (Kookana et al. 2010).

Rainfall is a key factor influencing the transport of pesticides in the environment. The onset of the rainy season is around October to November in most parts of Malawi, with the highest rainfall occurring around February to March or early April, especially in the north. The rains tail off in late April and May when winter begins. The amount and timing of rainfall in relation to pesticide application is a much more important factor than average annual rainfall and temperature (Wang et al. 2018). For Malawi, the observed and predicted increases in the proportion of rainfall that falls in heavy events during the wetter months of January and February affect the following pesticide pathways: leaching to surface and groundwater, runoff and erosion. There is a high probability of pesticide movement to surface and groundwater at higher rainfall intensities since wetter soils have higher hydraulic conductivities (Bloomfield et al. 2006). Hydraulic conductivity varies with soil type and the water content of the soil. The soils in the main sugarcane-growing areas of Malawi are chiefly alluvial in the Nkhotakota District and alluvials and vertisols in the Chikwawa District. The water holding capacity of vertisols is high compared to alluvials. This implies that there will be a higher likelihood of pesticide-rich water percolating to groundwater in areas with vertisols in situations of higher rainfall intensities. On the other hand, a higher soil water content will result in increased degradation rate of pesticides (Jebellie et al. 1996) and thus lower the pesticide risk estimate.

Increased rainfall intensities may also result in flooding and runoff. Runoff will directly influence the fate of pesticides through the increased erosion of soil particles and transport of sorbed pesticides (Bloomfield et al. 2006). Increased precipitation may enhance runoff contamination by pesticides (Silburn et al. 2013; Probst et al. 2005). Rainwater and floodwater runoff account for the transport of a quarter of the diuron applied yearly to sugarcane in Australia (Camenzuli et al. 2012).

Approximately 19% of the rainfall received in Malawi is lost through surface runoff (GoM 2008). It is therefore possible that a significant proportion of pesticides currently used in agriculture in Malawi are lost through this pathway. In the event of increased precipitation and floods, the concentration of pesticides such as acetamiprid and metolachlor would therefore need to be high if these episodes occur immediately after their application. About 33% of Malawians do not have access to potable water (WHO and UNICEF 2015). They depend on surface- and groundwater for drinking and household chores (Chidya et al. 2016; Chimphamba and Phiri 2014) and are consequently at greater risk of pesticide exposure.

3.5 Pesticide Sorption

Soil management practices influence sorption – the distribution or partitioning of a pesticide in an environment. Sorption not only reduces the risk of pesticide leaching but can also reduce the pesticide degradation rate, as the pesticides are not available to microorganisms. Dinisio and Rath (2016) reported high sorption of abamectin occurring in soils rich in organic matter. In another study, metolachlor and atrazine sorption increased in soils amended with biochar (Deng et al. 2017; Trigo et al. 2016). Biochar has some of the same effects as sugarcane burning after harvest and increases sorption. Adsorption of atrazine and endosulfan was better in soils covered with rice husks (Rojas et al. 2014). Leaching of MCPA significantly reduced in Mediterranean agricultural soils amended with olive oil mill wastes (Peña et al. 2015). These results show that efforts to improve soil fertility have a significant influence on the exposure of pesticides to the environment through the enhancement of pesticide degradation and sorption.

Crop management is also an important factor in pesticide sorption. In Malawi, as in many of the sugarcane producing countries, the burning of sugarcane fields prior to harvesting is practiced. Whilst much ash from burnt sugarcane residue is blown away from the fields by the wind, some ash remains, significantly influencing pesticide adsorption (the adhesion of pesticide molecules to surfaces of particles). Effective pesticide adsorption by burned sugarcane residues would lead to reduced availability of pesticides that typically target soil dwelling pests or those pesticides that have to be absorbed by plants through the roots (Yang and Sheng 2003). Farmers practicing residue burning may therefore report reduced efficacy of pesticides such as chlorpyrifos, dimethoate and clomazone (Xu et al. 2008; Kamm and Montgomery 1990). In addition, the practice negatively affects the population of microbes and total organic matter (Souza et al. 2012). Microbes are essential components for the degradation of pesticides. Thus, burning reduces pesticide risk through increased pesticide sorption. At the same time, it may also increase pesticide exposure risk due to an increased demand for inputs (fertilizers and herbicides).

Increases in rainfall coupled with intensive farming using nitrogen fertilizers and the burning of crop residues can result in the acidification of soils. The pH of a soil and the ionic state of the pesticide influence pesticide fate. For example, at pH 4, part of ametryn (pKa = 4.10) exists as a positively charged conjugate acid (de Paula et al. 2016). The electrostatic interaction between the ametryn conjugate and the ionized soil particles is enhanced resulting in a more stable structure. As a result, ametryn is more persistent in acidic soils (de Paula et al. 2016). According to Meyer and Heathman (2015), the soils under intensive sugarcane production in Chikwawa, southern Malawi, have become acidic. Very high temperatures (above 35 °C) coupled with frequent irrigation or flooding may also have contributed to soil acidification through soil mineral leaching in Chikwawa. In addition, burning crop residues removes excess positively charged ions contained in plant material, which are necessary for balancing negatively charged ions in organic molecules. This could neutralize the soil acidity upon decomposition (Rengel 2011). This scenario would increase the probability of soil contamination and negatively affect soil-dwelling, non-target organisms.

4 Conclusion

In summary, temperature and the timing and amount of rainfall will continue to influence degradation, sorption and transport of pesticides used in sugarcane production. Higher temperature will negatively affect pesticide toxicity, prompting farmers to use more and/or change pesticides. There is greater risk of pesticides contaminating water bodies through the runoff and erosion of sorbed pesticides. The persistence of pesticides such as ametryn and glyphosate may be higher in acidic soils. This study highlights the need to determine the occurrence of pesticide residues in sugarcane cropping and aquatic systems surrounding sugarcane plantations.

Acknowledgements This review was funded through the program 'Capacity Building for Managing Climate Change in Malawi (CABMACC)'. CABMACC is a completed collaborative program supported by the Norwegian Government and the Government of the Republic of Malawi and implemented by the Department of International Environment and Development Studies (Noragric) at the Norwegian University of Life Sciences (NMBU) and Lilongwe University of Agriculture and Natural Resources (LUANAR), Malawi.

References

Adhikari U, Nejadhashemi AP, Woznicki SA (2015) Climate change and eastern Africa: a review of impact on major crops. Food Energy Secur 4(2):110–132

Aktar MA, Sengupta D, Chowdhury A (2009) Impact of pesticides use in agriculture: their benefits and hazards. Interdiscip Toxicol 2(1):1–12

Bloomfield JP, Williams RJ, Goody DC, Cape JN, Guha P (2006) Impact of climate change on the fate and behavior of pesticides in surface and groundwater – a UK perspective. Sci Total Environ 369:163–177

Bonnett GT, Hewitt ML, Glassop D (2006) Effects of high temperature on the growth and composition of sugarcane internodes. Australian J Agric Res 57(10):1087–1095

Brooks N, Neil Adger W, Mick Kelly P (2005) The determinants of vulnerability and adaptive capacity at the national level and the implications for adaptation. Global Environ Change A 15:151–163

Camenzuli L, Scheringer M, Gaus C, Ng CA, Hungerbühler K (2012) Describing the environmental fate of diuron in a tropical river catchment. Sci Total Environ 440:178–185

Chai L, Wong M, Hansen HCB (2013) Degradation of chlorpyrifos in humid tropical soils. J Environ Manag 125:28–32

Challinor A, Wheeler T, Garforth C, Craufurd P, Kassam A (2007) Assessing the vulnerability of food crop systems in Africa to climate change. Clim Chang 83:381–399

Chandiposha M (2013) Potential impact of climate change in sugarcane and mitigation strategies in Zimbabwe. Afr J Agric Res 8:2814–2818

Chidya RCG, Matumula S, Nakoma O, Chawinga CBJ (2016) Evaluation of groundwater quality in rural-areas of northern Malawi: case of Zombwe extension planning area in Mzimba. Phys Chem Earth Parts A/B/C 93:55–62

Chimphamba JB, Phiri OL (2014) Borehole water pollution and its implication on health on the rural communities of Malawi. Malawi J Sci Technol 10(1):32–39

Dasgupta P, Morton JF, Dodman D, Karapinar B, Meza F, Rivera-Ferre MG, Toure Sarr A, Vincent KE (2014) Rural areas. In: Field CB, Barros VR, Dokken DJ, Mach KJ, Mastrandrea MD, Bilir TE, Chatterjee M, Ebi KL, Estrada YO, Genova RC, Girma B, Kissel ES, Levy AN, MacCracken S, Mastrandrea PR, White LL (eds) Climate change 2014: impacts, adaptation, and vulnerability. Part A: Global and sectoral aspects. Contribution of Working Group II to the Fifth Assessment Report of the Intergovernmental Panel on Climate Change. Cambridge University Press, Cambridge, UK/New York, NY, pp 613–657

de Beeck LO, Verheyen J, Olsen K, Stoks R (2017) Negative effects of pesticides under global warming can be counteracted by a higher degradation rate and thermal adaptation. J Appl Ecol 54:1847–1855

de Paula RT, de Abreu ABG, de Queiroz MELR, Neves AA, da Silva AA (2016) Leaching and persistence of ametryn and atrazine in red–yellow latosol. J Environ Sci Health B 51(2):90–95

Delcour I, Spanoghe P, Uyttendaele M (2015) Literature review: impact of climate change on pesticide use. Food Res Int 68:7–15

DAFF (2014) Sugarcane. http://www.daff.gov.za/docs/Brochures/sugarcane.pdf. Accessed 28 June 2018

Deng H, Feng D, Hec J, Li F, Yu H, Gea C (2017) Influence of biochar amendments to soil on the mobility of atrazine using sorption-desorption and soil thin-layer chromatography. Ecol Eng 99:381–390

Deressa T, Hassan R, Poonyth D (2005) Measuring the impact of climate change on South African agriculture: the case of sugarcane growing regions. Agrekon 44(4):524–542

Dinisio AC, Rath S (2016) Abamectin in soils: analytical methods, kinetics, sorption and dissipation. Chemosphere 151:17–29

Donahoe M (2014) Temperature extremes can affect pesticides. https://nwdistrict.ifas.ufl.edu/phag/2014/01/31/temperature-extremes-can-affect-pesticides/

Dong X, Sun H (2016) Effect of temperature and moisture on degradation of herbicide atrazine in agricultural soil. Int J Environ Agric Res 2(7):150–157

EU (2004) Review report for the active substance S-Metolachlor. S-Metolachlor SANCO/1426/2001 – rev. 3. http://ec.europa.eu/food/plant/pesticides/eu-pesticides-database/public/?event=activesubstance.ViewReview&id=381. Accessed on 24 June 2017

EXTOXNET (1994) Abamectin. http://pmep.cce.cornell.edu/profiles/extoxnet/24d-captan/abamectin-ext.html#6. Accessed16 July 2017

FAOSTAT (2018) Sugar cane. http://www.fao.org/faostat/en/#search/Sugar%20cane. Accessed on 5 March 2018

Fishel F (2015) Temperature effect on storage of agricultural pesticides. http://extension.missouri.edu/p/G1921. Accessed on 3 Oct 2017

Gawander J (2007) Impact of climate change on sugar-cane production in Fiji. WMO Bull 56:34–39

GoM (2008) Integrated water resources management and water efficiency (IWRM/WE) plan, 2008–2012, Malawi Water Partnership/CIDA, Lilongwe

Gonczi M (2016) Climate change and pesticides. Swedish University of Agricultural Sciences. https://www.slu.se/en/Collaborative-Centres-and-Projects/centre-for-chemical-pesticides-ckb1/information-about-pesticides-in-the-environment-/-climate-change-and-pesticides/. Accessed 3 Oct 2017

Hess TM, Sumberg J, Biggs T, Georgescu M, Haro-Monteagudo D, Jewitt G, Osdogan M, Marshall M, Thenkabali P, Daccache A, Marin F, Knox JW (2016) A sweet deal? Sugarcane, water and agricultural transformation in Sub-Saharan Africa. Glob Environ Chang 39:181–194

IPCC (2007) Working Group II: impacts, adaptation and vulnerability. Conceptual framework for the identification and assessment of key vulnerabilities. IPCC Fourth Assessment Report: Climate Change 2007. https://www.ipcc.ch/publications_and_data/ar4/wg2/en/ch19s19-1-2.html. Accessed 7 Dec 2017

Jebellie SJ, Prasher SO, Clemente RS (1996) Role of soil moisture content in reducing environmental pollution from pesticides. Can Water Res J 21(4):393–402

Jegede OO, Owojori OJ, Römbke J (2017) Temperature influences the toxicity of deltamethrin, chlorpyrifos and dimethoate to the predatory mite *Hypoaspis aculeifer* (Acari) and the springtail *Folsomia candida* (Collembola). Ecotoxicol Environ Saf 140:214–221

John EM, Sreekumar J, Jisha MS (2016) Optimization of chlorpyrifos degradation by assembled bacterial consortium using response surface methodology, soil and sediment contamination. Soil Sediment Conatm 25(6):668–682

Jones MR, Singels A, Ruane AC (2015) Simulated impacts of climate change on water use and yield of irrigated sugarcane in South Africa. Agric Syst 139:260–270

Kamm JA, Montgomery ML (1990) Reduction of insecticide activity by carbon residue produced by burning grass seed fields after harvest. Abstract. J Econ Entomol 83(1):55–58

Kasambala Donga T, Eklo M (2018, unpublished) Environmental load of pesticides currently used in sugarcane production in Malawi

Kerle EA, Jenkins JJ, Vogue PA (2007) Understanding pesticide persistence and mobility for groundwater and surface water protection. EM 8561-E. http://enhs.umn.edu/current/5103/pesticide/fate.html Accessed 23 Oct 2017

Knox JA, Rodríguez Díaz JW, Nixon DJ, Mkhwanazi M (2010) A preliminary assessment of climate change impacts on sugarcane in Swaziland. Agric Syst 103:63–72

Koloko M (2016) Red spider mite on sugarcane at Nchalo estate. ILLOVO Mzimbe Magazine 17(12):29–31

Kookana R, Holz G, Barnes C, Bubb K, Fremlin R, Boardmane B (2010) Impact of climatic and soil conditions on environmental fate of atrazine used under plantation forestry in Australia. J Environ Manag 91:2649–2656

Mahoney DJ, Gannon TW, Jeffries MD, Matteson AR, Polizzoto ML (2015) Management considerations to minimize environmental impacts of arsenic following monosodium methylarsenate (MSMA) applications to turfgrass. J Environ Manag 150:444–450

Marin FR, Ribeiro RV, Marchiori PER (2014) How can crop modeling and plant physiology help to understand the plant responses to climate change? A case study with sugarcane. Theor Exp Plant Phys 26:49–63

Matthieson L (2007) Climate change and the Australian sugarcane industry: impacts, adaptation and R&D opportunities. Technical Report, Sugar Research and Development, Brisbane, Australia

McSweeney C, New M, Lizcano G (2010) UNDP climate change country profiles: Malawi. http://country-profiles.geog.ox.ac.uk/. Accessed 14 Oct 2017

Mengistie BT, Mol APJ, Oosterveer P (2015) Pesticide use practices among smallholder vegetable farmers in Ethiopian Central Rift Valley. Environ Dev Sustain 19(1):301–324. https://doi.org/10.1007/s10668-015-9728-9

Meyer JH, Heathman WZ (2015) Report of further outcomes from the reconnaissance soil suitability survey of five estates in the Nchalo sugarcane supply area. Prepared for ILLOVO Sugar Malawi. 61pp

Mwale FD, Adeloye AJ, Beevers L (2015) Quantifying vulnerability of rural communities to flooding in SSA: a contemporary disaster management perspective applied to the Lower Shire Valley, Malawi. Int J Disast Risk Re 12:172–187

Noyes PD, McElwee MK, Miller HD, Clark BW, Van Tiem LA, Walcott KC, Erwin K, Levin ED (2009) The toxicology of climate change: environmental contaminants in a warming world. Environ Int 35(6):971–986

Peña D, López-Piñeiro A, Albarrán Á, Becerra D, Sánchez-Llerena J (2015) Environmental fate of the herbicide MCPA in agricultural soils amended with fresh and aged de-oiled two-phase olive mill waste. Environ Sci Pollut Res 22:13915–13925

Phiri MGI, Saka RA (2009) The impact of changing environmental conditions on vulnerable communities in the Shire Valley, Southern Malawi. In: Lee C, Schaaf T (eds) The future of drylands. Springer, Dordrecht, pp 545–559

PPDB (2017). http://sitem.herts.ac.uk/aeru/ppdb/en/atoz.htm. Accessed on 3 Oct 2017

Probst M, Berenzen N, Lentzen-Godding A, Schulz R (2005) Scenario-based simulation of runoff-related pesticide entries into small streams on a landscape level. Ecotoxicol Environ Saf 62(2):145–159

Rasheed R, Wahid A, Farooq M, Hussain I, Basra SMA (2011) Role of proline and glycine betaine pretreatments in improving heat tolerance of sprouting sugarcane (Saccharum sp.) bud. Plant Growth Regul 65:35–45

Rengel Z (2011) Soil pH, soil health and climate change. In: Singh B, Cowie A, Chan K (eds) Soil health and climate change. Soil biology, vol 29. Springer, Berlin, Heidelberg, pp 65–89

Rojas R, Vanderlinden E, Morillo J, Usero J, El Bakouri H (2014) Characterization of sorption processes for the development of low-cost pesticide decontamination techniques. Sci Total Environ 488–489:124–135

SADC/VAC (2016) Malawi vulnerability assessment committee results 2016. https://reliefweb.int/report/malawi/malawi-vulnerability-assessment-committee-results-2016. Accessed 7 Dec 2017

Sebaï TE, Devers M, Lagacherie B, Rouard N, Soulas G, Martin-Laurent F (2010) Diuron mineralisation in a Mediterranean vineyard soil: impact of moisture content and temperature. Pest Manag Sci 66:988–995

Silburn DM, Foley JL, deVoil RC (2013) Managing runoff of herbicides under rainfall and furrow irrigation with wheel traffic and banded spraying. Agric Ecosyst Environ 180:40–53

Souza RA, Telles TS, Machado W, Hungria M, Filho JT, Guimarães MD (2012) Effects of sugarcane harvesting with burning on the chemical and microbiological properties of the soil. Agric Ecosyst Environ 155:1–6

Stadlinger N, Mmochi AJ, Dobo S, Gyllbäck E, Kumblad L (2010) Pesticide use among smallholder rice farmers in Tanzania. Environ Dev Sustain 13(3):641–656

The World Bank Group (2017) Malawi data. https://data.worldbank.org/country/malawi. Accessed 7 Dec 2017

The World Fact Book (2016) Malawi. https://www.cia.gov/library/publications/the-world-factbook/geos/print_mi.html. Accessed 7 Dec 2017

Trigo C, Spokas KA, Hall KE, Cox L, Koskinen WC (2016) Metolachlor sorption and degradation in soil amended with fresh and aged biochars. J Agric Food Chem 64:3141–3149

USAID (2007) Adapting to climate variability and change. A guidance manual for development planning. USAID, Washington, 31pp

Vela N, Fenoll J, Navarro G, Garrido I, Navarro S (2017) Trial of solar heating methods (solarization and biosolarization) to reduce persistence of neonicotinoid and diamide insecticides in a semiarid Mediterranean soil. Sci Total Environ 590–591:325–332

Wang Q, Pang C, Wen Z, Zheng H, Cheng R, Ma J, Que X (2018) Effect of grass hedges on runoff loss of soil surface-applied herbicide under simulated rainfall in Northern China. Agric Ecosyst Environ 253:1–10

WHO and UNICEF (2015) Progress on sanitation and drinking water – 2015 Update and MDG assessment. WHO, Switzerland. 90pp. http://www.unicef.org/publications/index_82419.html. Accessed 17 Jan 2018

Wilson C, Stover E, Boman B (2004) Minimizing direct deposition of pesticides into waterways associated with Indian River citrus production. HortTechnology 14(4):545–550
Wood L, Moreniere L (2013) Malawi climate change vulnerability assessment. United States Agency for International Development, Prosperity, Livelihoods, and Conserving Ecosystems (PLACE) Indefinite Quantity Contract Core Task Order (USAID Contract No. AID-EPP-I-00-06-00008, Order Number AID-OAA-TO-11-00064). 102 pp
Xu C, Liu W, Sheng GD (2008) Burned rice straw reduces the availability of clomazone to barnyardgrass. Sci Total Environ 392:284–289
Yang Y, Sheng G (2003) Enhanced pesticide sorption by soils containing particulate matter from crop residue burns. Environ Sci Technol 37:3635–3639
Zhao D, Li Y (2015) Climate change and sugarcane production: potential impact and mitigation strategies. Int J Agron 2015:10. https://doi.org/10.1155/2015/547386
Zulu E, Ciera J, Musila N, Mutunga C, De Souza R (2012) Population dynamics, climate change, and sustainable development in Malawi. http://pai.org/wp-content/uploads/2013/01/Malawi_Country_Report_Final.pdf. Accessed 16 Sept 2017

Yield and Profitability of Cotton Grown Under Smallholder Organic and Conventional Cotton Farming Systems in Meatu District, Tanzania

T. N. Bwana, Nyambilila A. Amuri, E. Semu, J. E. Olesen, A. Henningsen, M. R. Baha, and J. Hella

Abstract Agronomic practices have a large effect on the yield and profitability of low-input smallholder cotton farming in Africa. A two-season field experiment was conducted in a semi-arid cotton growing area in Meatu District, Tanzania, to compare the yield and profitability of various conventional and organic cotton production practices. Besides the currently applied low-input conventional and organic cotton production practices, higher-input and innovative farming practices as well as control treatments (without fertiliser or pesticides) were tested. While season 1 had weather conditions that were very suitable for cotton production, much less

AH, NAA and JEO formulated the general research question and initiated the research project; TNB, NAA, ES and JEO formulated the specific research questions and designed the field experiment; TNB, supervised by NAA, ES, and JEO, conducted the field experiment and collected the data; TNB and AH conducted the economic analysis; AH conducted the statistical analysis; JH and MRB provided information about economics and politics of cotton production in Tanzania; TNB drafted the chapter and AH added and rewrote some paragraphs. All authors contributed to the revision of the chapter.

T. N. Bwana · N. A. Amuri (✉) · E. Semu
Department of Soil and Geological Sciences, Sokoine University of Agriculture, Morogoro, Tanzania
e-mail: namuri@sua.ac.tz

J. E. Olesen
Department of Agroecology, Aarhus University, Tjele, Denmark

A. Henningsen
Department of Food and Resource Economics, University of Copenhagen, Copenhagen, Denmark

M. R. Baha
Department of Food and Resource Economics, University of Copenhagen, Copenhagen, Denmark

Department of Food and Resource Economics, Sokoine University of Agriculture, Morogoro, Tanzania

J. Hella
Department of Food and Resource Economics, Sokoine University of Agriculture, Morogoro, Tanzania

B. R. Singh et al. (eds.), *Climate Impacts on Agricultural and Natural Resource Sustainability in Africa*, https://doi.org/10.1007/978-3-030-37537-9_10

rainfall in season 2 severely reduced the yield and land rent in both conventional and organic cotton production. In general, conventional and organic practices have similar cotton yields, but organic practices often generate higher land rents than conventional practices due to a higher price for organic cotton and lower production costs. In both seasons, the innovative organic practice generated the highest land rent of all conventional and organic practices, and it is statistically significantly higher than the land rents of all conventional farming practices.

1 Introduction

Cotton *(Gossypium hirsutum* L.) is the second largest export crop in Tanzania (after coffee) and largely contributes to national export earnings. Furthermore, it provides employment and income for over 500,000 households in rural Tanzania (TCB 2010). Cotton growing in Tanzania is dominated by smallholder farmers with farm sizes ranging from 0.5 to 10 hectares, with an average of 1.5 hectares. The production is characterised by manual operation under rain-fed conditions with minimal use of inputs such as fertilisers and pesticides. Two cotton production systems, conventional and organic, are practised in Tanzania, with the majority of cotton farmers practicing the conventional system (TCB 2010). Conventional cotton farming is based on the use of inorganic fertilisers and synthetic pesticides (Pimentel et al. 2005). However, most conventional smallholder farmers use little or no fertiliser, but they appreciably use pesticides for controlling the common cotton pests like aphids [*Aphis gossypii* (Glover)], American bollworm [*Heliothis armigera* (Hubner)] and cotton stainer [*Dysdercus* spp.].

The recommended application rates for fertiliser in cotton production vary depending on the soil type. For the Western cotton growing area (WCGA) of Tanzania, the revised fertiliser recommendations show application rates of 20–30 kg N ha^{-1}, 10–15 kg P ha^{-1} and 5 Mg ha^{-1} farm yard manure (FYM) (Mowo et al. 1993). However, a general national recommended rate of 40 kg N ha^{-1} and 18 kg P ha^{-1} is also reported by IFDC (2014). No specific national recommendations for pesticide use in cotton are available, and the application rates are based on guidelines from the pesticide manufacturers, which suggest four to six sprays per growing season.

Organic agriculture is a farming system that does not use genetically modified organisms (GMO), inorganic fertilisers, synthetic pesticides or any other agrochemicals (FAO 1998; Gold 2007; IFOAM 2014) but largely relies on soil fertility management for nutrient supply and natural pesticides for pest control. In recent years, organic cotton production in Tanzania has increased. In the season 2017/18, for example, Tanzania produced 3525 Mg of organic cotton, which placed the country in the seventh position of the world's leading organic cotton producers and in the second position in the world regarding the land area that is under conversion to organic cotton production (Textile Exchange 2018). Organic cotton production is practised in some areas in the Western cotton growing area, like Meatu and Maswa districts in Simiyu Region and in Singida Region. The production is contract based, where farmers enter production contracts with private companies. The contracting company provides organic seeds and biopesticides and offers training and extension services. In turn, the company is entitled to purchase the entire crop. The organic cotton production practices for nutrient management include the use of FYM, crop rotation and intercropping with legumes. For pest management, the practices also include trap crops (e.g. intercropping with sunflower) and the use of organic pesticides (neem-leaf extract or pyrethrum). However, there is no local documented information on agronomic, economic and environmental performance of these practices.

The average seed cotton yields in Tanzania fluctuated between 427 and 766 kg ha^{-1} 5–10 years ago but have decreased to 199–367 kg ha^{-1} in 2015/16 to 2018/19 (see Table 1; TCB 2019). This low yield is associated with major yield-limiting factors such as rain-fed growing conditions with infrequent rainfall, use of low yielding varieties and insufficient use of fertiliser and pesticides (TCB 2010). The Tanzanian government aims at increasing cotton production by 30% every year, initially aiming at increasing cotton yield from 750 kg ha^{-1} seed cotton (260 kg ha^{-1} of lint) in 2016/17 to 1500 kg ha^{-1} (520 kg ha^{-1} of lint) by 2020/21 (TCB 2016). The strategy for achieving this target includes several initiatives, e.g. building research capacity in various aspects of cotton farming, especially the breeding of new varieties and control of diseases and pests; establishing a participatory planning process for all stakeholders in the cotton sector; ensuring that all cotton production in the WCGA is done as contract farming; and establishing links among all relevant actors within the cotton value chain (e.g. extension service, input suppliers, farmers, cotton traders, ginneries) (TCB 2016). However, extensive use of inputs (fertilisers and pesticides) in these strategies is also associated with negative environmental impacts, such as nutrient losses causing eutrophication in water bodies, biodiversity loss, greenhouse gas (GHG) emissions, soil acidification and land degradation (Gomiero et al. 2011).

Table 1 Cotton production in Tanzania: area, production and yield

Season	Area (ha)	Production (Mg)	Yield (Mg ha^{-1})
2009/10	411,065	267,004	0.650
2010/11	382,934	163,518	0.427
2011/12	481,719	225,938	0.469
2012/13	465,996	357,133	0.766
2013/14	389,733	242,138	0.621
2014/15	455,272	202,312	0.444
2015/16	447,328	149,913	0.335
2016/17	423,341	122,362	0.289
2017/18	668,685	132,961	0.199
2018/19	607,029	222,725	0.367

Source: TCB (2019) and own calculations based on TCB (2019)

Among agricultural practices, organic farming practices are perceived to be more environmentally benign than conventional farming because of the avoidance of inorganic fertilisers and synthetic pesticides and the reliance on organic nutrient cycles (Tuomisto et al. 2012; Lorenz and Lal 2016). However, it remains unclear how the crop yields and economic performance compare between organic and conventional systems for crops grown under smallholder production systems in sub-Saharan Africa (SSA). Some authors argue that organic farming systems have lower yields than conventional farming and hence would not be able to meet the world's growing food demand (de Ponti et al. 2012). They also argue that organic farming is associated with low labour productivity, high production risks and high costs due to additional costs of certification (Borlaug 2000; Trewavas 2001; Nelson et al. 2004; Makita 2012). Others argue that under good management, yields from organic farming can be similar to or greater than those of conventional farming (Cavigelli et al. 2009; Seufert et al. 2012).

Most studies comparing yields of organic and conventional farming practices have been done in temperate climates (Rosenstock et al. 2013), leaving a wide data gap for tropical and subtropical conditions. This advocates for comparisons of the two farming systems in SSA in order to be able to give recommendations for sustainable cotton production in the region (Richards et al. 2016). As farmers often choose the production method according to their profitability, it is important that these comparisons extend beyond agronomic performance such as yield and also include profitability indicators such as the land rent. This study assesses the yields and land rents of cotton grown under low-input smallholder conventional and organic production systems.

2 Materials and Methods

2.1 Study Site Description

The field experiment was conducted in the Meatu District, Simiyu Region in Tanzania. The area is between latitudes 3°–4° S and longitudes 34°8′–34°49′ E at an altitude of 1000–1500 m.a.s.l. The area is within the semi-arid zone; rainfall ranges from 900 mm per year in the north to 400 mm in the south. The soil type at the experimental site was described to family level by Bwana (2019) as almost flat, moderately deep, clayey, moderately to strongly alkaline isohyperthermic, Pachic Calciustolls as per the USDA Soil Taxonomy (Soil Survey Staff 2014) and as Sodic Pellic Vertisols (Hypereutric, Mazic, Mesotrophic) according to the World Reference Base for Soil Resources (WRB) (IUSS Working Group WRB 2015). Crop and livestock farming are the major economic activities (URT 2017). Cotton is the main cash crop in the area where both conventional and organic cotton production is practised. Green gram (*Vigna radiata* (L.) R. Wilczek) is widely grown in the area and used in cotton-legume rotations and to some extent for intercropping with cotton. The study was done at BioRe Tanzania Ltd.'s demonstration farm, Mwamishali

Village, located at 3°31′11″ S and 34°14′05″ E, for two consecutive cotton growing seasons, season 1 (2015/16) and season 2 (2016/17).

2.2 Experimental Plot Initial Soil Properties

A composite soil sample was collected prior to our experiment at a depth of 0–20 cm, air-dried and sieved through a 2-mm sieve and analysed to determine initial soil properties. The composite sample was used to determine soil texture by the hydrometer method (Day 1965), and textural classes were determined using the USDA textural triangle (Soil survey staff 2014). The soil pH and electrical conductivity were determined by the potentiometric method (Okalebo et al. 2002). Organic carbon was determined by the Walkley and Black wet oxidation method (Nelson and Sommers 1982), total nitrogen by the Kjeldahl wet digestion-distillation method, extractable P by the Bray-1 method (Olsen and Sommers 1982), cation exchange capacity for basic cations (Ca, Mg, K and Na) by the NH_4OAc saturation method (Thomas 1982) and micronutrients (Cu, Fe, Zn, Mn) by the diethylenetriaminepentaacetic acid (DTPA) method (Motsara and Roy 2008). Undisturbed soil samples were also collected by using a core ring at the depth of 0–20 cm to determine soil bulk density and porosity by the core method (Blake and Hartge 1986).

In summary, initial soil properties of the experimental field consisted of a sandy clay texture (38% clay) and a soil pH of 9.0 and 7.2 in H_2O and $CaCl_2$, respectively; organic carbon was 1.03%; total nitrogen was 0.14%; extractable phosphorus was 16.0 mg kg^{-1}; and exchangeable bases in cmolc kg^{-1} were 17.87, 3.47, 1.54 and 0.11 for Ca^{2+}, Mg^{2+}, K^+ and Na^+, respectively. Micronutrient concentrations were 1.44, 1.11, 0.40 and 9.12 mg kg^{-1} for Cu, Fe, Zn and Mn, respectively. The bulk density was 1.36 g cm^{-3} and the soil is rated marginally suitable for cotton production due to low soil fertility (Bwana 2019). The manure used had an organic carbon content of 8%, total nitrogen of 1.03%, extractable phosphorus of 103 mg kg^{-1} and a C/N ratio of 7.8.

2.3 Weather Data

Weather data were recorded hourly using an automatic weather station (AWS) installed at the experimental site at the beginning of the experiment. The AWS collected data on precipitation, measured with an ECRN-100 high-resolution rain gauge (Decagon Devices Inc.). Air temperature and humidity were measured by a VP-4 sensor (Decagon Devices Inc.), solar radiation was measured by a PAR sensor, and soil temperature and moisture were measured by a 5TM sensor (Decagon Devices Inc.) at 20-cm depth. Hourly averages were logged in an Em50 data logger (Decagon Devices Inc.). The soil water-filled pore space (WFPS) was calculated from the measured volumetric soil moisture (VWC) using the relation WFPS (%) = 100 ∗ VWC ∗

$(1–BD * PD^{-1})^{-1}$, where BD = bulk density and PD = particle density, where the particle density (PD) was assumed to be 2.65 g cm^{-3} (Brady and Weil 2014).

2.4 Field Experimental Design and Treatments

A field experiment was conducted over two consecutive growing seasons, 2015/16 and 2016/17. The agronomic and economic performance of cotton was tested under a range of fertilisation and pest control strategies for both conventional and organic farming. As described in Tables 2 and 3, the current low-input conventional and organic cotton farming practices were tested against higher application rates and innovative practices, as well as against control treatments without fertilisers or pesticides. For conventional fertilisation management practices, two application rates of nitrogen (N) from the inorganic fertilisers diammonium phosphate (DAP) and urea and an innovative practice of combining inorganic fertilisers and organic manure were tested. For organic farming, the fertility management practices include two different levels of organic manure and an innovative practice of intercropping cotton and green gram.

Table 2 Soil fertility management treatments

		Description	
Management	Treatments	At planting	Later in the season
Conventional	*CF-0* – No fertilisers	No fertilisation	No fertilisation
	CF-30 – Currently practised fertilisation: 30 kg N ha^{-1}	75 kg ha^{-1} DAP (13.5 kg N ha^{-1}, 15 kg P ha^{-1})	At squire formation, top-dress with 35.9 kg ha^{-1} urea (16.5 kg N ha^{-1})
	CF-60 – Higher fertilisation rate: 60 kg N ha^{-1}	100 kg ha^{-1} DAP (18 kg N ha^{-1} + 20 kg P ha^{-1})	At squire formation, top-dress with 91.4 kg ha^{-1} urea (42 kg N ha^{-1})
	CF-30 + M – Innovative fertilisation: 3 Mg ha^{-1} FYM + 30 kg N ha^{-1}	3 Mg ha^{-1} FYM	At squire formation, top-dress with 65.2 kg ha^{-1} urea (30 kg N ha^{-1})
Organic	*OF-0* – No fertilisers	No fertilisation	No fertilisation
	OF-3 – Currently practised fertilisation: 3 Mg ha^{-1} FYM	3 Mg ha^{-1} FYM	No fertilisation
	OF-5 – Higher fertilisation rate: 5 Mg ha^{-1} FYM	5 Mg ha^{-1} FYM	No fertilisation
	OF-3 + L – Innovative fertilisation: 3 Mg ha^{-1} FYM + intercropping with green gram	3 Mg ha^{-1} FYM	At 2–4 weeks after planting cotton, plant green gram (*Vigna radiata*) between cotton rows

Table 3 Pesticide treatments

Management	Treatments	Description
Conventional	*CP-0* – No pesticides	No pesticide application
	CP-3 – Currently practised pesticide application: 3 sprays with synthetic pesticide	Each application: 371 ml ha^{-1} Ninja (50 g l^{-1} lambda-cyhalothrin) (in total: 1.112 l ha^{-1} Ninja)
	CP-6 – Higher pesticide application rate: 6 sprays with synthetic pesticide	Each application: 371 ml ha^{-1} Ninja (50 g l^{-1} lambda-cyhalothrin) (in total: 2.224 l ha^{-1} Ninja)
	CP-3-N + CU – Innovative pesticide application: 3 sprays with neem-leaf extract + cow urine	Each application: 24.8 kg ha^{-1} fresh neem leaves +3.5 l ha^{-1} sunflower oil +5.0 l ha^{-1} cow urine (in total: 74.3 kg ha^{-1} fresh neem leaves +10.4 l ha^{-1} sunflower oil +14.9 l ha^{-1} cow urine)
Organic	*OP-0* – no pesticides	No pesticide application
	OP-P – currently practised biopesticide application: Pyrethrum according to scouting	Scouting determines number and date(s) of sprays (season 2015/16: 1 application; season 2016/17: 1 application). Each application: 272 ml ha^{-1} natural Pyrethrin extract
	OP-N – innovative biopesticide application: Neem-leaf extract according to scouting	Scouting determines number and date(s) of sprays (season 2015/16: 1 application; season 2016/17: 1 application). Each application: 24.8 kg ha^{-1} fresh neem leaves +3.5 l ha^{-1} sunflower oil
	OP-N + CU – innovative biopesticide application: Neem-leaf extract + cow urine according to scouting	Scouting determines number and date(s) of sprays (season 2015/16: 1 application; season 2016/17: 1 application). Each application: 24.5 kg ha^{-1} fresh neem leaves +3.5 l ha^{-1} sunflower oil +5.0 l ha^{-1} cow urine

For conventional farming, the tested pest management practices included the currently practised three sprays with a synthetic pesticide (Ninja, 50 g l^{-1} lambda-cyhalothrin), a higher application rate of six sprays with the same synthetic pesticide and an innovative practice with three sprays with neem-leaf extract and cow urine. For organic farming, the pest management practices included the currently practised application of natural pyrethrum extract, application of neem-leaf extract and application of a mixture of neem-leaf extract and cow urine. The neem-leaf extract was prepared by soaking pounded fresh neem leaves in water at a rate of 12.5% w/w for 24 hours. The mixture was then sieved and sunflower oil added to improve adhesion. For the application of a mixture of neem-leaf extract and cow urine, cow urine was added. All applications of pesticides and biopesticides were done using knapsack sprayer at a rate of 198 litres per ha.

The experiment was arranged as split-split-plot in a randomised block design, where production systems (organic versus conventional) were the main plots with pest management as a sub-plot and nutrient management as a sub-sub plot. Treatments were replicated in three blocks, with a test plot size of 10 × 5 m as shown in Fig. 1. For organic treatments, a rotation was made in season 2 where organic plots were shifted to plots planted with sole legume in season 1, which is a

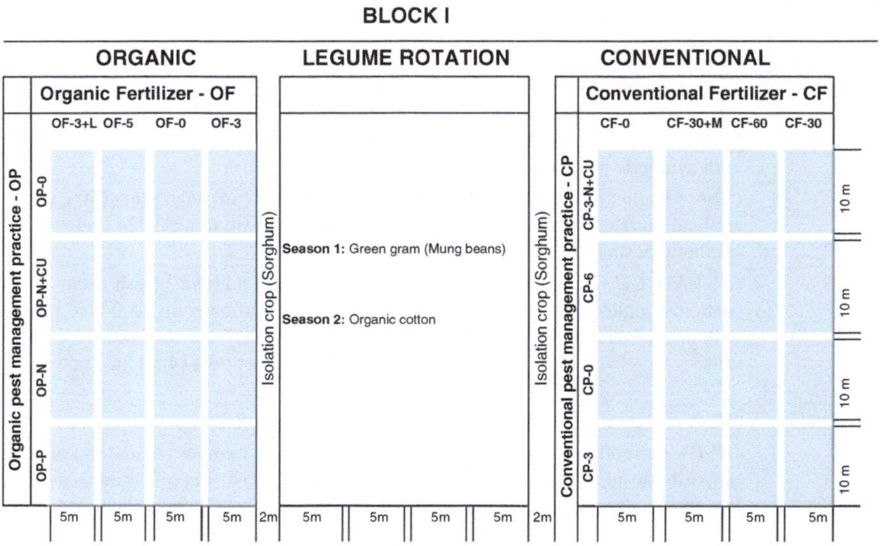

Fig. 1 Experimental plot layout (the abbreviations for the treatments are explained in Tables 2 and 3)

standard procedure for organic farmers in the area. Except for the innovative practices, the management of treatments followed the standard practices and recommended practices of farmers in the region. Organic and conventional plots were isolated by 2-m border rows planted with sorghum as indicated in Fig. 1.

2.5 *Determination of Yield*

Green gram yield was measured by harvesting the beans at the entire gross plot (10 × 5 m) during the third weeding. Cotton yield was measured by harvesting the net plot area (9 × 4 m) of each plot by hand picking, with the first round of harvesting done when approximately 60% of the cotton balls were open, which for seasons 1 and 2 was 142 and 135 days after sowing (DAS), respectively. A total of three rounds of harvesting cotton were conducted. The harvested green gram beans and seed cotton from each plot were weighed using a digital balance.

2.6 *Assessment of the Profitability*

We assess the economic performance of the analysed organic and conventional farming practices by the land rent, which is total revenues minus total production costs, except for the (opportunity) cost of the land. The total production costs include expenditures for purchased inputs and services as well as labour costs. The production costs were obtained by recording the costs of those inputs and activities that were done during the plot experiment and which would also be done by typical cotton producers in the area. These costs are similar to the costs that typical cotton producers in the area would have, whereas we assume that the opportunity costs of unpaid household labour are equal to the costs of hired labour for the same activity. Total revenue was calculated by multiplying the harvested quantities by the selling prices at the time of the harvest. The production costs and output prices that were used for calculating the land rents are presented in Tables 4 and 5, respectively. All costs and revenues were converted to US$ using an exchange rate of 2167.32 TZS US$$^{-1}$.

2.7 Statistical Analysis

Three different outcome variables were used in the statistical analysis: seed cotton yield (in Mg ha^{-1}), total revenue (in US$ ha^{-1}) and land rent (in US$ ha^{-1}). For each of these three outcome variables, the effects of the soil fertility treatments, the effects of the pesticide treatments and the effects of selected combinations of soil fertility and pesticide treatments were investigated. The selected combinations of soil fertility and pesticide treatments that were included in our analysis are conventional farming without fertilisation or pesticides (CF-0 & CP-0), currently practised conventional farming (CF-30 & CP-3), conventional farming with higher fertilisation and pesticide application rates (CF-60 & CP-6), innovative conventional farming practices (CF-30 + M & CP-3-N + CU), organic farming without fertilisation or pesticides (OF-0 & OP-0), currently practised organic farming (OF-3 & OP-P), organic farming with higher fertilisation rate (OF-5 & OP-P) and innovative organic farming practices (OF-3 + L & OP-N + CU). As the conventional and organic no-input treatments (CF-0 & CP-0 and OF-0 & OP-0) were identical in the first season but different in the second season (due to different pre-crops), the statistical analysis of the selected combinations of treatments considered these two treatments as the same treatment in season 1 but as two different treatments in season 2.

In order to investigate the effect of the soil fertility and pesticide treatments, we used the ordinary least squares (OLS) method to test effects on each of the three outcome variables using the soil fertility treatments, the pesticide treatments, the interaction terms between the soil fertility treatments and the pesticide treatments, and the block in which the plot was located as explanatory variables. We calculated the least squares mean values for each soil fertility and pesticide treatment (Searle et al. 1980) and conducted pairwise tests of equal means, where we present the results of these tests as "compact letter display" (Piepho 2004). We investigated the effects of the selected combinations of soil fertility and pesticide treatments in a similar way, but we used only the selected combinations of treatments and the block in which the plot was located as explanatory variables.

All calculations and statistical analyses were conducted with the statistical software "R" (R Core Team 2019) using the add-on packages "emmeans" (Lenth 2019), "multcomp" (Hothorn et al. 2008) and "ggplot2" (Wickham 2016).

Table 4 Production costs (in growing season 2015/16 and 2016/17)

Description	Treatments	Unit	Costs
Labour, oxen and plough for land preparation	All	US$ ha^{-1}	34.20
Manure 3 Mg ha^{-1} (10,000 TZS Mg^{-1})	CF-30 + M, OF-3, OF-3 + L	US$ ha^{-1}	13.84
Manure 5 Mg ha^{-1} (10,000 TZS Mg^{-1})	OF-5	US$ ha^{-1}	23.07
Labour for manure application 3 Mg ha^{-1}	CF-30 + M, OF-3, OF-3 + L	US$ ha^{-1}	3.42
Labour for manure application 5 Mg ha^{-1}	OF-5	US$ ha^{-1}	5.70
Labour for land pulverisation/harrowing	All	US$ ha^{-1}	17.10
Cotton seed	All	US$ ha^{-1}	7.98
Labour for cotton sowing	All	US$ ha^{-1}	22.80
Green gram seed	OF-3 + L	US$ ha^{-1}	11.97
Labour for green gram sowing	OF-3 + L	US$ ha^{-1}	17.10
DAP at planting (75 kg ha^{-1}, 1200 TZS kg^{-1})	CF-30	US$ ha^{-1}	41.53
DAP at planting (100 kg ha^{-1}, 1200 TZS kg^{-1})	CF-60	US$ ha^{-1}	55.37
Labour for DAP application at planting	CF-30, CF-60	US$ ha^{-1}	17.10
Urea as top-dress (35.9 kg ha^{-1}, 1200 TZS kg^{-1})	CF-30	US$ ha^{-1}	19.86
Urea as top-dress (91.4 kg ha^{-1}, 1200 TZS kg^{-1})	CF-60	US$ ha^{-1}	50.61
Urea as top-dress (65.2 kg ha^{-1}, 1200 TZS kg^{-1})	CF-30 + M	US$ ha^{-1}	36.11
Labour for urea application as top-dress	CF-30, CF-60, CF-30 + M	US$ ha^{-1}	17.10
Labour for first weeding	All	US$ ha^{-1}	22.80
Labour for second weeding	All	US$ ha^{-1}	20.52
Labour for third weeding	All	US$ ha^{-1}	15.96
Ninja (3 applications: 2000 TZS (150 ml)$^{-1}$)	CP-3	US$ ha^{-1}	6.84
Ninja (6 applications: 2000 TZS (150 ml)$^{-1}$)	CP-6	US$ ha^{-1}	13.68

(continued)

Table 4 (continued)

Description	Treatments	Unit	Costs
Natural Pyrethrin extract (1 application: 3500 TZS (110 ml)$^{-1}$)	OP-P	US\$ ha^{-1}	3.99
Neem leaves for biopesticide	CP-3-N + CU, OP-N, OP-N + CU	US\$ ha^{-1}	0.00
Sunflower oil for biopesticide (1 application: 3.5 l ha^{-1}, 2500 TZS l^{-1})	OP-N, OP-N + CU	US\$ ha^{-1}	3.99
Sunflower oil for biopesticide (3 applications: 10.4 l ha^{-1}, 2500 TZS l^{-1})	CP-3-N + CU	US\$ ha^{-1}	11.97
Cow urine for biopesticide (1 application: 4.94 l ha^{-1}, 500 TZS l^{-1})	OP-N + CU	US\$ ha^{-1}	1.14
Cow urine for biopesticide (3 applications: 14.83 l ha^{-1}, 500 TZS l^{-1})	CP-3-N + CU	US\$ ha^{-1}	3.42
Labour for scouting for pests	OP-P, OP-N, OP-N + CU	US\$ ha^{-1}	10.26
Labour for preparation of biopesticide (1 application)	OP-N, OP-N + CU	US\$ ha^{-1}	5.70
Labour for preparation of biopesticide (3 applications)	CP-3-N + CU	US\$ ha^{-1}	17.10
Labour for 1 application of (bio-)pesticides	OP-P, OP-N, OP-N + CU	US\$ ha^{-1}	4.56
Labour for 3 applications of (bio-)pesticides	CP-3, CP-3-N + CU	US\$ ha^{-1}	13.68
Labour for 6 applications of (bio-)pesticides	CP-6	US\$ ha^{-1}	27.36
Bags for harvested cotton	All	US\$ ha^{-1}	11.40
Labour for green gram harvesting	OF-3 + L	US\$ ha^{-1}	5.70
Labour for cotton harvesting	All	US\$ kg^{-1}	0.0277
Labour and vehicle for transport of cotton to selling point	All	US\$ ha^{-1}	5.70

Note: The abbreviations of the treatments are explained in Tables 2 and 3
Source: own recording of production costs, information obtained from experts, own calculations

Table 5 Output prices (at the end of growing season 2015/16 and 2016/17)

Output	Unit	Price
Cotton, conventional	US\$ kg^{-1}	0.461
Cotton, organic	US\$ kg^{-1}	0.554
Green gram	US\$ kg^{-1}	0.646

Source: observations at cotton selling points and local markets

3 Results

3.1 Weather Conditions in the Growing Seasons

The total rainfall recorded in the cotton growing season 1 (November 2015 to June 2016) was 759 mm, while that in season 2 (October 2016 to June 2017) was 522 mm (Table 6). The rainfall recorded in season 1 was above the long-term (1994–2011) annual rainfall mean of 668 mm (Kabote et al. 2013), while the rainfall in season 2 was below average.

The mean daily soil and air temperatures for seasons 1 and 2 were rather similar (Fig. 2). There was slightly higher mean solar radiation in season 2 (226 W m^{-2}) than in season 1 (211 W m^{-2}) and less soil moisture in season 2 than in season 1.

3.2 Yield and Economic Performance

The results regarding the yield and economic performance are summarised in Table 7, whereas the most important results are visualised in Figs. 3, 4 and 5. The weather conditions in the study area in season 1 (2015/16) were well suited for cotton production and resulted in relatively high cotton yields and a high profitability of cotton production. In season 2, however, moisture stress at the end of the growing season resulted in low yields and very poor economic performance.

The seed cotton yields in the two respective seasons in the currently practised low-input conventional (CF-30 & CP-3: 1.27 and 0.37 Mg ha^{-1}) and organic (OF-3 & OP-P: 1.37 and 0.63 Mg ha^{-1}) farming were not significantly different. However, the low-input organic practice had a higher land rent (LR) (524 and 139 US$ ha^{-1}) than the low-input conventional practice (278 and −114 US$ ha^{-1}) in both seasons. Higher-input conventional farming had a significantly higher cotton yield (CF-60 &

Table 6 Summary of mean daily values for weather parameters at the experiment site for the two cotton growing seasons (Nov 2015 to Jun 2016, Oct 2016 to Jun 2017)

Season		RH (%)	Temp (°C)	Precipitation (mm)	Solar radiation (W m^{-2})	Soil VWC (m^3 m^{-3})	Soil WFPS (%)	Soil temp (°C)
1	Mean	71	23.8		247	0.11	32.4	28.0
	Min	23	20.7	0	211	0.05	15.3	23.8
	Max	103	27.4	45	339	0.18	51.3	31.5
	Total			759				
2	Mean	54	24.6		259	0.11	32.7	29.4
	Min	39	20.4	0	226	0.05	13.8	22.6
	Max	75	28.5	48	438	0.21	63.0	35.5
	Total			522				

RH Relative humidity, *VWC* volumetric water content, *WFPS* water-filled pore space

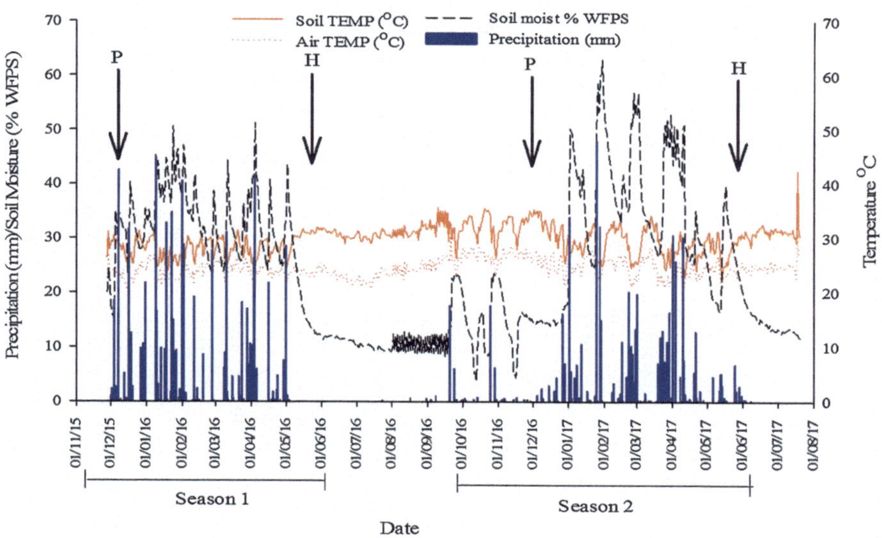

Fig. 2 Variation in rainfall, soil moisture (% water-filled pore space, WFPS) and soil and air temperature during the experiment period for season 1 (Nov 2015 to June 2016) and season 2 (Oct 2016 to June 2017). *Note: Arrows show planting (P) and harvesting (H) time*

CP-6: 1.76 and 0.50 Mg ha^{-1}) than higher-input organic farming (OF-5 & OP-P: 1.36 and 0.46 Mg ha^{-1}) in season 1, but the difference is insignificant for season 2, while the differences in the land rent between the higher-input conventional (422 and −121 US$ ha^{-1}) and organic (509 and 36 US$ ha^{-1}) practices are statistically insignificant for both seasons. The innovative conventional practice (CF-30 + M & CP-3-N + CU: 1.69 and 0.47 Mg ha^{-1}) had a significantly higher yield in season 1 than the innovative organic practice (OF-3 + L & OP-N + CU: 1.32 and 0.66 Mg ha^{-1}), but the difference is statistically insignificant for season 2. However, the innovative conventional practice (460 and −69 US$ ha^{-1}) had a statistically significantly lower land rent than the innovative organic practice (615 and 227 US$ ha^{-1}) both in seasons 1 and 2.

The highest seed cotton yields in season 1 were obtained in the higher-input conventional farming practice (CF-60 & CP-6: 1.76 Mg ha^{-1}) and the innovative conventional farming practice (CF-30 + M & CP-3-N + CU: 1.69 Mg ha^{-1}), whereas there are no statistically significant differences in the seed cotton yields between any of the treatments in season 2.

Given that organic cotton farming requires lower production costs and receives a higher output price than conventional cotton farming, it generally generates notably higher land rents than conventional cotton farming in spite of similar or lower seed cotton yields. Particularly the innovative organic practice (OF-3 + L & OP-N + CU: 615 and 227 US$ ha-1) generates the highest land rent both in seasons 1 and 2, and these land rents are significantly higher than the land rents of all conventional treatments.

Table 7 Least squares means of production costs, cotton yields, total revenue and land rents for different treatments in two growing seasons

Treatment	Costs (US$ ha^{-1})		Cotton yield (Mg ha^{-1})				Total revenue (US$ ha^{-1})				Land rent (US$ ha^{-1})			
Season	S1	S2	S1		S2		S1		S2		S1		S2	
	Fertility management practices													
CF-0	219	198	1.23	ab	0.45	a	566	a	208	a	347	a	10	bcd
CF-30	320	293	1.39	bc	0.43	a	643	ab	200	a	324	a	--93	ab
CF-60	369	339	1.56	c	0.50	a	722	bc	230	a	353	a	−109	a
CF-30 + M	298	269	1.52	c	0.49	a	701	bc	226	a	403	ab	−43	abc
OF-0	209	189	1.19	a	0.48	a	657	ab	268	a	448	b	79	de
OF-3	232	205	1.42	c	0.44	a	789	c	242	a	556	c	37	cd
OF-5	237	215	1.17	a	0.38	a	651	ab	212	a	414	ab	-3	abcd
OF-3 + L	266	242	1.39	bc	0.50	a	930	d	396	b	664	d	155	e
	Pesticide management practices													
CP-0	269	248	1.22	a	0.48	a	564	a	222	ab	295	a	−27	ab
CP-3	294	267	1.39	abc	0.40	a	640	ab	183	a	346	ab	−83	a
CP-6	319	290	1.53	bc	0.49	a	708	bc	224	ab	389	bc	−65	a
CP-3-N + CU	324	295	1.56	c	0.51	a	720	bc	235	ab	396	bc	−60	a
OP-0	219	195	1.29	a	0.44	a	747	c	280	ab	528	d	84	c
OP-P	239	217	1.36	ab	0.53	a	802	c	322	b	562	d	106	c
OP-N	241	218	1.22	a	0.38	a	716	bc	241	ab	474	cd	23	abc
OP-N + CU	245	221	1.31	a	0.45	a	762	c	275	ab	518	d	54	bc
	Selected combinations of fertility and pesticide management practices													
CF-0 & CP-0	188	171	1.08	a	0.44	a	545	a	203	a	357	ab	32	abc
CF-30 & CP-3	310	285	1.27	ab	0.37	a	588	a	171	a	278	a	−114	a
CF-60 & CP-6	388	354	1.76	c	0.50	a	810	b	232	Aab	422	abc	−121	a

CF-30 + M & CP-3-N + CU	322	288	1.69	c	0.47	a	782	b	219	ab	460	bc	−69	ab
OF-0 & OP-0	188	172	1.08	a	0.47	a	545	a	261	ab	357	ab	90	abc
OF-3 & OP-P	232	212	1.37	b	0.63	a	756	b	351	ab	524	cd	139	bc
OF-5 & OP-P	244	219	1.36	ab	0.46	a	752	b	255	ab	509	cd	36	abc
OF-3 + L & OP-N + CU	273	254	1.32	ab	0.66	a	888	b	481	b	615	d	227	c

Notes: S1 = growing season 2015/16, S2 = growing season 2016/17. The letters on the right-hand side of the mean values present the results of pairwise tests for equal means, where mean values in the same column and in the same panel of the table that have the same letter are not statistically significantly different at a significance level of 5%. The abbreviations of the treatments are explained in Tables 2 and 3

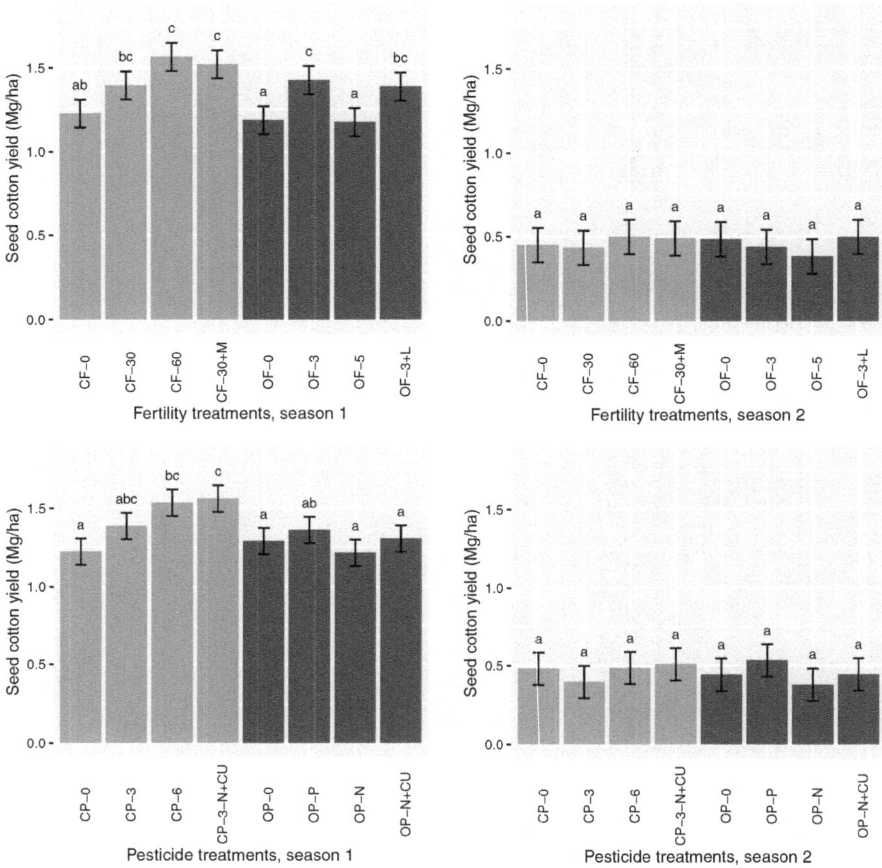

Fig. 3 Effect of conventional and organic fertility and pesticide treatments on seed cotton yields in growing seasons 1 (2015/16) and 2 (2016/17). *The bars indicate the least-squares means, while the vertical lines indicate the 95% confidence intervals of these mean values. The abbreviations of the treatments are explained in* Tables 2 and 3

3.3 Yield Compared to Potential Yield

The seed cotton yield ranged from 1.19 to 1.42 Mg ha^{-1} for organic farming practices and from 1.22 to 1.56 Mg ha^{-1} for conventional farming practices in season 1 (Table 7). For season 2, the yields ranged from 0.40 to 0.51 Mg ha^{-1} for conventional and from 0.38 to 0.52 Mg ha^{-1} for organic practices. As Fig. 3 indicates, both fertility and pest management practices had statistically significant effects on the seed cotton yield in season 1. However, in season 2, neither fertility management

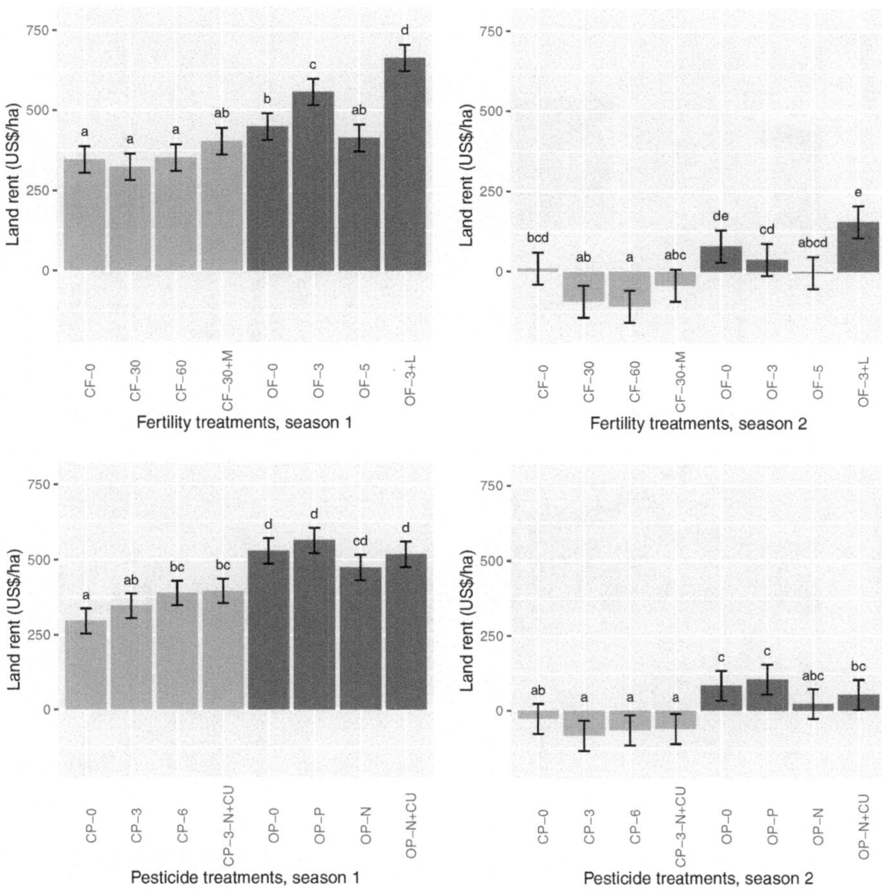

Fig. 4 Effect of conventional and organic fertility and pesticide treatments on the land rent in growing seasons 1 (2015/16) and 2 (2016/17). *The bars indicate the least-squares means, while the vertical lines indicate the 95% confidence intervals of these mean values. The abbreviations of the treatments are explained in* Tables 2 and 3

nor pest management practices had a significant effect on cotton yield. The yields observed in this study were considerably lower than the potential yield of the test variety (UK MO8) of 2.5 Mg ha^{-1} (Lukonge et al. 2007) for both seasons. For season 1, the yield was 52% to 43% below the potential yield in the organic system and 51% to 40% below the potential yield in the conventional system. For season 2, the yield was 84% to 80% and 85% to 79% below the potential yield in the organic and conventional system, respectively.

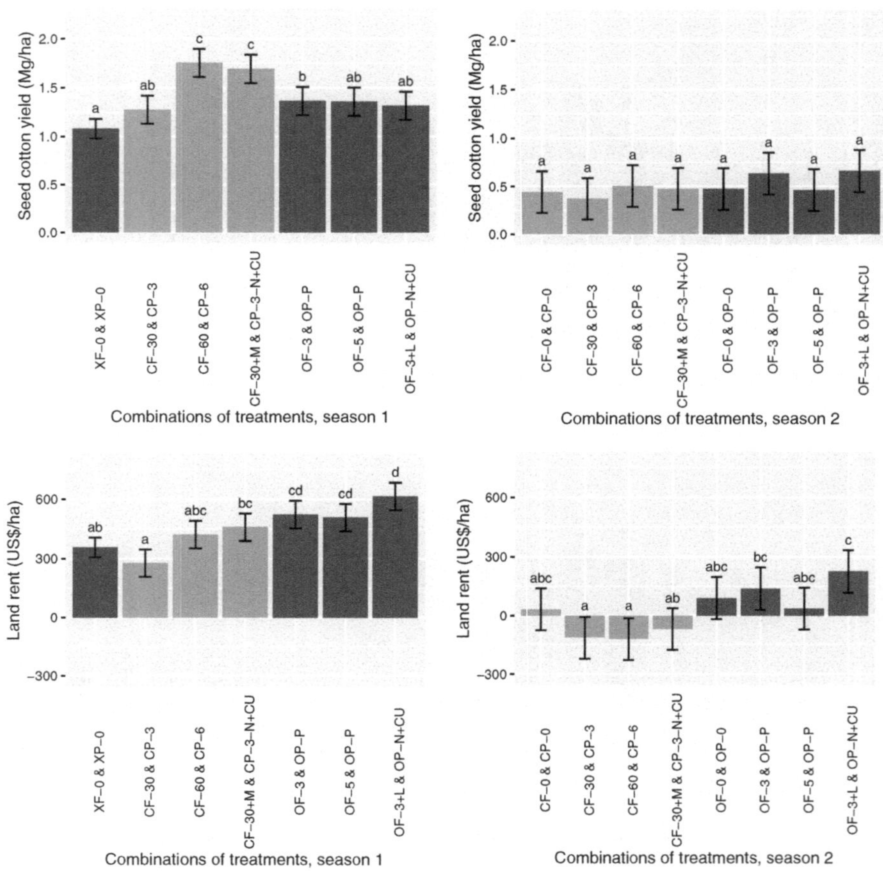

Fig. 5 Effects of selected combinations of fertility and pesticide treatments on the cotton yield and the land rent in growing seasons 1 (2015/16) and 2 (2016/17). *The bars indicate the least-squares means, while the vertical lines indicate the 95% confidence intervals of these mean values. The abbreviations of the treatments are explained in* Tables 2 and 3, *whereas treatment "XF-0 & XP-0" indicates treatments "CF-0 & CP-0" and "OF-0 & OP-0"*

4 Discussion

4.1 Yield and Economic Performance

Current Practices

The similarity of cotton yields between current organic and conventional practices can be linked to the fact that the current input levels in organic and conventional cotton farming are rather similar. For instance, the N input in the currently practised conventional fertilisation treatment (30 kg N ha^{-1}) is similar to the N input from 3 Mg ha^{-1} FYM in the currently practised organic fertilisation treatment (30.9 kg N ha^{-1}, given an N content of 1.03% in the applied FYM). This result aligns with Cavigelli et al.

(2009) who found that wheat yield was similar in organic and conventional systems. However, such comparisons are likely to be highly dependent on soil and climate conditions. Therefore, many studies indicate lower yields in organic than in conventional farming (e.g. Forster et al. 2013; Lee et al. 2015; Ponisio et al. 2015; Kniss et al. 2016; Suja et al. 2017; de Ponti et al. 2012; Seufert et al. 2012). These studies were done in high-input farming systems under suitable climatic conditions with high fertiliser levels in conventional farming. Our result of a better economic performance of organic production compared to conventional production is linked to the lower input costs, especially the use of manure instead of inorganic fertilisers, and a higher price of organic cotton compared to conventional cotton.

Higher-Input Scenario

For the higher-input scenarios, the higher yield of conventional farming compared to organic farming in season 1 is mainly an effect of the higher rainfall in this season, which favours higher N input in conventional (60 kg N ha^{-1}) than in organic farming (51.5 kg N ha^{-1}), and more effective pest management in the conventional treatments than in the organic treatments. The higher input of FYM in organic farming (OF-5) resulted in a significant reduction in crop yield, for which the reason is not known. An increase in N input in cotton farming increases seed cotton yield (Bell et al. 2003; Prasad and Siddique 2004) because the increased N rate increases the leaf photosynthetic rate (Cadena and Cothren 1995). This leads to higher boll weight and seed cotton yield. This result is in line with most studies which show higher yields in conventional than in organic systems (Seufert et al. 2012). For instance, Forster et al. (2013) found in a cotton-wheat-soybean rotation in India a higher yield in a conventional treatment receiving 105 kg N ha^{-1} than in an organic treatment receiving 65 kg N ha^{-1}.

In season 2, yields were considerably lower with no significant differences in yields between the two systems and treatments. This is linked to low soil moisture availability in season 2 (the rainfall received in season 2 was 26% less than in season 1), such that soil moisture was the most limiting factor that resulted in stunted plants which were then unable to utilise the higher N input. Due to lower production costs and a higher cotton price of organic cotton compared to conventional cotton, higher-input organic farming (OF-5 & OP-P) had a somewhat better economic performance than higher-input conventional farming (CF-60 & CP-6), but these differences are statistically insignificant in both growing seasons.

Innovative Practices

Manure-Fertiliser Combination

The innovative conventional practice of applying FYM in addition to the currently practised rate of inorganic fertilisers (CF-30 + M) gives a similar yield as the higher-input conventional fertility treatment (CF-60), which is likely caused by similar N inputs in these two treatments (CF-60: 60 kg N ha^{-1}; CF-30 + M: 60.9 kg N ha^{-1}).

The innovative conventional practice of applying FYM gives a slightly higher yield than the currently practised fertilisation with inorganic fertilisers only (CF-30), but this (small) difference in the yield is statistically insignificant. Hence, in contrast to many existing studies (e.g. Khaliq et al. 2006; Hulihalli and Patil 2008; Kumari et al. 2010; Anwar-ul-Haq et al. 2014; Moe et al. 2017; Rao et al. 2017), we do not find that combining inorganic fertilisers with manure gives significantly higher yields than applying inorganic fertilisers alone. For instance, Rao et al. (2017) find a significant synergistic interaction effect between FYM and inorganic fertilisers and argue that FYM acts as a source of additional nutrients and supports moisture retention. Kumari et al. (2010) argue that FYM increases microbial activity and, hence, nutrient availability to cotton plants. Moe et al. (2017) argue that combining inorganic fertilisers with manure results in higher yields due to continuous supply of nutrients throughout the growing season, given that inorganic fertilisers release nutrients rapidly during the early growth stages followed by gradual release of nutrients from organic manure at a later stage.

The economic performance of the innovative conventional fertility treatment is slightly better than the economic performance of the current practice and the higher-input fertility treatment, but these differences are statistically insignificant. This is in contrast to results from other studies which indicate significantly higher economic performance of combining inorganic fertilisers and manure than applying inorganic fertilisers alone. For instance, Anwar-ul-Haq et al. (2014) report higher yield and economic performance of cotton from combining 20 Mg ha^{-1} manure and NPK (at 88 kg N ha^{-1}) than sole NPK application (at 175 kg N ha^{-1}), which are very high application rates compared to those of this study (3 Mg ha^{-1} manure, 30 or 60 kg N ha^{-1} from inorganic fertilisers).

Three Sprays of Neem-Leaf Extract and Cow Urine

Spraying three times with neem-leaf extract in combination with cow urine has a similar effectiveness as spraying six times with a conventional pesticide containing lambda-cyhalothrin and likely has a slightly higher effectiveness than spraying three times with a conventional pesticide containing lambda-cyhalothrin, but this difference is statistically insignificant. The leaves of the neem tree (*Azadirachta indica)* contain biologically active components, and their potency is increased when mixed with cow urine (Gupta 2005). The biologically active components in the neem-leaf extract act broadly as toxicant, repellent, anti-feedant and growth-disrupting substances on insect pests (Gujar 1992) and also act as powerful insect growth regulators (IGR) (Subbalakshmi et al. 2012). The combinations of cow urine and neem-based products have shown significant synergistic effects to enhance product toxicity resulting in pest mortality (Gahukar 2013) but are safe to insect predators, particularly beetles (Gupta 2005). The land rent when spraying three times with neem-leaf extract and cow urine is similar to the one of spraying six times with a conventional pesticide and may be slightly higher than the land rent when spraying three times with a conventional pesticide.

Cotton-Legume Intercrop

In spite of potential competition for soil moisture and light between the cotton plants and the green gram plants, the cotton yield in the cotton-legume intercrop was similar to the cotton yield of the currently practised organic fertility practice, which applied the same amount of FYM but had no intercropping. Hence, our results contradict the results of several other studies that report lower cotton yields in cotton-legume intercropping (e.g. Khan and Khaliq 2004; Nandini and Chellamuthu 2004; Reddy and Shaik 2009; Hallikeri et al. 2007; Mankar and Nawlakhe 2009; Sankaranarayanan et al. 2012; Khargkharate et al. 2014; Jayakumar and Surendran 2017; Singh et al. 2017). For instance, Jayakumar and Surendran (2017) associate the lower cotton yields of intercrops with the early, vigorous growth of the intercrop that result in a smothering effect on the cotton crop. Similarly, Singh et al. (2017) report a significant reduction in seed cotton yield in cotton-mung bean and cotton-cowpea intercrop as compared to sole cotton. The higher yield of the intercrop in our case may be due to beneficial effects of the legume intercrop on soil fertility and nitrogen supply (Thilakarathna et al. 2016). Given the rather high cotton yield in the intercrop and the additional revenue from green gram production, cotton-green gram intercropping gives the highest land rent of all fertility treatments in both seasons. This result is in line with results reported by Jayakumar and Surendran (2017) and Singh et al. (2017) who also reported higher economic performance of cotton-legume intercrop compared to cotton without intercrop.

4.2 *Yield as Compared to Potential Yield*

The lower yield compared to the potential yield of the cotton variety UK MO8 in seasons 1 and 2 in this study for all treatments and their combinations is linked to the low rainfall in season 2 and soil fertility limitations. Low rainfall in season 2 severely affected the yield and, hence, masked the effects of the fertility and pesticide treatments. The rainfall in season 2 (522 mm) was on the lower side of the minimum water required for cotton growth (500 mm) (OECD 2008). With the same level of nutrient and pest management in the two seasons, soil moisture was the major limiting factor to primary productivity and biomass production. A series of intra-season dry spells were experienced in both seasons due to intermittent rain events (Fig. 2). The cotton yield in season 1 was higher than the cotton yield in season 2 but still 36% less than the potential yield of UK MO8, which is narrower than the average yield gap of 43% for cotton in semi-arid Africa as reported by Hengsdijk and Langeveld (2009). A similar study in India reported lower than potential yield in cotton in one season with poor growing conditions due to low rainfall and waterlogging in the conventional but not in the organic system (Forster et al. 2013). Hengsdijk and Langeveld (2009) show that water is the main contributor to the yield gap of up to 30% in semi-arid Africa regions to various crops including cotton, and they reported an average actual yield of 2.0 vs. a potential yield of 3.3 Mg ha^{-1}. The soil properties in the study area affecting soil fertility, including high soil pH, might also have reduced cotton yield. As indicated in Bwana (2019), the soil was classified as marginally suitable for cotton production due to soil limitations.

5 Conclusions and Recommendations

Notwithstanding the different results between the two seasons, we conclude that for the current cotton farming practices, there is no significant difference in the seed cotton yields between smallholder organic and conventional production practices. A difference would only occur under good rainfall conditions if the rate of nitrogen fertilisation is increased to 60 kg N ha^{-1} or more in the conventional farming system. Application of manure in combination with a low rate of inorganic nitrogen fertilisers had a similar effect on the yield than applying a higher rate of inorganic fertilisers. Applying neem-leaf extract and cow urine as pesticide gives a similar yield and land rent as applying pyrethrum (in organic farming) or as applying a high rate of a synthetic pesticide (in conventional farming). Under conditions of limited rainfall as in the second growing season of our experiment, moisture stress becomes limiting, and hence, fertility and pest management practices have no significant effects on the yield. Under the prevailing semi-arid conditions, smallholder farmers are rational to apply only low rates of fertilisers and pesticides.

Based on our study, we conclude that smallholder farmers in our study region can improve their economic situation by intercropping cotton with grain legumes and applying neem-leaf extract and cow urine for pest management. Within the bounds of farming practices currently practised in the study area, our results also indicate that individual farmers with interest in organic production methods could economically benefit by adopting organic cotton production. Further research could investigate various agronomic, environmental and economic aspects of cotton-legume intercropping (e.g. effects of different legume species) and of using neem-leaf extract and cow urine as biopesticide.

Acknowledgements The authors acknowledge financial support by the Ministry of Foreign Affairs of Denmark (Grant: 14-02KU) and the Government of the United Republic of Tanzania. The authors are grateful to BioRe Tanzania for availing the research site, to Bo Markussen for advice on the statistical analyses, and to the technical staff at Aarhus University, Denmark, and the Soil Science lab, Sokoine University of Agriculture, Morogoro, Tanzania, for their technical assistance.

References

Anwar-ul-Haq M, Akhtar J, Saqib M, Hussain A, Hussain T (2014) Integrated use of farm manure and mineral fertilizers to maintain soil quality for better cotton (*Gossypium hirsutum L.*) production. Pak J Agric Sci 51(2):413–420

Bell PF, Boquet DJ, Millhollon E, Moore S, Ebelhar W, Mitchell CC, Varco J, Funderburg ER, Kennedy C, Breitenbeck GA, Craig C (2003) Relationships between leaf-blade nitrogen and relative seed cotton yields. Crop Sci 43(4):1367–1374

Blake GR, Hartge KH (1986) Bulk density. In: Klute A (ed). Methods of soil analysis. Part 1. Physical and Mineralogical Methods. 2nd edn. SSSA Book Ser. 5.1. SSSA and ASA, Madison, WI. pp 363–375
Borlaug NE (2000) Ending world hunger. The promise of biotechnology and the threat of anti-science zealotry. Plant Physiol 124:487–490
Brady NC, Weil R (2014) The nature and properties of soils, 14th edn. Pearson Education Limited. Essex. 1046pp.
Bwana TN (2019) Assessment of environmental performance of smallholder organic and conventional cotton production systems in Meatu, Tanzania. PhD Thesis, Sokoine University of Agriculture, Morogoro, Tanzania
Cadena J, Cothren JT (1995) Yield response of cotton to nitrogen, irrigation, and PGR-IV regimes. In: Proceedings of Beltwide Cotton Conference, San Antonio, TX. 4–7 Jan 1995. National Cotton Council of America, Memphis, TN, pp 1142–1150
Cavigelli MA, Hima BL, Hanson JC, Teasdale JR, Conklin AE, Lu YC (2009) Long-term economic performance of organic and conventional field crops in the mid-Atlantic region. Renew Agr Food Syst 24:102–119
de Ponti T, Rijk B, van Ittersum MK (2012) The crop yield gap between organic and conventional agriculture. Agric Syst 108:1–9
Day PR (1965) Particle fractionation and particle size analysis. In: Black CA, Evans DD, White JL, Ensminger LE, Clark FE (eds) Methods of soil analysis, Part 1. American Society of Agronomy, Madison, WI, pp 545–566
FAO (1998) Evaluating the potential contribution of organic agriculture to sustainability goals. FAO's technical contribution to IFOAM's Scientific Conference, Mar del Plata, Argentina, 16–19 Nov 1998. http://www.fao.org/docrep/003/ac116e/ac116e02.htm
Forster D, Andres C, Verma R, Zundel C, Messmer MM, Mäder P (2013) Yield and economic performance of organic and conventional cotton-based farming systems–results from a field trial in India. PLoS ONE 8(12):e81039
Gahukar RT (2013) Cow urine: a potential biopesticide. Indian J Entomol 75(3):212–216
Gold MV (2007) Organic production and organic food: information access tools. Alternative Farming Systems Information Center. USDA National Agricultural Library (2007) Available in: http://www.nal.usda.gov/afsic/pubs/ofp/ofp.shtml. 18 May 2015
Gomiero T, Pimentel D, Paoletti MG (2011) Environmental impact of different agricultural management practices: conventional vs. organic agriculture. Crit Rev Plant Sci 30(1–2):95–124
Gujar GT (1992) Neem *(Azadirachta indica A Juss)* a natural insecticide: status and need for bioassay standards. Pestic Res J 4(2):69–79
Gupta MP (2005) Efficacy of neem in combination with cow urine against mustard aphid and its effect on coccinellid predators. Nat Prod Rad 4(2):102–106
Hallikeri SS, Aladakatti YR, Patil BC, Patil SS (2007) Intercropping of cotton with vegetables as the profitable cropping system under rainfed conditions of north Karnataka. University of Agricultural Sciences, Dharwad, Karnataka
Hengsdijk H, Langeveld JWA (2009) Yield trends and yield gap analysis of major crops in the world. Wageningen University, the Netherlands, p 170
Hothorn T, Bretz F, Westfall P (2008) Simultaneous inference in general parametric models. Biom J 50(3):346–363
Hulihalli UK, Patil VC (2008) Yield and fiber quality of cotton as influenced by fertilizer levels and organic manure. Ann Arid Zone 47(2):201–204
IFDC (2014) Report on consumption of fertilizers and fertilizer use by crop in Tanzania. International Fertilizer Development Centre (IFDC), 23pp
IFOAM (2014) The potential contribution of organic agriculture to the realization of the objectives of the Comprehensive Africa Agriculture Development Programme (CAADP) – a guide for stakeholders. IFOAM, Bonn, Germany. 60 pp
IUSS Working Group WRB (2015) World Reference Base for soil resources 2014. International soil classification system for naming soils and creating legends for soil maps. Update 2015. World Soil Resources Report 106. FAO, Rome: 188 pp

Jayakumar M, Surendran U (2017) Intercropping and balanced nutrient management for sustainable cotton production. J Plant Nutr 40(5):632–644

Kabote SJ, Mamiro D, Synnevåg G, Urassa JK, Mattee AZ, Mbwambo JS, Nombo CI, Masolwa LW, Chingonikaya EE (2013) Inter-annual anomaly and seasonal variability of rainfall and temperature in selected semi-arid areas of Tanzania. J Contin Educ Ext 4(2):295–317

Khaliq A, Abbasi MK, Hussain T (2006) Effects of integrated use of organic and inorganic nutrient sources with effective microorganisms (EM) on seed cotton yield in Pakistan. Bioresour Technol 97(8):967–972

Khan MB, Khaliq A (2004) Study of mungbean intercropping in cotton planted with different techniques. J Res (Sci) 15(1):23–31

Khargkharate VK, Kadam GL, Pandagale AD, Awasarmal VB, Rathod SS (2014) Studies on kharif legume intercropping with Bt cotton under rainfed conditions. J Cotton Res Dev 28(2):243–246

Kniss AR, Savage SD, Jabbour R (2016) Commercial crop yields reveal strengths and weaknesses for organic agriculture in the United States. PloS ONE 11(8):e0161673

Kumari AL, Veeraiah K, Kumari SR (2010) Long term effects of manures and fertilizers on productivity of rainfed cotton and soil fertility in Vertisols. J Cotton Res Dev 24(2):200–204

Lee J, Hwang S, Ha I, Min B, Hwang H, Lee S (2015) Comparison of bulb and leaf quality, and antioxidant compounds of intermediate-day onion from organic and conventional systems. Hortic Environ Biotechnol 56(4):427–436

Lenth R (2019) emmeans: Estimated Marginal Means, aka Least-Squares Means. R package version 1.3.5.1. https://CRAN.R-project.org/package=emmeans

Lorenz K, Lal R (2016) Environmental impact of organic agriculture. Adv Agron 139:99–152

Lukonge EP, Maryke L, Liezel H (2007) Combining ability for yield and fibre characteristics in Tanzanian cotton germplasm. Euphytica 161:383–389

Makita R (2012) Fair Trade and organic initiatives confronted with Bt cotton in Andhra Pradesh, India: A paradox. Geoforum 43:1232–1241

Mankar DD, Nawlakhe SM (2009) Yield attributes and yield of cotton (main crop) and greengram (intercrop) and quality of cotton as influenced by INM under cotton+ greengram intercropping. J Soils Crops 19(2):315–319

Moe K, Mg KW, Win KK, Yamakawa T (2017) Combined effect of organic manures and inorganic fertilizers on the growth and yield of hybrid rice (Palethwe-1). Am J Plant Sci 8:1022–1042

Motsara MR, Roy RN (2008) Guide to laboratory establishment for plant nutrient analysis. FAO fertilizer and plant nutrition bulletin 19

Mowo JG, Floor J, Kaihura FBS, Magoggo JP (1993) Review of fertilizer recommendations in Tanzania: Part 2, Revised fertilizer recommendations for Tanzania. Soil Fertility Report No. 6. National Soil Service, ARI–Mlingano, Tanga, Tanzania

Nandini S, Chellamuthu V (2004) Relative performance of cotton cultivars under sole and intercropping situations in the coastal region of Karaikal. In: Proceeding of International Symposium on Strategies for Sustainable Cotton Production-A Global Vision 2. Universty of Agricultural Sciences, Karnataka, India. pp 235–237

Nelson L, Giles J, MacIlwain C, Gewin V (2004) Organic frequently asked questions (FAQs). Nature 428:796–798

Nelson DW, Sommers LE (1982) Total carbon, organic carbon, and organic matter. In: Page AL, Miller RH, Keeney DR, Baker DE, Roscoe E, Ellis J, Rhodes JD (eds) Methods of soil analysis. Part 2. Chemical and Microbiological Properties. 2nd edn. Agronomy Monograph 9.2. ASA and SSSA, Madison, WI. pp 539–581

OECD (2008) Consensus Document on the Biology of Cotton *(Gossypium spp.)*. OECD Series on Harmonisation of Regulatory Oversight in Biotechnology Paris, OECD. ENV/JM/MONO(2008)38 No. 45: 64

Okalebo JR, Gathua KW, Woomer PL (2002) Laboratory methods of soil and plant analysis a working manual, 2nd edn. TSBF-CIAT and SACRED Africa, Nairobi

Olsen SR, Sommers LE (1982) Phosphorous. In: Page AL, Miller RH, Keeney DR, Baker DE, Roscoe E, Ellis J, Rhodes JD (eds) Methods of soil analysis. Part 2. Chemical and

Microbiological Properties, 2nd edn. Agronomy Monograph 9.2. ASA and SSSA, Madison, WI. pp 403–430
Piepho H-P (2004) An algorithm for a letter-based representation of all-pairwise comparisons. J Comput Graph Stat 13(2):456–466
Pimentel D, Hepperly P, Hanson J, Douds D, Seidel R (2005) Environmental, energetic, and economic comparisons of organic and conventional farming systems. AIBS Bull 55(7):573–582
Ponisio LC, M'Gonigle LK, Mace KC, Palomino J, de Valpine P, Kremen C (2015) Diversification practices reduce organic to conventional yield gap. Proc R Soc B 282(1799):20141396
Prasad M, Siddique MRB (2004) Effect of nitrogen and mepiquat chloride on yield and quality of upland cotton (Gossypium hirsutum). Indian J Agric Sci 74:560–562
R Core Team (2019) R: A language and environment for statistical computing. R Foundation for Statistical Computing, Vienna, Austria. https://www.R-project.org/
Rao PJ, Prasad PRK, Kumari AL, Rani PP, Rao CP (2017) Long-term effect of manures and fertilizers on nutrient status under cotton mono-cropping in vertisol. Int J Curr Microbiol Appl Sci 6(7):2084–2094
Reddy KR, Wetzel RG, Kadlec RH (2005) Biogeochemistry of phosphorus in wetlands. In: Sims JT, Sharpley AN (eds) Phosphorus: agriculture and the environment. Agron. Monogr. 46. ASA, CSSA, and SSSA, Madison, WI, pp 263–316
Reddy PRR, Shaik M (2009) Evaluation of cotton (Gossypium hirsutum)-based intercropping system through different approaches under rainfed conditions. Indian J of Agric Sci 79(3):210-214
Richards M, Metzel R, Chirinda N, Ly P, Nyamadzawo G, Vu QD, De Neergaard A, Oelofse M, Wollenberg E, Keller E, Malin D (2016) Limits of agricultural greenhouse gas calculators to predict soil N_2O and CH_4 fluxes in tropical agriculture. Sci Rep 6:26279. https://doi.org/10.1038/srep26279
Rosenstock TS, Rufino MC, Wollenberg E (2013) Toward a protocol for quantifying the greenhouse gas balance and identifying mitigation options in smallholder farming systems. Environ Res Lett 8:021003. https://doi.org/10.1088/1748-9326/8/2/021003
Sankaranarayanan K, Nalayini P, Praharaj CS (2012) Multi-tier cropping system to enhance resource utilization, profitability and sustainability of Bt cotton (*Gossypium hirsutum*) production system. Indian J Agric Sci 82(12):1044–1050
Searle SR, Speed FM, Milliken GA (1980) Population marginal means in the linear model: an alternative to least squares means. Am Stat 34(4):216–221
Seufert V, Ramankutty N, Foley JA (2012) Comparing the yields of organic and conventional agriculture. Nature 485(7393):229
Singh G, Choudhary P, Saini D, Lal Jat B (2017) Effects of legume intercrops on growth, yield and economics of hybrid american cotton under controlled condition. Glob J Bio Sci Biotechnol 6(1):61–82
Soil Survey Staff (2014) Keys to soil taxonomy, 12th edn. United States Department of Agriculture. NRCS, Washington DC, p 360
Subbalakshmi L, Mhukrishnanan P, Jeyaraman S (2012) Neem products and their agricultural applications. J Biopest 5:72–76
Suja G, Byju G, Jyothi AN, Veena SS, Sreekumar J (2017) Yield, quality and soil health under organic vs conventional farming in taro. Sci Hortic 218:334–343
TCB (2010) The cotton industry implementation plan, CIIP: 2010–2015. Tanzania Cotton Board, Dar es Salaam
TCB (2016) Corporate strategic plan 2016/17–2020/2021. Tanzania Cotton Board, Dar es Salaam
TCB (2019) Area under cotton and seed cotton production statistics for 2009/10–2018/19. Document sent by the Tanzania Cotton Board to Michael R. Baha on 7 Aug 2019
Textile exchange (2018) Organic cotton market report 2018. https://textileexchange.org/wp-content/uploads/2018/11/2018-Organic-Cotton-Market-Report.pdf. Site visited on 24 July 2019
Thilakarathna MS, McElroy MS, Chapagain T, Papadopoulos YA, Raizada MN (2016) Belowground nitrogen transfer from legumes to non-legumes under managed herbaceous cropping systems. A review. Agron Sustain Dev 36:58

Thomas GW (1982) Exchangeable cations. In: Page AL, Miller RH, Keeney DR, Baker DE, Roscoe E, Ellis J, Rhodes JD (eds) Methods of soil analysis. Part 2. Chemical and Microbiological Properties, 2nd edn. Agronomy Monograph 9.2. ASA and SSSA, Madison, WI. pp 159–166

Trewavas AJ (2001) The population/biodiversity paradox. Agricultural efficiency to save wilderness. Plant Physiol 125:174–179

Tuomisto HL, Hodge ID, Riordan P, Macdonald DW (2012) Does organic farming reduce environmental impacts? – a meta-analysis of European research. J Environ Manag 112:309–320

URT (2017) Simiyu investment guide. United Republic of Tanzania (URT). Retrieved from http://esrf.or.tz/docs/guide-simiyu.pdf on 11 Aug 2019

Wickham H (2016) ggplot2: elegant graphics for data analysis. Springer, New York

In Search of Climate-Smart Feeds: The Potential of Pearl Millet (*Pennisetum glaucum*, L.) to Replace Maize as an Energy Feed Ingredient in Broiler Diets in Malawi

Andrews C. L. Safalaoh and Edith Kavala

Abstract Feed is an important element in poultry production. To ensure sustainability, there is a need to explore nontraditional climate-resilient feed ingredients in poultry diets. This is necessary because of the increasing demand for key ingredients such as maize which is also a staple food crop and due to low supplies attributed to several factors including climate change effects. A study was conducted to evaluate the effects of replacing maize with varying inclusion levels of pearl millet on growth performance of broilers, carcass weight, and cost-effectiveness. A total of 260-day-old chicks were allocated to four treatments comprising pearl millet inclusion levels of 0%, 10%, 20%, and 100% in isocaloric and iso-nitrogenous starter and finisher diets. Each treatment had five replicates of 13 broilers each, in a completely randomized design (CRD), for 56 days. Data on feed intake and body weights were collected at weekly intervals. At the end of the experiment, five broilers from each treatment were randomly selected and evaluated for carcass weights and digestibility of diets, respectively. The results showed that there were no significant differences ($P < 0.05$) in mean final body weights, overall weight gain, carcass weight, and digestibility in all the treatment diets except for feed conversion ratio. The highest gross margin was obtained from the 0% pearl millet inclusion level, followed by 10% and 20% pearl millet inclusion levels. Use of pearl millet as the sole energy feed ingredient gave negative returns. It is concluded that replacing maize with pearl millet from 10% to 20% inclusion level as percent of the diet has no detrimental effect on broiler growth performance. The implication of these findings is that millet, which is more drought tolerant than maize, can potentially reduce the sole dependency on maize as an energy - feed source in poultry diets. Use of pearl millet can therefore contribute to adaptive capacity of broiler producers and feed compounders during droughts when maize supply is low.

A. C. L. Safalaoh (✉) · E. Kavala
Animal Science Department, Lilongwe University of Agriculture and Natural Resources (LUANAR), Lilongwe, Malawi
e-mail: asafalaoh@luanar.ac.mw

B. R. Singh et al. (eds.), *Climate Impacts on Agricultural and Natural Resource Sustainability in Africa*, https://doi.org/10.1007/978-3-030-37537-9_11

1 Introduction

Feed is an important element in livestock production, including poultry production. Many factors undermine food availability. They include the increasing livestock population due to increased demand for animal source proteins and climate-related disasters such as droughts and floods. Rojas-Downing et al. (2017) reported that climate change, which impacts cereal production such as maize (which is also used as livestock (or poultry) feed ingredient), poses a threat to food security due to competition between humans and animals/poultry. At the global level, livestock population is growing very rapidly. For example, Food and Agriculture Organization (FAO 2018) reported increases of 29% and 6% for chickens and cattle between 2006 and 2016. Malawi Government (2016) reported that poultry population increased by 24% between 2013 (28,542,805) and 2016 (35,439,568). These statistics imply that there is a need for a corresponding increase in food supplies. This is so because in livestock, including poultry (broiler or layer) production, feed is the most vital element as it can comprise up to 60–80% cost of production (Ibitoye et al. 2012; Panda et al. 2010). The issue of feed and exploration of alternative feedstuffs is of particular importance in countries which depend on grains for both human food and animal feed.

In Malawi, maize is the staple food grain and crop used as an energy source food in human diets. However, maize is also used as the major energy feed ingredient in poultry diets. But in recent years, production of maize has been jeopardized by the negative effects of climate change such as prolonged droughts and erratic rainfall. For example, FEWS NET (2017) reported that while some districts in central and southern Malawi reported an increase in maize crop production, maize production in the Lower Shire Valley for the 2017/2018 season was 36% lower than the 5-year average due to dry spells and infestation by the fall armyworm.

In this chapter, it is argued that the competition with humans and the resultant high costs can result in the formulation of expensive rations, thereby reducing profitability of the poultry enterprise. Hence, there is a need to find locally available alternative feedstuffs as replacements for maize to reduce the food-feed competition. It is further argued that because of climate change, there is a need to explore other climate-smart alternative energy feed ingredients to maize that are not only resilient to climate change but are also nutritious (Batonon-Alavo et al. 2015). Hence, there is a need for crop diversification as an adaptation measure to ward off negative impacts of climate change, with respect to food and feed supply (Rojas-Downing et al. 2017).

1.1 Pearl Millet as an Alternative Energy Feed Ingredient

One grain which is underutilized as an energy source in poultry diets in Malawi is pearl millet (*Pennisetum glaucum*). Several attributes of millet make it a suitable alternative to maize. While millet is widely grown as a major source of carbohydrates and proteins in the semiarid tropics of Africa and Asia (Saleh et al. 2013), it is not the major staple food crop in Malawi, apart from few areas such as the Lower Shire Valley. As such, it does not prompt competition between poultry and humans, who largely depend on and prefer maize to millet as an energy source food. Other uses of millet include brewing beverages such as sweet beer. Despite its availability in Malawi, millet is not used as an energy feed source when formulating and compounding poultry diets for the Malawi poultry industry. Globally, it has been reported that apart from Asia, very little of millet is used as feed, especially in Africa (Léder 2004).

In terms of agronomy, pearl millet grows well in areas characterized by low rainfall, and it is known for having the highest yield potential of all available types of millets under drought and heat stress conditions. In general, millets can grow in areas with seasonal rainfall of as little as 300 mm or less, while the minimum water requirement for maize is 500–600 mm (ibid.) in areas with erratic rains (Baurhoo et al. 2011a, b). In addition to being more drought and disease resistant than maize, pearl millet also grows well in soil that is acidic and low in natural fertility (Saleh et al. 2013; Léder 2004; Davis et al. 2003). Rao et al. (2004) reported that pearl millet could totally replace yellow maize in broiler diets without compromising performance. Under circumstances of low maize supply such as during droughts, the implication is that pearl millet could reduce dependency of broiler producers and feed compounders on the use of maize as an energy feed ingredient in broiler diets. These attributes make millet a potential "climate-smart" feed ingredient that requires further exploration.

In general, millet has a superior nutritional quality to maize grain. For example, Davis et al. (2003) reported that millet had a comparable true metabolizable energy (TME) value (3300–3448 kcal/kg) and a higher protein content (12–14%) than maize. The protein, essential amino acid, linoleic acid, and oil content of millet are also higher in millet than maize. Léder (2004) reported that pearl millet had a lysine content which was 21% greater than that of maize and 36% greater than sorghum. Inclusion of millet to up to 50% in broiler diets has been shown to result in growth performance and carcass yield of broilers that were equal or better than the broilers that were fed typical maize–soybean diets (ibid.). However, others have reported that lower inclusion levels of 5% and 10% whole millet in broiler diets gave similar results on performance and carcass yields of broilers compared to those fed maize-soybean-based diets (Batonon-Alavo et al. 2015; Hidalgo et al. 2004). Other studies (Ibitoye et al. 2012) have reported that the relative growth performance of broilers fed a pearl millet-based diet was higher than that of broilers fed a maize-based diet.

However, despite its potential use as a feed ingredient, millet should be used with some caution as it contains anti-nutritional factors such as tannins, phytic acid, and polyphenols which may limit energy utilization. The content of these anti-nutritional factors is lower compared to other cereal grains (Choct 2006). In addition, the concentration of anti-nutritional factors in pearl millet also depends on the variety or type of millet used due to genetic and agronomic factors.

In the preceding sections, it has been shown that pearl millet can be a potential replacement for maize as an energy feed ingredient in broiler diets. However, in Malawi, there is no information regarding the use of millet as an energy feed ingredient in poultry diets. This information is particularly necessary because Malawi is increasingly experiencing dry spells and severe droughts resulting in low production of maize, which is the staple food crop and is also used as a feed ingredient, especially in monogastric animal diets. It is argued that in order to increase its adaptive capacity, Malawi requires to diversify and expand its feed ingredient resource base. Alternative ingredients such as millet could also open retail and value adding avenues for millet growers and feed compounders respectively.

This study therefore set out to evaluate the impact on the growth performance of broilers if maize was replaced with pearl millet as an energy feed ingredient. Specifically, this study aimed to evaluate the nutrient composition of pearl millet, assess the effects of varying inclusion levels of pearl millet on broiler growth performance and carcass weights of broilers, and assess the effect on gross margins. This study hypothesized that there are no significant differences in growth performance and carcass weights of broilers when fed a millet-based diet or a maize-based diet.

2 Materials and Methods

Location: The experiment was conducted at the Bunda College Campus Animal Science Student's Farm at the Lilongwe University of Agriculture and Natural Resources.

2.1 Feed Ingredients and Chemical Analysis

All feed ingredients used in formulating the rations were purchased locally in Lilongwe, with the exception of the local variety pearl millet grain which was purchased in the Lower Shire Valley District of Chikwawa. Soybean was roasted on open fire before milling. All the three ingredients (maize, soybean, and millet) were then ground in a hammer mill to enable them to pass through a 2-mm sieve. Mixing of rations was done manually and repeated several times to ensure homogenous mixing of ingredients. Samples from each feed ingredient and rations formulated were collected and analyzed for dry matter, nitrogen, ether extracts, crude fiber, and ash using AOAC (2012) procedures. Gross energy was measured using a bomb calorimeter (ibid.). Diets were formulated to be iso-nitrogenous and isocaloric.

2.2 Dietary Treatments

Dietary treatments were prepared to contain 22% CP in starter and 18% in finisher rations. The treatments were as follows:

1. Treatment 1 (T1): Maize as sole energy source with no pearl millet
2. Treatment 2 (T2): 10% of the diet as pearl millet
3. Treatment 3 (T3): 20% of the diet as pearl millet
4. Treatment 4 (T4): 0 maize – 100% pearl millet as energy source

Before feed formulation, samples of maize, pearl millet, and soybean meal were analyzed in the laboratory for proximate composition analysis (AOAC 2012) as shown in Table 1. The results of the analysis were then used to formulate starter and finisher diets as presented in Table 2.

The proportion of ingredients in starter and finisher rations is presented in Table 2.

2.3 *Broiler Management and Experimental Design*

A total of 260-day-old Cobb broiler chicks were raised up to 56 days. On day 1, broilers were individually weighed and assigned to each of the four treatments with five replicates of 13 broilers each. The pens and equipment were disinfected 2 weeks before chick arrival. Litter comprised rice husks on which were placed one drinker and one feeder per pen. Artificial lighting of a 24-hr period was provided during the three first weeks of the brooding stage. Thereafter, the broilers were kept under natural conditions typical of smallholder poultry farmers in Malawi.

Table 1 Nutrient composition of the feed ingredients: pearl millet, maize, and soya bean (%)

Ingredient	DM	CP	ASH	CF	EE	Ca	GE (kcal/kg DM)
Pearl millet	91.99	13.3	4.2	7.6	7.31	0.07	3810.6
Maize	89.71	9.38	2.24	7.1	4.5	0.03	3765.2
Soya bean	89.99	35.05	4.45	4.6	19.9	0.37	4213.12

DM Dry matter, *CP* Crude protein, *CF* Crude fiber, *EE* Ether extract/crude fat, *Ca* Calcium, *GE* Gross energy

Table 2 Proportion of ingredients in starter and finisher diets of broilers as percentage of diet

Ingredients	Starter	Finisher	Starter	Finisher	Starter	Finisher	Starter	Finisher
Maize	43.4	59.61	34.86	51.07	26.32	42.53	0	0
Soya bean	51.15	35.40	49.69	33.94	48.23	32.48	43.72	25.2
Pearl millet	0	0	10	10	20	20	50.82	69.80
Salt	0.3	0.3	0.3	0.3	0.3	0.32	0.3	0.3
VMP	0.3	0.3	0.3	0.3	0.3	0.3	0.3	0.3
MCP	1.5	1.5	1.5	1.5	1.5	1.5	1.5	1.5
Methionine	0.77	0.56	0.77	0.56	0.77	0.56	0.77	0.56
Lysine	1	0.75	1	0.75	1	0.75	1	0.75
Lime	1.58	1.58	1.58	1.58	1.58	1.58	1.58	1.58
Total	**100**	**100**	**100**	**100**	**100**	**100**	**100**	**100**

T1 = 0%; T2 = 10%; T3 = 20% pearl millet and T4 = 0% maize
VMP Vitamin-mineral premix, *MCP* Monocalcium phosphate

From 3 weeks to the end of the experiment, the initial temperature in the brooder house was maintained at 32 degrees and gradually reduced to 21 degrees and monitored using a thermometer. Heat was provided using electricity light bulbs, supplemented with charcoal burner as necessary due to frequent electricity blackouts. Broilers were vaccinated against infectious bursal disease (Gumboro) on 14th and 35th day and against Newcastle disease (HB1) on seventh and 28th day through drinking water. Water was provided ad libitum.

Determination of Feed Intake, Body and Carcass Weight, and Mortality Rates

Measured quantities of feed were given on ad libitum basis to the broilers. Feed intake was determined by subtracting the amount of feed refusal collected from the amount of feed offered. Weekly body weights were taken individually for computation of weight gains. In week 7, five broilers per treatment were randomly selected and slaughtered for determination of carcass weights. Records on mortality were taken as they arose and feed intake was adjusted accordingly to reflect number of pens per pen.

Digestibility

Five broilers per treatment were randomly selected for a digestibility trial for three consecutive days at 6 weeks of age. The broilers were put on off feed for 24 hours but had access to water before commencement of the digestibility trial in order to empty the intestines. Feed intake was determined every day by subtracting the amount of leftovers collected from the amount of feed offered. Total fecal output was obtained on a daily basis from individual broilers, weighed, and stored in plastic bags until the end of the trial.

Ethical Considerations

All broilers were handled with proper care following recommended animal welfare ethics. Facilities including pens were accurately designed, constructed, equipped, and maintained to ensure that broilers were raised in comfort without undue stress and form of cruelty. Appropriate procedures were used in disease control, such as vaccination, isolation, and quarantine. Broilers were fed according to the minimum requirements (National Research Council 1994).

2.4 Data Analysis

The data collected was analyzed as completely randomized design following the procedures by Genstat. The data was subjected to one-way analysis of variance (ANOVA). Where the analysis of variance indicated the existence of significant differences among treatment means, Fisher's protected least significant difference (LSD) test was used to test and locate the treatment means that were significantly different from each other. The model used was as follows:

$$Y_{ij} = \mu + \tau_i + \varepsilon_{ij}$$

where

Y_{ij} denotes the growth response (broiler performance)
μ is the overall mean
τ_i is the effect of the *i*th replacement level (diet) or treatment ($i = 1, 2, 3, 4$)
ε_{ij} is the random and independent errors

3 Results

3.1 Nutrient Composition of Pearl Millet and Maize

The results on nutrient composition of pearl millet, maize, and soybean were presented in Table 1. The results showed that pearl millet was superior to maize in terms of crude protein, calcium, ether extract (fat), energy content, and crude fiber.

3.2 Effects of Different Pearl Millet Inclusion Levels on Growth Performance of Broilers

Table 3 shows the effects of pearl millet inclusion levels on performance of broilers from 1 to 8 weeks of age. The results show that there were no significant differences ($P < 0.05$) on the feed intake and weight gain of broilers fed either pearl millet or maize-based diet. However, the body weight gain of broilers fed the diet at 10% pearl millet inclusion level was higher than that of the other three treatments. The feed conversion ratios (FCR) for 10% and 20% inclusion levels were better than broilers fed diets where only maize (T1) or pearl millet (T4) was the sole energy feed ingredient used in the basal diet.

3.3 Effect of Different Pearl Millet Inclusion Levels on Carcass Characteristics of Broilers

The results in Table 4 show that different pearl millet inclusion levels had no significant ($P < 0.05$) effect on carcass characteristics of broilers. However, the relative values of carcass weight and dressing percentage were higher for the broilers fed 10% pearl millet-based diets.

Table 3 Effect of different pearl millet inclusion levels on growth performance of broilers

Parameter	T1	T2	T3	T4	SEM	*P*-value
Weight gain, g	2095	2323	2257	2154	82.2	0.066
Initial weight, g	46.75	47.50	46.53	48.22	1.018	0.375
Final weight, g	2142	2372	2304	2203	82.6	0.067
FCR, g:g	2.16^{a}	2.05^{b}	2.08^{b}	2.18^{a}	0.0008	0.001
ADG, g	37.41	41.51	40.31	38.47	1.468	0.066
DFI, g	80.84^{b}	85.2^{a}	83.68^{a}	83.69^{a}	2.288	0.001

$^{a–c}$Means that do not share same superscript within same rows are significantly different ($P < 0.05$)
FCR Feed conversion ratio, *ADG* Average daily gain, *DFI* Daily feed intake, T1 = 0%, T2 = 10%, T3 = 20% pearl millet, T4 = 0% maize

Table 4 Effect of different pearl millet inclusion levels on carcass characteristics

Parameter	T1	T2	T3	T4	SEM	*P*-value
Live weight, g	2116	2115	2088	2089	142.6	0.995
Carcass weight, g	1537	1594	1519	1528	129.4	0.942
Dressing percentage, %	72.51	75.96	73.03	73.43	2.744	0.613

$^{a–c}$Means that do not share same superscript within same rows are significantly different ($P < 0.05$)
T1 = 0%; T2 = 10%; T3 = 20% pearl millet; T4 = 0% maize

3.4 Effect of Different Pearl Millet Inclusion Levels on Feed Digestibility

Digestibility results showed that the diet with pearl millet as the sole energy source has significantly ($P < 0.001$) lower digestibility (64.73%) than the rest of the other treatments which were similar (0% pearl millet = 82.34%; 10% pearl millet = 81.83%; and 20% pearl millet = 81.5%). In general, digestibility tended to reduce with increasing inclusion levels of pearl millet.

3.5 Effect of Pearl Millet on Productivity Cost[1]

Taking variable costs into consideration, the average cost per broiler was MK1492.5 ($2.06), MK1512.20 ($2.09), MK1531.90 ($2.11), and MK1619.70 ($2.23) for treatments 1, 2, 3, and 4, respectively. Cost per broiler tended to increase with increasing levels of pearl millet used. Using an average selling price of MK2800 ($3.86) per whole carcass as farm gate price, the results show that the sole maize-based diet (T1) had the highest gross margin.

[1] 1 US $ = MK725

4 Discussion

4.1 Comparative Nutrient Composition of Pearl Millet and Maize

Results on nutrient composition of maize, pearl millet, and soybean meal used in this experiment fall within ranges reported in literature. For example, McDonald et al. (2011) reported that crude protein contents of maize and pearl millet were 9.8% and 12.1%, respectively. In this study, the crude protein content of maize was 9.38%, while that of pearl millet was 13.13%, which was higher than that reported by McDonald et al. (2011). However, Baurhoo et al. (2011a) reported a crude protein content of 14.48% for Canadian pearl millet which is higher than the crude protein content reported in this study. These differences could be attributed to the variety of millet used as a function of genetic differences. Energy content of pearl millet was also higher than that of maize. More interesting is the observation that the crude protein of pearl millet is higher than that of maize. This implies that apart from being a source of energy, pearl millet can contribute more to the dietary crude protein than maize. These results are in tandem with those reported in other studies where pearl millet was reported to contain higher amounts of crude protein and energy than maize (Saleh et al. 2013; Davis et al. 2003). The higher amount of energy in pearl millet than maize could be attributed to the higher fat content observed where pearl millet had 7.31% ether extract compared to 7.1% for maize. With these attributes on nutrient composition, some have argued that pearl millet can totally replace maize in broiler diets (Rao et al. 2004).

However, one limitation of using pearl millet could be its high crude fiber content which was higher than that of maize. This high crude fiber content may have negatively contributed to the low digestibility of the diet which had millet as the sole energy source, as reported in other studies (McDonald et al. 2011; Davis et al. 2003). On the other hand, Baurhoo et al. (2011b) reported no differences in apparent ileal digestibility of broilers fed maize versus pearl millet-based diets. These differences could be attributed to varieties used or how the grains were processed.

4.2 Effects of Different Pearl Millet Inclusion Levels on Growth Performance and Carcass Yields of Broilers

This study was about assessing the potential of including and using the local pearl millet variety as an energy source feed ingredient in broiler diets to spare maize for human consumption in Malawi. The results have shown that pearl millet can successfully be included in broiler diets within the range of 10–20% as percentage of the total diet, without negatively affecting growth and carcass yields. Earlier studies have reported inconsistent results. For example, Hidalgo et al. (2004) reported that pearl millet could be included at lower levels of

5–10%, while inclusion levels of up to 50% of the diet were reported in other studies (Léder 2004) without compromising growth performance and carcass weights (Davis et al. 2003). However, in this study, the diet with 50.82% pearl millet in starter diets and 69.8% in finisher diets had the lowest growth performance and carcass weights. This difference could be due to types and varieties of millet used.

In this study, the high growth performance for 10% and 20% pearl millet-based diets were also associated with improved feed conversion ratio and higher average weight gain than the other two treatments as has been reported in earlier studies (Baurhoo et al. 2011a, b). Positive results have also been reported by Batonon-Alavo et al. 2015; Ibitoye et al. 2012).

4.3 Effects of Different Pearl Millet Inclusion Levels on Cost of Production

Reducing costs and hence increasing profits is an important element of any profit-making business enterprise. In poultry, feed comprises the major cost in a broiler enterprise. Results showed that lowest cost of feed per broiler was on broilers fed the maize-based diet, followed by the 10% pearl millet-based diet. From the results, cost tended to increase with increasing levels of pearl millet inclusion. This was so because the cost of millet was higher than that of maize. Key to this finding is that while pearl millet can be included in broiler diets within the range of 10–20% of the diet, the element of costs should be seriously considered as it affects profitability of the enterprise. In this study, the higher cost per broiler for pearl millet based-diets was increased due to transport costs. These results are at variance with those of Medugu et al. (2010) who reported that the cost of feed per kilogram of live weight was lower for pearl millet-based diets than maize-based diets.

One other interesting observation in this study was that increasing levels of pearl millet tended to lower the amount of soya bean meal used. For example, the maize–based starter diet had 51.15% soya bean meal while the 10% and 20% pearl millet-based diets had 49.69% and 48.23% soya bean meal respectively. This is probably due to the pearl millet, which contributed more crude protein (13.13%) to the diet than maize with crude protein of 9.38%. Ceteris paribus, this could lower costs since the price of soybean meal was MK250 ($0.34)[2] per kilogram while millet was less than half of this price. The implication of this finding is that under ideal conditions, pearl millet can reduce costs associated with protein source feed ingredients, which are generally more expensive than energy source feed ingredients.

[2] 1 US $ = MK725

5 Conclusion

Based on the findings of this study, it is concluded that the local variety of pearl millet grown in Malawi can successfully replace maize in broiler diets at inclusion levels of up to 20% of the diet without compromising growth performance, carcass weights, dressing percentage, and feed conversion ratio. However, issues of costs should be taken into consideration when using pearl millet as an energy feed ingredient. Considering that Malawi's supply of maize is not reliable due to the changing weather patterns and competition with humans, pearl millet, which is less popular as human food and is more drought tolerant, offers an alternative energy feed ingredient for commercial feed compounders and provides a value addition pathway for farmers. Further studies are recommended to validate the results of this study, compare different varieties of millet used and available in Malawi and processing methods, and assess levels of anti-nutritional factors in different millet types.

Acknowledgments This study was supported through the "Towards 'Feed' Security and Climate Smart Feed Ingredients: Assessment of sorghum and millet as replacement for maize in poultry diets" Project with funding from the Directorate of Research and Outreach at Lilongwe University of Agriculture and Natural Resources (LUANAR) for which we are very grateful. The support of Ms. Thokozani Holla in data collection is also gratefully acknowledged.

References

AOAC (Association of official Analytical Chemists) (2012) In: 19th (ed) Official methods of analysis of AOAC international. AOAC, Washington, DC

Batonon-Alavo DI, Faruk MU, Lescoat P, Weber GM, Bastianelli D (2015) Inclusion of sorghum, millet and cottonseed meal in broiler diets: a meta-analysis of effects on performance. Animal 9(7):1120–1130

Baurhoo N, Baurhoo B, Mustafa AF, Zhao X (2011a) Comparison of corn-based and Canadian pearl millet based diets on performance, digestibility, villus morphology, and digestive microbial populations in broiler chickens. Poult Sci 90(3):579–586

Baurhoo N, Baurhoo B, Zhao X (2011b) Effects of exogenous enzymes in corn-based and Canadian pearl millet-based diets with reduced soybean meal on growth performance, intestinal nutrient digestibility, villus development, and selected microbial populations in broiler chickens. J Anim Sci 89(12):4100–4108

Choct M (2006) Enzymes for feed industry: past, present and future. Worlds Poult Sci J 62:5–15

Davis AJ, Dale NM, Ferreira FJ (2003) Pearl millet as an alternative feed ingredient in broiler diets. J Appl Poult Res 12(2):137–144

FAO (2018) World Food and Agriculture – Statistical Pocketbook 2018. Rome. 254 pp. Licence: CC BY-NC-SA 3.0.INGO available at http://www.fao.org/documents/card/en/c/CA1796EN. Accessed 20 June 2019

FEWS NET (2017) Malawi Food Security Outlook. https://fews.net/sites/default/files/documents/reports/MW_FSO_2017_06.pdf. Accessed Sept 2018

Hidalgo MA, Davis AJ, Dale NM, Dozier WA III (2004) Use of whole pearl millet in broiler diets. J Appl Poult Res 13(2):229–234

Ibitoye EB, Olorede BR, Jimoh AA, Abubakar H (2012) Comparative performance and organ relative weight of broiler chicken fed three sources of energy diet. J Anim Prod Adv 2(5):233–238

Léder I (2004) Sorghum and millets. In: Cultivated plants, primarily as food sources, 1st edn. EOLSS Publishers, Oxford, pp 66–84
Malawi Government (2016) Livestock population estimates. Department of Animal Health and Livestock Development (DAHLD), Lilongwe, Malawi
McDonald P, Edwards RA, Greenhalgh JFD, Morgan CA, Sinclaire LA, Wilkinson RG (2011) Animal nutrition. Prentice Hall, Pearson, Essex
Medugu CI, Kwari ID, Igwebuike J, Nkama I, Mohammed ID (2010) Performance and economics of production of broiler chickens fed sorghum or millet as replacement for maize in the semi-arid zone of Nigeria. Agric Biol J N Am 11:445–450
National Research Council (1994) Nutrient requirements of poultry. National Research Council, National Academy of Sciences, USA
Panda AK, Raju MVLN, Rama SVR, Lavanya G, Reddy EPK, Sunder GS (2010) Replacement of normal maize with quality protein maize on performance, immune response and carcass characteristics of broiler chicken. J Anim Sci 23(12):1626–1631
Rao SV, Raju MVLN, Reddy MR, Panda AK (2004) Replacement of yellow maize with pearl millet (Pennisetum typhoides), foxtail millet (Setaria italica) or finger millet (Eleusine coracana) in broiler chicken diets containing supplemental enzymes. Asian Australas J Anim Sci 17(6):836–842
Rojas-Downing MM, Nejadhashemi AP, Harrigan T, Woznicki SA (2017) Climate change and livestock: impacts, adaptation, and mitigation. Clim Risk Manag 16:145–163
Saleh AS, Zhang Q, Chen J, Shen Q (2013) Millet grains: nutritional quality, processing, and potential health benefits. Compr Rev Food Sci Food Saf 12(3):281–295

Climate Change and Weather Variability Effects on Cattle Production: Perception of Cattle Keepers in Chikwawa, Malawi

Janet Nanganga and Andrews C. L. Safalaoh

Abstract Climate change and associated weather variability are some of the factors that negatively affect agriculture production in the world. Despite the fact that livestock production is key livelihood and resilience asset of smallholder farmers, more emphasis is placed on the effects of climate change and weather variability on crop production than on livestock production. In Malawi, cattle act as a moving bank and are sold as a coping strategy to mitigate negative effects of climate change and weather variability such as prolonged droughts. However, there is paucity of information on smallholder cattle keepers' understanding and perception of the effects of climate change and weather variability on cattle production. The aim of this study was to assess the perceived direct and indirect effects of climate change and weather variability on cattle production. Data was collected through interviews using semi-structured questionnaires administered to 100 purposively selected cattle keepers in Ngabu, Chikwawa. Four focus group discussions were also held. Majority (99%) of the respondents indicated that they had some knowledge about the meaning of climate change and weather variability indicators or events such as drought (25.8%), erratic rainfall (22.7%), floods (22.7%), high temperature (18.6%) and strong winds (10.3%). Effects of weather variability were identified as including increased disease incidences (38.4%), heat stress (31.3%), drying up of water sources and reduced water unavailability (32.3%) and reduction in supply of fodder (49.5%). Noteworthy, 88% of the respondents mentioned that they were aware of the causes of weather variability including environmental degradation. The study shows that most cattle keepers understand weather variability and its causes and effects such as feed and water unavailability, which negatively impinge on cattle productivity. It recommends that farmers should adopt adaptation and mitigation measures such as growing pasture in wetlands, fodder preservation and water harvesting to ensure sustainable cattle production amid climate change.

J. Nanganga · A. C. L. Safalaoh (✉)
Animal Science Department, Lilongwe University of Agriculture and Natural Resources, Lilongwe, Malawi
e-mail: asafalaoh@luanar.ac.mw

B. R. Singh et al. (eds.), *Climate Impacts on Agricultural and Natural Resource Sustainability in Africa*, https://doi.org/10.1007/978-3-030-37537-9_12

1 Introduction

Agriculture is the mainstay of economies of most developing countries and mainly comprises crop and livestock production. Livestock production, including cattle, small ruminants and poultry farming, forms an important integral component of agriculture in the world that significantly contributes to food and nutrition security (Godber and Wall 2014). Livestock contributes to food and nutrition security through provision of animal-source foods such as meat, milk and eggs and as source of income and manure (Smith et al. 2013).

For marginalised rural poor smallholders, livestock is a source of income and acts as a four-legged bank account with vital functions, such as paying for household expenses like school fees, social functions, and serves as insurance against shocks like crop failures, accidents and illnesses (Thornton 2010; Njuki et al. 2013; Kristjanson et al. 2014). Livestock is also an important source of manure and income for the poor in agro-based economies like Malawi, which is dominated by crop and livestock production. Not surprisingly, livestock can be considered as the mainstay and key livelihood asset of most rural poor in the world, especially women.

In Malawi, cattle production is one of the agricultural activities practised by both smallholder and large-scale farmers as a source of livelihood (Chingala et al. 2017). Malawi has a population of 1, 470, 895 herds of cattle, supplying about 41,615 tonnes of beef (Malawi Government 2016). Like in other developing countries, cattle production is primarily based on an extensive free range system, with little or no supplementation apart from a few cattle feedlots scattered throughout the country.

In general, agriculture is highly vulnerable to the negative impacts of climate change and weather variability (Vermeulen et al. 2012), with negative implications on agriculture, affecting livestock production systems (Thornton and Gerber 2010). In Africa, negative impacts of climate change are worsened by overreliance on low-input rain-fed agriculture production and the continent's low adaptive capacity (FAO 2010). According to Thornton et al. (2007), there are poor prognosis and misunderstanding of localised impacts of climate change on crop and livestock agriculture. With a proper prognosis and a better understanding, aggressive exploration of mitigation and adaptation measures through which smallholder farmers can be supported to ensure food and nutrition security in the wake of climate change (Vermeulen et al. 2012) and weather variability would be required.

Malawi is not spared from the negative impacts of climate change. According to a report by a floods post-disaster needs assessment conducted in Malawi, it was shown that changes in weather variability were characterised by significant changes in temperature and rainfall patterns and increased frequency and intensity of droughts, floods, strong winds, strong rains, hailstorms, earthquakes, landslides and disease outbreaks (Malawi Government 2015). These extreme events are associated with negative impacts on livestock production, including meat quality (Gregory 2010), and impairment of production, reproductive performance, health status and immune response (Nardone et al. 2010). Other negative impacts attributed to climate change and variability-induced droughts and extreme weather events on livestock production include reduction in productivity of pastoral herds, reduction in water availability, reduced quantity and quality of forage and subsequent reduction in carrying capacity of rangelands (Megersa et al. 2014; Thornton and Herrero 2014), and incidences of livestock deaths in extreme cases (Gaughan and Cawsell-Smith 2015; Megersa et al. 2014). However, a key barrier to developing our understanding of climate-sensitive diseases is the paucity of information such as epidemiological and ecological observations on animal disease in developing countries (Grace et al. 2015).

2 Significance of Livestock and Climate Change Interface

This study is informed by and undertaken under the premise that the contribution of livestock such as cattle to food security and livelihoods in Malawi can be enhanced if adaptation and mitigation to climate change are well appropriated. Porter et al. (2014) indicated that the studies on observed impacts of climate change on livestock are fewer compared to those on crop production and fisheries. Thornton and Herrero (2014) argued that there is a need to better understand how mixed crop-livestock systems are impacted upon by climate change. In Malawi, there is limited information on the influence of socio-economic factors on farmers' perception of the impacts of climate change on smallholder beef production (Chingala et al. 2017). We argue in this chapter that to develop appropriate interventions and/or related policies, there is a need to have a clear and holistic understanding on how smallholder farmers perceive impacts of climate change and weather variability on cattle production. This understanding is particularly important because the effects of climate change mainly impact on the poor who are characterised by low adaptive capacity (Tol et al. 2004).

This study was therefore undertaken to assess the farmers' perception on the effects of weather variability on cattle production in Traditional Authority Ngabu, Chikwawa District, Southern Malawi. Specifically, this study aimed to identify and document the perceived impacts of weather variability on cattle production practices and productivity and to seek solutions and adaptation measures to address identified weather variability-related challenges that negatively impact smallholder cattle production.

3 Methodology

3.1 Location

The study was conducted in Traditional Authority Ngabu in Chikwawa District under the Shire Valley Agricultural Development (SVADD). Ngabu lies on latitude 16°26′71" S and longitude 34°52′9.53″ E. It is in the Shire River valley and is considered to be one of the hottest and driest districts in Malawi, except for the areas along the Shire River. This area was purposively chosen because it is one of the areas that have a high concentration of cattle population in Malawi. According to Malawi Government (2016), the area has a cattle population of 199,138, of which 175,966 (88%) are Malawi Zebu. Chikwawa District lies in a rain-shadow area along the Shire River Basin that is vulnerable to climate-related shocks and with low resilience abilities (Coulibaly et al. 2015). It is increasingly experiencing climate change and weather variability-induced events such as prolonged droughts and flooding (Malawi Government 2015).

3.2 Data Collection

The study used both qualitative and quantitative methods of data collection. A total of 100 cattle keepers were purposively selected from a list of members of the Chikwawa Livestock Association (CLA). The data was collected through face-to-face interviews held in farmers' homes using a pre-tested semi-structured questionnaire. As part of triangulation, additional data was also collected through four focus groups that comprised eight cattle keepers (four females and four males) and through interviews with key informants including the four members of the CLA and five livestock extension workers in the study area. Secondary data was collected through review of published and grey literature. The data collected included household demographics, farmers' understanding of weather variability and perceived weather variability-related impacts on livestock water supply (both quantity and quality), pasture/fodder availability, grazing or feeding systems or patterns, quality and quantity of feeds, incidences of heat stress, livestock vectors and disease prevalence, livestock-based livelihoods and coping or adaptation strategies employed.

3.3 Data Analysis

The data collected was analysed using IBM SPSS version 18 where descriptive statistics such as percentages and frequencies were used for summarising and presenting the results.

4 Results

4.1 Demographic Characteristics of Respondents

Out of the 100 respondents, 69% were male and 31% were female. Majority of the respondents (47%) were aged 33–49 years followed by those aged between 16 and 32 years (23%). The rest were 55–66 years old (16%) or older (14%). Sixty-one percent of the respondents were married, while 19%, 16% and 4% were single, divorced and widowed, respectively. Religion-wise, 71% of the respondents were Christians, and the rest were either Muslims or belonged to other religions.

4.2 Awareness of Weather Variability: Impacts and Causes

The majority (99%) of the farmers indicated that they were aware and had limited understanding of the meaning of the terms climate change and weather variability. In their responses, respondents indicated that climate change and weather variability are associated with and characterised by extreme events such as increasing frequency and intensity of droughts (25.8%), floods (22.7%) and erratic and shorter periods of rainfall (22.7%) as shown in Fig. 1. These were corroborated through the focus group discussions and interviews with key informants who also mentioned that weather variability is also associated with incidences of higher than average temperatures and strong dusty winds.

Several factors can contribute to climate change and weather variability. Interestingly, majority of the respondents (88%) indicated that they as individuals have in one way or another contributed to climate change and weather variability. The rest (12%) of the respondents indicated that they were neither sure of the actual meaning of climate change and weather variability nor did they know if their actions contribute to weather variability or climate change. Factors that contribute to climate change as mentioned by respondents include overgrazing (38%), poor manure disposal and management (34%) and deforestation (16%). These factors were also mentioned during focus group discussions and during interviews with key informants such as extension workers. Although extension workers would be expected to have technical know-how as resource persons for smallholder farmers, some extension workers indicated that they are not very conversant with the climate change and weather variability phenomena.

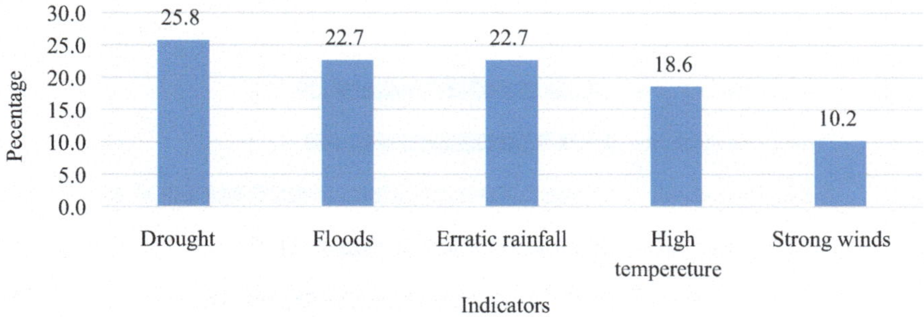

Fig. 1 The indicators of climate change and weather variability as perceived by respondents

4.3 Perceived Effects and Impacts of Climate Change and Weather Variability on Cattle Production

As shown in Fig. 2, almost half of the respondents (49.5%) indicated that the major impact of climate change and weather variability with respect to cattle production is reduced or total fodder unavailability due to extreme droughts and scorching high temperatures. Due to lack of fodder, respondents indicated that they are compelled to move cattle to hilly areas or wetlands where there may be some grass; sometimes they have to move more than 10 kilometres away from original kraals and traditional grazing areas. On the other hand, all respondents indicated that when the rains are very good, pasture or fodder is readily available. Fodder in form of pasture is also readily available in post-flood periods, with grass flourishing due to residual moisture in wetlands. However, both farmers and extension workers reported that the majority of cattle keepers (more than 70%) do not preserve fodder (such as hay or silage) during periods of good rains when grass is plentiful. In addition, it also came out clearly during focus group discussions that smallholder farmers and livestock keepers in general do not practice controlled or planned grazing in the area.

Other negative effects associated with weather variability were mentioned as increases in disease incidences (38.4%), drying up of water supply points and inadequate water supply (32.3%) and high temperatures (31.3%). From interviews with respondents and focus group discussions, it was learnt that inadequate water supply was due to decreased precipitation and unreliable rainfall patterns and periods. High temperatures were also associated with high frequency of wild bushfires which tended to destroy pasture, grazing lands and fodder trees. The bushfires are most often caused by people who go around hunting mice and other wild animals for food.

During focus group discussions, it was mentioned that over time, farmers experience increased cost of production arising from purchasing inputs such as feed and drugs to treat disease and pests. Feed costs included hiring people to cut grass from wetlands (locally known as dambos) or previously flooded areas or grass found along perennial river banks (29.3%).

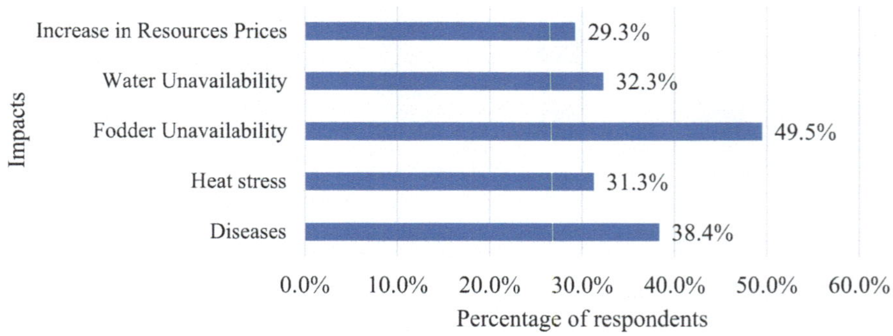

Fig. 2 Perceived effects and impacts of weather variability on cattle production

4.4 Interventions to Address Negative Impacts

Respondents mentioned several interventions that are employed to address the negative impacts of climate change and weather variability as indicated in Fig. 2. In terms of water inadequacy or unavailability, farmers draw water from boreholes or travel long distances in search of water points along previously known perennial rivers, which can be more than 10 kms away. Feed scarcity is addressed by moving cattle to faraway grazing areas found along perennial river banks or wetlands, grazing cattle in protected forests or purchasing grass from vendors (17% of respondents). On a positive note, it was noted that extension workers are now advising and encouraging farmers to adopt feed preserving methods such as hay or silage for use during the dry season or periods of drought; but only few have adopted this technology of feed preservation. The few that practise feed preservation do so during the rainy period of October to January, although the period is also characterised by changing rainfall patterns.

The commonest feeds preserved were hay (77%) and silage (23%), with others using natural perennial pasture grasses such as *Hyparrhenia rufa* (locally known as tsekera or kambumbu grass). Only 7% of the farmers indicated that they deliberately cultivate or grow pasture such as elephant grass (*Pennisetum purpureum*) as feed for their animals. Most of the farmers that grow pasture are feedlot owners that also supplement their animals with molasses in addition to grass they cut or gather along roads. Apart from feed preservation, some farmers also plant trees (24%), construct drainage ditches (17%), construct soil or stone bunds (10%), make terraces (10%) and erect fences (4%) as a way of mitigating effects of weather variability and climate change.

Some farmers do not practice fodder preservation at all. Reasons mentioned for not practising feed preservation include lack of knowledge and difficulty in practising feed preservation (36%); others are satisfied with communal extensive grazing (12%) which requires less labour. The majority (52%) had no specific reason for not practising feed preservation.

4.5 Extension Services and Other Interventions

One would ideally expect farmer advisory services to be readily available for smallholder farmers. However, 24% of the respondents indicated they have never received any government or non-governmental organisation advisory services on what intervention to employ in order to address issues related to weather variability or climate change. But it was encouraging to note that many farmers have had the privilege of being taught or have received support in the form of drilling of water wells and boreholes (24%) for water supply. Some have been taught how to do fodder preservation (15%), while others have received support with early disease outbreak alerts (23%). Sadly, there are no deliberate efforts to build structures specifically for capturing and storing water for livestock during periods of water scarcity.

5 Discussion

5.1 Demographic Characteristics of Respondents

Livestock ownership is a critical and important element in livestock production. In most parts of the developing world, livestock, especially large animals, are owned by men. Njuki et al. (2013) and Kristjanson et al. (2014) argued that livestock ownership, especially women's ownership of livestock, is important for rural households because it contributes to household food security by increasing household dietary diversity and food adequacy. From the results, it is evident that cattle ownership is dominated by males (69%), with only 31% of the respondents suggesting that even women are involved as key players in raising large animals such as cattle. This is against the general myth and arguments that smaller animals such as chickens and goats are the ones suitable for husbandry by women. This finding implies that women should be an integral component of livestock-based interventions, and there is a need to ensure their inclusion in household decision-making processes such as disposal of animals and animal products and general control over livestock assets and income derived therefrom (Njuki et al. 2013). In the context of climate change, poor women experience gender-specific constraints that hinder their ability to cope with and adapt to climate change (Terry 2009).

Coulibaly et al. (2015) reported that women in Chikwawa, Malawi, were the biggest victims in terms of their susceptibility to hazards induced by climate change compared to their male counterparts. In some instances, women have to walk long distances to fetch water for livestock, especially young calves which cannot walk long distances to access water. It was also noted that more than 50% of the respondents were less than 50 years old. This means that development of interventions should also consider all categories of people within the social strata. In this regard, gender- and youth-sensitive approaches are required when developing climate change-based policies and cattle production-related interventions aimed at empowering and enhancing participation of women as much as men. This would help support and promote gender equality in cattle production and to ensure that women benefit from livestock as livelihood assets.

5.2 Knowledge on Issues of Climate Change and Weather Variability

It has been argued that farming, especially crop production, is a risky endeavour for smallholder farmers in the semi-arid tropics, where season-to-season variability and uncertainty in rainfall determine productivity and profitability. As such, farmers tend to make decisions based on their experience and observations (Rao et al. 2011). In this study, it was noted that a considerable proportion of farmers indicated that they had some understanding of the climate change and weather variability phenomena, which are associated with events such as drought, floods, erratic rainfall, high temperature and strong, sometimes violent, winds. This finding agrees with Rao et al. (2011) who indicated that farmers mostly place more emphasis on negative impacts, which leads to high risk perception. An example of such negative impacts is high temperatures, which are associated with heat stress and manifest through heavy panting and sweating. Heat stress negatively affects livestock production and welfare as it causes reduction in feed intake and increases in water consumption, as reported in other studies (Coulibaly et al. 2015; Chingala et al. 2017; Gaughan and Cawsell-Smith 2015; Malawi Government 2015).

However, some respondents were not conversant with the climate change and weather variability phenomena. This implies that there is a need to enhance farmers' knowledge in terms of climate change and weather variability and their effects. In a study conducted in Kenya, Rao et al. (2011) found out that although farmers were aware of the general climate in their location, its variability, the probabilistic nature of the variability and the impacts of this variability on crop production, they had limited ability to synthesise the knowledge they had gained from their observations on climate change issues. It is argued that lack of understanding and attitude of "nonbelievers in climate change" can lead to scepticism in the farmers' attitudes to climate change, and this can ultimately become a significant barrier in their engagement and involvement in adaptation or mitigation interventions to climate change. In order to address the challenges associated with the impacts of weather variability on livestock such as cattle, there is a need for concerted efforts to help livestock keepers increase their adaptive capacity and gain better understanding of issues related to localised impacts of climate change (Thornton et al. 2007). This is particularly important because farmers themselves also acknowledge that their actions and activities, such as overgrazing and poor manure disposal and management, contribute to the negative effects of climate change. It has been reported that 65% of total greenhouse gas emissions by livestock is from beef and dairy cattle, where cattle dung or manure is one of the sources of nitrous oxide (Bailey et al. 2014).

5.3 Fodder and Water Availability and Climate Change

The impact of climate change on livestock production includes effects on availability of feed in the form of grain, pasture and forage crop production and quality (Rust 2018). In this study, overgrazing was mentioned as one of the factors that affect both quantity and quality of feed in terms of pasture. Overgrazing is also exacerbated by lack of planned or controlled grazing, which is communal in the study area. Droughts also affect vegetative growth of grasses and pasture in general, including shrubs. In addition, wild fires triggered by people hunting for mice also reduce feed availability. It has been reported that changes in temperature and rainfall negatively affect grassland productivity and species composition and dynamics, resulting in changes in animal diet and reduced nutrient availability for animals (Izaurralde et al. 2011; McKeon et al. 2009).

Reduced availability of fodder in the communal grazing lands results in a number of animals that can be supported per management unit or area of grazing land (carrying capacity), thereby affecting the grazing system (McKeon et al. 2009), which in turn affects feed intake of animals. With reduction and changes in feed intake, which impinge on nutritional needs of cattle, changes in ration formulation or feeding strategies and approaches may be required to meet the nutrient requirements of cattle. This requires interventions that will enhance the capacity of farmers to properly manage the available feed resources, including the need to adopt fodder preservation in the form of either hay or silage.

Water is a life's necessity for both humans and livestock. The results show that climate change and weather variability also negatively affect water supplies. This is probably due to erratic rains and droughts where the water table goes down, and the rivers, water wells and boreholes get dry. Cattle are therefore moved around in search of water, further increasing their water requirements. Thornton and Herrero (2014) reported that increases in temperature by 20° from 10 to 30 °C can increase water intake by 67% from 3–5 kg per kilogramme dry matter in *Bos indicus* cattle. Where cattle walk for long distances in search of fodder or water further implies that more energy is expended in the process.

5.4 Disease Incidences

There are many factors that contribute to incidences of livestock diseases, including climate change. In this study, respondents indicated that they have observed increased incidences of cattle diseases over the past years, which could be attributed to climate change or weather variability. However, this is merely based on anecdotal evidence. Bett et al. (2017) indicated that there is limited knowledge on the key processes involved in disease causation. This is mainly because of the lack of reliable long-term climate and disease data and information on other mechanisms affecting disease transmission dynamics, such as herd immunity. Aydinalp and Cresser (2008) argued that this could be due to the fact that most livestock diseases, including nematode infestations, are transmitted by vectors whose developmental stages are influenced by climatic conditions such as increased temperatures.

In addition, due to lack of fodder in traditional communal grazing areas, animals are moved from one area to another in search of feed and in the process may come into contact with diseased animals. Grace et al. (2015) reported that human behaviour can affect the way animals are raised. This may be due to vulnerability of animals to pathogenic organisms or exposure to environments with animal health risks to which they have not been previously exposed (Thornton 2010). Climate change–induced flooding may exacerbate diseases spread through water (ibid.). Bett et al. (2017) also reported that climate change in the form of increased air temperatures could be associated with epidemiology of infectious diseases and an increase in exposure to infectious pathogens such as arthropod vectors like *Culicoides imicola*, which transmits the bluetongue virus. Outbreaks of diseases such as Rift Valley fever in East Africa have also been associated with extreme weather events such as droughts (ibid.). The impact of diseases is that they reduce animal productivity and increase production costs as farmers are required to purchase vaccines and drugs to prevent and control diseases and vectors. Mitigation and adaptation interventions to minimise the negative impacts of livestock diseases and pests associated with climate change and weather variability are therefore required.

Overall, the impacts of climate change on livestock health, well-being, production and reproduction are well recognised worldwide, and hence, there is a need for defining and implementing relevant and contextualised adaptation and mitigation strategies (Bernabucci 2019).

6 Conclusion

The study results show that cattle production involves different categories of people, including women and youth, implying that target-specific interventions are required. The study has also revealed that cattle keepers feel that they are affected by climate change and weather variability, as observed by incidences of erratic rainfall and unreliable patterns of precipitation, which increase the frequency and intensity of droughts, strong winds and floods. Perceived effects on cattle production include water scarcity, increases in incidences of diseases and reduction in supply of fodder, which is exacerbated by increasing wild bushfires. These effects also negatively affect profitability in that farmers' expenses are increased in the form of purchasing drugs, vaccines and feed and hiring of extra labour to fetch water and feed. There is therefore a need for the government and other relevant actors to help cattle farmers to better understand the climate change and weather variability phenomena and enhance their adaptive capacity building to deal with negative impacts of climate change.

7 Recommendations

There is a need for development of mitigation and adaptation measures to address negative impacts of climate change on cattle production including, but not limited to:

(i) Enhancing the adaptive capacity of cattle keepers to better understand climate change and weather variability, their causes and effects, and the adaptation and mitigation measures that are required.
(ii) Encouraging farmers to start growing pasture and grasses, such as Rhodes grass and multipurpose tree species, in wetlands and along boundaries of their fields for those with limited land to address issues of feed scarcity.
(iii) Farmers should adopt and practise controlled and planned grazing, including rotational grazing, to address issues of overgrazing and to preserve fodder biodiversity.

(iv) Farmers should embrace fodder conservation like hay or silage for use during dry periods when there is no pasture. Fodder preservation should be accompanied by strategic ration formulations and feeding system to meet animal nutrient requirements.
(v) Deliberate efforts should be made to construct strategically placed water points for livestock (using solar power where possible), to ensure that livestock do not walk distances in search of water.
(vi) The community should be advised on the dangers of bushfires, with the aim of stopping this practice henceforth.
(vii) Farmers should provide shed for their animals to reduce impacts of heat stress.

Acknowledgements The authors wish to thank the Norwegian Government-funded Capacity Building for Managing Climate Change (CABMACC) Programme for providing financial support to conduct this research study in Chikwawa, Malawi. The cooperation and support of cattle owners, extension workers and members of the Chikwawa Livestock Association to willingly participate in this research study are highly appreciated.

Conflict of Interest The authors declare that there is no conflict of interest.

References

Aydinalp C, Cresser MS (2008) The effects of global climate change on agriculture. Am Eurasian J Agric Environ Sci 3(5):672–676

Bailey R, Froggatt A, Wellesley L (2014) Livestock–climate change's forgotten sector. Global Public Opinion on Meat and Dairy Consumption Chatham House. Available online at: https://gastronomiaycia.republica.com/wp-content/uploads/2014/12/estudio_consumo_carne.pdf. Accessed 10 July 2018

Bernabucci U (2019) Climate change: impact on livestock and how can we adapt. Anim Front 9(1):3–5

Bett B, Kiunga P, Gachohi J, Sindato C, Mbotha D, Robinson T, Lindahl J, Grace D (2017) Effects of climate change on the occurrence and distribution of livestock diseases. Prev Vet Med 137:119–129

Chingala G, Mapiye C, Raffrenato E, Hoffman L, Dzama K (2017) Determinants of smallholder farmers' perceptions of impact of climate change on beef production in Malawi. Clim Chang 142(1–2):129–141

Coulibaly JY, Mbow C, Sileshi GW, Beedy T, Kundhlande G, Musau J (2015) Mapping vulnerability to climate change in Malawi: spatial and social differentiation in the Shire River Basin. Am J Clim Chang 4(03):282

FAO (2010) 'Climate-smart agriculture': policies, practices and financing for food security, adaptation and mitigation. Food and Agriculture Organization of the United Nations, Rome. http://www.fao.org/docrep/013/i1881e/i1881e00.htm. Accessed 1 Mar 2019

Gaughan JB, Cawsell-Smith AJ (2015) Impact of climate change on livestock production and reproduction. In: Sejian V, Gaughan J, Baumgard L, Prasad CS (eds) Climate change Impact on livestock: adaptation and mitigation. Springer, New Delhi, pp 51–60

Grace D, Bett B, Lindahl J, Robinson T (2015) Climate and livestock disease: assessing the vulnerability of agricultural systems to livestock pests under climate change scenarios. CCAFS Working Paper no. 116. Copenhagen, Denmark. CGIAR Research Program on Climate

Change, Agriculture and Food Security (CCAFS). Available online at: www.ccafs.cgiar.org. Aaccessed 10 Aug 2018
Godber OF, & Wall R (2014) Livestock and food security: vulnerability to population growth and climate change. Global change biology, 20(10):3092–3102
Gregory NG (2010) How climatic changes could affect meat quality. Food Research International 43(7):1866–1873
Izaurralde RC, Thomson AM, Morgan JA, Fay PA, Polley HW, Hatfield JL (2011) Climate impacts on agriculture: implications for forage and rangeland production. Agron J 103(2):371–381
Kristjanson P, Waters-Bayer A, Johnson N, Tipilda A, Njuki J, Baltenweck I et al (2014) Livestock and women's livelihoods. In: Gender in Agriculture. Springer, Dordrecht, pp 209–233
Malawi Government (2016) Livestock Agriculture production estimates (APES), Department of Animal Health and Livestock Development, Lilongwe, Malawi
Malawi Government (2015) Malawi 2015 Floods Post Disaster Needs Assessment (PDNA). http://reliefweb.int/sites/reliefweb.int/files/resources/Malawi-2015-Floods-Post-Disaster-Needs-Assessment-Report.pdf. Accessed 8 Dec 2018
McKeon GM, Stone GS, Syktus JI, Carter JO, Flood NR, Ahrens DG, Bruget DN, Chilcott CR, Cobon DH, Cowley RA, Crimp SJ (2009) Climate change impacts on northern Australian rangeland livestock carrying capacity: a review of issues. Rangel J 31(1):1–29
Megersa B, Markemann A, Angassa A, Ogutu JO, Piepho HP, Zaráte AV (2014) Impacts of climate change and variability on cattle production in southern Ethiopia: perceptions and empirical evidence. Agric Syst 130:23–34
Nardone A, Ronchi B, Lacetera N, Ranieri MS, Bernabucci U (2010) Effects of climate changes on animal production and sustainability of livestock systems. Livest Sci 130(1–3):57–69
Njuki J, Waithanji E, Lyimo-Macha J, Kariuki J, Mburu S (eds) (2013) Women, livestock ownership and markets: bridging the gender gap in eastern and southern Africa. New York:Routledge
Porter JR, Xie L, Challinor AJ, Cochrane K, Howden SM, Iqbal MM, Lobell DB, Travasso MI (2014) Food security and food production systems. In: Field CB, Barros VR, Dokken DJ, Mach KJ, Mastrandrea MD, Bilir TE, Chatterjee M, Ebi KL, Estrada YO, Genova RC, Girma B, Kissel ES, Levy AN, MacCracken S, Mastrandrea PR, White LL (eds) Climate change 2014: impacts, adaptation, and vulnerability. Part A: Global and sectoral aspects. Contribution of Working Group II to the Fifth Assessment Report of the Intergovernmental Panel on Climate Change. Cambridge University Press, Cambridge, UK/New York, NY, pp 485–533
Rust JM (2018) The impact of climate change on extensive and intensive livestock production systems. Anim Front 9(1):20–25. https://doi.org/10.1093/af/vfy028
Rao KPC, Ndegwa WG, Kizito K, Oyoo A (2011) Climate variability and change: Farmer perceptions and understanding of intra-seasonal variability in rainfall and associated risk in semi-arid Kenya. Experimental agriculture, 47(2):267–291
Smith J, Sones K, Grace D, MacMillan S, Tarawali S, Herrero M (2013) Beyond milk, meat, and eggs: Role of livestock in food and nutrition security. Animal Frontiers 3(1):6–13
Thornton PK, Herrero M (2014) Climate change adaptation in mixed crop–livestock systems in developing countries. Glob Food Sec 3(2):99–107
Thornton PK, Gerber PJ (2010) Climate change and the growth of the livestock sector in developing countries. Mitig Adapt Strateg Glob Chang 15(2):169–184
Thornton PK (2010) Livestock production: recent trends, future prospects. Philos Trans R Soc B Biol Sci 365(1554):2853–2867
Thornton PK, Herrero M, Freeman HA, Okeyo Mwai A, Rege E, Jones PG, McDermott JJ (2007) Vulnerability, climate change and livestock-opportunities and challenges for the poor. https://cgspace.cgiar.org/handle/10568/2205. Accessed 15 Aug 2019
Terry G (2009) No climate justice without gender justice: an overview of the issues. Gend Dev 17(1):5–18
Tol RS, Downing TE, Kuik OJ, Smith JB (2004) Distributional aspects of climate change impacts. Glob Environ Chang 14(3):259–272
Vermeulen SJ, Aggarwal PK, Ainslie A, Angelone C, Campbell BM, Challinor AJ, Hansen JW, Ingram JSI, Jarvis A, Kristjanson P, Lau C (2012) Options for support to agriculture and food security under climate change. Environ Sci Pol 15(1):136–144

A Cohort Study of Reproductive Performance, Associated Infections and Management Factors in Zebu Cows from Smallholder Farms in Malawi

M. A. Bhatti, W. Chanza, S. Klevar, L. A. Kamwanja, T. B. Klem, D. C. Jansen, H. Holm, M. Chipandula, G. Njunga, M. Stokstad, and O. Reksen

Abstract Smallholder Zebu cattle farming is of prime importance for the livelihood of Malawians, and optimising cattle reproductive efficiency represents one way to increase food production. The objective of this study was to determine the reproductive performance and associated factors in a cohort of Malawi Zebu cows (n = 100) across 36 smallholder farms for a 2-year period. The birth of a live calf was observed in 78 dams and stillbirths/late abortions in 5 cows, 14 cows were censored, and 3 cows did not calve throughout the study period.

Median time from calving to resumption of luteal activity was 61 days. Mean calving interval was 457 days with 32% of the variation residing at herd level. A significant association was found between time from calving to conception (open days) and the presence of a breeding bull in a herd (156 versus 235 days) and between multiparous and primiparous cows (170 versus 220 days). Cows were screened three times, at 6-month interval, for antibodies against selected infectious agents. None were positive for *Brucella* spp., six were positive for bovine viral diarrhoea virus and two for *N. caninum* antibodies. No blood smears were positive for *Babesia* spp. or *Theileria* spp., but three were positive for *Trypanosoma* spp. Conclusively, this study indicated a low prevalence of the studied infections and a

M. A. Bhatti (✉) · T. B. Klem · D. C. Jansen · H. Holm · M. Stokstad · O. Reksen
Norwegian University of Life Sciences (NMBU), Faculty of Veterinary Medicine, Department of Production Animal Clinical Sciences, Oslo, Norway
e-mail: muhammad.azher.bhatti@nmbu.no

W. Chanza · L. A. Kamwanja · M. Chipandula
Trustees of Agricultural Promotion Programme, LPMU Headquarters, Lilongwe, Malawi

S. Klevar
Norwegian Veterinary Institute, Veterinærinstituttet, Oslo, Norway

G. Njunga
Trustees of Agricultural Promotion Programme, LPMU Headquarters, Lilongwe, Malawi

Department of Animal Health, Lilongwe, Malawi

B. R. Singh et al. (eds.), *Climate Impacts on Agricultural and Natural Resource Sustainability in Africa*, https://doi.org/10.1007/978-3-030-37537-9_13

large potential for increased reproductive efficiency in Zebu cattle of this area. Particularly, the keeping of a herd-own bull, adequate breeding management and care of first parity cows seem to be key factors to increase reproductive efficiency and thereby food production from Malawi Zebu.

1 Introduction

The African population of 1.2 billion people is expected to double by 2050 (Groth and May 2017) which will require significant improvements in food production. In addition, climate changes represent new challenges in crop and livestock production. The Food and Agricultural Organization (FAO) reports that 14.5% of greenhouse gas emissions are caused by livestock, and intensification of the production is the most efficient way of reducing carbon footprint from this sector (Gerber et al. 2013). In sub-Saharan African countries, severe dry weather has affected the large grazing areas leading to less grazing, and livestock struggle to maintain body condition resulting in compromised reproductive efficiency (Kanuya et al. 2006b; Matiko et al. 2008). During the dry season, rivers remain the only source of available water for the majority of cattle herds, and cattle need to walk longer distances to get water, which could also potentially influence their reproductive performance.

Rural Malawians' (80% of total population) existence is largely dependent on smallholder agriculture, and Malawian Zebu cattle farming contributes significantly to food production (Dixon et al. 2004). Optimising cattle reproductive efficiency represents one way to increase milk and meat production (Kiwuwa et al. 1983). Previous studies from other countries on indigenous Zebu cattle reproduction have emphasised the importance of having a breeding bull at the farm, which minimises the incidence of transmissible venereal diseases among herds (Fernandez et al. 1993; Rekwot et al. 2004; Young et al. 2014).

Improvements in Malawi Zebu breeding programs have been called for, and management of bulls has been identified as a major limiting factor (Nandolo et al. 2015; Rekwot et al. 2000). In a study of Tanzanian Zebu cows, Matiko et al. (2008) reported a significant effect of parity (> second parity) on post-partum resumption of ovarian cyclicity. It may, therefore, be speculated that first parity Zebu cows, maintained in a pastoral system with no breeding management, are mated too early for optimal reproductive performance.

Infectious disease agents like bovine viral diarrhoea virus (BVDV), *Brucella* spp. and *Neospora caninum* have been identified to exert direct effects on the reproductive organs, whereas protozoan parasites such as *Theileria parva*, *Babesia bovis*, *Babesia bigemina* and *Trypanosoma* spp. have an indirect effect on reproductive performance through compromised general health (Mukasa-Mugerwa 1989; Sekoni 1994). High prevalence of BVDV and *Brucella* spp. in Tanzania (Mathew et al. 2017) and *N. caninum* in Ethiopia (Asmare 2014) has been reported to exert a negative impact on Zebu cattle reproductive performance in these regions. The occurrence of these infections and their association with reproductive disorders is unknown in Malawi Zebu cattle.

The objective of this study was to investigate reproductive performance and its association with selected factors in Malawi Zebu to gain knowledge on potential intensification. Specific objectives were (i) to determine key reproductive performance indicators (time from calving to onset of regular luteal function, calving interval and calf mortality), (ii) to determine the relative impact of herd and cow level factors on calving interval by variance components analyses, (iii) to study associations between likelihood of pregnancy and availability of breeding bulls and parity of the cows and (iv) to determine the prevalence of selected infectious agents known to have an impact on reproductive performance.

2 Materials and Methods

A cohort study of Malawi Zebu cows was conducted in northern Malawi. The cohort was followed by veterinary examinations, four times during a 2-year period from September 2015 to October 2017.

Malawi is divided into 187 Extension Planning Areas (EPA), which are further subdivided into agricultural sections comprising 5–15 villages (Kundhlande et al. 2014). This study was conducted across seven agricultural sections (Bolero, Bata, Chozoli, Chikwawa, Chirambo, Mjuma and Jalira) of Bolero EPA, Rumphi District. Bolero EPA was selected because smallholder livestock farming is the predominant agrarian practice in this area (MZADD 2009).

Three farmer meetings were conducted in Bolero EPA before the start of the study. The selection of cattle farms was based on the presence of newly calved cows in the herd (cows that calved after August 15, 2015, and four subsequent weeks onwards). Cows from these herds were selected based on the calving information provided by the farmers in these three meetings.

A single-visit-multiple-subject survey technique (ILCA 1990) was used to obtain smallholder cattle farm data including herd size, availability of breeding bulls as well as parity, pregnancy and reproductive status of cows. Initially, 109 clinically healthy Malawi Zebu cows from 39 smallholder farms were selected and ear tagged. Later on, three farmers with nine cows withdrew before April 2016, leaving a cohort of 100 cows belonging to 36 farms, which were followed throughout the study. In this study, female calves older than 8 months were defined as heifers until first calving took place.

In addition to the permanent presence of a field assistant during the study period, four experienced large animal veterinarians (Faculty of Veterinary Medicine, Department of Production Animal Clinical Sciences, Norwegian University of Life Sciences) conducted cattle rectal examinations for reproductive status and pregnancy diagnosis at four time points at approximately 6-month intervals: April 2016, October 2016, April 2017 and October 2017. Pregnancy diagnosis was conducted based on the following criteria:

- Membrane slip + asymmetry (>35 days pregnant).
- Balloting of the foetus (>70 days).
- Fremitus in one of the uterine arteries (>105 days).
- Fremitus in both uterine arteries (>180 days).
- Large parts of the foetus apparent close to the pelvic inlet (>240 days).

An early abortion was suspected if cows were not late pregnant or had not calved within 8 months after diagnosis of an early pregnancy (>35 days). The terms *late abortion* and *stillbirth* were used for the foetus of the last trimester or full born calves dead at birth. If a cow with an assumed early abortion later became pregnant, the successful calving of a live calf was included in the assessment of the calving interval. The herd record keeping was regularly maintained by the farmers and monitored by the field assistant. Calving dates were obtained from these herd records, which enabled the calculation of calving intervals. Days from calving to pregnancy were approximated by subtracting 280 days from the next calving date. Observations of calf mortality in live born calves were restricted to the period from the date the calf was born until the end of December 2015.

The animals were treated for ticks in April 2016 and 2017 (Taktic®, Utrecht, The Netherlands) and for helminths in October 2016 and October 2017 (Kepromec®, Deventer, The Netherlands). Treatments were conducted in accordance with the manufacturer's recommendations.

2.1 Milk Sampling

Altogether, 96 cows (four cows omitted due to sampling irregularity) were subjected to milk progesterone analyses. A semi-quantitative assessment of progesterone in milk was performed using a cow-side P4 Rapid test (Ridgeway Science, St Briavels, UK). Progesterone analyses were used to determine the time from calving to onset of luteal activity (OLA) defined as three consecutive positive reactions to elevated milk progesterone values within 21 days. Milk was collected every third to seventh day from October to the end of December 2015.

2.2 Blood Sampling

Blood was sampled on three occasions: October 2015, April 2016 and October 2016 from 108, 99, and 81 cows, respectively. Blood was collected from either the coccygeal vein or the jugular vein into sterile vacutainers and immediately cooled in the field cooler until centrifugation at 1200 × g at room temperature. Sera were pipetted into sterile tubes and kept frozen at approximately −20 °C. The material was shipped on ice to the National Veterinary Institute in Oslo, Norway, and submitted to the Biosafety level 3 laboratories and kept frozen at −20 °C until analyses. EDTA mixed blood was sampled at two occasions (April 2016 and April 2017) and blood smears were prepared on site.

2.3 Laboratory Analyses

All sera were analysed for antibodies against BVDV, *Brucella* spp. and *N. caninum* using enzyme-linked immunosorbent assay (ELISA) from Boehringer Ingelheim Svanova in Uppsala, Sweden. The optical density (OD) was measured in a LabSystems Multiskan EX spectrophotometer (LabX, Ontario, Canada). All OD values were correlated to positive control sera referred to as corrected OD values, and percent positivity (PP) of each sample was calculated following the manufacturer's instructions. Anti-BVDV antibodies were detected using an indirect ELISA commercial kit (SVANOVIR® BVDV-Ab I-ELISA) following the manufacturer's instructions. Serum samples with PP ≥10% were considered positive and those with PP <10% were negative. The presence of antibodies to *Brucella* spp. was analysed using the SVANOVIR® *Brucella*-Ab-I-ELISA antibody test following the manufacturer's instructions. Serum samples with PP ≥40% were considered positive, and samples with PP <40% were considered negative. *N. caninum* specific antibodies were detected using an indirect SVANOVIR® *N. caninum* ISCOM ELISA following manufacturer instructions. Sera were considered positive when PP ≥20%, while samples with PP <20% were considered negative.

The air-dried blood smears were fixed in methanol, stained with Giemsa stain and examined for evaluation of *Theileria* spp., *Babesia* spp. and *Trypanosoma* spp. in microscopy at 100 × magnification by a trained technician from Central Veterinary Laboratory, Lilongwe, Malawi.

3 Statistical Analyses

The statistical software package STATA 14 (StataCorp LLC, Texas, USA) was used in all statistical analyses. Statistical significance was considered with a *P*-value less than 0.05 in all models. Variance components mixed model analyses using the 'xtmixed' procedure were performed allowing REML (restricted maximum likelihood) estimation to discriminate between variance of calving interval at cow, farm and section level. Agricultural sections and farm were included as random effect variables both separately and jointly in three models.

Univariable Kaplan–Meier (K-M) estimators were used to assess relationships between time from calving to pregnancy for the explanatory variables; presence of a breeding bull in the herd (1/0) and first parity versus multiparous cows (1/0). The time variable describing days from calving to pregnancy was calculated for cows that delivered a live calf within the period June 5, 2016, to October 6, 2017. All other cows were censored at the end of the study, at culling or death or when a late abortion/stillbirth had occurred. Statistical significance was assessed both by log likelihood and Wald chi-square statistics. Both approaches were used, as the Wald chi-square test is more sensitive to early observation than the default log-likelihood statistic (Dohoo et al. 2009). Explanatory variables with *P*-values of 0.10 and less for the association with likelihood of conception from the K-M estimators were included in a multivariable Cox proportional hazards model (Cox regression), with shared frailty at farm level.

Parity was included in the Cox model both as an ordinal variable (parity: 1, 2 and >2) and as first parity versus multiparous cows (1/0). Number of bulls in the herd was evaluated both as an ordinal variable (number of bulls: 1, 2 and >2) and to whether a breeding bull was present in the herd or not (1/0). Interaction terms between the bull and parity variables were tested and subsequently omitted for non-significance by backward elimination. As the likelihood ratio test for frailty was not significant, it was also removed from the final models. The proportional hazard assumption was evaluated by plotting the negative logarithm of the hazard function against the logarithm of survival time ($-\ln(H)$ versus $\ln(t)$) separately for the parity and the bull effects. The assumption of independent censoring was explored in sensitivity analyses of complete negative and positive correlations between censoring events and the event of interest. The overall fit of the model was assessed by Cox-Snell residuals and Moreau goodness of fit tests; outliers were evaluated by deviance residuals, whereas the lack of influential points was identified by scaled score residuals (Dohoo et al. 2009).

4 Results

The background information of herds from where the cohort of 100 cows were recruited is given in Table 1. The information regarding parity and milking status of cows was collected during the initial survey meetings.

Table 1 Background information of herds (with respect to agricultural sections) that included the 100 Malawi Zebu cows of the cohort study

	Adult cows					
	Milking	Dry	Heifers	Calves	Bulls	Total
Agricultural sections	(*n*)	(*n*)	(*n*)	(*n*)	(*n*)	(*n*)
Bata	52	28	38	68	10	196
Bolero	7	2	6	11	1	27
Chikwawa	25	13	16	27	6	87
Chirmabo	24	9	20	23	7	83
Chozoli	21	12	16	23	11	83
Jalira	4	3	4	4	1	16
Mjuma	5	2	3	6	1	17
Total	138	69	103	162	37	509

After three farmers had withdrawn, the number of bulls per farm ranged from zero to four, and 30 (out of 100) cows originated from 17 farms with no breeding bull in the herd. The primiparous cows (n = 19) of the cohort study originated from 13 farms.

The mean calving interval in 78 cows that subsequently delivered a live calf was 457.4 days with a standard deviation (SD) of 163.0 days, which corresponds to a mean period from calving to pregnancy of 177.4 days. The variance component analysis of the calving intervals from these 78 cows showed that 32% of the variance in calving interval resided at the herd level, with significant variation in calving interval between herds (P = 0.008), indicating an important contribution of herd management to reproductive outcome. The variance in calving interval between agricultural sections was not significant, indicating minimal geographical differences in reproductive efficiency within the study area.

Results of the reproductive outcomes of the study are reported in Table 2. Late abortion/stillbirth was reported for five cows. No cases of early abortions were reported by the farmers, but long intervals from a positive diagnosis of pregnancy until subsequent calving indicated that five cases of early abortions had occurred. Nine calves died before they reached the age of 3 months, and one calf died at 4 months of age, making the total loss of calves of about 13% (10 out of 78) during the dry period of 2015. This figure increased to 18% (15 out of 83) when late abortions and stillbirths were included.

Milk progesterone tests showed that OLA had already started in 43 cows at the beginning of the milk-sampling period in October 2015. The median value of OLA after calving was 61 days. OLA did not start in 25 out of 100 cows before milk sampling ended in the last week of December.

In herds with a breeding bull, the mean number of days to pregnancy was 156 (SD = 155), and for cows without a bull in the herd, the mean number of days to pregnancy increased to 235 (SD = 175). Figure 1 displays the Kaplan–Meier plot of pregnant cows by days after calving between cows from a herd with a breeding bull and cows in herds that relied on mating during periods of common grazing. The

Table 2 Reproductive status of 100 Malawi Zebu cows from 36 farms obtained at 6-month intervals from calving in August and September 2015 until October 2017. Cows at risk were followed until calving, abortion, death or until the cow had left the farm

Status	Oct. 2015	Apr. 2016	Oct. 2016	Apr. 2017	Oct. 2017	Oct. 2015–2017
Not pregnant	100	35 (1[a])	20 (4[a])	3	2	2
Pregnant	–	56	18	23	1	1
Late abortion/stillbirth	–	–	5	–	–	5
Calved	–	–	47	8	23	78
Dead	–	5	–	2	–	7
Sold/slaughtered	–	4	1	2	–	7
Cows at risk of calving	100	91	38	26	3	–

[a]The number of non-pregnant cows that possibly had experienced early abortions

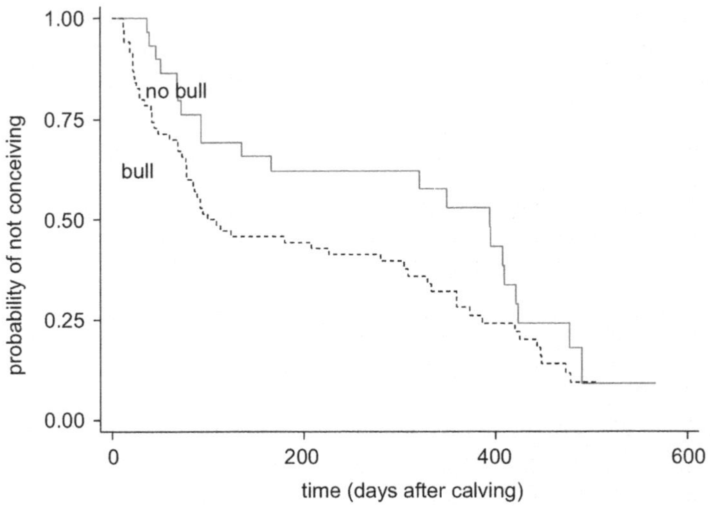

Fig. 1 Kaplan–Meier plot of the proportion of pregnant cows distributed by days after calving between cows from a herd with a breeding bull and cows in herds without a bull

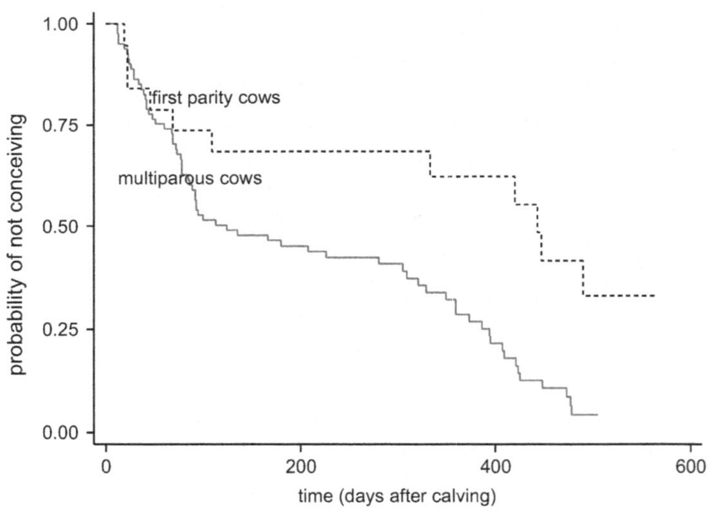

Fig. 2 Kaplan–Meier plot of proportion of pregnant cows distributed by days after calving between first parity and multiparous cows

difference between the two groups was not significant when assessed by the log-rank test for equality of survivor functions ($P = 0.097$) but significant according to the Wald chi-square statistic ($P = 0.032$).

The mean number of days from calving to pregnancy was 220 (SD = 202) for first parity cows and 170 (SD = 156) for multiparous cows. Figure 2 displays the Kaplan–Meier plot of first parity versus multiparous cows by days after calving.

Table 3 Multivariable Cox proportional hazards model of days from calving to pregnancy by the keeping of a bull in the herd (1/0) and first versus older parities of the cows (1/0)

Explanatory variables	Coeff. (*b*)	S.E	*P*	95% Confidence interval
Bull	0.549	0.259	0.034	0.042–1.055
First parity	−1.052	0.338	0.002	−1.715 to −0.390

The difference between first calving and older cows was significant when assessed by the log-rank test for equality of survivor functions ($P = 0.003$) but borderline non-significant according to the Wald chi-square statistic ($P = 0.051$).

The presence of a breeding bull in the herd and the effect of parity were both significantly related to pregnancy success when assessed in a multivariable Cox proportional hazards model of days from calving to pregnancy (Table 3). The hazard rate (HR) of pregnancy in primiparous cows compared to multiparous cows was 0.35 ($e^{-1.0524}$), indicating a significantly reduced likelihood of pregnancy in primiparous cows. The HR for pregnancy in cows from a herd with a breeding bull was 1.73 times higher ($e^{.0547}$) compared to cows in a herd without a bull.

An interaction term between the bull and parity effects was initially included but omitted for non-significance. The ordinal variables of number of bulls and parity were not significantly related to pregnancy rates in the multivariable models. The likelihood ratio of the multivariable Cox proportional hazards model was $\chi^2 = 14.8$ with 2 df ($P = 0.001$).

4.1 Antibody Results

Out of 288 serum samples collected during three occasions, no serum was positive in the *Brucella* spp. ELISA in either sampling. Only two cows, one sampled in October 2015 and another cow sampled in April 2016, were seropositive in the *N. caninum* ELISA. Four cows were found positive for BVDV in October 2016, whereas only one BVDV-positive cow was found in each of the samples obtained in October 2015 and April 2016 (Table 4). Of the blood smears, three cows were positive for *Trypanosoma* spp., but all were negative for *Babesia* spp. and *Theileria* spp. at all sampling events.

5 Discussion

This study shows that improved reproductive management and better care of first parity cows are key factors to increase meat and milk production from smallholder Zebu cattle farms of northern Malawi. In accordance with our findings, a recent study from Malawi (Nandolo et al. 2015) also reported that 64.6% farms were without a breeding bull in the herd. Furthermore, that farmers' decision on which bull to

Table 4 Number of seropositive animals for bovine viral diarrhoea virus (BVDV), *Brucella* spp. and *N. caninum* in Malawi Zebu herds

Infectious agents	Positive cows		
	October 2015 ($n = 105$)	April 2016 ($n = 95$)	October 2016 ($n = 81$)
BVDV	1	1	4
Brucella spp.	0	0	0
N. caninum	1	1	0

keep will affect the breeding in that region because cattle are mixed freely during the grazing period. In addition, a relatively high inbreeding rate was also estimated, and in conclusion, the authors suggested to establish community breeding programs (Nandolo et al. 2015). In accordance with the present study, parity has been reported to affect reproductive performance in Tanzanian Zebu cows (Matiko et al. 2008). This study also reported that the number of days open was significantly less in older cows as compared to the younger cows. Herd composition (keeping older cows) is a management strategy that may help in getting more calves and higher milk yield on a yearly basis. Also, avoiding premature mating of heifers may be a valuable strategy for increasing reproductive efficiency in first parity Zebu heifers (Madani et al. 2008).

The observed median days from calving to the onset of luteal function and mean calving interval were shorter than corresponding figures for Tanzanian Zebu cows (Kanuya et al. 2006a), indicating better reproductive performance in this cohort of Malawi Zebu cows.

N. caninum is one of the major causes of reproductive problems and abortions in dairy and beef cattle worldwide (Innes 2007). The two seropositive cows in our study indicate a very low seroprevalence in the study population and no impact on reproductive performance in our included herds. The finding of a *Brucella* spp.–free population in this study is important since human's brucellosis originates from animals. Previous studies from the neighbouring country, Tanzania, have reported differences in seroprevalence of BVDV and *Brucella* spp. in geographically very close areas (Mathew et al. 2015; Mathew et al. 2017; Swai et al. 2005). Further research is necessary to detect if the low occurrence is general or if the study area represents a free area surrounded by populations with higher prevalence. Such knowledge is necessary in order to choose the most cost-effective preventive measures to avoid introduction of these high-impact infections.

In addition, the occurrence of *Trypanosoma* spp., *Babesia* spp. and *Theileria* spp. has showed surprisingly lower prevalence than what has been found in other areas of neighbouring Tanzania. However, the results presented in the current study must be interpreted with care due to a relatively low sensitivity of the methods (Delgado et al. 1998).

The infections studied were chosen because they are widespread and generally cause substantial losses. Since the study population were free, or had very low prevalence, of all of them, it may be viewed as a population with low prevalence of infections. In such case, the reproductive outcome of cows from the best managed farms in this study may demonstrate a bench mark for reproductive efficiency in a low prevalence situation. However, a wide range of other infections could be present. Pathogens such as *Campylobacter fetus*, bovine herpesvirus-1 and *Leptospira* spp. that cause reproductive disorders could be further investigated. However, neither early nor late abortion rates were high in the current investigation. It is therefore plausible to assume that differences in herd management are of major importance for the observed variance in reproductive efficiency between the cows that were followed throughout the 2-year observation period. This is also demonstrated by a significant contribution of the herd effect in the variance component analysis of calving intervals. Furthermore, no variability in calving interval could be explained by agricultural section. This may indicate only minor geographical differences within the study area, with negligible impact on reproductive efficiency.

6 Conclusion

In conclusion, a large potential for increased reproductive efficiency by improving herd management was identified in the current investigation. Improved availability of bulls, better care of primiparous animals and higher calf survival in Malawi Zebu should be prioritised in order to mitigate the effects of climate change and improve farmers' economic situation and food security in the region.

Acknowledgement The authors would like to thank the smallholder farmers of Bolero EPA, Rumphi District, Malawi, who actively participated in this study. We are grateful to Dr. Liveness Jessica Banda and Dr. Fanny Chigwa, Animal Science Department, Lilongwe University of Agriculture and Natural Resources, for their help in laboratory analyses. This study was supported by the project: Capacity Building for Managing Climate Change in Malawi (CABMACC), Norwegian University of Life Sciences, Norway. The corresponding author is very much thankful to Prof. Peter Wynn (Graham Centre for Agricultural Innovation, Charles Sturt University, NSW 2650, Australia) for his thorough guidance and help in all the research activities.

Conflict of Interest The authors declare that they have no conflict of interest.

Funding This study was funded by Capacity Building for Managing Climate Change in Malawi (CABMACC).

References

Asmare K (2014) Neospora caninum versus *Brucella* spp. exposure among dairy cattle in Ethiopia: a case control study. Trop Anim Health Prod 46(6):961–966

Delgado J et al (1998) Low sensitivity of peripheral blood smear for diagnosis of subclinical visceral leishmaniasis in human immunodeficiency virus type 1-infected patients. J Clin Microbiol 36(1):315–316

Dixon J, Taniguchi K, Wattenbach H, TanyeriArbur A (2004) Smallholders, globalization and policy analysis, vol 5. Food & Agriculture Org, Rome, Italy

Dohoo I, Martin S, Stryhn H (2009) Veterinary epidemiologic research, 2nd edn. VER Chralottetwon, Prince Edward Island

Fernandez D, Berardinelli J, Short R, Adair R (1993) The time required for the presence of bulls to alter the interval from parturition to resumption of ovarian activity and reproductive performance in first-calf suckled beef cows. Theriogenology 39(2):411–419

Gerber PJ, Steinfeld H, Henderson B, Mottet A, Opio C, Dijkman J, Falcucci A, Tempio G (2013) Tackling climate change through livestock: a global assessment of emissions and mitigation opportunities: Food and Agriculture Organization of the United Nations. FAO, Rome, Italy

Groth H, May JF (2017) Africa's population: in search of a demographic dividend. Springer, Cham, Switzerland

ILCA (1990) Livestock systems research manual-Volume 1, ILCA working paper 1, International Livestock Centre for Africa Addis Ababa, Ethiopia. 287 pp

Innes EA (2007) The host-parasite relationship in pregnant cattle infected with Neospora caninum. Parasitology 134(13):1903–1910

Kanuya NL, Matiko MK, Kessy BM, Mgongo FO, Ropstad E, Reksen O (2006a) A study on reproductive performance and related factors of zebu cows in pastoral herds in a semi-arid area of Tanzania. Theriogenology 65(9):1859–1874

Kanuya NL, Matiko MK, Nkya R, Bittegeko SBP, Mgasa MN, Reksen O, Ropstad E (2006b) Seasonal changes in nutritional status and reproductive performance of Zebu cows kept under a traditional agro-pastoral system in Tanzania. Trop Anim Health Prod 38(6):511–519

Kiwuwa GH, Trail JC, Yousef M, Worku G, Anderson FM, Durkin J (1983) Crossbred dairy cattle productivity in Arsi region. Ethiopia, ILCA

Kundhlande G, Franzel S, Simpson B, Gausi E (2014) Farmer-to-farmer extension approach in Malawi: a survey of organizations, ICRAF working paper

Madani T, Yakhlef H, Marie M (2008) Effect of age at first calving on lactation and reproduction of dairy cows reared in semi-arid region of Algeria. Livest Res Rural Dev 20(6)

Mathew C, Stokstad M, Johansen T, Klevar S, Mdegela R, Mwamengele G, Michel P, Escobar L, Fretin D, Godfroid J (2015) First isolation, identification, phenotypic and genotypic characterization of *Brucella abortus* biovar 3 from dairy cattle in Tanzania. BMC Vet Res 11(1):156

Mathew C, Klevar S, Løken T, Mdegela R, Mwamengele G, Skjerve E, Godfroid J, Stokstad M (2017) Reproductive infections in cattle in Tanzania – lessons for control priorities. SOJ Microbiol Infect Dis 5(2):1–9

Matiko MK, Kanuya NL, Waldmann A, Ropstad E, Reksen O (2008) Environmental constrains on post-partum ovarian activity in Tanzanian Zebu cows. Theriogenology 69(7):896–904

Mukasa-Mugerwa E (1989) A review of a reproductive performance of female Bos Indicus (zebu) cattle, ILRI (aka ILCA and ILRAD)

MZADD (2009) Livestock production annual report, (unpublished reprot, Department of Animal Health and Livestock Development, Mzuzu)

Nandolo W, Gondwe T, Banda L (2015) Characterisation of breeding systems for Malawi Zebu cattle in Mzimba District, Northern Malawi. Malawian J Agric Natl Res Dev Stud 1(1)

Rekwot P, Ogwu D, Oyedipe E (2000) Influence of bull biostimulation, season and parity on resumption of ovarian activity of zebu (Bos indicus) cattle following parturition. Anim Reprod Sci 63(1):1–11

Rekwot P, Akinpelumi O, Sekoni V, Eduvie L, Oyedipe E (2004) Effects of nutritional supplementation and exposure to bulls on resumption of post-partum ovarian activity in Bunaji (Bos indicus) cattle. Vet J 167(1):67–71

Sekoni V (1994) Reproductive disorders caused by animal trypanosomiases: a review. Theriogenology 42(4):557–570

Swai E, French N, Karimuribo E, Fitzpatrick J, Bryant M, Brown P, Ogden N (2005) Spatial and management factors associated with exposure of smallholder dairy cattle in Tanzania to tick-borne pathogens. Int J Parasitol 35(10):1085–1096

Young J, Rast L, Suon S, Bush R, Henry L, Windsor P (2014) The impact of best practice health and husbandry interventions on smallholder cattle productivity in southern Cambodia. Anim Prod Sci 54(5):629–637

Effect of Dry Season Supplement Feeding of Malawi Zebu Cows on Reproductive Performance, Lactation and Weight Gain in Calves

M. A. Bhatti, W. Chanza, L. A. Kamwanja, S. Chikomola, M. Chipandula, A. Chikaonda, D. C. Jansen, S. Klevar, T. B. Klem, M. Stokstad, and O. Reksen

Abstract Climate change affects food security and livelihood in Malawi, and the Food and Agricultural Organization emphasises the intensification of milk and meat production to reduce greenhouse gas emissions. Measures to increase the robustness of small-holder dairy production to better cope with the new challenges are urgently needed. A field experiment was therefore conducted to investigate the effect of supplementing lactating Zebu cows on performance of the cows and their offspring in Malawi.

A total of 98 cows were included and allocated to experimental and control groups matched by parity, age of calves and geographical location of farm. The intervention was additional feeding during the dry season with 2 kg extra maize bran per cow per day and leguminous leaves (*Gliricidia sepium*). The experimental group in year one of the study ($n = 28$) received in total 114 kg additional leaves, and the experimental group ($n = 21$) received 240 kg extra leaves during the second year of study. The cows were followed for 15 consecutive months (experimental period). The outcome was reproductive performance, length of the lactation period and weight gain in calves.

The Kaplan-Meier survival estimator was used to compare calving interval and lactation period between experiment and control groups. Additional feeding did not stimulate reproduction efficiency in this trial. However, it promoted the length of the lactation period. Multivariable linear models predicted increased growth in calves of cows fed the higher quantity of leguminous leaves. Body girth was numerically lower in male compared to female calves. This study reveals a potential for intensified and more sustainable meat and milk production through changes in feeding regimes.

M. A. Bhatti (✉) · D. C. Jansen · T. B. Klem · M. Stokstad · O. Reksen
Norwegian University of Life Sciences (NMBU), Faculty of Veterinary Medicine, Department of Production Animal Clinical Sciences, Oslo, Norway
e-mail: muhammad.azher.bhatti@nmbu.no

W. Chanza · L. A. Kamwanja · S. Chikomola · M. Chipandula · A. Chikaonda
Trustees of Agricultural Promotion Programme, LPMU Headquarters, Lilongwe, Malawi

S. Klevar
Norwegian Veterinary Institute, Oslo, Norway

B. R. Singh et al. (eds.), *Climate Impacts on Agricultural and Natural Resource Sustainability in Africa*, https://doi.org/10.1007/978-3-030-37537-9_14

1 Introduction

Intensification of livestock production is necessary to decrease greenhouse gas emissions and improve farmers' economic situation and food security in Africa. The livestock sector contributes 14.5% of greenhouse gas emissions worldwide. Reducing the unproductive part of the herd by improving health, weight gain and reproductive performance is the most important contribution for reducing the carbon footprint from this sector (Gerber et al. 2013b).

Malawi has a cattle population of 1.5 million, dominated by the indigenous Zebu cattle (Nandolo et al. 2015). The Malawi Zebu cattle depend mainly on natural grass and crop residues, which most often have low nutritive value. Usually, there is no supplemental feeding with concentrates or legume crop. During the rainy season, cattle grazing is restricted to the hill and roadsides as the arable land is used for growing crops while the dambo (wetland) is unsuitable because of high prevalence of worms and liver flukes.

Malawi has a tropical climate with two distinct seasons, which are rainy (November to April) and dry (May to October) seasons, with annual rainfall ranging from 700 to 2400 mm (Banda et al. 2012). Climate change effects have become visible in Malawian livestock farming. Erratic rainfalls and prolonged dry season may further decrease fodder availability (Morton 2007), which severely affects production and reproduction of the livestock (Thornton et al. 2009). Since grazing lands are shrinking due to increased incidences of drought, farmers migrate into traditional grazing areas for cattle (Akinnagbe and Irohibe 2014). During the dry season, both the quality and quantity of pasture diminish, and the livestock struggle to maintain body condition; hence, reproduction may be compromised (Klinedinst et al. 1993). Particularly, the onset of ovarian activity after calving may be delayed, leading into longer calving intervals, fewer calves per cow and reduced herd growth and health (Chauhan and Ghosh 2014). Previous studies from Malawi have shown significant potential for increased reproductive efficiency by improved availability of bulls and better care of primiparous Zebu cows (Bhatti 2016). Also, improved calf survival and enhanced milk production will decrease the carbon footprint of milk and meat production (Gerber et al. 2013b).

Leguminous plants have been used to improve protein content in cattle feeding in several African countries. *Gliricidia sepium* leaves comprise of 21% crude protein, 23.8% crude fibre and 21–34% dry matter (Ministry of Agriculture and Fishries 2013). The high content of protein makes *Gliricidia* an excellent protein supplement to grass and other basal feeds. Numerous reports conclude that supplementing feeding with these leaves gives the cows increased weight gain and milk production (Aye and Adegun 2013). Therefore, the overall objective of this investigation was to study the effect of supplementing lactating Malawi Zebu cows during the dry season on reproductive performance and lactation in cows and weight gain in the offspring. Specifically, our first aim was to study the impact of supplemental feeding of leguminous leaves and maize bran on the calving interval. Our second aim was to study the association between supplemental feeding and the length of the lactation period and lastly to study the association between supplemental feeding and body girth circumference as a proxy for body weight attainment in the offspring.

2 Materials and Methods

In order to assess the association between the outcome variables and the intervention, a field experimental study was conducted in Bolero EPA, Rumphi district, Malawi. The outcomes were calving interval (days), length of the lactation period (days) and body weight attainment in calves (cm changes in body girth); and the intervention was supplemental feeding of the dam. Herds from seven agricultural sections were represented: Bolero, Bata, Chozoli, Chikwawa, Chirambo, Mjumba and Jalir.

A total of 98 cows were included in the study. They were enrolled at calving and followed for 15 months. The study was conducted for two consecutive years. The first trial ($n = 56$) commenced at calving between August 15th and September 12th in 2015 and ended on December 22nd in 2016, and the second trial ($n = 42$) started at calving between August 1st and September 19th in 2016 and ended on December 22nd in 2017. Half of the cows ($n = 49$) were assigned to a control group with no additional feeding, and the other half (28 in 2015 and 21 in 2016) were assigned to an experimental group receiving supplementary feeding. The extra feeding consisted of *Gliricidia sepium* leaves and maize bran and was given only during the driest months of the year: October, November and December. For both experimental groups, 2 kg maize bran was given per cow per day. In the first year of the study, each of the experimental cows was fed 114 kg leguminous leaves during the dry period (experimental group 1). The quantity of leaves was increased to 240 kg per cow for experimental cows in the second year of the study (experimental group 2). This gave us a total of four experimental groups: experimental group 1 (low-level additional feeding), experimental group 2 (high-level additional feeding), and two control groups for years 1 and 2, respectively (Table 1). Each pair of control and experimental cows was matched according to calving date (±15 days), parity of the dam (first, second and third parity and older) and within one out of the seven agricultural sections.

The feed was analysed for dry matter, ash (oven drying) and crude protein by Kjeldahl using the AOAC (2000) procedures at the Animal Science Department, Lilongwe University of Agriculture and Natural Resources, Malawi. Gross energy was calculated using 6100 Oxygen Bomb Calorimeter (Yan and Kim 2011). The acid detergent fibre was analysed by using an Ankom fibre analyser and Ankom (xt10) extractor as described by (Cao et al. 2009).

Reproductive outcome was measured as time from recruitment at calving in 2015 or 2016 and until next calving occurred within 15 months (calving interval). The follow-up period in each of the years ended on December 22nd the year after (2016 and 2017). Data from both years were collated ($n = 98$), and calving ($n = 58$)

Table 1 The study design for supplementary feeding in years 2015 and 2016

Year	Supplementary feed group (n)	Control group (n)	Supplementary feed offered (kg/cow)
2015	28	28	114
2016	21	21	240

was the event of interest (failure). Cows that had not delivered a calf before December 22nd 2016 or 2017 were right censored ($n = 28$). Cows that were excluded from the study (before the study period ended) were left censored ($n = 12$).

Milk samples were obtained two times weekly for progesterone assessment during the first year of the study (results reported in another publication included in this book), such that frequent visits by the field technician enabled an accurate determination of the lactation period in 56 cows, and cessation of milk production ($n = 42$) was the event of interest (failure). Cows that were still lactating at the end of the milk sampling period were right censored ($n = 14$). The last day of milk sampling was used as the day of censoring in these cows. None of the cows were left censored before cessation of milk yield.

The calves ($n = 98$) were subject to body girth measurements in October, November, December, January and March. The difference in body girth between October and January the following year, and between October and March the following year, was used to assess associations between body girth change (weight gain) and supplemental feeding of the dams. The age of the calf at the day of body girth measurement in October, January and March and sex of the calves were recorded for inclusion in statistical models.

3 Statistical Analyses

The statistical software package Stata 14 (StataCorp LLC, Texas, USA) was used in all statistical analyses. Statistical significance was considered with a *P*-value less than 0.05 in all models. Univariable Kaplan-Meier (K-M) estimators were used to assess relationships between calving interval and supplemental feeding (feeding, 1/0). Differences in survival functions between feeding groups were tested separately for each cohort and for the two cohorts joined. Similarly, the association between lactation period and feeding, 1/0, was assessed for the first year of the study. Difference between the groups was assessed by the log-rank test for equality of survivor functions.

Separate generalised multivariable linear regression models were run for the assessment of associations between body weight attainment and the explanatory variables: feeding, 1/0, gender (sex) and age of the calves in January and March. Initially, univariable associations between the two outcome variables, body girth attainment from October through January and October through March, and the explanatory variables were explored by analysis of variance (feeding, 1/0, and sex) or linear regression (age). Univariable associations with a *P*-value of less than 0.15 for the two-directional association with the outcome qualified for inclusion in a separate multivariable linear regression analysis. A backward selection procedure was employed, and explanatory variables with a *P*-value of less than 0.10 were kept in the final model. This resulted in two models explaining the body weight attainment for the two periods. The first model assessed body weight attainment for the period from October through January and included the explanatory variables: feeding, 1/0, and age of the calves in January. The second model estimated body weight attainment from October to March and included the explanatory variables: feeding, 1/0, and sex of the calves. Overall statistical significance of the models was assessed by the type III F-test in Stata. Homoscedasticity and normality of the residuals were assessed using plots of standardised residuals.

4 Results

The composition of feed shows that crude protein value of mixed supplement feed (*Gliricidia* and maize bran) is up to the recommended level (17–19%) for dairy cattle (Colmenero and Broderick 2006). The feed nutrient composition is presented in Table 2.

Altogether, 58 cows calved within the 15-month observation periods when the 2 years were collated. The reasons for left censoring were mortality ($n = 4$), sold for slaughter ($n = 5$), non-compliance ($n = 1$) and abortion ($n = 2$). Cows that had not delivered within the 15-month observation period were right censored ($n = 28$).

Feeding was not associated with calving interval in any of the three analyses conducted (separately for each experimental group and collated for both groups). For the assessment of both groups simultaneously (Fig. 1), the log-rank chi-square statistic was 0.21 ($P = 0.64$). The median number of days to the next calving (calving interval) was predicted to be 372 days with the upper and lower confidence interval (CI) of 322 and 415 days, respectively.

The length of the lactation period was associated with supplemental feeding for experimental group 1 of the study (log-rank statistic, 4.70, $P = 0.03$) (Fig. 2). The median number of days to cessation of milk yield was 133 days (CI: 125–158). This period was extended to a median value of 158 days in cows receiving supplemental feed versus 127 days in the control group that were subjected to pasture feeding only.

Supplemental feeding was significantly associated with body girth attainment in the calves in experimental group 2 (body weight) ($P < 0.01$) not only for the shorter period until January (Table 3), but also a prolonged effect of feeding was apparent until March in the second year of the study (Table 4).

These associations were not significant for experimental group 1 when cows were fed a markedly lower quantity of leguminous leaves (114 kg). *P*-values for the

Table 2 Nutritional composition of the supplemental feed

Sample	Crude protein (%)	Dry matter (%)	Ash (%)	Crude fibre (%)	Fat (%)	Energy (Cal/g)
Gliricidia sepium	19	87	8.5	16.7	11.2	6934
Maize bran	13	90	4.6	17.1	14.9	7258
Total supplemental feed	16	88.5	6.5	16.9	13.1	7096

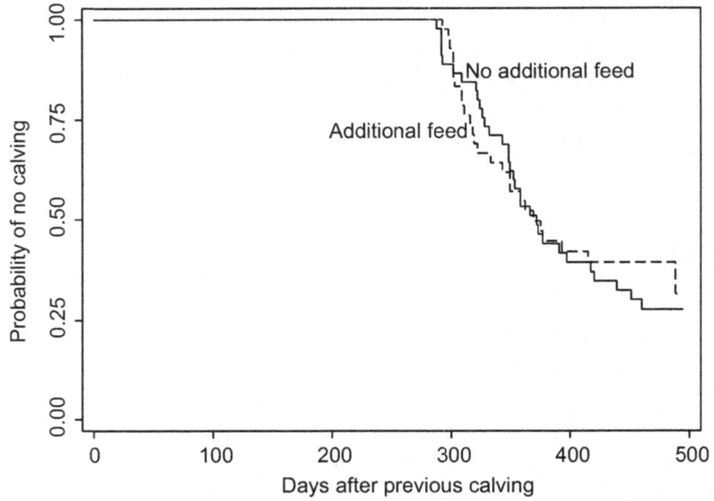

Fig. 1 Kaplan-Meier plot of proportion of calvings distributed by cows (n = 98) that received additional feed (n = 49) and a control group (n = 49) that were subjected to pasture feeding only. The time variable is days since the previous calving

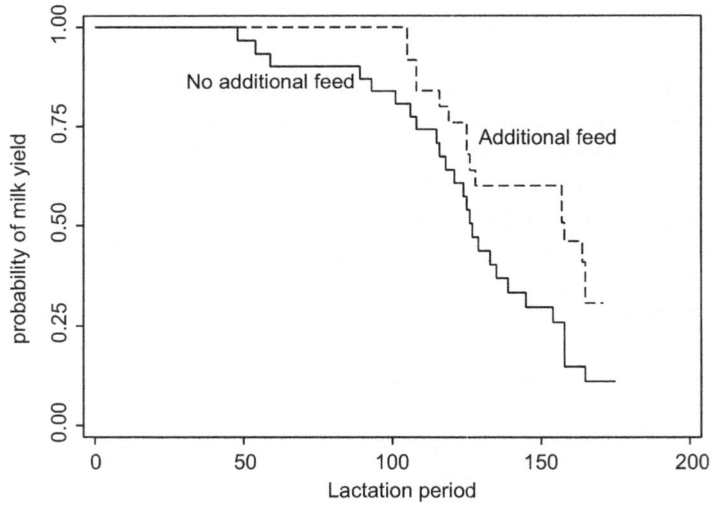

Fig. 2 Kaplan-Meier plot of the length of the lactation period distributed by cows (n = 56) that received additional feed (n = 28) and a control group (n = 28) that were subjected to pasture feeding only. The time variable is days since the previous calving

association between additional feeding and body weight gain in univariate tests during the October to January period were 0.37 and 0.49 for both years pooled and for the first year, respectively. Corresponding P-values for the October to March periods were 0.25 and 0.98.

Table 3 Associations between body girth attainment of offspring from October to January and additional feeding of maize bran and 240 kg leguminous leaves for the dams, adjusted for the difference in age (range 127–183 days)

Explanatory variable		Body girth attainment (cm)	S.E.	*P*-value
Feeding	No	-----	1.83	<0.01
	Yes	5.35		
Age in January		0.13/day	0.06	0.04
Intercept		−7.47	9.63	0.44

Table 4 Model assessing associations between body girth attainment from October to March and additional feeding of maize bran and 240 kg leguminous leaves for the dams adjusted for sex of the calves

Explanatory variable		Body girth attainment (cm)	S.E.	*P*-value
Feeding	No		1.94	<0.01
	Yes	5.42		
Sex	Female	------	2.10	0.08
	Male	−3.8		
Intercept		21.3	10.87	<0.01

The October–January model for experimental group 2 showed a positive linear relationship between body girth attainment and age of the calves, showing a more substantial body weight attainment for calves born early during the calving season. This model explained 24% of the variation ($R^2 = 0.24$) in body weight attainment (F-statistic: 6.04, $P < 0.01$) (Table 3). The mean age of the calves at the start of the study was 19.9 days in the group that received additional feeding (experimental group) and 19.2 days in the control group.

The effect of age was not apparent in the October to March model (Table 4). However, a numerical association between body weight attainment and sex becomes apparent when the calves were followed until March. The model assessing live weight from October through March explained 21% of the variation ($R^2 = 0.21$) in body weight attainment (F-statistic 5.27, $P < 0.01$) (Table 4).

5 Discussion

It has been stated that greenhouse gas emission from livestock production can be reduced by almost one-third if the practices currently applied by the producers with the lowest emission rate were widely adopted. This reduction can be achieved without changes in the production systems (e.g. from smallholder to industrial). The most significant decreases in the emission rate are associated with practices that improve efficiency at farm and animal level. They include better health and reproduction management and better feeding practices, such that the unproductive part of the herd decreases in number (Gerber et al. 2013a).

To our knowledge, it is the first Malawian study where nutritional regimes for cattle have been studied to meet the challenges that climate change represents for animal production in northern Malawi. The results in the present study show that supplementing lactating cows during the dry months when feed is scarce and of low quality is a manageable way of increasing body weight attainment in their calves. Calves from dams receiving supplemental feeding had 5 cm more body girth after adjustment for gender and time of birth in multivariable generalised linear models. These findings are in agreement with Maphane and Mutshewa (1999) who demonstrated that growth and body weight are profoundly affected by supplementation of feeding in dry season.

Body weight was greater in female calves in March. It is also reflected in previous investigations, revealing that female calves are better cared for than male calves. This is probably because they represent greater value for the farmer. These findings are in line with previous results (Chang'a et al. 2012; Young 1972) that found that predicting cattle live weights can be influenced by sex, environment, production system, animal husbandry practices and other factors. Increased food production and decreased environmental impact from livestock production require that the meat production potential in male calves is exploited in a more efficient matter than is currently the case.

Prolonged lactation periods may very well explain why the body weight attainment was greater in calves from experimental cows as compared to calves from cows in the control group in this study, although this was not measured during the same 2nd year. However, length of the lactation period was positively influenced by a relatively low quantity of leguminous leaves (114 kg) during experiment 1. Due to resource demanding work in the field, it was not possible to determine accurately when the lactation stopped during experiment 2. This was only possible during experiment 1 when the field assistant collected milk samples from all 56 cows twice weekly in 36 different farms. We were also not able to obtain accurate measures of milk yield, so lactation length was chosen as a measure for increased milk yield.

This study showed that feeding supplementation of lactating Malawi Zebu cows during the dry season did not influence reproductive outcome in this population of Zebu cattle. Another publication included in this book shows a relatively low incidence of some common infectious diseases in this population of Zebu cattle. The most important means of increasing reproductive efficiency in these 36 farms is therefore related to bull management and herd composition. Processing of feed ingredients and incorporation of leguminous leaves result in higher quality and digestibility of feed (Mekoya et al. 2008). It has previously been shown that an increase of feed digestibility has been associated with marked decreases in age at first calving. This is because feed digestibility is related to growth rate in animals (Keady et al. 2012), and the assumption is that growth rate and age at first calving are closely associated (Gerber et al. 2013a). However, the period of the current study did not allow for comparisons of age at first delivery.

6 Conclusion

Milk and meat production in indigenous Zebu cattle can be intensified by supplemental feeding of leguminous leaves during the dry period. Implementing such practices would reduce greenhouse gas emissions per unit of produced food and represents a potential for improved livelihood of smallholder farmers in Malawi.

Acknowledgements The authors would like to thank the Malawian smallholder farmers, who actively participated in this study. We are grateful to Dr. Liveness Jessica Banda and Dr. Fanny Chigwa, Animal Science Department, Lilongwe University of Agriculture and Natural Resources (LUANAR), for their help in laboratory analyses. This study was supported by the project: "Capacity Building for Managing Climate Change in Malawi (CABMACC)", Norwegian University of Life Sciences, Norway.

Conflict of Interest The authors declare that they have no conflict of interest.

Funding This study was funded by Capacity Building for Managing Climate Change in Malawi (grant number CABMACC).

References

Akinnagbe O, Irohibe I (2014) Agricultural adaptation strategies to climate change impacts in Africa: a review. Bangladesh J Agric Res 39(3):407–418

AOAC (2000) Official methods of analysis. AOAC International, Arlington

Aye P, Adegun M (2013) Chemical composition and some functional properties of Moringa, Leucaena and Gliricidia leaf meals. Agric Biol J N Am 4(1):71–77

Banda LJ, Kamwanja LA, Chagunda MGG, Ashworth CJ, Roberts DJ (2012) Status of dairy cow management and fertility in smallholder farms in Malawi. Trop Anim Health Prod 44(4):715–727

Bhatti MA (2016) Climate change resilience through enhanced reproduction and lactation performance in Malawian Zebu cattle. Masters Thesis: Norwegian University of Life Sciences, Ås, Norway, p 70

Cao Z, Anderson J, Kalscheur K (2009) Ruminal degradation and intestinal digestibility of dried or wet distillers grains with increasing concentrations of condensed distillers solubles. J Anim Sci 87(9):3013–3019

Chang'a JS, Løken T, Mdegela RH, Reksen O (2012) Factors associated with body weight attainment in calves on smallholder dairy farms in Tanzania. Open J Vet Med 2:66–73

Chauhan DS, Ghosh N (2014) Impact of climate change on livestock production: a review. J Anim Res 4(2):223

Colmenero JO, Broderick G (2006) Effect of dietary crude protein concentration on milk production and nitrogen utilization in lactating dairy cows. J Dairy Sci 89(5):1704–1712

Gerber PJ, Henderson B, Makkar HP (2013a) Mitigation of greenhouse gas emissions in livestock production. A review of technical options for non-CO2 emissions. FAO, Rome

Gerber PJ, Steinfeld H, Henderson B, Mottet A, Opio C, Dijkman J, Falcucci A, Tempio G (2013b) Tackling climate change through livestock: a global assessment of emissions and mitigation opportunities. Food and Agriculture Organization of the United Nations (FAO), Rome

Keady T, Marley C, Scollan N, Kuoppala K, Rinne M, Vanhatalo A (2012) Grass and alternative forage silages for beef cattle and sheep: effects on animal performance. In: Kuoppala K, Rinne M, Vanhatalo A (eds) Proceedings XVI international silage conference. Hämeenlinna, Finland, pp 152–165

Klinedinst PL, Wilhite DA, Hahn GL, Hubbard KG (1993) The potential effects of climate change on summer season dairy cattle milk production and reproduction. Clim Chang 23(1):21–36

Maphane GK, Mutshewa P (1999) Strategies for dry season feeding of animals in Botswana. In: Strategies for dry season feeding of animals in central and southern Africa. Proceedings of a joint ZSAP/FAO workshop held in Harare, Zimbabwe, pp 12–17

Mekoya A, Oosting S, Fernandez-Rivera S, Van der Zijpp A (2008) Multipurpose fodder trees in the Ethiopian highlands: farmers' preference and relationship of indigenous knowledge of feed value with laboratory indicators. Agric Syst 96(1–3):184–194

Ministry of Agriculture and Fishries (2013) Legumes an Ideal Fodder in the Dry Season. Division, L. R. a. I. Bodles Research Station Old Harbour, P. O. St. Catherine, Jamaica, p 2

Morton JF (2007) The impact of climate change on smallholder and subsistence agriculture. Proc Natl Acad Sci 104(50):19680–19685

Nandolo W, Gondwe T, Banda L (2015) Characterisation of breeding systems for Malawi Zebu cattle in Mzimba District, Northern Malawi. Malawi J Agric Nat Res Dev Stud 1:14–20

Thornton PK, van de Steeg J, Notenbaert A, Herrero M (2009) The impacts of climate change on livestock and livestock systems in developing countries: a review of what we know and what we need to know. Agric Syst 101(3):113–127

Yan L, Kim I (2011) Evaluation of dietary supplementation of delta-aminolevulinic acid and chitooligosaccharide on growth performance, nutrient digestibility, blood characteristics, and fecal microbial shedding in weaned pigs. Anim Feed Sci Technol 169(3):275–280

Young DL (1972) The estimation of liveweight from heart girth within specified age/sex groups of Kenya range cattle. East Afr Agric For J 38(2):193–200

Effects of Concentrate Supplementation on the Fatty Acid Composition of Fat Depots in Crossbred Goats

D. E. Mushi and L. O. Eik

Abstract The effects of concentrate diet supplementation on the fatty acid compositions of M. longissimus dorsi (LD), minced meat (MM) and omental fat (OF) in Small East African and Norwegian crossbred goats (9.5 months old, 17.1 kg live weight) were studied. Thirty-two castrated male goats were equally allotted into four levels of concentrate supplementation for 90 days: at 100% (T100), 66% (T66), 33% (T33) and 0% (T0) of ad libitum allowance. In LD and MM, proportions of linolenic acid and other n-3 polyunsaturated fatty acids (PUFA) decreased, whereas that of monounsaturated fatty acids (MUFA), conjugated linolenic acid (CLA) and total desirable fatty acids (DFA) increased with a higher level of supplementation. Concentrate supplementation had limited effect on the fatty acid composition of OF, marked by an increase in t-MUFA and CLA with a higher level of supplementation. Overall, concentrate supplementation beyond 66% of ad libitum feeding had limited effects on fatty acid composition, as depicted by the similarity between T66 and T100 goats in the proportions of oleic acid, CLA, total MUFA as well as n-6/n-3 PUFA ratio in LD and MM. Irrespective of the level of supplementation, LD was associated with higher proportions of total unsaturated fatty acids (UFA), MM with higher proportions of MUFA and OF with higher proportions of total saturated fatty acids (SFA). It is concluded that in order to optimise the healthfulness of goat meat, concentrate supplementation should be limited to 66% of ad libitum feeding. Consumers should be advised to refrain from high intake of internal fat depots due to the abundance of SFA and unfavourable n-6/n-3 PUFA ratios, a risk factor for cardiovascular diseases.

D. E. Mushi (✉)
Department of Animal, Aquaculture and Sciences, Sokoine University of Agriculture, Morogoro, Tanzania
e-mail: danim@sua.ac.tz

L. O. Eik
Department of International Environment and Development Studies, Noragric, Norwegian University of Life Sciences, Ås, Norway

B. R. Singh et al. (eds.), *Climate Impacts on Agricultural and Natural Resource Sustainability in Africa*, https://doi.org/10.1007/978-3-030-37537-9_15

1 Introduction

The increasing consumption of goat meat in developing countries and among ethnic groups in developed countries is attributed to the increasing proportion of individuals with high and middle income, the association of goat meat with religious festivals, smaller carcasses that allow for home slaughter and the low fat content of the meat (Beserra et al. 2004; McMillin and Brock 2005; Lee et al. 2008). To match supply with demand requires an increase in production, which can be achieved through improved feeding and genotypes. Goats deposit most of their fat internally around the viscera and less on the carcass and can thus provide a good source of lean meat (Babiker et al. 1990; Dhanda et al. 1999). Health-conscious consumers tend to prefer low-fat products. It is now known that it is the composition of fatty acids in meat and not fat content per se that determines the healthfulness of meat to consumers (Horcada et al. 2012; Warren et al. 2008). In view of this knowledge, the rating of goat meat for healthfulness should be based on its fatty acid profile and not only on its leanness.

Despite the potential of goats to contribute to the supply of healthy meat, there is limited research on the quality of goat meat, particularly regarding its fatty acid composition (Banskalieva et al. 2000). Wood et al. (2008) reported variations in the fatty acid composition of adipose tissue depots of pigs, sheep and cattle and in the factors affecting the composition. For instance, in order to speed up the growth rate and reduce the time taken for ruminants to reach slaughter weight, it is now common practice to finish such animals on energy-rich, concentrate diets. This practice may affect the composition of fatty acids in meat from ruminants. There is therefore a need to elucidate the effects of dietary intervention on the fatty acid composition of goat meat. A thorough study of fatty acid composition of meat animals should analyse the different fat depots to determine any depot-specific variation in fatty acid composition. This approach is particularly important in developing countries, where internal as well as subcutaneous fat depots are regularly consumed. Shirouchi et al. (2014) reported that each fat depot in an animal has unique characteristics and cannot be taken as representative for all other fat depots in the body. This study therefore seeks to determine the effect that supplementing the diets of crossbred goats with various levels of concentrate feed has on the fatty acid composition of minced meat (MM), muscle longissimus dorsi (LD) and omental fat (OF) in those animals.

2 Materials and Methods

2.1 Animals and Treatments

Thirty-two castrated male SEA and Norwegian crossbred goats (9.5 months old, 17.1 kg LWT) were allotted into eight weight blocks. Dietary treatments were assigned to each weight block in a completely randomised design. The dietary treatments were T0, where no concentrate supplementation was offered; T33 and T66, where the amount of concentrate on offer was equivalent to 33% and 66% of ad libitum concentrate intake, respectively; and T100, where concentrate was fed ad libitum allowing 10% refusals from the daily concentrate allowance.

2.2 *Feeding Management*

Animals were given a 3-week adaptation period during which they were treated with ivermectin for internal and external parasites. Goats were stall-fed in groups of four animals per pen. Grass hay was offered ad libitum allowing 20% refusals from the daily hay allowance. Concentrate was fed twice daily, at 9:00 and 15:00 hrs, while water was offered ad libitum. During an experimental period of 90 days, feed allowances and refusals were recorded daily for each group.

2.3 *Sampling of LD, MM and OF for Fatty Acid Analyses*

At the end of the experimental period, goats were fasted for 16 h before slaughter. At 45 minutes post-mortem, each carcass was dissected longitudinally into two equal halves through the median plane using a band saw. The carcasses were chilled at 0 °C for 24 hours before sampling. The M. longissimus dorsi (LD) was excised from each left-half carcass and split into blocks measuring approximately 7 cm in length. One block, at the posterior end of LD, was used for fatty acid analyses. The remaining left-half carcasses were dissected into muscle, fat and bone. The muscle and fat tissue from each half-carcass was thoroughly mixed together to represent the total edible tissue in a carcass and minced using a 5 mm sieve; three subsamples of the minced meat (MM) were then taken for fatty acid analyses. Approximately 10 g of omental fat (OF) was sampled for fatty acid analyses. All samples were packed in PVC bags and frozen at -25 °C until analyses were complete.

2.4 *Fatty Acid Analyses*

Analyses of fatty acids in the samples were carried out at the Department of Animal and Aquacultural Sciences at the Norwegian University of Life Sciences. Synthesis of fatty acid methyl esters (FAME) was conducted using the direct method of O'Fallon et al. (2007). Following the synthesis of FAME, 3 mL of hexane was added to each reaction tube after which the tubes were vortex-mixed for 5 min on a multi-tube vortex. The tubes were centrifuged for 5 min at 3000 rpm in a tabletop centrifuge (Wifug Ltd., England). The hexane layer containing the FAME was transferred from each tube into a gas chromatography (GC) vial. Each vial was capped and brought to −20 °C before fatty acid separation in a GC.

Fatty acid methyl esters were separated in a Thermo Finnigan Focus GC equipped with automatic injector, Restek capillary column (Rt – 2560; 0.25 mm internal diameter, 100 m long) and flame ionisation detector. Samples were injected in a split mode (1:40). The carrier gas was helium (He), which flowed at a pressure of 2.7 bars. Oven temperature programming was 70 °C for 2 min, 70 to 150 °C for 4 min, 150 °C for 34 min, 150 to 230 °C for 57 min and a final temperature of 230 °C for 10 min. The injection temperature was 230 °C, and the detection temperature was 255 °C. The identification of individual FAME in the samples was achieved by matching the retention time of the unknown FAME with that of known FAME in the standard mixture (FAME mix, 37 components, Supelco™). Fatty acid composition in the samples was calculated as the peak area percentage for each fatty acid, including unidentified fatty acids. Integration calibration software (Chromeleon, V6.7) connected to the chromatograph, which converts relative peak areas into weight percentages, was used for these calculations.

2.5 Physical and Chemical Compositions of Dietary Feeds

The grass hay consisted of *Brachiaria* spp. (70%) and *Bothriochloa* spp. (30%). Concentrate supplement was made of 28% sunflower seed cake, 70% maize bran, 1.3% lime, 0.2% salt and 0.5% mineral mix, with crude protein of 16.2% and estimated metabolisable energy of 13.4 MJ. Chemical and fibre compositions of both the grass hay and concentrate diet are reported in Table 1. Fatty acid composition of the concentrate diet is also indicated in Table 1.

2.6 Statistical Analysis

Experimental data were analysed using the general linear model (GLM) procedure of SAS (2001). Dietary treatments were considered as fixed effects and residual as random effects. Each individual animal served as an experimental unit for all fatty acids assessed. Due to small variation in the weight of animals within treatments, fatty acid composition was corrected for animal weight difference by using weight as a covariate in the GLM. In all analyses, when least square means were significant by ANOVA at $P < 0.05$, they were separated by the probability of difference (PDIFF) option of SAS.

In order to determine the distribution of fatty acids in different fat depots (LD, MM and OF), principal component analysis (PCA) was carried out on fatty acid compositions of the three depots using the Unscrambler version 9.2 (CAMO Process AS, Oslo, Norway) software. Full cross-validation method was employed in PCA.

Table 1 Chemical composition (g/kg DM) of concentrate diet and grass hay and fatty acid composition (% of total fatty acids) of concentrate diet fed

	Feeds	
Chemical composition	Concentrate	Grass hay
Dry matter (g/kg)	921.9	834.5
Organic matter	920.5	902.0
Crude protein	162.2	32.8
Ether extract	133.6	12.0
Crude fibre	145.8	353.3
Neutral detergent fibre	472.1	830.5
Acid detergent fibre	156.1	473.5
Ash	50.9	98.0
Nitrogen free extract	429.4	338.4
In vitro dry matter digestibility	545.6	395.5
In vitro organic matter digestibility	546.4	416.5
Metabolisable energy (MJ/Kg DM)	13.4	9.2
Fatty acids		
C16:0 Palmitic	12.30	–
C16:1n7 Palmitoleic	0.10	–
C18:0 Stearic	2.90	–
C18:1n9c Oleic	33.10	–
C18:2n6c Linoleic	46.80	–
C20:0 Eicosanoic	0.50	–
C18:3n3 Linolenic	0.40	–
C22:0 Behenic	0.32	–
C22:1n9 Erucic	0.01	–
C24:0 Lignoceric	0.30	–
C22:5n3 Docosapentaenoic	0.03	–
C22:6n3 Docosahexaenoic	0.01	–

3 Results

3.1 Fatty Acid Composition in M. Longissimus Dorsi (LD)

The main fatty acids in LD were palmitic (C16:0), stearic (C18:0), oleic (C18:1) and linoleic (C18:2) that together accounted for 52–76% of total fatty acid content (Table 2). The proportions of palmitic, oleic, elaidic, linolelaidic, rumenic (a conjugated linoleic acid (CLA)), total monounsaturated (MUFA), trans-MUFA and total desirable fatty acids (DFA – a total of all unsaturated fatty acids and stearic acid) increased ($P < 0.05$) with level of supplementation. Similarly, n-6/n-3 ratio increased ($P < 0.05$) with supplementation. The proportion of oleic, elaidic and trans-MUFA in T100 goats was twice as high ($P < 0.05$), whereas that of CLA was about four times higher than that recorded in T0 goats. However, T66 and T100 goats had comparable proportions of oleic, linolelaidic, CLA, total MUFA as well as n-6/n-3 polyunsaturated fatty acid (PUFA) ratio. On the other hand, the proportions of myristoleic, linolenic, eicosatrienoic, arachidonic, eicosapentaenoic, docosapentae-

Table 2 Least square means (±SE) for fatty acid compositions (percentage of total fatty acids) of LD muscle from castrated male SEA × Norwegian crossbred goats

		T0	T33	T66	T100	SE	*P*
Fatty acid	Structure	% of total fatty acids					
Capric	C10:0	0.03^{b}	0.06^{a}	0.06^{a}	0.05^{a}	0.01	**
Lauric	C12:0	0.06	0.06	0.06	0.05	0.01	NS
Myristic	C14:0	0.83	1.00	1.43	1.19	0.18	NS
Myristoleic	C14:1n5	0.23^{a}	0.12^{b}	0.09^{b}	0.10^{b}	0.02	**
Pentadecanoic	C15:0	0.30	0.27	0.31	0.20	0.04	NS
Palmitic	C16:0	13.02^{b}	16.70^{a}	17.84^{a}	16.95^{a}	0.80	**
Palmitoleic	C16:1n7	1.00^{a}	0.65^{bc}	0.67^{b}	0.55^{c}	0.04	***
Margaric	C17:0	0.60	0.66	0.90	0.87	0.10	NS
Stearic	C18:0	14.46^{c}	19.75^{a}	19.20^{a}	15.78^{bc}	0.73	***
cis-Vaccenic	C18:1n7	1.25^{a}	0.93^{bc}	0.92^{c}	1.10^{a}	0.06	**
Oleic	C18:1n9	17.70^{c}	27.34^{b}	33.34^{a}	32.48^{a}	1.60	***
Elaidic	C18:1n9t	0.18^{b}	0.34^{a}	0.40^{a}	0.35^{a}	0.02	***
Linoleic	C18:2n6	7.38^{bc}	10.67^{a}	7.30^{c}	10.31^{a}	0.90	*
Linolelaidic	C18:2n6t	0.11^{c}	0.23^{b}	0.35^{a}	0.34^{a}	0.02	***
Rumenic	C18:2c9t11	0.25^{c}	0.52^{b}	1.00^{a}	0.90^{a}	0.07	***
Linolenic	C18:3n3	0.92^{a}	0.31^{b}	0.16^{c}	0.16^{c}	0.03	***
Eicosanoic	C20:0	0.29^{a}	0.11^{b}	0.08^{b}	0.06^{b}	0.02	***
Gondoic	C20:1n9	0.12^{a}	0.06^{b}	0.03^{c}	0.05^{bc}	0.01	***
Eicosatrienoic	C20:3n3	13.65^{a}	5.00^{b}	2.63^{b}	4.05^{b}	1.00	***
Arachidonic	C20:4n6	0.11^{a}	0.07^{b}	0.04^{c}	0.06^{bc}	0.01	***
Heneicosanoic	C21:0	0.06^{a}	0.03^{b}	0.02^{b}	0.03^{b}	0.00	***
Behenic	C22:0	0.11^{a}	0.04^{b}	0.02^{b}	0.03^{b}	0.01	***
Erucic	C22:1n9	0.17^{a}	0.03^{b}	0.01^{b}	0.01^{b}	0.02	***
Eicosapentaenoic	C20:5n3	2.60^{a}	0.60^{b}	0.17^{b}	0.21^{b}	0.17	***
Docosapentaenoic	C22:5n3	3.74^{a}	0.96^{b}	0.36^{b}	0.41^{b}	0.22	***
Docosahexaenoic	C22:6n3	0.85^{a}	0.15^{b}	0.05^{c}	0.05^{c}	0.07	***
Lignoceric	C24:0	0.10^{a}	0.02^{b}	0.01^{b}	0.02^{b}	0.01	***
Total saturated	SFA	33.31^{c}	42.00^{ab}	43.62^{a}	39.37^{b}	1.34	***
Total unsaturated	UFA	52.70^{ab}	50.20^{b}	50.09^{b}	53.66^{a}	0.85	*
Total monounsaturated	MUFA	21.74^{c}	30.67^{b}	37.25^{a}	36.30^{a}	1.52	***
Total polyunsaturated	PUFA	30.95^{a}	19.50^{b}	12.83^{c}	17.36^{bc}	2.00	***
Total desirable	DFA	67.15^{b}	69.92^{a}	69.28^{a}	69.44^{a}	0.60	*
Total trans	t-MUFA	0.60^{b}	1.03^{ab}	1.35^{a}	1.25^{a}	0.20	*
Total n-6	n-6	7.64^{bc}	10.56^{a}	7.66^{c}	10.71^{ab}	1.00	*
Total n-3	n-3	21.26^{a}	7.03^{b}	3.30^{b}	4.80^{b}	1.25	***
Ratio (n-6:n-3)	n-6:n-3	0.36^{c}	1.54^{b}	2.26^{a}	2.23^{a}	0.12	***

SEA Small East African, *t-MUFA* sum of trans isomers ($\sum$ C18:1 trans^{4-16}). T0, T33, T66 and T100 refer to zero, 33%, 66% and 100% access to ad libitum concentrate allowance, respectively. ^{a,b,c}Least square means in the same row lacking a common letter differ ($P < 0.05$). *, ** and *** = $P < 0.05$, 0.01 and 0.001, respectively. *NS* not significant, *SE* standard error of the mean noic, docosahexaenoic, total n-3 PUFA and total PUFA decreased ($P < 0.05$) with supplementation. The proportion of linolenic acid, one of n-3 PUFA, in LD from T0 goats was six times higher than that of LD from T100.

3.2 Fatty Acid Composition in Minced Meat (MM)

The proportion of oleic, total MUFA, CLA, total unsaturated fatty acids (UFA) as well as n-6/n-3 PUFA ratio increased ($P < 0.05$) with higher levels of concentrate supplementation (Table 3). The proportion of CLA in T100 goats was more than two times higher than that recorded in T0 goats. Minced meat from concentrate-supplemented goats (T33, T66 and T100) had an almost 50% higher ($P < 0.05$) proportion of trans-MUFA than that of T0 goats, mainly due to trans-7-octadecenoic acid (C18:1 t7) (Table 3). However, MM from higher levels of concentrate supplementation, T66 and T100, had comparable proportions of UFA, MUFA and CLA as well as n-6/n-3 PUFA ratio. On the other hand, the proportion of stearic, linolenic, n-3 PUFA and total PUFA decreased with higher concentrate supplementation. The proportion of total PUFA in T100 goats was nearly half that recorded in T0 goats.

3.3 Fatty Acid Composition in Omental Fat (OF)

Concentrate supplementation had limited effects on the fatty acid composition in OF compared to other fat depots studied. The proportion of CLA and trans-MUFA increased ($P < 0.05$) with supplementation (Table 4). Omental fat from T66 and T100 goats had similar but 51% higher ($P < 0.05$) proportion of CLA than that of other dietary groups. On the other hand, the proportion of palmitic and total saturated fatty acids (SFA) decreased ($P < 0.05$) with supplementation.

3.4 Distribution of Fatty Acids in MM, LD and OF

Principal component one (PC1), which explained 63% of the variance in fatty acid composition, clearly separated LD from OF (Fig. 1). Minced meat was weakly separated from other depots along PC2, which explained 25% of the observed variance. Longissimus dorsi muscle was strongly associated with unsaturated fatty acids (UFA) especially PUFA, including n-3 and n-6 PUFA families (Fig. 2). On the other hand, OF was associated with SFA, whereas MM was associated with MUFA (Fig. 2). Specifically, LD muscle was associated with linolelaidic, oleic and heptadecanoic acids (Fig. 3). Omental fat was strongly linked with higher proportions of capric, octadecanoic and stearic acids. Minced meat was associated with CLA and palmitic and myristic acids along PC2 (Fig. 3).

Table 3 Least square means (±SE) for fatty acid compositions (percentage of total fatty acids) of minced meat (MM) from castrated male SEA × Norwegian crossbred goats

		T0	T33	T66	T100	SE	*P*
Fatty acid	Structure	% of total fatty acids					
Capric	C10:0	0.10	0.08	0.08	0.08	0.01	NS
Lauric	C12:0	0.14	0.11	0.11	0.11	0.02	NS
Myristic	C14:0	2.44	1.93	2.50	2.71	0.22	NS
Myristoleic	C14:1n5	0.05	0.05	0.08	0.09	0.02	NS
Pentadecanoic	C15:0	0.73	0.66	0.64	0.62	0.05	NS
Palmitic	C16:0	20.50	19.33	21.50	21.02	1.11	NS
Palmitoleic	C16:1n7	1.27[a]	0.90[b]	0.93[b]	0.91[b]	0.05	***
Margaric	C17:0	0.84[bc]	0.80[c]	1.23[a]	1.35[a]	0.11	**
Stearic	C18:0	23.73[ab]	26.28[a]	21.57[bc]	19.30[c]	1.00	***
cis-Vaccenic	C18:1n7	0.80[a]	0.61[b]	0.75[a]	0.81[a]	0.04	*
trans-Vaccenic	C18:1n7t	0.34	0.25	0.28	0.31	0.03	NS
Oleic	C18:1n9	26.51[c]	26.80[bc]	34.56[a]	36.20[a]	1.30	***
Elaidic	C18:1n9t	0.44	1.20	0.50	0.51	0.30	NS
Trans-7-octadecenoic	C18:1 t7	2.30[b]	3.44[a]	3.55[a]	3.20[a]	0.23	*
Linoleic	C18:2n6	4.16[a]	3.00[b]	3.60[ab]	3.91[a]	0.30	*
Rumenic	C18:2c9t11	0.62[c]	0.65[bc]	1.52[a]	1.58[a]	0.08	***
Linolenic	C18:3n3	0.56[a]	0.22[b]	0.20[b]	0.18[b]	0.03	***
Eicosanoic	C20:0	0.42[a]	0.15[b]	0.10[b]	0.10[b]	0.03	***
Gondoic	C20:1n9	0.10[a]	0.02[b]	0.02[b]	0.02[b]	0.01	***
Eicosatrienoic	C20:3n3	3.91[a]	0.52[b]	0.47[b]	0.62[b]	0.10	***
Arachidonic	C20:4n6	0.20[a]	0.04[b]	0.03[b]	0.03[b]	0.01	***
Eicosapentaenoic	C20:5n3	0.53[a]	0.04[b]	0.03[b]	0.03[b]	0.01	***
Docosapentaenoic	C22:5n3	1.30[a]	0.17[b]	0.12[b]	0.10[b]	0.10	***
Docosahexaenoic	C22:6n3	0.16[a]	0.02[b]	0.01[b]	0.02[b]	0.01	***
Total saturated	SFA	53.00	51.80	50.42	48.03	2.00	NS
Total unsaturated	UFA	43.31[bc]	40.21[c]	49.30[a]	51.43[a]	1.38	***
Total monounsaturated	MUFA	31.36[c]	35.20[bc]	43.00[a]	44.61[a]	1.34	***
Total polyunsaturated	PUFA	11.95[a]	5.00[b]	6.32[b]	6.82[b]	1.00	***
Total desirable	DFA	67.04	66.50	71.00	70.72	1.50	†
Total trans	t-MUFA	2.85[b]	6.25[a]	5.80[a]	5.70[a]	0.81	*
Total n-6	n-6	4.74[a]	3.07[b]	3.64[b]	4.00[ab]	0.34	*
Total n-3	n-3	6.43[a]	1.03[b]	0.80[b]	0.82[b]	0.63	***
Ratio (n-6:n-3)	n-6:n-3	0.96[c]	3.32[bc]	4.68[ab]	6.11[a]	0.84	**

SEA Small East African, *t-MUFA* sum of trans isomers ($\sum$ C18:1 trans$^{4-16}$). T0, T33, T66 and T100 refer to zero, 33%, 66% and 100% access to ad libitum concentrate allowance, respectively. [a,b,c]Least square means in the same row lacking a common letter differ ($P < 0.05$). *, ** and *** = $P < 0.05$, 0.01 and 0.001, respectively. † = $P < 0.1$. *NS* not significant, *SE* standard error of the mean

Table 4 Least square means (±SE) for fatty acid compositions (percentage of total fatty acids) of omental fat from castrated male SEA × Norwegian crossbred goats

		T0	T33	T66	T100	SE	*P*
Fatty acid	Structure	% of total fatty acids					
Capric	C10:0	0.12	0.12	0.12	0.12	0.01	NS
Lauric	C12:0	0.14	0.14	0.09	0.10	0.01	†
Myristic	C14:0	3.00[a]	2.10[c]	2.43[bc]	2.60[ab]	0.14	*
Pentadecanoic	C15:0	0.80[a]	0.80[a]	0.56[ab]	0.52[b]	0.10	*
Palmitic	C16:0	22.48[ab]	22.62[a]	20.94[bc]	20.11[c]	0.60	*
Palmitoleic	C16:1n7	1.43	1.17	1.20	1.07	0.07	†
Margaric	C17:0	0.52	0.42	0.48	0.48	0.03	NS
Stearic	C18:0	32.10	34.70	32.35	29.54	1.50	NS
cis-Vaccenic	C18:1n7	0.80	0.80	0.80	0.82	0.04	NS
trans-Vaccenic	C18:1n7t	0.21	0.31	0.28	0.30	0.03	NS
Oleic	C18:1n9	24.57	22.00	25.00	25.00	1.15	NS
Elaidic	C18:1n9t	0.35	0.51	0.52	0.68	0.11	NS
Linoleic	C18:2n6	3.51	3.18	3.08	3.26	0.20	NS
Rumenic	C18:2c9t11	0.90[bc]	0.84[c]	1.31[a]	1.34[a]	0.10	**
Linolenic	C18:3n3	0.22	0.20	0.15	0.12	0.04	NS
Eicosanoic	C20:0	0.18[ab]	0.22[a]	0.12[b]	0.12[b]	0.02	**
Eicosatrienoic	C20:3n3	0.22	0.16	0.20	0.20	0.02	NS
Arachidonic	C20:4n6	0.03[a]	0.02[b]	0.01[b]	0.01[b]	0.00	*
Docosanoic	C22:0	0.04[a]	0.03[ab]	0.01[bc]	0.02[c]	0.01	*
Docosapentaenoic	C22:5n3	0.13	0.10	0.05	0.06	0.02	NS
Docosahexaenoic	C22:6n3	0.05	0.02	0.03	0.02	0.01	NS
Total saturated	SFA	61.73[a]	63.32[a]	59.01[ab]	55.68[b]	2.00	*
Total unsaturated	UFA	34.83	31.85	36.00	36.70	1.43	NS
Total monounsaturated	MUFA	29.50	27.06	30.71	31.30	1.14	†
Total polyunsaturated	PUFA	5.33	4.80	5.16	5.38	0.32	NS
Total desirable	DFA	66.92	66.56	68.23	66.22	1.51	NS
Total trans	t-MUFA	1.34[c]	1.70[bc]	2.00[b]	2.50[a]	0.15	**
Total n-6	n-6	3.57	3.24	3.13	3.31	0.21	NS
Total n-3	n-3	0.60	0.47	0.42	0.40	0.05	NS
Ratio (n-6:n-3)	n-6:n-3	7.34	7.57	7.77	8.66	0.71	NS

SEA Small East African, *t-MUFA* sum of trans isomers ($\sum$ C18:1 trans$^{4-16}$). T0, T33, T66 and T100 refer to zero, 33%, 66% and 100% access to ad libitum concentrate allowance, respectively. [a,b,c]Least square means in the same row lacking a common letter differ ($P < 0.05$). *, ** and *** = $P < 0.05$, 0.01 and 0.001, respectively. † = $P < 0.1$. *NS* not significant, *SE* standard error of the mean

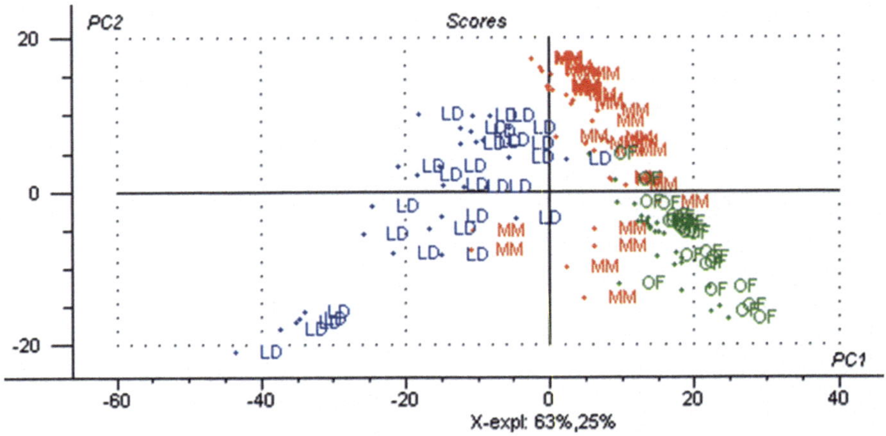

Fig. 1 Score plot (PCA) for longissimus muscle (LD), minced meat (MM) and omental fat (OF) from Small East African × Norwegian crossbred goats based on their fatty acid composition

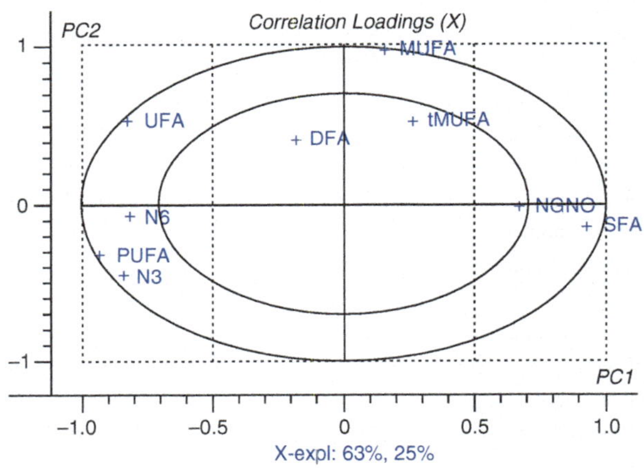

Fig. 2 Correlation loadings (PCA) for different groups of fatty acids with longissimus muscle, minced meat and omental fat from Small East African × Norwegian crossbred goats

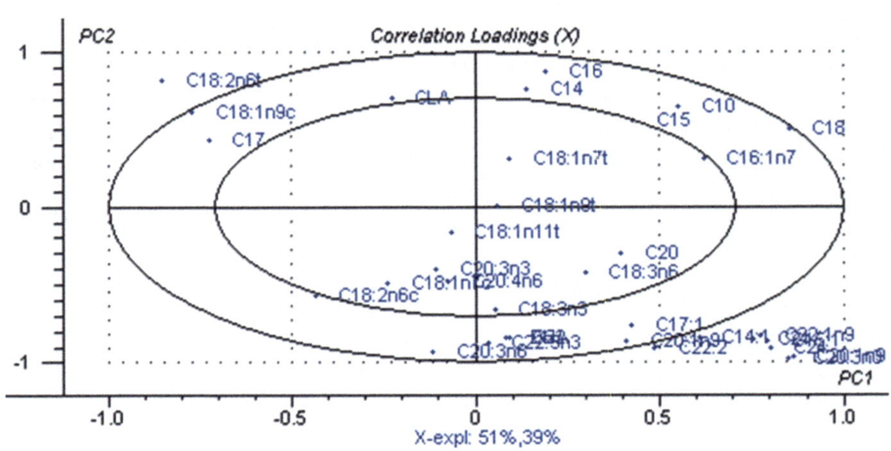

Fig. 3 Correlation loadings (PCA) for individual fatty acids with longissimus muscle, minced meat and omental fat from Small East African × Norwegian crossbred goats

4 Discussion

4.1 Fatty Acid Composition in M. Longissimus Dorsi (LD)

The higher proportion of palmitic (C16:0) and oleic (C18:1) acid observed in LD muscle from concentrate-supplemented goats may be due to the higher intake of the concentrate diet (Table 1). In agreement with our findings, Aurousseau et al. (2004) and Pordomingo et al. (2012) observed a rapid increase in the proportion of C16:0 in the longissimus muscle when the lambs and cattle were shifted from grazing to concentrate feeding. Saturated fatty acids such as palmitic, myristic (C14:0) and lauric (C12:0) raise LDL-cholesterol concentrations in blood, a risk factor for cardiovascular diseases (Lee et al. 2008). However, the health benefits associated with higher proportion of rumenic acid (a CLA), oleic acid as well as DFA recorded in LD from concentrate-supplemented goats partially compensate for the disadvantages associated with increased levels of C16:0. Conjugated linoleic acids (CLA) have been shown to have a number of health-promoting benefits including anti-carcinogenesis, improved immune system and lipid metabolism and prevention of diabetes and cardiovascular diseases (Schiavon et al. 2011; Carvalho et al. 2015; Bravo-Lamas et al. 2016).

The higher proportion of CLA observed in LD from goats on higher levels of concentrate supplementation (T66 and T100) in the present study agrees with that of Warren et al. (2008) and Horcada et al. (2012) and could be due to increase in fatness (Juarez et al. 2008). However, Juarez et al. (2008) reported an increase in CLA content as a result of switching from a concentrate-based diet to pasture. We have no plausible explanation for this discrepancy. In the rumen, linoleic (C18:2 n-6) and linolenic (C18:3 n-3) acids are hydrogenated to stearic (C18:0) acid (French et al. 2000; Warren et al. 2008; Bessa et al. 2008). However, increased levels of linoleic acid in ruminant diet tend to block complete hydrogenation of unsaturated fatty acid leading to accumulation of intermediate products such as vaccenic acid (C18:1 trans 11) and CLA in the rumen (Bas and Morand-Fehr 2000; Marinova et al. 2001; Daniel et al. 2004; Pordomingo et al. 2012). Part of these intermediates escape the rumen and are incorporated into body tissues and ruminant products. In addition, compared to forage diets, concentrate-based diets have a higher quantity of available carbohydrate, which shortens the retention time of feed in the rumen, consequently reducing the extent of biohydrogenation of polyenoic acids (Diaz et al. 2002; Demirel et al. 2006). Ruminant products are the major natural source of CLA. Juarez et al. (2008) reported a positive relationship between cis 9-trans11 CLA isomer content in meat and total fat content. Overall, the comparable proportions of CLA, linolelaidic, oleic, total MUFA as well as n-6/n-3 PUFA ratio in LD from T66 and T100 goats suggest that there is no gain in the deposition of such fatty acids beyond concentrate supplementation at 66% of ad libitum intake.

The decrease in n-3 in LD from goats on high levels of supplementation could be attributed to an increase in fatness. Increased fatness is associated with an increased proportion of SFA and MUFA in neutral lipids and a decline in the proportion of n-3 PUFA caused by dilution from SFA (Warren et al. 2008; Lee et al. 2008). In addition, fatness from high intake of concentrate diets is associated with increased deposition of n-6 PUFA. As shown in Table 1, concentrates are rich sources of n-6 PUFA, mainly due to linoleic acid (C18:2 n-6). The increase in deposition of n-6 PUFA with supplementation corresponds with the observed increase in n-6/n-3 ratio with supplementation. The higher proportions of n-3 PUFA observed in LD from non-supplemented goats (T0) agree with the findings of Nuernberg et al. (2005) and Horcada et al. (2012). The higher proportion of linolenic acid (C18:3 n-3) in non-supplemented (fed forage only) goats could be due to the low fatty acid concentration linked to their leanness, as well as due to forage consumption per se. Forages are rich sources of linolenic acid, a precursor for long chain n-3 PUFA (Daniel et al. 2004; Diaz et al. 2002). Demirel et al. (2006) found that level of linolenic acid in LD from lambs fed hay was 3.5 times higher than in those fed concentrate; in the present study, LD from goats with access to forage had a proportion of linolenic that was only six time higher than that of goats in ad libitum concentrate intake. To avoid various lifestyle diseases such as coronary heart disease, diabetes, cancer and atherosclerosis in consumers, the ratio of n-6/n-3 in meat is recommended to be below 4 (Aurousseau et al. 2004; Warren et al. 2008). In the present study, the ratio of n-6/n-3 in LD from goats on various levels of concentrate supplementation was below 4, indicating that irrespective of the level of concentrate supplementation, goats produce meat with favourable n-6/n-3 PUFA ratio in the lean.

4.2 Fatty Acid Composition in Minced Meat (MM)

With the exception of the lack of effect on the proportion of palmitic acid, concentrate supplementation affected fatty acid composition in MM in a similar way as in LD. In addition to intramuscular fat contained in LD, MM contained inter-muscular and subcutaneous fat depots, which we expected to be affected differently by concentrate supplementation. Deposition of fat in different depots in meat animals starts with internal fat, followed by intermuscular, subcutaneous depot and lastly intramuscular fat (Hausman et al. 2009; Joo et al. 2013). The lack of difference between MM and LD with respect to the proportion of major fatty acids following concentrate supplementation can be attributed to the uniqueness of goats in fat deposition. Unlike cattle and sheep, goats deposit most of their fat internally. Therefore, increased energy intake from concentrate supplementation will result in more internal fat than carcass fat in goats (Banskalieva et al. 2000). This means MM (representing carcass fat) from concentrate-supplemented goats was not significantly different from LD in terms of total fat content. Increased carcass fatness leads to the increase in total fatty acid concentration. This increase is characterised by higher concentration of triacylglycerol fatty acids with no change on the concentration of fatty acids in polar lipids (Webb et al. 1998; Juarez et al. 2008).

The observed higher proportion of oleic acid (C18:1 n-9) in MM from concentrate-fed goats is in accordance with Daniel et al. (2004), but not with Johnson and McGowan (1998). In addition to being produced de novo through the action of Δ-9 desaturase (Aldai et al. 2007), a substantial amount of oleic acid (C18:1 n-9) found in animal tissues is derived from the concentrate diet (Daniel et al. 2004). As seen in the present study (Table 1), concentrate diet contained considerable amount of oleic acid. On the other hand, the higher proportions of trans-MUFA observed in MM from concentrate-supplemented goats could be attributed to ruminal biohydrogenation of unsaturated fatty acids (UFA). The higher content of UFA, especially linoleic acid (C18:2), in the concentrate diet than in grass hay is a probable source of variation. Similar to the findings from the present study, Warren et al. (2008) found that meat from concentrate-fed animals has higher proportions of linoleic acid, CLA and vaccenic acid (C18:1 trans 11). However, Aurousseau et al. (2004) reported that the proportion of vaccenic acid was higher in grazing than in concentrate-fed lambs, findings that do not correspond with those of the present study. We have no plausible explanation for this discrepancy. Trans fatty acids have been linked to increased levels of plasma low-density lipoprotein (LDL) cholesterol, which has negative health effects (Mensink et al. 2003; Bas et al. 2007). The main trans isomer in meat (vaccenic acid), however, is not a significant risk factor for cardiovascular diseases compared to trans fatty acids formed by chemical hardening of vegetable oil (Nuernberg et al. 2005).

The observed increase in margaric acid (C17:0) with higher levels of concentrate supplementation may indicate changes to the rumen environment triggered by intake of a concentrate diet favouring amylolytic bacterial species and propionate production (Turner et al. 2012). Oliveira et al. (2015) observed a twofold increase in margaric acid in meat from goats with access to castor de-oiled cake compared with the control group. This indicates that margaric fatty acid content in goat meat can be influenced by diet alterations and is synthesised by rumen microorganisms.

In the present study, the ratio of n-6/n-3 in MM from goats on beyond 66% level of concentrate supplementation was beyond 4, indicating that the 66% of ad libitum feed is the maximum level of concentrate supplementation for goats in order produce meat with salubrious fatty acids (Horcada et al. 2012).

4.3 Fatty Acid Composition in Omental Fat (OF)

The observed limited effect of concentrate supplementation on the proportion of various fatty acids in OF depot disagrees with the conclusion of Barber et al. (2000) and Aldai et al. (2007) that the fatty acid profile of internal fat depots is affected by the preferential deposition of absorbed dietary fat into these depots. We have no plausible explanation for this discrepancy. Irrespective of the level of supplementation, n-6/n-3 PUFA ratio was above 4, indicating potential health risks associated with the consumption of internal fat depots such as omental fat. The increase in the proportion of rumenic acid with concentrate supplementation may stem from an increased supply of substrates for biohydrogenation, mainly linoleic acid. However, the comparable proportion of rumenic acid in T66 and T100 may indicate that the highest level of concentrate supplementation required for maximisation of rumenic acid deposition in OF is 66% of ad libitum intake. The higher proportion of SFA observed in OF from T0 and T33 could be linked to the higher extent of biohydrogenation due to relatively higher retention time of feed in the rumen caused by higher forage intake (Diaz et al. 2002; Marinova et al. 2001). Incidence of coronary diseases has been linked with intake of meat with high ratios of n-6/n-3 PUFA or SFA (Avilés et al. 2016).

4.4 Distribution of Fatty Acids in MM, LD and OF

The separation of intramuscular fat (LD) from other fat depots observed in the present study partly corroborates the findings of Aldai et al. (2007) whereby LD displayed different fatty acid composition (higher PUFA, lower SFA and MUFA) from that of both intermuscular and subcutaneous depots. On the other hand, the higher proportion of MUFA and CLA in MM may partly be due to the higher activity of Δ-9 desaturase in subcutaneous fat (Daniel et al. 2004; Aldai et al. 2007). This is an endoplasmic reticulum enzyme that is not only responsible for the conversion of SFA into MUFA but is also involved in the desaturation of trans-vaccenic acid into CLA (Avilés et al. 2016). Monounsaturated fatty acids of cis-configuration are hypocholesterolemic and do not reduce HDL cholesterol, which protects against coronary heart diseases (Webb et al. 1998).

The observed strong association between OF and SFA indicates that internal fats are more saturated than external fat depots. This might indicate effects of genetic programming that dictate a lower expression of Δ-9 desaturase in this depot. A higher level of saturation in internal fat might be an adaptation to the higher temperature in the core of the body than in the periphery (Monziols et al. 2007; Barton et al. 2007). Similar results were reported by Webb et al. (1998), who found that the level of saturation increases from external fat depots towards internal depots, with kidney and intermuscular fat depots being the most saturated while intramuscular and subcutaneous fat are the least saturated. Webb et al. (1998) noted significant effects of anatomical location on the proportion of C14:0, C16:0 and C18:1 in fat depots of Belgium Blue bulls, but there are limited reports on the effects of the same factor on fat acid composition of fat depots in goats.

5 Conclusion

It is concluded that feeding crossbred goats with concentrate diets increases the overall content of DFA, n-6 PUFA, trans-MUFA and CLA but lowers the proportion of n-3 PUFA in meat. Reasonable amounts of fresh forage should thus be offered to such goats with access to concentrate diets to balance for the proportion of n-3 PUFA, with known health-promoting benefits. Concentrate supplementation beyond 66% of ad libitum feeding has limited effects on fatty acid composition, as depicted by similarity between T66 and T100 goats in the proportions of oleic acid, CLA, total MUFA as well as n-6/n-3 PUFA ratio in LD and MM. Consumers are advised to refrain from high intake of internal fat due to its richness in SFA and unfavourable n-6/n-3 PUFA ratio, a risk factor for cardiovascular disease.

Acknowledgement Authors are grateful to the financial support provided by the SUA-EPINAV and Building Stronger Universities (BSU) projects.

Ethical Standard This experiment complies with the current laws of Tanzania.

References

Aldai N, Najera AI, Dugan MER, Celaya R, Osoro K (2007) Characterisation of intramuscular, intermuscular and subcutaneous adipose tissues in yearling bulls of different genetic groups. Meat Sci 76(4):682–691

Aurousseau B, Bauchart D, Calichon E, Micol D, Priolo A (2004) Effect of grass or concentrate feeding systems and rate of growth on triglyceride and phospholipid and their fatty acids in the M-longissimus thoracis of lambs. Meat Sci 66(3):531–541

Avilés C, Horcada A, Polvillo O, Membrillo A, Anaya G, Molina A, Alcalde MJ, Panea B (2016) Association study between variability in the SCD gene and the fatty acid profile in perirenal and intramuscular fat deposits from Spanish goat populations. Small Rumin Res 136:127–131

Babiker SA, Elkhider IA, Shafie SA (1990) Chemical-composition and quality attributes of goat meat and lamb. Meat Sci 28(4):273–277

Banskalieva V, Sahlu T, Goetsch AL (2000) Fatty acid composition of goat muscles and fat depots: a review. Small Rumin Res 37(3):255–268

Barber MC, Ward RJ, Richards SE, Salter AM, Buttery PJ, Vernon RG, Travers MT (2000) Ovine adipose tissue monounsaturated fat content is correlated to depot-specific expression of the stearoyl-CoA desaturase gene. J Anim Sci 78(1):62–68

Barton L, Marounek M, Kudrna V, Bures D (2007) Growth performance and fatty acid profiles of intramuscular and subcutaneous fat from Limousin and Charolais heifers fed extruded linseed. Meat Sci 76:517–523

Bas P, Morand-Fehr P (2000) Effect of nutritional factors on fatty acid composition of lamb fat deposits. Livest Prod Sci 64:61–79

Bas P, Berthelot V, Pottier E, Normand J (2007) Effect of level of linseed on fatty acid composition of muscles and adipose tissues of lambs with emphasis on trans fatty acids. Meat Sci 77(4):678–688

Beserra FJ, Madruga MS, Leite AM, da Silva EMC, Maia EL (2004) Effect of age at slaughter on chemical composition of meat from Moxoto goats and their crosses. Small Rumin Res 55(1–3):177–181

Bessa RJB, Lourenco M, Portugal PV, Santos-Silva J (2008) Effects of previous diet and duration of soybean oil supplementation on light lambs carcass composition, meat quality and fatty acid composition. Meat Sci 80(4):1100–1105

Bravo-Lamas L, Barron LJR, Kramer JKG, Etaio I, Aldai N (2016) Characterization of the fatty acid composition of lamb commercially available in northern Spain: emphasis on the trans-18:1 and CLA content and profile. Meat Sci 117:108–116

Carvalho VB, Leite RF, Almeida MTC, Paschoaloto JR, Carvalho EB, Lanna DPD et al (2015) Carcass characteristics and meat quality of lambs fed high concentrations of crude glycerin in low-starch diets. Meat Sci 110:285–292

Daniel ZCTR, Wynn RJ, Salter AM, Buttery PJ (2004) Differing effects of forage and concentrate diets on the oleic acid and conjugated linoleic acid content of sheep tissues: the role of stearoyl-CoA desaturase. J Anim Sci 82(3):747–758

Demirel G, Ozpinar H, Nazli B, Keser O (2006) Fatty acids of lamb meat from two breeds fed different forage: concentrate ratio. Meat Sci 72(2):229–235
Dhanda JS, Taylor DG, Murray PJ, McCosker JE (1999) The influence of goat genotype on the production of Capretto and Chevon carcasses. 4. Chemical composition of muscle and fatty acid profiles of adipose tissue. Meat Sci 52(4):375–379
Diaz MT, Velasco S, Caneque V, Lauzurica S, de Huidobro FR, Perez C, Gonzalez J, Manzanares C (2002) Use of concentrate or pasture for fattening lambs and its effect on carcass and meat quality. Small Rumin Res 43(3):257–268
French P, Stanton C, Lawless F, O'Riordan EG, Monahan FJ, Caffrey PJ, Moloney AP (2000) Fatty acid composition, including conjugated linoleic acid, of intramuscular fat from steers offered grazed grass, grass silage, or concentrate-based diets. J Anim Sci 78(11):2849–2855
Hausman GJ, Dodson MV, Ajuwon K, Azain M, Barnes KM, Guan LL et al (2009) Board-invited review: the biology and regulation of preadipocytes and adipocytes in meat animals. J Anim Sci 87:1218–1246
Horcada A, Ripoll G, Alcalde MJ, Sañudo C, Teixeira A, Panea B (2012) Fatty acid profile of three adipose depots in seven Spanish breeds of suckling kids. Meat Sci 92:89–96
Johnson DD, McGowan CH (1998) Diet/management effects on carcass attributes and meat quality of young goats. Small Rumin Res 28(1):93–98
Joo ST, Kim GD, Hwang YH, Ryu YC (2013) Control of fresh meat quality through manipulation of muscle fiber characteristics. Meat Sci 95(4):828–836
Juarez M, Horcada A, Alcalde MJ, Valera M, Mullen AM, Molina A (2008) Estimation of factors influencing fatty acid profiles in light lambs. Meat Sci 79(2):203–210
Lee JH, Kouakou B, Kannan G (2008) Chemical composition and quality characteristics of chevon from goats fed three different post-weaning diets. Small Rumin Res 75(2–3):177–184
Marinova P, Banskalieva V, Alexandrov S, Tzvetkova V, Stanchev H (2001) Carcass composition and meat quality of kids fed sunflower oil supplemented diet. Small Rumin Res 42(3):219–227
McMillin KW, Brock AP (2005) Production practices and processing for value-added goat meat. J Anim Sci 83(13_suppl):E57–E68
Mensink RP, Zock PL, Kester ADM, Katan MB (2003) Effects of dietary fatty acids and carbohydrates on the ratio of serum total to HDL cholesterol and on serum lipids and apolipoproteins: a meta-analysis of 60 controlled trials. Am J Clin Nutr 77(5):1146–1155
Monziols M, Bonneau M, Davenel A, Kouba M (2007) Comparison of the lipid content and fatty acid composition of intermuscular and subcutaneous adipose tissues in pig carcasses. Meat Sci 76(1):54–60
Nuernberg K, Dannenberger D, Nuernberg G, Ender K, Voigt J, Scollan ND, Wood JD, Nute GR, Richardson RI (2005) Effect of a grass-based and a concentrate feeding system on meat quality characteristics and fatty acid composition of longissimus muscle in different cattle breeds. Livest Prod Sci 94(1–2):137–147
O'Fallon JV, Busboom JR, Nelson ML, Gaskins CT (2007) A direct method for fatty acid methyl ester synthesis: application to wet meat tissues, oils, and feedstuffs. J Anim Sci 85(6):1511–1521
Oliveira CHA, Silva AM, Silva LM, van Tilburg MF, Fernandes CCL, Moura AA, Moreno FBMB, Monteiro-Moreira ACO, Moreira RA, Bezerra FJ, Rondina D (2015) Meat quality assessment from young goats fed for long periods with castor de-oiled cake. Meat Sci 106:16–24
Pordomingo AJ, Garcia TP, Volpi Lagreca G (2012) Effect of feeding treatment during the backgrounding phase of beef production from pasture on: II. Longissimus muscle proximate composition, cholesterol and fatty acids. Meat Sci 90:947–955
SAS (2001) Statistical analysis system. User's guide, version 8. SAS Institute, INC, Carry, NC, p 2
Schiavon S, De Marchi M, Tagliapietra F, Bailoni L, Cecchinato A, Bittante G (2011) Effect of high or low protein ration combined or not with rumen protected conjugated linoleic acid (CLA) on meat CLA content and quality traits of double-muscled Piemontese bulls. Meat Sci 89(2):133–142
Shirouchi B, Albrecht E, Nuernberg G, Maak S, Olavanh S, Nakamura Y, Sato M, Gotoh T, Nuernberg K (2014) Fatty acid profiles and adipogenic gene expression of various fat depots in Japanese Black and Holstein steers. Meat Sci 96:157–164

Turner TD, Karlsson L, Mapiye C, Rolland DC, Martinsson K, Dugan MER (2012) Dietary influence on the m. longissimus dorsi fatty acid composition of lambs in relation to protein source. Meat Sci 91:472–477

Warren HE, Scollan ND, Enser M, Hughes SI, Richardson RI, Wood JD (2008) Effects of breed and a concentrate or grass silage diet on beef quality in cattle of 3 ages. I: Animal performance, carcass quality and muscle fatty acid composition. Meat Sci 78(3):256–269

Webb EC, De Smet S, Van Nevel C, Martens B, Demeyer DI (1998) Effect of anatomical location on the composition of fatty acids in double-muscled Belgian Blue cows. Meat Sci 50(1):45–53

Wood JD, Enser M, Fisher AV, Nute GR, Sheard PR, Richardson RI, Hughes SI, Whittington FM (2008) Fat deposition, fatty acid composition and meat quality: a review. Meat Sci 78(4):343–358

Goat Milk Quality and Possible Dairy Products from Rural Households of Tanzania and Malawi Under the Farmer-Processor Partnership

I. A. Ketto, G. Msalya, F. Chigwa, R. K. Abrahamsen, L. O. Eik, G. C. Kifaro, and R. L. Kurwijila

Abstract Dairy goat farming in Tanzania and Malawi, promoted through collaboration between Norwegian University of Life Sciences, Sokoine University of Agriculture and Lilongwe University of Agriculture and Natural Resources, has improved income and nutrition of the poor households of both countries. In Tanzania, the successful introduction of dairy goats can be traced in Mgeta division, Uluguru mountains, Morogoro region, where goat milk has improved nutrition in the farm families. Surplus milk in Mgeta is processed into yoghurt by the farmers' association (TWAWOSE). Milk and yoghurt are sold in local markets within the division with potential for reaching distant markets, including Morogoro Municipality (50 km) and Dar es Salaam (250 km) where market promotions for goat milk and yoghurt were successfully conducted. However, the progress of dairy goats' milk industry is constrained by lack of reliable milk records, limited knowledge in hygienic milk handling at farm level and poor infrastructure which lead to the production of poor-quality raw milk that does not meet minimum standards for milk processing. The objective of this review is to present current knowledge on dairy goat milk quality and possible dairy products to be considered in rural households of Tanzania and Malawi, under the farmer-milk processor partnership. The potential and available technologies for processing goat milk are suggested including fluid milk beverage, yoghurt and cheese from goat milk. Further, we discuss the possible partnership between dairy goat farmers and the private sector (e.g. milk processors), a vital link for supporting innovations

I. A. Ketto (✉)
Asas Dairies Limited, Iringa, Morogoro, Tanzania

G. Msalya · G. C. Kifaro · R. L. Kurwijila
Sokoine University of Agriculture, Chuo Kikuu, Morogoro, Tanzania
e-mail: msalya@sua.ac.tz; kifaro@sua.ac.tz

F. Chigwa
Lilongwe University of Agriculture & Natural Resources, Lilongwe, Malawi

R. K. Abrahamsen · L. O. Eik
Norwegian University of Life Sciences, Aas, Norway
e-mail: roger.abrahamsen@nmbu.no; lars.eik@nmbu.no

B. R. Singh et al. (eds.), *Climate Impacts on Agricultural and Natural Resource Sustainability in Africa*, https://doi.org/10.1007/978-3-030-37537-9_16

and commercialisation of the dairy goat industry. Farmers benefit from the technologies and skills in good farming practices and milk handling, while processors gain by obtaining good quality milk and milk products. As a result, increased income and better nutrition from dairy goats will be achieved.

1 Introduction

A large number of poor households in sub-Saharan Africa (SSA) depend on the livestock for their livelihood. However, their production systems are characterised by a low input-output system, insufficient to meet the food demands of the fast-growing population. Low productivity of indigenous livestock and low purchasing power are the major causes of food insecurity in the region. In particular, there is limited access to animal protein sources such as meat, milk and eggs in poor households who mainly depend on cereals (Otte and Chilonda 2003). There is a need to improve the existing production systems, particularly for households in the rural areas which depend entirely on agricultural activities for their livelihood. In Tanzania, about 50% of agricultural households keep livestock species mainly local cattle, small ruminants and local chicken; these provide reliable pathways out of poverty and help to achieve animal protein nutrition requirements (MALF 2016). Although indigenous livestock species are preferred due to their adaptive characteristics to the tropical settings, they are characterised by low productivity; therefore, it is difficult to achieve increased animal protein production based on these animals alone.

The introduction of Norwegian dairy goats, also known as Norwegian Landrace (NL), is one example of the successful projects intended to improve the availability of protein products of animal origin in rural communities, where it was initially difficult to obtain them. For example, in the Uluguru mountain areas, in Morogoro, Tanzania, the NL goats were shown to improve nutrition and income status of the poor households of Mgeta division (Eik et al. 2008). Consequently, household milk consumption in the division increased from 0 in 1988 to 1.6 L/day in 2012 (Kifaro et al. 2012). Since their introduction in 1993 through collaborative research between Norwegian University of Life Sciences (NMBU) and Sokoine University of Agriculture (SUA), dairy goat farming has been significantly sustained, leading to increase in the numbers of both goats and farmers (Msalya et al. 2017). In addition, dairy goat farmers association (TWAWOSE) was established with a purpose of ensuring the sustainability of the project through improved management of breeding bucks, organisation of the sales of live animals and facilitation of extension services by farmers themselves (Lie et al. 2012). Recently, a processing centre was established by TWAWOSE and aimed at collecting milk, processing it into yoghurt and selling it in the local markets of Mgeta and beyond. In Malawi, dairy goats have been promoted through various programmes including University Development Linkages Project (UDLP) (UDLP 1997) through which different dairy goat breeds were introduced by Bunda College of Agriculture, the Salima Research Station and non-governmental organisations, including Small Scale Livestock Promotion Programme (SSLPP) (Chigwa 2011). Unlike Tanzania, there have been more challenges in dairy goat farming in Malawi including those reared under on-station experiments at Bunda College and farmer households with regard to survival, poor management and utilisation of goat milk (Chigwa 2011). For these and other reasons, the dairy goats and goat milk value chains have performed relatively poorer.

The major challenge limiting the increased productivity of dairy goats in both countries is the low level of awareness on the importance of goat milk in human health and nutrition. It is therefore not surprising that there have been almost no new innovations in dairy goat husbandry, the handling of milk and the level of milk processing and packaging to meet the potential demand. In addition, the local farming communities have low working capital and low capacity for handling raw milk, as well as several constraints with regard to milk processing. Furthermore, the farmers have low entrepreneurship skills and lack stable markets for raw or processed milk and thus fail to run dairy goat enterprises efficiently. These factors have contributed to the difficulties in scaling up and improving production to meet the larger market demands (Lie et al. 2012). In a paper published by Msalya et al. (2016), the possibility of a private-public partnership (PPP) involving SUA, TWAWOSE and a private dairy processing company within Morogoro Municipality, namely, Shambani Graduates Limited (SGL), was floated. It was suggested as a sustainable way of providing a stable and reliable market for milk from TWAWOSE and other dairy goat farmers; it would improve storage and processing and also enable goat milk to reach a larger market such as Morogoro, Dodoma and Dar es Salaam. To achieve such collaborations or partnerships, understanding of the quality aspects of milk as a raw material for processing and improved milk handling practices at the farm level are essential prerequisites (Msalya et al. 2016). In the present chapter, we present and discuss these aspects and highlight the importance of farmers adopting the relevant technologies for processing liquid milk beverages, yoghurt and cheese, as well as improving packaging and shelf life of goat milk in the rural settings. Firstly, the handling of goat milk, its chemistry, processing and nutritional possibilities are discussed. Attention, if duly given to these issues, will lead to improvements in the management of goats, milk productivity and handling, and overall high quality of milk delivered to the collection centres. Consequently, consumption is expected to increase, while more farmers will turn to goat farming because of stable and reliable markets as well as increasing profits.

2 Goat Milk Quality, Handling and Processing

2.1 Goat Milk Quality

Milk as a raw material used in food production should be from healthy animals, free from antibiotics, and should have a low somatic cell count (SCC) and bacteria count (Park and Guo 2006). According to García et al. (2014), milk quality could be defined as the potential of milk to withstand different technological treatments such as heat and pressure, have good renneting properties and be easy to acidify using lactic acid bacteria in order to get a product with the preferred nutritional, health and organoleptic properties (colour, odour and texture) to the consumer. Variables like milk composition, SCC and the presence of inhibiting substances such as antibiotics and pesticides determine the processability, nutritional, health and organoleptic properties of the milk products. The presence of antibiotics in milk slows down or prevents the lactic acid fermentation of milk (Raynal-Ljutovac et al. 2005) and consequently affects the production of fermented milk. It is important to understand how these variables can be influenced to improve the quality of milk as a raw material for processing. At the farm, milk should be stored at 4 °C after milking; cold storage of milk reduces the growth of bacteria and improves the keeping quality of milk. However, cold storage of milk changes the structure and physicochemical properties of casein (CN) micelles by the dissolution of β-CN and micellar calcium (Raynal-Ljutovac et al. 2005). Furthermore, it creates growth possibilities for psychrotrophic bacteria such as *Pseudomonas* in milk (Raynal-Ljutovac et al. 2005). These bacteria produce heat-resistant lipolytic and proteolytic enzymes, which hydrolyse lipids and proteins, respectively, and may, therefore, induce the development of rancid and bitter taste in milk and milk products (Paludetti et al. 2018).

In developing countries, especially in the rural areas where dairy goat farming is practised, cold storage of milk to control the bacterial growth is a challenge because of the power outage and lack of facilities and equipment for rapid cooling of milk at the farm. The use of hydrogen peroxide and thiocyanate, which activates the lactoperoxidase system (LPS) in milk, could be an option in tropical countries. The LPS is the most important natural antibacterial system in milk, which prevents the growth of both pathogenic bacteria and psychrotrophs, hence prolonging the keeping quality of raw milk. If LPS is the option adopted in enhancing the keeping quality of the raw milk, understanding of its possible effect on the processing and quality properties of fermented milk is important. Nakada et al. (1996) reported that LPS treated milk gave yoghurts made from mixed cultures (*Lactobacillus delbrueckii* ssp. *bulgaricus* and *Streptococcus thermophilus* (ratio 1:1)) the desired acidity at the same time as the control yoghurts. The LPS treated yoghurts showed limited post-acidification, and this contributed to yoghurts with a favourable balance between sweetness and sourness. The effects of LPS on the rheological properties of yoghurts were evaluated by Özer et al. (2003). This study showed that viscosity and storage modulus of the products were decreased by the increase in the concentration of LPS in yoghurt milk.

The high level of somatic cell count (SCC) in milk is another factor which limits the quality and processability of goat milk. The SCC in dairy goats varies between parity, breed and stage of lactation, with the highest levels being towards the end of lactation (Skeie 2014). Higher SCC in goat milk compared to cow milk is due to the higher basal levels of SCC in bacteria-free udder in goat milk compared to cow milk (300,000 vs. 70,000 ml^{-1}) (Paape et al. 2007). A different set of standards for udder health using SCC is practised between countries. For example, legal limit for SCC varies between countries; for example, most of the European countries have adopted lower levels (400,000 ml^{-1}) for both cow and goat milk, while in the United States of America, higher levels (750,000 and 1,000,000 ml^{-1}) are used for goat and cow milk, respectively (Paape et al. 2007). Higher SCC is associated with impaired processability of milk (Raynal-Ljutovac et al. 2005; Skeie 2014) and may lead to undesirable effects on the health of the consumer, due to the toxins produced by pathogenic bacteria, as a result of udder infection (Skeie 2014). To commercialise goat milk, it is important to set the acceptable level of SSC, and it should be included in the routine recording of the milk quality as a measure for reducing the increase in bacteria load in milk and to improve the welfare of milk-producing animals. One of the means for reducing SCC is to control the most contagious diseases in dairy goats, including caprine arthritis encephalitis virus (CAEV), caseous lymphadenitis (CLA) and Johne's disease (Skeie 2014). Elevated SCC has been associated with these diseases (Sanchez et al. 2001).

Milk lipids are the main contributors to flavour in goat milk and milk products (Chilliard et al. 2003). Lipids are enclosed within milk fat globule (MFG) in the form of triglycerides. The MFG is surrounded by the milk fat globule membrane (MFGM), which prevents lipolysis of milk fat by endogenous lipoprotein lipase (LPL). When MFGM is disrupted by either agitation or temperature fluctuations, the triglycerides are exposed to LPL and hydrolysed to release free fatty acids (FFA). The LPL activity is controlled by both environmental and genetic factors (Chilliard et al. 2003). Higher content of FFA was linked to the tart or rancid flavour of the milk (Eknæs and Skeie 2006) and impaired rennet coagulation properties (Skeie 2014). The negative effects of high contents of FFA on the flavour of the milk and on cheesemaking properties were a serious issue for the NL goat milk. The Norwegian dairy company (TINE SA) has therefore introduced results from the analysis of FFA as one of the parameters in the payment system for the goat milk. Efforts made by the NL goat industry on feeding and selective breeding have led to an improvement in flavour and cheesemaking properties of milk from these animals in recent years (Skeie 2014). A well-established milk quality control is needed during the commercialisation of goat milk from the smallholder dairy goat farmer in rural areas. Training program on the hygiene of milk production, milk handling and dairy recording of milk quality should be established together with a well-established disease control and treatment program. The joint collaboration with dairy processing companies would be the possible means to enforce quality standards of goat milk from rural areas of Tanzania and Malawi.

2.2 *Chemical Composition of Goat Milk*

Milk components occur in three phases. One phase is the true solution of lactose, organic and inorganic salts, vitamins and small molecules in water. Proteins are dispersed in an aqueous phase of the milk at a molecular level (whey proteins, α-lactalbumin, β-lactoglobulin, serum albumin, etc.) and larger colloidal particles (casein micelles) with a diameter of 50–600 nm. The last phase is the milk lipids, which occur in the emulsified state as fat globules with a diameter of 0.1–20 μm (Fox et al. 2015). Both nutritional and technological properties of milk are determined by its chemical composition. The gross milk composition varies within species (due to genetics, stage of lactation, milking intervals and feeding regimes) and between species. The gross milk composition of goat, cow and human milk is presented in Table 1. According to the data presented by Park et al. (2007), goat milk is richer in total protein content and fat content than cow milk; however, the content of lactose and total casein is lower in goat milk than in cow milk. There is a considerable variation in protein composition between species; the content of α_{s1}-casein (CN) in goat milk is lower than in cow milk, while the contents of α_{s2}-CN, β-CN, κ-CN and non-protein nitrogen (NPN) are higher in goat milk compared to the cow milk (Park 2010a). Having a higher content of κ-CN, goat milk has a smaller casein micelle size than cow milk. This is because the mean diameter of the casein micelle size is negatively correlated to the contents of κ-CN (Ketto et al. 2017).

The amount of oligosaccharides in goat milk differs from those in other domesticated mammals, including cow milk (23 vs. 6 mg/100 g) (Raynal-Ljutovac et al. 2008). Higher amounts of oligosaccharides in goat milk, together with the comparable structure to the oligosaccharides in human milk, would make goat milk a good source of oligosaccharides compared to those from other domesticated mammals (Kiskini and Difilippo 2013).

Goat and sheep milk fat are composed of higher contents of the medium-chain fatty acids (C6:0, C8:0, C10:0) (the capra fatty acid) and C18:1 fatty acids compared to cow milk (Park et al. 2007). Medium-chain fatty acids give the specific aroma in the goat and sheep milk compared to cow milk (Markiewicz-Kęszycka

Table 1 Average chemical composition of milk from goat, cow and human

	Goat	Cow	Human
Gross composition, %			
Protein	3.4	3.2	1.2
Fat	3.8	3.6	4.0
Lactose	4.1	4.7	6.9
Ash	0.8	0.7	0.3
Casein (CN) composition (g100 g^{-1})			
α_{s1}-CN	0.27[a]	1.00	0.06
α_{s2}-CN	0.41	0.37	ND
β-CN	1.09	1.00	0.31
κ-CN	0.50	0.35	0.10
Whey composition (g100 g^{-1})			
β-Lactoglobulin	0.22	0.33	ND
α-Lactalbumin	0.12	0.12	0.18
Lactoferrin	0.01	0.01	0.15
Serum albumin	-	0.04	0.04
Lysozyme	Traces	Traces	0.03
Immunoglobulins	0.05	0.07	0.12

Source: Park et al. (2007)
ND = Not determined
[a]The content varies depending on the genetic polymorphism on the gene which codes for α_{s1}-CN

et al. 2013). Higher levels of conjugated linoleic acid (CLA) (C18:2 *cis* 9, *trans* 11) were found in goat and sheep milk than in cow milk (Markiewicz-Kęszycka et al. 2013). The content of CLA in milk will vary depending on the feeding regimes; for instance, milk from animals fed on pasture was found to contain higher levels of CLA compared to the indoor fed animals (Zervas and Tsiplakou 2011).

It is also of interest for processors of goat milk to know that the composition may vary considerably during the lactation period and according to the feeding regime. Brendehaug and Abrahamsen (1986) investigated the variation in the chemical composition of milk from a herd of NL goats during a lactation period including periods of variations in feeding regime. They found, for instance, that the fat content decreased over the first 4 months of lactation and increased again during the mountain pasture period. The protein content decreased during the first 4 months and then increased until the end of lactation, while the lactose concentration decreased throughout the lactation. Variation in content of the individual fatty acids and in the mineral content was also observed.

Milk is a good carrier of minerals, essential for the growth and development of infants. However, there is a difference in mineral contents as well as vitamins in milk from different species (Table 2). The contents of calcium, phosphorus, magnesium and potassium are higher in goat milk compared to cow milk. On the other hand, cow milk has higher contents of sodium and sulphur compared to goat milk (Park 2010b). Goat milk has a higher content of vitamin A compared to cow milk because goats are efficient converters of β-carotene (pro-vitamin A) to vitamin A (this explains why goat milk is whiter than cow milk). However, goat milk has lower contents of pyridoxine, vitamin B12 and folate compared to cow milk (Turck 2013). If goat milk is used as a substitute for human milk in the infant diet, the supplementation of a folate-rich ingredient is important in order to reduce the risk of anemia.

2.3 *Goat Milk Processing*

Technological Properties of Goat Milk

In dairy processing, milk is subjected to heat treatment to reduce bacterial load, prolong shelf life and to improve the technological properties of milk. Understanding the technological properties of milk, for example, heat stability, is important in the manufacturing of dairy products. From a scientific point of view, there has been a limited focus on the heat stability of goat milk (Zadow et al. 1983), probably because it is less commercialised than cow milk. Goat milk is less stable to heat compared to cow milk, this could be explained by a higher content of ionic calcium in goat milk compared to cow milk (Zadow et al. 1983). Heat stability of goat milk for UHT could be improved by pH adjustment, the addition of calcium sequestrant (2% disodium phosphate) and preheating of milk before processing (Zadow et al. 1983).

Table 2 Mineral and vitamin composition (amount in 100 g) in goat, cow and human milk

	Goat	Cow	Human
Minerals			
Calcium, Ca (mg)	134	122	33
Phosphorus, P (mg)	121	119	43
Magnesium, Mg (mg)	16	12	4
Sodium, Na (mg)	41	58	15
Potassium, K (mg)	181	152	55
Chlorine, Cl (mg)	150	100	60
Sulphur, S (mg)	28	32	14
Iron, Fe (mg)	0.07	0.08	0.20
Copper, Cu (mg)	0.05	0.06	0.06
Manganese, Mn (mg)	0.03	0.02	0.07
Zinc, Zn (mg)	0.56	0.53	0.38
Iodine, I (mg)	0.02	0.02	0.01
Selenium, Se (μg)	1.33	0.96	1.52
Vitamins			
Vitamin A (IU)	185	126	190
Vitamin D (IU)	2.3	2.0	1.4
Thiamine (mg)	0.07	0.08	0.01
Riboflavin (mg)	0.21	0.16	0.02
Niacin (mg)	0.27	0.08	0.17
Pantothenic acid (mg)	0.31	0.32	0.20
Vitamin B6 (mg)	0.05	0.04	0.01
Folic acid (μg)	1.0	5.0	5.5
Biotin (μg)	1.5	2.0	0.4
Vitamin B12 (μg)	0.07	0.36	0.03
Vitamin C (mg)	1.29	0.94	5.0

Source: Park et al. (2007)

To commercialise goat milk, it is important to establish quality control measures of the raw milk to ensure the quality of the final product (Park and Guo 2006). Rennet coagulation properties (clotting time, curd firming rate and curd firmness) of the milk should be observed as a part of the milk quality control procedure to evaluate the milk quality for cheesemaking. Favourable rennet coagulation properties (short clotting time and high curd firmness) are associated with the improved texture of the cheese and improved cheese yield. Higher contents of total Ca and P, protein, fat and κ-CN have been shown to be associated with favourable cheesemaking properties (Ketto et al. 2017). Cow milk has a higher α_{s1}-CN and total casein content than goat milk. This explains the differences in texture between the goat and cow cheeses (Skeie et al. 2014). Through selective breeding, it was possible to reduce the frequency of goats which produce milk with low or without α_{s1}-CN (α_{s1}-CN "null" variant) in Norwegian dairy goats (Skeie et al. 2014). The elimination of such goats over the years improved the flavour of the milk and the textural properties of the goat cheeses in Norway (Skeie et al. 2014). The proportion of casein to the total protein (casein index, %) is considered to be crucial for efficient cheesemaking.

Milk quality for fermented milk (yoghurt) can be monitored by determining its acid coagulation properties (gelation time, the speed of gelation and gel firmness), using glucono-δ-lactone. Good acid coagulation properties would be short gelation time and high final gel firmness. Goat milk, compared to sheep and cow milk, showed rapid changes in pH, probably because of a lower casein content in goat milk (Salaün et al. 2005). Rysstad and Abrahamsen (1987) found, however, only a slightly faster decrease in pH of goat milk than in cow milk when fermenting the milk with a yoghurt starter. After 8 hours of incubation at 43 °C, the pH reached 4.2 in both types of milk. One of the quality attributes of goat milk for yoghurt making is the higher protein content (Tamime et al. 2006). Desired protein content and total solids (TS) could be achieved by ultrafiltration (UF) technique to achieve the desired properties for both set and stirred style yoghurts made from goat milk such as higher curd tension and viscosity, respectively (Abrahamsen and Holmen 1981).

Fluid Milk Beverages

Fluid milk for consumption is a popular product in some western countries. It can either be low fat (skimmed) or full fat, flavoured or plain. There is limited research on liquid milk beverages made from goat milk. The standard procedure for the fluid milk processing of cow milk applies for goat milk, i.e. receiving, filtering, standardisation, homogenisation, pasteurisation, cooling, packaging and distribution (Park and Guo 2006). Before processing, milk is filtered to remove the foreign materials, for example, body cells from the udder. Filtration can be achieved by mechanical filtration (using a clarifier or bactofuge) at a commercial scale (Park and Guo 2006). In such cases, also a high number of spore-forming bacteria and somatic cells will be removed from the milk. The aim of homogenisation is to reduce the size of fat globules. After homogenisation, a new fat globule membrane is formed from the original material in the MFGM and micellar casein and some whey proteins. This reduction in fat globule size will change the colour (whitening) of the milk compared to the unhomogenised milk (Park and Guo 2006). After homogenisation, milk is pasteurised to reduce the bacteria load and improve the shelf life of the milk. As mentioned earlier, goat milk has lower heat stability than cow milk, and higher temperature treatment (UHT) might, therefore, lead to the coagulation or sedimentation of the milk. The low-temperature long time (LTLT) pasteurisation such as 63 °C for 30 min could be ideal batch pasteurisation conditions for goat milk, while high-temperature short-time (HTST) pasteurisation (72 °C for 15 s) is applied in automated processing lines (Park and Guo 2006). Cream separation is an option, depending on market demand, for example, low-fat milk beverage (2%). The cream after separation could be used in the production of high-fat products such as butter, ghee and ice cream. Before such use, the cream should be pasteurised. Figure 1 shows a process chart for the liquid milk beverage as adopted from Park and Guo (2006). Development of off-flavour is one of the

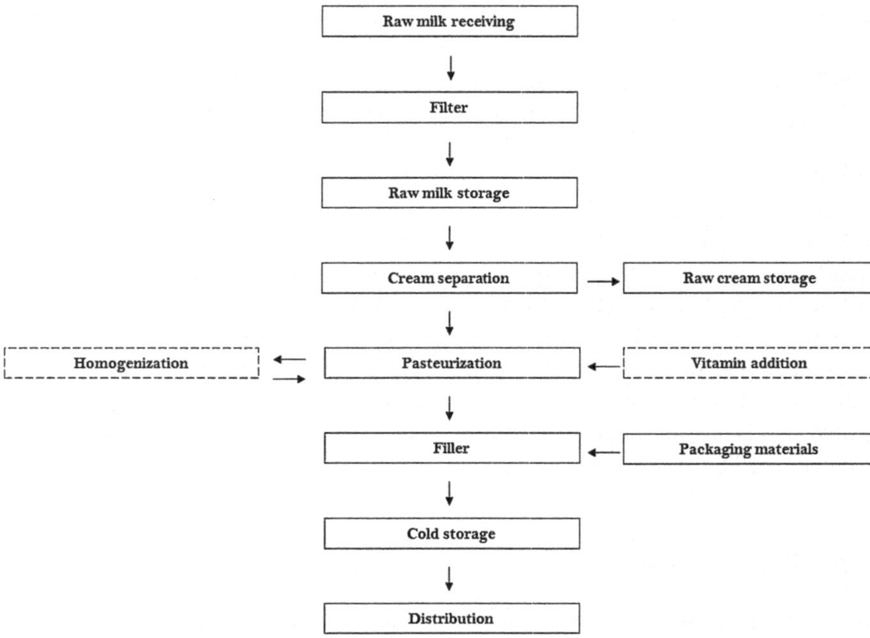

Fig. 1 Process chart for the goat liquid milk beverage. (Source: Park and Guo (2006))

challenges for liquid milk. However, improvements in milk handling would reduce the possible development of off-flavours in the final milk product for consumption (Loewenstein et al. 1980).

Yoghurt

Yoghurt is one of the fermented milk products made from goat milk. It has a long tradition in many countries (Tamime and Robinson 2007a). The same process used to produce yoghurt from cow milk could be used to produce goat milk yoghurt. A standard procedure for set-style yoghurt from goat milk is shown in Fig. 2. As mentioned before, homogenisation is normally used to reduce the size of the fat globules and enhance the technological properties of milk and to reduce the creaming effect. It is mostly practised in cow milk processing because of the large fat globule size in this milk. Homogenisation of goat milk, before yoghurt production, was however found to improve curd tension and viscosity of yoghurt (Abrahamsen and Holmen 1981). Heat treatment of milk during yoghurt processing is very essential to the technological properties of goat milk before processing. Heat treatment of milk, i.e. pasteurisation, restores the structure of casein micelles through recovering of the soluble β-CN and micellar calcium, dissolved to the serum phase during cold storage of milk (Raynal-Ljutovac et al. 2005). During yoghurt making, high heat treatment is employed (for instance, 80 °C for 30 minutes). At this temperature,

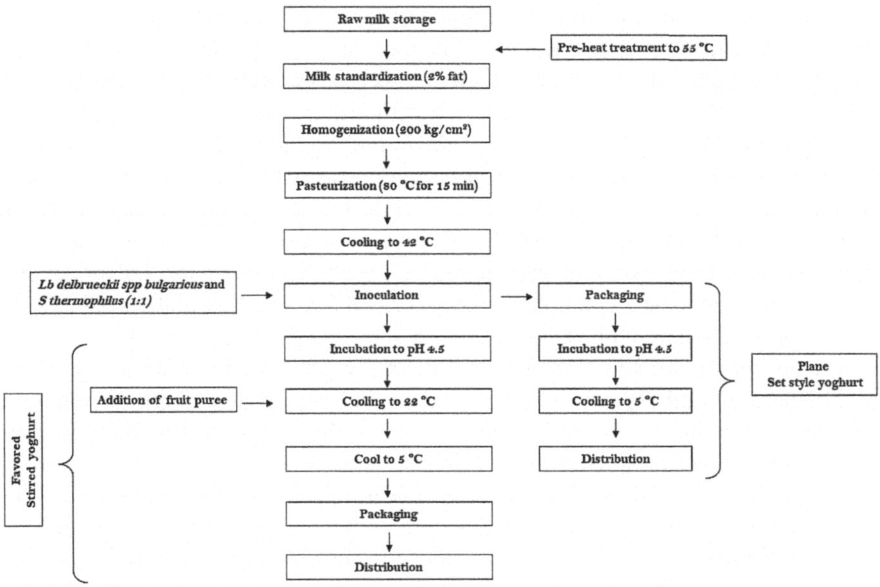

Fig. 2 Flowchart for set- and stirred-style yoghurt. (Source: Modified from Park and Guo (2006) and Walstra et al. (2006))

whey proteins (especially β-LG) are denatured and attached to the casein micelles, by intermolecular disulphide bonds and hydrophobic interactions. This reaction between whey proteins and casein micelles is associated with the improved texture of yoghurts (Tamime and Robinson 2007a).

After heat treatment, milk is cooled to approximately 42 °C, which is a favourable temperature for the growth of yoghurt starter cultures (*Lb. delbrueckii* spp. *bulgaricus* and *S. thermophilus*). In most cases, the combination of two starters is used during yoghurt processing, because they grow efficiently when they are together (Walstra et al. 2006). Yoghurt starter bacteria ferment lactose to lactic acid (homofermentative). In some cases, other bacteria, for example, probiotic bacteria (for instance, *Lb. acidophilus*, *Lb. paracasei* ssp. *casei* and *Bifidobacteria* ssp.) and mesophilic LAB, are mixed to modify the physicochemical and nutritional properties of the fermented milk products (Park and Guo 2006; Tamime et al. 2006). The properties of yoghurt are largely influenced by the type and composition of the starter cultures used. For example, the use of exopolysaccharide-producing starter (EPS) modifies the textural properties of yoghurt compared to yoghurts prepared by using the non-EPS LAB. The use of EPS seems to be a cheaper and acceptable natural thickening agent used to modify the textural properties of yoghurts compared to the use of stabilisers, which is discouraged by many countries (Tamime et al. 2006). In addition, EPS has been postulated to be associated with some health benefits, for example, lowering the cholesterol levels, prebiotic effect and modulation of the immune system (Ruas-Madiedo et al. 2010). Another factor which determines the textural properties of yoghurt is the amount of inoculation and the incubation temperatures. Stronger gel firmness at the same incubation temperature was obtained at higher inoculation rate compared to the lower inoculation rates (Tamime and Robinson 2007b). Lower incubation temperature triggers slower acidification, which gives more time for the colloidal calcium phosphate (CCP) to dissolve from the casein micelles before pH 5.1. A larger proportion of CCP would be dissolved from the casein micelles, and a firmer gel is expected at pH 4.6 (iso-electric pH for casein) (Tamime and Robinson 2007b). The events occurring during acidification of milk are described in two main categories, i.e. the changes of the pH due to acidification, which alters the ζ-potential on the surface of CN micelles, and the solubilisation of colloidal calcium phosphate from CN micelles (Heertje et al. 1985).

A review of the fermentation of goat milk with yoghurt starter bacteria was given by Abrahamsen and Rysstad (1991). One of the most characteristic differences between metabolites produced in cow and in goat milk yoghurt was found to be a considerable higher production of CO_2 in goat milk yoghurt than in yoghurt from cow milk (Rysstad and Abrahamsen 1987). The other typical difference was the production of the most characteristic flavour compound in yoghurt, namely acetaldehyde. Far more acetaldehyde was produced in the cow milk yoghurt than in the goat milk yoghurt. This observation was further studied and elucidated by Rysstad et al. (1990), who concluded that the reason for lower production of acetaldehyde in goat milk yoghurt was the relatively high amount of glycine in this milk compared to the content in cow milk. Acetaldehyde is mainly produced by the yoghurt culture by fermenting of the amino acid threonine. By the help of the enzyme threonine, threonine aldolase is transformed directly to acetaldehyde and glycine. A feedback inhibition of threonine aldolase produced by the presence of higher amounts of free glycine in the goat milk took place. This indicates that it may be rather difficult to obtain the same level of typical yoghurt flavour in yoghurt from goat milk as normally can be obtained from the fermentation of cow milk with yoghurt cultures.

Cheeses

The history of goat cheese manufacturing started back during the Mesopotamia era, and then further developments were made along the Mediterranean basin and some EU countries. Today, goat cheese is an important dairy product in many parts of the world. In most countries in SSA, there is a limited cheese consumption in general, but urbanisation and the growth of tourism in East and Southern African countries, including Tanzania and Malawi, will most likely stimulate the consumption of cheese in the near future.

Principally, cheeses can be classified into three main groups: hard, semi-hard, and soft and acid fresh cheeses; in each group, cheese can either be flavoured with spices or be plain (Park and Guo 2006). In some traditions, some of the goat cheese varieties are made from unpasteurised milk, and they have often unique texture and flavour compared to the cheeses made from pasteurised milk (Delgado et al. 2011). However, due to public health concerns (EFSA 2015), pasteurisation of cheese milk is encouraged so as to meet the food safety regulations, especially in the EU. Additives, for example, salts and/or cultures like a mould of *Penicillium* genera, in addition to LAB cultures, in some cheese varieties will influence the flavour of the finished cheese (Loewenstein et al. 1980). Innovations in cheese processing have led to a vast number of cheese varieties in the market; some are made exclusively from goat milk or from goat milk mixed with milk from other species.

The processing parameters during cheese processing, including preheat treatment of cheese milk, type, and dose of starter bacteria and the ageing conditions and time, modify the physical and biochemical properties of the finished cheese. Modifications of these parameters have led to the different cheese varieties with distinct flavour (Loewenstein et al. 1980). A unique cheese variety in Norway (Brown whey cheese) is made by evaporation of whey (a by-product from cheesemaking) combined with cream and whole milk. Depending on the type of cheese milk used (i.e. goat and/or cow milk), degree of browning and fat content, different varieties of brown whey cheese exist (Skeie and Abrahamsen 2017). Based on the international definition of cheese, the Brown whey cheese is, however, not a cheese since only the whey from cheesemaking is used and mixed with some milk and cream without any cheesemaking taking place.

Despite low consumption of dairy products in Africa, various dairy products including cheeses are produced in many African countries (Mattiello et al. 2018). Processing of milk into cheese improves both the nutritive value and shelf life of milk. However, the quality of cheese milk in terms of protein composition, bacteria load and SCC levels is important for the cheese yield and the quality and texture of the finished product. Brined cheese (goat cheese ripened in brine) would be ideal for the tropical environment. White brined cheeses would keep longer than other cheese varieties (Hayaloglu 2017). Many types of brine-ripened cheeses exist in many countries, mainly along the Mediterranean Sea, Middle East, Egypt and some Balkan countries (Hayaloglu 2017).

Halloumi cheese is a semi-hard well-known example of brined cheese. It originated in Cyprus; however, it has gained popularity in other parts of the world (Papademas and Robinson 1998). The flow chart for Halloumi cheese production is presented in Fig. 3. Halloumi cheese is made without the addition of starter bacteria; hence, the cheese microflora originates from the milk and from infections during cheesemaking (Anifantakis and Kaminarides 1983). Another unique feature during Halloumi cheese production is the high temperature for cooking of the cheese curd (94–96 °C) in the whey. This gives the cheese a compact and elastic texture (Hayaloglu 2017). During cooking of the cheese curd, whey proteins are denatured, precipitated and collected as a raw material for the production of whey cheeses such as Anari cheese. Finally, the cheese is ripened in brine for over 30 days to achieve the desired sensorial properties. The biochemical processes during the ripening of Halloumi cheese are summarised by Hayaloglu (2017).

Traditionally, Halloumi cheese is made from either ovine milk or goat milk or a mix of the two milks (1:1) (Papademas and Robinson 1998). In recent years, bovine milk is also used (alone) or mixed with ovine and goat milk under large-scale production of Halloumi cheese to meet the large market demand. The types of milk

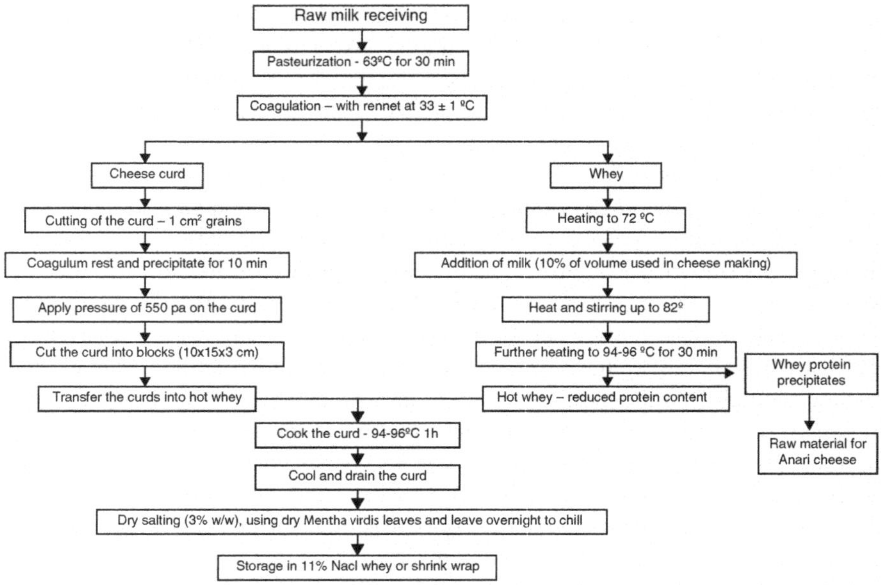

Fig. 3 Process chart for the production of traditional Halloumi cheese. (Source: Anifantakis and Kaminarides (1983) and Hayaloglu (2017))

used, milk composition, processing parameters and amount of salt explain the differences in the ripening of the cheese and the physical quality of the cheese (Elgaml et al. 2017; Milci et al. 2005; Miloradovic et al. 2017). Studies were made to compare the properties of Halloumi cheese made from the ovine, goat and bovine milk or a mix of these types of milk (Elgaml et al. 2017; Milci et al. 2005). According to Kaminarides et al. (2000), Halloumi cheese made from ovine milk, goat milk and the mix of the two did not differ in moisture content (MC), fat in dry matter (FDM) and salt contents, with values close to the recommended amount (46 and 43% for MC and FDM, respectively) according to Cypriot (Cyprus) standards for Halloumi cheese (Papademas and Robinson 1998). However, a higher cheese yield and harder Halloumi cheese were obtained from the ovine milk compared to the cheese made from goat milk. This was attributed to the differences in chemical composition between the two milks (Theophilou and Wilbey 2007). Egyptian studies showed that Halloumi cheese (fresh cheese) made from goat milk exhibited a higher yield compared to Halloumi cheese made from bovine milk (Elgaml et al. 2017).

3 Nutritional Aspects of Goat Milk and Goat Milk Products

Milk hypersensitivity is one of the challenges affecting the consumption of dairy foods globally. This is due to the body reaction towards a certain type of food (e.g. milk) after consumption. If the food is of milk origin, these reactions can be grouped into two, i.e. lactose intolerance (non-immunological) and milk protein allergies (immunological). About 75% of the world population are unable to digest lactose due to the lack of lactase (β-galactosidase), the enzyme which hydrolyses lactose. The gut microorganisms ferment the indigested lactose into gases, and due to its salt osmolality, it draws water into the lumen, which triggers diarrhoea. However, lactose intolerance is not synonymous with milk intolerance. Many lactose-intolerant people are able to consume reasonable amounts of milk, especially when consumed concurrently with other foods or when consumed in the form of fermented milk products. The second form of milk hypersensitivity is milk allergy, which accounts for 1–2% and 2–8% of adults and children, respectively (Helm and Burks 2000). This condition leads to several clinical symptoms, for example, diarrhoea, vomiting, rhinitis and anaphylaxis. According to Park (1994), over 40% of individuals suffering from cow milk allergy can tolerate goat milk proteins. Goat milk can possibly be a good substitute for some people with allergies to cow milk.

The small-sized fat globules in goat milk provide a large surface area for gastric lipase; hence, they are easier to digest compared to the larger fat globules in the cow's milk, especially for the infants (Raynal-Ljutovac et al. 2008). Goat milk is, as already mentioned, richer in medium-chain fatty acids (C6:0, C8:0 and C10:0) than cow milk, which is considered as essential for the human health, i.e. they provide a rapid energy source for malnourished individuals, inhibit the growth of bacteria and virus and dissolve the cholesterol deposits (Markiewicz-Kęszycka et al. 2013). Medium-chain fatty acids in goat and sheep milk contribute to the special organoleptic properties (aroma) in the products from these milks such as goaty and muttony aroma, respectively (Markiewicz-Kęszycka et al. 2013). Goat milk is also richer in CLA, in particular, C18:0 *cis* 9, *trans* 11, than cow milk. The benefits of CLA to human health have been studied using animal model experiments. It has been claimed that CLA exhibits anti-carcinogenic properties, enhances growth and plays part in disease prevention (Lehnen et al. 2015). The higher contents of niacin and riboflavin in the milk from goats compared to cow's milk are sufficient to meet the needs of the infants according to the FAO and WHO guidelines (Jenness 1980; Park et al. 2007).

4 Possible Farmer-Processor Partnerships

Initiatives by different organisations and research institutes in Tanzania and Malawi, for example, SUA and LUANAR, backed up by NMBU have successfully enabled the introduction and sustenance of production systems of NL and other goats in rural communities of the two countries (Chigwa 2011; Msalya et al. 2016). As a result, diets and livelihoods of the poor people in these communities have been diversified significantly (Eik et al. 2008). Msalya et al. (2016) suggested collaborations between farmers, private-milk-processing enterprises such as Shambani Graduates Ltd. (SGL) and a research institution (SUA) as an option feasible for the pro-poor value chains for climate-smart goat's milk processing and dairy goat maintenance. In Tanzania, other large processing plants can help the farmer increase production by providing assistance on optimum feeding practices and education in milk handling and entrepreneurship (Msalya et al. 2016). Other available options in Tanzania include the model of operation adopted by the Tanga Dairy Cooperative Union (TDCU), a smallholder dairy cattle farmer union in Tanga, northern Tanzania. In this model, Tanga Fresh Limited (TFL), one of the major milk processors in the country, created a reliable market for the fresh milk from the smallholder dairy farmers within the TDCU (Nell et al. 2014). TFL processes mainly ultra-high-temperature (UHT) treated milk, sour milk and yoghurts (plain and flavoured) which are sold in the major cities within the country. This and other commercial processors both in Tanzania and Malawi can work with farmers, whereby farmers send milk to nearby collection centres for bulking and further transportation to milk processing plants.

5 Conclusions

Innovations in the dairy goat management including feeding, milk handling, and disease control and milk processing are the key steps for the improvements of dairy goat productivity, marketing and the income of the poor dairy goat farmers in rural communities of Tanzania and Malawi. Goat milk possesses characteristics which make it a suitable substitute for cow milk with respect to human nutrition and raw material for the processing of various milk products. However, smallholder dairy farmers in developing countries lack sufficient capital to purchase agricultural inputs (i.e. animal feeds, veterinary services, etc.) and knowledge in the production of quality milk as the raw material for milk processing. The involvement of the private sector (i.e. milk processors) is an important strategy in the implementation of the innovations within the dairy goat sector in Tanzania and Malawi. Through these improvements of dairy goat husbandry and marketing strategies, dairy goat farmers will deliver quality milk to the milk collection centres and to the dairy factory for the processing of high-quality milk products, which will be available to the distant and larger markets. This will improve the living standards of dairy goat farmers and will add to the animal protein source for the undernourished infants in Tanzania and Malawi.

Acknowledgements The Royal Norwegian Government through NORAD supports various research and development (R&D) projects for Norwegian dairy goats in Tanzania and Malawi. A significant piece of information to write this paper has been drawn from these projects. Until recently NORAD was supporting the programme for Enhancing Pro-poor Innovations in Natural Resources and Agricultural Value Chains (EPINAV), including the project for upscaling dairy goats' technologies at Sokoine University of Agriculture (SUA). In Malawi, NORAD supports the programme for Capacity Building for Managing Climate Change in Malawi (CABMACC) at the Lilongwe University of Agriculture and Natural Resources (LUANAR). SUA and LUANAR collaborate with the Norwegian University of Life Sciences (NMBU) in implementing the programme.

References

Abrahamsen RK, Holmen TB (1981) Goat's milk yoghurt made from non-homogenized and homogenized milks, concentrated by different methods. J Dairy Res 48(3):457–463

Abrahamsen RK, Rysstad G (1991) Fermentation of goats' milk with yogurt starter bacteria: a review. Cult Dairy Prod J 26(8):28

Anifantakis EM, Kaminarides S (1983) Contribution to the study of Halloumi cheese made from sheep's milk. Aust J Dairy Technol 28:29–31

Brendehaug J, Abrahamsen RK (1986) Chemical composition of milk from a herd of Norwegian goats. J Dairy Res 53(2):211–221

Chigwa FC (2011) Strategies for improving dairy goat kid rearing systems for milk and meat production in Norway, Tanzania and milk utilization in Malawi. PhD thesis, 2011:50. Norwegian University of Life Sciences, Ås, Norway. ISBN: 978-82575-1013-8

Chilliard Y, Ferlay A, Rouel J, Lamberet G (2003) A review of nutritional and physiological factors affecting goat milk lipid synthesis and lipolysis. J Dairy Sci 86(5):1751–1770

Delgado FJ, González-Crespo J, Cava R, Ramírez R (2011) Proteolysis, texture and colour of a raw goat milk cheese throughout the maturation. Eur Food Res Technol 233:483–488

EFSA (2015) Scientific opinion on the public health risks related to the consumption of raw drinking milk. EFSA J 13(1):1–95

Eik LO, Kifaro GC, Kiango SM, Nordhagen ØM, Safari J, Mtenga LA (2008) Productivity of goats and their contribution to household food security in high potential areas of East Africa: a case of Mgeta, Tanzania. Afr J Food Agric Nutr Dev 8(3):278–290

Eknæs M, Skeie S (2006) Effect of different level of roughage availability and contrast levels of concentrate supplementation on flavour of goat milk. Small Rumin Res 66(1–3):32–43

Elgaml N, Moussa MAM, Saleh AE (2017) Comparison of the properties of Halloumi cheese made from goat milk, cow milk and their mixture. J Sust Agric Sci 43(2):77–87

Fox PF, Uniacke-Lowe T, McSweeney FP, O'Mahony JA (2015) Dairy chemistry and biochemistry. Springer, New York

García V, Rovira S, Boutoial K, López MB (2014) Improvements in goat milk quality: a review. Small Rumin Res 121(1):51–57

Hayaloglu AA (2017) Cheese varieties ripened under brine. In: McSweeney PLH, Fox PF, Cotter PD, Everett DW (eds) Cheese chemistry, physics and microbiology. Academic Press, San Diego, pp 997–1040

Heertje I, Visser J, Smits P (1985) Structure formation in acid milk gels. Food Microstruct 4(2):267–277

Helm RM, Burks AW (2000) Mechanisms of food allergy. Curr Opin Immunol 12(6):647–653

Jenness R (1980) Composition and characteristics of goat milk: *review*. J Dairy Sci 63(10):1605–1630

Kaminarides S, Rogotie E, Mallatou H (2000) Comparison of the characteristics of halloumi cheese made from ovine milk, caprine milk or mixtures of these milks. Int J Dairy Technol 53(3):100–105

Ketto IA, Knutsen TM, Øyaas J, Heringstad B, Ådnøy T, Devold TG, Skeie SB (2017) Effects of milk protein polymorphism and composition, casein micelle size and salt distribution on the milk coagulation properties in Norwegian Red cattle. Int Dairy J 70:55–64

Kifaro GC, Msalya G, Nziku ZC, Kimbita EK (2012) Status of Norwegian dairy goats in two districts of Morogoro region at the start of an up-scaling project. In: 1st annual conference of the EPINAV programme, Morogoro, Tanzania

Kiskini A, Difilippo E (2013) Oligosaccharides in goat milk: structure, health effects and isolation. Cell Mol Biol (Noisy-le-grand) 59(1):25–30

Lehnen TE, da Silva MR, Camacho A, Marcadenti A, Lehnen AM (2015) A review on effects of conjugated linoleic fatty acid (CLA) upon body composition and energetic metabolism. J Int Soc Sports Nutr 12:36

Lie H, Rich KM, Kurwijila LR, Jervell AM (2012) Improving smallholder livelihoods through local value chain development: a case study of goat milk yogurt in Tanzania. Int Food Agribusiness Manag Rev 15(3):55–85

Loewenstein M, Speck SJ, Barnhart HM, Frank JF (1980) Research on goat milk products: a review. J Dairy Sci 63(10):1631–1648

MALF (2016) The Tanzania smallholder livestock sector: a review based on the 2012/13 National Panel Survey. Ministry of Agriculture, Livestock and Fisheries (MALF). Dar-es-Salaam, Tanzania

Markiewicz-Kęszycka M, Czyżak-Runowska G, Lipińska P, Wójtowski J (2013) Fatty acid profile of milk – a review. Bull Vet Inst Pulawy 57(2):135–139

Mattiello S, Caroprese M, Matteo CG, Fortina R, Martini A, Martini M, Parisi G, Russo C, Zecchini M (2018) Typical dairy products in Africa from local animal resources. Ital J Anim Sci 17(3):740–754

Milci S, Goncu A, Alpkent Z, Yaygin H (2005) Chemical, microbiological and sensory characterization of Halloumi cheese produced from ovine, caprine and bovine milk. Int Dairy J 15(6):625–630

Miloradovic Z, Kljajevic N, Miocinovic J, Tomic N, Smiljanic J, Macej O (2017) High heat treatment of goat cheese milk. The effect on yield, composition, proteolysis, texture and sensory quality of cheese during ripening. Int Dairy J 68:1–8

Msalya G, Lie H, Mfinanga V, Ringheim AS, Sandvik R, Åsli M, Christophersen OA, Haug A, Mushi DE, Mwaseba D et al (2016) Public-private partnership for sustainable production and marketing of goat's milk in light of climate change. In: Lal R, Kraybill D, Hansen DO, Singh BR, Mosogoya T, Eik LO (eds) Climate change and multi-dimensional sustainability in African agriculture: climate change and sustainability in agriculture. Springer, Cham, pp 505–524

Msalya G, Sonola VS, Ngoda P, Kifaro GC, Eik LO (2017) Evaluation of growth, milk and manure production in Norwegian dairy goats in one highland of Tanzania 30 years after introduction. South Afr J Anim Sci 47:202–212

Nakada M, Dosako S, Hirano R, Oooka M, Nakajima I (1996) Lactoperoxidase suppresses acid production in yoghurt during storage under refrigeration. Int Dairy J 6(1):33–42

Nell AJ, Schiere H, Bol S (2014) Quick scan dairy sector in Tanzania. Dutch Ministry of Economic Affairs, The Hague

Otte J, Chilonda P (2003) Classification of cattle and small ruminant production systems in sub-Saharan Africa. Outlook Agric 32(3):183–190

Özer B, Grandison A, Robinson R, Atamer M (2003) Effects of lactoperoxidase and hydrogen peroxide on rheological properties of yoghurt. J Dairy Res 70(2):227–232

Paape MJ, Wiggans GR, Bannerman DD, Thomas DL, Sanders AH, Contreras A, Moroni P, Miller RH (2007) Monitoring goat and sheep milk somatic cell counts. Small Rumin Res 68(1):114–125

Paludetti LF, Kelly AL, O'Brien B, Jordan K, Gleeson D (2018) The effect of different precooling rates and cold storage on milk microbiological quality and composition. J Dairy Sci 101(3):1921–1929

Papademas P, Robinson R (1998) Halloumi cheese: the product and its characteristics. Int J Dairy Technol 51(3):98–103

Park YW (1994) Hypo-allergenic and therapeutic significance of goat milk. Small Rumin Res 14(2):151–159

Park YW (2010a) Improving goat milk. In: Griffiths M (ed) Improving the safety and quality of milk. Woodhead Publishing, Cambridge, UK, pp 304–346

Park YW (2010b) Goat milk: composition and characteristics. In: Pond WG, Baer CK (eds) Encyclopedia of animal science. CRC Press, Boca Raton

Park YW, Guo M (2006) Goat milk products: types of products, manufacturing technology, chemical composition, and marketing. In: Park YW, Haenlein GFW (eds) Handbook of milk of non-bovine mammals. Blackwell Publishers, Hoboken

Park YW, Juárez M, Ramos M, Haenlein GFW (2007) Physico-chemical characteristics of goat and sheep milk. Small Rumin Res 68(1–2):88–113

Raynal-Ljutovac K, Gaborit P, Lauret A (2005) The relationship between quality criteria of goat milk, its technological properties and the quality of the final products. Small Rumin Res 60(1):167–177

Raynal-Ljutovac K, Lagriffoul G, Paccard P, Guillet I, Chilliard Y (2008) Composition of goat and sheep milk products: an update. Small Rumin Res 79(1):57–72

Ruas-Madiedo P, Salazar N, de los Reyes-Gavilán CG (2010) Exopolysaccharides produced by lactic acid bacteria in food and probiotic applications. In: Brennan PJ, Itzstein MV, Moran AP (eds) Microbial glycobiology. Academic Press, San Diego, pp 885–902

Rysstad G, Abrahamsen RK (1987) Formation of volatile aroma compounds and carbon dioxide in yogurt starter grown in cows' and goats' milk. J Dairy Res 54(2):257–266

Rysstad G, Knutsen WJ, Abrahamsen RK (1990) Effect of threonine and glycine on acetaldehyde formation in goats' milk yogurt. J Dairy Res 57(3):401–411

Salaün F, Mietton B, Gaucheron F (2005) Buffering capacity of dairy products. Int Dairy J 15(2):95–109

Sanchez A, Contreras A, Corrales JC, Marco JC (2001) Relationships between infection with caprine arthritis encephalitis virus, intramammary bacterial infection and somatic cell counts in dairy goats. Vet Rec 148(23):711–714

Skeie SB (2014) Quality aspects of goat milk for cheese production in Norway: a review. Small Rumin Res 122(1–3):10–17

Skeie SB, Abrahamsen RK (2017) Brown whey cheese. In: McSweeney PLH, Fox PF, Cotter PD, Everett DW (eds) Cheese chemistry, physics and microbiology, vol 4. Academic Press, San Diego, pp 1117–1132

Skeie SB, Inglingstad RA, Brunborg LJ, Eknæs M (2014) The influence of the deletion in exon 12 of the gene encoding αs1-casein (CSN1S1) in the milk of the Norwegian dairy goat breed on milk coagulation properties and cheese quality. Small Rumin Res 122(1–3):50–58

Tamime AY, Robinson RK (2007a) Historical background. In: Tamime AY, Robinson RK (eds) Tamime and Robinson's yoghurt, 3rd edn. Woodhead Publishing, Cambridge, UK, pp 1–12

Tamime AY, Robinson RK (2007b) Background to manufacturing practice. In: Tamime AY, Robinson RK (eds) Tamime and Robinson's yoghurt, 3rd edn. Woodhead Publishing, Cambridge, UK, pp 13–161

Tamime AY, Skriver A, Nilsson LE (2006) Starter cultures. In: Tamime AY (ed) Fermented milks. Blackwell Science Ltd, Oxford, UK, pp 11–52

Theophilou P, Wilbey RA (2007) Effects of fat on the properties of halloumi cheese. Int J Dairy Technol 60(1):1–4

Turck D (2013) Cow's milk and goat's milk. World Rev Nutr Diet 108:56–62

UDLP (1997) Improving the nutritional well being of infants and children through small ruminants milk and meat. University Development Linkage Project: Bunda College of Agriculture, L., Malawi.: USAID-funded Bunda/Lincoln/Minnesota Research Programme

Walstra P, Woulters JTM, Geurts TJ (2006) Dairy science and technology, 2nd edn. CRC, Taylor & Francis, Boca Raton

Zadow JG, Hardham JF, Kocak H, Mayes JJ (1983) The stability of goat's milk to UHT processing. Aust J Dairy Technol 38(1):20–23

Zervas G, Tsiplakou E (2011) The effect of feeding systems on the characteristics of products from small ruminants. Small Rumin Res 101(1):140–149

The Need for Farmer Support and Record Keeping to Enhance Sustainable Dairy Goat Breeding in Tanzania and Malawi

G. Msalya, Z. C. Nziku, T. Gondwe, G. C. Kifaro, L. O. Eik, and T. Ådnøy

Abstract Improved dairy goats were introduced in Tanzania and Malawi to uplift nutritional standards and income of rural poor households. Breeding and management interventions were applied to ensure successful introgression of exotic genotypes into local breeds and environment. In Tanzania, 2% of goats are crossbreds of local and foreign goats. In Malawi, the exact proportion of purely local and crossbreds is not known due to poor recording. Norwegian Landrace (NL) and Saanen are the dominant breeds in Tanzania and Malawi, respectively, and their demand is increasing. In both countries, the greatest challenges are to supply goats with high genetic potential due to lack of proper breeding policy, lack of replacement bucks, and fast-generation turnovers. Increasing demand leads to high cropping rate and forces farmers to breed goats without systematic recording. A higher input/output system that requires more control is needed in order to produce better animals and obtain long-term benefits. Simple, manageable, and sustainable dairy goat breeding programs should be designed. The objective of this review is to recommend capacity building strategy needed to sustain the already seemed working technology under small-scale farming systems. For example, creation of community breeding structures and upgrading of local crossbreds through artificial insemination (AI) using elite bucks are greatly recommended. The AI was shown to be a promising genetic gain strategy for NL x local goats in Tanzania, after testing through simula-

G. Msalya (✉) · G. C. Kifaro
Department of Animal, Aquaculture and Range Sciences (DAARS), Sokoine University of Agriculture(SUA), Morogoro, Tanzania
e-mail: msalya@sua.ac.tz

Z. C. Nziku
Tanzania Livestock Research Institute (TALIRI), Tanga Research Centre, Tanga, Tanzania

T. Gondwe
Lilongwe University of Agriculture and Natural Resources (LUANAR), Lilongwe, Malawi

L. O. Eik
Department of NORAGRIC, Norwegian University of Life Sciences (NMBU), Ås, Norway

T. Ådnøy
Department of Animal and Aquacultural Sciences, Norwegian University of Life Sciences (NMBU), Ås, Norway

B. R. Singh et al. (eds.), *Climate Impacts on Agricultural and Natural Resource Sustainability in Africa*, https://doi.org/10.1007/978-3-030-37537-9_17

tion with 1000 does. It has been calculated that with records from 6 daughters per 30 young test bucks per year and selecting 3 best bucks as future elite bucks, a considerable genetic gain could be achieved. African environment conditions ensure genetically adapted future generations. For now, genetic improvement and community-based breeding program with buck rotation is suggested. Furthermore, needs for selection of replacement stock, animal identification, and performance records have been highlighted.

The demand of dairy goats is increasing each year in many African countries including Tanzania and Malawi. Their introduction has mainly been through importing exotic bucks and does, and through semen, which led to crossbreeding with local breeds. However, due to lack of support structures and capacity, breeding of these animals has not been appropriate; and for this reason, the performance has not been as expected. Breeding programs and protocols for dairy goats in both countries are lacking as most farmers breed goats based on their own knowledge. Scientific and farmer-based recording are lacking, and this makes it difficult to track the genetic composition of over 90% of the animals, which would have provided information about the breeding value (BV) of the dairy goats and enabled the institution of in-country programs to sustain imported and improved genotypes. Modern breeding practices are difficult for farmers to carry out themselves, without technical back-up from the government or research institutions. In this chapter, we review and discuss the dairy goat breeding options for developing countries using the cases of Tanzania and Malawi. We have considered the methods which can be adopted by the farmers with less financial investments.

1 Introduction

Goats are important animals in developing countries in Africa and Asia and provide a great potential for increased benefits with respect to food, nutrition, and financial security. In Tanzania, these animals are ranked second in terms of importance after cattle, and the total number in 2018 was estimated to be 18.8 million (MLFD 2018). From the Malawi census reports, there are more goats (8.4 million) compared to cattle which were about 1.7 million in 2017 (DAHLD 2018). Goats have fast-generation rate, are cheap to keep, and are popular and sometimes the only form of economic activity in most rural poor communities of African countries where they are nicknamed the “poor man’s cows” (Devendra 2013). However, goats in both countries are mainly indigenous, with low productivity due to poor genetic potential and low-input management (Chenyambuga et al. 2004; Nandolo et al. 2015; DAHLD 2018). For example, over 98% of the goats in Tanzania belong to the Small East African (SEA) breed, which has been shown to have a maximum growth rate of 5–7 g/day, mature size of 20–25 kg, low carcass weights between 6 and 13 kg (Chenyambuga et al. 2004), and milk production of less than 0.7 L/day, with very short lactation periods (Ryoba and Hansen 1988). Similar values were reported in Malawi from a previous study (Banda et al. 1993). Low productivity of the majority of livestock species, including goats, is probably the main reason for low income, poverty, and increasing animal protein deficiencies particularly among children and women in the rural communities of Tanzania and Malawi, which depend on livestock as their major economic activity (MLFD 2018). In addition, the production systems for these goats are poor and challenged by lack of sufficient nutritious feeds and prevalence of diseases and recently face the negative effects of climatic variability (Nziku et al. 2016). It is therefore difficult to sustain production of sufficient animal protein based on the indigenous livestock species. Interventions to improve trait production and production through genetics and management are required in developing countries.

Dairy goats were introduced and have shown to improve nutrition and income status of the poor households in developing countries including Tanzania (Eik et al. 2008). Household milk consumption in one division of Tanzania increased from 0 in 1988 to 1.6 (L/day) in 2012 after introduction of Norwegian Landrace (NL) dairy goats in 1993 (Kifaro et al. 2012). In addition, the income of households keeping dairy goats in the same community increased considerably (Msalya et al. 2017). In different communities in Tanzania, Chenyambuga et al. (2014) showed that household income obtained from dairy goat farming was between 25% and 30%. In Malawi, the importance of dairy goats was perceived, and dairy goats have been distributed to farmers under projects such as University Development Linkage Project (UDLP) and NORAD ARDEP Project (UDLP 1997; Lungu 2012) and through non governmental organizations (NGOs) including Small Scale Livestock Promotion Program (SSLPP) (Chigwa 2011). Moreover, dairy goats are appropriate for combating the challenges of climatic impacts while providing considerable income and nutritional needs of the poor households compared to large ruminants (Zervas and Tsiplakou 2013). Compared to cows, goats are relatively inexpensive to keep, require less fodder and a smaller plot of land for grazing while producing sufficient milk, and are hence suitable for rural poor households (Peacock 2007).

Additional attributes of goats include their adaptation to a wide range of climates and management practices, as well as their ability to thrive in arid and semiarid lands that are unfavorable for crop production and other purposes (Zervas and Tsiplakou 2013). In addition, they produce fewer emissions than cattle and fit well into integrated and climate-friendly farming systems in countries negatively hit by the impacts of climate change (Nziku et al. 2016). For these reasons, dairy goats have become popular in Tanzania and Malawi, and their number and demand are increasing. It has been estimated that dairy goats form 2% of all goats in Tanzania, while the number of these goats was estimated at about 4% of the total goats of Malawi (Chigwa 2011; MLFD 2018). The NL and Saanen are the dominant dairy breeds in Tanzania and Malawi, respectively. Mgeta division in Tanzania is the major breeding and distribution center for NL, while Babati district in Manyara region is the major area for breeding of Toggenburg goats. Mgeta has been the main center in the country that supplies goats to the farmers; and for this reason, the cropping rate is high, leading to fast-generation turnovers (Kifaro et al. 2007). In Malawi, dairy goat breeding was established at Bunda College of Agriculture (BCA), now Lilongwe University of Agriculture and Natural Resources (LUANAR), where mainly the Saanen breed was tested. At different times, the NL, Damascus, and

Alpines breeds were also tested at the center. Despite the introduction of superior dairy breeds, the situation on the ground was not satisfactory, with milk productivity declining. Efforts to exploit and sustain imported genetics and ideal management were not considered at inception (Kifaro et al. 2007). A nucleus dairy goat flock for pure NL animals was established at Mulbadaw, Manyara, Tanzania, to respond to this need and assist dairy goat breeding in both Tanzania and Malawi. However, the management of animals at nucleus center has been reported to be poor and not fully useful (Internal Reports of the Project for Up-scaling of pro-poor innovative dairy goat technologies for improved livelihood security and human capacity in selected highland areas, abb. Innovative Dairy Goats Project under the Programme for Enhanced Pro-poor Innovations in Agricultural value chain, EPINAV, Unpublished; Lungu 2012). Strategies to facilitate the introduction and sustenance of dairy goats are required to assist farmers to exploit the maximum potential of dairy goats. It is for this reason that this paper was written. First, we discuss the breeding practices for dairy goats in both countries, and thereafter, we show the options suitable for designing the sustainable breeding programs for dairy goats. Finally, a conclusion is drawn based on the current practices and trend and the expected need for sustainable dairy goat breeding.

2 The Review Methods and Data Searching

For about 8 years, the Innovative Dairy Goat Project operated in six villages of Mgeta Division, Mvomero District, Morogoro region in Tanzania to assist farmers with support services of keeping records and improving breeding of NL goats financed by NORAD. In Malawi, a UDLP project established a nucleus dairy goat breeding center at BCA and supplied mainly Saanen bucks to villages in Mkwinda and Mitundu Extension Planning Areas (EPAs) in Lilongwe. Previously, NORAD was also involved in a project to introduce dairy goats in Malawi. Furthermore, another NORAD project, namely, NUFU, integrated dairy goats into human health interventions in some communities of Lungwena EPA, Southern part of Lake Malawi district. One of the major objectives of these projects was to ensure sustainable supply of dairy goats for the current and prospective farmers. This paper is based on interdisciplinary dairy goats' research and observations from these projects. Literature search regarding dairy goat breeding in both countries was done to supplement data from the projects. Evaluation and use of literature for dairy goat breeding followed the systematic search of literature (SSL) and the scientific steps highlighted in it as suggested by O'Connor et al. (2014) and used by various previous authors including Msalya (2017). The keywords involved for searching literature in relation to the subject were dairy goats breeding and dairy goats in Tanzania and Malawi, respectively. At least five search engines were used including Google, PubMed, Web of Science/Web of Knowledge, CAB Direct, and African Journal Online (AJOL). We only picked papers from reliable sources and with complete references. From these our review results are summarized.

3 Review Results

3.1 Practices and Challenges Related to Dairy Goats Breeding in Tanzania and Malawi

In Tanzania, bucks were imported from Norway, bred either with pure NL does at Sokoine University of Agriculture (SUA), and /or crossbred with the local does of SEA goats mainly in the farming communities of Mgeta (Mtenga and Kifaro 1993). Also, pure NL bucks were sent to Mgeta where the farmers were trained to manage and rotate them in the project villages (the smallest units in the division). Therefore, a buck was transferred from one village to another when it has stayed in such a village for 2 or 3 years (Fig. 1). Buck replacement was done by SUA researchers. A similar approach was implemented in Malawi between 1992 and 2003, with Saanen pure breeding taking place at BCA, and pure bucks distributed to breeding centers with holding pens in the villages, for purposes of crossbreeding with local does (Lungu 2012). In Malawi, Saanen breed was preferred due to its superior performance compared to other breeds including Alpine, Toggenburg, and Damascus that were evaluated (Lungu 2012).

Rotation of bucks was difficult among farmers, and records were not in place to facilitate identification of parents and offspring for genetic parameter evaluation. The problem of recording arose due to increasing numbers of goats and, subsequently, lack of replacement strategy of pure breeds. These circumstances forced researchers to import additional pure NL bucks from Norway (for Tanzania) as well as pure Saanen from South Africa (for Malawi). In 2010, pure NL bucks and does were imported to Malawi from Tanzania, and these were included in the nucleus flock at BCA. Management of pure goats was performed with support from projects but has been shown to be difficult for the farmers (Kifaro et al. 2007). These problems provided experiential learning for researchers and had to look for alternative strategies. In Tanzania, the Mulbadaw nucleus flock for pure NL goats was estab-

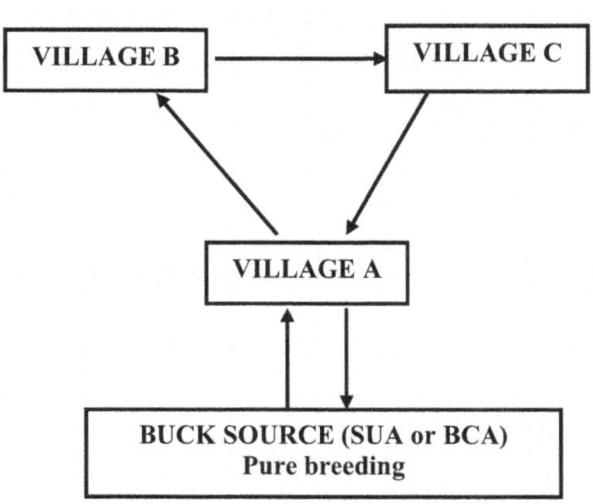

Fig. 1 Example of buck rotation in dairy goat farming villages (Adapted from a handbook of farmers (in Swahili) prepared by Sokoine University of Agriculture, The Programme for Enhancing Pro-poor Innovations in Natural Resources and Agricultural Value Chains, 2013)

lished to address this challenge and was a plan to avoid future importation expenses for farmers. At Mulbadaw, pure bucks were to be bred and supplied to farmers in Tanzania and neighboring countries (Kifaro et al. 2007). A second option for Tanzania included the use of artificial insemination (AI) in goat breeding, and in this case, semen was imported from Norway stored at SUA, and experts were sent to Mgeta and Mulbadaw to synchronize the does respectively, obtaining conception rates (CR) of 54% and 55% (Kifaro et al. 2007). The bucks resulting from the AI and nucleus herd were used for breeding in the project villages as well as in neighboring and distant villages. In Table 1, we present results of a second AI in Tanzania conducted in similar areas by the Innovative Dairy Goat Project in 2012.

Five years after the first AI in Tanzania, we report a little higher CR in the nucleus flock at Mulbadaw farm and poor CR in farming villages of Mgeta division (Table 1). Both in Mulbadaw and Mgeta, cases of abortions and misconception were reported, and most of these were associated to lack of good records and poor selection of does, some of which were synchronized while they were pregnant. Furthermore, poor results can also be associated with poor timings and technical know-how of the AI experts.

In Malawi, initially dairy goats were tested at research station before goats were taken to the farmers. In this case, the dairy goats in Malawi were tested at BCA and were distributed through various projects including the UDLP and NGOs (Chigwa 2011). Some of the breeding bucks were obtained from Mulbadaw farm in Tanzania. As another goat strategy, community-based goat breeding (CBGB) programs were implemented, and the approach for these has been reported by Nandolo et al. (2016). The focus was to improve productivity of Malawi local goats through community integrated selection of bucks for breeding. The local goats were improved to increase supply of meat to households and to provide does for crossing with exotic dairy and meat breeds at the same time. The CBGB programs in Malawi mainly relied on the choice of traits appreciated by the farmers (Nandolo et al. 2016). The authors showed that farmers in different communities have different considerations before making their choices of breeding goats. For example, farmers with more than 7 does preferred the ones with high qualities for productive and reproductive traits as well as good morphometric data (Nandolo et al. 2016). Other considerations in the breeding choices were higher growth rates, higher mature weights, and survival traits, which are directly related to off-take rates. The authors finally suggested that it was important to take on board different selection strategies for different sites (Nandolo et al. 2016).

Table 1 Results of the second artificial insemination conducted in dairy goats in Tanzania

Center	Inseminated does	Kidded does	Conception rate (%)
Mulbadaw	25	15	60
Mgeta	18	4	20
Total	43	19	44.2

3.2 *Dairy Goat Crossbreeding Programs in Tanzania and Malawi*

The main breeding practice in both Tanzania and Malawi has been crossbreeding to upgrade the low-producing indigenous goats into high-yielding animals (50–75% exotic blood). In Tanzania, various goat breeds were imported into the country for this purpose since early 1960s, and they mostly included Saanen, Toggenburg, Anglo-Nubian, and Alpine (Mtenga and Kifaro 1993). Since that time, research on dairy goats in the country has tended to concentrate on breeding for improved reproductive performance and milk yield after crossbreeding with the indigenous goats (Mtenga and Kifaro 1993). Preliminary performance for these programs was reported by these authors (Table 2).

In Malawi, both dairy and meat breeds of goats have mainly been obtained from South Africa, Tanzania, and other neighboring countries, and the approach was also crossbreeding. The main breed used in dairy crossbreeding with the local goats was Saanen; an earlier report on the performance was produced by Karua and Banda (1990), and it mainly concentrated on growth traits (Table 3). The results reported by these authors clearly showed that the crossbred goats were higher in performance compared to the local goats.

Recently, end surveys were carried out in Malawi to evaluate performance of the 10-year crossbreeding program, and results are reported in Table 4. It was shown that there was a declining trend on the number of crossbred goats in communities and a reduced usage of bucks for crossbreeding. It was also observed that breeding was difficult to control in the rural communities. For goats in the communities, indiscriminate mating took place, leading to unwanted backcrossing toward local breeds. These observations defeated the crossbreeding program in the communities. Also, the number of bucks at BCA has declined. It was concluded that the crossbreeding was not accompanied by a sustainable breeding program.

Table 2 Performance of crossbred dairy goats during the early years of introduction (early 1990s)

Measured trait	Anglo-Nubian	Toggenburg	Saanen	Blended
Kidding rate (%)	62	52	45	63
Twinning rate (%)	15	18	16	14
Preweaning mortality (%)	52	24	48	–
Birth weight (kg)	2.7	2.8	3.8	2.5
Birth to 16 weeks (kg)	83	84	77	62
Birth to 72 weeks (kg)	41	43	45	–
Lactation length (days)	154	143	129	268
Yield (total kg)	105	85	142	123
Yield (kg/day)	0 7	0.6	1.1	0.5

Source: Mtenga and Kifaro (1993); goats recorded were crosses of 50%

Table 3 Birth weights of Saanen × local (SL) and pure local (PL) at Bunda College farm (kg)

Breed	BW (kg)	WW (kg)	Weight gain (g/day) until weaning	W364 (kg)	Weight gain (g/day) until 364 days
Saanen crosses (M)	2.61 ± 0.39[a]	12.80 ± 5.14	92.32	48.41	25.00 ± 3.00
Saanen crosses (F)	2.48 ± 0.54[a]	11.96 ± 2.77	82.67	41.82	22.50 ± 1.78
Local (M)	1.93 ± 0.53b	16 7.60 ± 1.53	50.00	29.32	14 99 ± 1.66
Local (F)	1.84 ± 0.48b	8.33 ± 2.07	56.16	31.07	16.16 ± 1.55

Source: Karua and Banda (1990)

BW birth weight, *WW* weaning weight, *W112* weight at 112 day, *W364* weight at 364 days, *M* male, *F* female

Table 4 Results from a survey to evaluate crossbreeding after project closed

Parameter	Unit	Value	Remarks
Households benefitted from program	%	56	Number of participating households within the study sites
Households that had crosses	%	14	Declining prevalence of crossbred goats
Kids observed by breed being local	%	87	Increasing number of local goats
Proportion of half crossbred kids	%	7.4	Fewer crossbreds compared to local goats
Proportion of local back crossed kids	%	5.8	Shows there was indiscriminate mating taking place
Proportion of local goats over crosses	%	94	Declining prevalence and nonsignificant contribution of crossbreeding

Source: Lungu (2012)

3.3 Available and Suitable Dairy Goat Breeding Options for Tanzania and Malawi

3.3.1 Scheme for Breeding Dairy Goats Based on Simulation (Determining Number of Bucks)

One way through which breeding can be planned is through simulation. In Tanzania, Nziku et al. (2017b) showed a progeny testing and selection breeding program with an acceptable model fitting into the rural farming communities of Mgeta where NL goats are being farmed. Briefly, the authors used 1000 female goats out of the 2000 population of goats which were considered available (Nziku et al. 2017c). The 1000 does were considered in generation zero as having known parents, and they were assumed to be mated to 100 bucks (10 does per each buck); and assuming a twining rate of 1.1, about 1100 kids per generation (550 male and 550 female) would be expected (Nziku et al. 2017c). Further assumptions were 10% mortality, 30% culling of old does, and various buck replacement rates. Each time kids are born, recording for traits of interest is taking place, until four generations. As a result, the value for important reproductive traits is being determined, including age at first mating, age at first kidding, adult productive life, and generation interval among other traits. The simulation assumed that all goats come from the same breeding population and have the same conception and survival rates after weaning. Furthermore, the authors determined productive potential in their model, including the estimated 210 kg of milk in 210 days (Nziku et al. 2017b, c). Finally, considerable accuracy in selection intensity is gained when many bucks are tested for traits such as daily milk yield, than when only few of them are tested. However, the authors noted the importance of revising the breeding programs from time to time, due to the fact that one breeding may not fit in all production systems. This confirms the ideas in Falconer and Mackay (1996) that the best animal selected under temperate environment may not be the best under tropical conditions. In Mgeta's situation, testing 30 young bucks per year may be the best current option. However, the proposed breeding program may not be perfect in the future due to constant shifts in the available knowledge and practical options. It is therefore necessary to revise breeding programs from time to time.

3.3.2 Use of Records to Design a Breeding System for Dairy Goats

Recently, Nziku et al. (2017c) used data obtained from 62 stallholder dairy goat farms in Mgeta (Mvomero district) to suggest a breeding program for this community. The authors based their program on the little data available from the farmers regarding traits in dairy goats, including milk yield per doe per day, and weights at different stages of growth (growth performance). In the analyses, the authors also took into account other information that they considered important in designing breeding program in dairy goats. This included the individual identities (ID), parents of individual goats (buck and doe IDs), birth date, sex, herd dynamics (kids born, animals sold, animals slaughtered, number of deaths), inseminations (date of insemination and semen ID), and health of animals (e.g., type of diseases and treatment effected). Based on the analyses, it was concluded that a simplified breeding plan using buck rotation and occasional AI as the one initially planned at Mulbadaw nucleus flock was beneficial to farmers (Nziku et al. 2017a, b, c).

4 Discussion

In the above data presentation, we have highlighted the practices of farmers and other stakeholders, including researchers in dairy goat breeding. Although there have been successful results from various programs, we note the chain stops at some points; and we also note that farmers are responsive to instructions of researchers when projects are ongoing. However, goats have short generation interval, and it is imperative to strictly supervise breeding. Norway is one of the countries in the world with an advanced and comprehensive recording and breeding scheme for dairy goats, particularly the NL goats (Ådnøy 2014). Over the years, the goal of breeders of NL goats has been to improve production of the goats while maintaining their identity; and for these reasons, their breeding program comprised of over 30,000 dairy goats in 340 herds, which are controlled in terms of records (Blichfeldt 2013). Over 90% of the farmers in the country willingly participate in the goat recording scheme operated by the Farmers Dairy Cooperative (TINE), in which eight important traits are recorded daily and are included in the breeding goals. These traits are daily milk yield, milk content such as percentage dry matter (DM), crude protein (CP), fat, lactose, udder/teat conformation, milk speed evacuation, free fatty acid contents, and occurrence of mastitis. The records are analyzed to obtain annual values for each of these traits per animal or per herd; and in all these cases, individuals' pedigree information is recorded (Ådnøy et al. 2000). Farmers are encouraged to test best buck kids within own herds and to use semen for introducing new genes. This is possible since farmers keep records and have large herd sizes per farm, and they control the animals. Also, the infrastructure necessary for goat recording and performance evaluation is well articulated from the farm to the institutional level (Blichfeldt 2013). To keep genetic variation high and increase milk quality, Norway occasionally imports superior genes from other countries. For example, between 2007 and 2011, French Alpine semen were imported and made available to farmers (Ådnøy 2014). Every year, semen from elite bucks are stored for possible future use (Blichfeldt 2013). In addition, genotypes from special populations, i.e., wild goats, are secured in the same manner. By doing so, it is possible to safeguard biodiversity and adapt the breeding goals if needed. The NL breed is adapted to and bred under Norwegian conditions that are very different from climatic and other environmental conditions in Tanzania and Malawi. NL and other exotic dairy genotypes will normally have a higher milk production potential than local SEA goats, but they will also be more vulnerable to diseases and be less robust to African conditions. Therefore, we propose that goats descending from imported AI should be tested and selected for breeding under conditions similar to where they will be used in the future. A blend of such imported stock and local goats should be tested, and the best should be selected to have a more robust dairy goat breed in the future.

It is important for developing countries including Tanzania and Malawi to maintain sustainable replacement animals and ensure that these are raised carefully and have better genetic merit compared to their parents. It is therefore recommended that the breeding program for dairy goats becomes continuous and the resulting populations are evaluated (Kifaro et al. 2007). This has been shown to be the starting point of the goat milk value chain which determines the success of subsequent value chain nodes. Recruiting breeding bucks from local communities within one locality such as Mgeta is accepted as having greater genetic potential for traits such as adaptability than importing from temperate countries. In Tanzania, some of the NL bucks imported from Norway did not perform as expected, and some died before mating for different reasons (Kifaro et al. 2007). However, it is of great importance to bring in new genes as shown in the example of NL breeding in Norway (Ådnøy 2014). There is greater potential of planning sustainable breeding in the local dairy goat farming communities of Tanzania and Malawi, and the populations of these animals are considered sufficient for conducting selection and achieve recommended genetic gains (Nziku et al. 2017c). Higher genetic gain and scheme sustainability can be expected if farmers and their organizations are able to play major roles in the scheme implementation while working closely with research and academic institutions. Although it may be hard to pick all lessons from developed countries such as Norway, Tanzania and Malawi can keenly supervise the accuracy of records and invest in either AI or a stricter buck circle program to improve the genetic potential and overall productivity of the goats. Moreover, farmers in these countries will need continued support for some years, accompanied by formation of stronger farmers' organizations. In Tanzania, organizations such as Twawose as well as the NL and Toggenburg Breeders Associations (NOBRA and TOBRA) need to be strengthened.

5 Conclusions

Farmers and other key beneficiaries along the dairy goats' market chain in Tanzania and Malawi could benefit more if they are capacitated through sustainable collaborative research and development with different developmental partners. This will enable them to participate to their best level to work along with the proposed breeding strategies. Faster genetic progress for dairy goats in Tanzania and Malawi can be taped from any best running breeding program while continuing to develop own capacity to run animal breeding programs. We note also that capacity building for important people such as the extension officers is needed, and they should be equipped with the important and basic techniques for dairy goat breeding such as goat AI technologies and recording techniques, if sustainable and improved results are to be obtained. We emphasize the importance of records, without which successful dairy goat breeding programs cannot be achieved. Technical assistance from the governments of these countries or research institutions is still needed as farmers still lack the important techniques for dairy goats breeding.

Acknowledgments The Royal Government of Norway through NORAD supports various research and development (R&D) projects for Norwegian dairy goats in Tanzania and Malawi from which information to write this paper was drawn. Until recently, NORAD was supporting the Programme for Enhancing Pro-poor Innovations in Agricultural Value chains (EPINAV), including the project for upscaling dairy goats' technologies at Sokoine University of Agriculture (SUA). In Malawi, NORAD is supporting the Program for Capacity Building for Managing Climate Change in Malawi (CABMACC) at The Lilongwe University of Agriculture and Natural Resources (LUANAR). The SUA and LUANAR collaborate with the Norwegian University of Life Sciences (NMBU) in implementing the programs.

References

Ådnøy T (2014) The dairy goat industry in Norway: challenges in a historical perspective. Rev Small Rumin Res 122:4–9

Ådnøy T, Nævdal I, Svendsen M (2000) In: Gruner L, Chabert Y (Eds) Buck circles for dairy goat breeding in Norway (Circles des books pourl'amélioration génétique des chèvres laitières en Norvège). Institut del'Elevage/INRA, p. 1047

Banda JW, Ayoade JA, Karua SK, Kamwanja LA (1993) The local Malawi goat. World Anim Rev 74/75:49–57. http://www.fao.org/docrep/U9550T/u9550T0j.htm

Blichfeldt T (2013) Breeding for goats' milk quality in Norway. Goat milk quality. Regional IGA Conference 2013 in Tromsø, Norway

Chenyambuga SW, Hanotte O, Hirbo J, Watts PC, Kemp SJ, Kifaro GC, Gwakisa PS, Petersen PH, Rege JEO (2004) Genetic characterization of indigenous goats of Sub-Saharan Africa using microsatellite DNA markers. Asian Australas J Anim Sci 17:445–452

Chenyambuga SW, Jackson M, Ndemanisho EE, Komwihangilo DM (2014) Profitability and contribution of small-scale dairy goat production to income of smallholder farmers in Babati and Kongwa districts, Tanzania. Livest Res Rural Dev 26(27). http://www.lrrd.org/lrrd26/2/chen26027.html

Chigwa FC (2011) Strategies for improving dairy goat kid rearing systems for milk and meat production in Norway, Tanzania and milk utilization in Malawi. PhD. Thesis, Department of Animal and Aquacultural Sciences, Norwegian University of Life Sciences, Ås, Norway, Thesis Number 2011:50, pp 14–15

DAHLD (2018) Agricultural productivity estimate report for 2017 2018. Department of Animal Health and Livestock Development, Ministry of Agriculture, Irrigation and Water Development (MOAIWD), Lilongwe

Devendra C (2013) Investments on pro-poor development projects on goats: ensuring success for improved livelihoods. Asian Australas J Anim Sci 26:1–18

Eik LO, Kifaro GC, Kiango SM, Nordhagen ØM, Safari J, Mtenga LA (2008) Productivity of goats and their contribution to household food security in high potential areas of East Africa: a case of Mgeta, Tanzania. Afr J Food Agric Nutr Dev 8:278–290

Falconer DS, Mackay TFC (1996) Introduction to quantitative genetics, 4th edn. Harlow, London, pp 131–134

Karua SK, Banda JW (1990) Dairy goat breeding in Malawi: Gestation length, birth weights and growth of the indigenous Malawi goats and their Saanen crosses. Proceedings of the First Biennial Conference of the African Small Ruminant Research Network. ILRAD, Nairobi, Kenya10–14 December 1990. Available in: In IIRI 2002. Animal Genetic Resources Virtual Library. CD ROM. Small Ruminant Research and Development in Africa

Kifaro GC, Eik OL, Mtenga LA, Mushi DE, Safari J, Kassuku AA, Kimbita EN, Maeda-Machangu AD, Kanuya NL, Muhikambele VRM, Ndemanisho E, Ulvund MJ (2007) The potential use of AI in sustainable breeding of dairy goats in developing countries. A case study of Norwegian dairy goats in Tanzania. Tanzan J Agric Sci 8:19–24

Kifaro GC, Msalya G, Nziku Z, Ndemanisho EE, Kimbita EN, Mushi DE, Max RA, Mtenga LA (2012) Status of Norwegian dairy goats in two districts of Morogoro region at the start of an up-scaling project. Oral presentation, 1st annual conference of the EPINAV program, Morogoro, Tanzania, 18–19 Dec 2012

Lungu G (2012) An assessment of prevalence, productivity and usage of crossbred goats in villages of Lilongwe. BSc report, University of Malawi

MLFD: Ministry of Livestock and Fisheries Development (2018) Budget speech, pp 4–5, http://www.mifugouvuvi.go.tz/uploads/publications/en1526555624-Hotuba%20%20ya%20Waziri%20wa%20Mifugo%20na%20Uvuvi%20kwa%20mwaka%202018-19%20%20tarehe%2014%20%20Mei%202018%20_Final.pdf

Msalya G (2017) Contamination levels and identification of bacteria in milk sampled from three regions of Tanzania, evidence from literature and laboratory analyses. Veterinary Medicine International 2017:9096149, pp 1–10. https://doi.org/10.1155/2017/9096149

Msalya G, Sonola VS, Ngoda P, Kifaro GC, Eik LO (2017) Evaluation of growth, milk and manure production in Norwegian dairy goats in one highland of Tanzania 30 years after introduction. South African J Anim Sci 47:202–212 https://doi.org/10.4314/sajas.v47i2.12

Mtenga LA, Kifaro GC (1993) Dairy goat research and extension at Sokoine University of Agriculture (lowlands) and Mgeta (highlands) areas of Tanzania. In: Kategile JA, Mubi S (eds) Future of livestock industries in East and Southern Africa, Proceedings of a Workshop held at Kadoma Ranch Hotel, Zimbabwe, 20–23 July, 1992. ILCA (International Livestock Centre for Africa), Addis Ababa, p 227. Also available at http://www.fao.org/wairdocs/ILRI/x5485E/x5485e0d.htm

Nandolo W, Gondwe T, Banda L (2015) Characterization of indigenous goat breeds and their production systems for purposes of designing community-based breeding programs in Malawi. Project Report submitted to USDA under African Goat Improvement Network

Nandolo W, Wurzinger M, Mészáros G, Van Tassell C, Gondwe T, Mulindwa H, Lamuno D, Sölkner J (2016) Identification of breeding objectives in community-based goat breeding programs in Malawi. Acta Argic Slov Suppl 5:103–108

Nziku ZC, Asheim LJ, Eik LO, Mwaseba D, Kifaro GC (2016) Climate change adaptation in vulnerable crop and livestock production systems in Mgeta. Tanzan Afr J Food Agric Nutr Dev 16:10835–10865. https://doi.org/10.18697/ajfand.74.15605

Nziku ZC, Kifaro GC, Eik LO, Steine BCT, Ådnøy T (2017a) Reasons for keeping dairy goats in Tanzania, and possible goals for a sustainable breeding program. Anim Prod Sci 57:338–346. https://doi.org/10.1071/AN15423

Nziku ZC, Kifaro GC, Eik LO, Steine T, Ådnøy T (2017b) Use of simulation to examine a dairy goat breeding program in Tanzania. Livest Res Rural Dev 29(43). http://www.lrrd.org/lrrd29/3/czab29043.html

Nziku ZC, Kifaro GC, Eik LO, Steine T, Msalya G, Ådnøy T (2017c) Situation analysis and prospects for establishing a dairy goat breeding program in Tanzania. Livest Res Rural Dev 29:223. http://www.lrrd.org/lrrd29/12/Cza29223.html

O'Connor AM, Anderson KM, Goodell CK, Sargeant JM (2014) Conducting systematic reviews of intervention questions I: writing the review protocol, formulating the question and searching the literature. Zoonoses Public Health 61:28–38

Peacock C (2007) The goats model. A proven approach to reducing poverty among smallholder farmers in Africa by developing profitable goat enterprises and sustainable support services. Working papers, Farm-Africa

Ryoba R, Hansen KK (1988) Composition of goat's milk compared to cow's milk. In: Mgheni M, Ekern A (eds) Improved dairy production from cattle and goats in Tanzania. Reports from research projects 1982–1985. Noragric at the Agricultural University of Norway, As, pp 60–62

Sokoine University of Agriculture and The programme for Enhancing Pro-poor Innovations in natural resources & Agriculture Value chains (2013) Utunzaji bora wa mbuzi wa maziwa mtunze mbuzi akutunze. Kifaro GC, Msalya G (eds), p 36

UDLP (1997). Improving the nutritional well being of infants and children through small ruminants milk and meat. USAID-funded Bunda/Lincoln/Minnesota Research Programme. University Development Linkage Project: Bunda College of Agriculture, Lilongwe, Malawi. pp 1

Zervas G, Tsiplakou E (2013) Goat milk. In: Park YW, Haenlein GFW (eds) Milk and dairy products in human nutrition: production, composition and health. Wiley-Blackwell, London, pp 498–518

Stratified Livestock Production and Live Animal and Meat Export from Ethiopia: Lessons from the Experience of a Donor Funded Project

Adugna Tolera and Lars Olav Eik

Abstract This paper reviews the livestock production system and the performance of meat and live animal export from 2005/2006 to 2011 and the catalytic role played by Ethiopia Sanitary and Phytosanitary Standards and Livestock and Meat Marketing Program (SPS-LMM), a development project designed with objectives of improving the capacity of Ethiopian veterinary services and enhancing Ethiopia's competitive advantage for meat and live animal exports. The performance of formal live animal and meat export increased by 442% and 241%, respectively, from a base of \$27 million for live animal and \$18.5 million for meat in 2005/2006 to \$148 million for live animal exports and \$63 million for meat exports in 2010/2011, respectively. These increments were due to the increased volume of exports, increased selling price of meat and live animals in the importing countries, and increased formalization of the livestock trade. There is a strong linkage between the pastoral livestock production and the feedlot operations and export abattoirs operating in the central highland areas of the country. In general, coordinated and concerted efforts of the public and private sector actors, with catalytic support of development projects, are needed to increase the live animal and meat export performance and to fuel economic growth of the country. However, it is imperative that all actors along the value chain benefit fairly from the development initiatives.

1 Introduction

Ethiopia is endowed with a high livestock population, diverse animal genetic resources, and diverse agroecologies suitable for different livestock production systems. Livestock production significantly contributes to the livelihoods of millions of

A. Tolera (✉)
School of Animal and Range Sciences, College of Agriculture, Hawassa University, Hawassa, Ethiopia

L. O. Eik
Department of International Environment and Development Studies, Norwegian University of Life Sciences, Ås, Norway

B. R. Singh et al. (eds.), *Climate Impacts on Agricultural and Natural Resource Sustainability in Africa*, https://doi.org/10.1007/978-3-030-37537-9_18

people in the country. It serves as a means of employment, a source of income, a source of nutritious food (meat, milk, eggs etc.), and a source of vital inputs for crop production (draught power, manure, and cash income for the purchase of various inputs). Animals, especially equine and camels, serve as very important means of transport in rural areas. Manure from animals is also used to improve soil fertility for crop production, and dried dung and biogas is used as a source of fuel for rural communities. In addition, livestock plays important social functions and is also a means of saving and security.

Livestock production also makes significant contributions to the national economy, contributing about 16–19% of the overall gross domestic product (GDP) and about 45% of the agricultural GDP (Behnke 2010). Livestock products also contribute to the export earnings of the country and as sources of industrial raw materials. Livestock production also enables the utilization of over 60% of the land surface (the rangelands), which cannot be effectively utilized otherwise. In addition, in mixed crop-livestock production systems, integration of livestock production into the system enables the use of over 50% of the plant biomass (crop residues), which cannot be directly used by humans, to convert them into highly valuable products and increase the resource utilization efficiency of the overall system (Tolera et al. 2012).

In Ethiopia, there is a potential for expansion of crossbred dairy cattle production in cooler highlands and for stratified beef production, in which breeding and stocker animals are raised in pastoral areas, and finisher animals are fed in feedlots in the highlands. The demand for livestock products is increasing from time to time, both in the domestic and export markets. Moreover, there is increased interest in promoting intensive and market-oriented livestock production as a means of livelihood, income source, and economic growth. Intensification and market orientation of livestock production demand increased supply of high-quality feed.

In recent years, the Ethiopian Government has given special attention and consideration to the development of export-oriented meat industry. This value-addition strategy primarily focuses on supporting livestock producers, particularly pastoral communities, through creating market linkages to the feedlots and processing industries, job creation, saving hides and skins which are the major foreign currency generators as compared to any livestock products. In addition, Ethiopia enjoys proximity to high demand meat markets in the Middle East and North African countries, which has attracted investments in export abattoirs both by nationals and foreign companies.

However, despite the huge livestock resource potential, engagement of a good number of companies in the industry, proximity to strategic markets, and supportive Government policy environment, the export of meat remained very low. The project known as Ethiopia Sanitary and Phytosanitary Standards and Livestock and Meat Marketing (SPS-LMM) Program was designed and implemented by the Norman Borlaug Institute for International Agriculture of Texas A&M University in partnership with the Ethiopian Ministry of Agriculture and other public and private sector actors as an intervention strategy to develop the export meat industry. The project had two broad objectives of improving the capacity of Ethiopian veterinary service to undertake SPS and related activities in support of increased livestock and meat export and enhancing Ethiopia's competitive advantage for meat and livestock exports. The project was implemented from the beginning of 2006 up to September 30, 2011 (SPS-LMM 2011a).

This paper reviews the performance of the meat and live animal export industry of Ethiopia during the SPS-LMM project period, lessons learned, and the way forward to enhance live animal and meat export and improve the livelihoods of livestock producers in general and the pastoral communities in particular. The objective is to review livestock production systems, livestock and meat export performances, and the role played by SPS-LMM Program. The information included in this paper is based on a review of published and unpublished reports, export data collected from the Revenue and Customs Authority of Ethiopia, and information from other sources.

2 A Brief Review of the Livestock Production Systems

There are four major livestock production systems in the country. These include pastoral and agropastoral system, mixed crop-livestock production system, small-scale urban and peri-urban (landless) livestock production, and relatively large-scale commercial livestock production. The different livestock production systems differ in resource endowments, available feed resources, and livestock feeding practices.

Pastoral and agropastoral production system Grazing and browsing (foliages and pods of trees and shrubs) of rangelands are the main source of feed in the pastoral and agropastoral livestock production systems, with occasional use of crop residues, stubble grazing, and conserved forages. Seasonal mobility tracking forage and water is a common practice of coping with seasonal feed and water shortage. During periods of severe drought, pastoral herds become dependent on emergency feeding on feeds sourced from highlands (hay, straw, and agroindustrial by-products). Recently, irrigated fodder production along river basins is increasing. Seasonal variation in fodder availability and quality, encroachment of invasive species, restricted livestock mobility, land use change, recurrent droughts, weakening of customary institutions, and lack of sustained investment in rangeland improvement are among the major challenges affecting livestock feed supply and pastoral livestock production (Gizachew 2012).

Mixed crop-livestock systems The major sources of feed in this system include grazing of poor-quality natural pasture (private or communal), including road sides and crop boundaries as well as crop residues and stubble grazing. These are rarely supplemented with green forages, tree leaves, household wastes, and purchased feeds. Natural pastures used to be a major source of feed in this system. However, the area of grazing land is decreasing from time to time due to the expansion of cropping into grazing lands causing shrinkage of grazing land. The existing grazing lands exhibit decreased productivity due to heavy use and other factors (prolonged and excessive use). Crop residues are becoming increasingly important sources of animal feed. They make up >50% of the biomass of crops and contribute about 50%

of feed supply. However, they are characterized by low nutritive value, wide variability in quality, seasonal availability, and bulkiness to transport over long distance (Tolera 2007). Food-feed crops such as maize, sorghum, cassava, sweet potatoes, etc. offer promising opportunities for dual-purpose use as a source of human food and animal feed (Tolera et al. 2012). In general, there is marked seasonal fluctuation of feed supply and quality.

Urban and peri-urban livestock operations These include small- to large-scale dairy farms, feedlots, and poultry farms. Most of such operations do not have land for grazing or feed production, and as a result, they are dependent on purchased feed (both roughage and concentrate). They are located around the major urban centers of the country where they have better access to agroindustrial by-products, compound concentrate feeds, and conserved forages in the form of hay or straw. They are affected by unreliable supply and quality and increasing price of purchased feed. The unavailability and very high price of vitamin and mineral supplements is a particular challenge (Tolera et al. 2012).

Large- scale commercial livestock production These include commercial dairy farms as well as cattle and small ruminant ranches and feedlots. They have better access to land, financial, and material inputs for feed production, conservation, and storage.

3 Feeding and Conditioning Systems for Slaughter Animals

Both large-scale (commercial feedlots) and small-scale (backyard) fattening operations are carried out in Ethiopia. Commercial feedlots feed a relatively large number of animals at a time. They keep from as few as about 20–50 animals to as many as 5000 heads of cattle at a time. Almost all commercial feedlots depend on purchased concentrates and roughage feeds as they do not have land for feed production. Most feedlots are located in the central part of the country, mostly around Mojo, Adama, Wonji, and Melkassa areas of East Shewa Zone of Oromia Regional State. This gives them easy access to agroindustrial by-products such as wheat bran, oilseed cakes, and molasses, which form a major portion of the concentrate mixture fed to the animals. Purchased native grass hay from Sululta, north of Addis Ababa, and tef or wheat straws make up the roughage component of the diet (Tolera 2008).

On the other hand, there are traditional and indigenous systems of cattle and small ruminant fattening practices in different parts of the country. These are typically carried out in the backyard using any feed resources produced on the farm. Notable examples of backyard fattening practices are carried out in Wolayita and Hararge areas. Farmers in Wolayita have a long tradition of fattening oxen using locally available feeds. They feed one or two oxen for about 3–4 months and sell during festive holidays such as *Meskel* and Christmas. The main feed resources are cut-and-carry grass and various agricultural by-products such as cereal and pulse straws, sweet potato vines and tuber, thinning or whole crop maize, enset supplemented with boiled maize and haricot bean, and household wastes such as *atella* (a by-product of home brewed or distilled beverages) and coffee residues. Similarly, the fattening practice in East and West Hararge zones of Oromiya Regional State is based on thinning and leaf stripping of maize and sorghum, grasses, and weeds from croplands and other agricultural by-products such as sweet potato vines. Backyard fattening in Arsi Negelle area is based on *areqe atella* (a residue resulting from home distilling of alcoholic liquor, *areqe*) and wheat straw supplemented with a small amount of wheat bran and linseed cake or any other oilseed cake. Similar small-scale backyard fattening of one or two animals is practiced in different parts of the country using different crop residues and/or cut grass with occasional supplementation with different agricultural and agroindustrial by-products (Tolera 2008).

4 Source of Animals for Meat and Live Animal Export

According to the traditional classification of livestock production systems, there are two distinct agroecologies and livestock production systems. The highland areas, whose altitude is over 1500 meters above sea level (m.a.s.l), cover 40% of the total area of the country (1.2 million km^2) and host about 60% of the total livestock population. On the other hand, lowland areas whose altitude is below 1500 m.a.s.l are covering 60% of the land mass of the country. Though the exact number is not known because of various reasons, it is indicated in Jabbar et al. (2007) that pastoral and agropastoral areas own 60% of the goats, 40% of sheep, 20% of cattle, and 100% of the camel population of the country. According to Ayele et al. (2003), cattle and sheep are also the major livestock in highland areas, whereas camels and goats are the prominent domestic animals in the pastoral lowlands below 1500 meter above sea level. The major pastoral lowland areas are located in four regional states: Somali (44 districts), Oromia (34 districts), Afar (29 districts), and SNNP (6 districts). These are the major sources of livestock for export market.

The peculiar nature of livestock marketing in Ethiopia is the source of animals. Though there is a huge resource in the highlands, the export quality animals are obtained from the pastoral lowlands. The feedlots and export abattoirs exclusively buy animals from lowlands. The rate of weight gain of livestock in feedlots and meat color of shoats are the major reasons that exclude highland livestock from the export market apart from the competition in the highland markets with domestic consumers. However, lowlands are prone to recurrent drought, and production is not market oriented. Thus, animals destined for meat and live export are produced in a stratified production system in which the breeding herd and the stocker animals are raised on rangelands in the pastoral and agropastoral areas. On the other hand, the final feeding and conditioning of slaughter and export animals is carried out in the central highlands where there is a better access to agroindustrial by-products and concentrate mixtures.

Over 90% of the animals conditioned in feedlots, and those destined for meat and live animal export are sourced from the pastoral areas. In the pastoral areas, the livestock largely depends on rangelands consisting of native vegetation (grasses, bushes, shrubs, as well tree leaves and pods), which may be augmented by crop residues in the agropastoral production systems. On the other hand, feedlot operations are carried out in the central highlands using concentrate feeds composed of different agroindustrial by-products (oilseed cakes and brans) and a small amount of dry roughages (hays or straws). Feeding and conditioning of the animals in the feedlots is carried out for about 2–3 months before slaughter or live export. In this typical stratified beef production system, the breeding herd and the young stock are kept in the pastoral areas while the final fattening and finishing is done in feedlots in the central highlands of the country.

The stratification of the production system is based on relative livestock and feed resource endowment of the pastoral areas vs. the highlands and access to markets for livestock production inputs and conditioned animals. Most feedlots are located in the central parts of the country, particularly around Mojo, Adama, Wonji, Melkassa, and Adami Tullu areas. This gives them easy access to agroindustrial by-products such as wheat bran, oilseed cakes, and molasses, which form a major portion of the concentrate mix fed to the animals. Purchased native grass hay and straws of tef and/or wheat make up the roughage component of the diet. Properly formulated and uniformly mixed best cost ration is recommended for rapid body weight gain at best cost.

The stratified beef production system linking the pastoral area with feedlots and export abattoirs plays beneficial roles to various actors along the value chain, which include the following:

- It creates market opportunities for the pastoral livestock keepers and increases the off-take rate of pastoral herds and flocks.
- It reduces grazing pressure on the rangelands by moving some animals out of the system to the feedlots.
- It creates employment and livelihood opportunities for various actors along the value chain (pastoralists, livestock traders, transporters, feedlot operators, live animal and meat exporters, abattoirs, butcheries, etc.).
- It increases the contribution of the livestock sector to the national economy and foreign currency earning of the country through the export of live animals and animal products.

5 The Influence of SPS-LMM on Livestock and Meat Export

5.1 The Status of Livestock and Livestock and Meat Export Prior to the Project

Livestock and livestock products are among the top five export commodities in Ethiopia (IGAD-ICPALD 2013). The annual meat export potential of Ethiopia was estimated at 72,000 metric tons (MT) of meat, valued at $136 million (LMA 2004). However, in 2005/2006 Ethiopian fiscal year, prior to the commencement of the

SPS-LMM activities, Ethiopia exported only 7917 MT of meat and 163,375 head of livestock, earning a total of $45.8 million (SPS-LMM 2011b). The export of both meat and live animals remained low and far below the potential for several years because of inadequate feeding and conditioning systems and fragility of the performance due to weak SPS certification system which made the country susceptible to disease-related trade bans. In addition, the trade relationship has been limited to very few trading partners and importing countries. Meat export was limited mainly to chilled shoat carcasses destined primarily to low-income guest workers in the Kingdom of Saudi Arabia (KSA) and United Arab Emirates (UAE).

However, official export of live animals represented a small fraction of the total number of live animals exported as a very large number of cattle, sheep, goats, and camels are exported through informal cross-border trade. For instance, according to Desta et al. (2011), the gross value added in the informal cross-border livestock trade on Berbera and Bosasso marketing corridors in the Somali region of Ethiopia was estimated to be as high as 144 million USD. Livestock are also informally exported through Kenya, Sudan, Djibouti, and Eritrea. These unofficial exports contribute to the welfare of Ethiopians by financing the importation of a wide range of consumer goods, including necessities such as clothing and staple food items. GebreMariam et al. (2013) indicated that informal cross-border trade in Ethiopia is about four times the volume of its formal exports. Thus, excluding informal exports, official figures often undervalue the real contribution of livestock to the national economy. The private sector operators in the live animal and meat export business often suffer from the insufficient and inconsistent supply of price competitive, export quality meat and live animals. In addition, they have limited knowledge of market preferences of type and quality of meat products. Another persistent problem experienced by the Ethiopian meat and live animal exporters is lack of established, reliable, and profitable contracts with traders from the importing countries.

5.2 *Activities Accomplished by the SPS-LMM Program*

The SPS-LMM activities were designed to address five principal constraints limiting the sustainable expansion of meat and livestock export from Ethiopia. These constraints include import restrictions based on SPS requirements imposed by importing countries; insufficient and inconsistent supply of price competitive, export quality meat and livestock; lack of capacity for cattle slaughter and for cold chain processing and packaging export quality beef products; lack of capability for cost-effective, cold chain transport of meat products by road and sea for delivery to MENA and other international markets; limited knowledge of market preferences for type and quality of meat products; and lack of established and profitable contracts with meat importers in international markets (SPS-LMM 2011a).

The activities of the Program were organized under two broad objectives, viz., the Sanitary and Phytosanitary Standards (SPS) and Livestock and Meat Marketing (LMM) objectives. The activities carried out under the SPS objective focused on increasing the capacity of the veterinary services, improving national capacity for diagnosing export limiting diseases, improving local vaccine production at the Ethiopian Veterinary Institute (NVI), improving Ethiopia's SPS certification capability to comply with the requirements and standards of importing countries, and establishing SPS certification system. Under the marketing objective, the SPS-LMM Program focused on improving Ethiopia's capacity and competitive advantage for meat and livestock exports through increasing cost-effective production of export quality meat and live animals, improving the efficiency and effectiveness of export marketing, and enhancing demand for Ethiopian livestock and meat in the target markets.

5.3 *The Contribution of SPS-LMM to Live Animal and Meat Export Performance*

As a result of the very close working relationship of the SPS-LMM Program with public institutions and private operators, significant increase was registered in meat and live animal exports from Ethiopia between 2005 and 2011. The value of formal live animal and meat exports increased from a base of \$27 million and \$18.5 million in 2005/2006, respectively, to \$148 million and \$63 million in 2010/2011, respectively (Figs. 1 and 2) an increment of 241% in the value of meat exports and 442% in the value of live animal exports. These sharp increases reflect an increase in the number and volume of meat exported (a doubling of meat exports and near tripling of live animal exports), an increase in the value per MT of meat and head of livestock, and, in the case of live animals, an increased formalization of the livestock trade. Meat exports were more than doubled (from 7917 MT to 16,877 MT), while official live animal exports nearly tripled (from 163,375 head to 472,045 head) between 2005/2006 and 2010/2011 (Figs. 1 and 2). In 2011/2012, a year after the termination of the project, the foreign currency generated from the export of live animals and meat further increased to \$207.1 million and 78.8 million, respectively, as a result of increased number of live animals (800,000 head) and more volume of

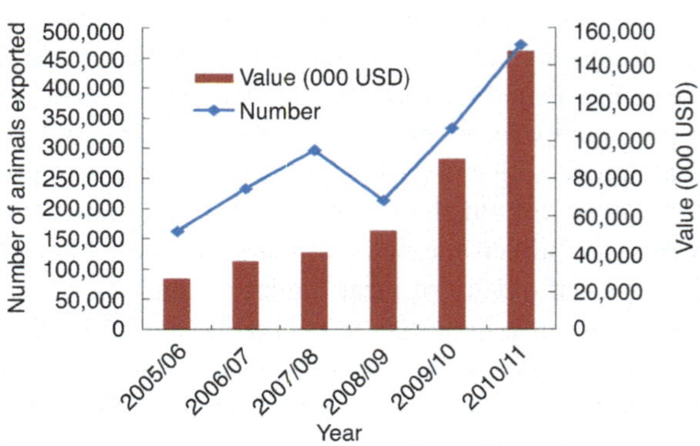

Fig. 1 Number and value of live animals exported from Ethiopia during 2005/2006–2010/2011 (SPS-LMM 2011b)

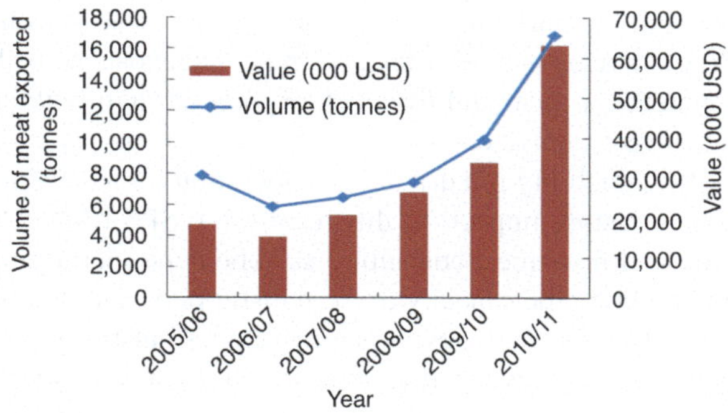

Fig. 2 Volume and value of meat exported from Ethiopia during 2005/2006–2010/2011 (SPS-LMM 2011b)

meat (17,800 tons) exported. Furthermore, the exports of meat, live animals, and animal products increased in importance from 11% of the total value of exports in 2004/2005 to 13% in 2015 (NBE 2017).

About 90% of the meat export comes from goat meat, whereas the over 60% of the live animal export comes from the export of cattle. The meat export is mainly dominated by chilled carcasses of sheep and goats. The main destination (about 95%) of meat export goes to United Arab Emirates (UAE) and Kingdom of Saudi Arabia (KSA), where the main customers are the low- to medium-income communities.

In addition to the influence on live animal and meat export, the SPS-LMM Program served as a gateway for investment attraction to the livestock sector. This has attracted more foreign and domestic investors in feedlots and meat industry sector. A typical example of such development is the vertically integrated Verde Beef that integrates feed production, feedlot operation, slaughter services, and meat processing with higher capacity for increasing the volume and quality of meat exported.

During the life of the Program, Ethiopia rapidly expanded its exports to traditional markets such as KSA, UAE, and Egypt while also diversifying into new markets. Between 2006 and 2011, Turkey, Jordan, Vietnam, China, Kuwait, Oman, Qatar, Angola, and Comoros imported Ethiopian meat products for the first time. During the same period, Bahrain, Kuwait, Oman, and Lebanon officially imported Ethiopian live animals for the first time. Some of these new importing countries rapidly ramped up volumes, such as Turkey, which became the third largest importer of Ethiopian meat products after the UAE and KSA at the end of the project period (SPS-LMM 2011a). Other new markets expanded more slowly, but even relatively small volumes of imports reflect a new willingness from importing country authorities to recognize Ethiopian SPS certification systems—a highly significant development.

The rapid increase in exports is the result of a combination of contributing factors, which include the commitment and support of the Ethiopian Government, the investments of private meat and live animal exporters, and the catalytic role played by SPS-LMM Program. SPS-LMM helped the public and private sectors to achieve their objectives by providing technical expertise to strengthen animal health and SPS certification systems and upgrading private operators' skills in animal feeding, processing, marketing, and exporting. The close working relationship between the Program team and relevant government agencies helped to create an enabling environment for meat and livestock exports through building the technical capacity of the government agencies.

Sustainability is a driving consideration for a development project or program to have a lasting impact. In this regard, the SPS-LMM Program invested significant time and resources, consulting stakeholders, identifying needs and capacity gaps, and building the capacity of local institutions and organizations to fulfill their mandate. The relevant government agencies, institution(s), and association(s) were involved as active participants or co-organizers in every training, every needs assessment, and every study tour organized by the Program.

6 Conclusion

There is a strong linkage between the pastoral livestock production and the feedlot operations and export abattoirs operating in the central highland areas of the country. Over 90% of the animals conditioned in feedlots and those destined for meat and live animal export are sourced from the pastoral areas. The performance of the live animal and meat export from Ethiopia showed a significant increase from 2005/2006 to 2011 because of the high attention given by the government to the sector, willingness of the private sector operators to make the necessary investments, and the catalytic support provided by development partners like SPS-LMM. The increased meat and live animal export through coordinated and concerted efforts of the public and private actors is important to fuel economic growth, fairly benefitting all actors along the value chain.

References

Ayele S, Assegid W, Jabbar MA, Ahmed MM, Belachew H (2003) Livestock marketing in Ethiopia: A review of structure, performance and development initiatives. Socio-economics and Policy Research Working Paper 52. ILRI (International Livestock Research Institute), Nairobi, Kenya. 35pp

Behnke R (2010) The contribution of livestock to the economies of IGAD Member States: study findings, application of the methodology in Ethiopia and recommendations for further work, IGAD Livestock Policy Initiative, Working paper no. 2–10. IGAD

Desta S, Berhanu W, Sebsibe A, Guerne-Bleich E (2011) Assessment of cross border informal livestock trade in Somali region. Food and Agriculture Organization Research Report, Addis Ababa, Ethiopia

GebreMariam S., Amare S., Baker D., Solomon A. and Davies R. 2013. Study of the Ethiopian live cattle and beef value chain.

Gizachew L (2012) Drought induced livestock feed crises and emergency responses. In: Tolera A, Yami A, Alemu D (eds) Livestock feed resources in Ethiopia: challenges, opportunities and the need for transformation. Ethiopian Animal Feed Industry Association, Addis Ababa

IGAD-ICPALD (2013) The contribution of livestock to the Ethiopian economy. ICPALD Policy Brief

Jabbar M., Negassa A. and Gidyelew T. 2007. Geographic distribution of cattle and shoats populations and their market supply sheds in Ethiopia, ILRI, October 2007

LMA (Livestock Marketing Authority) (2004) Meat exports market study. MoARD, Addis Ababa, Ethiopia

NBE (National Bank of Ethiopia) (2017) Annual report 2015/16. Addis Ababa: NBE

SPS-LMM (2011a) Ethiopia sanitary and phytosanitary standards and livestock & meat marketing program completion report

SPS-LMM (2011b) Focus on Ethiopia's meat and live animal export. SPS-LMM Trade Bulletin 5, July 2011

Tolera A (2007) Feed resources for producing export quality meat and livestock in Ethiopia: Examples from selected Woredas in Orimia and SNNP Regional States. SPS-LMM, Texas Agricultural Experimentation Station (TAES)/Texas A&M University System. Hawassa University, Addis Ababa

Tolera A (2008) Feed resources and feeding management: a manual prepared for feedlot operators and extension workers. Texas A & M University, SPS-LMM, Addis Ababa

Tolera A, Yami A, Alemu D (eds) (2012) Livestock feed resources in Ethiopia: challenges, opportunities and the need for transformation. Ethiopian Animal Feed Industry Association, Addis Ababa

Of 'White Elephant' in Fisheries: A Conflict Resolution Model Around the Usage of Climate-Smart Fish Postharvest Technologies in Lake Malawi

Fundi Wandisunga Kayamba-Phiri, Moses M. Limuwa, and Trond Storebakken

Abstract This study develops and analyses a conflict resolution model surrounding the rejection of a climate-smart fisheries technologies project in Chipala village, Nkhotakota district, Malawi. Solar tent driers and improved smoking kilns were introduced by the SEED Fish project in Chipala and Vinthenga villages. However, the technologies are not in use in Chipala due to conflicts among the different stakeholders involved. Qualitative approaches were used to collect and analyse data from different stakeholders within the district's decentralised structures. The targeted participants were involved in the planning and implementation of the SEED Fish project. The community scorecard (CSC) is an accountability tool used between service providers and service users and was adapted as a conflict resolution model to facilitate discussions on the roles that different stakeholders played in the run – up to the failure of the project. The resolution, in form of an action plan between the two parties, was made to discuss how the conflicts could have been avoided by jointly designing an ideal project plan that would ensure adoption and sustainable use of the technologies. This study concludes that there was need to incorporate social-cultural and social-political factors in planning and implementation, for it is the failure to incorporate these factors that led to and perpetuated existing conflicts in the study area. The recognition and involvement of fish processors was appreciated by the community and implementers as central for the success of such innovations and avoiding conflicts. Involving and reinforcing of decentralised structures not only reduces the risk of conflict but also contributes to the success of co-management of natural resources. This study contributes to the theoretical understanding of how to resolve conflicts while contributing towards good governance and sustainable development processes.

F. W. Kayamba-Phiri
Norwegian University of Life Sciences, Faculty of Landscape and Society, Department of International Enviroment and Development Studies (Noragric), Ås, Norway

M. M. Limuwa (✉) · T. Storebakken
Norwegian University of Life Sciences, Faculty of Biosciences, Department of Animal and Aquacultural Sciences, Ås, Norway
e-mail: mlimuwa@luanar.ac.mw

B. R. Singh et al. (eds.), *Climate Impacts on Agricultural and Natural Resource Sustainability in Africa*, https://doi.org/10.1007/978-3-030-37537-9_19

1 Introduction

This study develops and analyses the process of a conflict resolution model adapted for the community of Chipala village in Nkhotakota district, Malawi. The study was conducted as a result of the failure of a fisheries postharvest technologies project, where the project was rejected by the community and therefore deeming it a 'white elephant'. Failure of development interventions to meet their intended objectives has been researched since the Marshall Plan (Hogan 1989). However common, similar mistakes in development initiatives continue to be made. Development might fail due to failure to respect and recognise the local settings and mistargeting of participants among others (Klausen 1964), which are also the reasons for the conflicts in Chipala district.

Participation of local communities in identifying their pressing needs builds ownership and legitimacy. However, participation on its own is not enough, because there is also a tendency by local communities to jump into initiatives, which yield short-term and fast gains. Samndong (2018) highlights the importance of participation in forest governance as a tool to achieve REDD+ compliance status. However, even though the local communities participated, it was mostly tokenism.

The solar fish dryer intervention is one of the Capacity Building for Managing Climate Change in Malawi (CABMACC) funded projects in Malawi, implemented between 2015 and 2017. Under the project, a solar tent dryer and two improved smoking kilns were constructed in Chipala. Its overall objective was to improve incomes of rural communities through improved fisheries management and enterprise development for enhanced resilient livelihoods to climate change. This paper asks and answers questions related to participation: how conflicts resulted and how they could have been avoided to promote adoption and community participation.

Data were collected in Chipala village among project implementers, fish processors (FPs), the Beach Village Committee (BVC), district assembly officials, members of the area, and village development committees. Data were primarily collected by qualitative methods, using CARE International's community scorecard, which included focus group discussions and key informant interviews.

In a traditional sense, governance has been related to the government and its activities. However, interactive governance theory and other approaches to governance in fisheries argue that governance includes other actors besides government (Jentoft and Chuenpagdee 2009). In Iceland, an individual transfer quota system (ITQ) was introduced to achieve ecological stewardship, economic efficiency, and safety at sea (Sampson 2013). The system is a neoliberal solution to the control of fisheries resources. Furthermore, Pálsson (2006) finds that ITQs, a high modern regime, mostly benefit capital and boat owners and scientific experts while marginalising the small fishers, crews, and local knowledge. The emphasis on recovering fish stocks across the globe has seen economic and ecological components at the centre of policy and governance, while the sociocultural domain is either completely ignored or not prioritised (Urquhart and Acott 2013; Reed et al. 2013). Modernist governance often fails to consider nature and society together

(Pálsson 2006), thereby undermining the local place and its importance, whereas both are complex and diverse (Sampson 2013; Jentoft and Chuenpagdee 2009).

For Johnson (2006), governance is a process that should reinforce the ties among different stakeholders and foster dialogue, debate, and collaboration through interaction. McGoodwin (1990) criticises the capitalist way of governing resources by underpinning the exclusion of social and cultural considerations. However, the interactive governance theory merges the existing governing system and the system to be governed to a common platform where they interact to manage fisheries (Johnson 2010; Jentoft 2007), while paying attention to all the societal and cultural values of place (Jentoft and Chuenpagdee 2009).

The government of Malawi adopted the community-based natural resource management (CBNRM) initiative under a co-management arrangement, guided by the co-management of natural resources policy, which is currently operational. Co-management is aimed at increasing user communities' partnerships in managing resources that they depend on for livelihoods, to yield better management outcomes (Ngochera et al. 2017). This is appropriately aligned to the decentralisation policy and legislation (GoM 1998a, b), based on principles of bottom-up planning. This means that communities decide and prioritise issues and aspects of development directly impacting their livelihoods in their local areas. Co-management was also adopted because government lacked the capacity to manage natural resources across the country. The benefits of CBNRMs are nonfinancial: 'the empowerment of people in rural areas, conservation of biodiversity, and the development of more secure livelihoods and the reduction of risk' (Fabricius 2004: 3).

Like other natural resource-based sectors, the implementation of co-management regimes in the fisheries sector was assumed to benefit the resources and their users (Ngochera et al. 2017). However, recent empirical studies (Weyl 2008; Béné et al. 2009; Njaya et al. 2011; Hara, Donda, and Njaya 2002) highlight the potential problems that may arise from such natural resource governance reforms, due to lack of capacity and resources. These studies analysed co-management arrangements of fisheries in Malawi as having problems that arise particularly around power distribution, how to determine the responsibilities of the various role players in co-management arrangements, such as Beach Village Committees (BVCs). The prevalence of such problems is quite understandable, considering that BVCs consist of different community members, who represent their communities in various fisheries and related activities.

Thus, equally important are the relational interactions between levels of resource management in decision-making and in carrying out co-management prescriptions. Cleaver (2002) coins the term 'bricolage' which she defines as 'how mechanisms for resource management and collective action are borrowed or constructed from existing institutions, styles of thinking and sanctioned social relationships'. Caution is needed in ensuring that institutional vestiges relating to centralised management regimes do not conspire to frustrate mechanisms promoting participatory, co-management approaches. For instance, while the government has moved away from a top-bottom to bottom-up approach to natural resource management, there still are traits of top-bottom interactions between local organisations and resource users. At the same time, similar traits can still be traced within government sectors in their approach to co-management, where decisions are not entirely bottom-up. For instance, in Chimaliro Forest Area in the northern region of Malawi, the formation of Village Natural Resource Management Committees (VNRMCs) was done by the community during community meetings, but 'under the auspices of and with the advisory services provided by the forestry department' (Kayambazinthu 2000). The study therefore tries to understand the levels of control that the community had or can have in their own development process, specifically the introduction of innovative technologies such as the dryer and smoking kilns.

2 Participation in Fisheries Postharvest Management

Participation is highly advocated by development practitioners as a means of achieving inclusion within communities, which include decision-making related to planning and resource control. There are many conceptual frameworks that have been employed to understand participation (Chambers 1997; Agarwal 2001). This study used the framework (Fig. 1) developed by Arnstein (1969), which has also been used to understand natural resources conservation and many anti-poverty development interventions (Arnstein 1969; Samndong 2018). In Fig. 1, the bottom two rungs,

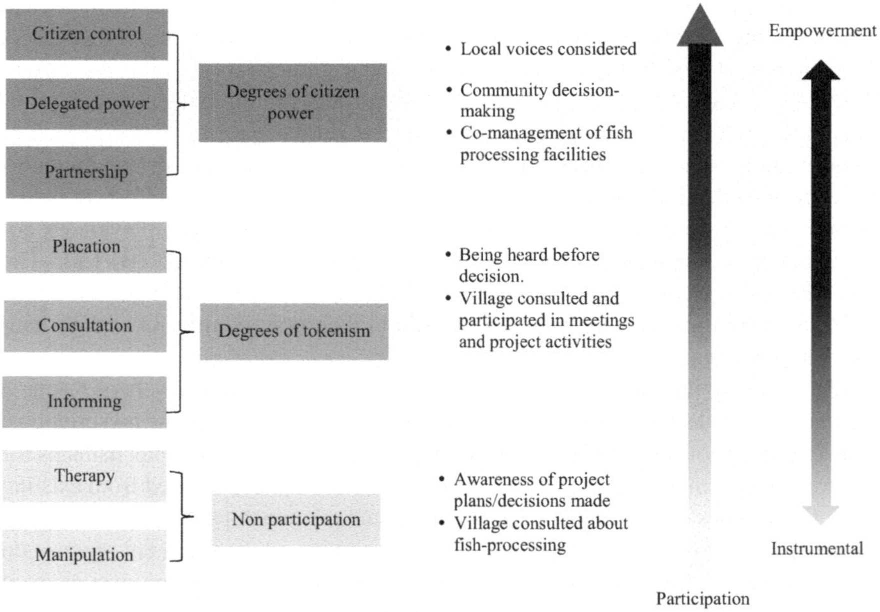

Fig. 1 Characterising community participation in solar drying fish processing technology. (Adapted from Arnstein (1969) as modified by Samndong (2018))

therapy and manipulation, reflect nonparticipation. Under this level of control, power holders merely 'educate' or 'cure' the intended participants without evidence of empowerment to influence the process (Arnstein 1969).

This study established that the fishers and some community structures were consulted and knew about decisions to plan and implement the project in their area. Rungs 3, 4, and 5 (informing, consultation, and placation) manifest some degree of tokenism. These three rungs illustrate improved participation as beneficiaries are not only informed but also voice their opinions. However, this does not mean that their opinions will be taken on board. The three top most levels of control reflect characteristics of citizen power, where participants can influence decisions which affect their day-to-day welfare. Contextualising this framework assists in avoiding homogenous interventions for stakeholders, which Cleaver (1999) also discusses in her paper on paradoxes of participation (Brown 2002). Plans that carefully consider aspects of community sustainability, participation, and empowerment result in high-impact development initiatives (Brown 2002).

The involvement of local communities varies, as influenced by institutions and power. For example, those who seem to benefit or participate are deemed powerful and suppress the powerless. Therefore, instead of the less powerful communities benefiting, the interventions are captured by the local elites. Other social institutional factors like gender have also been found to influence development (Haraldsdottir 2002; Chiwaula et al. 2018). In fishing communities, the local elites are those who own fishing equipment, as well as the BVC. Individuals who own equipment exploit those without to work for them. However, other studies on fishing communities of Malawi report that it is actually the fishing crew who dictates the sharing of the proceeds compared to the owners of the fishing equipment (Hara and Jul-Larsen 2003). The BVC, at times, undermines the participation of community stakeholders in the fish value chain by dominating decision-making. Such tendencies affect inclusive development interventions within fishing communities, which reflects how power dictates who participates.

Participation in development initiatives has also been conceptualised on the basis of improving effectiveness in resource use and empowerment of the beneficiaries (Cleaver 1999). In essence, participation without power to effect necessary changes is not new in development interventions and is often called the 'empty ritual of participation'. Participation without power paints a picture that all stakeholders were consulted and they agree to the decisions made (Arnstein 1969).

3 The Community Scorecard

Since the data (Kayamba-Phiri 2018) showed no evidence of a feedback meeting between the SEED Fish implementers and the community, we decided to use the community scorecard to resolve the conflict by first evaluating the project, without placing emphasis on the conflict itself, to facilitate discussions of both the positive and negative aspects of the project.

The scorecard can be used in different sectors, as long as there are two parties involved: service users and service providers. For service users, CSC helps to give feedback to service providers about their performance. For service providers, CSC helps government institutions/NGOs to learn directly from communities about their level of satisfaction with services and/or programs. Table 1 below summarises the benefits and challenges of using the scorecard (CARE Malawi 2013).

Figure 2 describes the community scorecard process. The groundwork involves selection of local facilitators, participants, and a neutral location for the exercise. From this stage, the process is managed by the facilitators who assist service users and providers to create and score the input tracking, performance, and self-evaluation scorecards. In this study, before the interface meeting, the community focus groups met to resolve the conflict within the community. The interface meeting involved both service users and providers who gave feedback to each other's scorecards (the performance scorecard and self-evaluation scorecard, respectively). In the end, an action plan was developed which would ensure accountability, transparency, development, efficiency, and empowerment for future development initiatives.

Using the CSC for conflict resolution in this study was even more time-consuming than a normal CSC process, as the CSC was adapted to emphasise or highlight occurrences where a focus group or combined groups mentioned reasons or statements pertaining to the conflicts surrounding the project. We observe from this study that the length of the process might vary,

depending on the magnitude and number of conflicts identified. For instance, there were three levels of conflicts arising from the SEED Fish project, which required an additional mediation session between the BVC and the community.

The CSC process requires participants in each focus group to elect representatives who form the group to consolidate the community scorecards into one community scorecard. Because the aim was to resolve conflicts, we decided to have all community participants in a mediation session concerning the conflict between the BVC and the community. This way there was more input as more community mem-

Table 1 Benefits and challenges of using the CSC

Benefits	Challenges
Promotes dialogue and improves relationships with service providers	Requires time (training facilitators and introducing a new tool to the community)
Facilitates a common understanding of issues and solutions	Can lead to conflict if not facilitated well
Clarifies the roles and responsibilities of the service users in service diversity	It requires good facilitation as the CSC deals with behaviour and personalities
Can expose corrupt officials	Sometimes individuals can be targeted (finger pointing)
Can show the service providers how to be accountable and responsible	If not well facilitated, it can raise expectations that cannot be fulfilled by service providers
Can be used to monitor progress and service quality together with the community	
Promotes a common understanding of issues and solutions to problems	

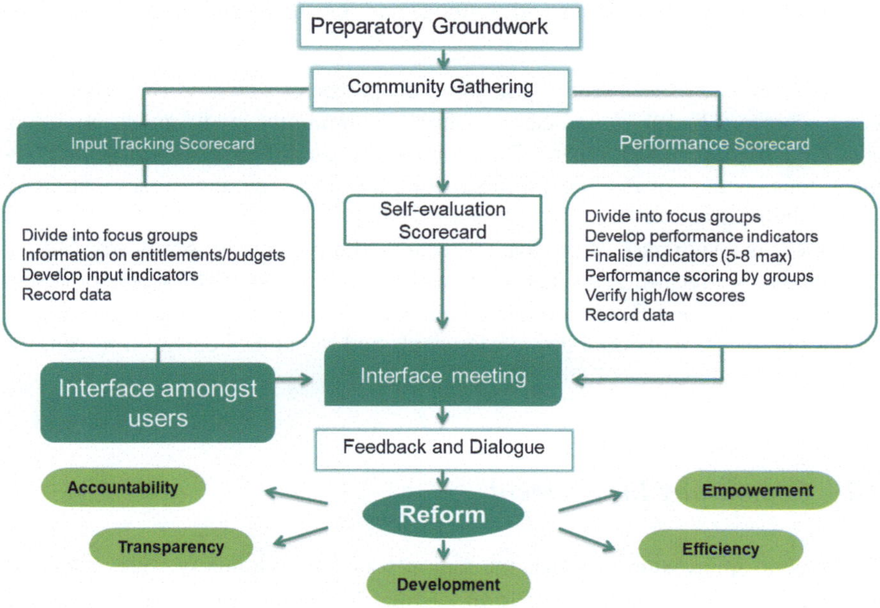

Fig. 2 Conflict resolution model. (Adapted from CARE International's Community Scorecard (CARE Malawi 2013))

bers were able to voice their concerns. Further, the participants were able to discuss their ideas on how the conflict started, which varied among the different focus groups. This was done to ensure that the final interface meeting between service users and providers did not include unresolved issues from within the community.

Despite the district being matrilineal dominated (Phiri 1983), the involvement of women in the actual fishing is not common, but women own fishing equipment. In addition, women largely command postharvest activities, which include processing of fish. Limuwa and Synnevåg (2018) noted that the increased involvement of Nkhotakota women in fisheries did not result in power to control fisheries-related incomes; however, in this study, we observed that through the scorecard process, the female participants were central in terms of arguing and defending scores given.

3.1 The Scorecard

The successes and challenges of the project were similar across all the groups, with a few differences especially from the service provider group discussion. The issues that were identified in all the community focus groups were:

- The location and management of the facilities
- Conflicts between the fisheries department and the community, as well as between the community and the BVC

- The lack of access to loans and trainings and the small holding capacity of the solar dryer tent

The community focus groups prioritised the issues respectively. The service provider scorecard had three indicators, namely, community involvement, adoption, and infrastructure development. The community identified community involvement, management of facilities, infrastructure development, and capacity building as their indicators.

Table 2 is a summary of the five scorecards, the community scorecard, and the consolidated scorecard. The Likert scale was between 1 and 5. The scores represent 5, very good; 4, good; 3, not so bad; 2, bad; and 1, very bad. The data shows that in all but one of the individual group scorecards and the consolidated scorecard, the scores remained under 3 (Table 2).

3.2 Community Involvement

The general impression was that the majority of the community were not made aware of the project. Participants from all four community focus groups reported that there were no awareness meetings organised by the project to inform the community at large about the project. Similarly, the Area and Village Development Committee (ADC/VDC) participants stressed the lack of consultation of development committees in the planning phase of the project. As one participant put it, the ADC and VDC are regarded as the 'custodians of rural development' in their areas. According to procedure, the ADC and VDC are meant to participate in all development projects in a targeted area, regardless of the sector, and, in this case, fisheries.

A related institutional challenge confronting co-management of natural resources appears to be associated with competing and often overlapping leadership roles at the local community level. At this level, the roles of different local committees, as well as chiefs, in co-management arrangements are unclear (Kamoto 2014;

Table 2 Summary of community, service provider, and consolidated scorecards

Indicator	Female	Male	BVC	ADC/ VDC	Community	Service providers	Consolidated scorecard
1. Community involvement	3	1	1	1	1	3	3
2. Management of facilities	1	1	1	3	1		**1**
3. Infrastructure development	3	3	2	4	3	4	**3**
4. Capacity building	2	1	1	2	1		**2**
5. Adoption						1	

Hara et al. 2014). This is evident from the statements of the project implementers who stressed that they had followed procedure as instructed by the fisheries extension office in Nkhotakota. Competition for power arises between committees such as the BVC and development committees as vehicles for user participation in co-management of natural resources. The confusion arose because the roles of the different committees were not defined on the onset, which affected the management process and community participation. The BVC assumed all management, and as the ADC and VDC were not involved, there was no accountability at the local level.

The service providers reported that the project staff conducted an inception meeting with the fisheries department and the District Executive Committee (DEC), after which they organised a similar meeting in the community, which was attended by BVC members who would later manage the facilities. The rest of the community, however, felt that the inception meeting should have involved others in addition to the BVC, such as the ADC, VDC, local chiefs, and, most importantly, the fish processors themselves, none of whom are members of the BVC. The argument of the community members was that if the inception meeting did not involve them as the main users of the facilities, there was bound to be conflicts around management of the facilities. This was a learning point for the service providers, as they acknowledged that they could have considered including the ADC/VDC and other community members in the initial meetings. From the advice given from the fisheries department, it seemed that an inception meeting with the DEC and later the BVC would suffice.

The fish processors complained that they had not been included in any of the processes concerned with the planning of the project; they were only informed about the usage and payment regulations for the facilities. Their further argument was that the conflict was a result of assumptions made by the BVC about fish processing and, without consultation, setting prices that were too high for fish processors – a top-down approach. Thus, there was no co-management between the BVC and the fish processors, as intended by the project.

During the FGD, the implementers were of the view that the Chipala community was a complex community to work with, and this affected community involvement. They compared Chipala and Vinthenga villages and the level of commitment that the community members had to construction and management of the tent dryers. The members of Chipala community did not take part in construction of the tent, to which the community responded that they were not made aware that they needed to get involved; they thought it was the responsibility of the BVC. The mediation session between the fish processors and the BVC confirmed the implementers' concerns about the community in Chipala having a complex governance structure, which had adopted a top-down approach with regard to community participation.

The implementers also expressed shock at the fact that the area where the facilities were constructed in Chipala had previously been a busy fish market, and this was used as justification for constructing the facilities in Chipala. In response, the community explained that the drastic change was a result of the decision by the fisheries department to stop all trade on the premises for security reasons.

A female FP further expressed that the top-down approach used by the project and the BVC made them assume that the tent was either constructed for the fisheries department or the BVC and not for the community at large. A female member of the ADC stated that some of them were not sure of the use of the facilities; some thought the tent dryer was for storing animals. The community group thought that the tent was built for the fisheries department, as it was constructed on their premises, rather than on a site that would have been favourable for the fish processors themselves. As explained in their focus group, one of the service providers explained to the community that selection of the site was based on technicalities such as the soils and wind patterns. The implementers endeavoured to explain the technical justification to the community, but a female member of the BVC responded by stating that they had potential sites within their village that fit the mentioned site requirements. The problem was that they had not been asked to suggest options.

Co-management is aimed at increasing user communities' partnerships in managing resources that they depend on for livelihoods to yield better management outcomes (Ngochera et al. 2017). This is appropriately aligned to the decentralisation policy and legislation (GoM 1998a, b), based on principles of bottom-up planning, which means that communities decide and prioritise issues and aspects of development directly impacting their livelihoods in their local areas. The discussions between implementers and the community indicated that the community had the capacity to understand technicalities of development initiatives based on their local knowledge and were able to not only contest decisions but also make suggestions that would increase adoption and participation.

3.3 Management of Facilities

The community reported that the project had not monitored progress, which resulted in reduced commitment to managing the tent. The service providers however explained that once they had handed over to the BVC, the community assumed ownership and thus management of the facilities. However, conflicts arose in the Chipala community. First, the fisheries department forbade fish processing activities on their premises. This did not include the dryer and kilns; but the disagreements between the BVC and the fish processors made the dryer dysfunctional. The fish processors complained that the prices were too high, compared to the neighbouring project site. Since there was no revenue, the BVC could not maintain the structure. One of the female fish processors expressed her willingness to renovate the tent dryer by replacing the sheets and door that was stolen. This was rejected by the BVC, who in turn claimed that maintenance should have been sponsored by the project.

The BVC also found the price to be fair, noting that the fish processors made large profits from their business. In response, the fish processors explained the process and the fluctuation of prices on the market. The community interface meeting was the first time that the BVC and fish processors as well as the implementers and different local development stakeholders had a formal discussion about the management of the tent. All parties agreed that ownership of the facilities was unclear due to the top-down approach used both by the project staff and the BVC. Issues of management should have been addressed at the onset of the project, and this would have ensured that different actors were involved and could be held accountable, both at local and district level.

3.4 *Infrastructure Development*

The indicator had the highest score (3) on the consolidated scorecard (Table 2), as both the service providers and the community members acknowledged that the facilities were a good development in the area. The community viewed the solar drying method as time efficient and hygienic and produced high-quality dried fish. Thus, despite the fact that none of the facilities were in use and the tent dryer was dilapidated, the community thought the project made some achievements. All four community focus groups stressed that the facilities were too far from the fish landing site, which was a negative effect on development. The service providers, however, scored themselves a 4 (Table 2). To justify the score, they explained that they had succeeded in conducting an inception meeting and meeting the BVC, to whom they also explained how the technology worked. Further, they had succeeded in constructing the tents and handing them over to the community through the BVC. The community felt that the tent was too small. One of the male fish processors posed a hypothetical question to the service providers concerning how many fish processors would have been able to use the tent dryer if management of the tent had been successful. The service providers explained that the holding capacity of the tent dryer was adequate as the project was a trial in Nkhotakota. They further explained that the holding capacity was based on estimations of catches received from the fisheries department. They could not have constructed bigger tent dryers as in other fishing communities in Malawi whose catches were large than those in Nkhotakota.

The female fish processors also complained that no toilets had been constructed at the site, which affected the hygiene of the area. In response, the service providers explained that construction of toilets should have been the community's contribution to the project, as a way of reducing dependency on assistance from projects. This point would have been valid if the community had been involved in the planning. Participatory planning of the project would have clarified the expectations and contributions from all parties involved.

3.5 *Capacity Building*

The community's main concern was the lack of training on construction and usage of the facilities. They complained that they were not involved in the construction and were not trained to maintain the facilities; and the challenge was that they would have to depend on experts from outside their community at a higher cost than would be required to train a person from Chipala. This had the effect of limiting capacity building within their community as no one had been taught new skills. The BVC and fish processors' groups particularly pointed out that the project did not honour their promise of loans to fish processors. This complaint was quickly rejected by the service providers during an interface meeting that was neither planned nor communicated. The community asserted that the project had not achieved anything by way of value addition. The community had expected linkages to markets, exchange visits to areas where the facilities were being successfully used, as well as packaging and branding of their products. Other activities such as value addition to the products were not achieved because the dryer was not used for long and thus issues of packaging or marketing were not explored. One of the service providers even showed pictures of branded products from Vinthenga where the dryer is in use.

3.6 Adoption

The indicator was only identified by the service providers, under which they gave a score of 1. The issues raised under this indicator however were raised by the community under other indicators: consultation, ownership, value addition, technology testing, and tent location.

4 Action Plan: The Resolution

Following the discussion of scores, the two parties were asked to draw up an action plan that would best rectify the problems discussed, to prevent such conflicts from recurring. The action plan is presented in Table 3. The action plan even though hypothetical was planned as realistically as possible.

We find that the community at Chipala would have benefited from using the facilities had the CSC been used during the project, ideally midterm. The resolution would have allowed for reallocation of funds to activities that the community and the implementers, through the CSC process, realised were crucial in introducing technological innovations in a community. In as much as participatory management planning is time-consuming and costly, we note that such processes result in building the management capacity of community members, as well as implementers. The prioritisation of issues in the resolution (Table 3) reveals that the community considers participatory management planning as key for them to assume ownership of development initiatives. We consider participatory approaches as value for money, unlike creating a 'white elephant'. The data shows that despite the enactment of policies and legislation on devolution of authority and decentralisation, the norms of centralised management remain deep-rooted in government departments (Chinsinga 2005), as well as in rural communities. Thus, participatory approaches

Table 3 Action plan

Priority theme	Action	Who will lead it	With whom	Completion date
Procedures (inception phase)	Dissemination meetings with community leaders such as VDC and ADC	SEED fish project	DEC	July to December 2018
Setting up solar dryer committee	Community meeting	VDC and local chiefs	BVC and fisheries department extension workers	December 2018 to January 2019
Setting rules and regulations	Discussion of project objectives and plan	Newly formed solar dryer committee (SDC)	BVC	December 2018 to January 2019
Capacity building	Trainings for fish processors and SDC	SEED fish project	Fisheries department and district council	July 2018
Participation	Community meeting chaired by the SDC	SDC and BVC	Fisheries department extension workers	July 2018
Research	Dissemination meeting between SEED fish project staff and the community	Fisheries department extension workers	VDC	Quarterly

are effective if different stakeholders within the community are accorded equal power over the process for accountability and sustainability. Further, regardless of their nature, whether purely for development or research, projects that affect livelihoods require an understanding of sociopolitical context, as well as contingencies for different outcomes. Policy makers should adopt integrated management planning that addresses the diverse interests in the natural resources and the ecological, socioeconomic, and external factors that threaten sustainability of ecosystems and livelihoods of dependent communities (Jamu et al. 2011).

5 Conclusion

This study managed to bring different actors of this project together in resolving conflict. The study concludes that the participation in the project activities by the community was not legitimate and did not empower them as intended. Instead, participatory approaches empowered the community and have the potential to increase community participation in innovations that directly affect their livelihoods. We therefore find that using the CSC tool midterm of the project would have resolved conflicts and allowed for adoption and community participation. Furthermore, the paper recommends understanding of sociopolitical context, in this case local institutions as a prerequisite for any development intervention. The top-down approaches used by the BVC indicated traces of centralised management, which can also be seen at district level. Thus, the empowerment of the community also depends on community members being awarded equal power in participatory decision-making, which also ensures integrated management planning as varied interests are incorporated into the development process.

References

Agarwal B (2001) Participatory exclusions, community forestry, and gender: an analysis for South Asia and a conceptual framework. World Dev 29(10):1623–1648

Arnstein SR (1969) A ladder of citizen participation. J Am Inst Plann 35(4):216–224

Béné C, Belal E, Ousman MB, Ovie S, Raji A, Malasha I, Njaya F, Andi MN, Russell A, Neiland A (2009) Power struggle, dispute and alliance over local resources: analyzing 'Democratic' decentralization of natural resources through the lenses of Africa inland fisheries. World Dev 37(12):1935–1950

Brown K (2002) Innovations for conservation and development. Geogr J 168(1):6–17

CARE-Malawi (2013) The Community Score Card (CSC): a generic guide for implementing CARE's CSC process to improve quality of services. Cooperative for Assistance and Relief Everywhere, Inc

Chambers R (1997) Whose reality counts?: putting the first last. Intermediate Technology Publications Ltd (ITP)

Chinsinga B (2005) District Assemblies in a fix: the perils of the politics of capacity in the political and administrative reforms in Malawi. Dev South Afr 22(4):529–548. https://doi.org/10.1080/03768350500322891

Chiwaula LS, Chirwa GC, Binauli LS, Banda J, Nagoli J (2018) Gender differences in willingness to pay for capital-intensive agricultural technologies: the case of fish solar tent dryers in Malawi. Agric Food Econ 6(1):1–15

Cleaver F (1999) Paradoxes of participation: questioning participatory approaches to development. J Int Dev 11(4):597

Cleaver F (2002) Reinventing institutions: bricolage and the social embeddedness of natural resource management. Eur J Dev Res 14(2):11–30. https://doi.org/10.1080/714000425

Fabricius C (2004) The fundamentals of community-based natural resource management. In: Fabricius C, Koch E, Magome H, Turner S (eds) Rights, resources and rural development: community-based natural resource management in Southern Africa. Earthscan, London

GoM (1998a) Local Government Act, 8 March 1999. Lilongwe, Government of Malawi

GoM (1998b) Malawi National Decentralization Policy. Government of Malawi, Decentralization Secretariat, Lilongwe

Hara M, Donda S, Njaya (2002) Lessons from Malawi's experience with fisheries co-management initiatives. In: Geheb K, Sarch MT (eds) Africa's inland fisheries: the management challenge. Fountain Publishers, Kampala, pp 31–48

Hara M, Jul-Larsen E (2003) The "lords" of Malombe: an analysis of fishery development and changes in fishing effort on Lake Malombe, Malawi. FAO Fish Tech Pap 2:179–200

Hara M, Donda S, Ngochera, Berge E (2014) Fragmentation of natural resource management on the Southeast Arm of Lake Malawi and the Conceptual Framework. In: Fragmentation of resource management on the South East Arm of Lake Malawi. LIT Verlag, Zürich, pp 1–15

Haraldsdottir G (2002) Cooperation and conflicting interests: an ethnography of fishing and fish trading on the shores of Lake Malawi. Monography. University of Iowa, Anthropology, Iowa. 253 pp

Hogan MJ (1989) The Marshall Plan: America, Britain and the Reconstruction of Western Europe, 1947–1952. Cambridge University Press

Jamu D, Banda M, Njaya F, Hecky R (2011) Challenges to sustainable management of the lakes of Malawi. J Great Lakes Res 37, Supplement 1(0):3–14. https://doi.org/10.1016/j.jglr.2010.11.017

Jentoft S (2007) In the power of power: the understated aspect of fisheries and coastal management. Hum Organ 66(4):426–437

Jentoft S, Chuenpagdee R (2009) Fisheries and coastal governance as a wicked roblem. Mar Policy 33(4):553–560

Johnson DS (2006) Category, narrative, and value in the governance of small-scale isheries. Mar Policy 30(6):747–756

Johnson, D. (2010). Institutional adaptation as a governability problem in fisheries: Patron-client relations in the Junagadh fishery, India. Fish and Fisheries, 11, 264–277

Kamoto J (2014) Forest management in Mangochi District, with emphasis on the South East Arm of Lake Malawi catchment. In: Donda S, Hara M, Ngochera M, Berge E (eds) Fragmentation of resource management on the South East Arm of Lake Malawi. Dynamics around fisheries. LIT Verlag, Zürich, pp 92–117

Kayamba-Phiri FW (2018) Between the sun and fish are people: a socio-economic study of Solar dryers for fish processing in Malawi. Master's thesis. Ås: Norwegian Unversity of Life Sciences (NMBU), International Environment and Development Studies. 131 pp

Kayambazinthu D (2000) Empowering communities to manage natural resources: where does the power lie? The case of Malawi. In: Shackleton S, Campbell B (eds) Empowering communities to manage natural resources: case studies from Southern Africa. SADC Wildlife Sector-Natural Resources Management Programme and CSIR, Lilongwe

Klausen AM (1964) Technical assistance and social conflict: a case study from the Indo-Norwegian fishing project in Kerala, South India. J Peace Res 1(1):5–18

Limuwa MM, Synnevåg G (2018) A gendered perspective on the fish value chain, livelihood patterns and coping strategies under climate change – insights from Malawi's small-scale fisheries. Afr J Food Agric Nutr Dev 18(2):13525–13544

McGoodwin JR (1990) Crisis in the world's fisheries: people, problems, and policies. Stanford University Press

Ngochera M, Donda S, Hara M, Berge E (2017) Defragmenting resource management on the Southeast Arm of Lake Malawi: case of fisheries. Centre for Land Tenure Studies Working Paper 12/17, Norwegian University of Life Sciences, Ås, Norway

Njaya F, Donda S, Béné C (2011) Analysis of power in fisheries co-management: experiences from Malawi. Soc Nat Resour 25(7):652–666. https://doi.org/10.1080/08941920.2011.627912

Pálsson G (2006) Nature and society in the age of postmodernity. In: Biersack A, Greenberg JB (eds) Reimagining political ecology. Duke University Press, Durham

Phiri KM (1983) Some changes in the matrilineal family system among the Chewa of Malawi since the nineteenth century. J Afr Hist 24(02):257–274

Reed M, Courtney P, Urquhart J, Ross N (2013) Beyond fish as commodities: understanding the socio-cultural role of inshore fisheries in England. Mar Policy 37:62–68

Samndong R (2018) The participation illusion: questioning community participation in a REDD+ pilot project in the Democratic Republic of Congo. Int For Rev 20(3):390–404

Sampson YA (2013) Artisanal fishing and livelihoods in Kwahu-North District, Ghana. Unpublished master's thesis, Kwame-Nkrumah University of Science and Technology, Accra

Urquhart J, Acott T (2013) Constructing 'The Stade': fishers' and non-fishers' identity and place attachment in Hastings, South-East England. Mar Policy 37:45–54

Weyl O (2008) Lessons learnt from 10-years of co-management in Lake Malombe, Malawi, Africa and their applicability to Sri Lanka's perennial reservoirs. Participatory Approaches to Reservoir Fisheries management: Issues Challenges and Policies, Dambulla, Sri Lanka, 3–6 Oct 2004

Part IV
Policy and Institutions for Sustainable Agriculture and Natural Resource Management

Policy and Action for Food and Climate Uncertainties in Malawi

Ruth Haug and Ola T. Westengen

Abstract The purpose of this paper is to assess lessons learned from past and present policies and actions related to creating a conducive environment for a sustainable food system transformation in Malawi. We take a food system approach because it provides the opportunity for a holistic and integrated analysis of the drivers and constraints that influence the Malawian food system such as climate change and high levels of food insecurity and rural poverty. Lessons learned include the importance of national ownership and recognition of what is politically feasible and institutionally implementable in efforts towards a more sustainable food system. Finding solutions to food and climate uncertainties lies in the interface between policy and action and requires small-scale farmers' creativity and ability to adjust to change, without transferring the responsibility of climate adaptation to small-scale farmers with limited adaptation capacity. An integrated approach to food and climate uncertainties will call for socioeconomic and technological innovations that mutually strengthen each other whilst ensuring that the technology works for small-scale men and women farmers in accordance with the intentions set out in the sustainable development goals.

1 Introduction

Climate change is a serious threat to future food and livelihood security in Malawi (Thurlow et al. 2014; Challinor et al. 2016; IPCC 2018). The country has already experienced numerous disasters triggered by adverse weather. Between 1967 and 2015, Malawi went through 7 severe droughts and 20 floods, including the floods of 2015 in which Malawi received the highest rainfall ever recorded (GoM 2015). In 2016, the disaster continued with an El Niño-induced drought that contributed to hunger and a renewed need for humanitarian assistance (GoM 2016). There is a close relationship between vulnerability to climate change and poverty (Hallegatte et al. 2016). Countries that are heavily dependent upon agriculture, such as Malawi where agriculture constitutes around one third of BNP and employs the majority of the population (WB 2018), are particularly vulnerable to climate change. Malawi is

R. Haug (✉) · O. T. Westengen
Noragric, Landsam, Norwegian University of Life Sciences (NMBU), Ås, Norway
e-mail: ruth.haug@nmbu.no

B. R. Singh et al. (eds.), *Climate Impacts on Agricultural and Natural Resource Sustainability in Africa*, https://doi.org/10.1007/978-3-030-37537-9_20

among the poorest countries in the world, and poverty has slightly increased during the last years from 50.7% in 2010/2011 to 51.5% in 2016/2017 (NSO 2017; WB 2018). However, extreme poverty decreased in the same period from 24.5% to 20.1% (NSO 2017). The food security situation has also worsened; IFPRI (2018) reports a decline in household food security from 32% food insecure households in 2010/2011 to 61% in 2016/2017. The seriousness of the current food insecurity and poverty situation combined with expected adverse impacts of climate change in the future paint a worrisome scenario for Malawi in the years to come. The Government of Malawi (GoM) and various international actors have responded to these tremendous challenges by investing in agricultural input subsidies and promoting climate-smart agriculture and through the implementation of safety nets such as the direct provision of cash or food. Whilst climate-smart agriculture (CSA) is currently considered a helpful adaptation approach, it is also criticized for transferring the climate burden to small-scale farmers with already low adaptation capacity (Karlsson et al. 2018). CSA is also criticized for its narrow focus on technological fixes at the level of production at the expense of a focus on important political, sociopolitical and economic drivers of change (Taylor 2018). The purpose of this paper is to assess lessons learned from past and present policies and actions related to creating a conducive environment for a sustainable food system transformation in Malawi. A food system approach is used to analyse how various drivers influence the social, economic and environmental sustainability of the Malawian food system, with the aim of contributing to a better understanding of the challenges faced and the opportunities for action.

2 Approach

To assess lessons learned from policies and actions aimed at transformation to a more sustainable food system in Malawi, we ask the following questions: What are the main reasons for the lack of sustainability in the Malawian food system? What measures have been put in place by local, national and international actors to contribute to a more sustainable food system? What lessons can be learned from various efforts to a more sustainable food system? We use the food system framework illustrated in Fig. 1 to analyse the food system in Malawi (HLPE 2017). This analytic framework is applied because it provides opportunities to map the food system situation, to identify challenges, to understand different drivers and to assess to what degree the measures included in the sustainable development goals (SDGs) will be able to address the identified challenges. In addition, the framework illustrates how various factors relate to each other in a holistic way. Particular emphasis is put on SDG 2.3, which states that agricultural productivity and incomes of small-scale food producers should be doubled by 2030 (UN 2018). The reason for focusing on SDG 2.3 is the importance of agricultural productivity, food security and poverty in climate change adaptation and contributions to a sustainable food system in a typical agrarian economy such as that in Malawi. Our approach also draws on a livelihood

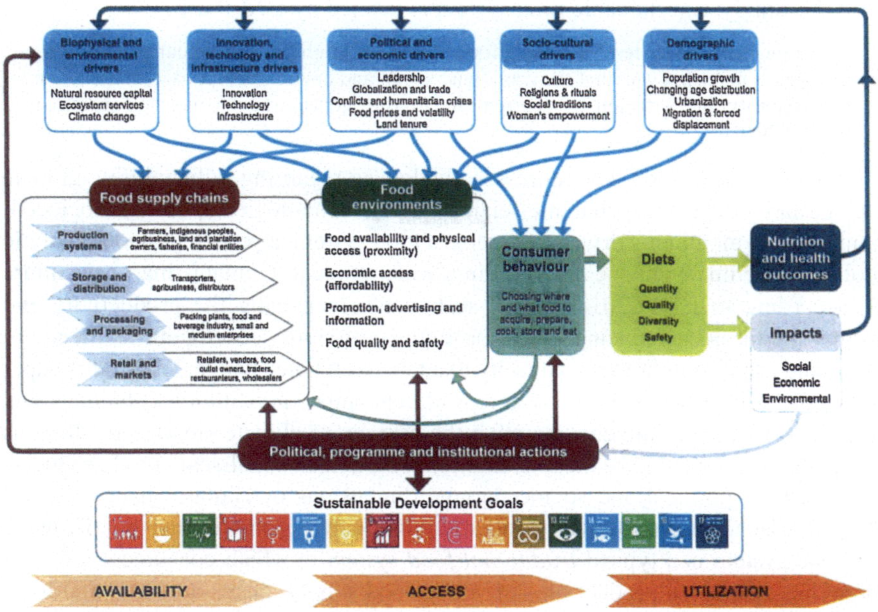

Fig. 1 Conceptual framework for food systems for nutrition. (HLPE 2017)

framework (Ellis and Freeman 2004; Ellis et al. 2003) in the context of the Poverty Reduction Strategies (PRSP) implemented in Malawi and many other countries supported by the International Financial Institutions in the late 1990s and early 2000s. In Ellis et al. (2003), the livelihood framework links macrolevel efforts to develop comprehensive poverty reduction policies (such as the PRSPs and the SDGs) with a microlevel understanding of the conditions faced by the rural poor.

A review of the available literature, along with data from the fourth *Integrated Household Survey* (IHS4), is used to undertake the food system analysis (NSO 2017). In addition, a contextual understanding of the food system in Malawi is provided through our participation in the Norwegian-funded project *Capacity Building for Managing Climate Change in Malawi* (CABMACC).

3 Sustainability of the Food System in Malawi

The food system in Malawi is far from sustainable and is at risk of becoming even less so in the years ahead due to climate change along with other factors. In this context, we define a food system as follows:

> A food system gathers all the elements (environment, people, inputs, processes, infrastructures, institutions, etc.) and activities that relate to the production, processing, distribution, preparation and consumption of food, and the outputs of these activities, including socio-economic and environmental outcomes (HLPE 2017:11).

Accordingly, a sustainable food system is defined by UNEP as:

> A sustainable food system (SFS) is a food system that delivers food security and nutrition for all in such a way that the economic, social and environmental bases to generate food security and nutrition for (current and) future generations are not compromised (UNEP 2018).

A food system approach addresses challenges regarding vulnerability, climate change and food in a way that interrelates natural, technological, political, socioeconomic and demographic drivers. Thereby, a food system approach provides analytic tools to put climate change and food in a wider context. Climate-smart agriculture is, according to FAO (2018), about sustainably increasing the productivity and income of farmers, reducing greenhouse gas emissions, developing resilience in crops and improving farmers' capacity to adapt to climate change. Whilst this paper does not go into detail on the different climate-smart agriculture techniques and results, it does keep into consideration the critique of climate-smart agriculture as being about technical fixes that put additional burden on small-scale food producers and overlook political and socioeconomic dimensions of climate change (Taylor 2018; Karlsson et al. 2018). In an agrarian-dominated economy such as Malawi's, the food system is a typical *traditional food system* in which consumers rely to a large degree on own production and where food supply chains are short and local (HLPE 2017). Hence, an analysis of the Malawian food system entails a strong focus on primary food production, agricultural productivity and rural poverty. In this paper, each of the five drivers included in the HLPE (2017) framework is reviewed and discussed in a Malawian context.

Biophysical and environmental drivers include natural resource capital, ecosystem services and climate change. In Malawi, the natural resource capital is limited by land scarcity. Malawi is one of the countries in Africa with the smallest average farm size, and average farm size is further declining. Median land area cultivated per household declined from 0.57 hectares in 2010/2011 to 0.45 hectares in 2016/2017 (NSO 2017). Land scarcity is more severe in the central and southern parts of the country than in the northern part. The natural resource capital is also effected by deteriorating soil quality, contributing to low yield levels (Carr 2014; WB 2018). Inputs such as chemical fertilizers are out of reach for most small-scale farmers who do not receive subsidized input vouchers (Haug and Wold 2017). Soil improvement measures such as conservation agriculture, mulching and use of organic manure are technological measures promoted by extension officers and NGOs. Land scarcity and poor soil fertility are important factors in the assessment of natural resource capital, though other factors that contribute to food loss such as pests (e.g. the armyworm) should also be mentioned.

Adverse weather and climate change are included in the biophysical and environmental drivers. Since 1967, Malawi has been hit hard by severe disasters amounting to 8 serious droughts and 20 floods (GoM 2015, 2016). In the future, climate change is expected to increase both the frequency and severity of drought and flooding. IPCC (2014) found that maize-based farming systems such as that in Malawi may experience a decline in yield of around 18–22%. The USAID climate change risk profile for Malawi (2017: 2) includes:

...higher average temperatures of 1–3 °C by 2050, increase in the number of days and nights considered "hot" by 2060, overall increases or decreases in rainfall difficult to project, later onset/earlier cessation of rainy season, increase in average monthly rainfall from Dec– Jan and a decrease from Feb–April, increases in the proportion of rainfall during extreme events of up to 19% annually by 2090.

Climate change will most likely result in reduced crop yields followed by hunger unless effective measures are put in place (Arndt et al. 2014).

The second category of drivers includes *innovation, technology and infrastructure.* Infrastructure such as the provision of roads is still a challenge in many places in the country and a constraint for the efficient marketing of crops and delivery of inputs. Small-scale farming is, to a large degree, traditional although input subsidies have helped in making chemical fertilizer and improved seed more accessible for farmers receiving input coupons (Pauw et al. 2014; Haug and Wold 2017). One important way of adapting to climate change is through switching to crop varieties that are able to cope with climate-related stress. Both public extension systems and actors such as international NGOs promote climate-smart agriculture measures such as mulching, water harvesting and the use of basins, organic material and manure (Kamangira et al. 2016). According to Kakota et al. (2017), extension officers lack knowledge and skills regarding the dissemination of climate-smart agriculture practices. Technological efforts have focused on productivity gains and climate adaptation. Value addition in the food value chain has received less attention. Input subsidy programs have consumed a large proportion of public spending to agriculture and have played a social security role in the country (Dorward and Chirwa 2014). There are different views on how successful the subsidized input programs have been in relation to agricultural development and food security. In the 10-year period 2005/2006 to 2014/2015, the country was self-sufficient in maize, even producing enough to export it to neighbouring countries (Arndt et al. 2014b; Pauw et al. 2014; GoM 2016). However, fertilizer subsidies are contested in relation to market distortion, climate adaptation, resilience building and long-term sustainability (Chinsinga and Chasukwa 2016; Haug and Wold 2017). The input subsidies contributed to productivity gains and kept hunger at bay for a large proportion of Malawian smallholders, before the country was hit by serious flooding in 2015 and drought in 2016. In recent years, input subsidies have become out of reach for many small-scale farmers as the proportion of voucher receivers sharply declined from 60% in 2010/2011 to 36% in 2016/2017 (NSO 2017).

The third category of drivers are *political and economic* drivers, including leadership, globalization and trade, conflict and humanitarian crises and food prices and volatility. Globalization and trade are important for Malawi both in relation to high economic dependency on tobacco export and regarding the influx of low-cost consumer goods that outcompete national industrial efforts (Collier and Dercon 2009). The World Bank (2018) describes Malawi as a country with weak governance, poor institutions, limited ability to implement policy and a high level of dependency on international aid. The agro-food sector is heavily dependent upon support from actors outside the country and is vulnerable to donor fatigue and a lack of trust in the government from the donors (Kamangira et al. 2016). Corruption scandals such as *cashgate* in 2013 have contributed to donors withdrawing from Malawi or using alternative channels for their support other than the public sector (WB 2018). The input subsidy program has also been criticized for leakage and insufficient targeting of the poorest categories of small-scale farmers (Chinsinga and Poulton 2014). However, the input subsidy program is an area in which the GoM has been able to implement policy that has given results in productivity gains in agriculture (Haug and Wold 2017). *The National Agricultural Policy* (NAP) in Malawi is formulated with the *National Agricultural Investment Plan* (NAIP) as its main implementation vehicle, including areas such as productivity, value addition, resilience and trade (MAIWD 2016, 2018). A revised seed policy was launched in May 2018 after having been contested regarding a lack of recognition of the importance of the informal seed system (Nation 2016). The revised seed policy aims to ensure effective seed regulations, high seed quality, internationally acceptable seed certification and growth in the domestic seed industry (MAIWD 2018b). The policy promotes a conducive environment for the private seed sector, but maintains legal space for *Quality Declared Seed* (QDS) and an integrated seed system (MAIWD 2018b).

Malawi also has policies and plans for climate change and resilience such as the *National Resilience Plan: Breaking the cycle of food insecurity in Malawi* (GoM 2016). The availability and affordability of food is core in many of the Malawian policies after serious hunger has repeatedly afflicted the country for decades. Food prices in Malawi have a history of being volatile, for example, low at harvest time and then rising with a peak in the hunger period before the next season's harvest. Ad hoc decisions on the use of export bans to control food prices have contributed to uncertainty and a lack of predictability in relation to investment in farming (WB 2018). Malawi's history of strategically managing grain reserves has proved disastrous, an example being the drought of 2001 during which the grain reserves were gone just when they were needed for price stabilization, and prices hiked to five times their seasonal average (HLPE 2012). In Malawi, input subsidies have historically been the largest social security program, but in 2016/2017 following two seasons of bad harvest and a substantial reduction in the input subsidy program, in-kind food assistance became the largest safety net program, supporting 6.7 million people (Duchoslav and Kenamu 2018). Whether to subsidize farm inputs or provide direct food/cash relief, and to what degree, or how to find the right balance between these two measures has been an ongoing discussion between donors to Malawi over the last two decades (Haug and Wold 2017).

The fourth category of drivers are *sociocultural drivers*, including culture, religion, social traditions and gender inequality. Female-headed households tend to be poorer than male-headed households, with lower agricultural productivity, fewer assets and less access to services (WB 2018). Malawi scores poorly on various gender inequality indices, and the country is ranked number 148 out of 160 countries by UNDP (2018). About half of the Malawian population lives below the poverty line, with a slight increase in poverty from 50.7% in 2010/2011 to 51.5% in 2016/2017 (WB 2018). However, ultra-poverty decreased in the same period from 24.5% to 20.1% (NSO 2017). The decrease in ultra-poverty is explained by an expansion in the coverage of social safety nets such as the *Food Security Response*

Program (FIRP) and improved ability to target the poorest categories of people (Duchoslav and Kenamu 2018). Gender inequalities in access to services such as agricultural extension contribute to women smallholders having poorer access to knowledge and advice relating to climate-smart agriculture than male farmers (Kakota et al. 2017). The World Bank (2018) places gender inequality as a foundational issue that needs to be addressed, along with weak governance, in order to break the cycle of low growth and slow poverty reduction in Malawi.

Demographic factors such as population growth, urbanization and migration are the fifth category of drivers in the food system framework. According to WB (2018), the Malawian population is expected to increase from 17 million people in 2017 to around 34 million people in 2038, in spite of a reduction in the total fertility rate. Urbanization has been rather slow, and migration is occurring on a rural-to-rural basis (NSO 2017). The poverty level in urban areas is as low as 17% compared with 57% in rural areas (WB 2018). Urbanization will probably increase in the years ahead as land scarcity escalates and farming is perceived more as a poverty trap than a prosperous livelihood opportunity.

The five different categories of drivers reviewed above shape the food supply chain, the food security situation and the nutritional and health outcomes in Malawi. The food supply chain is characterized by low farm-level productivity, a high degree of farm-level self-consumption of food, limited sales outside of local markets, high marketing costs, few storage facilities, limited processing and value addition and the limited role of retail outlets/supermarkets in the food system. However, the agricultural sector plays an important economic role, as tobacco, tea and sugar constitute most of the Malawian export market. The fourth *integrated household survey* provides recent data and the possibility for comparison over time, regarding the food and nutrition security situation (NSO 2017). Based on the IHS4 data, IFPRI found a dramatic decline in household food security in Malawi from 2010/2011 to 2016/2017, as more than half of the households fall into the category *very low food security* in 2016/2017, up from 32% to 61% (IFPRI 2018). The proportion of households experiencing food shortage in the 12 months preceding the survey amounted to 73% (NSO 2017). The food insecurity situation is more serious for female-headed households than male-headed households (NSO 2017). Different mechanisms to cope with food insecurity such as the consumption of less preferred foods, limiting portion sizes and reducing the number of daily meals were used more frequently in 2016/2017 than in 2010/2011 (IFPRI 2018). In the *Global Hunger Index*, which is based on undernourishment, child wasting, child stunting and child mortality, Malawi is ranked as number 87 out of 119 countries (von Grebmer et al. 2018). The proportion of undernourished people in the population has, according to the Global Hunger Index, increased from 21.8% in 2010 to 26.3% in 2016 (von Grebmer et al. 2018). IFPRI (2018) reports that the percentage of stunted children declined between 2010/2011 and 2016/2017, whilst the proportion of children who were underweight or wasting increased. The general decline in food and nutrition security from the third to the fourth integrated household survey is indeed worrisome. This unfortunate trend can be partly explained by the 2015 flooding and 2016 drought and, to some degree, by the decline in the input subsidy program. As the food system analysis indicates, however, political, economic, social and demographic drivers are also contributing factors.

4 Measures Towards a More Sustainable Food System

So far, we have analysed the Malawian food system and identified the drivers that influence it and that contribute to its lack of sustainability. There are a myriad of interrelated factors that result in worrying nutrition and health outcomes and unfavourable social, economic and environmental impacts of the food system. From the food system analysis, a set of priority areas appears to be of utmost importance in order to move towards a more sustainable food system:

- *Biophysical and environmental drivers*: Land scarcity, climate change, low soil fertility and food loss due to pests
- *Innovation and technology*: Affordability of inputs and the efficiency of climate-smart agriculture as a sufficient tool for climate adaptation as well as value addition along the food chain
- *Political and economic*: Weak governance, corruption, lack of ability to implement policy, a high level of donor dependency and uncertainty related to shifting donor priorities
- *Sociocultural*: Gender inequality and discrimination of women that hinders climate change adaptation and food security
- *Demographic:* High rural population growth that needs to be addressed in a land scarcity and climate change context, as well as the urgent need for non-farm job creation in order to enable diversification of rural livelihoods

These challenges are by no means new and have been addressed by numerous efforts of men and women farmers, extension officers, researchers, NGOs, the private sector, GoM and donors. Farmers are the most important investors in agriculture (RAI 2014) and do whatever they can within labour, health, capital, knowledge and market constraints to adapt to climate change and unfavourable natural and political frame conditions. Malawi is one of the countries in Africa whose government gives the proportionally highest priority to agriculture in the national budget (AU 2018). In addition, donors have substantially supported the agricultural sector in Malawi, although in a rather unpredictable manner in which varied donor preferences and a lack of trust in the Government have influenced the type of support given and the channels used. Weak governance, corruption and poor implementation capacity result in donors opting to manoeuvre around the state and find alternative solutions to secure efficiency and results. Such behaviour might contribute to weakening the role of the state in the long term.

The GoM's main agro-food sector priority has been to boost production by heavily subsidizing inputs such as fertilizer and improved seed. After serious droughts in 1991/1992 and 1994/1995, input subsidies started, but were then abandoned again after pressure from donors (Harrigan 2003). Production went down and food insecurity went up, contributing to new input subsidy programs being established such as the *Starter Pack Program* and the *Extended Targeted Input Program* (Sjaastad et al. 2007). The current *Farm Input Support Program* (FISP) started in the 2005/2006 without direct donor support in the first year, though donors did contribute later with 5% and 32% of the annual cost (Dorward and Chirwa 2014). Various actors such as donors and NGOs have criticized FISP for not being economically viable, distorting the market, poor targeting, leakage, not being environmentally sustainable in relation to long-term soil productivity and crowding out other necessary measures in relation to longer-term resilience building and adaptation to climate change (Chinsinga and Poulton 2014; Chinsinga and Chasukwa 2016; Haug and Wold 2017).

Whilst FISP is far from perfect, particularly regarding targeting and leakage, the GoM has the proven ability to implement policy and actions, such as those that resulted in maize production that exceeded the national requirements in the period 2005/2006, until floods and drought hit in 2015 and 2016 (Pauw et al. 2014; GoM 2016). In situations of floods and drought, input subsidies will fall short of fixing the problem. In recent years, FISP has drastically declined in terms of farmers receiving vouchers, whilst other programs have increased such as efforts towards promoting climate-smart agriculture and direct social security programs such as school feeding and FIRP. In 2016/2017, FIRP provided either in-kind food relief or direct cash transfers to vulnerable households, reaching 6.7 million people (Duchoslav and Kenamu 2018). The recent trend of declining support to small-scale farmers for direct agricultural production and increased direct assistance with food or cash is worrisome in relation to future climate threats in an agrarian economy such as Malawi's. Regarding services, climate change adaptation is also about institutional support and innovation. The Ministry of Agriculture, Irrigation and Water Development has launched a new *National Agricultural Extension and Advisory Services Strategy* that aims to strengthen this service (MAIWD 2018c). A new element of the strategy is to set minimum standards for service providers and approval mechanisms for type and content of advice, as well as the registration of providers and better coordination of the different actors involved in extension and advisory services for farmers (MAIWD 2018c). The strategy is ambitious and will need both funding and capable institutions in order to be properly implemented. The regulatory and coordinating role of the public sector at different levels will be strengthened if the strategy is put into effect.

Climate change adaptation related to the agro-food system is also highlighted in the *National Climate Change Investment Plan* for 2013–2018 (NCCIP). This plan highlights institutional capacity, environmental conservation and productivity increase through adaptation investment and climate-smart agriculture (MECCM 2013). Whilst the funding and implementation of NCCIP have been challenging, international initiatives such as the *Green Climate Fund* (GCF) have provided support to a large program working on the upscaling of the use of modernized climate information and early warning systems (M-CLIMES) (GCF 2018). M-CLIMES provides early warning information whilst strengthening preparedness and local capacity regarding the ability to respond to weather-related disasters. Preparedness is a priority area for the GoM, managed through its *Department of Disaster Management Affairs* (DoDMA) and *National Disaster Preparedness and Relief Committee* (NDPRC). After the 2015 floods, planned measures by the Government included more forecasting and early warning, better stakeholder coordination, the establishment of emergency response operation centres and a shift towards more preparedness and a reduced need for emergency response (GoM 2015; DoDMA 2015). The actions implemented by the GoM to strengthen preparedness include input subsidies and emergency responses including social protection such as school meals and direct transfers of food and cash. Preparedness is defined to include the three activities: *anticipate, respond to and recover from* a disaster/hazard (UNISDR 2016). Preparedness includes many different measures such as risk analysis, vulnerability assessment, crop insurance, food storage, technical change in agriculture (e.g. climate-smart agriculture and drought-resistant cultivars), direct livelihood support, subsidizes, food relief, support to local markets and service provision (HLPE 2012). Although the GoM is giving priority to preparedness through its various policies, weak institutions and lack of ability to implement policy are hampering effective actions and results. Resilience building and, in turn, a more sustainable food system are thus also hampered.

In the HLPE (2017) food system framework, the sustainable development goals are given as policy and programs that will influence and contribute to a sustainable food system. The GoM has signed up to the SDGs. SDG2 on zero hunger is of particular interest in relation to a sustainable food system, though most of the SDGs are relevant in this regard. Since Malawi has a typical traditional food system, SDG2.3 is highly relevant in relation to food and nutrition security:

By 2030, double the agricultural productivity and incomes of small-scale food producers, in particular women, indigenous peoples, family farmers, pastoralists and fishers, including equal access to land, other productive resources and inputs, knowledge, financial services, markets and opportunities for value addition and non-farm employment (UN 2018).

In this regard, small-scale food producers are those who farm land (or livestock) in the bottom 40% of land size (or livestock number) and *obtain annual economic revenue from agricultural activities falling in the bottom 40% of economic revenues* (UN 2018). The two indicators used to measure progress include the volume of production per labour unit by classes of farming/pastoral/forestry enterprise size, and the average income of small-scale food producers, by sex and indigenous status (UN 2018). To double the agricultural productivity, the required growth rate is 4.62% per year. Malawi is one of the countries with the smallest average farm size that has previously managed to double both land and labour productivity over a 15-year period (1992–2007 and 1994–2009). This was mainly due to input subsidies and conducive weather (Mikecz and Vos 2016). Whilst SDG2.3 can be a realistic target regarding the doubling of the agricultural productivity of small-scale food producers in general, special efforts will be needed to ensure that women and the bottom 40% are included. It will likely be particularly challenging to double the *income* of the bottom 40% of small-scale food producers, including women.

5 Lessons Learned Regarding Efforts Towards a More Sustainable Food System

Numerous efforts towards a more sustainable food system in Malawi have been initiated from the local level by small-scale men and women farmers and up to the government and international levels. However, the results have not yielded the necessary effects, as food insecurity appears to be increasing whilst climate uncertainties are threatening the crops and livelihoods of small-scale farmers. There are many lessons that can be learned from past and current policies and actions. At the international and national level, there is no lack of appropriate policies, strategies and plans, but the capability for implementation is weak. The description of the implementation of PRSP processes in Ellis et al. (2003:1496) is still relevant today: *it is an agenda promoted from the top by donors and pursued with varying degrees of enthusiasm, or lack of it, by national governments*. Benson et al. (2018) underline disconnect between what they regard as *reasonably high-quality* policies and policy processes and the limited outcome of these policies, stating that the quality of policy implementation does not meet the intentions of the policies. At the local level, the climate adaptation capacity appears to be low, and efforts tend to be limited to technological solutions in a way that transfers the burden of adaptation to small-scale producers (Karlsson et al. 2018). The lessons regarding weak policy implementation at the national level and a low capacity for local climate adaptation indicate that more focus should be given to the formulation of implementable policies and revisit the way climate uncertainties are being addressed through climate-smart technology at the small-scale farmer level.

The analysis of the food system drivers provides an overview of the different types of challenges that need to be addressed to achieve a more sustainable food system with less hunger, poverty and climate uncertainty. Regarding political and economic drivers, weak governance, corruption, lack of ability to implement policy, and a high level of donor dependency do not provide the best frame conditions for making the food system more sustainable. The analysis shows that *biophysical and demographic factors* such as small farm size, land scarcity and high rural population growth need to be acknowledged in climate adaptation measures in a way that goes beyond current climate-smart agriculture techniques. High levels of rural poverty and low profitability in farming require measures such as alternative job creation, both within and outside the agricultural sector including value addition along the food chain. Value addition could be sought in certain crops such as groundnuts and potatoes, though the marketing of this could be challenging. Value addition from seed producers might be a possibility, if international actors are willing to invest and actively commission Malawian small-scale farmers to produce seed for a bigger market than Malawi. With climate change, the production of climate-resilient cultivars could become a Malawian specialty both for domestic and other markets, including climate-smart seed and income opportunities. Seed production is given as a possible example of the type of value addition along the food chain that can improve profitability in agriculture and thereby reduce rural poverty and build resilience towards shocks. However, climate-smart seed and value addition from seed production is just one possible example of a niche activity that could be pursued by both the GoM and small-scale farmers. More and better value addition options may be pursued through bottom-up initiatives in which farmers are active partners in identifying ways of reducing poverty and food insecurity whilst at the same time contributing to climate change adaptation and a more sustainable food system.

Donor dependency and influence are high in Malawi, and donor priorities are important drivers in relation to policy and action. The WB (2018: x–xii) underlines that the foundational issues of weak governance, lack of ability to implement policy and gender inequality need to be addressed and suggests four different strategic pathways for Malawi for the years ahead:

- *Increasing agricultural productivity, commercialization and diversification*
- *Diversifying the economy and creating jobs, including structural transformation, private sector and service delivery*
- *Harnessing the demographic dividend and building human capital*
- *Building resilience against shocks and enhancing environmental sustainability including strengthening of social protection programs*

Earlier in this paper, we reviewed GoM's policies and strategies that, to a large degree, already cover these four pathways, though policy implementation and thereby the delivery of results have been challenging. However, regarding the first pathway of agricultural productivity, the input subsidy program is an example of national ownership followed by ability to implement policy that, in a specific period (2005/2006–2013/2015), yielded good results and kept serious hunger at bay (Pauw et al. 2014; Haug and Wold 2017). Although input subsidies alone are far from enough to ensure a sustainable food system in Malawi, FISP provides useful lessons regarding implementation capability. Simultaneously prioritizing all four pathways will be a challenging strategy for Malawi within the current situation of weak governance, inadequate monetary resources and limited institutional capacity, as highlighted by the WB. It is of course vital to know *what* to do, but a lesson learnt from Malawi is that it is equally important to know *how* to do it – how can Malawi implement its policies?

6 Conclusion

A food system approach provides the opportunity for an integrated analysis of drivers and constraints that influence the food system, as well as a holistic understanding of how policies, programs and institutional actions can contribute to a more sustainable food system. Malawi relates to a broad range of national and international policies aiming to improve the food security and poverty situation in the country; however, only a few of these policies are implemented with visible improvements at the rural household level. Lessons learned from policies and actions in the Malawian food system include the importance of national ownership and recognition of what is politically feasible and institutionally implementable in efforts towards a more sustainable food system. Finding solutions to food and climate uncertainties lies in the interface between policy and action and requires small-scale farmers' creativity and ability to adjust to change, without transferring the responsibility of climate adaptation to small-scale farmers with limited adaptation capacity. In order to build resilience to climate change, there is a need to move beyond weather forecasting and early warning systems, to provide realistic livelihood opportunities for small-scale farmers to move out of poverty and towards improved food security. The tendency to render food security as a problem that can be dealt with through technical solutions alone, such as climate-smart agriculture, must be complemented by thorough analyses of the economic, political and socio-political drivers of food system transformation. An integrated approach to food and climate uncertainties will call for socioeconomic and technological innovations that mutually strengthen each other whilst ensuring that the technology works for small-scale men and women farmers in accordance with the intentions set out in the sustainable development goals.

Acknowledgments The Norwegian-funded project on *Capacity Building for Management of Agriculture and Climate Change Program* (CABMACC) and HLPE CFS for permission to include Fig. 1 from HLPE (2017) in the chapter.

References

Arndt C, Schlosser AC, Strzepek K, Thurlow J (2014) Climate change and economic growth prospects for Malawi: an uncertainty approach. J Afr Econ 23(Supplement 2):25

Arndt C, Pauw K, Thurlow J (2014b) The economy wide impacts and risks of Malawi's farm input subsidy programme. Working paper 2014/099. UNU-WIDER, Helsinki

AU (African Union) (2018) Inaugural biennial review report of the African Union commission on the implementation of the Malabo declaration on accelerated agricultural growth and transformation for shared prosperity and improved livelihoods. http://www.resakss.org/sites/default/files/BR%20English%20Draft%20Print.pdf. Accessed Apr 2018

Benson T, Nyirenda Z, Nankhuni F, Maredia M (2018) The quality of agriculture and food security policy processes at national level in Malawi, results from the 2017/18 Malawi agriculture and food security policy processes, Feed the Future Innovation Lab for Food Security Policy Research Paper 109. East Lansing: Michigan State University.

Carr S (2014) The challenge of Africa's Nitrogen drought: Some indications from the Malawian experience. Policy note 19. International Food Policy Research Institute (IFPRI), Washington, DC

Challinor AJ, Koehler AK, Ramirez-Villegas J, Whitfield S, Das B (2016) Current warming will reduce yields unless maize breeding and seed systems adapt immediately. Nat Clim Chang 6:954–958

Chinsinga B, Chasukwa M (2016) Searching for a holy grail? The nexus between agriculture and youth unemployment. In: Banik D, Chinsinga B (eds) Political transition and inclusive development in Malawi. Routledge, London, pp 112–131

Chinsinga B, Poulton C (2014) Beyond technocratic debates; the significance and transience of political incentives in the Malawian farm input subsidy program (FISP). Dev Policy Rev 32:123–150

Collier P, Dercon S (2009) African agriculture in 50 years: smallholders in a rapid changing world. Expert meeting on how to feed the world in 2050. FAO, Rome

DoDMA (2015) Malawi: flood. Situation report. Department of Disaster Management Affairs, Malawi, p 12

Dorward A, Chirwa E (2014) Evaluation of the 2013/14 farm input subsidy program, Malawi. Program implementation analysis. Ministry of Agriculture Irrigation and Water Development, Malawi

Duchoslav J, Kenamu E (2018) Are social safety nets and input subsidies reaching the poor in Malawi? Strategy and support program. Working paper 26. IFPRI, Malawi

Ellis F, Freeman HA (2004) Rural livelihoods and poverty reduction strategies in four African countries. J Dev Stud 40:1–30

Ellis F, Kutengule M, Nyasulu A (2003) Livelihoods and rural poverty reduction in Malawi. World Dev 31:1495–1510

FAO (2018) Being climate smart in agriculture. http://www.fao.org/fao-stories/article/en/c/1172370/. Accessed Dec 2018

GoM (2015) Malawi 2015 floods post disaster needs assessments report. Ministry of Disaster Management Affairs, Malawi

GoM (2016) The national resilience plan. Breaking the cycle of food insecurity in Malawi. Office of the Vice President, Malawi

Green Climate Fund (2018) Projects funded. https://www.greenclimate.fund/projects/fp002. Accessed Dec 2018

Hallegatte S, Bangalore M, Bonzanigo L, Fay M, Kane T, Narloch U, Rozenberg J, Treguer D, Vogt-Schilb A (2016) Shock waves, Report. World Bank, Washington, DC

Harrigan J (2003) U-turns and full circles: two decades of agricultural reform in Malawi 1981–2000. World Dev 31(5):847–863

Haug R, Wold BKG (2017) Social protection or humanitarian assistance: contested input subsidies and climate adaptation in Malawi. IDS-Bulletin 48(4):93–110

HLPE (2012) Social protection for food security. A report by the High Level Panel of Experts on Food Security and Nutrition, Rome

HLPE (2017) Nutrition and food systems. A report by the High Level Panel of Experts on Food Security and Nutrition of the Committee on World Food Security, Rome

IFPRI (2018) Key facts series: Malawi – food and nutrition security June 2018. International Food Policy Research Institute, Washington, DC

IPCC (2014) Climate change 2014. Impact, adaptation and vulnerability. Working Group II contribution to the fifth assessment report of the intergovernmental panel on climate change. IPCC, World Meteorological Organization, Geneva

IPCC (2018) Summary for policymakers. In: Masson-Delmotte V, Zhai P, Pörtner HO, Roberts D, Skea J, Shukla PR, Pirani A, Moufouma-Okia W, Péan C, Pidcock R, Connors S, Matthews JBR, Chen Y, Zhou X, Gomis MI, Lonnoy E, Maycock T, Tignor M, Waterfield T (eds) Global warming of 1.5 °C. An IPCC special report on the impacts of global warming of 1.5 °C above pre-industrial levels and related global greenhouse gas emission pathways, in the context of strengthening the global response to the threat of climate change, sustainable development, and efforts to eradicate poverty. World Meteorological Organization, Geneva, p 32

Kakota TV, Maonga BB, Synnevag G, Chonde C, Mainje M (2017) Harmonization of extension messages on climate smart agriculture in Malawi: do we speak with one voice, and to whom? J Agric Ext Rural Dev 9(11):255–261

Kamangira D, Makoko K, Munthali GT, Pungulani L (2016) Status of agricultural innovations, innovation platforms and innovation investment – Malawi. PARI project country report: Republic of Malawi. FARA, Accra

Karlsson L, Naess LO, Nightingale A, Thompson J (2018) 'Triple wins' or 'triple faults'? Analysing the equity implications of policy discourses on climate-smart agriculture (CSA). J Peasant Stud 45(1):150–174

MAIWD (2016) National Agriculture Policy (NAP). Ministry of Aagriculture, Irrigation and Water Development, Malawi
MAIWD (2018) National Agriculture Investment Plan (NAIP). Ministry of Agriculture, Irrigation and Water Development, Malawi
MAIWD (2018b) National seed policy. Ministry of Agriculture, Irrigation and Water Development, Malawi
MAIWD (2018c) The national agricultural extension and advisory services strategy. Ministry of Agriculture, Irrigation and Water Development, Malawi
MECCM (2013) National climate change investment plan 2013–2018. Ministry of Environment and Climate Change Management, Malawi
Mikecz O, Vos R (2016) Can smallholders double their productivity and incomes by 2030? ESA working paper no. 16–05. FAO, Rome. http://www.fao.org/3/a-i5959e.pdf. Accessed Nov 2018
Nation (2016) Draft seed policy risks marginalizing smallholder farmers. https://mwnation.com/draft-seed-policy-risks-marginalising-smallholder-farmers/. Accessed Dec 2018
NSO (2017) Integrated household survey (IHS4) household socio-economic characteristics report, Zomba. National Statistics Office (NSO), Malawi
Pauw K, Beck U, Mussa R (2014) Did rapid small-holder-led agricultural growth fail to reduce rural poverty? Making sense of Malawi's poverty puzzle. Wider working paper 2014/123. UNU-WIDER, Helsinki
RAI (2014) Responsible investment in agriculture and food systems. http://www.fao.org/cfs/home/activities/rai/en/. Accessed Dec 2018
Sjaastad E, Kaarhus R, Vedeld P, Wold BKG (2007) Privatization and liberalization in the agricultural sector: an examination of processes and outcomes in three African cases, Noragric report no 39. Norwegian University of Life Sciences (NMBU), Noragric
Taylor M (2018) Climate-smart agriculture: what is it good for? J Peasant Stud 45(1):89–107
Thurlow J, Arndt C, Schlosser A, Strzepek K (2014) Climate change and economic growth prospects for Malawi: an uncertainty approach. J Afr Econ 23:83–107
UN (2018) UNSTAT SDG indicators. https://unstats.un.org/sdgs/metadata/. Accessed Nov 2018
UNDP (2018) Gender inequality index. hdr.undp.org/en/composite/GII. Accessed Nov 2018
UNEP (2018). https://www.unenvironment.org/explore-topics/resource-efficiency/what-we-do/sustainable-lifestyles/food-and-food-waste. Accessed Dec 2018
UNISDR (2016) UNISDR terminology – preparedness definition https://www.unisdr.org/files/7817UNISDRTerminologyEnglish.pdf. Accessed Sep 2016
USAID (2017) Malawi – climate change risk profile. https://www.climatelinks.org/sites/default/files/asset/document/2017_USAID_Climate%20Change%20Risk%20profile%20-%20Malawi.pdf. Accessed Dec 2018
von Grebmer K, Bernstein J, Patterson F, Sonntag A, Klaus LM, Fahlbusch J, Towey O, Foley C, Gitter S, Ekstrom K, Fritschel H (2018) Global hunger index 2018. Forced migration and hunger. Welthungerhilfe and Concern Worldwide, Berlin/Dublin
WB (2018) Malawi. Systematic country diagnostic. Breaking the cycle of low growth and slow poverty reduction. Malawi country team report. Africa Region. World Bank, Washington, DC

Need for Personal Transformations in a Changing Climate: Reflections on Environmental Change and Climate-Smart Agriculture in Africa

Bishal K. Sitaula, Ognjen Žurovec, Bal Chandra Luitel, Anne Parker, and Rattan Lal

Abstract In this paper, we discuss the need for personal transformation in the rapidly changing environmental context. We also provide a short reflective and analytical framework for climate-smart agriculture (CSA) and its linkages with vulnerability, resilience, and livelihoods and reflect on how CSA can impact positively on the livelihoods of smallholder farmers in Africa. Furthermore, we stress the importance of reflective thinking on interdependence and the linkages between agricultural and environmental problems that are deeply rooted in human wants/greed and that manifest in various forms such as unsustainable agriculture, biodiversity losses, climate change, and land/soil degradation. To address these problems at their source, we attempt to introduce the idea of personal transformation, which is lacking in current approaches to education and agricultural development. We present a few educational perspectives, where a transformation refers to a change or shift of meaning on dominant ideas, beliefs, and assumptions. We argue that transformation takes place within the self after internalizing the transformative individual views or ideas from different perspectives. In other words, we cannot transform ourselves through the lens of beliefs and assumptions that are taken for granted. It is being aware of one's own

B. K. Sitaula (✉) · O. Žurovec
Department of International Environment and Development Studies, Norwegian University of Life Sciences, Ås, Norway
e-mail: bishal.sitaula@nmbu.no

B. C. Luitel
School of Education, Kathmandu University, Dhulikhel, Nepal

A. Parker
Department of Environmental Sciences, Naropa University, Boulder, CO, USA

R. Lal
School of Environment and Natural Resources, Carbon Management and Sequestration Center, The Ohio State University, Columbus, OH, USA

B. R. Singh et al. (eds.), *Climate Impacts on Agricultural and Natural Resource Sustainability in Africa*, https://doi.org/10.1007/978-3-030-37537-9_21

context and development of consciousness that brings changes within the self. We anticipate that education promotes and enhances awareness of personal transformation in various arenas of development, including agriculture.

1 Introduction and Objectives

In the rapidly changing world, an apparent failure of humans and institutions to come to grips with climate change is being increasingly felt. We may have learned how humans contribute to climate change at the proximal level, but we rarely ask *why* we leave a larger than necessary carbon footprint. The reality is that, due to the urgency of emerging climate crises, incremental change in technology may not address the problems at their source. We have also learned through many years of repeated research that dramatic changes in agriculture and land use have taken place in larger parts of Africa, South Asia, and other parts of the world, as a result of economic growth and increasing demands for production. Forestlands are converted into agricultural land and intensified for food production. Agricultural systems in countries across the African continent have to undergo substantial change in order to address the pressing challenges of population growth and climate change. There is a great need for a transition to agricultural production systems that are more productive, use inputs more efficiently, have less variability and greater stability in their outputs, and are more resilient to risks, shocks, and long-term climate variability. Smallholder farmers are particularly vulnerable due to their dependence on agriculture for food security and livelihoods. Given the importance of agriculture in the context of rural livelihoods, food security, and poverty reduction, meaningful climate-smart agriculture (CSA) has the potential to increase the livelihoods and resilience of millions of smallholder farmers in Africa. In addition, there is an urgent need for transformative change in our individual consumption behaviours and lifestyles that are linked to food demands that have important implications for future pathways for agricultural development. The objectives of this paper are therefore to (1) reflect on the need for transformation in environmental change context; (2) discuss the changing context and provide a short, analytical framework for CSA and its linkages with vulnerability, resilience, and livelihoods and reflect on how CSA can impact positively on the livelihoods of smallholder farmers in Africa; and (3) illustrate educational strategies in an adult learning context for personal transformation that may eventually permeate to farming communities.

2 Transformation in a General Environmental Context

The failure of humans and institutions to come to grips with climate change is being increasingly felt. We may have learned how humans contribute to climate change at the proximal level, but we rarely ask *why* we leave a larger than necessary carbon footprint. The reality is that due to the urgency of emerging climate crises, incremental changes in technology may not address the problems at their source. Given the importance of agriculture in the context of climate change, rural livelihoods, food security, and poverty reduction, meaningful climate resilient agriculture, also known as climate-smart agriculture (CSA), has the potential to improve the livelihoods and resilience of millions of smallholder farmers in Africa. Our current framing of the agriculture- and climate-related problem is inadequate, both in the way the problem is articulated and in the solutions that are proposed. People are separated practically, cognitively, and emotionally from their environment and may lack an understanding of climate change and its impacts on agriculture. To counter this, the concept of transformation is increasingly used to advocate individual and systemic change for wider implications, not just in response to environmental or climate change (O'Brien 2012). We need to shift our cognition of the problem and increase global consciousness at large. The importance of transformation was already acknowledged in the IPCC Fourth Assessment Report (IPCC 2007). Transformation was included in a special report for managing the risks of extreme events and disasters to advance climate change adaptation (IPCC 2012). In this report, the IPCC defines transformation as 'fundamental changes to the attributes of a system, including value systems, regulatory, legislative or bureaucratic regimes, financial institutions, technology and biophysical systems' (p. 564).

In order to address climate change and agricultural sustainability in a changing climate, mindfulness and an awareness of interdependence are needed. This transformative awareness should be applied to higher education and research relevant for climate change adaptation and mitigation for sustainable agricultural development. This implies an increasing need for reflective thinking on interdependence and the linkages between agricultural and environmental problems deeply rooted in human wants/greed and that manifest in various forms, such as unsustainable agriculture, biodiversity losses, climate change, and land/soil degradation (Gisladottir and Stocking 2005). We need to look at our own assumptions about the innovations we make and have made (Jasanoff 2007) and transform our efforts to 'science with society'. Our present education system is mainly focused on what is *known* (as a third-person experience or information that is available publicly) and much less on the *process* of knowing or on who is the 'knower' (first-person experience). The knower, who internalizes the knowledge, will develop different perceptions based on their experiences, geographical location, cultural heritage, etc. In this regard, how agriculture education is presented and the motivation of scientists involved will influence transformation in individuals and in agriculture. Scientific work should be conducted with creativity and compassion rather than as purely mechanical work. If we believe that environmental and agricultural problems are largely a human construct, humans should then have the ability to deconstruct these problems. A human construct can be deconstructed with proper education, training, and increased understanding and awareness. We need to reflect on interdependence and explore who we are, where we come from, and what our role in this world is in a transformative context. If we only have a materialistic perception or worldview based on consumerism in the modern age, we cannot expect significant changes to happen in a sustainable direction. We need to establish universal connectedness and a spirit of wholeness and mindfulness in life while developing agriculture.

Given the number of ongoing global crises, transformation of individuals and of agriculture may be our only option, whether it is chosen or forced upon us. Innovations such as emerging technological solutions (incremental changes in technology) are not keeping pace with the extent and severity of global crises. Today's modernist civilization is developed on the paradigm of infinite growth, which encourages unscrupulous consumption of natural resources. Globally, negative impacts on agriculture and the environment are drastically increasing. Humankind is entering the third millennium with multiple global challenges concerning the provision of food, water, and energy for a growing population. How do we reduce global hunger (821 million hungry people in the world (FAO et al. 2018)), address a dysfunctional global food system (which allows 821 million hungry people and fosters 1.5 billion overweight people (Magdoff and Tokar 2010)), and expect to feed 9.8 billion people in 2050 as projected by UN (2017). Human relation to the environment is in imbalance and can only be restored by a fundamental shift in consciousness and, in our production and consumption patterns, through personal transformation.

Humans often live 'in denial' and make confronting environmental challenges difficult. From a climate change point of view, these denials can be grouped into three different forms: literal, interpretive, and implicatory denial (Cohen 2013). Literal denial is usually associated with commercial and ideological issues, for example, the use of fossil fuels. In this case, organized attempts and campaigns are made in order to undermine, deny, or dismiss the scientific consensus that the use of fossil fuels has a profound impact on climate change (Dunlap and McCright 2010). In the case of interpretive denial, the fact that climate change is happening is not denied, but is given a different interpretation, for example, that climate change is not anthropogenic, but is caused by a change in solar radiation or in the earth's orbit. Perhaps, the most significant form of denial is implicatory denial. Although we are living our everyday lives with sufficient information to confront global problems, we choose to turn a blind eye and not to take appropriate actions to change our lifestyle. The hard question remains on how to change human behaviour and alter our attitude and lifestyles in order to respond to ongoing climate change. It is difficult to expect behavioural changes to happen voluntarily, as individuals are attached to their habits and default behaviours (Norgaard 2009). We propose that this behavioural change must be met through globalization and personal transformation. We need to introduce a transformative wisdom at all levels of education and research including agriculture. If we really want to address climate change, it is not enough to have understanding and knowledge of climate change, associated risks, and impacts of this complex problem (cognitive aspect). We must also have emotional concern for perceived climate stresses (affective aspect) and the intention to behave in a way that helps in coping/adapting to climate change (conative aspect). We believe that only through the combination of these three aspects (cognitive, affective, and conative) can we create fertile ground for human-climate interaction and move towards a solution to the problem.

Transformative wisdom is required for humanity to deal effectively and ethically with the challenges and opportunities that are being presented before us. Wisdom is tentatively defined as a deep understanding of people, things, events, or situations that result in the ability to act to consistently produce optimum results using a minimum of time and energy. It also has a broad meaning in a spiritual context, which involves seeing things clearly (as they are); acting in prudent and effective ways with the well-being of 'the whole' in mind; knowing when to act and when not to act; being able to handle whatever arises with peace of mind and an effective, compassionate, holistic response; and being able to anticipate and avoid potential problems. In order to reach this level of wisdom, we need to boost our emotional balance and develop an open mental attitude, i.e. bring about a shift in cognition and consciousness.

In our educational processes, we were not equipped to realize the connection between what we experience (as first-person experiences) and deeper connectivity with everything else, key to life and our well-being. We propose that science needs to be developed in a way that creates linkages between the environment (where the source of both well-being and stress lie), human well-being, human existence and our relationship with other beings and systems, and the science of personal transformation, including wisdom traditions (integration of mind, body, and environment). A conceptual diagram for the proposed linkages is illustrated in Fig. 1.

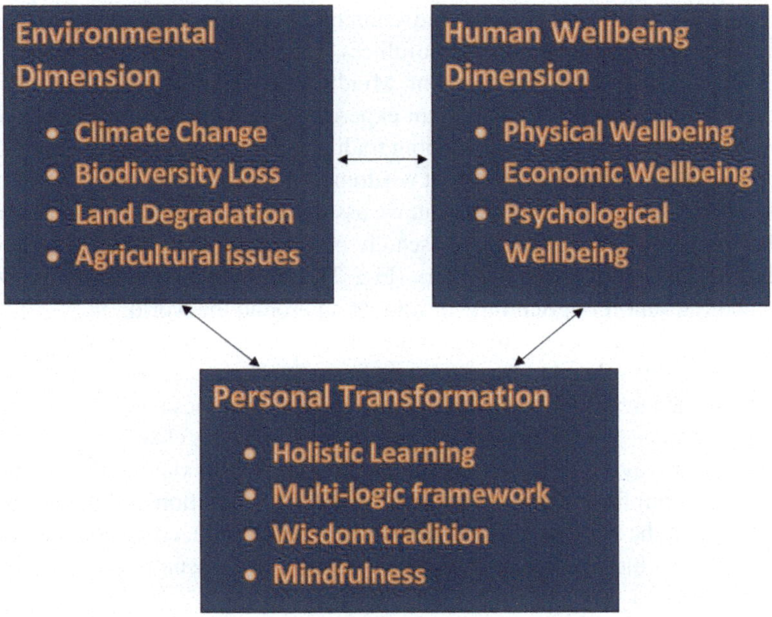

Fig. 1 Proposed linking environmental and human well-being dimensions with personal transformation

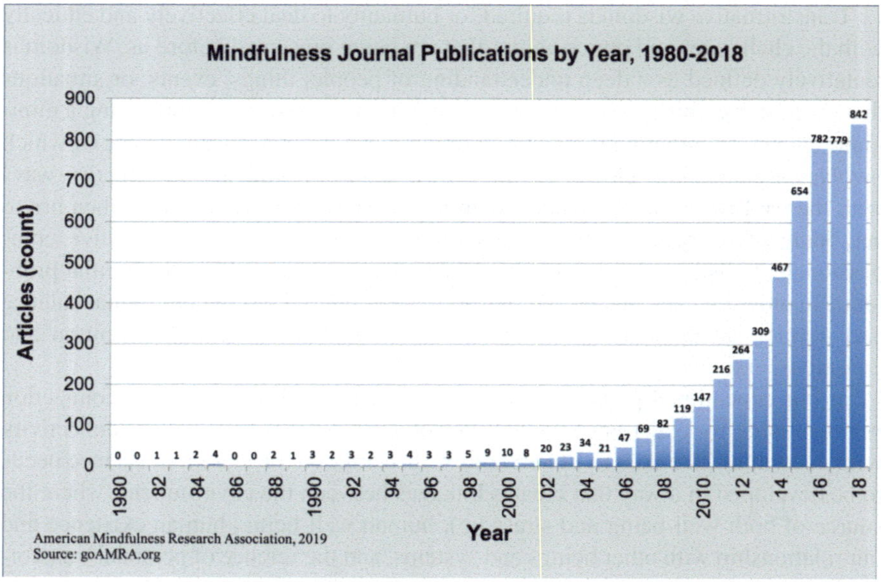

Fig. 2 Trend in mindfulness journal publications, by year (1980–2018)

Personal transformation with examples of wisdom traditions is studied at universities specializing in mindfulness and contemplative education, such as Naropa University,[1] where the concept of mindfulness is used in the context of the environment and natural resource management. Mindfulness is defined as a psychological quality of being attentive to the present experience on a moment-to-moment basis (Kabat-Zinn 2009). Based on the wisdom traditions of noetic sciences, mindfulness creates a basis for moment-to-moment wisdom in individuals, and that wisdom creates the basis for change. Only then can we assume that the change and transformation becomes truly meaningful. Research on personal transformation through mindfulness has grown over the years (Fig. 2), and we argue that this growth is timely and relevant for agricultural development around the world.

This study reflects on agriculture in the African context; we propose that new or emerging thoughts related to personal transformation are necessary to address both agricultural and environmental problems. It is evident that these problems cannot be solved on the same level at which they were created. There is clearly a need to move from a circular way of thinking (Young 2009) to an axial (axial revolution) way of thinking. This implies a more transformative shift in cognition and consciousness and a change in the way we look at our system of agriculture, education, and capacity building, including critical reflections on current assumptions and beliefs (O'Brien 2012).

[1] www.naropa.edu

3 Reflections on Agriculture in an African Climate Change Context

Almost half of the population in Africa lives in extreme poverty (FAO et al. 2015). While much has been done over the last two decades to reduce poverty and hunger, there are still about 230 million people who are deemed undernourished (FAO et al. 2015). At the same time, there has been a rapid increase in population on the African continent. It is estimated that half of the predicted 2.2 billion increase in the global population for the period 2013–2050 will occur in Africa alone (UN 2017). Increasing urbanisation occurs in parallel with increasing population, and urban dwellers are projected to account for more than a half of the population in Africa by 2030 (UN 2018). Projections based on population growth and food consumption patterns indicate that agricultural production will need to increase by 70–100% in order to meet food demands by 2050, in a ‘business-as-usual’ scenario (FAO 2017). This poses a serious challenge for agricultural production systems, which must undergo a major transformation in the coming decades in order to meet the intertwined challenges of achieving food security and reducing poverty, while adapting to climate change and preserving the existing natural resource base.

African farming systems are very diverse in terms of livelihood strategies, population pressures, access to markets, institutions, and agroecological conditions. What is common, however, is the domination of smallholder farming systems. It is estimated that more than 33 million farms are smallholdings, which constitute 80% of the total number of farms in Africa (Altieri et al. 2012). Furthermore, smallholder farms produce about 70% of the total food supply across the continent and up to 90% in some regions of sub-Saharan Africa (Wiggins 2009). Although they play a vital role in achieving food security and poverty reduction, the irony is that most of smallholder farmers are living in hunger and poverty. The main features of smallholder farmers are dependence on agriculture for livelihoods and subsistence, lack of access to assets, and labour-intensive production systems using simple, often outdated technologies. They are also often forced to farm degraded and infertile soils. Productivity of their lands is low, as shown by the low yields achieved. Consequently, smallholder farmers receive low economic returns for their production efforts. Most smallholder farmers located in rural areas lack both physical and institutional infrastructure, which limits their livelihood opportunities. These problems are mostly seen on farms in remote or isolated regions. Smallholders are disadvantaged in both domestic and international markets because of the small volume and often low quality of the products they trade in. Lastly, women often play a vital role in production. They are responsible for 30–60% of the agricultural labour supplied in Africa (Palacios-Lopez et al. 2015) yet seldom have any major role in political activities or decision-making processes in patriarchal communities.

Projected increases in annual surface temperatures and decreased rainfall in many regions of Africa and the increased incidence of extreme weather events such as droughts and floods will pose further constraints on already vulnerable smallholder farming systems. Future projections predict significant yield decreases of staple crops in sub-Saharan Africa, such as maize (*Zea mays*), sorghum (*Sorghum bicolor*), millet (*Pennisetum glaucum*), and cassava (*Manihot esculenta*). Yield reduction could be as much as 50% by 2020 in rain-fed croplands (Jones and Thornton 2009). Livestock systems will also be affected, as climate change is expected to reduce availability of forage for grazing animals and increase the prevalence of animal pests and diseases. The high dependence of smallholders on agriculture for their livelihoods and food security makes them very vulnerable to climate variability and change. Before looking into ways to enable smallholder farmers to be able to cope with the consequences of adverse weather and climate conditions, we need to have a common understanding of what comprises vulnerability and how it is linked to other concepts used in this chapter, most notably resilience, livelihoods, and transformation.

Agricultural production is the backbone of the rural economy in African countries. It employs more than 60% of the population and generates a substantial economic value, contributing to about 30% of the GDP. Despite its enormous significance in terms of economic value in rural regions, as well as closely linked implications on food security and poverty reduction, the main characteristics of these regions are widespread poverty and extensive areas of low agricultural productivity due to steadily degrading resource bases, weak markets, conflicts, and high climatic risks (Vermeulen et al. 2012). Climate change is expected to further exacerbate the risks that farmers face. The increased global human population, growth, and industrialisation along with the emission of greenhouse gases (GHG) caused by the burning of fossil fuels, deforestation, clearing of land for agriculture, and agricultural intensification are considered to be the main culprits of climate change. According to the IPCC 5th Assessment Report, the earths' surface temperatures have already increased by 0.5 °C in most parts of Africa in the past hundred years. The increase of global mean surface temperature by the end of this century is predicted to be between 2 and 4 °C according to most scenarios (Pachauri et al. 2014), while some parts of Sahel and tropical West Africa could experience surface temperature increases between 3 and 6 °C above the late-twentieth-century baseline (Niang et al. 2014). Temperatures in the African continent are likely to rise more quickly than in other land areas, particularly in more arid regions. While there is a lack of sufficient observational data to draw reliable conclusions about trends in annual precipitation and the projected changes in annual precipitation, most of the current findings show a likely decrease in northern, western, central, and southern Africa. The increase in annual precipitation and wetter climate is projected only for eastern Africa. Furthermore, the increasing frequency of extreme weather events, such as heat waves, flooding, and drought, has also been reported in some areas over the past few decades, and it is expected to increase in the future.

Agriculture, and therefore the livelihoods of those dependent on agriculture, is extremely vulnerable to climate change. Climate change affects food production directly through changes in agroecological conditions and indirectly by affecting growth and distribution of incomes and thus the demand for agricultural produce (Schmidhuber and Tubiello 2007). The impacts of climate change on agriculture are primarily reflected in changes in mean temperatures and precipitation, which subsequently lead to a reduction of crop yields and livestock productivity, the emergence of new pathogens and diseases, crop failure, and long-term production declines. Furthermore, the increased frequency and intensity of extreme weather events such as drought and floods threaten food security, erode the asset base, and jeopardize livelihoods, further entrenching poverty (IPCC 2014). Consequently, there is a dire need to improve agricultural productivity, while also building the resilience and adaptive capacity of those vulnerable populations to cope with increased climate change and variability. This should, however, be seen in a larger context of sustainability. More efforts are needed to create sustainable current food and farming systems to a future food, health, and well-being for all. Because agriculture is so fundamental, we believe that mindfully developing a new agriculture approach (such as climate smart agriculture) and adding a more 'human' dimension could be relevant in a changing climate. Climate-smart agriculture (CSA) is currently one of the advocated approaches to achieve food production in a changing climate in Africa. We briefly describe this approach in the following section.

4 Climate-Smart Agriculture (CSA)

CSA is a relatively recent concept. It was defined by the FAO in 2009 and has been reshaped through inputs and interactions of multiple stakeholders over time (Lipper and Zilberman 2018). The basic principle of the concept is that agriculture must become 'climate smart', meaning that agriculture and food systems must undergo significant transformation in order to meet the challenges of food security and climate change. This is to be accomplished through the achievement of national food security and development goals through three defined objectives (pillars): (i) sustainable increase in productivity, (ii) increased resilience through adaptation of different technologies and practices, and (iii) reducing GHG emissions where possible (FAO 2013). What distinguishes CSA from other sustainable agricultural approaches is a specific focus on climate change and the search for possible synergies and negotiation of trade-offs in the pursuit of increased food security, resilience, adaptation, and mitigation outcomes in a broader landscape or system perspective (Lipper et al. 2014). Given the importance of agriculture in the context of livelihoods, food security, and poverty reduction, CSA has the potential to increase the livelihoods and resilience of millions of smallholder farmers in Africa. However, if CSA is promoted without including the human dimension, this will not address the problem at its source.

In the following sections, we provide the analytical framework for CSA and its linkages with vulnerability, resilience, livelihoods, and transformation and highlight how CSA can impact positively on the transformation of agricultural production. Furthermore, we briefly discuss the main priorities in the way forward towards more productive and resilient smallholder farming systems and discuss the barriers that exist concerning CSA implementation.

Vulnerability, resilience, and adaptive capacity Vulnerability is a broad concept used in different research traditions and encompasses both biophysical and social systems. In the context of climate change, the IPCC defines vulnerability as 'The degree to which a system is susceptible to, or unable to cope with, adverse effects of climate change, including climate variability and extremes. Vulnerability is a function of the character, magnitude, and rate of climate variation to which a system is exposed, its sensitivity, and its adaptive capacity' (IPCC 2001, p. 21). This definition includes an external biophysical dimension, represented through exposure to climate change and variability, as well as an internal social dimension of a system, which comprises its sensitivity and adaptive capacity. When determining the vulnerability of a system (which can refer to either countries, ecosystems, societies, or individuals), one must consider both its ability to recover from disturbance (i.e. resilience) as well as its resistance to being disturbed (i.e. adaptive capacity). At the individual level, personal transformation leading towards better psychological resilience and adaptation is poorly studied among farming populations.

Resilience has roots in ecology and can be described as the capacity of biophysical and social systems to prevent, mitigate, or cope with risk and recover from shocks. The conceptual linkages between vulnerability and resilience are very complex and blurry, as they originate from different social scientific disciplines, but can be applied to both biophysical and social systems. At first approximation, resilience can be seen as an antonym of vulnerability because, to some, vulnerability refers to the capacity to preserve the structure of the system, while resilience refers to its capacity to recover from nonstructural changes in dynamics (Gallopín 2006). Others, however, view resilience as a component of vulnerability or use both terms interchangeably (Gallopín 2006). Without intention to engage further into this epistemological debate, we would like to point out the relations between the concepts of adaptive capacity (as an integral part of vulnerability) and resilience. In the context of climate change, adaptive capacity is defined as 'the ability of a system to adjust to climate change (including climate variability and extremes) to moderate potential damages, to take advantage of opportunities, or to cope with the consequences' (IPCC 2001, p. 21). This implies that resilience is a component of adaptive capacity in the context of mitigation or recovery from shocks, while the ability to adjust and respond to changes provides capacity to maintain resilience. For example, a cropping system that is drought tolerant implies the preparedness of the system to cope and mitigate the damage caused by the extended periods of drought, which is therefore drought-resilient. The adaptive capacity of this system, however, provides the means of increasing and maintaining resilience (e.g. conservation agriculture knowledge and skills via extension, access to drought-resistant crop varieties, or irrigation system). Therefore, systems with high adaptive capacity and resilience are able to reconfigure without significant declines in crucial functions in relation to primary productivity, natural resource-based sustainability, social relations, and well-being during and after stresses and shocks (Nyamwanza 2012).

Changes in climatic conditions will require different adaptation strategies, in terms of both livelihood strategies and adjustments in agricultural production to alleviate the severity of climate change impacts and increase in adaptive capacity. If enhancing the adaptive capacity of smallholders is a starting point of adaptation, it is essential to understand how their livelihoods are composed, accessed, and sustained within the local agroecological and socioeconomic conditions. Furthermore, it is also essential to boost psychological resilience through personal transformation (refer to the next section).

According to Ellis (2000, p. 10), 'A livelihood comprises the assets (natural, physical, human, financial and social capital), the activities, and the access to these (mediated by institutions and social relations that together determine the living gained by the individual or household'. Assets are the basis on which livelihoods are built. They are stocks of capital or the means of production available to a given individual, household, or community that can be used in their livelihood activities (Ellis 2000). Five types of livelihood assets have been described as follows (Carney 1998; DFID 1999): (i) natural capital (e.g. land, water, biological resources), (ii) social capital (networks and connectedness, membership in groups, kinship, trust), (iii) human capital (skills, knowledge, labour, health), (iv) (UN) physical capital (infrastructure and technologies), and (v) financial capital (stocks of cash or equivalent derived from income or savings). Limited access to livelihood assets often shapes poverty and consequently the lack of adaptive capacity. A focus on assets helps to establish what resources are available and accessible to aid in adaptation.

The ability of a livelihood to be able to cope with and recover from stresses and shocks is central to the definition of sustainable livelihoods (Scoones 1998). The range and combination of activities and choices that individuals and households undertake, depending on the range and the degree of utilisation and diversification of their available asset base in order to achieve their livelihood goals, result in different livelihood strategies. Livelihood strategies are composed of activities that generate the means of household survival (Ellis 2000). They are dynamic and respond to changing pressures and opportunities, resulting in adoption and adaptation over time. In the context of smallholders, households with access to agricultural means of production have the choice between both agriculture and non-agricultural economic activities. Different adaptation strategies within agriculture may be also considered, such as extensification versus intensification, as well as 'exit options' such as off- and non-farm activities, migration, and remittance strategies. It should be mentioned, however, that if the extensification approach increases the land area needed to feed the world, then it is not a viable option.

Most rural households depend on a diverse portfolio of activities and income sources, where agricultural production is featured alongside a range of different activities, which, together, contribute to survival and increased well-being (Ellis 2000). Livelihood diversification is pursued for a mixture of motivations, and these vary according to context: a desire to accumulate or to invest, a need to spread risk or maintain incomes, a requirement to adapt to survive in eroding circumstances, or a combination of these (Hussein and Nelson 1998). The core aim is to achieve sustainability of livelihoods, which implies that livelihoods are stable, resilient, resistant, and robust when it comes to external and internal stresses (Scoones 2009). The stronger, more resilient, and more varied the asset base, the greater the people's adaptive capacity and the level of security and sustainability of their future livelihoods (Cooper et al. 2008).

The role of CSA in agricultural adaptation to climate change The promoters of CSA (Lipper and Zilberman 2018) have perceived this as an approach with the goal to develop the technical, policy, and investment conditions to achieve sustainable agricultural development for food security under climate change (FAO 2013). It is meant to integrate the four dimensions of sustainable development (i.e. economic, social, environmental, and institutional) by jointly addressing the food security, ecosystems management, and climate change challenges, both at the farm level and beyond. At the farm level, CSA aims to strengthen livelihoods and food security, especially of smallholders, by improving the management and use of natural resources and adopting appropriate approaches and technologies for the production, processing, and marketing of agricultural commodities. Beyond the farm level, CSA seeks to support countries by putting in place the necessary policy, technical, and financial mechanisms to mainstream climate change adaptation and mitigation into agricultural sectors and provide a basis for operationalizing sustainable agricultural development under changing conditions.

CSA appears not to represent a set of practices that can be universally applied, but is rather an approach that involves different elements embedded in local contexts. CSA relates to actions both on-farm and beyond the farm and incorporates technologies, policies, institutions, and investment. Different elements that can be integrated in CSA approaches include (i) improved management of farms, crops, livestock, aquaculture, and capture fisheries; (ii) produce more with less while increasing resilience; (iii) restoration of degraded lands for productive agriculture and forestry; (iv) ecosystem and landscape management to conserve ecosystem services, which are key to increasing resource efficiency and resilience; and (v) UN services for farmers and land managers to enable them to implement the necessary changes (Goyal and Nash 2017). These aspects only include transformation in the technical sphere, but do not address the need for transformation in the personal sphere (O'Brien and Sygna 2013).

In order to achieve the objectives of CSA to enhance food security while contributing to climate change mitigation, as well as preservation of the natural resource base and vital ecosystem services, a shift from single-objective production systems to the management of the entire landscape is required to meet CSA's multiple objectives. This also requires a transition to agricultural production systems that are more productive, use inputs more efficiently, have less variability and greater stability in their outputs, and are more resilient to risks, shocks, and long-term climate variability. More productive and more resilient agriculture requires a major shift in the way land, water, soil nutrients, and genetic resources are managed to ensure that these resources are used more efficiently (FAO 2013). In recent decades, we witnessed some similar concepts that either have failed or had limited success in African countries. What makes CSA different is that it is defined as a concept built upon a technical foundation that largely already exists and in which a range of sustainable agricultural approaches are the cornerstones of implementing CSA in practice. CSA also suffers from a common problem of externalizing our relationship to our participation in the activity; that is, a tendency to take it without examining its context specificity lies at the root of many of our challenges. It is easy to separate ourselves, our inner attitudes, and our psychological state, from our work. CSA is thus lacking the inclusion of skills for inner resilience, such as appreciation of interdependence, mindfulness, and other psychological aspects of the human dimension that enable transformation in individuals practicing CSA.

Current priority actions for CSA towards the transformation of smallholder farming systems Agriculture is the main source of livelihood for most rural households in Africa. It provides food, income, and jobs and can thus be an engine of economic growth in rural areas and beyond (Dethier and Effenberger 2012). As climate change continues to threaten agriculture and smallholder livelihoods, it is important that actions are taken to reduce risks and capitalize on opportunities. Unlocking the potential of smallholder farmers through CSA can potentially play a vital role in transforming agriculture in African countries.

Smallholder farmers are facing numerous challenges, especially in terms of access to essential factors of production such as financial capital, natural capital (soils, water) inputs (seeds, fertilizers, pesticides), knowledge, technology, infrastructure, and services, and also have to cope with poor market access (Hertel and Rosch 2010). Despite the many challenges they face, smallholders have proven to be champions of adaptation, able to cope and survive in the changing environmental and socioeconomic conditions throughout history. Thus, smallholder farmers can be the agents of change in the transformation of agricultural systems in Africa. In order to drive sustainable agricultural development that builds resilience and increases adaptive capacity, however, they need to be provided with knowledge and incentives to adapt. In this section, we present some of the priorities that need to be addressed by policymakers and other stakeholders in order to facilitate adaptation of climate-smart agricultural practices.

While CSA takes into account the four dimensions of food security, namely, availability, accessibility, utilization, and stability, the increased productivity in agriculture is the main entry point (FAO 2013). In this regard, CSA borrows heavily from the concept of sustainable intensification (Campbell et al. 2014). Sustainable agricultural intensification is defined as producing more output from the same area of land while reducing the negative environmental impacts and at the same time increasing contributions to natural capital and the flow of environmental services (Pretty 2008). One of the key points of increased productivity is the closing of the yield gap (Pradhan et al. 2015). The best yields that can be obtained locally depend on the capacity of farmers to access and use, among other things, seeds, water, nutrients, pest management, soils, biodiversity, and knowledge (Godfray et al. 2010). Low yields occur because of technical constraints that prevent farmers from increasing productivity, or for economic reasons arising from market conditions. Productive and sustainable agricultural systems make the best of both crop varieties and livestock breeds and their agroecological and agronomic management. Improving agricultural productivity is central to reducing poverty and food security. According to Irz et al. (2001), every 10% increase in farm yields leads to a 7% reduction in poverty in Africa. More food produced on farms enhances food security and may lead to a transition from subsistence to commercial agriculture, given that the functioning markets and infrastructure are in place. Higher incomes from sales of agricultural products may lead to more investments in on-farm livelihood assets and other sectors beside agriculture. An increasing number of studies demonstrate the potential of adopted CSA and sustainable agricultural practices for improving food security, adapting to climate change, and/or mitigating GHG emissions. Practices such as crop improvements, agroforestry and soil conservation, conservation agriculture (CA), integrated soil fertility management (ISFM), integrated pest management (IPM), horticulture, livestock and fodder crops, aquaculture, and novel policies and partnerships are some of the emerging climate-smart options for increased farm productivity, rural livelihoods, and adaptive capacity of farmers and production systems in African countries (Pretty et al. 2011).

Enabling policies are crucial for the adoption of CSA practices by smallholder farmers in Africa Policy decisions have the potential to impact smallholder farmers, and policy instruments can be used to subsidize agriculture and enable access to essential inputs, which in turn can impact food production, food security, and livelihoods (Pretty 2008). Policies to enable adoption of CSA should focus on equitable mechanisms for transferring resources and should not create barriers to adaptations to the new systems and transformations towards more ethical practices. Indeed, it may not be necessary to create new policies but to prudently communicate the basic principles and targets/benefits of CSA. A strong framework and linkages between all players must be in place to ensure guidance of local CSA initiatives (Ampaire et al. 2015). Policy interventions to enable the adoption of CSA should (i) provide access to information, (ii) take into account the fact that smallholder farmers have limited assets and capacity for risk-taking, and (iii) provide interventions that can assist risk management and productivity without increasing inappropriate land-use and environmental degradation. Furthermore, if policies are to be successful in enabling the adoption of CSA, the policy formulation process, which in Africa is to a large extent unidirectional and top-down, must expand to include other organizations including cooperative movements, the private sector, and the opinions of smallholder farmers themselves (Ampaire et al. 2015). In addition, the role of women in African agriculture must not be ignored, and policy formation should address gender imbalances, which could help to decrease poverty and food insecurity. One should also study and reflect on the competence of scientists and farmers and the policymakers' perceptions and attitudes regarding the incorporation of CSA into various contexts based on the need for fulfilment not just achievement of crop yield.

More research at the national, regional, and local levels is required in many cases, given the context-specific nature of the CSA approach and the diversity of farming systems, political and socioeconomic conditions, and agroecological zones in Africa. The current evidence base is inadequate to support effective decision-making at the national and local levels, which impedes the ability of decision makers to identify what is climate smart in the given biophysical and socioeconomic context (Lipper et al. 2014). The barriers to adaptation in changing climatic conditions and the means to overcome them are largely unknown. The barrier may be at the personal sphere of individuals (belief, values, worldviews, and paradigm. Refer to O'Brien and Sygna 2013), which requires personal transformation. Furthermore, what is climate smart in one place may not be that in another. Thus, transdisciplinary research at the landscape level and the assessment of synergies and co-benefits, as well as the development of metrics to quantify the benefits and trade-offs of different options based on stakeholder objectives, are needed to derive locally acceptable and feasible solutions.

Education and knowledge transfer of CSA practices is needed for African farmers to facilitate new mitigation and adaptation agendas to positively impact sustainable intensification. Extension services have played a key role in promoting agricultural productivity and dissemination of knowledge, and their role in promoting adaptation measures to climate change will certainly have the same importance (Bryan et al. 2009; Maponya and Mpandeli 2013). It is necessary to strengthen or enhance the capacity of both public and private agricultural extension services, enabling them to take leading roles in terms of strengthening innovation process, building linkages between farmers and other agencies, and institutional development. As leaders in knowledge generation, research institutes and agricultural universities have an important role to provide means and tools for reflective practices such as first person inquiry by educating mindful, context responsive, and economically viable agricultural practices among to students, extension workers, and policymakers. This can be accomplished on different ways, such as through the use of demonstration plots, joint research projects, field trials, training seminars, etc. (Asopa and Beye 1997), and should include aspects of transformative learning and personal transformation (described in more detail in the following sections).

It is evident from the discussion above that much more work is needed in order to achieve sustainable livelihoods and resilient communities and individuals in Africa, both in technical and political contexts, and including aspects of personal transformation. One of the greatest challenges is to mobilize sufficient technical support and financial resources for the necessary actions at the local level and to ensure that resources reach the most vulnerable target groups. While the estimated costs of adaptation and mitigation in Africa and worldwide largely exceed the funds available, CSA opens up new funding opportunities for agricultural development by allowing the sector access to climate funds to enable adaptation and mitigation in agriculture. A number of strategies, policies, partnerships, and investments have been initiated in recent years to put the CSA concept into practice, such as the Adaptation Fund, Least Developed Countries Fund, Special Climate Fund, Clean Development Mechanism, Global Environment Facility Trust Fund, and Future Green Climate Fund. These initiatives may address problem on the proximal level, but do not address the problem of human desire and greed.

Farmers are faced with many barriers to adaptation, including but not limited to financial, sociocultural, institutional, and technological challenges. They are also constrained by the lack of weather and climate data. There is a serious need to determine the drivers and barriers to sustained adoption of CSA practices in order to understand the local context within which adaptations take place and to thoroughly evaluate the suitability of target practices and how their widespread adoption might best be facilitated. There is also a need to provide supportive incentives and policy reforms for farmers, such as access to modern inputs and favourable credit terms, investments in infrastructure, tenure security, and functional markets.

Recent partial success stories in CSA implementation show its potential to help African countries to increase agricultural production and incomes, adapt to climate change, build resilience, and reduce greenhouse gas (GHG) emissions. African policymakers and stakeholders have to show commitment by mainstreaming CSA into national development programmes and policies that support sustainable agricultural development, to decrease food insecurity and poverty. Public institutions and their strategic partners, most notably the private sector and agro-advisory service providers, need to work together to scale up CSA practices, disseminate the knowledge, and stimulate farmers to invest in CSA practices and technologies. As stated above, a transformative approach is required for this, consisting of a 'powerful unleashing of human potentials to commit, care and affect change for better life' (O'Brien 2012), including the personal transformation of agricultural professionals such as farmers. In the next section, we describe the key concepts of personal transformation in an educational context.

Primarily, as far as upholding sustainable agricultural practices are concerned, intervention is required at the individual level of agricultural professionals, as sustainable practices will be initiated and carried out by human agents. With appropriate training and educational processes, humans have the potential to change themselves and their agricultural practices. Such change processes take longer periods of time to come about due to the required shift in personal and collective consciousness. For such changes to occur, a carefully developed educational strategy for personal transformation is required that will eventually permeate farmers' behaviour. The next section explains such strategies in an educational/adult learning context.

5 Personal Transformation: Key Concepts and Ideas

From an educational perspective, transformation refers to a change or shift of meaning on dominant ideas, beliefs, and assumptions. A critical approach developed in the mind of individuals may help ideas to be viewed differently. Thus, by becoming critical and reflective, an individual can bring change within the self and others. Transformation takes place within the self after the individual views the ideas from different perspectives. In other words, we cannot transform ourselves by viewing a belief or assumption in a 'taken-for-granted' manner. Being aware of one's own context and the resulting development of consciousness brings about change within the individual.

The concept of personal transformation concerns a shift or change in the thoughts and associated behaviours of individuals. It pertains to the self-awareness that takes place after gaining knowledge through specific learning procedures such as critiquing, imagining, envisioning, and reflecting. Education promotes and enhances awareness for personal development and thus personal transformation. Moreover, it enhances the level of consciousness of individuals (Natanasabapathy et al. 2011). In other words, personal transformation is a 'multidimensional concept, a dynamic, uniquely individualized process of expanding consciousness whereby an individual becomes critically aware of old and new self-views and chooses to integrate these views into a new self-definition' (Wade 1998, as cited in Devine and Sparks 2014, p. 713). Thus, personal transformation is a shift of visions and thoughts (Miller et al. 2005) in individuals.

An individual undergoes personal transformation through critical vision (Taylor and Cranton 2012). Personal transformation is associated with transformative learning, and behavioural change takes place when an individual goes through an education system and acquires knowledge. 'Changes in behaviour depend on how well learning takes place and what and how we give value to the knowledge and what we become' (Shapiro 2005, as cited in Natanasabapathy et al. 2011). Thus, personal transformation is concerned with 'becoming'. The process of 'becoming' is a lifelong activity that involves reflective thinking and transformation of perspectives. Personal transformation concerns ethical actions, thereby developing empathic attitude towards the land, other species, and people. Here it is argued that the transformative learning is a means of personal transformation.

5.1 Transformative Learning: A Mean of Personal Transformation

Transformative learning theory provides understanding of the processes of adult learning. It concerns adaptive learning in the sociocultural context according to the needs and aspiration of the learners (Mezirow 1990). Moreover, transformative learning encourages participatory and experiential learning approaches. The learners enhance their knowledge, skills, and abilities through an interactive and experiential basis, involving themselves in the learning processes. Transformative educators do not adopt traditional pedagogical processes of depositing content through lectures and considering the learners' mind as 'banks' (Freire 2005). Instead, they participate in the learning activities, guiding the learners towards their learning needs/goals. In doing so, transformative educators gain knowledge and skills through reflection, enabling self-transformation.

With regard to transformative learning, Freire (2005) has coined the term 'conscientization' or *consciousness raising* (Dirkx 1998). Frere's adult education focuses on raising critical awareness or consciousness among adults. His notion of transformative learning is to liberate oppressed individuals and groups (e.g. farmers) through developing ethical actions. Freire (2005) believes that education for adults should develop the ability to analyse, raise questions, and take ethical actions on social, political, cultural, and economic issues that are directly related to shaping their lives (Dirkx 1998). This learning process enables adults to move along the path of personal transformation. Likewise, transformative learning develops the ability of learners to view their world through dialectical ways of actions and reflections.

Transformative learners are able to think reflectively and foster freedom from suppression and oppression.

According to Freire (2005), learners should develop critical views against existing power dynamics and hierarchies (such as between large companies and farmers). They need to understand how power structures contribute to suppress or oppress them. The ability to contest and challenge existing ideas or thoughts that lead to suppressive practices in society (such as Western countries exploiting natural and human resources in Africa) can result in transformation. Freire (2005) does not believe in the concept of learners accumulating expertise in a 'knowledge bank' through the traditional pedagogical approach; rather, he believes in the development of critical thinking and enhancing consciousness for ethical actions.

5.2 *Path of Transformation*

The transformative learning theory focuses on personal change. This change/shift can be emotionally challenging for individuals maintaining a continuous pursuit to achieve a sustainable outcome. Transformation occurs in different stages. An individual moves from a stage of ignorance to one of knowing and then to a stage of awareness (Natanasabapathy et al. 2011). She/he gains self-awareness with increasing knowledge. The increased awareness through acquisition of knowledge leads to shared vision. Transformative learning does not exist between the stages of 'ignorance' and 'aware'; the 'aware' stage is thus the transitional stage in the path of transformation (Fig. 3). The stage of 'conscious' is the stage in which transformed action is applied as a conscious learner, demonstrating that transformative learning (which involves reflection, connection, and realization) takes place (Fig. 3).

The process of transformation becomes visible when one's understanding is applied in action. Thus, the path of transformation is not merely knowing the world or 'becoming'; rather, it is to be applied to the subject's behaviour. Transformation includes the behavioural change of learners in a meaningful manner. This concept

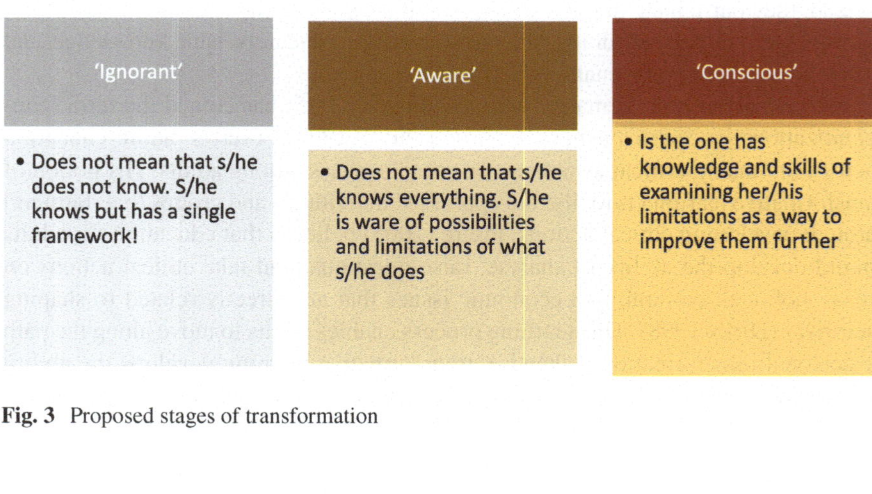

Fig. 3 Proposed stages of transformation

of 'being' and then 'doing' brings change or a shift at the individual level. Therefore, transformation focuses on a continuous process of reflecting, connecting acting in all stages.

5.3 Transformation Through Holistic Learning

Holistic learning is an alternative way of thinking or perspective regarding the education of the whole person rather than imparting specific aspects of knowledge, to prepare learners to compete in the global market. It attempts to develop the whole person, including 'intellectual, emotional, physical, social, aesthetic, and spiritual' elements (Miller et al. 2005). Holistic learning theory is defined as knowledge being a social construct with three different facets: explicit, implicit, and emancipatory knowledge (Yang 2004).

According to Yang (2004), knowledge is viewed as the understanding of one's realities through mental correspondence, personal experiences, and emotional affection towards the outside world. He argues that explicit knowledge is the cognitive aspect that represents the understanding of realities. Moreover, he believes that implicit knowledge is not openly expressed or articulated and is personal and context specific. It is hard to formalize and communicate as it exists in one's behaviour, actions, and experiences. Yang (2004) further views that emancipatory knowledge is an affective reaction to the outside world. It is an understanding based on emotional affection and defines the point of view of the world outside.

These three facets of knowledge manifest during the learning process and are intrinsically different, though complementary to each other. The three knowledge facets interact and are therefore dynamic, thereby creating a holistic perspective and holistic human knowledge. Holistic learning theory believes not only in these three knowledge facets but also three knowledge layers: foundation, manifestation, and orientation.

Holistic learning theory believes that knowledge is created and transformed after the interaction of three different facets of knowledge. These relations are reflected in seven modes of learning: participation, conceptualization, contextualization, systematization, validation, interpretation, and transformation (Fig. 4).

According to Yang (2004), participation is the process in which the learners participate in the learning process and generate implicit knowledge. Conceptualization is the process of articulating implicit knowledge into explicit forms. Contextualization is the process of using explicit knowledge at a practical level. The process of systematization helps to incorporate explicit knowledge in developing logic and reasoning. The validation process enables us to examine and modify existing values, beliefs, and assumptions. Legitimization is the process of justifying explicit knowledge based on emancipatory knowledge. According to Yang (2004), transformation is the process of converting 'old' meanings of attitudes, beliefs, values, feelings, aspirations, and ethics into another form. Materialization is the process of transforming emancipatory knowledge into tacit/implicit knowledge. Interpretation is the process of garnering meaning from tacit learning.

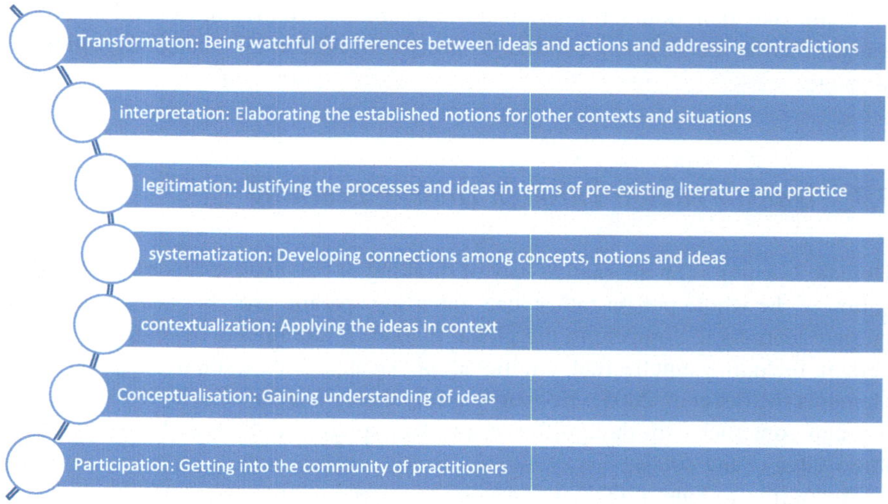

Fig. 4 Synthesized seven modes of holistic learning

5.4 Transformation Through a Multilogic Framework

In the literature of transformation, a host of logics has been discussed as ways of incorporating transformation into the natural and social sciences. A logic framework illustrates the complexity of our worldviews. Three forms of logic are defined, namely, poetic, metaphorical, and dialectical logics, with a focus on how they may facilitate the personal transformation of professionals working in the field of natural and social science.

Creative arts such as poetry can bring positive changes to an individual's worldviews. Writing poetry may help to bring about meaningful change in the mindset of the individual. In other words, writing poems serves as catalyst in one's journey of revisioning the self and the way one chooses to be in this world (Kreuter and Reiter 2014). Furthermore, creative poetic writing may enable us to transcend our limitations of thinking. It encourages the imagination, fuels our dreams, and plants seeds of hope that are vital for our renewal. Thus, poetic logic is an integral part of personal transformation.

5.5 Transformation Through Wisdom Traditions

Wisdom traditions of the East, West, North, and South have preserved various perspectives of the transformation of individuals and communities. Be it the meditation practices of the East, the poetic ruminations of the Middle East, 'we-ness' wisdoms of Africa, or contemplative traditions of the West, there are immense potentialities for transforming our hearts, hands, and minds in the pursuit of greater good for humanity. These wisdom traditions include those arising from major world religions and the vernacular traditions of Asia, Africa, and the Americas.

Wisdom traditions encourage individual transformation from conditions and conditionalities, advocating a deeper connection with ourselves, with each other, and with the world around us. Wisdom encourages this larger view of life and emphasizes the interconnectedness of all things and beings. 'The value of meditation or contemplative prayer is undeniable and it creates greater acceptance and spaciousness in our relationship to ourselves and the people and situations we encounter' (Bower 2014, p. 7). Meditation fosters the helpful ability to pause in our response to the world, enabling us to connect with our highest aspirations and make wiser, more life-affirming choices.

6 Summary and Conclusions

We conclude that personal transformation is required to address ongoing agriculture and environmental challenges in a changing world. There is a need to include knowledge and skill development activities that enable personal transformation in various arenas of development, including agriculture. Personal transformation is necessary in the current state of affairs, where human wants and greed are becoming an underlying planetary force, shaping the planet's future, including the long-term future of agriculture. Such deeply rooted psychological human dimensions are inadequately linked to the agricultural and environment-related problems that threaten sustainability. At this juncture of time, what is required most is personal transformation, integrated in all types of education including agricultural development.

Agricultural education and personal transformation should be combined to promote food security, peace-building, and harmony and to balance development with sustainability.

A number of strategies, policies, partnerships, and investments have been initiated in recent years to put CSA into practice. Despite the few successes of select initiatives in agriculture, much more work is needed to achieve sustainable livelihoods and resilient communities in Africa, for individuals and in technical and political contexts.

Our definition of transformation is a 'powerful unleashing of human potentials to commit, care and affect change for better life' (Sharma 2007; O'Brien 2012) and implies the same in the context of environmental change and agriculture. In addition, transformation implies change in the way we perceive the world and the systems we have created. There is a need for transformation in ongoing practices, for example, in the practice of transforming research into policy and actions. However, without unleashing the human potential to commit, care, and affect change for a better life (in agriculture or otherwise), any measure taken will be less affective in implementing environmental and agriculture-related programmes. Wisdom traditions encourage individual and societal transformation, advocating a deeper connection with ourselves, with each other, and with the world around us.

References

Altieri MA, Funes-Monzote FR, Petersen P (2012) Agroecologically efficient agricultural systems for smallholder farmers: contributions to food sovereignty. Agron Sustain Dev 32(1):1–13

Ampaire E, Happy P, Van Asten P, Radeny M (2015) The role of policy in facilitating adoption of climate-smart agriculture in Uganda. CGIAR Research Program on Climate Change, Agriculture and Food Security (CCAFS), Copenhagen

Asopa VN, Beye G (1997) Management of agricultural research: a training manual. Food & Agriculture Organisation of the United Nations, Rome

Bower P (2014) Leadership and the wisdom traditions. http://www.garyhubbellconsulting.com/media/40600/leadership_and_wisdom.pdf. Accessed 19 Dec 2018

Bryan E, Deressa TT, Gbetibouo GA, Ringler C (2009) Adaptation to climate change in Ethiopia and South Africa: options and constraints. Environ Sci Pol 12(4):413–426. https://doi.org/10.1016/j.envsci.2008.11.002

Campbell BM, Thornton P, Zougmoré R, van Asten P, Lipper L (2014) Sustainable intensification: what is its role in climate smart agriculture? Curr Opin Environ Sustain 8:39–43. https://doi.org/10.1016/j.cosust.2014.07.002

Carney D (1998) Sustainable rural livelihoods: what contribution can we make? Department for International Development, London

Cohen S (2013) States of denial: knowing about atrocities and suffering. Polity Press, Malden

Cooper PJM, Dimes J, Rao KPC, Shapiro B, Shiferaw B, Twomlow S (2008) Coping better with current climatic variability in the rain-fed farming systems of sub-Saharan Africa: an essential first step in adapting to future climate change? Agric Ecosyst Environ 126(1):24–35. https://doi.org/10.1016/j.agee.2008.01.007

Dethier J-J, Effenberger A (2012) Agriculture and development: a brief review of the literature. Econ Syst 36(2):175–205. https://doi.org/10.1016/j.ecosys.2011.09.003

Devine C, Sparks W (2014) Defining moments: toward a comprehensive theory of personal transformation. Int J Humanit Soc Sci 4(5.1):31–39

DFID (1999) Sustainable livelihoods guidance sheets. Department for International Development, London

Dirkx JM (1998) Transformative learning theory in the practice of adult education: an overview. PAACE J Lifelong Learn 7:1–14

Dunlap RE, McCright AM (2010) Climate change denial: sources, actors and strategies. In: Lever-Tracy C (ed) Routledge handbook of climate change and society. Routledge, Abingdon, pp 240–259

Ellis F (2000) Rural livelihoods and diversity in developing countries. Oxford University Press, Oxford

FAO (2013) Climate smart agriculture sourcebook. Food and Agriculture Organisation of the United Nations, Rome

FAO (2017) The future of food and agriculture – trends and challenges. Food and Agriculture Organisation of the United Nations, Rome

FAO, IFAD, & WFP (2015) The state of food insecurity in the world 2015. Meeting the 2015 international hunger targets: taking stock of uneven progress. Food and Agriculture Organisation of the United Nations, Rome

FAO, IFAD, UNICEF, WFP & WHO (2018) The state of food security and nutrition in the world 2018: building climate resilience for food security and nutritio. Food and Agriculture Organisation of the United Nations, Rome

Freire P (2005) Pedagogy of the oppressed. Bloomsbury Publishing, New York

Gallopín GC (2006) Linkages between vulnerability, resilience, and adaptive capacity. Glob Environ Chang 16(3):293–303. https://doi.org/10.1016/j.gloenvcha.2006.02.004

Gisladottir G, Stocking M (2005) Land degradation control and its global environmental benefits. Land Degrad Dev 16(2):99–112

Godfray HCJ, Beddington JR, Crute IR, Haddad L, Lawrence D, Muir JF et al (2010) Food security: the challenge of feeding 9 billion people. Science 327:812. https://doi.org/10.1126/science.1185383
Goyal A, Nash J (2017) Reaping richer returns: public spending priorities for African agriculture productivity growth. The World Bank, Washington, DC
Hertel TW, Rosch SD (2010) Climate change, agriculture and poverty. The World Bank, Washington, DC
Hussein K, Nelson J (1998) Sustainable livelihoods and livelihood diversification. Institute of Development Studies, Brighton
IPCC (2001) Climate change 2001: impacts, adaptation, and vulnerability: contribution of Working Group II to the third assessment report of the Intergovernmental Panel on Climate Change. Cambridge University Press, Cambridge, NY
IPCC (2007) In: Core Writing Team, Pachauri RK, Reisinger A (eds) Climate change 2007: synthesis report. Contribution of Working Groups I, II and III to the fourth assessment report of the Intergovernmental Panel on Climate Change. IPCC, Geneva, p 104
IPCC (2012) Glossary of terms. In: Field CB, Barros V, Stocker TF, Dahe Q (eds) Managing the risks of extreme events and disasters to advance climate change adaptation: special report of the intergovernmental panel on climate change. Cambridge University Press, New York, p 564
IPCC (2014) Climate change 2014: synthesis report (Contribution of Working Groups I, II and III to the fifth assessment report of the Intergovernmental Panel on Climate Change). IPCC, Geneva
Irz X, Lin L, Thirtle C, Wiggins S (2001) Agricultural productivity growth and poverty alleviation. Dev Policy Rev *19*(4):449–466. https://doi.org/10.1111/1467-7679.00144
Jasanoff S (2007) Technologies of humility. Nature 450(7166):33
Jones PG, Thornton PK (2009) Croppers to livestock keepers: livelihood transitions to 2050 in Africa due to climate change. Environ Sci Pol 12(4):427–437
Kabat-Zinn J (2009) Wherever you go, there you are: mindfulness meditation in everyday life. Piatkus Books, London
Kreuter EA, Reiter S (2014) Building resilience in the face of loss and irrelevance: poetic methods for personal and professional transformation. J Poet Ther 27(1):13–24
Lipper L, Zilberman D (2018) A short history of the evolution of the climate smart agriculture approach and its links to climate change and sustainable agriculture debates. In: Lipper L, McCarthy N, Zilberman D, Asfaw S, Branca G (eds) Climate smart agriculture : building resilience to climate change. Springer, Cham, pp 13–30
Lipper L, Thornton P, Campbell BM, Baedeker T, Braimoh A, Bwalya M et al (2014) Climate-smart agriculture for food security. [Perspective]. Nat Clim Chang 4:1068. https://doi.org/10.1038/nclimate2437.
Magdoff F, Tokar B (2010) Agriculture and food in crisis: conflict, resistance, and renewal. Monthly Review Press, New York
Maponya P, Mpandeli S (2013) The role of extension services in climate change adaptation in Limpopo province, South Africa. J Agric Ext Rural Dev 5(7):137–142
Mezirow J (1990) Fostering critical reflection in adulthood: a guide to transformational and emancipatory practice. Jossey-Bass, San Francisco
Miller JP, Karsten S, Denton D, Orr D, Kates IC (2005) Holistic learning and spirituality in education: breaking new ground. SUNY Press, New York
Natanasabapathy P, Bourke J, Joshi R (2011) Exploring transformative learning in the open and distance learning environment. In: Philosophy of Education Society of Australasia conference
Niang I, Ruppel OC, Abdrabo MA, Essel A, Lennard C, Padgham J, Urquhart P (2014) Africa (climate change 2014: impacts, adaptation, and vulnerability. Part B: regional aspects. Contribution of Working Group II to the fifth assessment report of the Intergovernmental Panel on Climate Change). Cambridge University Press, Cambridge, NY
Norgaard KM (2009) Cognitive and behavioral challenges in responding to climate change: the World Bank, Policy Research Working Paper 4940

Nyamwanza AM (2012) Livelihood resilience and adaptive capacity: a critical conceptual review. Jàmbá J Disaster Risk Stud 4(1):1–6
O'Brien K (2012) Global environmental change II: from adaptation to deliberate transformation. Prog Hum Geogr 36(5):667–676
O'Brien K, Sygna L (2013) Responding to climate change: the three spheres of transformation. In: Proceedings of transformation in a changing climate. University of Oslo, Oslo, pp 19–21
Pachauri RK, Allen MR, Barros VR, Broome J, Cramer W, Christ R et al (2014) Climate change 2014: synthesis report. Contribution of Working Groups I, II and III to the fifth assessment report of the Intergovernmental Panel on Climate Change. IPCC, Geneva
Palacios-Lopez A, Christiaensen L, Kilic T (2015) How much of the labor in African agriculture is provided by women? The World Bank, Washington, DC
Pradhan P, Fischer G, van Velthuizen H, Reusser DE, Kropp JP (2015) Closing yield gaps: how sustainable can we be? PLoS One 10(6):e0129487. https://doi.org/10.1371/journal.pone.0129487
Pretty J (2008) Agricultural sustainability: concepts, principles and evidence. Philos Trans R Soc London B Biol Sci 363(1491):447–465
Pretty J, Toulmin C, Williams S (2011) Sustainable intensification in African agriculture. Int J Agric Sustain 9(1):5–24. https://doi.org/10.3763/ijas.2010.0583
Schmidhuber J, Tubiello FN (2007) Global food security under climate change. Proc Natl Acad Sci 104(50):19703–19708. https://doi.org/10.1073/pnas.0701976104
Scoones I (1998) Sustainable rural livelihoods: a framework for analysis (IDS Working Paper 72). Brighton
Scoones I (2009) Livelihoods perspectives and rural development. J Peasant Stud 36(1):171–196. https://doi.org/10.1080/03066150902820503
Sharma M (2007) Personal to planetary transformation. Kosmos Fall/Winter:31–35
Taylor EW, Cranton P (2012) The handbook of transformative learning: theory, research, and practice. Wiley, San Francisco
UN, U. N., Department of Economic and Social Affairs, Population Division (2017) World population prospects: the 2017 revision, key findings and advance tables (Working Paper No. ESA/P/WP/248)
UN, U. N., Department of Economic and Social Affairs, Population Division (2018) World population prospects 2017. http://population.un.org/wpp/. Accessed 21 Dec 2018
Vermeulen SJ, Aggarwal PK, Ainslie A, Angelone C, Campbell BM, Challinor AJ et al (2012) Options for support to agriculture and food security under climate change. Environ Sci Pol 15(1):136–144. https://doi.org/10.1016/j.envsci.2011.09.003.
Wiggins S (2009) Can the smallholder model deliver poverty reduction and food security. Future Agricultures Consortium, Brighton
Yang B (2004) Holistic learning theory and implications for human resource development. Adv Dev Hum Resour 6(2):241–262
Young M (2009) A meta model of change. J Organ Chang Manag 22(5):524–548

Part V
Value Added Options for Smallholder Market Access and Integration

Between the Sun and Fish Are People: A Socio-economic Study of Solar Dryers for Fish Processing in Malawi

Fundi Wandisunga Kayamba-Phiri, Gry Synnevåg, and Moses M. Limuwa

Abstract The study examined how the use of solar tent dryers has improved the livelihoods of fish processors in Chipala and Vinthenga villages in Nkhotakota, Malawi. The study used the sustainable livelihoods framework to analyse the findings. A mixed methods research design was employed in the study. To analyse the qualitative data, a thematic analysis was used, and for the quantitative data, inferential and descriptive statistics were used such as logistic regression model and t-tests. The findings showed that the solar tent dryers have improved fish processing; however, the impact of the dryers on livelihoods is minimal and not well accounted for. Some key sustainability measures were lacking and therefore posed a threat to the continued use of the method. As a result, the dryer was not in use in Chipala which indicated no impact on livelihoods. The logistic regression model indicated the village of a fish processor as a determining factor for using the dryer, due to the difference in governance structures, which affected the management of the tent dryers. Adoption was directly affected by poor governance: a top-down approach employed by different stakeholders. Adoption increased in Vinthenga as the fish processors appreciated the solar drying method as faster and less involving. However, small holding capacity of the dryer in Vinthenga resulted in an average usage of 3 times a week (n, 19). Profits from solar-dried fish were slightly insignificant at 20% during peak season (P) and 11% during off-peak season (O), as compared to the traditional methods of which the most profitable were the frying (25% (P) and 21% (O)) and smoking (20% (P) and 19% (O)) methods. Women were involved in all activities in the fisheries' value chain except for catching fish, which is restricted to men. Gender roles and perceptions affected the socio-economic status of fish processors, as gender equality was contextualised as a monetary responsibility shift to women who were involved in fish processing and other enterprise. All factors considered, the impact of solar tent dryers has been low on the livelihoods of fish processors.

F. W. Kayamba-Phiri (✉) · G. Synnevåg · M. M. Limuwa
Norwegian University of Life Sciences, Faculty of Landscape and Society, Department of International Environment and Development Studies (Noragric), Ås, Norway
e-mail: kayamba-phirif@ei-malawi.org

B. R. Singh et al. (eds.), *Climate Impacts on Agricultural and Natural Resource Sustainability in Africa*, https://doi.org/10.1007/978-3-030-37537-9_22

1 Introduction

Lake Malawi is home to the largest number of fish species in a fresh water body, in the world, with an estimated around 1000 fish species, most of which are indigenous to Lake Malawi (MBERU 2001). The lake has attracted tourists, but most importantly, its fish has been a major source of protein for most of the population in Malawi, as fish has been the cheapest source of protein in the country (Matiya et al. 2005). However, while the lake attracts a lot of glory, fish populations are dwindling (USAID 2015). Furthermore, there are post-harvest losses incurred. These changes have consequences for the country's fish industry, but also for those who mainly depend on fish for their livelihood: the fisherfolk. Thus, the livelihoods of the fisherfolk form the basis of this investigation, more especially fish processors who are involved in all post-harvest activities.

This study analyses the contributions of a research project titled 'Sustainable Environment and Enterprise Development for Climate Change Adaptation in Fisheries (SEED-Fish)' towards improved livelihoods. Funded by the Capacity Building for Managing Climate Change (CABMACC), SEED-Fish was implemented in Linga Extension Planning Area (EPA) in Traditional Authority (TA) Malengachanzi, in Nkhotakota district in Malawi. The project was managed and coordinated by the Department of Fisheries and Aquatic Science at Mzuzu University (Mzuni) in partnership with Lilongwe University of Agriculture and Natural Resources (LUANAR, Bunda College), the Malawi Department of Fisheries Research Unit at Monkey Bay in Mangochi district (FRU) and WorldFish – Malawi Centre in Zomba district.

The project was implemented with the main objective to improve incomes and livelihoods as well as environmental management of fisherfolk communities, to enhance their resilience to the effects of climate change. The project has three specific objectives:

(i) To evaluate, test and adopt the use of clean energy and sustainable processing technologies on small fishes
(ii) To develop and test small-scale fisherfolk entrepreneurial model
(iii) To develop tools or models for building capacity and governance for improved income and environmental conservation

During implementation of the SEED-Fish project, two solar tent dryers were constructed in Chipala and Vinthenga Beach Village Committees (BVC) areas in 2015 and 2016 for the implementation of the SEED-Fish project. The reports, since installation of the dryers, show immense appreciation of the technology by the fish processors (FPs) such that there is need for more solar tent dryers as demand has overstretched supply (SEED-Fish 2016) in one of the villages. Following this development, the project intended to extend the holding capacity of the solar tent dryer at Vinthenga Beach Village Committee (BVC) to include more shelving for drying the fish.

Fig. 1 Solar tent dryer constructed at Chipala in 2016. (Source: Fundi Kayamba-Phiri)

With one-fifth of Malawi comprising of lake bodies, fishing is a widespread trade around the surrounding, mainland areas. Fish are sold fresh or dried, but especially dried because only 10% of the Malawian population have access to electricity (Kambewa et al. 2007) for storage. Dried fish is thus one of the most common and cheaply available fish on the market. Some of the processing methods require firewood to smoke or dry the fish, and these have contributed to deforestation around fishing areas. The SEED-Fish project under CABMACC introduced improved fish processing solar tent dryers and improved smoking kilns to serve as a climate change adaptation strategy by reducing deforestation that results from processing fish using firewood (LUANAR 2013). However, it must be noted that innovative technologies, such as solar tent dryers, are expensive and require strategic consideration on how the communities can sustainably maintain and finance the introduced technologies.

Therefore, this study examines how the usage of solar tent dryers (Fig. 1) improved the livelihoods of fish processors in Chipala and Vinthenga. In this regard, the study addresses the question: how does the usage of solar tent dryers contribute towards building sustainable livelihoods for FPs? Specifically, the paper assesses the sustainability and adoption of using solar tent drying method. The research study uses a mixed method of research. Specific data collection methods included focus group discussions (FGDs), key informant interviews, participant observation, and household survey using a questionnaire administered to a hundred respondents.

2 Conceptualising the Sustainable Livelihoods Approach

A livelihood comprises capabilities, tangible and intangible assets and activities required as a means of living. These components of a livelihood jointly determine the living gained by individuals or households (Ellis 1999). A livelihood is sustainable when it can cope with and recover from stresses and shocks and maintain or enhance its capabilities and assets, both now and in the future, while not undermining the natural resource base (DFID 1999). Further, sustainability here is defined as a requirement that the use of resources today does not reduce real incomes in the future. To this end, little research has been done on livelihood sustainability amongst fish processors (FPs) in Nkhotakota (Allison and Mvula 2002). The study therefore tries to understand how the introduction of innovative processing methods affects the livelihoods of FPs, that is, both tangible and intangible assets and resources.

The sustainable livelihoods approach (SLA) is founded upon the notion that intervention must be based upon an understanding of what underpins livelihoods and on what is known as intentional development (Morse and McNamara 2013). Cowen and Shenton (1998) categorise development into two basic forms: immanent and intentional development. Immanent development denotes a broad process of advancement of human societies that is driven by factors that include advances in science, medicine, the arts, communication, etc. Immanent development is thus a long-term and continuous process for governments with investments in infrastructure, health and education. However, immanent and intentional development take place simultaneously, with the former providing the basis for areas where intentional development is needed.

Morse and McNamara (2013) argue that intentional development has not been successful. There are several reasons for this argument. First, intentional development is based on a construct of what is and what is not developed and thus also what development means. Secondly, due to the construct of what is developed, a top-down approach to development is applied, with rich countries setting the development agenda and determining how it should be implemented. This is in turn viewed as a reconfiguration of colonialism, which has also made the beneficiaries of the interventions dependent on aid and thereby creating a lack of agency to become self-sufficient (Moyo 2010). It is therefore necessary to examine the working of the technologies that are on offer in order to determine whether the intervention process will result in sustainable livelihoods for the poor.

Poverty has been defined as multidimensional, because of the inability of one dimension alone, income, for instance, to reflect the living situation of an individual. Other measures, such as access to basic needs, education, health facilities as well as assets like housing and livestock, are included to determine if one is poor or not (Alkire et al. 2015). The poor need to survive every day as well as accumulate assets to become resilient to shocks and long-term stresses. The framework model is therefore divided into people, basic needs, resources, assets, shocks and trends, access and control.

Vulnerability context, thus shocks and trends, is not always negative and may include the introduction of innovative technologies, giving an opportunity to assess whether the technologies are being utilised by the targeted group and their impact on the socio-economic status of the FPs. The study assesses the FPs' sources of income, to determine what other shocks may affect their flow of income. The vulnerability context thus draws attention to the complexity of factors that contribute directly or indirectly towards the overall vulnerability of the poor. Given this complexity, the poor may not always be able to change or manipulate their situation, because of lack of assets or structures and procedures to adapt to their situation.

The sustainable livelihoods framework (SLF) model divides livelihood assets into five groups: human, natural, financial, physical and social capital. Within the framework and this study, there is a close link between the human and social capital and the vulnerability context, particularly shocks. This is because some of the shocks are the result of systems of natural resource management in a community. The perceptions and behaviours towards the natural resources, especially woodfuel, are assessed in the study. Under financial capital, the income levels of the FPs are assessed. Access to microfinancing institutions or other alternatives such as village banks gives insight into how the technologies have improved their income, as well as how much the FPs can contribute towards maintaining and upscaling construction of solar tent dryers. The physical capital assessed are: access to the dryer and other fish processing resources, transportation as well as information. Linga EPA is accessible by road and is closest to the trading centre, which entails having better access to markets than other communities along Lake Malawi shoreline do. The social capital looks at the family dynamics of the FPs and how these affect their trade. Access and control of resources, especially income, were crucial, especially because most FPs are women, some of whom may not have control over their income. Involvement in different community groups is also assessed, especially the way it contributes to the sustainable management of the dryers.

The transformational structures and procedures in the SLA relate to the levels of governance, the role of the private sector, the laws, policies, culture and institutions within the community. In this regard, the local governance structures are assessed in terms of how they serve the local communities, how information flows and how decisions are made between the local, district and national levels. Thus, the assessment includes the laws and policies governing the trade of fish in Malawi and how the local communities inform policy processes.

Governance is also assessed in terms of how the laws and policies are followed through in the district. Similar assessment is made for the private sector, of which SEED-Fish is part. The project is assessed in terms of how it has worked with both the local government as well as the communities themselves. Only men catch fish out in the lake, but women are involved in inland processes, including owning gear, processing and marketing of the products. The culture of the community is also assessed in terms of how the fish value chain is organised and by assessing the perceptions of the fisherfolk towards the roles that men and women have.

3 Results and Discussion

3.1 Fish Processing in Chipala and Vinthenga: Value Chain and Value Addition

The results (n, 99) show that only about 28% of the fish bought from fishers is sold by wholesalers, while 72% is sold by retailers. The proportions per processing method are based on the method that a respondent used the most, which indicated that though fresh fish is preferred, there are few FPs who solely sell fresh fish. In this case, all retailers and some wholesalers are fish processors. The analysis also shows that wholesalers sell their fish in markets outside of Nkhotakota, while the retailers sell their fish either in the local market or their homes. Only 3% of the fish processed using the solar drying method is sold by wholesalers, which shows a minimal exposure of the method to markets outside of Nkhotakota. Within Nkhotakota, solar-dried fish constituted 10% of fish sold processed. The bulk (56%) of the fish caught was sold sun-dried.

The study focussed on the methods used to process different fish species. The data revealed that the two most common fish species sold by the respondents are *Copadichromis virginalis* and *Engraulicypris sardella,* locally known and hereafter referred to as *Utaka* and *Usipa* respectively. For Utaka and Usipa, the methods used by fish processors are illustrated in Figs. 2 and 3. In general, the figures show that fresh and sun-dried fish are the most commonly sold from Chipala and Vinthenga. Solar-dried fish was the third most commonly sold fish at 12%.

The low rate of solar-dried fish can partly be attributed to the small number of respondents using the solar-dried method. However, this represents half of the population in Vinthenga, upon which assumptions are based. The high percentage of fresh Utaka and Usipa corresponds with the qualitative data which indicates that FPs would prefer more ice machines that would improve the quality and preservation of fresh fish.

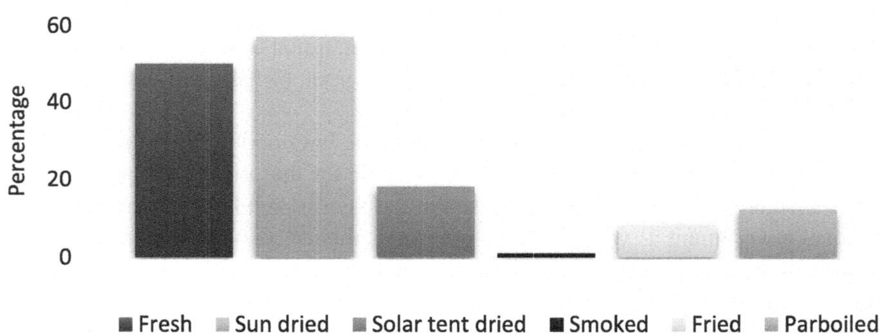

Fig. 2 Distribution of processing method used for *Engraulicypris sardella* (Usipa) (n, 57)

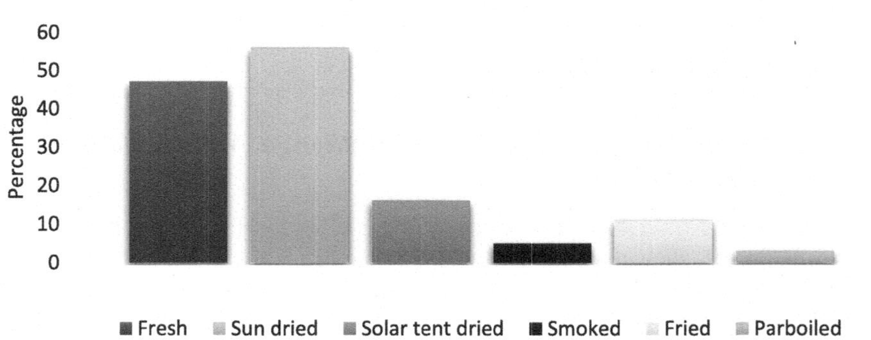

Fig. 3 Distribution of processing method used for *Copadichromis virginalis* (Utaka) (n, 56)

Benefits of the Solar Drying Method

Fish processors in both villages appreciated the improved quality of dried fish using the solar drying method. The participants indicated that the customers will always buy the solar-dried fish first because of the appealing clear colour. Other studies revealed related disadvantages of using the open sun-drying method, as the fish is exposed to contamination as well as insect attack (Yean et al. 1998; Banda et al. 2017). Furthermore, they state that open sun-drying is seasonal and affects the quality of the product. Within the dry seasons, the fish must be exposed to enough heat from the sun to dry evenly, which seldom happens, as noted by the participants. The use of the solar dryers ensures that the heat captured in the tent is evenly distributed. The participants also appreciated the difference in taste, noting that the sun-dried fish had a mild sour taste due to uneven drying. The better-quality solar-dried fish attracts an additional K500, normally representing a 13–15% increase in the price of a bucket of fish. Related findings on the cost of solar-dried fish in other districts along Lake Malawi found it to be twice more than sun dried (Chiwaula et al. 2018).

Actual fish processing times per method could not be estimated and compared within the two villages as FPs in Vinthenga do not use the smoking or frying methods, which are methods used in Chipala where there were no adopters of the solar drying method. However, observations in Chipala showed that the frying method was the most time consuming as it took 30 minutes on average to process 50 fish (Utaka).

The cost for firewood was the same in both villages, at MK100 per bundle of three small pieces of wood. The dimensions of the bundle were estimated to be 2 cm^2 and 40 cm long. For 50–200 fish (Utaka), the FPs estimated 10 bundles of firewood were required for using the frying method. The smoking method used double the amount of firewood for the same amount of fish. However, for species such as *Oreochromis* spp. (Chambo), the FPs expressed that more firewood is used, even though they could not estimate, as none of them sold smoked Chambo. Processing methods were preferred by FPs based on the best-quality fish (44%), time efficiency (31%), minimal cost of inputs (19%) and customer's preference (6%). Respondents who preferred the solar drying method attributed it to time efficiency. However, the analysis also shows that not all solar dryer users perceived the solar drying method as a preferred method.

Another advantage of using the dryer is reducing post-harvest losses during the rainy season. Using the dryer achieves the project's objective of reducing post-harvest losses that are associated with climate variability, with prolonged rainy seasons which affect fish processors' income. One female FP who had used the dryer in Chipala appreciated the fact that while FPs might not be able to use the dryer consistently, it still reduces their post-harvest losses during the rainy season.

3.2 *Adoption of Technology Versus Limited Use of Fuelled Processing Methods*

Adoption of the Solar Drying Method

There were 19 respondents who indicated using the solar drying method, all from Vinthenga village. The number of fish processors (n, 99) per method are shown in Fig. 4. The data shows that the majority (62%) of the fish processors in Chipala and Vinthenga were below the age of 30, indicating a high level of participation of young fish processors (FPs). The large percentage of young FPs has positively affected adoption, as 68% of solar dryer users were below the age of 30, while only 5% were above the age of 40. As the project staff noted, it was expected that the older fish processors, due to their many years of experience using traditional methods, may not be as willing to adopt new methods as compared to their younger counterparts.

The FGD participants indicated that their reasons for adopting the technology were the possibility of improving the quality of dried fish as well as time efficiency, affording them more time for other duties. None of them indicated using fuelled methods; however, they continued to use the sun-drying method after adopting the solar drying method. The reasons for continuing to use the sun-dried method were lack of space in the tent and customer preference for sun-dried fish.

The respondents adopted the technology between 2016 and 2017, with most adopting in 2016 when the dryer was constructed. Respondents who did not adopt the method (n, 80) indicated that it was because they did not have access to a dryer (35%), were not aware of the technology (26%), lacked capital (5%), preferred fresh fish (3%), were not interested (9%) and processed a small amount of fish (5%).

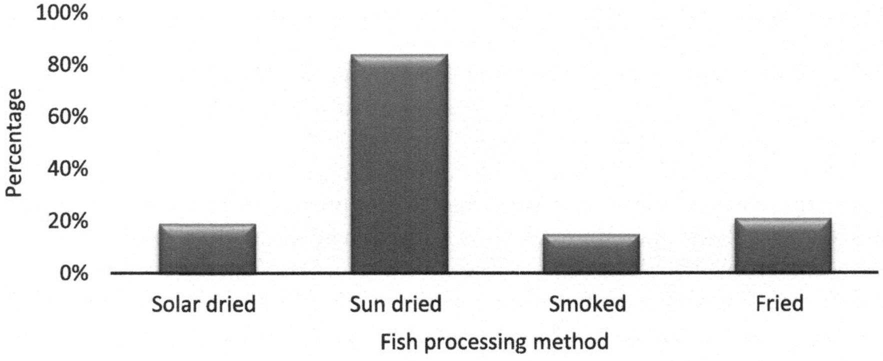

Fig. 4 Number of fish processors per method

Implementation Strategies

The implementation strategies, especially the involvement of the community in each stage of the project, differed in Chipala and Vinthenga. For instance, the community in Vinthenga were more involved from the onset of the project by selecting the construction site and supporting the artisans in the construction process. Implementation was also affected by conflicts within the community, which were unresolved throughout the project period. Adoption was halted, and thus also management of the facilities.

Various levels of governance were in conflict: between the Fisheries extension office and the BVC, between the community and the SEED-Fish project and, third, between the BVC and fish processors. The choice of the site came from the Fisheries Department in Chipala, thus using a top-down approach, causing the BVC and other members of the community to conclude that the dryer was constructed for the Fisheries Department, rather than the community. Governance, as an interaction between different stakeholders or levels of governance, has been a challenge; however, as Johnson (2006) states, governance as a process should reinforce ties between stakeholders. There was lack of harmonisation of the levels of governance, and that was observed in the differences between the two communities, even though they are neighbouring villages. The loop diagram (Fig. 5) shows how governance has affected livelihood outcomes. The situation in Chipala showed that if the governance structures are not transformed through development initiatives, projects will not yield the expected outcome, as the community goes through a circle of low adoption and thus minimal impact due to poor governance. The findings here confirm that in value addition in fisheries, economic and ecological components are at the centre of governance while side-lining or totally ignoring the sociocultural domain (Urquhart and Acott 2013; Reed et al. 2013).

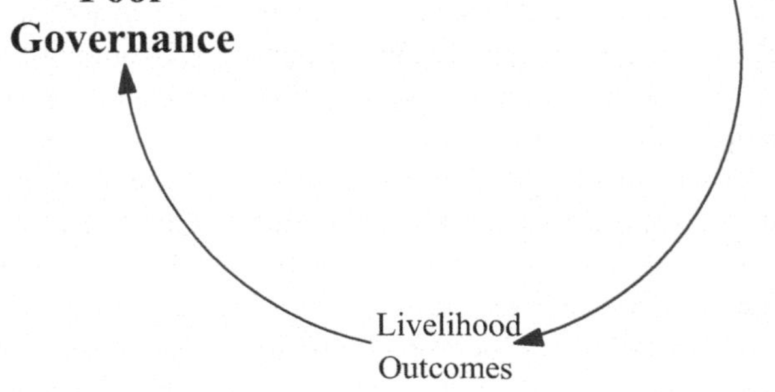

Fig. 5 Illustration of the effect of governance on livelihood outcomes

The lack of skill and knowledge transfer in Chipala and Vinthenga underscores the need for information as a stand-alone capital in the SLA. Odero (2006) recommends that the attachment of information to social capital tends to reduce the visibility and importance of information and knowledge sharing. Ideally, such knowledge sharing would then feed into evidence-led policymaking processes at village level. This would ensure that the different stakeholders, such as the BVC and fish processors, are well equipped to lobby for policies that improve their livelihoods, as well as empowering them to question policies or projects that are not in their best interest.

A male participant in the FGD at Vinthenga observed that selling fresh fish would be a suitable alternative to selling solar-dried fish. He also stated that the demand for solar-dried fish is not vividly high now because of the inconsistent use of the dryer, which can be attributed both to climate variability as well as the holding capacity of the dryer. In all the discussions, it was communicated that most people preferred fresh fish, and that other processing methods are mainly used when FPs return from the market with fresh fish. Access to ice for people to preserve fresh fish is the only barrier. One of the challenges with intentional development is the assumption of the needs of the poor by project implementers (Morse and McNamara 2013). The findings showed emphasis on the preference for fresh fish as compared to any processing method. Some participants also stressed the need for ice-making machines instead of improved processing technologies such as the dryer and kilns. However, due to customer preferences there is a demand for processed fish.

3.3 Implications of Gender Perspectives on the Livelihoods of Fish Processors

Women are the main fish processors in Nkhotakota. With exception of a few male fish processors, men are more involved in catching fish. There are more than 150 female fish processors in Chipala and Vinthenga, but the figures of male FPs were not clear amongst the fish processors. Some estimates put them to less than 30 in both villages. The BVCs could not provide actual figures of fish processors in their respective villages.

The female FPs explained that at times they hire men to help them transport the fish from the boat to the dryer or drying racks, as well as to dry their fish. These men are not their husbands. The participants agreed amongst themselves that the response from their husbands is normally that they cannot spend their time drying fish, or helping their wives doing the same.

Understanding the gender perspective of the fish value chain gave insight into the different processes and roles that men and women play in the value chain and why. The responses to why women cannot catch fish, for example, gave understanding of the discourse and who holds power of it. The male participants explained that fishing was forbidden for women due to the heavy workload of fishing, because men and women are different in physical and mental strength. The perspectives of the female participants were similar, and they seemed to be inspired by the general male perspective about the women's role and limitations in fisheries. The women reiterated what the men had stated: that a woman cannot be a fisher. The women went further to explain that from what they hear their husbands narrating about their fishing experiences, they are sure they cannot go through it. This discourse is acculturated in the beliefs of both men and women, so much that examples of women who fish are unbelievable and unwelcome. Because recurring instances show the vulnerability of women on the lake, such as women drowning and men surviving, women are convinced that they cannot fish. Thus, the fishing activities that males and females adopt in the fish value chain become their identity, and transformation of such roles in the future is a challenge (Thompson et al. 1983).

Men who are fishers will sometimes contribute to the necessities in the household, but most times they will not do so on the basis that the wife has income from her FP business to take care of the household. Women decide how to spend the money from their business. From their FP income, the women buy most necessities for the household, as well as housing materials and education expenses such as tuition fees and school uniform. Thus, as explained:

> When the fish processing business is not going well, then there are serious problems in the household. Female participant, solar drying FGD, Vinthenga

The fish processors in the study were members of village banks, as well as *chihandi*, or members of micro-savings and loan groups. Some members indicated that they had savings with formal banks (Fig. 6). *Chihandi* is a simpler system as compared to village banks, as members of a group agree to contribute a certain amount per day and the total amount is allocated to one member of the group per day. The village bank requires more organisation, as every week the group meets to save money in the form of shares and take out loans as well as discuss other group investments.

Fig. 6 Fish processors' saving methods

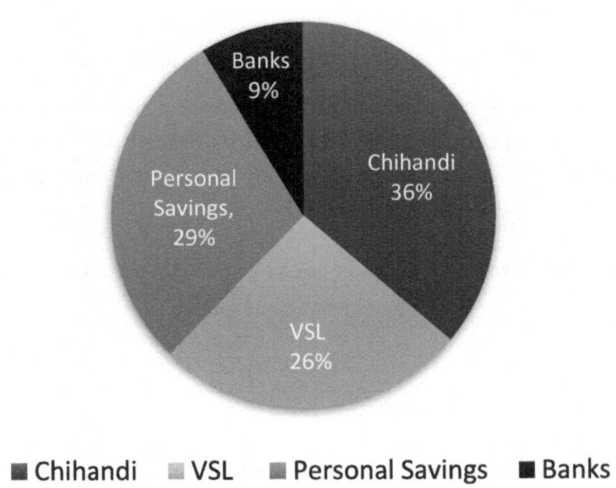

The fish processors stated that the village banks serve as safety nets when they have lost their capital from their processing business. The money saved in the village banks was mainly from their fish processing business. They however expressed the desire for microfinance institutions where they could access bigger loans to invest in other businesses, as an alternative to fish processing, especially when fish is scarce.

A 'savings culture' rarely exists in fishing communities due to two main factors. First, there is a perception that fish is in abundance in the lake if money is needed, which is a culture element. Secondly, options to save through financial organizations are limited (Ardjosoediro and Neven 2008). Though there are microfinance opportunities, the options available in the two communities are few and designed for communities or individuals with small incomes, whereas FPs desire access to bigger loans to invest in their enterprise, as well as potentially invest in more solar tent dryers. The average income in the sample, of about half the population of FPs, was MK 67,328.28. Although village banks serve to protect rural communities from taking out big loans and thus failure to repay, the regulations require that persons are restricted to a certain share and loan limit.

The data (Table 1) indicated that gender affected participation in three of the four saving methods. The data further revealed that women engage more in saving activities than men.

While fish processing is their main source of income, female FPs told the research team that being involved in one or more microfinance groups assisted them in recovering from loss of capital in their fish processing enterprise, as well as easing the pressure of providing for their families. The village headman in Vinthenga stated that gender equality is a buzz phrase these days, and thus women are empowered to have enterprises; however, once women begin earning money, there is a tendency for the men to abandon their financial responsibility for the household to the women. But the women told the researchers that their aim in engaging in fish processing was to make a contribution towards what the husband already provides. One female FP in Vinthenga put it as follows:

> For us business is just to enables us to help the man take care of the household, and that's it.

The discussion about how FP income is managed within the household quickly turned emotional as the female FPs voiced their frustrations about the household dynamics and how these dynamics affect their enterprise. The women told the

Table 1 Cross-tabulations of gender and saving methods

	Gender		Significance
	Male	Female	
Village bank savings	0	18	**
Institutional bank savings	3	3	*
Chihandi savings	2	23	
Personal savings	7	13	**

Notes: $N = 99$; * and ** show cross-tabulation at significance levels of 0.10 and 0.05, respectively researchers that most women are only married so that they can be known as 'so and so's wife'. The only man in the group laughed at this, but he confirmed that sometimes the men may offer some money to their wives as 'help'.

The experience in both villages reveals that while the introduction of development initiatives, especially value addition innovations, has the potential to improve communities, the failure to address gender attitudes may result into unforeseen burdens for some sections of the society. In this particular context, the husbands of female fish processors feel less obligated to support their families because their wives are economically empowered. Thus, the higher the income of the female fish processor, the higher the financial burden they assume over the family.

It should be kept in mind that although women are involved in all activities except the catching of fish, they make less money than the fishermen, especially those who own the fishing business themselves (Ngochera et al. 2017). The data showed that there were no male processors within the lowest-income brackets of between MK 2000 and MK 19000; there were 10 female FPs within this bracket. Out of the male FPs in the sample, 4 were within the highest-income bracket of between MK 100,000 and MK 300,000. Thus, male fish processors have higher income due to more access to capital and mobility, which is limited for women because they typically have more responsibilities within the household (Allison and Mvula 2002).

3.4 Sustainability Measures

Participation

The introduction of innovative technologies that add value to a product ought to take advantage of already existing commitment to the trade, thereby expanding the market of the produce and improving income (Ngochera et al. 2017). The management of the dryer in Vinthenga revealed communal ownership of the development initiative, as per design of the project. Rental fees collected from usage of the dryer were used for maintenance of the same. The solar dryer committee (SDC) spent K100,000 to maintain the tent dryer, and this cost covered plastic sheets, wood planks and nails. The committee members at Vinthenga also suggested that using old sugar plastic bags could be an alternative to the plastic sheets purchased at Arkay Plastics Ltd., which showed the innovation within the community to support the introduced technology. After the tent was constructed in 2017, it was handed over to the BVC, which then handed the tent over to a subcommittee formed at the tent. The SDC consists of fish processors and a representative from the BVC (Fig. 7).

Contrary to the community in Vinthenga, the BVC members in Chipala took a more laid-back approach to their duty of managing the dryer. Co-management between the government and BVCs over natural resources presents conflicting lines of authority between local institutions (Ngochera et al. 2017). The BVC assumed leadership over the technology albeit with limited knowledge about fish processing

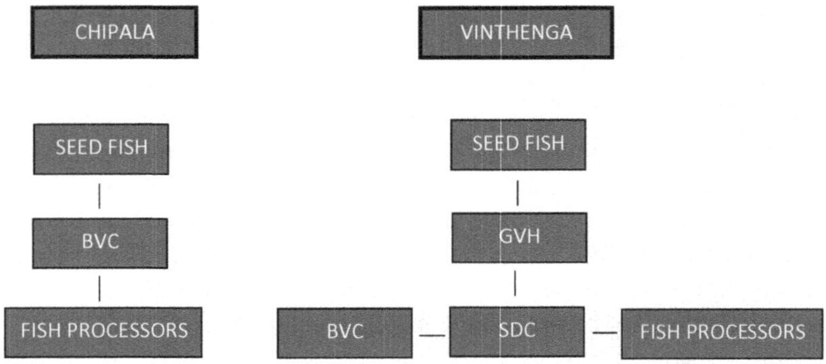

Fig. 7 Illustration of levels of participation in Chipala and Vinthenga

and marketing, as the majority of them are only fishermen. The delegation to a subcommittee (the SDC) in Vinthenga ensured that experienced fish processors were part of the decision-making processes concerning the dryer and thus empowering the fish processors to use and advocate for using the solar tent dryer. For example, of the 19 respondents using the dryer, 15 were members of the SDC. Thus, the improvement of social capital in Vinthenga has the potential to yield the expected outcome of improved income, if enough tent dryers are made available for the processors.

The discussions in Chipala revealed that the the location of the dryer was the main reason for the low participation in solar drying activities. The construction of the dryer away from the landing site, and on the Fisheries Department's premises, caused the community to reject the project and therefore not to assume ownership, as they were not involved in the choice of the site. The District Fisheries Officer (DFO) explained that the reason for having the dryer on Fisheries Department property was for research that would be carried out while the community use it for their trade.

We used a logistic regression model to further understand what factors influenced the fish processors to participate in the project or fail to do so (Table 2). The results show that the location or community which a FP was from, represented by the village variable in the model, is the only determining factor for participation in solar drying activities, with those in Chipala having a less likelihood of participating. This corresponds with the data from the interviews which shows the differences in commitment to the project, with the community in Chipala showing the least interest in taking part in solar drying activities.

The lack of participation can also be attributed to the conflicts within the community, which prohibited FPs from using the tent dryer, as well as the prices set by the BVC in Chipala, which were too high for FPs if they were to use the dryer. Prices in Chipala were pegged at MK500 per 5 kg bucket, as compared to MK250 per day in Vinthenga, regardless of the amount of fish processed. The BVC members' insistence in Chipala that the price does not need to be adjusted is evidence

Table 2 Logistic regression model for determining factors of participation in solar drying activities

	B	S.E.	Sig.	Exp(B)
Age	0.08	0.069	0.25	1.08
Gender (male)	−0.61	1.243	0.62	0.54
Village (Chipala)	−3.32	1.185	0.01***	0.04
Ethnicity			1	
Ethnicity (1)	16.76	40192.95	1	18990159
Ethnicity (2)	−2.09	44250.63	1	0.12
Ethnicity (3)	38.36	56841.43	0.99	4.58
Ethnicity (4)	−0.40	46000.44	1	0.67
Ethnicity (5)	16.90	40192.95	1	21959568
Ethnicity (6)	0.01	56841.43	1	1.01
Years of education	0.14	0.10	0.18	1.15
Number of children in household	0.21	0.48	0.65	1.24
Household size	−0.09	0.50	0.85	0.91
Constant	−21.02	40192.95	1	0

***$p < 0.01$

that the dryer is seen more as an enterprise, rather than a development initiative for the community to improve their livelihoods. When livelihood strategies are mismatched with the incomes of targeted groups or communities, the livelihood outcomes are anything but transformed (DfID 1999). Thus, based on the data, the main challenge was participation (and ownership, in the recoded dependent variable), which affected the level of adoption, especially in Chipala, as the logistic regression model also explains.

Capacity Building

Training and educating communities have been defined as means to widen the capital base and thus enlarging people's choices and continued involvement in development initiatives. The average years of education that fish processors had attained was 5.84, which indicates that most of them (77%) had only attained primary education. In as much as the fish processors' level of education did not significantly relate to participation in solar drying activities ($\chi 2 = 10.76$, $df = 13$, $p = 0.63$), the average years of education, and experience in the trade, indicate potential for training in the construction of dryers.

The lack of training in the project means that committees, and any other leadership involved in the fish processing value chain, have not gained technical know-how about dryers and thereby have limited knowledge about a technology that they are supposed to promote in their communities and policy processes. It also limits their innovativeness to improve the service by combining their theoretical and technical know-how with their already existing capital base. Thus, the lack of capacity building here is a hindrance for increased access to and use of solar dryers and thereby limiting transformation of the communities' livelihood assets and political and social structures.

The project developed an entrepreneurial model for the communities to use; however, it was finalised towards the end of the project and was not disseminated to the communities. The project staff in Nkhotakota also had plans to expose the FPs in Vinthenga and Chipala to other communities that are also using the solar drying method, so that they can appreciate the experiences of other FPs; however, these were also not conducted. Exchange visits are described as one of the most effective tools for increasing adoption of innovative technologies because of the encouragement that participants gain from the experiences and viewpoints of their peers (Matras et al. 2013). The closest community from Nkhotakota is in Salima, which is the neighbouring district.

The project staff were of the view that the communities would not be able to construct the tent dryer on their own, precisely because they were not trained in facility construction and economic management. Accordingly, trainings conducted within the project were only for the contracted artisans who constructed the tent dryers. The communities were merely informed about the technology and the benefits of solar drying. While the project staff believed that the community was capable of financing the construction of a tent dryer, he identified the lack of collective financial initiatives amongst fisherfolk as a challenge for construction.

Supply and Demand for Smoked and Dried Fish

Despite the benefits of solar dryers, the participants told the research team that the profit was marginal because of the inconsistent usage of the dryer and the demand for smoked and fried fish for its distinct taste. Figure 8 illustrates the customer-preferred methods, as perceived by the respondents (n, 99). This shows that the demand for smoked and fried fish was not as high as that of sun-dried, fresh and solar-dried fish, which underscores the greater importance of the latter methods.

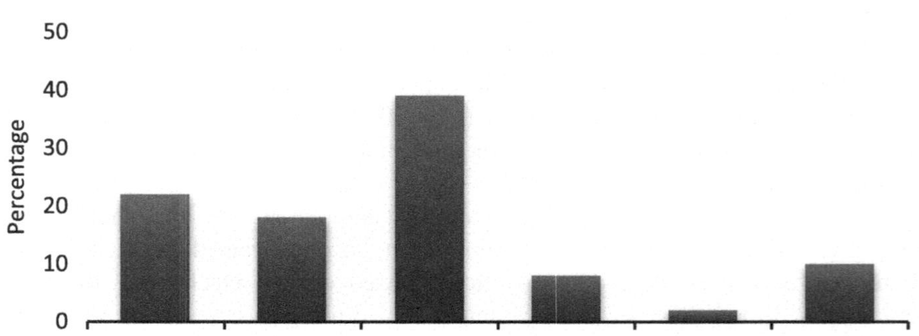

Fig. 8 Customer-preferred methods of fish processing

Table 3 Means of processed fish prices (MK) during peak season

Selling prices/method during peak season				
	N	Mean	Std. deviation	Std. error mean
Sun dried	84	3952.38	1183.77	129.16
Smoked	15	5333.33	3045.29	786.29
Fried	20	6960.00	2115.21	472.98
Solar dried	18	4183.33	1862.72	439.05

Table 4 Means of processed fish prices during off-peak season

Selling prices/method during off-peak season				
	N	Mean	Std. deviation	Std. error mean
Sun dried	84	5625.00	1426.67	155.66
Smoked	15	7133.33	2837.67	732.68
Fried	20	8535.00	2008.73	449.16
Solar dried	18	5500.00	1823.05	429.69

To understand the supply and demand of smoked and fried fish, selling prices were also assessed. Tables 3 and 4 are results from one sample t-tests comparing the average selling prices for the four processing methods. Sun-dried prices were included because the method is closest to the solar drying method and requires similar resources. For both peak (Table 3) and off-peak periods (Table 4), the analysis shows that, respectively, the most expensive bucket of fish is fried, with a mean of MK 6960 when fish catches are high and MK 8535 when fish catches are low. Second to fried fish was smoked fish, followed by solar-dried and lastly sun dried. Solar-dried fish is shown to be more profitable than sun-dried fish.

In terms of demand, with the smoked and fried fish being the most expensive, this implies that there is greater demand for use of the two methods. Since the inputs of such methods are higher as compared to all the other processing methods, the higher prices are determined by and are reflective of the inputs. However, after deducting the cost of processing, fried fish was still the most profitable at 25% during peak season and 21% during off-peak season. Solar-dried fish had a profit of 20% and 11% during peak and off-peak seasons, respectively.

Value Addition

Of the solar dryer users (n, 19), 21% sold their solar-dried fish outside of Nkhotakota, while 79% sold their fish locally. Some FPs (21%) revealed that the solar drying method had improved their fish processing business. The majority attributed the improvement to better-quality fish and profit (75%), while others attributed the improvement to lower costs, high demand of solar-dried fish and efficiency during rainy season. The lack of improvement was attributed to the lack of space in the dryer.

Value for Money

Apart from the fact there are times when the tent dryer is not in use due to scarcity of fish, when fish is available the holding capacity of dryer is not enough for all FPs. Thus, tracking of income improvements is not consistent. Furthermore, it was discovered that the project had not tracked the incomes of the fish processors who were benefiting from the solar dryer, implying that there were no assessments conducted by the project to appreciate the monetary value of the investment in the communities. There are no figures of how many people are currently using the dryer, how much they earned before the project and how much they earned after its introduction. The committee at the tent dryer were instructed to keep a record; however, they were not consistent in this task and estimated that 35 FPs were using the dryer.

The study showed that 44% of the FPs' profit was attributed to the usage of solar dryers during the peak season and 31% during off-peak season. During the peak season, the attributed proportion is higher which assumes more frequent usage of the dryer, and the opposite is true for the off-peak season. The data corresponds with the field observations that the dryer was empty during the week of data collection. The FGD participants expressed that fish catches were low at the time, and thus proportions during data collection were lower.

It is worth noting that some of the main activities of the SEED-Fish project were not fulfilled, largely because these activities were dependent on research conducted under the project. The disadvantage of implementing research projects that directly affect livelihoods is the overlooking of challenges that the research process might face during the implementation period. The emphasis on research thus overshadowed the impact of the technologies on the livelihoods of fish processors.

Diversification and Resilience

The fish processors said they also engaged in some small-scale businesses, but these did not compare to the fish processing business in terms of income. To verify this, respondents' income from fish processing and that from their other sources of income were compared. The average contribution of fish processing to the total income of the respondents was 83%, though it ranged between 2% and 100%. In determining income diversification, Simpson's diversity index (Simpson 1949) was used to measure the extent to which sources of income were diversified. The index ranges between 0 and 1, with 0 indicating no diversification of the total number of sources of income from one entity and 1 being the highest level of income source diversification. Diversity was measured at 0.98 which showed that the sources of income were highly diversified, making the FPs' income activities sustainable for their livelihoods.

4 Conclusion

The solar dryers have improved fish processing, and thus there is potential for the method to be adopted if more dryers are constructed, especially in Vinthenga, where the demand to use the dryer is high. But this assessment is based on one dryer which was in use. In Chipala, where the community rejected the development, the initiative is a white elephant project, and its impact with regard to improving livelihoods of FPs is minimal and not well accounted for. Likewise, the impact of the technologies on the reduction of wood-fuelled methods and their impact on deforestation could not be assessed because the community was not using the kilns installed by the project in Chipala.

The main reason for adopting the technology was the quality of fish produced in the tent dryer, as there was a need for a more hygienic and profitable processing method other than sun-drying. Adoption increased as the FPs realised that the method reduced their time spent and activity while processing fish. However, the capacity of the dryer in Vinthenga has meant that increases in adoption limit the impact of the dryer on income, as fish processors cannot use the dryer consistently because of the demand.

Women were involved in all activities in the fisheries' value chain except for catching fish, which is restricted to men. Fish processing profits are affected by the fact that there are no female fishers because they have little bargaining power over the prices set by fishermen whose knowledge of the market is limited as they are only involved in catching fish. The study concluded that, all factors constant, the current fish value chain meant that even with the solar drying method, post-harvest losses during the off-peak seasons would prevail due to the fluctuation of prices set by fishers who have little knowledge of the market. The gender roles and perceptions affected the socio-economic status of fish processors, in that gender equality was contextualised as a responsibility shift to women who were involved in fish processing and other enterprise. The discussion revealed that this perception of gender equality was not recognised as affecting or rather stalling the improvement of women's assets, though the more money women made, the less responsible their husbands became.

The lack of key sustainability measures poses a threat to the continued use of the method. The lack of skills and knowledge transfer, on how the technology works or how to construct dryers to increase usage, meant that the communities had been empowered to manage a technology that they do not entirely understand, let alone know how to replicate. In Chipala, the abandonment of the dryer as well as the conflicts surrounding it meant that unless governance structures are transformed the method would not be sustainable as the community has rejected the project, even though they have been exposed to the benefits through the success story of their neighbours in Vinthenga.

All factors considered, the impact of the solar tent dryers on the livelihoods of fish processors has been insignificant. The solar drying method was adopted in one village and rejected in the other. Amongst the causes of this rejection are the management methods, including the pricing of the services that was not arrived at by popular participation. Making improvements to existing systems and methods is a major development challenge that will continue to affect efforts to improve livelihoods. As described in the SLA, and reiterated throughout this study, at the centre of such technological innovations should be people, especially the poor. Understanding local communities and institutions is recommended as a prerequisite for any development intervention; this precaution will help to foster good governance and sustainable usage of technological innovations.

References

Alkire S, Foster JE, Seth S, Santos ME, Roche JM, Ballon P (2015) Multidimensional poverty measurement and analysis. Oxford University Press, Oxford, UK

Allison EH, Mvula PM (2002) Fishing livelihoods and fisheries management in Malawi. LADDER Working Paper No 23, Overseas Development Group, Norwich. http://www.odg.uea.ac.uk/ladder/

Ardjosoediro I, Neven D (2008) The Kenya capture fisheries value chain: An AMAP-FSKG Value Chain Finance Case Study. United States Agency for International Development (USAID), USA

Banda et al., (2017). Nutrition, microbial and sensory quality of solar tent dried (Samva Nyengo) and open sun dried C. virginalis (Pisces; cichlidae). International Journal of Marine Sciences, 7(11):96–101

Chiwaula LS, Chirwa GC, Binauli LS, Banda J, Nagoli J (2018) Gender differences in willingness to pay for capital-intensive agricultural technologies: the case of fish solar tent dryers in Malawi. Agric Food Econ 6(1):1

Cowen M, Shenton R (1998) Agrarian doctrines of development: Part 1. J Peasant Stud 25:49–76

DFID (1999) Sustainable livelihoods guidance sheets. Department for International Development, London

Ellis F (1999) Rural livelihood diversity in developing countries: evidence and policy implications. Overseas Development Institute Natural Resource Perspectives, No. 40: Overseas Development institute

Johnson DS (2006) Category, narrative, and value in the governance of small-scale fisheries. Mar Policy 30(6):747–756

Kambewa P, Mataya B, Sichinga K, Johnson T (2007) Charcoal: the reality-a study of charcoal consumption, trade and production in Malawi. International Institute for Environment and Development, London

LUANAR (2013) Capacity building for managing climate change in Malawi (CABMACC). Programme proposal document. LUANAR, Lilongwe

Matras F, Sidi F, Treinen S (2013) Exchange visits: advice for improving the impact. Good Practice Fact Sheet: West Africa. Food and Agriculture Organisation (FAO)

Matiya G, Wakabayashi Y, Takenouchi N (2005) Factors Influencing the Prices of Fish in Central Region of Malawi and its Implications on the Development of Aquaculture in Malawi. Journal of Applied Sciences, 5:1424–1429

MBERU (2001) Fish in Lake Malawi and other Lakes in Malawi. http://www.est.ryukoku.ac.jp/est/yuhma/FLM-pamph/index.html. Accessed on 10 June 2017

Morse S, McNamara N (2013) Sustainable livelihood approach: a critique of theory and practice. Springer, New York

Moyo D (2010) Dead aid: why aid is not working and how there is another way for Africa. Farrar, Straus and Giroux, New York

Ngochera M, Donda S, Hara M, Berge E (2017) Defragmenting resource management on the southeast arm of Lake Malawi: case of fisheries. Centre for Land Tenure Studies Working Paper 12/17, Norwegian University of Life Sciences, Ås

Odero K (2006) Information capital: 6th asset of sustainable livelihood framework. Discov Innov 18(2):83–91

Reed M, Courtney P, Urquhart J, Ross N (2013) Beyond fish as commodities: understanding the socio-cultural role of inshore fisheries in England. Mar Policy 37:26–68

SEED Fish (2016) Quarterly Report. Received through email (30 May 2017)

Simpson EH (1949) Measurement of diversity. Nature 163:688

Thompson P, Wailey T, Lummis T (1983) Living the fishing, vol 50. Routledge & Kegan Paul, London

Urquhart J, Acott T (2013) Constructing 'The Stade': fishers' and non-fishers' identity and place attachment in Hastings, South-East England. Mar Policy 37:45–54

USAID (2015) The importance of wild fisheries for local food security: Malawi. file:///C:/Users/Bruker/Downloads/Malawi_fisheries_profile_cleared_9_29_15_508.pdf.Accessed7June2017. https://rmportal.net/biodiversityconservation-gateway/projects/closed-global-projects/measuring-impact/mi-project-resources/the-importance-of-wild-fisheries-for-local-food-security/the-importance-of-wild-fisheries-for-local-food-security-malawi

Yean SY, Pruthiarenun R, Doe P, Motohiro T, Gopakumar K (1998) Dried and smoked Fish products. In: Doe E (ed) Fish drying and smoking: production and quality. CRS Press, New York

Profitability of Supplementary Feeding of Indigenous Cattle in Dry Areas of Tanzania

Alexander Solstad Ringheim, Daniel Mushi, Ephraim Mtengeti, Ismail Selemani, Magnus Åsli, Lars Olav Eik, Elikira Kimbita, and Leif Jarle Asheim

Abstract By 2050, global food consumption is expected to rise by 60% compared to the 2005–2007 level. In sub-Saharan Africa (SSA), the population increase may be as much as 250% by the same period. Hence, there is an urgent need to increase food production and introduce productivity-enhancing measures in SSA agriculture, including the livestock sector, which is the main focus of this article. The current productivity of the Tanzanian livestock sector is low due to seasonal variations in the availability and quality of pasture and other feeds. The cattle gain weight during the rainy season and lose weight in the subsequent dry season. Additionally, pastoralists face challenges due to the conversion of grazing areas into cropland, overgrazing, and the increasingly frequent droughts. Although the optimum age for slaughter is 3.5–4.5 years, farmers in Tanzania slaughter their cows at 5–6 years. This article argues that this may be an unhelpful economical management practice. To study the effects of improved feeding on economic performance, we collected data on on-farm supplementation experiments with indigenous Zebu cattle, in collaboration with pastoral communities and a large-scale commercial wheat farm in Hanang, Tanzania. The study compared the income and costs associated with traditional cattle keeping (TS) for 6 years at slaughter, with that of two levels of concentrate supplementation, low (LSS) and medium (MSS), allowing for slaughtering at 4.5 and 3.5 years, respectively. Adjusted net margins for the three systems were 199, 911 and 978 USD, respectively. Our results strongly suggest that farmers should

A. S. Ringheim (✉)
International Environment and Development Studies, Noragric, Norwegian University of Life Sciences (NMBU), Ås, Norway

D. Mushi · E. Mtengeti · I. Selemani · M. Åsli · E. Kimbita
Sokoine University of Agriculture (SUA), Morogoro, Tanzania

L. O. Eik
Department of International Environment and Development Studies, Norwegian University of Life Sciences, Ås, Norway

L. J. Asheim
Norwegian Institute of Bioeconomy Research (NIBIO), Ås, Norway

B. R. Singh et al. (eds.), *Climate Impacts on Agricultural and Natural Resource Sustainability in Africa*, https://doi.org/10.1007/978-3-030-37537-9_23

supplement the feeding of their young stock regularly, in times when the animals cannot sustain themselves on grazing alone. The primary explanations for the recommendation were that supplementation would lead to increased production of meat and reduced variable costs, that is, feeds and drugs. Our study was limited to steers. Future studies should include supplementation of cows to obtain annual calving and use of crop by-products instead of concentrates.

1 Introduction

By 2050, global food consumption is expected to rise by 60% compared to 2005–2007 levels (FAO 2009). In sub-Saharan Africa (SSA), including Tanzania, this figure is projected to be even higher due to an estimated population increase of 250% and a three-fold increase in cereal demand (Van Ittersum et al. 2016). Tanzania's total land base is 88.6 million hectares. Approximately 60 million hectares of this land base is rangeland suitable for livestock production (Mushi et al. 2015). Tanzania has a cattle population of 26 million (FAO 2019a), predominantly consisting of the indigenous Zebu breed. In 2012, the overall livestock and beef sector's contributions to the national GDP were 4.6% and 2%, respectively (MLDF 2013). Meat consumption in developing countries like Tanzania is low in comparison to western countries (FAO 2005), amounting to around 10 kg per capita at the millennium, where beef constituted approximately 7 kg.

Tanzanian livestock sector dominated by indigenous cattle (Zebu) display low productivity levels (Michael et al. 2018). Significant seasonal variations affect the availability and quality of feeds, resulting in substantial differences in weight gains and losses for cattle during the rainy and dry seasons. Older animals characterise the cattle stock, with age at slaughtering around 5–6 years (Selemani et al. 2015; Mushi et al. 2015). Moreover, available grazing land has been decreasing due, in part, to increase in areas for crop production. In the study area, the Hanang District in Manyara region, 40,000 ha of grazing areas were converted into wheat fields in the 1980s (Lane and Pretty 1990). Also, the role of climate change as well as the varied and suboptimal management practices seems to be key factors contributing to reduced productivity of Tanzania rangelands (Sangeda and Maole 2014).

In contrast, over the last two decades, the Ethiopian livestock sector has significantly increased its productivity, in part due to improved feeding and management. One of the productivity-enhancing measures has been that bulls and steers are bought from nomadic areas and transported to grain-producing areas and fattened for approximately 90 days in feedlots before slaughtering (Tolera and Eik 2019). These animals were fed concentrates and roughages, mostly derived from crop by-products. Such feedlots vary in size, ranging from a few cattle to several thousand. This approach has led to increased economic activity throughout the entire value chain. Pastoralists improve their income by selling animals, crop farmers generate revenue from the sale of crop by-products, and feedlot system administrators derive profit when 'finished' animals finally make their way into local or export markets.

There is a strong linkage between Ethiopia's pastoral livestock production, the feedlot operations and the export abattoirs in the central-highland areas. For the period from 2005 to 2011, the joint efforts of governmental agencies, private sector operators and NGOs resulted in significant increase of live-animal and meat exports from Ethiopia (Tolera and Eik 2019).

From 2013 onwards, a similar value chain approach as in Ethiopia was introduced among agropastoralists in the Hanang District of Tanzania. In this paper we aim to discuss, based on these experiences, the feasibility and economics of supplementation of young stock with concentrate. We also consider possible positive synergies between a large wheat farm, the HLH Farm and Development Ltd. (HLH Farm) and nearby pastoralists.

2 Material and Methods

2.1 Area, Climate and Production Systems

The study area was in the Hanang District of the Arusha Region in Northern Tanzania (Fig. 1) at the HLH Farm and among surrounding agropastoralist communities from the three neighbouring villages of Gawidu, Dirma and Mulbadaw/Hiriri. The climate in Hanang is semi-arid and characterised by two distinct dry seasons, the first starting in May and lasting to early November and the second from December to February. Temperature ranges from 20 to 30 Celsius, and annual rainfall from 408 to 802 mm (Safari et al. 2011; Selemani et al. 2015).

The HLH Farm is a commercial, non-profit company owned by the foundation 'Friends of Haydom', a Norwegian-based NGO. Since its establishment in 2006, the farm has hosted numerous research projects undertaken by SUA and NMBU aiming to improve the livelihood of smallholder farmers in the country. We used information from the data pool of the HLH Farm to develop a model for the current and improved livestock production systems in the area. Three livestock-keeping practices were set up using different levels of concentrate supplementation for the young stock (Table 1).

Effects of concentrate supplementations were modelled based on responses from feeding experiments conducted at the HLH Farm (Mushi et al. 2015). Cost of the concentrates was calculated based on local prices for the different ingredients: maize bran, cotton seed cake, minerals and salt.

2.2 Method and Model

According to Malcolm (2004), farm management is about making choices, and economics is the core discipline of choice to achieve a mix of goals of varying degrees in the face of many unknowns through the alternative use of resources. The logic is: 'What has been and is the situation?' 'What is likely to be the new situation if I do

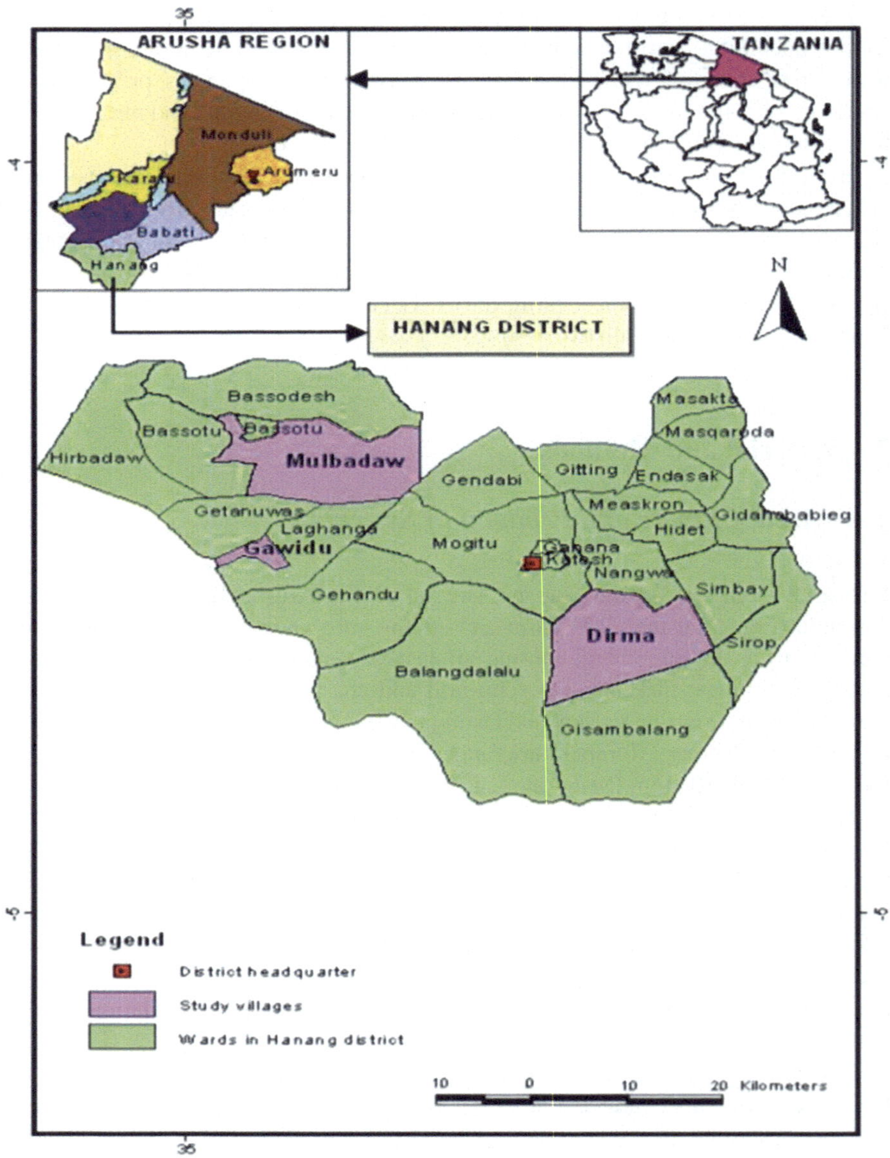

Fig. 1 Map of the study area and villages

this, or that, or nothing different?' and 'Am I likely to be sufficiently better off, all things considered, for it to be worthwhile doing this instead of that or doing nothing different?' The changes in the various measures are the policy options. Modelling farm systems using the whole farm approach can be very useful, though farm models can only be partial representations of reality, and the results need to be tempered by consideration of the unmeasurable aspects, not covered in the models.

Table 1 Overview of modelled production systems

Livestock system	*Description*
Traditional system (TS)	Mainly grazing for all animals, occasional supplementation with crop by-products; mostly wheat straw, bean halm and maize Stover
Low supplementation system (LSS)	Similar to TS plus daily supplementation of young stock with a minimal amount of concentrates
Moderate supplementation system (MSS)	Similar to TS with higher levels of concentrates offered

He claimed that most good answers could be captured with a few key numbers in a few essential budgets, and that whole farm budgeting methods that focus on risky elements as well as case studies can be very useful and enlightening.

Based on the above, we have carried out a simple calculation of a few key measures of profitability to compare the economic situation of the three systems. We base the profitability assessment on farm budget analysis modelling and on the farm business income (FBI), sometimes referred to as farm business profit (FBP). This value is derived by subtracting the production costs, that is, both fixed costs (FC) and variable costs (VC), from the gross production value (PV) (DEFRA 2010):

$$\mathrm{FBI} = \mathrm{PV} - \mathrm{VC} - \mathrm{FC} \tag{1}$$

The variable costs, that is, costs that depend strongly upon the size of the heard such as feed and medicine, are generally the most important in pastoralist production systems. Fixed costs include the cost of holding pen construction, storage facilities and traditional pastoralist tools. In the case of supplementary cattle feeding in Mulbadaw, the most commonly used equipment (fixed costs) are regularly maintained and repaired. We aggregated these costs as fixed maintenance costs, using Tanzanian shilling (TZS) based on the price level for the first quarter of 2014.

2.3 Data Material and Model Assumptions

We did 29 interviews with farmer/pastoralists in the collaborating villages of Gawidu, Dirma and Mulbadaw/Hiriri, between March 30 and April 1, 2015. Respondents were selected at random by farm group leaders within each of the villages. Participants included: farmers practising both improved and traditional agro-pastoral beef cattle keeping, solely traditional pastoral beef cattle keepers and extension officers. The interviews were conducted using a semi-structured questionnaire that covered socio-economic and farming system characteristics. The focus of the survey was mainly on economic and husbandry figures related to the cattle keeping, that is, herd size, diseases and mortality, feeding and perceived limitations to production. We used Excel to calculate means, frequency distributions or percentages describing the production characteristics (i.e. farm and family) of the respondents or farming systems.

We based our modelling on an average cattle herd, which included a total of 48.5 animals out of which 18 were cows. Although pastoralists practice random mating, we assumed a peak calving season at the beginning of the dry season on December 1. We estimated the birth weight of calves to 20 kg and daily weight gains of 500 g during the suckling period, enabling calves to reach 50 BW kg when weaned at 5 months at the end of the rainy season. Figures for weight gains according to feeding intensity (Table 2) were obtained from feeding experiments undertaken both at the HLH Farm (Selemani et al. 2015) and elsewhere, in Tanzania and Kenya. The weight gains were assumed to be similar for steers, bulls and heifers. Steers and bulls kept for meat were presumed slaughtered at 480 kg live body weight (LBW), and dressing percentage was 50. The bottom line in Table 1 shows the age at slaughter of the three systems.

We used information from pastoralists, extension worker and researchers to estimate age at first calving, mortality, calving intervals and herd replacement rates. Time for first calving can be lowered from three to two years for heifers, with regular concentrate supplementation as in the LSS and MSS alternatives. Although some heifers calve at 2 years, average calving age in the TS is 3 years. For cows in all treatments, we modelled biannual calving due to low feed intake. Across treatments, the replacement rate for cows and animal mortality was 14 and 15%, respectively.

In Table 3, based on the information from the pastoralists, we present some essential price estimates used for the modelling. The cost of concentrates was calculated from a standard feed mixture, corresponding with the composition of the feed used in the feeding experiments and consisting in 70% maize bran, 27% sunflower seed cake, 2% mineral mix and 1% salt.

Traditionally, livestock graze rangelands and also consume after-harvest crop residues on agropastoralists' fields. These crop residues have limited market value due to their low energy contents and high transaction costs in bringing them to feeding places. The crop residues may have an alternative economic benefit to be included when modelling. Since manure is naturally left behind when cattle graze on residues or for feedlots, manure could be relocated to the field; we have not differed between systems.

Table 2 Parameters for weight gains and supplementation levels for females, steers and bulls

System	TS[a]	LSS[b]	MSS[b]
Supplementary feeding, concentrates, kg/animal/day:			
Dry season	0	0.3	0.4
Rainy season	0	0.2	0.3
Body weight gain (BWG), g/animal/day:			
Dry season	0.12	0.22	0.28
Rainy season	0.34	0.38	0.52
Average	0.22	0.29	0.39
Age at slaughter, years:	6.0	4.5	3.5

[a]*Traditional system: basal diet, mainly grazing for all animals, occasional supplementation with crop by-products*

[b]*Low and medium supplementation systems: basal diets + different levels of concentrate supplementation*

Table 3 Selected price estimates for economic modelling

Items	US dollar[a]
Value of the carcass (240 kg slaughter weight)	324.9
Drugs, annual costs per animal	10.8
Price of concentrates, per kg	0.2
Market fee, per animal	1.4
Cost of transport to the auction area, per animal	1.3
Fencing and feeding arrangements, annual expenses per unit	33.6

[a]*Original costs in Tanzanian shillings (TZS), converted 1 USD = 2300 TZS*

3 Results and Discussion

3.1 Effects of Concentrate Supplementation on Gross and Net Margins

In Table 4, we present the effects of concentrate supplementation on gross and net margins among agriculturalists keeping cattle in the Hanang District, Tanzania.

The results of our modelling indicate positive effects of concentrate supplementation on farmer's income. In addition to the cost of concentrates, buying drugs was a significant expense for the agropastoralists in our study. The most common diseases in the area were anaplasmosis, contagious bovine pleuropneumonia (CBPP), East Coast fever (ECF) and diarrhoea. Since these costs were annual and per animal, the number of years the animals are kept before slaughtering will affect the total cost of medicine throughout the life cycle.

Pers. Communication, 2019, the HLH Farm Manager T. Rugland stated his assessment of the cattle keeping and concentrate supplementation practice as follows. The fact that the HLH Farm during the period from 2014 to 2019 has continued feedlotting of cattle, a new farm enterprise is a further indication of successful innovation. Mostly steers are purchased at local auctions, fattened for 3 months and sold at a profit to local butchers. Previously, we transported the animals 300 km to Arusha for slaughter at an established abattoir, allowing for better handling of the carcass, but the pastoralist communities could not assess valuable plus-products (blood, inner organs). The collaborating agropastoralists have also continued to buy bundled untreated wheat straw from the farm for dry season feeding. T. Rugland notes that this collaboration has nurtured a positive relationship between the parties.

3.2 Sustainability of the Current Pastoralist Production System

Recently, a rapid global increase in production and consumption of animal products has taken place. According to (FAO 2019a), it is expected that the demand for animal products will continue to grow.

Table 4 Effects of concentrate supplementation on gross and net margins among cattle owners in the Hanang District

System	TS[a]	LSS[b]	MSS[b]
Livestock enterprise output[c]:			
American dollar (USD)	706	1846	1981
Variable costs, USD:			
Veterinary drugs	469.6	429.5	403
Concentrates	0	462.7	556.8
Market fees	3.1	8.2	8.7
Gross margins from livestock[c]: USD	234	946	1013
Fixed costs, fencing and transport, USD	34.9	34.9	34.9
Adjusted net margins[d]: USD	199	911	978

[a]*Traditional system: basal diet, mainly grazing for all animals, occasional supplementation with crop by-products*

[b]*Low and medium supplementation systems: basal diets + different levels of concentrate supplementation*

[c]*Income calculation and exchange rate based on 2014 and 2019 figures, respectively. 1 USD = 2300 TZS*

[d]*Extra work for supplementing concentrates was not included*

Feed for livestock production currently accounts for one-third of the global cropland (FAO 2019b), competing for land, water, energy and labour. The vagaries of climate change and socio-economic pressures make the production of animal feedstuffs increasingly challenging.

Increasing productivity and more efficient use of inputs throughout the whole livestock sector will be fundamental to meet the growing demand for quality livestock products while minimising its impact on the environment and the world's natural resources (FAO 2019b). For small- to medium-scale livestock production systems, increased productivity is currently constrained by skill level, knowledge and appropriate technologies, compounded by insufficient access to markets, goods and services and weak institutions. The result is that both production and productivity remain well below potential, and losses and wastage can be high. However, adapted breeds, local feed resources and animal health interventions are available, along with improved and adapted technologies that include sound animal husbandry, on- and off-farm product preservation and value-adding product processing. When combined with supportive policies and institutional development in respective countries, innovative technologies may substantially improve productivity and income generation and contribute to poverty reduction (FAO 2019a).

The importance of ruminants is highlighted by the fact that Tanzania has 50 million inhabitants, 26 million cattle, 17 million goats and nine million sheep (FAO 2019c). Kurwijila and Boki (2003) estimated that only 2.7% of livestock are dairy animals, mostly crossbreeds. The country has the third largest cattle population in Africa after Ethiopia and Sudan. About 90% of the livestock population is of indigenous types, known for their low genetic potential in milk and meat production (Agricultural Sector Development Program 2016). The animals are kept under pastoral systems and are dispersed in most rangelands of the country. The method involves herders moving with high numbers of livestock animals. They do so to access the scattered, ecologically specialised and seasonally varied grazing lands and watering points. The system is vulnerable and may be among the drivers of deforestation and land degradation. Some of the efforts to make pastoralism more sustainable included making pastoralists more sedentary by providing watering points in 'rural villages. However, most of the rangeland is in a poor state due to overgrazing and wildfires. Productivity is low and significant areas of pastures are required to meet pastoral needs. During the dry season, the rangeland grasses lose nutritive value to such a degree that it may not be suitable for supporting any cattle production at all.

Cultural norms may in part explain why Tanzania's pastoralists continue to keep their cattle beyond the optimum point for slaughter. Cattle serve several roles as they represent the accumulation of assets, security, prestige and status, defined as wealth (Doran et al. 1979). To optimise income, efficiently rearing and selling cattle at the optimum time is paramount. Livestock may also serve as a form of savings account, which can be 'cashed in' to meet future planned or unexpected expenditures.

3.3 *The Feasibility of Feeding Concentrates and Options for Roughage-Based Supplementation*

The results of this study support the conclusion that concentrate supplementation has positive effects on farm income. One of the primary positive outcomes is the shortening of the time taken to produce a steer of given carcass weight. Furthermore, when we supplement young stock, heifers may give birth at a younger age, thereby increasing efficiency. Improved feeding could also lead to reduced mortality rates. Unfortunately, our dataset was not strong enough to validate such a claim. Currently, calving occurs typically every second or third year at no specific time of the year. Practising supplementation could also allow for annual calving when combined with a fixed calving season, which would lead to increased productivity.

In the recent review publication, 'Grazed and confused', Garnett et al. (2017) summarise current knowledge on red meat and its effects on global warming. The report indicates an adverse impact of the high consumption of red meat in industrialised countries. On the other hand, options for improved sustainable management of the vast global rangelands do exist. The natural pastures and crop residues, which are the primary feed resources in most developing countries, including Tanzania, cannot sustain effective animal production or even animal maintenance because of their inherent nutrient deficiencies, low digestibility and limited intake capacity of animals. Supplementation of the low-quality roughages with small quantities of high-quality supplements could enhance productivity or at least avoid body weight losses during critical feed shortage periods of the year. Supplementary forage legumes significantly improve the voluntary feed intake and digestibility of low-quality roughages. Moreover, forage legume supplements would improve N metabolism and utilisation and increase energy availability to the animals, leading to improved production and reproductive performance of animals. These measures may also enhance soil organic carbon and other soil qualities, improve rangeland carbon sequestration and thereby contribute to reducing climate change.

In our study area, the Hanang District, authorities reallocated large communal grazing areas for wheat farming, which in turn led to conflicts. However, a fruitful partnership between the HLH Farm and surrounding agropastoralists has now been established. Previously, to utilise the straw, pastoralists came in with the herd after the wheat was harvested, and this resulted in low utilisation of the feed as well as damage to the farm contours, which in turn led to erosion. Farmworkers in the study collected part of the straw and stored it for use by cattle feed in the dry period. We are still working on identifying the sustainable ratio for dividing the straw between animal feed and mulching for sustained soil productivity. Currently, we recommend a rate of equal use, that is, half the amount to remain in the field and the other removed and stored as dry season feed. To avoid depleting the soil fertility, farmers should monitor levels of soil nutrients. Interestingly, appropriate use of fertiliser results in increased yields of both grain and straw, benefitting both humans and cattle. Co-utilisation between large-scale farms and pastoralists in this way may lead to increased food production, both directly from the grain and indirectly through ruminants utilising the straw.

4 Conclusions

The results described in this study indicate that farmers should endeavour to supplement their young stock regularly, at least in times when the animals cannot sustain themselves on grazing alone. We recommend that future studies model the effect of supplementing cattle in feedlots before slaughter. Concentrate supplementation has predominantly been discussed in this paper; further studies should also assess the use of roughages and treated/untreated straw and stovers. Moreover, future studies should also address the supplementary feeding of cows.

References

Agricultural Sector Development Programme Phase Two (ASDP II) (2016) Tanzanian Government Programme Document, pp. 204. Visited 1 Apr 2019, www.tzdpg.or.tz/fileadmin/documents/external/national_development_frameworks/ASDP2--

DEFRA (Department for Environment, Food and Rural Affairs) (2010) Definition of terms used in Farm Business Management. Third edition. Published by the Department for the Environment, Food and Rural Affairs. Visited 14 Feb 2019, www.neat-network.eu/sites/neat-network.eu/files/Definition%20of%20terms.pdf

Doran MH, Low ARC, Kemp RL (1979) Cattle as a store of wealth in Swaziland: implications for livestock development and overgrazing in Eastern and Southern Africa. Am J Agric Econ 61(1):41–47

FAO (2005) Livestock sector brief. United Republic of Tanzania. Livestock information, sector analysis and policy branch. Visited 16 2019, www.fao.org/ag/againfo/resources/en/publications/sector_briefs/lsb_TZA.pdf

FAO (2009) How to Feed the World in 2050. Visited 16 2019, www.fao.org/fileadmin/templates/wsfs/docs/expert paper--

FAO (2019a) Livestock and animal production. Visited 16 Jan 2019, www.fao.org/ag/againfo/themes/en/animal_production.html

FAO (2019b) CountrySTAT, Tanzania. Visited 28 Jan 2019, www.fao.org/countryprofiles/en/

FAO (2019c) FAO's role in animal production. Visited 28 Jan 2019, www.fao.org/animal-production/en/

Garnett T, Godde C, Muller A, Röös E, Smith P, de Boer I, zu Ermgassen E, Herrero M, van Middelaar C, Schader C, van Zantenhttps H (2017) Grazed and confused? Ruminating on cattle, grazing systems, methane, nitrous oxide, the soil carbon sequestration question – and what it all means for greenhouse gas emissions. Food Climate Research Network, Oxford Martin Programme on the Future of Food Environmental Change Institute, University of Oxford. Visited 28 Jan 2019, www.fcrn.org.uk/sites/default/files/project-files/fcrn_gnc_report.pdf

Lane C, Pretty JN (1990) Displaced pastoralists and transferred wheat technology in Tanzania. Visited 22 Mar 2019, https://coady.stfx.ca/displaced-pastoralists-and-transferred-wheat-technology-in-tanzania/

Kurwijila LR, Boki KJ (2003) A review of the small scale dairy sector –Tanzania. Visited 16 Jan 2019, www.fao.org/fileadmin/templates/ags/docs/dairy/P1assessmenttanzania.pdf

Malcolm LR (2004) Farm Management analysis: a core discipline, simple sums, sophisticated thinking. Aust Farm Bus Manag J 1(1):45–56

MLDF (2013) Ministry of livestock development and fisheries. Budget Speech for 2013/2014. Visited 16 Feb 2019, www.parliament.go.tz/docs/83a5d-WIZARA-YA-MIFUGO-NA-UVUVI.pdf

Mushi DE, Eik LO, Bernues A, Ripoll-Bosch R, Sundstøl F, Mo M (2015) Reducing GHG emissions from traditional livestock systems to mitigate changing climate and biodiversity. I: Sustainable intensification to advance food security and enhance climate resilience in Africa. Springer Publishing Company 2015 ISBN 978-3-319-09359-8: 343-366

Safari J, Mushi DE, Mtenga LA, Kifaro GC, Eik LO (2011) Growth, carcass and meat quality parameters of Red Maasai sheep fed wheat straw- based diets. Trop Anim Pro Health 43(1):87–89

Sangeda AZ, Maole JL (2014) Tanzanian rangelands in a changing climate: impacts, adaptions and mitigation. Net J Agric Sci 2(1):1–10

Selemani IS, Eik LO (2015) The effects of concentrate supplementation on growth performance and behavioral activities of cattle grazed on natural pasture. Trop Anim Health Prod 48:229. https://doi.org/10.1007/s11250-015-0935-z

Selemani IS, Eik LO, Holand Ø, Ådnøy T, Mtengeti E, Mushi DE, Sørheim O (2015) Feeding strategies for improved beef productivity and reduced GHG emission in Tanzania: effect of type of finish-feeding on carcass yield and meat quality of zebu steers. I: Sustainable intensification to advance food security and ehance climate resilience in Africa. Springer Publishing Company 2015 ISBN 978-3-319-09359-8:367-382

Stephen M, Mbwambo N, Mruttu H, Dotto MM, Ndomba C, da Silva M, Makusaro F, Nandonde S, Crispin J, Shapiro B, Desta S, Nigussie K, Negassa A. Gebru G (2018) Tanzanian Livestock Master plan, Bill & Melinda Gates Foundation. Visited 2 Apr 2019, www.cgspace.cgiar.org/bitstream/handle/10568/92405/livestockMasterPlan_Tanzania.pdf

Tolera A, Eik LO (2019) Stratified livestock production and live animal and meat export from Ethiopia: lessons from the experience of a donor-funded project. Submitted for publication in "The Springer Book-Project"

Van Ittersum MA et al (2016) Can sub-Saharan Africa feed itself? PNAS, vol. 113. No 5, Visited 16 Jan 2019, www.pnas.org/cgi/doi/10.1073/pnas.1610359113

Integrating Smallholder Farmers to Commodity Value Chains in Sub-Saharan Africa: Challenges, Prospects and Policy Issues

Fredy Timothy Mlyavidoga Kilima and Lusato Revocatus Kurwijila

Abstract The participation of smallholder farmers in commodity value chains in sub-Saharan Africa has generally been low. Commodity and location-specific factors in favour of and against farmers' participation in agricultural value chains have been documented in contemporary literature. However, the analysis of farmers' prospects to integrate into such value chains has predominantly focused on specific commodities or groups of related commodities, leading to recommendations that apply to studied commodities only. There is a need to compare and contrast farmers' prospects to integrate into different agricultural commodities and generate comprehensive lessons on what works and under what circumstances. This chapter has adopted a multi-commodity approach to assess such prospects for smallholder dairy and vegetable farmers in selected countries in sub-Saharan Africa. The approach is deemed appropriate because it identifies challenges and reveals opportunities for a wider group of actors within the value chains. The main objective is to identify several productivity, quality and market linkage-enhancing interventions that can increase farmers' participation in value chains and their incomes.

1 Introduction

1.1 The Sub-Saharan Africa Context

Agriculture in sub-Saharan Africa (SSA) comprises predominantly smallholder farming (Staatz et al. 2007; Wiggins 2009), which is highly diversified to allow farmers meet basic demands for food and diversify their income portfolios (IFAD 2010).

F. T. M. Kilima (✉)
Moshi Co-operative University, Moshi, Tanzania
e-mail: ftmkilima@gmail.com

L. R. Kurwijila
Sokoine University of Agriculture, Morogoro, Tanzania
e-mail: kurwiji@sua.ac.tz

B. R. Singh et al. (eds.), *Climate Impacts on Agricultural and Natural Resource Sustainability in Africa*, https://doi.org/10.1007/978-3-030-37537-9_24

Empirical evidence from SSA reveals that agricultural growth and pro-poor performance depend on the participation of smallholder farmers in commodity value chains (Diao et al. 2007), which is currently low (Weinberger and Pichop 2009; Lie et al. 2012). The percentage of smallholder dairy farmers in countries such as Kenya, Tanzania, Rwanda, Uganda, Zimbabwe and Zambia who are linked to milk processing is estimated to be around 20%, 4%, 10%, 11%, 32% and 17%, respectively (Akwabi-Ameyaw 1997; de Janvry et al. 1993; Kilima et al. 2015; Sikawa and Mugisha 2016). There are both optimistic and pessimistic views regarding the participation of smallholders in high-value chains.

Pessimistic views revolve around four key issues: Firstly, the small-scale nature of smallholder farming may result in farmers' inability to supply consistent amounts of quality products, which would limit farmers' prospects to engage in commodity trade, where buyers seek to avert risks associated with supply disruptions and quality problems (Poirier and Quinn 2006). Evidence from Tanzania suggests that smallholder dairy goat farmers were not as keen as large processors to comply with milk quality standards, as they tend to focus on maximizing short-term rather than long-term gains and the sustainability of commodity value chains (Lie et al. 2012). There have been isolated incidences where smallholder dairy farmers have recorded significant productivity gains but failed to integrate into high-value milk chains owing to their inability to observe milk quality and standards (Kilima et al. 2015).

The second pessimistic view concerns the poor development of most value chains in SSA, due to inadequate investment in infrastructure and business support services. These deficiencies lead to poor coordination mechanisms and ineffective exchange of agricultural commodities (Barrett 2008; Poulton et al. 2010). This underdevelopment justifies continued subsistence farming in isolated, remote areas where there are limited incentives for smallholder farmers to engage in markets, because transaction costs for trade are generally high and prohibitive for farmers' engagement (Jayne et al. 2010; Fafchamps et al. 2005; Fafchamps and Minten 2012).

Thirdly, there exists a sustained mistrust between smallholder farmers and traders, rooted in unfair exchange in determining volumes and product quality, especially when accurate scientific measures are needed to determine value during the exchange process. Kirsten and Sartorius (2002) report that contract manipulation by agribusiness firms, and farmers' distrust of these firms, is rampant in SSA. In connection to this problem, a significant majority of illiterate smallholder farmers suspect that buyers attempt to gain through illicit practices such as understating actual volumes, using unobservable quality attributes of agricultural produce or both (Eaton and Shepherd 2001; Minot and Ngigi 2010; Mansur et al. 2009). Buyers and sellers are usually ill-informed about their interdependency in the context of the value chain, and both may have short-term incentives to misrepresent the truth and violate contract terms. The resulting mistrust impedes smallholder farmers in the upgrading of their practices, at a time when upgrading is critical for coping with the ongoing and rapid changes (transformation and modernization) of global food systems (Maertens et al. 2012). The mistrust is also associated with additional transaction costs for physically inspecting commodities before exchange, thereby eroding gains for sellers, buyers or both.

The fourth pessimistic view is linked to the joint effect of policy-related interventions and the existing (inadequate) infrastructure, which together distort markets and discourage long-term investment in crop and market development (Barrett 2008; Smale et al. 2012; Van Donge et al. 2012). Some scholars acknowledge that such policy distortions could have a greater negative impact on crop production and market development than the problems associated with the development of new technologies (Smale et al. 2012).

In contrast, optimistic views on enhancing smallholder farmers' participation in commodity value chains revolve around ongoing efforts to promote chain development in SSA. These efforts have primarily focused on identifying and piloting business models to overcome specific challenges while at the same time ensuring traceability and reducing exposure to risks such as inconsistent volume and quality of produce. Broadly speaking, these initiatives have attempted to address problems such as the high dispersion of producers, diseconomies of scale among producers and poor access to information, technology and finance. Actions that appear promising in overcoming these problems include tapping into the support of the public sector to improve market linkages among smallholder farmers and knowledge management through local and global innovation systems (Fischer and Qaim 2011; Kirsten and Sartorius 2002; Mojo et al. 2017; Pietrobelli and Rabellotti 2011; Riisgaard et al. 2008; World Bank 2006).

While varied forms of business models have been piloted, the participation of smallholder farmers in commodity value chain has persistently been low with the exception of isolated commodities such as high-value crops for export. This low participation is a result of the greater focus on subsidies or grants, rather than focus on the farmers themselves, or resources from other local actors to finance market and social development. This creates overdependence on external support and makes the initiatives unsustainable (Humphrey 2008). Moreover, literature citing the successful integration of smallholder farmers, such as Abdulai and Birachi (2009) and Nhodo and Changa (2013), does not fully account for variations in commodity markets, particularly regarding the structure of the value chain in relation to others. Consequently, factors such as the composition of actors and their behaviour, the level of chain development, the intensity of competition and the nature of the commodity have rarely been compared. In general, little is known about the efficacy of different business models in delivering benefits to smallholders and other actors along different value chains, implying that drivers of success or failure are not adequately understood to generalize the results. This chapter reviews relevant business models that have been deployed for similar purposes in SSA, so as to draw lessons from smallholder milk and vegetable farmers in selected countries in SSA and to guide future endeavours. The chapter seeks to:

(i) Map the application of different models in selected commodity value chains
(ii) Assess the performance of these models as applied to commodity value chains
(iii) Make recommendations to promote the integration of smallholder farmers to specific value chains

Practitioners are now seeking to identify appropriate means to promote smallholder farmers' participation in commodity value chains. Such means should rely on market-based approaches to ally value chain actors, so that they concentrate on particular activities that allow them to capture maximum benefits. This could be achieved through identifying the distribution channels, pricing mechanisms, product attributes, selling propositions and chain configurations that yield the greatest competitive advantage and livelihood outcomes for chain actors. It is thus crucial to focus on actor-specific market interventions rather than generic interventions that do not always help the poorest, because they do not usually have the capacity (human, physical resources or social) to participate effectively in market exchange. Poorer chain actors can potentially seize opportunities in local (farm-level) markets, which are easier to penetrate, less competitive and less stringent in terms of volume and quality requirements than larger markets.

A blending of interventions that are either solely private sector or solely public sector may ensure harmonized and dedicated support in promoting the integration of smallholder farmers in agricultural value chains locally and beyond. In cognisance of this requirement, this chapter reviews country-specific applications of the selected forms of vertical coordination to pinpoint the salient features and effects of such applications on farmers' participation. The ultimate goal is to document the key lessons learnt in order to inform future technical and policy interventions.

2 Methods

This chapter relies on existing literature to identify products and markets for which contract farming and collective action seem most suitable. This is achieved through the description of selected value chains and a desk review of empirical literature to map the applications of the models in selected commodities while justifying why farmers need to engage in exchange processes, pinpointing current and future prospects for their inclusion and revealing policy implications.

3 Description of Commodity Value Chains

3.1 Milk Value Chain

Studies conducted in SSA have identified huge growth potential for the dairy subsector that can benefit many value chain actors. The potential for a viable dairy industry in SSA stems from the ever-growing population in the region (currently estimated at 2.69% vis-à-vis the world average of 1.12%), along with a growth in consumers' incomes (Delgado 2003; Speedy 2003) and increased investment in small- to medium-scale milk processing plants, which both process their own milk and source from other dairy farmers (Kilima et al. 2015). These investments provide incentives for further growth in urban milk production. Milk marketing by individual smallholder dairy farmers is largely determined by their actual consumption needs at the household level, consumption by people in the neighbourhood, their potential to produce surpluses over their household requirements and their ability to access external markets, particularly urban centres. Research findings in Kenya and Nigeria reveal that milk demand is higher in urban areas than in rural areas, and urban dwellers consume more milk products than rural consumers in the same income bracket (Jansen 1992; Staal and Mullins 1996).

Milk production systems vary greatly within SSA, with a mix of smallholder dairy farmers and larger dairy farms (small, medium and large scale). Milk is produced by indigenous cows, crossbreeds and improved breeds of cow, with indigenous cattle being the most significant source of milk in the region (Tambi et al. 1997; Kurwijila 2002; Njombe and Msanga 2009). Milk production systems include grazing systems, mixed farming systems and industrial (or landless) systems.

In general, milk production and marketing systems are classified as either urban, peri-urban or rural systems (Tsehay 2002). The majority of consumers drink fresh liquid milk, and its marketing is dominated by a traditional-informal value chain, with only a small proportion of the total production being marketed through the formal chain.[1] Most of the rural systems are integrated crop–dairy systems, to exploit synergies between dairy enterprise, staple food crops and other cash crops (Kurwijila 2002; Staal 2002). In SSA, there are pastoralist and agropastoralist milk production systems, which are both considered part of the rural system. Access to milk markets by producers in these systems depends on the cost of collection and transportation of milk, which is determined by the distance to processing centres and consumption areas (Omiti and Staal 1996; Omiti 2002).

The way milk is produced and marketed raises several concerns regarding the possibility and capacity of 'ordinary' smallholder dairy farmers to participate in the formal value chain. There is evidence suggesting that milk supplied by smallholders is not consistent in its availability, with more milk produced during the rainy season and less in the dry season. The informal milk market cannot absorb the surplus milk during the flush season. This surplus could be channelled to buyers in the formal channel, who have the capacity to prolong the shelf life of milk and even out the supply between low and high seasons. However, many farmers have been unable to market their milk in formal channels because of their inability to differentiate milk products and identify market niches, communicate product quality and safety to immediate buyers and reduce costs of production while maintaining quality and high levels of production.

---

[1] In the context of the milk value chain, a formal value chain is defined as a chain whose actors are licensed to handle milk after complying with defined minimum standards for competence in hygienic milk handling and processing to safeguard consumers' health. This chain is operated as an organized system of milk collection, using well-established mechanisms of bulking (e.g. through cooperative societies or agents) and transportation in insulated tankers to factories where it is processed and packaged before marketing.

Farmers' continued participation in the informal value chain is associated with several problems. First, they are exposed to market risks (e.g. price shocks) because forward contracts cannot be enforced effectively, partly due to uncertainties over milk quality and partly due to weak institutional capacity to enforce contracts. Second, the prospect for increasing earnings is limited as farmers supply unprocessed (low-value) milk. This is also likely to happen to formal chain participants if there is poor price transmission between upstream and downstream actors. Third, milk buyers in the informal chain tend to buy small quantities of milk, as they have limited capacities to store the milk or distribute it to distant markets, which, in turn, limits growth in milk production and market development.

Experience from East Africa shows that farmer's institutions are important vehicles for market integration. Farmer groups, associations and primary co-operative societies enable collective action, economies of size and increased bargaining power vis-à-vis corporate buyers. In Kenya, *Kenya Co-operative Creameries Ltd.* (KCC) has been the bedrock of dairy enterprise development over many decades (Muriuki 2011). In Tanzania, the *Tanga Dairies Co-operative Union* (TDCU) has anchored more than 6000 smallholder farmers firmly in the formal dairy value chain linked to *Tanga Fresh Ltd.*, the biggest dairy processor in the country (Kilima et al. 2015). In the southern highlands of Tanzania, *Njombe Livestock Farmers Association* (NjoLiFa), a farmers' co-operative in the Njombe region, has amalgamated the individual work of farming families into a formidable player in milk processing and the production of high-quality cheese, linked to the distant Dar es Salaam market, the commercial capital of Tanzania. Similar stories abound in Uganda and Rwanda, the rising stars in the dairy hub of East and Southern Africa (Technoserve 2008a, b).

3.2 Vegetable Value Chain

Historically, vegetables have played a vital role for farming communities in many SSA countries. In the context of resource-poor smallholders, these crops are particularly important because they can be successfully farmed in small areas and are usually intercropped with other crops to enhance resource-use efficiency. The perception that vegetables are subsistence crops has led to the neglect of these crops in terms of crop and market development initiatives (Weinberger and Lumpkin 2005). Consequently, the production and marketing of vegetables in SSA has been a modest affair, with produce being grown in small quantities for local consumption, resulting in less investment in processing. In general, the aggregate demand for horticultural crops in SSA shrunk between 1971 and 2000 (Weinberger and Lumpkin 2005).

A significant change in production and marketing was noted after the 1990s, when African countries changed their trade policies from protected to more open economies, which paved the way for new trade agreements that created new avenues for trade in high-value products. Since then, vegetable production became one of the opportunities for farmers in sub-Saharan African countries to increase their export revenues (Barrett and Browne 1996). A significant portion of this trade was accounted for by export to Europe under contract farming, or hybrid forms of this vertical coordination. This coordination was through cold chains for handling the perishable products.

In summary, a larger proportion of total vegetable production is still intended for local consumption in spot markets that are located in urban centres, without contractual agreements. However, there are often informal arrangements that allow farmers and buyers to perform some of the functions that would otherwise exist under formal contracts.

Literature reveals several challenges for smallholder vegetable growers to participate in lucrative markets including:

(i) Their limited ability to afford the costs required to align their farms and post-harvest practices to meet specific requirements (e.g. product variety, standards, delivery methods and food safety), which are gaining increasing importance in export markets (Dolan and Humphrey 2000).
(ii) Negative effects of demand shocks; unlike milk, which is largely consumed locally, some of the vegetables produced under contract farming (e.g. baby corn and green beans) rarely have high domestic demands in SSA. This can result in high exposure to risk from demand shocks for farmers, exporters or both (Ashraf et al. 2009).
(iii) Sudden change in contract terms; most of vegetables exported to Europe have been certified products (e.g. organic, EurepGAP). Certified products are produced under standard practices and processes that may change when growers are unable to adjust their practices and processes (Averill 2007).
(iv) Women who are normally disadvantaged in terms of resources (land and valuable assets) may not participate fully to derive the same benefits as men (Carney 1992).

3.3 Common Challenges and Key Analytical Issues

Milk and vegetables are high-value products, but are characterized by a high ratio of transaction costs to final value because they are perishable and require heavy investment in processing to prolong shelf-life (Hobbs and Young 2000; Sinja et al. 2006). Production of these high-value products is an opportunity for smallholder farmers to increase returns to improve family livelihoods, but poorer farmers may lack the necessary resources to engage in production. The real gain that poorer farmers obtain from participation in production and marketing may be much lower than the nominal price in the market. The high transaction costs arise from market imperfections that limit farmers' access to services, such as credit. This limits participation of poor smallholders in activities that can potentially enhance their access to alternative (formal) markets.

The problem of poor participation of small-scale dairy and vegetable farmers in the formal value chain is also due to the challenges farmers face in local production and marketing. The problem is compounded further by the new competitive environment of liberalized and globalized food markets, where free trade and food safety are receiving ever-increasing global attention (Danielou and Ravry 2005; Swinnen and Maertens 2007; Maertens et al. 2012). Liberalization and globalization usually lead to an influx of imported products, which in turn leads to stiffer price competition and the adoption of new strategies of chain management. These changes constitute new challenges for small dairy and vegetable farmers in SSA, where standards are not strictly enforced in the market place. In many SSA countries, food standards for local markets are below those in export markets and there are no effective means to ensure countrywide compliance (Jayasuriya et al. 2006). It is worth noting that developed and developing countries in SSA face different economic, technological, institutional and political challenges. A change in standards in a developed country may only result in a slight change within the industry and require a strengthening of the domestic regulations to ensure that new standards are met. However, for emerging agricultural value chains in SSA, particularly in countries where institutional failures are rampant, a new standard may require major investment in infrastructure and significant legal and/or organizational changes to redress such failures. Furthermore, the costs of compliance can vary markedly across firms and countries because they are accustomed to different national standards and ways of life and they face unique challenges.

Economic literature identifies intra-firm and inter-firm coordination and management of chain activities as the best way to improve chain efficiency and gains to chain actors (Mazacco 1996). Various forms of chain governance structures (vertical coordination forms) have been developed by lead actors in a bid to reduce transaction costs so as to gain competitive advantage (Hobbs 1996; Hobbs et al. 1998; Hobbs and Young 2001). In situations where quality control is important, the trend has been a shift from loosely coordinated chains to closely coordinated ones and, in extreme cases, to full vertically integrated chains. The loosely coordinated chains are mostly characterized by spot procurement, whereas the relationship between actors in closely coordinated chains are forged through either contractual arrangements, joint ventures, strategic alliances or vertical integration. The informal and formal milk and vegetable value chains in SSA are likely to have their own unique governance structures, falling either within the categories discussed here or variants of the same. It is important to identify these structures and assess their effect on smallholder farmers' entry, the chain efficiency and participants' benefits.

3.4 Integration of Smallholder Farmers to Commodity Markets

The rationale for integrating smallholder farmers into commodity value chains in SSA and other developing countries is deeply rooted in the desire to promote the adoption of new institutional economic (NIE) theory, with the purpose of matching governance forms to market failures and transaction characteristics in the region (Kirsten and Sartorius 2002). Almost all countries in SSA are witnessing major changes in terms of property rights and market set-up following the previous and on-going economic changes including market liberalization (Poulton and Hanyani-Mlambo 2004) and transformation and the modernization of global food systems (Swinnen and Maertens 2007; Maertens et al. 2012). These changes have caused new private vertical coordination systems to emerge to address market failures resulting from the collapse of state-managed systems and other market shocks. The most significant challenge is that liberalization, and technological and other market changes, can hardly allow smallholder farmers to gain market access and realize increased productivity and income growth unless coupled with appropriate institutional innovations that fuel market development and address other market entry barriers.

The predominant forms of vertical coordination are contract farming and collective action. However, most of the means of vertical coordination are biased towards producers of export crops (e.g. cotton), estate crops for large-scale processing and export (e.g. sugarcane), highly perishable products (e.g. milk) and selected high-value crops (e.g. vegetables and flowers), also for export, but not smallholder farmers' food crops.

Contract Farming

In the context of agricultural commodities, contract farming entails a farmer committing to produce according to a prior agreement and a buyer to purchase the produce from the farmer. These agreements normally specify volume, quality and time of delivery as well as the type of input to be used and the price or pricing formula to be adopted. Contract farming may particularly be important when spot purchase is ineffective to facilitate commodity exchange and assure compliance with requirements such as volume, codified standards and traceability (Narrod et al. 2009; Kirsten and Sartorius 2002). Formal contractual agreements have been reported in countries such as Zimbabwe (Nhodo and Changa 2013), Kenya (Abdulai and Birachi 2009), Senegal (Swinnen and Maertens 2007), Madagascar (Minten et al. 2009), Malawi (Kumwenda and Madola 2005) and Ghana (Eaton and Shepherd 2001), while informal contracts have been reported in Uganda (Bolwig et al. 2009) and Zimbabwe (Masakure and Henson 2005). Moreover, variants that lie between the 'loosely coordinated' and 'closely coordinated' forms of contracting, herein generally referred to as 'hybrid forms', have also been reported in countries such as Kenya (Strohm and Hoeffler 2006). The hybrid forms of coordination entail one or more intermediary agents, where famer co-operatives fit in pertinently.

An evaluation of contracts involving smallholder farmers in SSA reveals both good and bad prospects. There have been concerns with respect to the emergence of captive markets, where a wider adoption of legally binding contractual agreements might depress spot market prices and reduce farmers' earnings, which has been the root cause of side selling in some countries (Eaton and Shepherd 2001; Minot and Ngigi 2010; Mansur et al. 2009). In addition, there has been concern regarding abrupt changes in demand or supply side factors that can cause a flourishing contract scheme to collapse, leading to a disruption in production (Reardon and Barrett 2000). Other concerns relate to deliberate moves by some exporters to use contracts as interim measures to stabilize supply while contemplating the adoption of their own long-term measures to meet such an objective or contracting large-scale producers (Minot and Ngigi 2010). Equally, sudden changes in standards may be detrimental to smallholder farmers who are incapable of upgrading practices instantly to comply with new requirements (Minot and Ngigi 2010; Narrod et al. 2009).

Moreover, selectivity bias has been noted across almost all contracts within SSA. This is because buyers (processors and exporters) usually prefer to work with medium- and large-scale farmers, and not smallholders who require more support in terms of accessing credit, inputs and technical assistance. Socio-economic characteristics of smallholder farmers have also been identified to influence agricultural contracts. Minten et al. (2009) found that contract growers in a vegetable value chain in Madagascar were relatively better off than non-contracted farmers in terms of their education and welfare levels, as well as income stability. Maertens and Swinnen (2009) found that producers who were contracted to produce vegetables and fruits for export were either independent farmers or employed within the export sector and had significantly higher incomes than those not working in the sector. Moreover, the literature also revealed that farmer contracts have been effective in selected commodities (Abdulai and Birachi 2009; Barrett and Bowne 1996).

Some scholars have identified that inclusion of smallholder farmers could be achieved through intermediation involving a wide range of agencies such as co-operatives, village leaders and local traders, especially local buyers and sellers (Coulter et al. 1999; Narrod et al. 2009). The rationale for this is that outsourcing services such as input control, field visits and quality measurement can be more effectively performed by an agent than by a foreign buyer. While possible, this arrangement raises concerns with respect to the way agencies generate their revenues to finance contract-specific activities and the implications for farmers' earnings.

An evaluation of contract farming involving grower schemes in Tanzania by West (2017) reveals several challenges with respect to: (1) harmonizing the interests and motives of actors in the public and private sector, (2) variation in levels of investment and market set-ups, (3) variation in agroecosystems (e.g. rainfall amount and distribution along with soil conditions) where commodities are farmed and (4) differences between small- and large-scale producers with respect to their exposure to risks, ability to hedge against such risks and rewards. Small producers' incentives to engage in these schemes seems to be diversifying and securing their livelihoods; this discourages long-term investment and commercialization in agricultural enterprises and threatens the sustainability of contract farming and the participation of small-scale producers.

Collective Action

Collective action refers to actions jointly undertaken by a group of people whose goal is to enhance their status and achieve a common objective. New institutional economics (NIE) identifies this form of action as being governed by institutional rules, including behavioural norms, by which economic agents interact to achieve desired outcomes (World Bank 2002). The theory asserts that formal rules and regulations, norms of collective action and enforcement mechanisms play key roles in agricultural markets. These means of governance are vital for dissemination of information to members and business partners, mediating transactions, facilitating the transfer and enforcement of property rights as well as managing competition (Peacock et al. 2004; Valentinov 2007). These instruments and enforcement mechanisms can be used to remedy market imperfections in rural markets. Institutional innovations that reduce transaction costs and enhance market coordination through producer organizations and marketing groups are vital in overcoming some of these problems.

Farmer organizations, particularly co-operatives, play a key role in coordinating agricultural and production services in SSA (Bingen et al. 2003; Kydd and Dorward 2004; Shiferaw and Muricho 2011). Historically, earlier co-operatives were introduced by colonialists to promote the production of cash crops. These institutions were inherited by decolonizing governments, intensively promoted even in areas where the concept was new or unpopular, highly subsidized and involved in many aspects of the production, marketing and trade of agricultural commodities (Baffes 2003; Shiferaw et al. 2009). During this period, the performance of co-operatives was generally impressive because of the dedicated government support through intensive market regulations and interventions. However, over time, governments in SSA incurred huge expenses to sustain these institutions, leading to economic hardships and fiscal deficits that required major reforms to restore macroeconomic balance and efficiency. Thus, many sub-Saharan countries were forced to adopt structural adjustment policies and liberalized their economies in the 1990s. One of the preconditions to liberalize the economies was an immediate withdrawal of government support to co-operatives (Develtere et al. 2008). This withdrawal created a void that was not immediately filled by the private sector. The private sector emerged slowly in areas of high agricultural potential where commercial crops were grown, and there was a tendency to purchase produce from commercial farmers rather than smallholder farmers in low agricultural potential areas (Jayne et al. 2002). To date, the prevailing institutional environment and the internal governance structures of co-operatives have not fully adjusted to accommodate the changing economic conditions. The most significant challenges for co-operatives in SSA include the adoption of ineffective means to promote members' participation and independence, along with the absence or incompleteness of legal instruments to ensure adequate oversight on co-operative affairs (Kydd and Dorward 2004; Develtere et al. 2008).

An assessment of the adoption of this means of vertical coordination, and the effectiveness of the adoption, reveals that co-operatives in SSA may not offer all of the required support to smallholder farmers (Shiferaw et al. 2009) because they tend to have multiple roles while facing several operational and institutional challenges. One critical challenge regarding the role of co-operatives in facilitating farmers' access to inputs and markets is the institutional means to reduce operating costs, especially when compared with trading partners who tend to be investor-oriented firms. Costs for co-operatives are often high, owing to operation inefficiencies that result from centralized decision-making involving all members or their representatives (Chevallier 2011). Consequently, without external influence, farmer co-operatives in SSA rarely contract outputs and provide inputs on credit to members, but are largely seen as means to enhance competition in procurement of agricultural produce. Many of these organizations have not been able to broaden their functions/services, instead of performing core functions such as milk collection, and thus usually fail to seize new opportunities.

As in contract farming, issues related to differential impacts also apply to farmer organizations. A study conducted in Rwanda reveals larger income effects of membership for members with larger farms (Verhofstadt and Maertens 2014). Bernard and Spielman (2009) found that poorer farmers tend not to participate in these organizations although they may derive indirect benefit(s), while Tesfay and Tadele (2013) found limited prospects for women in Ethiopia to engage in marketing co-operatives. Fischer and Qaim (2012) echo these concerns with an example of the commercialization of agricultural produce through a co-operative system in Kenya that lead to men taking over the marketing responsibility from women.

The marketing structure for milk and vegetables tends to involve several intermediary agents. For example, milk marketing co-operatives may involve primary societies that deliver milk to processors through contracts, signed by their respective apex organizations. The intermediation is likely to be associated with high transaction costs that could be minimized through allowing primary societies to directly enter into agreement with the processor on their own account. In Tanzania, for example, dairy farmers were charged Tshs. 20–30, 10, and 10 per litre of milk as primary society levies, union levies and milk transport costs to the plant, respectively (Kilima et al. 2015).

A brief analysis of vertical coordination forms in the milk value chain in selected countries within SSA reveals small market shares for smallholder farmers (4–20%) where, in isolated incidences (e.g. in Zimbabwe), a larger proportion is supplied by large-scale dairy farms (Table 1). The value chain faces various challenges, ranging from farmers' inability to meet quality requirements and participate effectively in marketing to agency problems (mismanagement of co-operatives) and moral hazard issues (side selling and loan default).

A similar analysis of vertical coordination mechanisms in the vegetable value chain (Table 2) reveals that farmers' market shares are relatively larger (>27) than shares for smallholder dairy farmers. In Ghana and Madagascar, for example, smallholder farmers are the major suppliers of exported vegetables. However, this seems to occur where: exporters have limited choices when contracting suppliers (Minten et al. 2009) or are willing to diversify their supply (Swinnen 2005); labour

Table 1 Key features of selected milk value chains from SSA countries

Business model	Country	Driving force	Rationale for smallholder milk producers	Precondition	Approximate market share (%)	Challenges encountered	Source of information
Collective action	Kenya	Perishability.	• Easing bulking and transportation of milk to processing plant. • Ensuring adequate oversight on milk quality and economies of size during transportation, processing and marketing. • Group as platform to access other business support services.	Previous experience with contract farming.	20	Difficulties in holding management accountable to members leading to: • Mismanagement of resources; • Overinvestment in fixed assets; and capacity underutilization.	de Janvry et al. (1993); Akwabi-Ameyaw (1997)
Contract and collective action	Tanzania	• Perishability. • Market uncertainty.	• Easing bulking and transportation of milk to processing plant. • Ensuring adequate oversight on milk quality and economies of size during transportation, processing and marketing. • Group as platform to access other business support services. • Securing marketing. • Stabilizing earnings.	• Establishment of milk processing plant. • Governance system involving co-operatives.	4	• Difficulties in holding management accountable to members. • Non-compliance with terms of contract (side selling).	Kilima et al. (2015)
Contract	Rwanda	Price risk.	Hedging season low price during milk flush season.	Established milk collection centres.	10	Frequent milk rejection	http://www.value-chains.org/dyn/bds/docs/772/SNVRwandaDairy.pdf

(continued)

Table 1 (continued)

Business model	Country	Driving force	Rationale for smallholder milk producers	Precondition	Approximate market share (%)	Challenges encountered	Source of information
Contract and collective action	Uganda	Price risk.	• Securing earning through contract. • Accessing credit. • Accessing information and advisory services.	• Establishment of milk processing plant. • Governance system involving co-operatives.	10.5	Participation much lower for dairy farmers producing less milk.	Sikawa and Mugisha (2016)
Contract	Zimbabwe	Market assurance.	[a]	Well-established dairy sector.	32	Limited participation of smallholder.	http://www.snv.org/public/cms/sites/default/files/explore/download/rarp_2016-dairy-subsector-study.pdf
Contract	Zambia	[a]	[a]	[a]	17	No milk is sourced from smallholder dairy farmers owing to limited data on farmers' location and production levels.	http://www.lrrd.cipav.org.co/lrrd25/4/mumb25073.htm

[a]Means information not available

Table 2 Key features of selected vegetable value chains from SSA countries

Business model	Country	Driving force	Rationale for smallholder vegetable producers	Precondition	Approximate market share (%)	Challenges encountered	Source of information
Contract and collective action	Ghana	Market assurance.	• Overcoming problems related to poor access to credit and farm inputs.	• Tight relationships with exporters. • Effective use of smallholder farmer groups organized by exporters.	85	• High transaction cost in dealing with less educated and inexperienced smallholders.	Gorman and Webber (2010)
Contract and collective action	Kenya	Market assurance.	• Reducing market transaction cost. • Easing market access.	• Rise in supermarkets. • Duty-free access to European markets.	27	• Difficulties in ensuring compliance with food safety and quality requirements. • Traceability requirements make smallholder farmers' participation difficult.	Minot and Ngigi (2010)
Collective action	Senegal	Market assurance.	Joint effort to achieve certification and quality control, access information and gain market access.	Effective engagement of exporters to ensure farmers' access to inputs, logistical services and technology.	52	• Difficulties in meeting stringent food quality and safety standards. • Framers' inability to meet supply requirements. • Inadequate experience of farmers' federations to work with private sector.	Swinnen et al. (2013)
Contract	Madagascar	• Easing market access. • Hedging against seasonal price fluctuation.	Assurance of export market.	High level of transparency. Micro-contract with farmers and rigorous requirement for all suppliers. Attractive benefits to contracted farmers that are stipulated in the contracts. Good quality and standard oversight.	90–100	• Contract enforcement problems including side selling. • Weak domestic demand for high-quality products.	Minten et al. (2005)

for production is cheap; natural resources to support vegetable production, particularly land and water, are abundant (Swinnen et al. 2013); and exporters have their own means to mobilize producers into effective farmers' groups. However, there are uncertainties regarding farmers' continued participation in both milk and vegetable value chains because their participation is predicted to decline with the increase in quality and standard requirements (Henson and Hamphrey 2010; Danielou and Ravry 2005).

4 Conclusions and Policy Implications

4.1 Conclusion

The chapter sought to map the application of contract farming and the co-operative model of business to assess smallholder farmers' prospects for integrating into agricultural commodity value chains, using milk and vegetable value chains from selected countries in SSA as case studies. The analysis has revealed complex dynamics within the value chains that result in limited scope to achieve the desired outcome of promoting the linkage of smallholder farmers to formal markets, using the identified means of vertical coordination.

In terms of the application of the selected forms of vertical coordination, the major findings are:

(i) We found crop-specific biases in terms of the application of the two business models, where there is keener application of the hybrid forms of vertical coordination in the vegetable value chain than in the milk value chain. This bias is obviously meant to strengthen oversights through effective engagement of intermediary agents.
(ii) Most of the studied coordination mechanisms are buyer (exporter)-driven. Buyers' main incentive to engage smallholder farmers seems to be to seize opportunities in foreign markets, accorded by specific trade agreements that are bound to change over time.
(iii) We consistently found lower farmers' participation prospects where there are no economies of size, implying limited participation for smallholders producing less produce and those who are deficient in key competencies and resources, with women being potentially the most affected.
(iv) There are no effective institutional mechanisms to allow participating farmers to hedge against potential risks from supply disruption and abrupt changes in contract terms.
(v) Smallholder farmers' participation in high-value chains seems to be favourable when critical resources (labour, land and water) that are needed for production are readily available and cheap. This raises concerns for farmers' future prospects, as climate shocks to agricultural production become more eminent.

4.2 Policy Implications

The findings from this analysis have important policy ramifications. The following recommendations are made to address the challenges undermining product and value chain development:

(i) Enhancing production efficiency at the farm level and other nodes of the value chains is the best way to enhance productivity within the value chains, promote business linkage between small-scale farmers and processors or exporters and ensure diversified income opportunities and gains to actors involved
(ii) Developing effective grades and standards, devising effective strategies to monitor and regulate production and handling processes, encouraging actors to uphold the existing quality and standard requirements and assess compliance thereof through dedicated regulatory authorities
(iii) Ensuring that there is efficient and effective management of marketing information to strengthen the ability of co-operatives to bargain and assume greater roles in marketing and, where possible, eliminate unnecessary intermediation. This intervention might help in reducing price erosion through deductions arising from the engagement of many intermediaries
(iv) Extending technical support to co-operatives to identify an ideal business model that can potentially entail the restructuring/reorganization of the current system based on objective assessment of operational and managerial efficiency
(v) Enhancing farmers' resilience to agricultural risks, especially climate change and other market shocks
(vi) Strengthening private and public sector collaboration in all endeavours aiming at raising agricultural productivity, product quality and farmers' access to agricultural technology, input and output markets

References

Abdulai A, Birachi EA (2009) Choice of coordination mechanism in the Kenyan fresh milk supply chain. Rev Agric Econ 31(1):103–121

Akwabi-Ameyaw K (1997) Producer cooperative resettlement projects in Zimbabwe: lessons from a failed agricultural development strategy. World Dev 25:437–456

Ashraf N, Giné X et al (2009) Finding missing markets (and a disturbing epilogue): evidence from an export crop adoption and marketing intervention in Kenya. Am J Agric Econ 91(4):973–990

Averill V (2007) African trade fears carbon footprint backlash. BBC News, 21. http://www.beatricespadacini.com/pdf/Media%20Work/African%20trade%20fears%20carbon%20footprint%20backlash.pdf. Accessed 31 Dec 2018

Baffes J (2003) Tanzania's coffee sector: constraints and challenges in a global environment. World Bank, Washington, DC

Barrett CB (2008) Smallholder market participation: concepts and evidence from eastern and southern Africa. Food Policy 33(4):299–317

Barrett H, Browne A (1996) Export horticultural production in sub-Saharan Africa: the incorporation of the Gambia. Geography 81(1):47–56
Bernard T, Spielman DJ (2009) Reaching the rural poor through rural producer organizations? A study of agricultural marketing cooperatives in Ethiopia. Food Policy 34(1):60–69
Bingen J, Serrano A et al (2003) Linking farmers to markets: different approaches to human capital development. Food Policy 28(4):405–419
Bolwig S, Gibbon P et al (2009) The economics of smallholder organic contract farming in tropical Africa. World Dev 37(6):1094–1104
Carney JA (1992) Peasant women and economic transformation in the Gambia. Dev Chang 23(2):67–90
Chevallier M (2011) The cooperatives' sources of efficiency: a catalyst for the emergence of localized norms. J Coop Stud 44(1):31–40
Coulter JA, Goodland A et al (1999) Marrying farmer cooperation and contract farming for service provision in a liberalizing Sub-Saharan Africa. Overseas Development Institute, London. Natural Resources Perspectives, No. 48: 1–4. https://www.odi.org/sites/odi.org.uk/files/odi-assets/publications-opinion-files/2869.pdf. Accessed 10 Dec 2018
Danielou M, Ravry C (2005) The rise of Ghana's pineapple industry. Africa Region Working Paper Series 93. The World Bank, Africa Region, Washington, DC. http://documents.worldbank.org/curated/en/788441468035966030/pdf/349970PAPER0GH0Pineapple0AFRwp93.pdf. Accessed 01 Mar 2019
de Janvry A, Sadoulet E et al (1993) Introduction to a special section, state, market, and civil organizations: new theories, new practices, and their implications for rural development. World Dev 21:565–575
Delgado CL (2003) Rising consumption of meat and milk in developing countries has created a new food revolution. J Nutr (Special Edition) 133(11):3907–3910
Develtere P, Pollet I et al (2008) Cooperating out of poverty: the renaissance of the African cooperative Movement. https://www.ilo.org/public/english/employment/ent/coop/africa/download/coop_out_of_poverty.pdf. Accessed 10 Dec 2018
Diao X, Hazell PB et al (2007) The role of agriculture in development: implications for Sub-Saharan Africa, vol 153. International Food Policy Research Institute. https://ageconsearch.umn.edu/record/55405/files/dsgdp29.pdf. Accessed 14 Dec 2018
Dolan C, Humphrey J (2000) Governance and trade in fresh vegetables: the impact of UK supermarkets on the African horticulture industry. J Dev Stud 37(2):147–176
Eaton C, Shepherd A (2001) Contract farming: partnerships for growth. http://www.fao.org/3/y0937e/y0937e00.pdf. Accessed 14 Dec 2018
Fafchamps M, Minten B (2012) Impact of SMS-based agricultural information on Indian farmers. World Bank Econ Rev 26(3):383–414
Fafchamps M, Gabre-Madhin E et al (2005) Increasing returns and market efficiency in agricultural trade. J Dev Econ 78:406–442
Fischer E, Qaim M (2011) Linking smallholders to markets: determinants and impacts of farmer collective action in Kenya. World Dev 40(6):1255–1268
Fischer E, Qaim M (2012) Gender, agricultural commercialization, and collective action in Kenya. Food Secur 4(3):441–453
Gorman M, Webber M (2010) Understanding the value chain and integrating information into strategy—Nigerian domestic catfish. In: Webber CM, Labaste P (eds) Building competitiveness in Africa's agriculture. A guide to value chain concepts and applications. World Bank, Washington, DC, pp 49–54
Henson S, Humphrey J (2010) Understanding the complexities of private standards in global agri-food chains as they impact developing countries. J Dev Stud 46(9):1628–1646
Hobbs JE (1996) A transaction cost approach to supply chain management. J Supply Manag 1(2):15–27
Hobbs JE, Young LM (2000) Closer vertical co-ordination in agri-food supply chains: a conceptual framework and some preliminary evidence. Supply Chain Manag 5(3):131–143

Hobbs JE, Young LM (2001) Vertical linkages in agri-food supply chains in Canada and the United States. http://publications.gc.ca/collections/Collection/A22-226-2001E.pdf. Accessed 20 Dec 2018

Hobbs JE, Kerr WA et al (1998) Creating international competitiveness through supply chain management: Danish pork. Supply Chain Manag 3(2):68–78

Humphrey J (2008) Private standards, small farmers and donor policy: EUREPGAP in Kenya. Working paper series, 308. Brighton: IDS. https://opendocs.ids.ac.uk/opendocs/handle/123456789/4167. Accessed 03 Mar 2019

IFAD (2010) Rural poverty report 2011 – new realities, new challenges: new opportunities for tomorrow's generation. IFAD, Rome

Jansen HGP (1992) Dairy consumption in northern Nigeria: implications for development policies. Food Policy 17(3):214–226

Jayasuriya SD, MacLaren D et al (2006) Meeting food safety standards in export markets: issues and challenges facing firms exporting from developing countries. A paper presented at the IATRC summer symposium on food regulation and trade: institutional framework. Concepts of analysis and empirical evidence. Bonn, Germany, 28–30 May 2006

Jayne TS, Govereh J et al (2002) False promise or false premise? The experience of food and input market reform in Eastern and Southern Africa. World Dev 3011:1967–1985

Jayne, TS, Sitko N et al (2010) Malawi's maize marketing system. Report commissioned by the World Bank and Government of Malawi/Ministry of Agriculture, Lilongwe. https://ageconsearch.umn.edu/record/62162/. Accessed 13 Dec 2019

Kilima FTM, Mvena ZSK et al (2015) Challenges of value chain development: experiences from milk production and marketing in Iringa Municipality and Tanga City. J Contin Educ Ext 6(2):1053–1068

Kirsten J, Sartorius K (2002) Linking agribusiness and small-scale farmers in developing countries: is there a new role for contract farming? Dev South Afr 19(4):503–529

Kumwenda I, Madola M (2005) The status of contract farming in Malawi. https://www.fanrpan.org/archive/documents/d00096/Contract_Farming_Malawi_May2005.pdf. Accessed 13 Dec 2018

Kurwijila LR (2002) An overview of dairy development in Tanzania. In: Rangnekar D, and Thorpe W (eds) Smallholder dairy production and marketing—opportunities and constraints. Proceedings of South-South workshop held at National Dairy Development Board, Anand, India, 13–16 March 2001, pp 174–184

Kydd J, Dorward A (2004) Implications of market and coordination failures for rural development in least developed countries. J Int Dev 16(7):951–970

Lie H, Karl M et al (2012) Improving smallholder livelihoods through local value chain development: a case study of goat milk yogurt in Tanzania. Int Food Agribusiness Manag Rev 15(2):55–85

Maertens M, Swinnen JFM (2009) Trade, standards and poverty: evidence from Senegal. World Dev 37(1):161–178

Maertens M, Minten B et al (2012) Modern food supply chains and development: evidence from horticulture export sectors in sub-Saharan Africa. Dev Policy Rev 30(4):473–497

Mansur K, Tola M et al (2009) Contract farming system: a tool to transforming rural society in Sabah. https://www.researchgate.net/publication/23991622_Contract_Farming_System_A_Tool_to_Tranforming_Rural_Society_in_Sabah. Accessed 15 Dec 2018

Masakure O, Henson S (2005) Why do small-scale producers choose to produce under contract? Lessons from nontraditional vegetable exports from Zimbabwe. World Dev 33(10):1721–1733

Mazacco MA (1996) HACCAP as business management tool. Am J Agric Econ 78(3):770–774

Minot N, Ngigi M (2010) Are horticultural exports a replicable success story? Evidence from Kenya and Côte d'Ivoire. In: Haggblade S, Hazell P (eds) Successes in African agriculture: lessons for the future. John Hopkins University Press, Baltimore

Minten B, Randrianarison L et al (2005) Supermarkets, international trade and farmers in developing countries: evidence from Madagascar. Cornell University, Ithaca, NY

Minten B, Randrianarison L et al (2009) Global retail chains and poor farmers: evidence from Madagascar. World Dev 37(11):1728–1741

Mojo D, Fischer C et al (2017) The determinants and economic impacts of membership in coffee farmer cooperatives: recent evidence from rural Ethiopia. J Rural Stud 50:84–94

Muriuki HG (2011) Dairy Development Report. Kenya Report, FAO Rome. SNV/PPD Consultants Ltd, June 2013. Dairy Sector Policy Study and Capacity Needs Assessment of Stakeholder Associations. Kenya

Narrod C, Roy D et al (2009) Public-private partnerships and collective action in high value fruit and vegetable supply chains. Food Policy 34:8–15

Nhodo L, Changa M (2013) Contestations surrounding contract farming in rural Zimbabwe: a case of Mukosi cotton growers in Masvingo South. Int J Polit Good Gov 4(1):1–16

Njombe AP, Msanga YN (2009) Livestock and dairy industry development in Tanzania. Department of Livestock production and Marketing Infrastructure Development, Ministry of Livestock Development, Tanzania

Omiti J (2002) Impact of liberalisation in Kenya's dairy sector. In: Rangnekar D, Thorpe W (eds) Smallholder dairy production and marketing—Opportunities and Constraints. Proceedings of South-South workshop held at National Dairy Development Board, Anand, India, 13–16 March 2001, pp 368–379

Omiti J, Staal S (1996) The Ugandan dairy subsector: a rapid appraisal. ILRI/NARO/MAAIF Research Report. ILRI (International Livestock Research Institute)/NARO (National Agricultural Research Organization of Uganda)/MAAIF (Ministry of Agriculture, Animal Industry and Fisheries of Uganda), Kampala, Uganda, p 137

Peacock C, Jowett A et al (2004) Reaching the poor – a call to action: investment in smallholder agriculture in sub-saharan Africa. Farm-Africa, London

Pietrobelli C, Rabellotti R (2011) Global value chains meet innovation systems: are there learning opportunities for developing countries? World Dev 39(7):1261–1269

Poirier CC, Quinn FJ (2006) Survey says: solid gains. Supply Chain Manag Rev 10(1):34–41

Poulton C, Hanyani-Mlambo B (2004) Competition and coordination in liberalized African cotton market systems. World Dev 32(3):519–536

Poulton C, Dorward AR et al (2010) The future of small farms: new directions for services, institutions and intermediation. World Dev 38(10):1413–1425

Reardon T, Barrett CB (2000) Agroindustrialization, globalization and international development: an overview of issues, patterns and determinants. Agric Econ 23:195–205

Riisgaard L, Bolwig S et al (2008) Integrating poverty and environmental concerns into value-chain analysis. Dev Policy Rev 28(2):195–216

Shiferaw B, Muricho G (2011) Farmer organizations and collective action institutions for improving market access and technology adoption in Sub- Saharan Africa: review of experiences and implications for Policy. https://www.researchgate.net/publication/281321739_Farmer_Organizations_and_Collective_Action_Institutions_for_Improving_Market_Access_and_Technology_Adoption_in_Sub-_Saharan_Africa_Review_of_Experiences_and_Implications_for_Policy. Accessed 24 Feb 2019

Shiferaw B, Obare G et al (2009) Leveraging institutions for collective action to improve markets for smallholder producers in less-favored areas. Afr J Agric Res Econ 3(1):1–18

Sikawa GY, Mugisha J (2016) Factors influencing south-western Uganda dairy farmers' choice of the milk marketing channel: a case study of Kirihura District- south western Uganda. https://www.mocu.ac.tz/upload/researches/SIKAWA%20AND%20MUGISHA.pdf. Accessed 12 Mar 2019

Sinja J, Njoroge L et al (2006) Milk market access for smallholders: a case of informal milk trader groups in Kenya. In: A research workshop on collective action and market access for smallholders, Cali, Colombia, pp 2–6

Smale M, Byerlee D et al (2012) Maize revolutions in sub-Saharan Africa. In: Keijiro O, Larson DF (eds) An African green revolution: finding ways to boost productivity on small farms. Springer, Dordrecht, pp 165–195

Speedy AW (2003) Global consumption of animal source foods. J Nutr (Special Edition) 133(11):4048–4053

Staal SJ (2002) The competitiveness of smallholder dairy production: evidence from sub-Saharan Africa, Asia and Latin America. In Rangnekar D, and Thorpe W (eds) Smallholder dairy production and marketing—Opportunities and Constraints. Proceedings of South-South workshop held at National Dairy Development Board, Anand, India, 13-16 March 2001, pp 250–264

Staal SJ, Mullins G (1996) Dairy consumption and its determinants in coastal Kenya. Kenya. Agricultural Research Institute-KARI/ILRI Collaborative Research Report. Nairobi, Kenya. ILRI. http://agris.fao.org/agris-search/search.do?recordID=QM9700012. Accessed 03 Nov 2018

Staatz J, Dembele N et al (2007) Agriculture for development in sub-Saharan Africa. Background paper for the World Development Report 2008. http://citeseerx.ist.psu.edu/viewdoc/download?doi=10.1.1.649.7875&rep=rep1&type=pdf. Accesses 13 Dec 2018

Strohm K, Hoeffler H (2006) Contract farming in Kenya: theory, evidence from selected value chains and implications for development cooperation. PSDA Main Report. http://www.fao.org/uploads/media/PSDA_CFKenyaSelectedVCs_Main%20Report_final.pdf. Accessed 13 Dec 2018

Swinnen JFM (2005) The Dynamics of Vertical Coordination in Agrifood Chains in Eastern Europe and Centra Asia. ECSSD Case studies working paper no. 41. World Bank, Washington, DC

Swinnen JF, Maertens M (2007) Globalization, privatization, and vertical coordination in food value chains in developing and transition countries. Agric Econ 37:89–102

Swinnen J, Colen L et al (2013) Constraints to smallholder participation in high-value agriculture in West http://www.fao.org/3/i3222e/i3222e09.pdf. Africa. In: Elbehri A (ed) Rebuilding West Africa's food potential. FAO/IFAD, pp 287–313

Tambi E, Mukhebi AW et al (1997) Privatisation of animal health services in Kenya: an evaluation of the Kenya Veterinary Association Privatisation Scheme (KVAPS). Organization of African Unity (OAU), Inter-African Bureau for Animal Resources (IBAR), Pan-African Rinderpest Campaign (PARC) and International Livestock Research Institute (ILRI) Report. ILRI, Nairobi, Kenya, p 84

Tesfay A, Tadele H (2013) The role of cooperatives in promoting socio-economic empowerment of women: evidence from multipurpose cooperative societies in south-eastern zone of tigray, ethiopia. International Journal of Community Development 1(1):1–11

Technoserve (2008a) The dairy value chain in Rwanda. https://cgspace.cgiar.org/bitstream/handle/10568/2410/Dairy%20Value%20Chain%20Rwanda%20Report.pdf;sequence=1. Accessed 13 Mar 2019

Technoserve (2008b) The dairy value chain in Uganda. https://cgspace.cgiar.org/bitstream/handle/10568/2406/Dairy%20Value%20Chain%20Uganda%20Report.pdf?sequence=1. Accessed 13 Mar 2019

Tsehay (2002) Small-scale milk marketing and processing in Ethiopia. In: Rangnekar D, Thorpe W (eds) Smallholder dairy production and marketing—Opportunities and Constraints. Proceedings of South-South workshop held at National Dairy Development Board, Anand, India, 13–16 March 2001, pp 352–367

Valentinov V (2007) Why are cooperatives important in agriculture? An organizational economics perspective. J Inst Econ 3(1):55–69

Van Donge JK, Henley D et al (2012) Tracking development in South-East Asia and sub-Saharan Africa: the primacy of policy. Dev Policy Rev 30(special edition):5–24

Verhofstadt E, Maertens M (2014) Can agricultural cooperatives reduce Poverty? Heterogeneous impact of cooperative membership on farmers' welfare in Rwanda. Appl Econ Perspect Policy 37(1):86–106

Weinberger K, Lumpkin TA (2005) Horticulture for poverty alleviation—the unfunded revolution. AVRDC working paper No. 15. http://203.64.245.61/fulltext_pdf/EB/2001-2010/eb0013.pdf. Accessed 13 Mar 2019

Weinberger K, Pichop G (2009) Marketing of African indigenous vegetables along urban and peri-urban supply chains in sub-Saharan Africa. In: Shackleton MP, Drescher AW (eds) African indigenous vegetables in urban agriculture, 1st edn. EarthScan, London, pp 225–244

West JJ (2017) Agricultural investments for development in Tanzania: reconciling actors, strategies and logics? Dissertation, Norwegian University of Life Sciences

Wiggins S (2009) Can the smallholder model deliver poverty reduction and food security for a rapidly growing population in Africa? FAC Working Paper No. 08. https://opendocs.ids.ac.uk/opendocs/bitstream/handle/123456789/2338/FAC_Working%20_Paper_08.pdf?sequence=1. Accessed 28 Feb 2019

World Bank (2002) World development report 2002: building institutions for markets. Oxford University Press, New York

World Bank (2006) Enhancing agricultural innovation: how to go beyond the strengthening of research systems. World Bank, Washington, DC

Economic Rationale of Using African Weaver Ants, *Oecophylla longinoda* Latreille (Hymenoptera: Formicidae) for Sustainable Management of Cashew Pests in Tanzania

Gration M. Rwegasira, Maulid M. Mwatawala, Rozalia G. Rwegasira, Abdullah N. Rashidi, Nene Wilson, and William George

Abstract Cashew nut, *Anacardium occidentale* Linnaeus, is an important crop to the African economy, contributing to livelihood of many families. It is among the top foreign-currency-earning crops in East and West African countries. The crop is attacked by several sucking pests that survive on multiple host plants. These include: *Pseudotheraptus wayi*, *Helopeltis anacardii*, *H. schoutedeni*, and *Selenothrips rubrocinctus*. Management of these pests is usually a challenge which triggers irrational use of synthetic pesticides. Excessive use of pesticides concerns human health and pollution in the environment. The use of African weaver ants, *Oecophylla longinoda*, has presented an alternative to pesticides for sustainable cashew pest management. The predator was found to be as highly effective ($P < 0.0001$) as the recommended insecticide (Lambda cyhalothrin) ($P < 0.0001$) in controlling cashew pests. The several economic analyses on the profitability of the technology over the recommended pesticides by partial budgeting, marginal rate of returns, benefit-cost ratio and net present value proved *O. longionda* to be superior to pesticides. The partial budgeting indicated a net benefit of US$7.72 per cashew tree by changing from insecticides to *O. longinoda* within two seasons.

G. M. Rwegasira (✉) · M. M. Mwatawala
Department of Crop Science and Horticulture, Sokoine University of Agriculture, Morogoro, Tanzania
e-mail: grwegasira@sua.ac.tz; mwatawala@sua.ac.tz

R. G. Rwegasira
Morogoro Region Administrative Secretariat, Morogoro Regional Secretariat, Morogoro, Tanzania

A. N. Rashidi
Research Ministry of Agriculture, Natural Resources, Livestock and Fisheries, Pemba, Tanzania

N. Wilson
TARI Naliendele, Mtwara, Tanzania

W. George
University of Dodoma (UDOM), Dodoma, Tanzania

B. R. Singh et al. (eds.), *Climate Impacts on Agricultural and Natural Resource Sustainability in Africa*, https://doi.org/10.1007/978-3-030-37537-9_25

The dominance analysis for marginal rate of returns indicated a net profit of US$11.39 per tree in two seasons compared to US$5.74 per tree gained from Lambda cyhalothrin. If adopted for use, the predator will provide a sustainable solution to cashew pest management and overcome the pesticide residue threats in marketed cashew from Africa.

1 Introduction

Management of cashew pests, wherever cashew nut is grown, has been a long-term challenge due to the perennial nature of the crop that ensures continued availability of the host plant, season after seasons. In addition, most of the notorious cashew pests are known to survive in multiple host plants which sustain their continued existence and subsequent reoccurrence on cashew plants whenever the suitable plant parts such as tender growing leaves, flower buds, inflorescences, fruits and nuts become available. Cashew nut growers in Africa rely heavily on pesticides (Martin et al. 1997; NARI 2010) whose advantages in controlling pests leave a lot of undesirable effects to human, the rest of fauna in agroecosystems and the environment. The limited knowledge on pesticide science and application among farmers in Africa further compounds the problems associated with pesticide use (Dwomoh et al. 2009). The farmers tend to use the same insecticide season after season provided it has proven to work, without worrying about the potential development of resistance by the target pest. In Tanzania, one of the major cashew-producing countries in sub-Saharan Africa (SSA), the use of Lambda cyhalothrin (Karate®) has been on preference by most farmers for more than a decade. Non-use of protective devices, poor dosages, wrong timing of application and poor handling of insecticides after use aggravate the negative effect of insecticides to farmers' health and the environment. Accidental or intentional poisoning is not uncommon among cashew growers in Africa. Insecticide abuse in cashew nut production has for long been a part and parcel of common practices, such that consumers in export markets have continuously called for safer products produced through organic farming. Developing methods for sustainable pest control that would provide alternatives to the chemical control of cashew pest would not only ensure healthy safety among consumers, it would but also provide protection to growers and the environment. The use of African weaver ants, *Oecophylla longinoda* Latreille (Hymenoptera: Formicidae), has for a long time been recommended for sustainable management of cashew pests, but knowledge on effective exploitation of the biocontrol agent was lacking. The comparative effectiveness of *O. longinoda* against commonly used insecticides has never been explored. The objectives of the current study were (i) to determine the comparative effectiveness of *O. longinoda* against major cashew pests and (ii) to undertake economic analyses of the *O. longinoda*-based management to quantify the economic rationale of using the biocontrol agent as an alternative to recommended insecticides. In the present report, *O. longinoda* has proven very effective against major cashew pests and promises a great breakthrough in cashew pest management strategies. The use of weaver ants was found to be comparatively as effective as insecticides, yet cheaper in cost and very sustainable. Detailed below are insights into the application of weaver ant technology on cashew pests based on the experiences in Tanzania.

2 Insect Pests as Cashew Production Constraints

Yield statistics suggest Africa to contribute half of the world production of cashew (FAOSTAT 2018). But comparatively, the yield per unit area attained in Asian countries is ten times that of Africa. The low production per unit area in SSA is attributed to damage and losses caused by insect pests and diseases at different growth stages (Maruthadurai et al. 2012). The compensatory value attained in the African continent is attributed to the extensiveness of the areas under production of the crop. Biotic constraints, particularly saps sucking insect pests and the fungal diseases (powdery mildew disease), *Oidium anacardii* (Noack), have largely contributed to the low yields (NARI 2010). These pests can lead to 60% to 100% yield losses, depending on the variety, location and season (Agboton et al. 2013; ICIPE 2011).

Insect pests that attack cashew nut crops in SSA are varied with complex ecologies (Hammed et al. 2008). The crop is attacked by more than 60 different insect species throughout its growth period (Dwomoh et al. 2008). Although there are different species of cashew pests in different SSA countries, variations are limited due to similarities in ecology and weather parameters. Thus, common pests in any cashew-growing country in Africa would be the same. The most important insect pests in SSA include the coreid coconut bug, *Pseudotheraptus wayi* Brown (Hemiptera: Coreidae); the mirid bugs, *Helopeltis anacardii* Miller and *H. schoutedeni* Reuter (Hemiptera: Miridae) and thrips, *Selenothrips rubrocinctus* (Giard). These are often the dominant species which suck and feed on new tender shoots, flower panicles and young developing fruits (NARI 2010). The extent of damage caused by these pests varies with location, variety and management practices (Dwomoh et al. 2009).

3 Management of Cashew Pests

The high devastation level and the knowledge that existed before 2016 necessitated a total reliance on calendar spray of insecticides to control cashew pests. Lambda cyhalothrin and its generics as well as endosulfan have been the lone tool that cashew growers depend on (CABI 2005; Martin et al. 1997; Mitchell 2000; NARI 2010). Chemical control remains one of the well-known control measures among growers, but the recently confirmed effectiveness of *O. longinoda* on cashew pests provides the best alternative (NARI 2010). Smallholder farmers are seeking alternative technologies like the use of weaver ants that reduce production costs and/or increase yields. The negative effects associated with irrational use of insecticides include: adverse effects on non-target species, resistance to repeatedly used insecticides, pest resurgence, secondary pest outbreaks, environmental contamination and negative effects on the health of the farmers, who often lack the adequate protective gears (Frank et al. 2015; Peng et al. 2014; Saini et al. 2014). In addition, costs to repeatedly purchase insecticides would also be reduced. Unfortunately, knowledge of the use of *O. longinoda* as biological control agent against cashew pests is limited to few individuals, particularly researchers, and has not been accessed by cashew growers.

4 The Use of *Oecophylla longinoda* in Controlling Cashew Pests

African weaver ants, *Oecophylla longinoda* Latreille, are arboreal predatory insects which virtually attack any pest and whatever creature that would intrude their territory (Fig. 1). They are territorial (Van Mele and Cuc 2007) and usually attack a wide range of organisms ranging from insects and reptiles to large mammals. The nonselective predatory nature is what makes them the best fit for multiple pest control, particularly in orchard and field crops such as cashew, mango, citrus and coconut. The effective use of *O. longinoda* in African cashew stemmed from the long-time application of the closely related species that belongs to the same genus, *Oecophylla smaragdina* Fabricius (Christian et al. 2008; Cole and Jones 1948; Lokkers 1986; Crozier et al. 2009), in Asia, particularly China, since 304 A.D. (Hölldobler and Wilson 1990; Van Mele 2008). Green ants, *O. smaragdina,* are geographically limited to Southeast Asia and Northern Australia, including many tropical western Pacific Islands (Lokkers 1986), while *O. longinoda* is found in Africa (Van Mele and Cuc 2007; Way 1954) with a wide-band presence across equatorial Africa

Fig. 1 The predator, *O. longinoda*, on cashew. (**a**), *O. longinoda* patrolling the cashew nuts; (**b**), *O. longinoda* attacking *P. wayi*, one of the key cashew pests

(Crozier et al. 2009). *O. smaragdina* and *O. longinoda* can control more than 50 different pests in multiple crops and forest trees (Way and Khoo 1992). Their ability to control pests can be similar or more than chemical pesticides (Peng and Christian 2005; Dwomoh et al. 2009; Offenberg et al. 2013).

The first report on *O. longinoda* in East Africa was made in the early 1950s (Way 1954) when the predator was found colonizing plants in natural forests, particularly along the coasts of Kenya and Tanganyika and islands of Pemba, Unguja and Mafia. In West Africa, the use of *O. longinoda* was first reported in Benin in the 1980s by Majer (1986) and thereafter Way and Khoo (1992). In different African countries, *O. longinoda* has been reported as an effective biocontrol agent against citrus (Ativor et al. 2012) and cashew pests (Dwomoh et al. 2009) in Ghana; coconut (Way 1953; Varela et al. 2012; Seguni et al. 2011) and cashew pests in Tanzania (Olotu et al. 2013) and Mozambique (Peng et al. 2010) and cashew and mango pests in Benin (Adandonon et al. 2009; Anato et al. 2015; Sinzogan et al. 2008; Van Mele et al. 2007). The missing link in all the reports was how to make use of the predator and critical analyses of its comparative efficacy and economic advantages, over the commonly used cashew pest control methods which are mainly pesticide-based.

The knowledge gaps that would otherwise hinder the effective exploitation of this predator as an alternative means in controlling pests of diverse crops was recently addressed through collective undertaking by researchers at Sokoine University of Agriculture in Tanzania and University of Abomey-Calavi (Université d'Abomey-Calavi) in Benin, in collaboration with scientists at Aarhus University of Denmark, who undertook detailed studies on how to effectively exploit *O. longinoda* in managing cashew pests as detailed in the sections that follow. It is imperative to realize that *O. longinoda* is a generalized predator for many insect pests, and once introduced into the crop, it takes care of almost all kind of insect pests, regardless of the species, except for a few with which they survive symbiotically such as aphids.

5 Methodology

5.1 Oecophylla longinoda *Colony Identification and Mapping*

Weaver ants survive in natural forest and other bushes whose plant leaves are suitable for their nest construction. During a four-year experimentation with *O. longinoda* in Benin and Tanzania, the initial colonies of weaver ants used in cashew were either obtained from bushes around the target experimental plants or collected elsewhere at a distance and transported to the experimental units (Abdulla 2016). Identification of the existence of weaver ants were by spotting of nests on trees, followed by tracking of the adult weaver ant trails as they moved for foraging. Each weaver ant colony was mapped by testing for enmity among insects of the neighbouring trees, since weaver ants of different colonies tend to fight and kill each

Plate 1 Battle among *O. longinoda* of different colonies that leads to deaths of many

Fig. 2 Harvesting of *O. longinoda*, packaging for transportation and introduction by hanging

other (Peng et al. 2010). Thus, weaver ants which belonged to the same colony liked each other and did not fight, while those of different colonies fought to death (Plate 1). Trees occupied by members of the same colony were marked with tags of the same colour. All colonies existing in the same area were mapped.

5.2 Oecophylla longinoda *Colony Collection and Transportation to Experimental Fields*

After mapping the colonies, searches for a nest containing weaver ant queen for each colony were done using the procedures developed by Nene et al. (2017). The queen nest was harvested alongside other workers' nests, kept in aerated cloth bags and transported to experimental cashew fields for release (Fig. 2). It is recommended that colony harvesting should consider a mandatory inclusion of the queen because a colony without a queen may not be established, and if they do, they would last for only 6 months (Peng et al. 2010). Colonies with *O. longinoda* queen usually live for up to 7 years.

5.3 Introduction and Rearing of Oecophylla longinoda *Colonies*

Experiments were conducted in already established cashew trees of variety AC4 aged 12 years and planted at spacing of 12 m by 12 m at Naliendele Agricultural Research Institute in Tanzania. Harvested nests for each colony were introduced into the canopy of experimental cashew trees following the already established experimental design. Trees introduced with members of the same colony were connected with Manila ropes to aid movements of the weaver ants among trees for increased patrol to protect the target plants (Peng et al. 2010). Generally, using *O. longinoda* as biocontrol agents requires a simple ten-step protocol summarized below:

1. Identifying colony source.
2. Mapping the colonies and marking boundaries for each colony.
3. Identifying the queen nest.
4. Harvesting the nests and separating colonies. Queen nest must be harvested first.
5. Packaging the harvested nests in perforated or aerated bags and tying them to avoid escape.
6. Carefully transporting nests to the target farm area. Transportation time should not exceed 3 hours.
7. Preparing the field to which new release is intended. This would in a sequence come in as step number 2, but it is included at this stage for convenience.
8. Locating the right recipient tree and opening loose the containing bags to release the ants.
9. Post-release care, which includes protection from enemies, and supplementary feeding.
10. Promotion for further establishment, patrol of the trees by aid of ropes for pest control and networking among members of the same colony as well as maintaining colony separators.

In the current experiment, a randomized complete block design (RCBD) was used with three control options as treatments, namely, (1) use of *O. longinoda*, (2) use of recommended insecticides and (3) control (where nothing was applied). The experiments were replicated four times. Each experimental unit covered 96 cashew trees with the total experimental areas with 384 trees. After introduction, *O. longinoda* were supplied with ground fish protein and sugar until colonies were well established and ready to patrol the cashew (Abdulla 2016). The recommended insecticide, Karate® (50 g/L of Lambda cyhalothrin), was sprayed on cashew trees after every 2 weeks at the recommended rate of 5 mL/L per tree (NARI 2010), during the active growth of cashew using a motorized backpack sprayer (*M 225–20 Motor-Rückensprühgerät*). Sulphur (falcon dust) was also prophylactically sprayed at a rate of 250 g/tree at flowering period, supplemented with systemic fungicide, Bayfidan (triadimenol 250 g/L), to control powdery mildew disease (PMD; *Oidium anacardii*). The number of major cashew pests was scouted every week and scores of inflicted injuries on shoots, inflorescence and nuts were recorded in each of the experimental units as recommended (Martin et al. 1997; NARI 2010).

5.4 *Effectiveness of* **Oecophylla longinoda** *in Controlling Cashew Pests*

The efficacy of tested methods in controlling cashew pests was established based on the data for insect records and injury scores. The number of cashew pests in each treatment was compared using pairwise multiple comparisons (Steel-Dwass method), whereby each of the tested pest control methods was compared with the rest of the methods. The nut yield collected from each tree was weighted in kilograms (kg) to obtain average yield from each treatment.

5.5 *Economic Analyses of Using* **Oecophylla longinoda**

The pest management costs (variable costs) incurred in each of the treatments were itemized. The harvested cashew nuts were categorized into grades, first, second and third, to help relate the market price with the obtained yield from each treatment. Apart from cashew grades, the market price considered the mode of crop management, whereby organically produced cashew nuts fetched higher market prices compared to inorganically produced ones. In order to perform dominance analysis, insect pest management practices (treatments) were arranged in order of increasing variable costs (William 2015). A treatment was dominated if its variable costs were higher than the proceeding treatments but its net benefit was lower. Such a treatment was termed as dominated treatment and was denoted by letter "D".

6 Study Findings

Among the research undertakings were the development of protocol on how to collect and introduce colonies in new orchards (Rwegasira et al. 2015), guidelines on identification of queen nest and subsequent collection for re-establishment in new fields (Nene et al. 2017) and techniques to massively rear *O. longinoda* queens (Rwegasira et al. 2017). Also undertaken was a detailed examination of the efficacy of the predator in controlling the cashew pests and the comparative efficacy with the commonly used insecticides (Abdulla et al. 2016) and economic analyses of the comparative use of *O. longinoda* and synthetic insecticides (William 2015).

6.1 Efficacy of Tested Insect Pest Control Methods

Studies on the comparative efficacy indicated *Oecophylla longinoda* to be equally effective in controlling cashew pests with the recommended insecticides (Table 1), with similar impacts on the yield quantity (Fig. 3) as shown below (Abdulla 2016). Despite limited details in qualitative measurements and analyses of obtained yield, the nuts from cashew kept under *O. longinoda* treatment looked shiner and bigger in size than the ones under conventional insecticide treatment. Both Karate and *O. longinoda* were significantly ($P < 0.0001$) highly effective against the cashew pests when compared with the control treatment. The trend suggests that the effectiveness of *O. longinoda* increased with season, with the predator inflicting more impact on pests in the second growing season compared to the first season. Thus, when using *O. longinoda*, more effective pest control should be expected with increased stay and colonization of the crop habitat by the predator.

6.2 Economic Relevance in Using **Oecophylla longinoda** *to Control Cashew Pests*

One major question that usually affects the conception and ultimate use of new technology however relevant it is would be its profitability (Evans 2015) and capital investment required. Thus, knowledge on these critical determinants is usually imperative to end users. According to William (2015), the knowledge developed and economic analyses made indicated that the use of *O. longinoda* is not only effective against major cashew pests but also very profitable.

Moreover, the long-term sustainability of pest management accrued from the established *O. longinoda* colonies adds to long-term benefit realized by the end

Table 1 Pairwise multiple comparisons (Steel-Dwass method) of the efficacy of recommended insecticide and *O. longinoda* against cashew pests during a two seasons' studies in Tanzania

Pests	Treatment comparisons	*P* value Season 1	*P* value Season 2
P. wayi	*O. longinoda* vs. Karate	0.223	0.208
	O. longinoda vs. control	<0.0001	<0.0001
	Karate vs. control	<0.0001	<0.0001
Helopeltis species	*O. longinoda* vs. Karate	0.181	0.103
	O. longinoda vs. control	<0.0001	<0.0001
	Karate vs. control	<0.0001	<0.0001
Thrips spp.	*O. longinoda* vs. Karate	0.104	0.062
	O. longinoda vs. control	<0.0001	<0.0001
	Karate vs. control	<0.0001	<0.0001

Source: Abdulla et al. (2016)

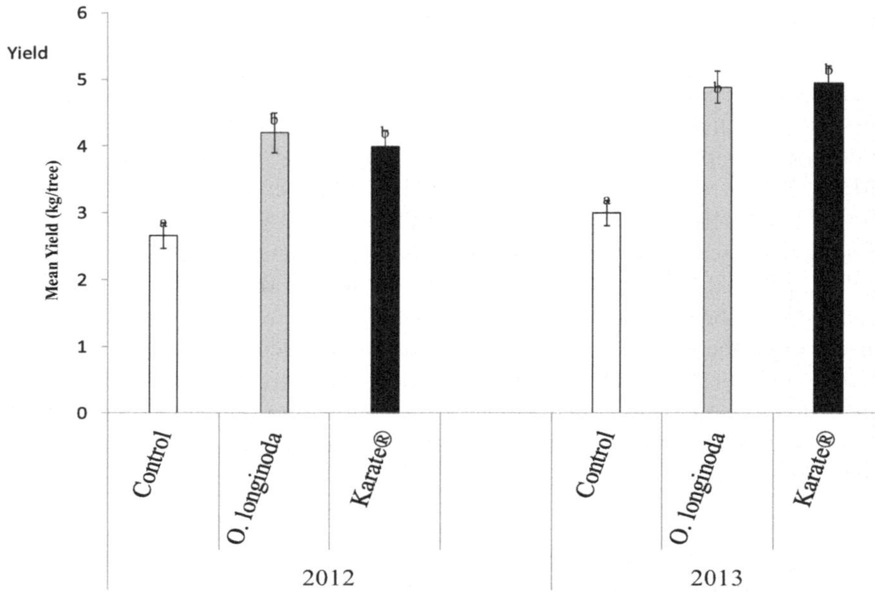

Fig. 3 Comparison of *O. longinoda* with recommended insecticide in controlling cashew pests as reflected on obtained economic yield. (Source: Abdulla 2016)

user. Very impressively, the use of *O. longinoda* leaves neither an undesired effect on the environment nor negative health effect to the users, as the case would be with insecticides. The economic analyses of the profits attainable by controlling cashew pests using the *O. longinoda* would be undertaken by varied methods such as partial budgeting analysis (PBA), marginal rate of returns and benefit-cost ratio (BCR).

Partial Budgeting

Partial budgeting is a tool for farm managers used to compare the profitability of a new project in comparison to the current, prior to effective implementation (Tigner 2006). It helps farm owners to evaluate the financial effect of incremental changes. It is a strong tool that provides insights into the potential investments and should thus be used as effective analytical tool before rational decisions can be made. The only limitations in using this method are that only resources that will be changed are considered, the resources in the business that are left unchanged are not considered and only the change or project under consideration is evaluated for its ability to increase or decrease income in the farm business (Tigner 2006). The practicability of the decision to opt for *O. longinoda* in cashew pest management instead of the currently recommended Lambda cyhalothrin insecticides was experimentally established by William et al. (2015) as detailed (Table 2).

Table 2 Partial budget (US$/tree) of African weaver ants on cashew in two seasons

Particulars	Based on season 1 crop budget (US$/tree)		Based on season 2 crop budget (US$/tree)	
	Switching from Karate to *O. longinoda*	Switching from control to *O. longinoda*	Switching from Karate to *O. longinoda*	Switching from control to *O. longinoda*
Incremental benefits				
Reduced costs	2.12	6.02	5.44	9.34
Additional benefits	3.79	0.00	2.39	0.00
Total incremental benefits	5.35	6.02	7.83	9.34
Incremental costs				
Additional costs	0.50	0.47	0.11	0.11
Reduced revenue	0.00	0.00	0.00	0.00
Total incremental costs	0.50	0.47	0.11	0.11
Net change in benefits	4.85	5.55	7.72	9.23

Source: Modified from William et al. (2015)

The presented findings on partial budget analysis suggested that the adoption of *O. longinoda* would increase income (positive net change in benefits) in both seasons. The incremental benefits in farming with *O. longinoda* exceeded the incremental costs incurred. Thus, using *O. longinoda* was an economically profitable management practice. It conforms to Evans (2015) observations that if a technology is relatively new, requiring some new skills, higher benefits associated with less costs may be appropriate to a farmer to change or shift from his/her old technology.

Marginal Rate of Returns

Limitations in using the partial budgeting (PB) method indicated in section "Partial Budgeting" is that the decisions to switch from one control method to the other may not be adequately guided if the other options of cost and benefit comparisons are not explored necessitating the need for additional methods of economic analyses. The marginal rate of returns (MRR) provides for such comparisons and addressed some weaknesses associated with the partial budgeting method. Marginal rate of returns refers to the amount of additional revenue that a business can expect to earn per each additional cost that it spends on production (Ryan 2016). The MRR is principally a dominance analysis of the results obtained using the partial budgeting

Table 3 Dominance analysis of cashew pests' management options

Pest management options	Costs that vary (US$/tree)		Net benefits (US$/tree)	
	Season 1	Season 2	Season 1	Season 2
No control	4.92	6.13	5.37	5.42
O. longinoda	6.72	7.61	9.59	11.39
Karate	8.34	9.73	4.18	5.74
MRR (%)	235 > 100		405 > 100	

Source: With modification from William et al. (2015)

method. Thus, a project would continue increasing the profit until its MRR value is equivalent to zero. Using the dominance analysis, it was established that using *O. longinoda* was superior (Table 3) to the insecticide, Lambda cyhalothrin (William 2015).

The dominance analyses indicated that *O. longinoda* was dominant over Lambda cyhalothrin and the no control option, confirming that the insecticide was less profitable in both seasons. The marginal rate of return calculated based on the profitability indices qualified that *O. longinoda* was superior and economically a better alternative (William 2015). The option to use *O. longinoda* would make the best choice because the percentage MRR was valued at 235% in the first season and 405% in the second season. This simple analysis suggested further that the profit realized by using the predator would increase season after season, leading to greater profit from using the technique. Preference of *O. longinoda* to other options would be in line with the recommendation by Das and Mandal (2010) that rational farmers adopt a new innovation that has a comparatively higher MRR.

Benefit-Cost Analyses

The decision to invest in any enterprise is often guided by the costs to be incurred and the accrued benefit from the respective venture. Several methods that compare two investment components, the cost and benefit, have been used by the researchers to assess the profitability of using *O. longinoda* over the insecticide Lambda cyhalothrin in controlling cashew pests (William et al. 2015). These included the net present value (NPV), benefit-cost ratio (BCR) and internal rate of return (IRR). On per tree basis analyses of the costs spent and benefit gained, it was proven through all these evaluation criteria that *O. longinoda* recorded highest values compared to the insecticide Lambda cyhalothrin (Table 4). Considering the fact that cost is vital to these methods, the no-control proved superior to the insecticide-based pest control option when Lambda cyhalothrin was used. The use of *O. longinoda* would therefore make a feasible alternative to the insecticides. Biological control agents have always proved profitable, recording highest value of the BCR. Van Den Berg (2010) reported the biological control of the spiny blackfly in Switzerland to be highly cost-effective with a BCR of 199:1.

Table 4 Net present value (US$/tree), benefit-cost ratio and internal rate of return (%) analyses of treatments in cashew production

Particulars	Treatments		
	O. longinoda	Lambda cyhalothrin	Control
Present value of benefits	30.53	24.16	18.91
Present value of cost	12.40	15.62	9.54
Net present value	18.13	8.54	9.37
Benefit-cost ratio	2.5:1	1.5:1	2.0:1
Internal rate of return	57	24	41
Ranking based on NPV	1	4	3
Ranking based on BCR	1	4	3
Ranking based on IRR	1	4	3

Source: Modified from William (2015)

7 Effective Use of *Oecophylla longinoda* on Cashew Pests

The effective use of the African weaver ants in controlling pests requires a thorough understanding of the biology, behaviour and habitat of *O. longinoda*. Weaver ants live in colonies with clear division of roles and responsibilities among members of castes. A colony is composed of one or more queens, males, large workers and small workers (Van Mele and Cuc 2007). The insects reproduce by oviparity, whereby the queen has the sole responsibility of laying eggs that are tended by small workers, which nurse them through subsequent juvenile stages until adulthood when they are capable of fending for themselves and play other roles in the colony (Crozier et al. 2009). The nursing workers also feed the queen. Large workers have roles of foraging and bringing food to the colony and also protecting, defending, maintaining, constructing nests and expanding the colony (Crozier et al. 2009; Way 1954). Males are rarely produced, usually to mate with newly produced queens when the queen mother senses the need for producing daughter queens. Principally the queen is a mother to all members of the colony; thus, she controls and sustains the colony. Therefore, the intension to use *O. longinoda* for pest management must inevitably ensure the presence of the queen.

African weaver ants are arboreal and aggressive (Van Mele and Cuc 2007). They construct nests using tree leaves and use them as shelters. In each colony, the queen would be accommodated in one nest, which is well constructed and always highly protected (Abdulla et al. 2016; Nene et al. 2017). The protocol for locating *O. longinoda* queens prior to nest collection has been detailed by Nene et al. (2017). Harvesting and subsequent introduction of *O. longinoda* colonies into new fields requires correct mapping of each individual colony, identification of the colony queen and harvesting the whole colony nests together with the queen and subsequent introduction into new fields. Details on the colony identification, mapping, harvesting to introduction and managing of *O. longinoda* for sustained predation of the cashew pests have been elaborated (Abdulla 2016).

8 Conclusion

The use of *O. longinoda* in managing cashew and other crop pests is currently the best and most sustainable alternative to insecticide-based pest management strategies in Tanzania, particularly in orchard trees, as the differences between using recommended insecticides and weaver ants were not statistically significant ($P < 0.001$). Moreover, using *O. longinoda* fits well with organic farming providing a reliable solution to the attempt by the world authorities to minimize pollution and protect human beings and the environment from agricultural produce with pesticide residues. The technology has been proven economically feasible by all analytical methods such that the benefits accrued from its use are invaluable. When partial budgeting method was used, the net change in benefit ranged from US$4.85 to US$7.72 per tree by switching from conventional insecticide (Karate) to *O. longinoda*, while switching from no control to using *O. longinoda* led to net change in benefit from US$5.55 to US$9.23 per tree. The marginal rate of return indicated net benefit of up to US$11.39 per tree when *O. longinoda* was used, compared to US$5.74 per tree when Karate® was used. Thus, the use of *O. longinoda* against cashew pests does not only protect crops and add economic gains from investments in production; it also ensures protection of farmers' health and the environment. It is an undeniable fact that *O. longinoda* technology will reduce the production costs, offer sustainable solution to cashew pests and open ways to markets that requires high-quality agricultural produce with no pesticide residuals.

Acknowledgement Authors are thankful to Dr. Joachim Offenberg and Prof. Mogen G. Nielsen of Aarhus University, Denmark, for technical collaboration in designing the research work. The study was financially supported by the Danish International Development Assistance (DANIDA) through grant DFC no. 10-025AU.

References

Abdulla NR (2016) Weaver ant, *Oecophylla longinoda* Latreille (Hymenoptera: Formicidae) as biocontrol agent on major insect pests of cashew and mango in Tanzania. PhD thesis, 206+i–xiv pp

Abdulla NR, Rwegasira GM, Jensen K-MV, Mwatawala MW, Offenberg J (2016) Potential of *Oecophylla longinoda* Latreille (Hymenoptera: Formicidae) in managing major insect pests in organic cashew production systems. J Int Soc Org Agric Res. https://doi.org/10.1007/s13165-016-0152-8

Adandonon A, Vayssieres J-F, Sinzogan A, Van Mele P (2009) Density of pheromone sources of the weaver ant *Oecophylla longinoda* affects oviposition behaviour and damage by mango fruit flies (Diptera: Tephritidae). Int J Pest Manag 55(4):285–292

Agboton BV, Salifu D, Segumi Z, Sijaona ME, Shomari S, Ekesi S, Maniania NK (2013) Bioecology of some key cashew insect pests and diseases in diverse habitats and landscapes in Tanzania. J Appl Entomol 137:782–789

Anato FM, Wargui RB, Sinzogan AAC, Offenberg J, Adandonon A, Vayssières JF, Kossou DK (2015) Reducing losses inflicted by insect pests on cashew, using weaver ants as a biological control agent. Agric For Entomol 17:285–291

Ativor IN, Afreh-Nuamah K, Billah MK, Obeng-Ofori D (2012) Weaver ant, *Oecophylla longinoda* (Latreille) (Hymenoptera: Formicidae) activity reduces fruit fly damage in citrus orchards. J Agric Sci Technol A 2(2012):449–458

CABI (2005) Crop protection compendium CD ROM. CABI. Available online: at www.cabicompendium.org/cpc. Visited 15th Sept, 2013

Christian K, Peng R, Lan LP (2008) Implementation of the IPM program using weaver ants as a major component for cashew growers in Vietnam. Proceedings of the good agronomic practices workshop, in Binh Thuan, Vietnam, 21–22 July 2008, 10pp

Cole AC, Jones JW (1948) A study of the weaver ant, *Oecophylla smaragdina* (Fabricius). Am Midl Nat 39:641–651

Crozier RH, Newey PS, Schluns EA, Robson SK (2009) A masterpiece evolution-*Oecophylla* weaver ants (Hymenoptera: Formicidae). Myrmecol News 13:57–71

Das AMB, Mandal M (2010) An economic analysis of Stevia (*Stevia rebaudiana*) cultivation through stem cutting and tissue culture propague in India trends. J Agric Econ 3:216–222

Dwomoh EA, Acknor JB, Afun JVK (2008) Survey of insect species associated with cashew (*Anacardium occidentale* Linn.) and their distribution in Ghana. Afr J Agric Res 3(3):205–214

Dwomoh EA, Afun JV, Ackonor JB, Agene VN (2009) Investigations on *Oecophylla longinoda* (Latreille) (Hymenoptera: Formicidae) as a biocontrol agent in the protection of cashew plantations. Pest Manag Sci 65:41–46

Evans E (2015) Marginal analysis: an economic procedure for selecting alternative technologies practices. Department of Food and Resource Economics (Inst. Food Agr. Sci. Publ., Univ. of Florida, Gainesville, FL), 29pp

FAOSTAT (2018). http://www.fao.org/faostat/en/#data/QC. Visited 17th May, 2018

Frank E, Tina R, Heiko S (2015) Reducing pesticide use and risks – what action is needed? Briefing paper. Helvetas Swiss Interco-operation, Switzerland, 31pp. Germany. IntechOpen publishers. www.intechopen.com, 498pp

Hammed LA, Anikwe JC, Adedaji AR (2008) Cashew nuts and production development in Nigeria. Am –Euras. J Sci Res 3(1):54–61

Hölldobler B, Wilson EO (1990) The ants. Belknap Press of Harvard University Press, Cambridge, MA, 732pp

ICIPE (2011) Integrated management of major insect pests and diseases of cashew in East and Western Africa. Deutsche Gesellschaft für Internationale Zusammenarbeit GmbH and The Advisory Service on Agricultural Research for Development, 2pp. http://www.giz.de/expertise/downloads/giz2011-en-factsheet-icipe-integrated-management-insect-pests.pdf. Visited 1st Oct, 2014

Lokkers C (1986) The distribution of the weaver ant, *Oecophylla smaragdina* (Fabricius) (Hymenoptera: Formicidae) in northern Australia. Aust J Zool 34:683–687

Majer JD (1986) Utilizing economically beneficial ants. In: Bradleigh S (ed) Economic impact and control of social insects. Vinson Praeger, New York, pp 314–331

Martin PJ, Topper CP, Bashiru RA, Boma F, De Waal D, Harries HC, Kasuga LJ, Katanila N, Kikoka LP, Lamboll R, Maddison AC, Majule AE, Masawe PA, Milanzi KJ, Nathaniels NQ, Shomari SH, Sijaona ME, Stathers T (1997) Cashew nut production in Tanzania: constraints and progress through integrated crop management. Crop Prot 16(1):5–14

Maruthadurai R, Desai AR, Prabhu HRC, Singh NP (2012) Insect pests of cashew and their management. Technical Bulletin No. 28, ICAR Research Complex for Goa, Old Goa, 18pp

Mitchell P (2000) Leaf-footed bugs (Coreidae). In: Schaefer CW, Panizzi AR (eds) Heteroptera of economic importance. CRC Press, Boca Raton\London\New York\Washington, DC, pp 337–403

NARI (2010) Agricultural research Annual Report. Government printer, Dar es Salaam, Tanzania, pp 23–25

Nene W, Rwegasira GM, Mwatawala M (2017) Optimizing a method for locating queen nests of the weaver ant, *Oecophylla longinoda* Latreille (Hymenoptera: Formicidae) in cashew, Anacardium occidentale L. Plantations in Tanzania. Crop Prot 102(2017):81–87. © 2017 Elsevier Ltd. https://doi.org/10.1016/j.cropro.2017.08.013

Offenberg J, Nguyen TB, Wiwatwitaya D (2013) The effectiveness of using weaver ant biocontrol in Southern Asian citrus and mango. J Asian Myrmecol 5:139–149

Olotu MI, Du Plessis H, Seguni ZS, Maniania NK (2013) Efficacy of the African weaver ant Oecophylla longinoda (Hymenoptera: Formicidae) in the control of Helopeltis spp. (Hemiptera: Miridae) and Pseudotheraptus wayi (Hemiptera: Coreidae) in cashew crop in Tanzania. Pest Manag Sci 69:911–918. https://doi.org/10.1002/ps.3451

Peng RK, Christian K (2005) Integrated pest management in mango orchards in the Northern Territory Australia, using the weaver ant, *Oecophylla smaragdina*, (Hymenoptera: Formicidae) as a key element. Int J Pest Manag 51:149–155

Peng RK, Christian K, Lan LP, Binh NT (2010) Integrated cashew improvement program using weaver ants as a major component – photo book for cashew growers in Vietnam. Agriculture Publisher, Ho Chi Minh City, Vietnam, pp 69–71

Peng R, Lan LP, Christian K (2014) Weaver ant role in cashew orchards in Vietnam. J Econ Entomol 107(4):1330–1338

Rwegasira GR, Mwatawala M, Rwegasira GM, Mogens GN, Offenberg J (2015) Comparing different methods for trapping mated queens of weaver ants (Oecophylla longinoda; Hymenoptera: Formicidae). Biocontrol Sci Tech 25(5):503–512. https://doi.org/10.1080/09583157.2014.992861

Rwegasira GR, Mwatawala MW, Rwegasira GM, Axelsen J (2017) Optimizing methods for rearing mated queens and establishing new colony of *Oecophylla longinoda* (Hymenoptera: Formicidae). Int J Trop Insect Sci 37(4):217–224. https://doi.org/10.1017/S1742758417000224

Ryan C (2016) Definition of Marginal Rate of Return. https://budgeting.thenest.com/definition-marginal-rate-return-25528.html. Visited 3rd May, 2018

Saini RK, Yadav GS, Kumari B (2014) Compendium of lectures delivered during the Advanced Training Course on "Novel Approaches in Pest and Pesticide Management in Agro-Ecosystem". 19th August to 8th September, 2014. Centre of Advanced Faculty Training (CAFT), Department of Entomology, CCS Haryana Agricultural University, Hisar, 338pp

Seguni ZSK, Way MJ, Van Mele P (2011) The effect of ground vegetation management on competition between the ants *Oecophylla longinoda* and *Pheidole megacephala* and implications for conservation biological control. Crop Prot 30:713–717

Sinzogan AAC, Van Mele P, Vayssiéres JF (2008) Implications of on-farm research for local knowledge related to fruit flies and the weaver ant *Oecophylla longinoda* in mango production. Int J Pest Manag 54(3):241–246

Tigner R (2006) Partial budgeting: a tool to analyze farm business changes. Iowa State University Extension. https://www.extension.iastate.edu/agdm/wholefarm/html/c1-50.html. Visited 3rd May, 2018

Van den Berg EC (2010) Detection, quantification and monitoring Prosopis spp. In the Northern Cape Province of South Africa using Remote Sensing and GIS. Dissertation for Award of MSc Degree at North-West University, 234pp.

Van Mele P (2008) A historical review of research on the weaver ant *Oecophylla* in biological control. Agric For Entomol 10:13–22

Van Mele P, Cuc NTT (2007) Ants as friends: improving your tree crops with weaver ants. African Rice Center (WARDA), Cotonou, Benin and CABI, Egham, UK, 72pp

Van Mele P, Vayssieres JF, Tellingen EV, Vrolijks J (2007) Effects of African weaver ant, *Oecophylla longinoda*, in controlling mango fruit flies (Diptera: Tephritidae) in Benin. J Econ Entomol 100(3):695–701

Varela AM, Seif A, Nyambo B (2012) A guide to IPM in mango production in Kenya. The International Centre of Insect Physiology and Ecology, 2nd edn. ICIPE Science Press, Nairobi, pp 34–36

Way MJ (1953) The relationship between certain ant species with particular reference to the biological control of the coreid, *Theraptus* sp. Bull Entomol Res 44:669–691

Way MJ (1954) Studies on the life history and ecology of the ant *Oecophylla longinoda* Latreille. Bull Entomol Res 45:93–112

Way MJ, Khoo KC (1992) Role of ants in pest management. Annu Rev Entomol 37:479–503
William JG (2015) Economic Value of Using African Weaver Ants as Biological Control in Fruit Production and Export in Tanzania. Sokoine University of Agriculture, PhD Thesis 142pgs
William JG, Hella J, Lars E, Offenberg J, Mwatawala M, Rwegasira G (2015) Partial budgeting analysis of different strategies for management of insect pests in cashew and mango orchards in Tanzania. Int J Sust Agric Res 2(4):98–110

Part VI
Upscaling Innovative Technologies on Smallholder Farms

Determinants of ISFM Technology Adoption and Disadoption Among Smallholder Maize Farmers in Central Malawi

Joseph S. Kanyamuka, Charles B. L. Jumbe, Jacob Ricker-Gilbert, Abdi-Khalil Edriss, and Wezi G. Mhango

Abstract This paper sought to address the following key research question: what drives some smallholder farmers to adopt integrated soil fertility management (ISFM) practices in a changing climate while others drop out? Data used in this study were collected from 200 farming households during the 2015/2016 agricultural season in Kasungu District of Central Malawi. This paper applied multivariate probit and bivariate probit model, respectively, to analyse the joint adoption and disadoption decisions regarding ISFM technologies and/or practices. Significant factors for both adoption and continued use of ISFM practices include access to legume seed, access to extension, secure land tenure, group membership and landholding size. Interestingly, the same factors discourage disadoption of ISFM practices. For instance, access to legume seed and access to extension discourage disadoption of maize-legume intercropping and rotation, while larger landholding sizes encourage adoption of legume-maize rotation. In addition, the study found that female farmers were more likely to continue the practice of maize-legume intercropping. We therefore recommend that ISFM technology packages that include the use of inorganic fertilizer should go along with other complementary interventions such as maize-legume intercropping and rotation. In addition, efforts in the promotion and scaling up of ISFM practices should focus on those households with secure land tenure rights, encourage access to extension and should place emphasis on the complementarities that exist in the adoption of decisions regarding ISFM technologies in maize-based farming systems.

J. S. Kanyamuka (✉)
New Alliance Policy Acceleration Support (NAPAS:Malawi), Lilongwe, Malawi

C. B. L. Jumbe · A.-K. Edriss · W. G. Mhango
Lilongwe University of Agriculture and Natural Resources, Lilongwe, Malawi

J. Ricker-Gilbert
Purdue University, West Lafayette, IN, USA

B. R. Singh et al. (eds.), *Climate Impacts on Agricultural and Natural Resource Sustainability in Africa*, https://doi.org/10.1007/978-3-030-37537-9_26

1 Introduction: Background and Context

Declining soil fertility due to low and inappropriate fertilizer use, continuous monocropping and inappropriate crop residue management are considered major constraints to increased crop productivity in sub-Saharan Africa (SSA) and Malawi in particular (Mhango et al. 2012; Ngwira et al. 2012; Vanlauwe 2015; Kanyamuka et al. 2018). To address the problem, the Government of Malawi has been implementing the Farm Input Subsidy Program (FISP) to enable farmers to access chemical fertilizers and improved seed to boost maize production (Holden and Lunduka 2010). Following successive implementation of the programme, Malawi was reportedly able to achieve significant food production and productivity (primarily in maize), contributing to increased food security, high real wages and poverty reduction (Dorward and Chirwa 2011).

However, FISP is largely implemented using government resources and takes up a large share of the national budget, thereby posing sustainability challenges. As noted by Dorward and Chirwa (2011), the later years of the programme have been accompanied by high international fertilizer prices and import costs. Future Agricultures Consortium (FAC 2010) also notes that input subsidies such as the FISP are difficult to remove and are badly targeted such that only affluent famers get much of the benefits at the expense of poor farmers. Therefore, there is a need to identify sustainable strategies that can make the FISP more profitable to smallholder farmers. One of the sustainable intensification strategies that has proven successful in raising the efficiency of fertilizer use and improving smallholder productivity is integrated soil fertility management (ISFM) technologies (Ollenburger 2012; Sommer et al. 2013; Vanlauwe 2015).

Fairhurst (2012) defined ISFM as a set of soil fertility management practices that necessarily include the use of inorganic fertilizer, organic inputs and improved germ plasm, combined with the knowledge of how to adapt these practices to local conditions aiming at optimizing agronomic use efficiency of the applied nutrients and improving crop productivity. Vanlauwe (2015), Mhango et al. (2012) and Ollenburger (2012) note that integrating legumes in farming systems is key to implementation of ISFM. Therefore, LUANAR, formerly known as Bunda College of Agriculture, with financial support from the McKnight Foundation Collaborative Crops Research Program, has been implementing a project entitled "Legume Best Bets to Acquire Phosphorous and Nitrogen and Improve Family Nutrition" to assist the smallholder farmers boost agricultural production. The project has been implemented in Northern Malawi at Ekwendeni in Mzimba District and in Central Malawi at Mkanakhoti Extension Planning Area (EPA) in Kasungu District since 2006/2007.

While the benefits of ISFM technologies in enhancing fertilizer use efficiency and improving maize productivity are widely acknowledged in the literature (Marenya and Barrett 2007; Fairhurst 2012; Lambrecht et al. 2014a), their adoption among smallholder farmers remains fairly low (Kassie et al. 2013; Kamau et al. 2013). This is despite several promotional efforts by various stakeholders that include government and non-governmental organizations. In addition, while the number of farmers adopting such ISFM technologies and practices in the initial stages looks promising, the trend of such numbers over time has generally been on the decline (Marenya and Barrett 2007). This is especially true with the Legume Best Bets Project as revealed by a reconnaissance survey undertaken by the authors in 2015. It is against this background that the present study was undertaken to address the following important research question: what drives some smallholder farmers to adopt ISFM technologies while others drop out? Addressing this question is vital for understanding the factors that influence both the adoption and disadoption of the ISFM technologies. Such knowledge is vital to improve the design of better strategies for promoting adoption and both up scaling and out-scaling of ISFM technologies. This may also help technology promoters know who to target with ISFM technologies. The rest of the paper is organized as follows: Sect. 2 addresses methodological challenges and Sect. 3 presents the results and discussion, followed by the conclusion and policy recommendations emanating from the study findings.

2 Methodology

2.1 Study Area, Data and Sampling Design

The study used primary data collected from 200 smallholder maize farmers in Central Malawi from Kasungu District, one of the major maize-producing areas, which fall under Kasungu Agricultural Development Division (KADD). Kasungu District and Mkanakhoti Extension Planning Area (EPA) in particular was purposively selected because it is where LUANAR (formerly Bunda College), in collaboration with the McKnight Foundation, has been implementing the Legume Best Bets Project since 2006/2007. From the EPA, six project villages were identified and individual farmers were randomly selected using population proportional to size (PPS) sampling procedure. The data were collected during the 2014/2015 growing season, focusing on four ISFM technologies: inorganic fertilizer (151 adopters and 49 non-adopters), improved seed (153 adopters and 47 non-adopters), maize-legume intercropping (92 adopters and 108 non-adopters) and maize-legume rotation (133 adopters and 67 non-adopters). Improved seed for maize included hybrid varieties and open-pollinated varieties (OPVs). Among others, data on the performance of the ISFM technologies, access to productive inputs (such as FISP coupons), farmer socio-economic characteristics and institutional factors were collected. Figure 1 presents the map of Malawi showing the study area.

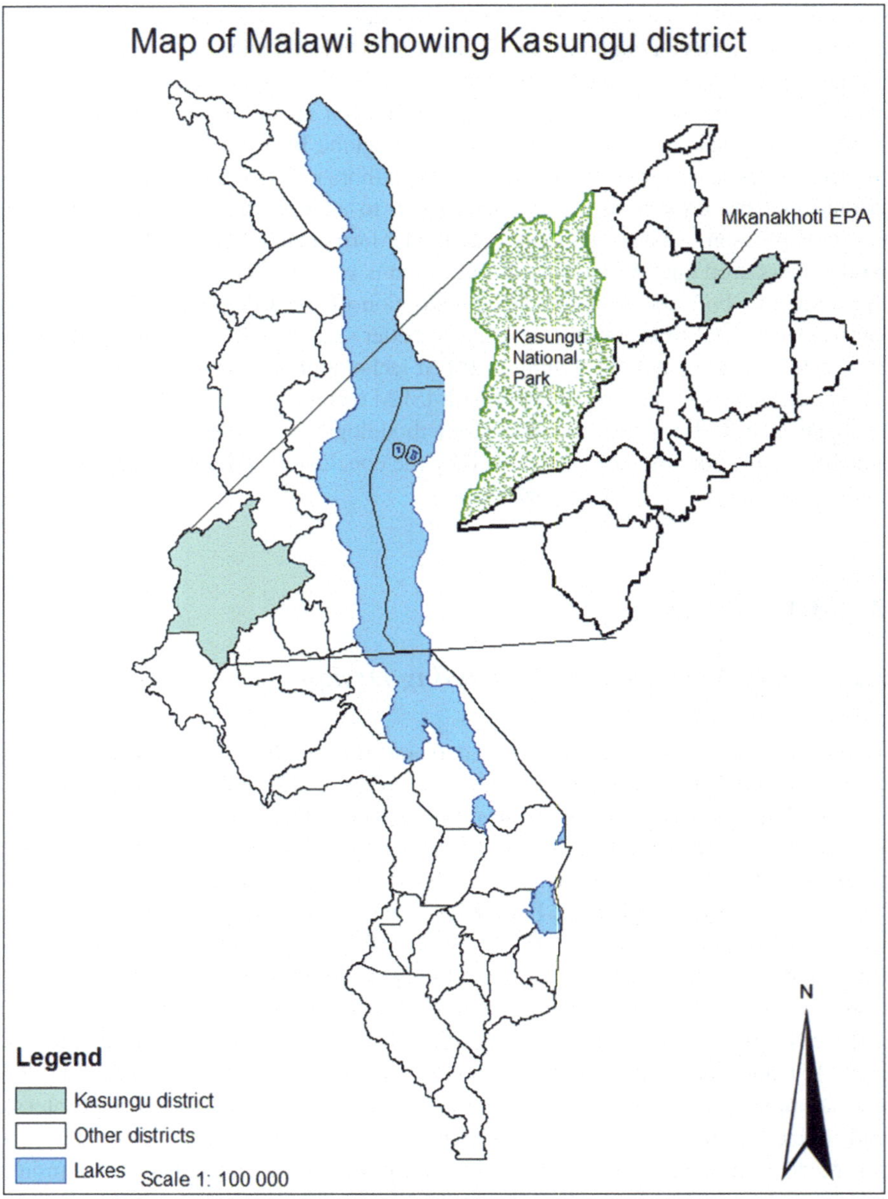

Fig. 1 Map of Malawi showing the study area – Kasungu District. (Source: Authors' own generation)

2.2 *Conceptual Framework and Empirical Strategy*

Conceptual Framework

This study postulates that adoption of ISFM technologies is influenced by household characteristics (age, gender, education level of the household head, household size, household assets and wealth status), institutional factors (access to credit, access to extension, farmer group membership and land tenure), farm and plot characteristics (farm size and soil fertility status) as well as technology attributes. While the effect of household characteristics such as age on adoption of ISFM technologies is indeterminate, male headship of the household is expected to have positive influence due to the comparative advantage in access to productive resources such as cash, which is important to purchase inorganic fertilizer and improved seed. However, the age of the household head may reflect farming experience and past investments in soil fertility-enhancing technologies. Education level of the household head is hypothesised to influence adoption of ISFM technologies positively due to higher management skills, better understanding of concept of ISFM and its implementation as well as the ability to source relevant information about the technologies.

Institutional factors such as access to extension increase farmers' access to information, which is a critical factor in agricultural technology adoption process, while access to credit services facilitates farmers' access to productive resources (Lambrecht et al. 2014b). In the context of ISFM, inorganic fertilizer and improved seed are externally purchased and are deemed unaffordable among some of the resource-constrained smallholder farmers in the study area. Thus, the expectation is that those farmers with access to credit are more likely to adopt ISFM technologies as compared to those farmers without access to credit. Land tenure is expected to influence adoption of ISFM technologies positively, in that those households with secure land tenure rights are more likely to invest in soil fertility-enhancing technologies because of the guaranteed benefits despite the length of the return period, while those with insecure land tenure are more likely to abandon the ISFM technologies.

Plot characteristics such as landholding size have, in the literature, been reported to have positive influence on adoption of agricultural technologies (e.g. Marenya and Barrett 2007 and Tizale 2007). However, in this study, the influence of landholding size on adoption of ISFM technologies is hypothesized to be either positive or negative depending on the type of technology under consideration. For instance, landholding size is expected to be negatively correlated with maize-legume intercropping and positively correlated with legume-maize rotation. Soil fertility status of the plot can influence adoption of ISFM technologies. For instance, farmers' perception of fertile soils is expected to influence adoption of ISFM technologies such as inorganic fertilizer and improved seed. Foster and Rosenzweig (2010) note that the marginal product of inorganic fertilizer tends to be higher on "better" soils than poor soils.

Empirical Strategy

While the individual components in the set of ISFM technology adopted by the farming households are mutually exclusive, farming households in real life apply a mix of two or more mutually exclusive soil fertility interventions bundle (Marenya and Barrett 2007). They do so to exploit the complementarities between alternative interventions such as improving soil fertility and reducing run-off, among others. Following Marenya and Barrett (2007) and Kassie et al. (2013), we employ a multivariate framework to model the joint adoption and disadoption of multiple ISFM technologies. Consistent with the random utility model, the likelihood of adopting a particular technology (or bundle) is such that the utility derived from that specific alternative is greater than or equal to the utilities of all other alternative technologies in the choice. Thus, the decision to (dis)adopt an ISFM technology (or package) is joint and can be modelled as the difference between the benefit and cost of adoption – or continuation of a practice/technology. The latent (dis)adoption decision is determined by:

$$Y^*_{ipj} = X_{ipj}\beta_j + \in_{ipj} \tag{1}$$

$$\in_{ipj} = \alpha_{pj} + \eta_{ij} \tag{2}$$

where Y^*_{ipj} reflects farm household's (dis)adoption on plot p of an ISFM technology (or package),j (j= inorganic fertilizer, improved seed, maize-legume intercropping and legume-maize rotation). X_{ipj} is a matrix of regressors representing household and plot-specific attributes of adoption of technology j that explain constraints and preferences, β_j is a vector of parameters to be estimated for the jth technology adoption equation and $\in_{ipj}$ is a composite error term comprising plot-specific unobserved characteristics (α_{pj}) and unobserved individual farmer characteristics (η_{ij}). Nevertheless, we do not observe the net benefit of adoption but only the choice of whether or not the farmer choses the technology/practice. Thus, the observed discrete choice of a technology/practice is related to latent (unobserved) variable, such that Eq. (1) is transformed indicating whether a farm household is adopting an ISFM technology or not:

$$Y_{ipj} = \begin{cases} 1 \text{ if } Y^*_{ipj} > 0 \\ 0 \text{ if } Y^*_{ipj} \leq 0 \end{cases} \tag{3}$$

2.3 Empirical Models

Determinants of ISFM Technology Adoption

In this study, adoption of a particular ISFM technology was defined as the use or application of a particular ISFM technology or practice by the farmer for at least two consecutive growing seasons. The 2 years were purposively chosen as it is considered minimum for the farmer to complete a full system (cycle) of maize legume intercropping, with rotation at the centre of each system.

While other studies on adoption of ISFM (Nata et al. 2014; Yirga and Hassan 2008) have ignored the possible interrelationships between various practices, this study applied a multivariate probit (MVP) to model (simultaneously) multiple (dis)adoption decisions regarding ISFM technologies and/or practices. This involved modelling the influence of a set of explanatory variables on each of the different technologies/practices while allowing the unobserved and measured factors, accounted for by error terms, to be freely correlated. The correlations between the technologies may be either positive or negative depending on whether farmers consider the technologies as complements or substitutes, respectively (Kassie et al. 2013).

Following Kassie et al. (2013), the multivariate econometric model is characterized by a set of binary-dependent variables (Y_{hpj}) such that:

$$Y^*_{hpj} = X'_{hpj}\beta_j + \mu_{hpj}, \quad j = 1,\ldots,4 \tag{4}$$

and

$$Y_{hpj} = \begin{cases} 1 & \text{if } Y^*_{hpj} > 0 \\ 0 & \text{otherwise} \end{cases} \tag{5}$$

where Y^*_{hpj} reflects both the latent adoption and disadoption decision for h^{th} household on plot p of jth technology (or package). $j = 1, \ldots, 4$ denotes the type of ISFM technology (inorganic fertilizer, improved seed, legume-maize rotation and maize-legume intercropping). Equation (5) is premised on the fact that a rational h^{th} farm household has latent variable, Y^*_{hpj}, which captures the unobserved preferences or demand associated with the jth adoption of ISFM technology. The latent variable is assumed to be a linear combination of observed characteristics (X_{hpj}), both household- and plot-level characteristics, that influences the adoption and disadoption of a jth ISFM practice, as well as unobserved characteristics contained in the disturbance term, μ_{hpj}. β_j is the vector of parameter estimates obtained through the maximum likelihood estimation procedure. The multivariate probit model assumes interdependency of the adoption decisions for the ISFM technologies and that the error terms in equation (4) jointly follow a multivariate normal (MVN) distribution, having zero conditional mean and unity-normalized variance where $\mu_{hpj} \sim MVN(0, \Sigma)$ and covariance matrix given by:

$$\Sigma = \begin{bmatrix} 1 & \rho_{12} \rho_{13} \cdots \rho_{1m} \\ \rho_{12} & 1 \quad \rho_{23} \cdots \rho_{2m} \\ \rho_{13} & \rho_{23} \quad 1 \quad \cdots \rho_{3m} \\ \vdots & \vdots \quad \vdots \quad 1 \ \vdots \\ \rho_{1m} & \rho_{2m} \rho_{3m} \cdots 1 \end{bmatrix} \tag{6}$$

The off-diagonal element in the matrix represents the unobserved correlation between the jth and m^{th} ISFM practice. It thus follows that equation (5) specifies a MVP model that jointly represents decisions to adopt a particular ISFM practice while allowing for correlation across the error terms of the unobserved equations. Table 1 presents the description of the specific variables hypothesized to influence (dis)adoption of ISFM technologies considered in this study.

Table 1 Description of the explanatory variables hypothesized to influence adoption of ISFM technologies

Variables	Description and measurement	Expected sign
Age	Age of the household head (years)	±
Gender	Gender of the household head (1 = male, 0 = female)	±
Education	Highest formal education level attained by the household head	±
Household size	Household size proxy for labour availability measured as total number of persons in a household	+
Landholding size	Total household farm size (hectares)	±
Extension access	Access to extension services (1 = yes, 0 = no)	+
Credit access	Access to credit (1 = yes, 0 = no)	+
Land ownership	Proxy for land tenure system (1 = own land, 0 = otherwise)	+
Distance from house to plot	Measured in kilometres (km)	−
FISP access	Whether the farmer received a FISP coupon or did not (1 = yes, 0 = otherwise)	+
Group membership	Whether the farmer belongs to any group/association or not (1 = yes, 0 = otherwise)	+
Farming experience	The number of years the farmer has been in farming	±
Soil fertility perception	1 = fertile, 0 = otherwise	±

3 Results and Discussion

3.1 Descriptive Statistics

Table 2 presents a summary of key demographic and socio-economic characteristics by technology and farmer category. In terms of gender composition, the sample was dominated by male-headed households across all the technologies and the pooled sample, with 172 (86.0%) households being headed by males while 28 (14.0%) were headed by females. The mean age across all the four technology adopters and non-adopters hovered around 40. The mean age of adopters and non-adopters for maize-legume intercropping was 44.5 and 43.8, respectively, while that of legume-maize rotation was 44.2 and 43.9 for adopters and non-adopters, respectively. The mean age for the total sample was 44.1 years. Landholding size did not vary much among the four technologies and farmer groups. However, it is worth noting that for legume-maize rotation, adopters had considerable higher landholding size of 2.1 ha against 1.8 ha for non-adopters, on average. For improved seed, both adopters and non-adopters had almost the same landholding size of 2.0 and 2.1 ha, respectively, while adopters of inorganic fertilizer had an average total landholding size of 2.0 as compared to non-adopters with 2.2 ha, while the total sample had an average landholding size of 2.0 ha.

Access to FISP coupon voucher reflects government support towards improving smallholder farmer access to productive resources. Table 2 shows that access to FISP coupon was higher among adopters for inorganic fertilizer, improved seed and legume-maize rotation. For instance, 95 (62.1%) inorganic fertilizer adopters had access to FISP coupons, while only 19 (40.4%) of the non-adopters had access to FISP coupons. On the other hand, 58 (37.9%) of inorganic fertilizer adopters and 28 (59.6%) inorganic fertilizer non-adopters had no access to FISP coupons. As for improved seed, 91 (60.3%) of the adopters had access to FISP coupons against 60 (39.7%) who did not. This means that having access to FISP coupon does not guarantee adoption of inorganic fertilizer and improved maize seed but rather redemption of that coupon as some farmers who accessed the FISP coupons had sold them to fellow farmers and vendors.

3.2 Factors Influencing Adoption of ISFM Technologies

Table 3 presents the results of the multivariate probit model for the factors influencing adoption of ISFM technologies. The model fits the data well with the Wald$x^2(37) = 258.25$; $p = 0.0000$, thereby rejecting the null hypothesis that all regression coefficients are jointly equal to zero. The results justify the application of the MVP model as shown by the likelihood ratio test for the overall correlation of error terms as indicated by correlation coefficients (rho) [$x^2(6) = 17$; $p = 0.0080$)]. This implies that the null hypothesis that the error terms are not correlated is rejected

Table 2 Summary of key demographic and socio-economic characteristics by farmer and technology category

Variables		Inorganic fertilizer		Improved seed		Maize-legume intercropping		Legume/maize rotation		Total
		Adopters $n = 153$	Non-adopters $n = 47$	Adopters $n = 151$	Non-adopters $n = 49$	Adopters $n = 92$	Non-adopters $n = 108$	Adopters $n = 133$	Non-adopters $n = 67$	$n = 200$
Age of the household head	(Years)	44.4	43.4	44.6	42.6	44.5	43.8	44.2	43.9	44.1
Household size		5.8	5.3	5.7	5.7	5.8	5.6	5.8	5.6	5.7
Number of adults		2.6	2.3	2.5	2.5	2.6	2.4	2.5	2.4	2.5
Farming experience	(Years)	19.6	21.1	20.0	19.8	20.9	19.1	20.3	19.2	20
Landholding size	(ha)	2.0	2.2	2.0	2.1	2.2	1.8	2.1	1.8	2.0
Plot distance	(km)	0.65	0.75	0.62	0.83	0.56	0.80	0.70	0.63	0.7
Sex of the household head (%)	Male	135 (88.2)	37 (78.7)	131 (86.7)	41 (83.7)	79 (85.9)	93 (86.1)	116 (87.2)	56 (83.6)	172 (86.0)
	Female	18 (11.8)	10 (21.3)	20 (13.3)	8 (16.3)	13 (14.1)	15 (13.9)	17 (12.8)	11 (16.4)	28 (14.0)
Education level of household head	Never attended	5 (3.3)	3 (6.4)	5 (3.3)	3 (6.1)	3 (3.3)	5 (4.6)	7 (5.3)	1 (1.5)	8 (4.0)
	Primary	111 (72.5)	33 (70.2)	104 (68.9)	40 (81.6)	70 (76.1)	74 (68.5)	92 (69.1)	52 (77.6)	143 (72.0)
	Secondary	36 (23.5)	11 (23.4)	41 (27.1)	6 (12.3)	18 (19.6)	29 (26.9)	33 (24.8)	14 (20.1)	47 (23.5)
	Tertiary	1 (0.7)	0 (0.0)	1 (0.7)	0 (0.0)	0 (0.0)	1 (1.0)	1 (0.8)	0 (0.0)	1 (0.5)
Access to FISP coupon	Yes	95 (62.1)	19 (40.4)	91 (60.3)	23 (46.9)	52 (56.6)	62 (57.4)	72 (54.2)	42 (62.7)	114 (57.0)
	No	58 (37.9)	28 (59.6)	60 (39.7)	26 (53.1)	40 (43.4)	46 (42.8)	61 (45.8)	25 (37.3)	86 (43.0)
Group membership	Yes	94 (61.44)	31(66.0)	100 (66.2)	25 (51.0)	73 (79.35)	52 (48.2)	80 (60.2)	45 (67.2)	125 (62.5)
	No	59 (38.56)	16 (34.0)	51 (33.8)	24 (49.0)	19 (20.65)	56 (51.8)	53 (39.8)	22 (32.8)	75 (37.5)

Access to credit	Yes	34 (22.2)	11 (23.4)	38 (25.2)	7 (14.3)	21 (22.83)	24 (22.2)	35 (26.3)	10 (14.9)	45 (22.5)
	No	119(77.8)	36 (76.6)	113 (74.8)	42 (85.7)	71 (77.17)	84 (77.8)	98 (73.7)	57 (85.1)	155 (77.5)
Access to extension	Yes	108 (70.6)	28 (59.57)	107 (70.86)	29 (59.2)	76 (82.6)	60 (55.6)	94 (70.7)	42 (62.7)	136 (68.0)
	No	45 (29.4)	19 (40.53)	44 (29.14)	20 (40.8)	16 (17.4)	48 (44.4)	39 (29.3)	25 (37.3)	64 (32.0)
ISFM training	Yes	73 (47.7)	18 (38.3)	76 (50.3)	15 (30.6)	57 (62.0)	34 (31.5)	57 (42.9)	34 (50.8)	91 (45.5)
	No	80 (52.3)	29 (61.7)	75 (46.7)	34 (69.4)	35 (38.0)	74 (68.5)	76 (57.1)	33 (49.2)	109 (54.5)

Figures in parentheses are percentages

Table 3 Results of the multivariate probit model for factors influencing adoption of ISFM technologies

Explanatory variables	ISFM technology			
	Inorganic fertilizer $n = 153$	Improved seed $n = 151$	Intercropping $n = 92$	Rotation $n = 133$
Sex of household head (1 = male, 0 = female)	0.1120 (0.3810)	0.5426 (0.3562)	0.0737 (0.3269)	0.4619 (0.3231)
Age of household head (years)	−0.0175 (0.0105)*	0.0107 (0.0091)	0.0030 (0.0077)	
Farm experience (years)				0.0126 (0.0090)
Land ownership (1 = own land, 0 = otherwise)	0.4282 (0.4750)	0.6390 (0.4524)	0.1020 (0.2622)	0.5732 (0.2711)**
Land size (ha)			0.0080 (0.0538)	0.1010 (0.0574)*
Number of adults	0.2322 (0.1452)			
Access to fertilizer (FISP) coupon (1 = yes, 0 = no)	1.0838 (0.2589)***	0.1805 (0.2306)		
Soil fertility perception (1 = average, 0 = otherwise)	0.3871 (0.2611)	0.6798 (0.2339)***		
Soil fertility perception (1 = fertile, 0 = otherwise)				0.4931 (0.2489)**
Education level of spouse (1 = senior primary, 0 = otherwise)	0.0603 (0.2695)			
Education level of spouse (1 = junior secondary, 0 = otherwise)		0.7057 (0.4680)		
Extension access (1 = yes, 0 = no)	0.6618 (0.2722)**		0.5257 (0.2380)**	
Log of household income (MK)	0.2153 (0.0930)**	0.2144 (0.0967)**		
Radio ownership (1 = yes, 0 = no)		0.3630 (0.2539)		
ISFM training (1 = yes, 0 = no)		0.5807595 (0.2450)**		−0.2590 (0.1979)
Seed access (1 = yes, 0 = no)			0.6659 (0.2067)***	−0.4047 (0.1984)**
Credit access (1 = yes, 0 = no)		0.3794 (0.2892)	−0.3425 (0.2566)	0.5254 (0.2479)**
Market access (1 = yes, 0 = no)			0.0021 (0.0005)**	0.2841 (0.2061)
Group membership (1 = yes, 0 = no)			0.6521 (0.2228)***	

(continued)

Table 3 (continued)

Explanatory variables	ISFM technology Inorganic fertilizer $n = 153$	Improved seed $n = 151$	Intercropping $n = 92$	Rotation $n = 133$
Constant	−2.4531 (1.2717)	−3.7936 (1.3068)	−1.4136 (0.5716)	−0.8797 (0.5406)

$N = 178$; Wald chi2(37) = 258.25; Prob > chi2 = 0.0000; Log pseudolikelihood = −330.66398
Likelihood ratio test of rho21 = rho31 = rho41 = rho32 = rho42 = rho43 = 0: chi2(6) = 17.3651; Prob > chi2 = 0.0080
Figures in parentheses are percentages
*significant at 10% ($p < 0.1$)
**significant at 5% ($p < 0.05$) and
***significant at 1% ($p < 0.01$)

and concludes that the error terms are correlated. As presented in Table 3, the pair-wise correlation error terms were significant, thus confirming complementary (positive correlation) and/or substitutable (negative correlation) between some of the ISFM technologies. This implies that the adoption of a given ISFM technology is dependent on the adoption of other ISFM technologies, and thus aggregating them into individual ISFM technologies (single-equation approach) would yield biased and inconsistent results.

Consistent with previous studies (e.g. Marenya and Barrett 2007; Yirga and Hassan 2008; Teklewold et al. 2013; Kamau et al. 2013; Kassie et al. 2015a), the following variables were significant in influencing adoption of ISFM technologies: age of the household head, land tenure, access to FISP coupons, access to extension, access to credit, access to markets, access to legume seed, household income, training in ISFM, soil fertility perception status, landholding size and group membership. The selection of these variables was guided by a thorough literature review based on relevant economic theories. Age of the household head was significant in influencing only adoption of inorganic fertilizer at 10% alpha level. This finding concurs with the finding of Marenya and Barrett (2007) who observed that as decision-makers age, their planning horizons shrink, and so the incentives for them to invest in the future productivity of their farms diminish. Land tenure positively and significantly influenced the adoption of legume-maize rotation at 5% alpha level ($p < 0.05$). This suggests that the propensity to adopt legume-maize rotation is high on owner-cultivated plots than on rented and/or leased plots due to tenure insecurity. Because the long-term soil fertility investment benefits (such as legume-maize rotation) accrue over time, this implies that secure land tenure will influence positively on adoption decision (Teklewold et al. 2013). Access to subsidized fertilizer coupons was positive and highly significant in influencing adoption of inorganic fertilizer at 1% alpha level with p-value of 0.000. Kassie et al. (2015b) noted that access to subsidy coupons has facilitated the adoption and use of inorganic fertilizer in Malawi. Access to extension also positively and significantly influenced adoption of inorganic fertilizer and maize-legume intercropping at 5% alpha level

($p < 0.05$), confirming the critical role played by extension in creating awareness as a prerequisite stage to technology adoption.

The results of the MVP model also revealed a positive and significant relationship between access to markets (both output and input markets) and adoption of maize-legume intercropping. The finding is consistent with the findings of other previous studies such as Lambrecht et al. (2014b). Legumes such as soybean, pigeon peas and groundnuts constitute second major cash crop after tobacco in the study area, and thus, despite soil fertility improvement and household food and nutritional benefits, legumes are grown for immediate cash purposes. This means that access to output markets for grain legumes is crucial in realizing this objective.

Mhango et al. (2012) also noted that limited access to both input (mainly for seed) and output markets was one of the major constraints to adoption of maize-legume intercropping. Thus, facilitating access to input and output markets among the smallholder farmers by either introducing these markets in their vicinities or reducing transaction costs that farmers face in accessing the markets by improving the network of road conditions and provision of affordable transportation means would greatly increase the adoption of maize-legume intercropping. Vanlauwe et al. (2014) also argue in support that increasing the adoption of sustainable intensification strategies such maize-legume intercropping will, among others, require access to profitable output markets for enhanced productivity growth. According to Sanginga and Woomer (2009), maximum benefits from ISFM technologies can only be obtained within an enabling environment, where factors such as input and output markets are in place, coupled with functional service delivery, institutions and progressive policies.

Access to legume seed was found to be positive and highly significant ($p = 0.001$) in influencing adoption of maize-legume intercropping. This finding is consistent with the findings by Amare et al. (2011) who, using a bivariate probit model, found a positive and significant relationship between access to legume seed and adoption of maize-pigeon pea intercropping. This is not surprising as most of the farmers in the sample size reported access to seed constraint as a major impediment to adoption of legume diversification in maize-based farming system. This was more pronounced among the non-adopters. The finding implies holding other things equal; access to legume seed increases the propensity to adopt maize-legume intercropping.

Landholding size is another important determinant of adoption of ISFM technologies, significantly influencing adoption of legume-maize rotation at 10% alpha level ($p < 0.1$). This implies that the likelihood of adopting legume-maize rotation is higher with households that have large landholding sizes than those households with smaller landholding size, other things being equal. Tizale (2007) found similar results in Ethiopia and argued that large landholding sizes make crop rotation such as legume-maize rotations more attractive than maize-legume intercropping, which is often practised to maximize land use and spread economic risk. Group membership, proxy measure of social capital, was found to be positive and highly significant in influencing adoption of maize-legume intercropping suggesting that households that belong to any farmer group or rural institution pose high probability of adopting maize-legume intercrop as opposed to those households that do not. Group membership was, among others, found to facilitate access to legume seed by linking farmers to external institutions such as Bunda College.

Complementarity and Substitutability of Adoption of ISFM Technologies

Table 4 presents the covariance matrix of the adoption equation error terms depicting the different types of relationships that exist among the technologies.

Table 4 indicates that the correlation coefficient between the adoption of inorganic fertilizer and improved seed varieties is positive implying the complementary relationship between the two technologies. This finding is in agreement with the findings of other studies such as of Hassen (2015) who found a significant complementary relationship between adoption of improved varieties and inorganic fertilizer in Ethiopia. Another complementary and significant relationship was observed between the adoption of inorganic fertilizer and maize-legume intercropping to an extent of about 24%. This suggests that the two practices have reinforcing complementary effects on soil fertility and thus on the productivity of maize, the staple food crop in the study area. On the other hand, a negative and significant relationship was found between adoption of legume-maize rotation and inorganic fertilizer (about 34%) indicating substitutability in the adoption decisions of the two technologies. This was common among wealthy households that would prefer to adopt inorganic fertilizer because of their ability to afford chemical fertilizers and practice continuous cultivation (monocropping) as compared to their poor counterparts.

The complementary relationship that exists between adoption of inorganic fertilizer and maize-legume intercropping raises important policy implications. The results suggest that inorganic fertilizer use must not be promoted as a stand-alone practice for the management of soil fertility because the recommended fertilizer application rates are often beyond the reach of resource-constrained smallholder farmers. Therefore, combining the FISP package with complementary interventions such as legume seed (accompanied by extension messages) with the aim of promoting maize-legume intercropping and rotations will help in raising the efficiency of

Table 4 Correlation coefficient estimates (covariance matrix) of the adoption equations of the joint MVP model

Technology/practice	Inorganic fertilizer	Improved seed	Maize-legume intercropping
Improved seed	0.091 (0.164)		
Maize-legume intercropping	0.239 (0.141)*	0.027 (0.144)	
Legume-maize rotation	−0.339 (0.146)**	−0.177 (0.126)	0.325 (0.113)***

Likelihood ratio test of rho21 = rho31 = rho41 = rho32 = rho42 = rho43 = 0: chi2(6) = 17.3651 Prob > chi2 = 0.0080

Figures in parentheses are standard errors

*significant at 10% ($p < 0.1$)

**significant at 5% ($p < 0.05$) and

***significant at 1% ($p < 0.01$); ρ (rho) = correlation coefficient fertilizer use by accelerating the release of nutrients into the soil while producing long-term benefits for sustainable crop production. The adoption of maize-legume intercropping and rotation also exhibited a significant and positive (complementary) relationship (32.5%). This may be explained by the fact that smallholder farmers are advised to follow a two- to three-year adoption cycle of legume diversification in the maize-based farming system as promoted by the Legume BB project than with crop rotation at the centre. This implies that the adoption of a particular cropping system in the current year was dependent on the cropping system practised by the farmer in the previous season, while the cropping system on the next growing system is conditional on the present cropping system employed by the farmer.

3.3 Determinants of Disadoption of Maize-Legume and Legume-Legume Intercropping and Rotation

Among the four technologies under consideration, only two of the technologies had significant number of disadopters to merit regression analysis. These are maize-legume intercropping and rotation. It is worth noting that these are the only practices that are being promoted by the Legume Best Bets project. Thus, bivariate probit regression model was applied to test whether the two disadoption decisions are correlated or not. The aim here is to test whether the same factors that encourage adoption also discourage disadoption of the ISFM technologies. Table 5 presents the results of the bivariate probit model for the determinants of ISFM technology disadoption, with particular focus on maize-legume intercropping and rotation.

The results of the bivariate probit regression show that the model is highly significant overall, with the Wald $x^2(20) = 53.57$; $p = 0.0001$. This implies that the model fits the data reasonably well and the null hypothesis that all regression coefficients are jointly equal to zero is rejected. The Wald test of rho = 0 with $x^2(1) = 9.82887$; $p = 0.0017$) supports the application of the bivariate probit model and conclude that the error terms in the two disadoption decisions are correlated. This is also confirmed by the positive and significant correlation coefficient (rho) of 0.5972 (about 60%) suggesting that a strong complementary relationship exists between the disadoption of maize-legume intercropping and legume-maize rotation. In other words, the two technologies are abandoned jointly. This may be so because of the central role of legumes in the two cropping system, and thus whatever happens to legumes will affect the sustainability of the two systems. This has a critical policy implication in that efforts to avert the abandonment of one practice will most probably reduce the likelihood of abandonment of the other.

Five of the explanatory variables were significant in explaining disadoption of maize-legume intercropping and legume-maize rotation. These are sex of the household head, farming experience, total land size, access to legume seed and access to extension. An interesting result was found here in that a symmetrical relationship exists whereby some of the variables that encourage adoption ISFM technologies

Table 5 Determinants of disadoption of ISFM technologies

Explanatory variables	Conditional marginal effects	
	Maize-legume intercropping	Legume-maize rotation
Number of adults	0.0069 (0.0348)	0.0078 (0.0426)
Sex of the household head (1 = male, 0 = female)	0.2276 (0.1124)**	−0.0513 (0.1128)
Farm experience (years)	0.0072 (0.0033)**	−0.0085 (0.0041)**
Land size (ha)	0.0203 (0.0105) **	−0.0321 (0.0154)**
Market access (1 = yes, 0 = no)	0.0302 (0.0787)	−0.0657 (0.0799)
Access to FISP coupon (1 = yes, 0 = no)	0.0898 (0.0765)	−0.0315 (0.0828)
Access to seed (1 = yes, 0 = no)	−0.3868 (0.0557)***	0.2599 (0.0894)***
Extension access (1 = yes, 0 = no)	−0.0984 (0.0873)	−0.0713 (0.0777352
Credit access (1 = yes, 0 = no)	0.0898 (0.1303)	−0.1561 (0.1538)
Group membership (1 = yes, 0 = no)	−0.0876 (0.0812)	0.0333 (0.0810)
Constant	−1.024282 (0.4379)**	−0.5011 (0.6712)
/athrho	0.6888 (0.2197)***	

$N = 190$; Wald chi2 (20) = 53.57; Prob > chi2 = 0.0001
Log pseudolikelihood = −134.78223
Wald test of rho = 0: chi2(1) = 9.82887; Prob>chi2 = 0.0017
Figures in parentheses are standard errors
*significant at 10% ($p < 0.1$)
**significant at 5% ($p < 0.05$) and
***significant at 1% ($p < 0.01$)

also discourage abandonment. Results of the bivariate probit model revealed that male headship of a household significantly ($p < 0.05$) increases the probability of disadopting maize-legume intercropping by almost 22.8%. In other words, female-headed households are more likely to continue with adoption of maize-legume intercropping than male-headed households. This may be explained by the fact that men are more active and always at the forefront in the main cash crop of the area, tobacco, than female farmers/women who are left to manage food and secondary cash crops like legumes (such as soya beans, pigeon peas and groundnuts). This finding may also attest to the important role that women play in food production.

Whereas years of farming experience had positive influence on adoption of maize-legume rotation, results of the bivariate probit revealed that years of farming experience is negatively and significantly ($p < 0.05$) correlated with disadoption of legume-maize rotation and positively correlated with disadoption of maize-legume intercropping. This implies that farmers with more years of experience in farming are less likely to abandon legume-maize rotation. This may be explained by the fact that farmers with more years of farming experience accumulate more knowledge on crop husbandry such as crop management and are thus better able to manage the two crops on the same plot as compared to farmers with little farming experience. The conditional marginal effects in Table 5 show that a 1-year increase in farming experience increases the propensity to abandon maize-legume intercropping by 0.75 and reduces the likelihood of disadopting legume-maize rotation by 0.85%.

Landholding size as an important productive asset was also negatively and significantly ($p < 0.05$) correlated with disadoption of legume-maize rotation and positively and significantly correlated with disadoption of maize-legume intercropping (see table on conditional marginal effects). Other things being equal, households with large land size are more likely to continue with adoption of maize-legume intercropping than those with small landholding size. This means that landholding size played a symmetrical role in encouraging both adoption and long-term use/adoption of maize-legume intercropping and legume-maize rotation. The positive correlation between landholding size and maize-legume intercropping was expected, since one of the major reasons why farmers practice intercropping is to maximize land use; thus, farmers with more landholding sizes are more likely to abandon maize-legume intercropping and switch to legume maize rotation. The conditional marginal effects show that a unit increase in landholding size increases the likelihood of abandoning maize-legume intercropping by almost 2% and reduces the likelihood of abandoning legume-maize rotation by 3%. This relationship was significant at 5% alpha level.

While access to legume seed highly encourages the adoption of maize-legume intercropping, results of the bivariate probit regression indicate that access to legume seed among smallholder farmers was significant and negatively correlated with disadoption of maize-legume intercropping. Seed access constraint was the second most frequently reported reason for disadoption of maize-legume intercropping after pests and diseases among smallholder farmers. This implies that the probability of both adoption and continued adoption of maize-legume intercropping is largely dependent on guaranteed access to legume seed among smallholder farmers. Access to legume seed has the largest conditional marginal effects on both intercropping and rotation and shows that access to seed reduces the probability to abandon maize-legume intercropping by almost 39% and significantly increases the probability to abandon legume-maize rotation by almost 26%. Access to extension was negative and significantly correlated with disadoption of both maize-legume intercropping and rotation at 5% and 10% alpha level, respectively. This implies that households with greater access to extension services are less likely to abandon the practices than those with no access to extension services, holding other things constant. This implies that continued adoption of these two technologies is dependent on a vibrant extension system to a certain extent. The marginal effects in Table 5 show that access to legume seed reduces the probability to disadopt maize-legume intercropping by almost 4%.

Farmer Perceptions of the Reasons for Disadoption of ISFM Practices

Farmers' perceptions on reasons for disadoption of ISFM practices, particularly maize-legume intercropping and rotation, were mixed, with some consistencies with our model results. Figure 1 shows that the most frequently reported reason for abandonment of the technologies is lack of access to legume seed with 67% of the responses. This is followed by lack of access to markets to incentivize continued adoption and labour constraints (8.6%). Labour constraints were mostly reported for maize-legume intercropping when it comes to crop residue incorporation. Other reasons for disadoption included limited land size (5.4%) especially for legume-maize rotation and pests and diseases (5.4%).[1]

4 Conclusion and Policy Recommendations

This study has analysed and assessed the determinants of both adoption and disadoption of ISFM technologies among the smallholder maize farmers in Central Malawi. This has been achieved through the application of the multivariate and bivariate probit models. Consistent with the literature in the area, results have revealed that adoption decisions regarding ISFM technologies are interdependent. The study has revealed that the key drivers of ISFM technology adoption are land tenure, access to extension, access to credit, access to markets, access to legume seed, landholding sizes and farmer group membership, among others. On the other hand, similar factors are found to discourage disadoption of ISFM technologies and/or practices. For instance, access to legume seed discourages disadoption of maize-legume intercropping, while larger landholding sizes encourage adoption of legume-maize rotation. Worth noting, females are more likely to continue with adoption of maize-legume intercropping.

The study has also established complementary and substitutable relationships in the (dis)adoption decisions regarding ISFM technologies. For instance, inorganic fertilizer and maize legume intercropping are adopted as complements (24%), while maize-legume intercropping and rotation are adopted as complements (33%). From the results, we recommend that ISFM technologies that include inorganic fertilizer should not be promoted as individual components but as a package based on their "best-bet" combinations. For instance, while the agricultural input subsidy in Malawi has increased adoption of inorganic fertilizer and improved maize seed, more should be done to scale up the legume component so to promote complementary interventions such as maize-legume intercropping and rotation. This may be achieved through increasing access to legume seed and both input and output markets among smallholder farmers. We also recommend that promotion of ISFM technologies should focus on those with secure land tenure rights, particularly for women, so that they are incentivized, as these have shown to have the higher likelihood of investing in ISFM technologies.

[1] Particularly, pigeon peas are heavily attacked by pests if not treated.

References

Amare M, Asfaw S, Shiferaw B (2011) Welfare impacts of Maize–Pigeon pea intensification in Tanzania. Agric Econ 2012:1–17

Dorward A, Chirwa E (2011) The Malawi agricultural input subsidy programme: 2005/06 to 2008/09. Int J Agric Sustain 9(1):232–247

FAC (2010) Raising agricultural productivity in Africa: options for action, and the role of subsidies. Africa progress panel policy brief

Fairhurst T (ed) (2012) Handbook for integrated soil fertility management. Africa Soil Health Consortium, Nairobi

Foster AD, Rosenzweig MR (2010) Microeconomics of technology adoption. Economic Growth Center. Discussion Paper No. 984. Yale University, USA

Hassen S (2015) Disadoption, substitutability, and complementarity of agricultural technologies: a random effects multivariate probit analysis. Environment for development. Discussion Paper Series. Washington D.C., USA

Holden S, Lunduka R (2010) Too poor to be efficient? Impacts of the targeted fertilizer subsidy programme in Malawi on Farm Plot Level Input Use, Crop Choice and Land Productivity. Noragric Report No. 55, Norway

Kamau M, Smale M, Mutua M (2013) Farmer demand for soil fertility management practices in Kenya's Grain Basket. Selected paper prepared for presentation at the agricultural and applied economics association (AAEA) and CAES joint meeting. Washington, D.C., USA. August 4–6

Kanyamuka JS, Jumbe CBL, Ricker-Gilbert J (2018) Making agricultural input subsidies more effective and profiatble in africa: the role of complementary intrventions. Book Chapter. Food systems sustainability and environmental policies in modern economics. https://doi.org/10.4018./978-1-5225-3631-4.ch008

Kassie M, Jaleta M, Shiferaw B, Mmbando F, Mekuria M (2013) Adoption of interrelated sustainable agricultural practices in smallholder systems: evidence from rural Tanzania. Technol Forecast Soc Chang 80(3):525–540

Kassie M, Teklewold H, Jaleta M, Marenya P, Erenstein O (2015a) Understanding the adoption of a portfolio of sustainable intensification practices in eastern and southern Africa. Land Use Policy 42(2015):400–411

Kassie M, Teklewold H, Marenya P, Jaleta M, Erenstein O (2015b) Production risks and food security under alternative technology choices in Malawi: application of a multinomial endogenous switching regression. J Agric Econ 66(3):640–659

Lambrecht I, Vanlauwe B, Maertens M (2014a) Integrated soil fertility management: from concept to practice in Eastern DR Congo. Bioeconomics Working Paper Series. KU Leuven, Belgium

Lambrecht S, Vanlauwe B, Merckx R, Miet Maertens M (2014b) Understanding the process of agricultural technology adoption: mineral fertilizer in Eastern DR Congo. Elsevier. World Dev 59:132–146

Marenya PP, Barrett CB (2007) Household-level determinants of adoption of improved natural resources management practices among smallholder farmers in western Kenya. Food Policy 32(2007):515–536

Mhango WG, Snapp SS, Phiri GYK (2012) Opportunities and constraints to legume diversification for sustainable maize production on smallholder farms in Malawi. Renewable Agric Food Syst 28(3):234–244

Nata JT, Mjelde JW, Boadu FO (2014) Household adoption of soil-improving practices and food insecurity in Ghana. Agric Food Sec 2014(3):17

Ngwira AR, Kabambe VH, Kambauwa G, Mhango WG, Mwale CD, Chimphero L, Chimbizi A, Mapfumo P (2012) Scaling out best-fit legume technologies for soil fertility enhancement among smallholder farmers in Malawi. Afr J Agric Res 7(6):918–928

Ollenburger MH (2012) Modelling integrated soil fertility management options in Malawi under variable soil and climate conditions. MSc thesis. Michigan State University, USA
Sanginga N, Woomer PL (2009) Integrated soil fertility management in Africa: principles, practices and development process. Tropical Soil Biology and Fertility Institute of the International Canter for Tropical Agriculture, Nairobi
Sommer R, Bossio D, Dester L, Dimes J, Kihara J, Koala S, Mango N, Rodriguez D, Thierfelder C, Winowiecki L (2013) Profitable and sustainable nutrient management systems for East and Southern African smallholder farming systems-challenges and opportunities: a synthesis of the Southern and eastern African situation in terms of the past experiences, present and future opportunities in promoting nutrient use in Africa. CYMMYT, Colombia
Teklewold H, Kassie M, Shiferaw B (2013) Adoption multiple sustainable agricultural practices in rural Ethiopia. J Agric Econ 64(3):597–623
Tizale CY (2007) The dynamics of soil degradation and incentives for optimal management in the Central Highlands of Ethiopia. PhD Thesis. University of Pretoria, South Africa
Vanlauwe B (2015) Sustainable agricultural resources management: unlocking land potential for productivity and resilience. Background paper. United Nations Economic Commission for Africa, Nairobi
Vanlauwe B, Coyne D, Gockowski J, Hauser S, Huising J, Masso C, Nziguheba G, Schut M, Asten PV (2014) Sustainable intensification and the African smallholder farmer. Curr Opin Environ Sustain 8:15–22
Yirga C, Hassan RM (2008) Multinomial logit analysis of farmers' choice between short and long-term soil fertility management practices in the Central Highlands of Ethiopia. Researchgate. Accessed 14 Oct 2015

Exploiting Arbuscular Mycorrhizal Fungi-Rhizobia-Legume Symbiosis to Increase Smallholder Farmers' Crop Production and Resilience Under a Changing Climate

Ezekiel Mugendi Njeru, Morris Muthini, Mercy Martha Muindi, Omwoyo Ombori, Shem Bonuke Nchore, Steve Runo, and John M. Maingi

Abstract Beneficial soil microbiota, such as arbuscular mycorrhizal fungi (AMF) and rhizobia, provide essential agroecosystem services in smallholder farming systems. Such microorganisms have great potential to promote crop production and resilience under a changing climate in sub-Saharan Africa. However, their function is affected by agronomic management practices, crop genotype and soil quality, among other factors. In this work, we sought to determine the effect of soil quality and crop genotype on nodulation, percentage mycorrhizal colonization and growth of maize and cowpea crops. Soil samples were obtained from ten smallholder farms with known management history in Embu and Kitui counties of Kenya and analysed for physicochemical parameters. Greenhouse bioassays were then carried out, where the samples were put in sterilized pots in four replicates and maintained in a completely randomized design. Four cowpea and maize genotypes (locally grown landraces and recommended genotypes from *Kenya Agricultural and Livestock Research Organization*) were grown in pots for 40 days. After harvesting, nodulation in the case of cowpea, shoot dry weights and mycorrhizal root colonization were determined. Remarkably, cowpea genotypes differed significantly ($p < 0.0001$) in nodule number. The locally cultivated landrace (C2) recorded the lowest nodulation with 30.4 nodules plant^{-1}, compared to the open pollinated varieties (OPVs): C1, 39.15; C3, 43.70; and C4, 40.6 nodules plant^{-1}. Among the maize genotypes, the locally cultivated landrace (M3) recorded a significantly ($p = 0.008$) higher percentage of mycorrhizal root colonization (68.9%) compared to the OPVs: M1 58.1% and M2 65.3%, while the hybrid (M4) had the lowest root colonization of 57.8%. Soil characteristics influenced nodulation and mycorrhizal colonization, where soil P was positively correlated to cowpea nodulation. Soil organic matter, nitrogen, pH and calcium positively correlated with AMF maize

E. M. Njeru (✉) · M. Muthini · M. M. Muindi · S. Runo · J. M. Maingi
Department of Biochemistry, Microbiology and Biotechnology, Kenyatta University, Nairobi, Kenya

O. Ombori · S. B. Nchore
Department of Plant Sciences, Kenyatta University, Nairobi, Kenya

B. R. Singh et al. (eds.), *Climate Impacts on Agricultural and Natural Resource Sustainability in Africa*, https://doi.org/10.1007/978-3-030-37537-9_27

root colonization. Our results demonstrate the strong effect of soil quality and crop genotype on AMF-rhizobia-legume symbiosis, which affects overall crop growth and production. These factors should therefore be critically considered during the development of efficient low-cost inocula for enhanced smallholder farmers' crop production.

1 Introduction

Over the last decade, agricultural productivity has continued to decline in many smallholder agroecosystems, which are the main source of livelihood in developing countries. This has been as a result of several underlying factors such as inherent poor soil fertility, pests and diseases and increased human population, which has consequently led to land fragmentation, degradation and water stress associated with the changing climate (Mwendwa and Giliba 2012; Cozzolino et al. 2013). Among these factors, low soil fertility presents one of the major challenges to sustainable crop production since most smallholder farmers cannot afford to invest in chemical fertilizers (St.Clair and Lynch 2010). Recently, most research to improve crop production has targeted plant breeding for crop cultivars that are high yielding, drought tolerant and resistant to pests and diseases (Boon et al. 2013). However, the widening gap between food production and increasing population has continued unabated, which calls for concerted efforts to enhance sustainable agriculture practices. In this context, key strategies such as the utilization of soil biota for improvement of soil fertility and land productivity through biological processes have been brought to the limelight (Oruru and Njeru 2016).

Nitrogen and phosphorus are the major limiting nutrients required for plant growth (Antunes and Goss 2005). Due to its involvement in crucial plant processes such as respiration, energy transfer, phospholipid and nucleic acid synthesis, plants require phosphorus in large amounts compared to other nutrients (Chen and Graedel 2016). Globally, phosphate-based fertilizer application has increased fivefold from 5 million tons of phosphorous (P) per year in 1961 to 20 million tons per year in 2013 (Chen and Graedel 2016). Most sub-Saharan Africa soils are deficient of P in both inorganic and organic forms since most of the P is poorly available for plant uptake, due to its immobilization and precipitation with other soil minerals such as iron (Fe) and aluminium (Al) in acid soils and calcium (Ca) in alkaline soils. Consequently, only a small percentage of P (<1%) is directly available for plant uptake (Battini et al. 2017), resulting in a major limitation to crop production.

Microorganisms associated with plant roots can be used sustainably in mobilizing P in the soil, thus making it available to the plants (Philippot et al. 2013). These include arbuscular mycorrhizal fungi (AMF) which constitute a key functional group of soil microbiota that can greatly contribute to crop productivity and ecosystem sustainability (Gianinazzi et al. 2010; Njeru 2018). They form symbiotic relationships with the roots of most agricultural crops (ca. 80%), improving the nutritional status of their hosts and protecting them from pathogenic microorganisms and abiotic stresses (Smith and Read 2008; Berta et al. 2013). AMF are also important because of their ability to improve the sustainability of agricultural systems and plant communities and enhance soil biodiversity. The underground hyphal networks formed by AMF can influence plant growth, nutrient acquisition as well as plant-plant interactions. Rondon et al. (2006) and Ondieki et al. (2017) reported that common beans and cowpeas colonized by AMF accumulated approximately 60% more phosphorus content compared to crops without mycorrhizal colonization. This highlights the importance of effectively managing AMF for maximum utilization of phosphates thus minimizing the use of inorganic phosphate fertilizers.

Plant genotypes differ in their dependency on AMF, as reflected by various mycorrhizal colonization and growth performance of host plants after mycorrhizal inoculation (Aquino et al. 2015; Yang et al. 2016; Oruru et al. 2018). Therefore, breeding programs should consider mycorrhizal responsiveness which varies among genotypes (Chu et al. 2013). Moreover, the influence of AMF root colonization by soil properties has also been reported. Van Tuinen et al. (2016) reported variation in AMF colonization between plots due to soil characteristics in each plot while working on interaction between AMF and *Vitis vinifera* (L). Therefore, there is a need to establish suitable soil characteristics for maximum utilization of AMF for improved maize production.

Nitrogen is one of the most essential nutrients for plants since it facilitates various metabolic processes that regulate plant growth and development (Ondieki et al. 2017; Razaq et al. 2017). This is because it is a major constituent of important organic compounds, proteins, enzymes, amino acids and nucleic acids. According to Leghari et al. (2016), nitrogen availability not only improves plant yield but also the quality of food produced. To enhance nitrogen content in the soil, inorganic nitrogen fertilizers have been utilized widely in the past (Otieno et al. 2009). However, nitrogenous fertilizers contribute to greenhouse gas (GHG) emissions globally (Argaw 2013; Erbas 2017), as well as being costly and beyond the reach of most small-scale farmers. There is therefore a need for cheaper and environmentally friendly alternative sources of nitrogen.

Biological nitrogen fixation (BNF), which utilizes the symbiotic interaction between diazotrophic bacteria (rhizobia) and legumes, is a viable alternative to inorganic nitrogen fertilizer use (Oliveira et al. 2010). BNF can sustainably improve nitrogen input in agroecosystems and reduce the need for inorganic fertilizers (Lindstrom et al. 2010). In turn, this will mitigate the negative impacts of inorganic fertilizer production and utilization on the global nitrogen cycle and global warming (Garg and Geetanjali 2007). Based on transcriptome analyses, the legume plant genotype has a significant influence on symbiosis with rhizobia other than the rhizobia strain. Therefore, the ability to have effective symbiotic interaction with rhizobia should be a suitable character for selection of cowpea genotypes for plant breeding programs (Gomes et al. 2015).

Maize (*Zea mays* L.) is one of the most essential cereal crops in Kenya. It is a staple food for above 90% of the Kenyan population (Wambugu et al. 2012). Although food security in Kenya usually depends on adequate supplies of maize to meet the local demand, maize production has declined over the years. The downward trend has been attributed to depleted soil fertility (mainly phosphorus, nitrogen, potassium) due to lack of application of adequate fertilizers and poor soil health management practices (Onono et al. 2013). A change in soil management practices can help reduce the excessive use of mineral fertilizers especially P and N. This includes inoculation with beneficial soil microbiota such as AMF and inclusion of leguminous crops such as cowpea in the cropping systems.

Cowpea (*Vigna unguiculata* (L.) Walp.) is an important multipurpose legume grown in Kenya mainly by smallholder farmers, among other legumes, such as common bean and pigeon peas (Ondieki et al. 2017). It is a major source of income and supplements nutrients such as dietary proteins, vitamins and fibre. Cowpea has the ability to establish symbiosis with rhizobia, which enhances nodulation and the uptake of nitrogen through BNF, promoting plant growth and productivity (Mwendwa and Giliba 2012). The crop can thus be grown in mixed intercropping systems without the need for application of chemical fertilizers (Ondieki et al. 2017).

One way of sustainably improving crop production in smallholder farms is through promotion of healthy plant-microbe interactions. To enhance such interactions in smallholder farming systems, three critical elements should be considered holistically within the scope of increased agroecosystem functionality. These include the soil environment, crop genotype and microbial isolates. In this study, we hypothesized that soil quality and crop genotype would significantly affect nodulation, AMF colonization and growth of locally cultivated maize and cowpea crops in semiarid areas of Eastern Kenya. Our specific objectives were (i) to isolate and characterize native cowpea rhizobia using morphological and biochemical characteristics, (ii) to determine the effect of genotype and soil characteristics on cowpea nodulation and growth and (iii) to determine the effect of maize genotypes and soil characteristics on AMF colonization and growth.

2 Materials and Methods

2.1 Study Site

The study was conducted in smallholder farms in the semiarid areas of Embu (near Machang'a Primary School, latitude 1°10′S, longitude 37°47′E) and Kitui (near Kathiani Primary School, latitude 0°46′S, longitude 37°39′E) counties in lower

Eastern Kenya. The areas are hot and dry with temperatures ranging between 14 and 34 °C. Annual precipitation ranges between 500 and 1050 mm, with an average of 900 mm per annum. The areas represent some of the typical semiarid areas in Kenya predominated by smallholder farmers where agriculture is the main stay.

Cowpea and maize are among the key crops that are grown for food and cash generation. The two crops can be intercropped and provide balanced nutrition with maize, a stable food in Kenya, providing carbohydrates and cowpea providing both vitamins (leaves used as vegetables) and proteins (grains used as a protein source). The selection of farms for the study depended on farm history, where only farms with active cultivation of the two experimental crops and with no inoculation history were selected. Greenhouse bioassays and laboratory experiments were carried out in Kenyatta University, Nairobi, Kenya.

2.2 Soil Sampling and Greenhouse Bioassays

This study involved greenhouse experiments and laboratory analyses. Soil samples (20 kg) were obtained from the 10 farms (5 farms each from Embu County and Kitui County) at a depth of 0–30 cm. The soil samples from each farm were then dried and separated into two portions. One portion was used for analysis of physico-chemical properties and the other portion used for rhizobia and AMF trapping, with cowpea and maize as the host plant, respectively. Four rain-fed maize (*Zea mays* L.) cultivars, comprising of two open pollinated varieties (M1 [OPV 1] and M2 [OPV 2]), M3 (a landrace) and M4 (hybrid maize), and four cowpea (*Vigna unguiculata* (L.) Walp.) cultivars (three open pollinated varieties (C1 [OPV 1], C3 [OPV 2], C4 [OPV 3]) and C2 (a landrace), were grown in 3-L pots in a split-plot arrangement in a completely randomized experimental design. Each treatment was replicated four times. The crops were irrigated as required and harvested after 40 days. The maize crop was examined for AMF root colonization, while cowpea plants were tested for the nodule dry weight and nodule number.

2.3 Isolation and Characterization of Rhizobia

Procedures described by Castellane et al. (2015) were used for the isolation of pure rhizobia isolates from nodules of cowpeas from trapping experiments in the greenhouse. All isolates were characterized based on selected morphological features: colony size, colour, shape, mucosity, border, elevation, transparency and capacity to produce exo-polysaccharides. The isolates were cultured in yeast extract mannitol agar (YEMA) media and later in YEMA containing Congo red dye and YEMA containing bromothymol blue.

2.4 Determination of Percentage Mycorrhizal Colonization of the Maize Genotypes

Fine roots were washed in running tap water and thereafter cleared with 10% KOH at 80 °C water baths. The cleared roots were rinsed thoroughly with distilled water, acidified with 2% HCl, followed by staining with 0.05% (w/v) trypan blue in lactic acid (Liu et al. 2016). The samples were then washed with 10% lactic acid and stored at 4 °C in a refrigerator until analysis. From each sample, random subsets of about 100 root segments were selected for AMF colonization assessment. The percentage of root length colonized by AMF was determined by observing the stained roots for arbuscules, vesicles and intraradical hyphae under a dissecting microscope using the gridline intersect method (Giovannetti and Mosse 1980).

2.5 Soil Analysis

A pH meter was used to determine soil pH in a 1:2.5 soil/water suspension. Soil organic matter was analysed via the wet combustion Walkley and Black oxidation method. Macro-Kjeldahl digestion, titration and distillation method was used to determine total nitrogen (Okalebo et al. 2002). Available P was determined using the Olsen extraction method (Okalebo et al. 2002). The soils were leached using ammonium acetate and the amounts of exchangeable cations (Ca^{2+} and Mg^{2+}) in the leachate analysed using atomic absorption spectroscopy. Flame emission spectroscopy was used to measure K^+ and Na^+ concentrations.

2.6 Statistical Analyses

Percentage data on maize root AMF colonization were arcsine transformed before analysis to fulfil the assumptions of ANOVA. The results were then analysed using two-way analysis of variance, considering the source of soil and crop genotypes as independent factors. Wherever feasible, post hoc testing was performed using Tukey's HSD test. For cowpea genotypes, data on the number of nodules and nodule dry weight were analysed using two-way analysis of variance (ANOVA). Means were separated by Tukey's HSD test at 5% probability level. All ANOVAs and post hoc tests were done using SAS software version 9.2. The bacteria isolate groups were clustered based on their morphological characteristics applying the Euclidean distance similarity and single linkage (nearest-neighbour) procedures using PAST software version 3.20. Redundancy analysis (RDA) to relate soil characteristics with plant growth characteristics (shoot and root dry weights), cowpea nodulation and maize AMF colonization potential was done using Canoco software version 5. RDA is a multivariate analysis that is utilized to determine the relationship between two different sets of data.

3 Results and Discussion

3.1 Soil Characteristics

Soil chemical properties varied across farms in the two counties. Soils were characteristically slightly acidic to medium alkaline, with pH ranging from pH 5.8 in farm 1 to pH 8.0 in farm 4 (Table 1). This was satisfactory for maize and cowpea production. The soil pH was above the critical value of 5.5 recommended for East African soils (Okalebo et al. 2002). Soil pH has been reported to influence P availability and should be maintained between 5.5 and 7.2 (Cerozi and Fitzsimmons 2016). At high pH, phosphorus forms calcium phosphate, an insoluble compound that diminishes P availability to the plants (Cerozi and Fitzsimmons 2016). The percentage soil organic matter (SOM) in soils obtained from the farms ranged from 0.27% to 1.31%, while soil N content ranged from 0.04% to 0.14% in farms 8 and 4, respectively (Table 1). Generally, SOM and % N in both Kitui and Embu counties were below the recommended minimum of 3% and 0.25%, respectively, for East African soil (Okalebo et al. 2002). This requires input of nitrogen into the soil from organic or inorganic sources, or through biological nitrogen fixation. Soils from farms 1, 7 and 10 recorded the highest P concentration of 20 mg kg^{-1}, while soils from farm 4 had the lowest P concentration of 1 mg kg^{-1} (Table 1). Farms 1, 7, 8 and 10 had P levels within or above the critical values of 15 mg kg^{-1} (Okalebo et al. 2002). All the soils had K^+ above the critical value of 0.22 cmol.kg^{-1}, which is adequate for crop production. Adequate K^+ levels are important for many biochemical processes in the plant, including efficiency in the use of nitrogen, and therefore significantly influence plant yield. The exchangeable Na^+ ions were adequate in all of the farms since they were within the recommended minimum of <1

Table 1 Soil chemical properties

SITE	County	Soil pH	SOM %	Total *N*	Av. P mg kg^{-1}	K (Exchangeable (cmol. kg^{-1}))	Ca	Na
Farm 1	Kitui west	5.83	0.78	0.09	20	1.16	3.6	0.12
Farm 2	Kitui west	7.80	1.03	0.11	5	0.24	58.4	1.1
Farm 3	Kitui west	7.98	0.85	0.10	8	0.6	57.4	0.24
Farm 4	Kitui west	8.00	1.31	0.14	1	0.28	58.5	1.56
Farm 5	Kitui west	7.96	1.19	0.13	5	0.58	66	0.44
Farm 6	Machang'a	7.72	0.67	0.08	5	0.78	10.3	0.16
Farm 7	Machang'a	6.29	0.69	0.08	20	0.64	4.5	0.16
Farm 8	Machang'a	6.30	0.27	0.04	15	0.24	2.8	0.06
Farm 9	Machang'a	7.72	0.67	0.08	5	0.78	10.3	0.16
Farm 10	Machang'a	6.87	0.35	0.05	20	0.3	20	3.83
Critical value[a]		5.5	3	0.25	15	0.22	4.0	<1

[a]Okalebo et al. (2002). Soil organic matter (SOM): N, nitrogen; K, potassium; Ca, calcium; Na, sodium; Av. P, average phosphorus

cmol.kg^{-1} for East African soils as recommended by Okalebo et al. (2002). Exchangeable calcium ions were above the critical value in all farms except farms 1 and 8 (Table 1).

3.2 *Effect of Cowpea Genotypes and Soil Characteristics on Rhizobia Nodulation*

There was significant ($p < 0.0001$) variation in nodule number and nodule dry weight between farms. Cowpea crops from farm 8 showed the highest nodulation, while those from farm 2 recorded the lowest nodulation (Table 2). The variation in nodulation among the farms may be due to soil characteristics such as available phosphorus and calcium, pH, salinity soil nitrogen (Agoyi et al. 2017) as well as cropping history (Mothapo et al. 2013).

Cowpea nodulation significantly ($p < 0.0001$) varied between cowpea genotypes. Interestingly, open pollinated genotypes showed higher nodule number compared to

Table 2 Rhizobia nodulation, shoot and root dry weight plant^{-1} in different cowpea genotypes (mean ± standard error [SE])

Treatment	Region	Nodule number	Nodule dry weight (g)	Shoot dry weight (g)	Root dry weight (g)
Farm					
Farm 1	Kitui west	38.44 ± 2.62^{abc*}	0.18 ± 0.01^{a}	1.30 ± 0.15^{ab}	0.26 ± 0.03^{ab}
Farm 2	Kitui west	30.94 ± 3.10^{c}	0.10 ± 0.01^{b}	0.90 ± 0.12^{bc}	0.17 ± 0.02^{bcd}
Farm 3	Kitui west	41.81 ± 2.99^{abc}	0.16 ± 0.02^{ab}	0.81 ± 0.07^{bc}	0.12 ± 0.01^{d}
Farm 4	Kitui west	33.81 ± 1.88^{bc}	0.16 ± 0.02^{ab}	1.06 ± 0.09^{abc}	0.17 ± 0.02^{bcd}
Farm 5	Kitui west	44.63 ± 3.66^{ab}	0.15 ± 0.01^{ab}	0.99 ± 0.12^{bc}	0.21 ± 0.02^{abcd}
Farm 6	Machang'a	34.44 ± 3.02^{bc}	0.15 ± 0.02^{ab}	1.11 ± 0.11^{abc}	0.16 ± 0.02^{bcd}
Farm 7	Machang'a	44.56 ± 4.43^{ab}	0.20 ± 0.02^{a}	1.03 ± 0.12^{abc}	0.21 ± 0.03^{abcd}
Farm 8	Machang'a	46.63 ± 3.56^{a}	0.16 ± 0.01^{ab}	1.50 ± 0.14^{a}	0.30 ± 0.05^{a}
Farm 9	Machang'a	32.44 ± 2.60^{c}	0.10 ± 0.01^{b}	1.11 ± 0.13^{abc}	0.24 ± 0.04^{abc}
Farm 10	Machang'a	36.75 ± 1.75^{abc}	0.11 ± 0.01^{b}	0.68 ± 0.09^{c}	0.13 ± 0.02^{cd}
Variety					
C1		39.15 ± 1.92^{a}	0.14 ± 0.01^{a}	1.13 ± 0.09^{a}	0.22 ± 0.021^{a}
C2		30.35 ± 1.51^{b}	0.14 ± 0.01^{a}	1.18 ± 0.09^{a}	0.22 ± 0.02^{a}
C3		43.70 ± 2.42^{a}	0.14 ± 0.01^{a}	0.97 ± 0.06^{a}	0.18 ± 0.02^{a}
C4		40.58 ± 1.73^{a}	0.16 ± 0.01^{a}	0.92 ± 0.05^{a}	0.17 ± 0.01^{a}
P values					
Farm		<0.0001	<0.0001	<0.0001	<0.0001
Variety		<0.0001	0.5649	0.0304	0.0437
Farm × variety		0.0191	0.2885	0.4487	0.0693

*Values with dissimilar letters along the columns are significantly different at $p < 0.05$ according to Tukey's HSD test. C1, OPV-1; C2, landrace-Kikamba; C3, OPV-2; C4, OPV 3

the local landrace (Table 2). This corresponds with previous findings that cowpea and soybean nodulation is greatly dependent on the crop genotype. Similar results were recently reported among common bean genotypes (Argaw and Muleta 2018; Motaroki et al. 2018) which could be could be attributed to differences in genetic makeup of the various genotypes.

There was a significant difference ($p < 0.0001$) in shoot and root dry weight between the farms. Cowpea crops grown in soils from farm 8 recorded the highest shoot and root dry weights, while those grown in soils from farms 10 and 3 had the lowest shoot and root dry weights, respectively. Moreover, root dry weights were significantly different ($p = 0.04$) between the cowpea varieties, where variety C1 and C2 had the highest shoot and root dry weights (Table 2). Although cowpea varieties C3 and C4 had a high number of nodules, this did not translate to higher shoot dry weights. This could be attributed to the presence of rhizobia isolates that are parasitic (deleterious) to the plant and do not fix nitrogen effectively. This is corroborated by Jida and Assefa (2011), who reported that rhizobia effectiveness cannot be sufficiently determined by nodule dry weight and nodule number; hence, in the rhizobia legume association, rhizobia may infect the host plant but may not be effective in nitrogen fixation.

Relationship Between Soil Characteristics, Cowpea Nodulation and Plant Growth Parameters

Soil characteristics varied across the farms. This may explain the differences observed in shoot dry weights, root dry weights, nodule number and nodule dry weights. Ulzen et al. (2016) reported that a deficiency of nutrients in soil, and other soil parameters such soil pH and temperature, affected nitrogen fixation and ultimately nodulation in cowpeas. This is also consistent with the findings by Slattery and Pearce (2002) who, after working with 6 legumes as trap hosts for rhizobia (field pea [*Pisum sativum*], faba bean [*Vicia faba*], lentil [*Lens culinaris*], vetch [*Vicia sativa*], chickpea [*Cicer arietinum*] and lupin [*Lupinus angustifolius*]), reported that soil characteristics such as soil pH, nutrients, temperature and osmotic stress affect the symbiotic relationships between rhizobia and legumes.

Based on RDA, available phosphates in the soil had a great influence on nodule number and nodule dry weight (Fig. 1). Nitrogen fixation process requires about 25–30 mole of ATP per mole of nitrogen fixed. As a result, nodules provide a strong sink for phosphorus. Moreover, phosphorus increases the overall nutrition of legume crops including nodule number and shoot and root dry weights. According to Bargaz et al. (2012), high available P in the soil stimulates nodulation in legumes, thus helping the plants overcome the inhibitory effects of high N on nodulation.

Potassium availability also correlated closely with high shoot and root dry weights. The shoot dry weights and root dry weights also closely correlated with available phosphates and the nodule number. However, the shoot and root dry weights were closely influenced by the additive effect of soil pH, calcium ions, sodium ions, nitrogen and soil organic matter (Fig. 1).

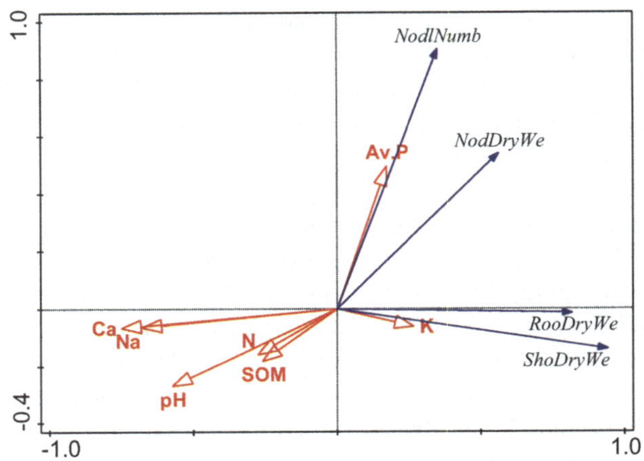

Fig. 1 *Redundancy analysis* (RDA) relating the variation in cowpea nodule number, shoot dry weight, root dry weight and the soil properties in each farm

3.3 Morphological and Biochemical Characteristics of Rhizobia Isolates

A large diversity of 22 nodule isolate groups were obtained from the greenhouse trapping experiment. The rhizobia isolates were characterized based on their growth characteristics in YEMA plain medium, YEMA with Congo red and YEMA with bromothymol blue. The colony elevation was convex, domed or flat. The colony consistency was either dry, mucoid firm gummy or mucoid soft gummy for colonies with excessive extracellular polysaccharide (EPS) production. Colony appearance was opaque, translucent or transparent. The colour was white, creamy and watery (Table 3). The nature of growth was confluent and shiny for all isolates. Only *group i* isolates were dry and opaque (Table 3).

Bromothymol Blue (BTB), YEMA and Congo Red Reaction of the Rhizobia Isolates

The native rhizobia isolates *groups i, x, xvii, xviii* and *xxii* turned BTB blue and were considered as alkaline producers and slow growers (Ondieki et al. 2017), while all the other groups acidified the media containing BTB turning it to yellow, a characteristic of fast-growing rhizobia (Jida and Assefa 2011). None of the nodule isolates absorbed Congo red when grown in YEMA media containing Congo red stain. Pure nodule isolate *groups ii, iv, vii, ix, xii, xiv, xv* and *xx* produced purple pigmentation on the media, while *groups v, xi, xiii, xvii, xviii, xxi* and *xxii* had pink colonies in YEMA media with Congo red (Table 3). A distinctive characteristic of rhizobia is their inability to absorb Congo red. However, some species of rhizobia may absorb

Table 3 Morphological and biochemical characteristics of native rhizobia isolates nodulating cowpea

		Isolate characteristics										
Group	Size (mm)	Elevation	Margin	BTB reaction	EPS production	Colour	Congo red reaction	YEMA reaction	Texture	Colony	Growth	Transparency
i	1.25	F	sm	B	Dry	w	crna	non	Mfg	shinny	cnt	o
ii	6.8	D	sm	Y	M	cr	crna + pp	non	Mfg	shinny	cnt	t
iii	2	D	sm	Y	Dry	crw	cna	non	Mfg	shinny	cnt	t
iv	4.25	D	sm	Y	M	cr	crna + pp	non	Mfg	creamy	cnt	t
v	3.5	D	sm	Y	M	crw + opc	crna + pk	non	Mfg	shinny	cnt	t
vi	3.8	D	sm	Y	M	crw + opc	crna	non	mfg	shinny	cnt	t
vii	6.1	Cnx	sm + irr	Y	M	crw + opc	crna + pp	non	mfg	shinny	cnt	t
viii	3.2	Cnx	sm	Y	M	cr	crna	non	mfg	shinny	cnt	t
ix	3.25	Cnx	sm	Y	M	crw + opc	crna + pp	non	mfg	shinny	cnt	t
x	1	f	sm	B	dry	w	crna	non	mfg	shinny	cnt	t
xi	3.7	d	sm	Y	m	w	crna + pk	non	mfg	shinny	cnt	tp
xii	4.5	cnx	sm	Y	m	cr	crna + pp	non	mfg	shinny	cnt	t
xiii	3.6	d	sm	Y	m	w	crna + pk	non	mfg	shinny	cnt	t
xiv	6	cnx	sm	Y	m	w	crna + pp	non	mfg	shinny	cnt	t
xv	3.6	Cnx	sm	Y	m	cr	crna + pp	non	mfg	shinny	cnt	t
xvi	7	d	sm	Y	C	cr	crna	non	mfg	shinny	cnt	t
xvii	3.5	d	sm	B	m	w	crna + pk	non	mfg	shinny	cnt	t
xviii	3.25	d	sm	B	m	w	crna + pk	non	mfg	shinny	cnt	t
xix	2.8	d	sm	Y	m	cr	crna	non	mfg	shinny	cnt	t
xx	2.9	Cnx	sm	Y	m	w	crna + pp	non	mfg	shinny	cnt	t
xxi	3.2	d	sm + irr	Y	m	w	crna + pk	non	mfg	shinny	cnt	t
xxii	4.4	Cnx	sm	B	m	w	crna + pk	non	mfg	shinny	cnt	t

f flat, *d* domed, *cnx* convex, *sm* smooth, *sm + irr* smooth irregular, *B* blue, *Y* yellow, *M* moderate, *w* white, *cr* creamy, *crw + opc* creamy white with opaque centre, *crna* Congo red non-absorbing, *crna + pp* Congo red non-absorbing with purple pigmentation on media, *crna + pk* Congo red non-absorbing with pink colour appearance, *mfg* mucoid firm gummy, *cnt* confluent, *t* translucent, *tp* transparent, *o* opaque

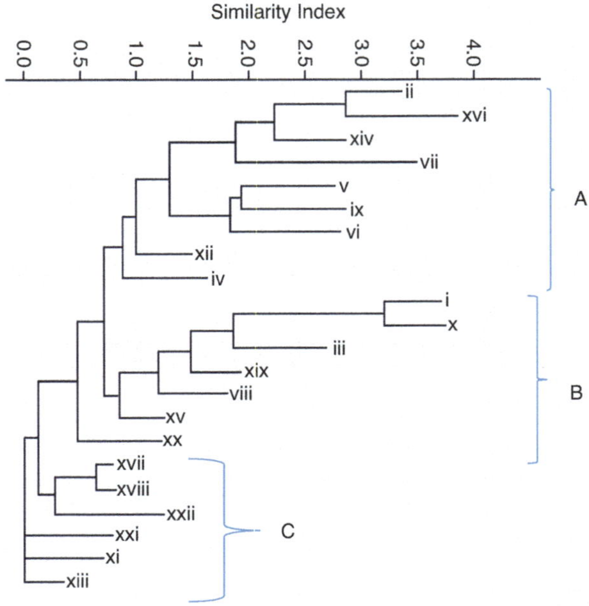

Fig. 2 Cluster analysis based on neighbour joining method and Euclidian similarity index showing morphological relationship of the cowpea nodule isolates

small amounts of the dye to give a pale pink to pink appearance. In addition, some species have been observed to produce orange to deep pink colonies due to absorption of Congo red stain (Ondieki et al. 2017).

Cluster Analysis of the Nodule Isolates Based on Morphological Characteristics

Using the neighbour joining method and Euclidian similarity index, bacteria isolates obtained from cowpea nodules were clustered, based on their morphological characteristics, into three main clusters (cluster A, B and C). Cluster A represented most of the nodule isolates groups (nine groups), while cluster C represented only six bacteria isolate groups (Fig. 2).

3.4 Effect of Soil Characteristics and Genotypes on Percentage AMF Colonization and Maize Growth

As observed in this study, there was significant ($p = 0.0417$) variation in maize root dry weight between farms (Table 4). Significant differences ($p = 0.0286$) in root dry weights were also observed between maize genotypes, where maize hybrid (M4) had the highest root dry weight (1.09 g plant^{-1}), while open pollinated maize genotype (M1) had the lowest root dry weight (0.71 g plant^{-1}) (Table 4). Maize

Table 4 Root and shoot dry weight in g plant^{-1} and AMF colonization of maize genotypes in different soils (mean ± standard error [SE])

Treatment	Region	Root dry weight (g)	Shoot dry weight (g)	Percentage AMF colonization
Farm				
Farm 1	Kitui west	$1.16 \pm 0.25^{a*}$	0.75 ± 0.10^{c}	57.9 ± 6.0^{d}
Farm 2	Kitui west	0.72 ± 0.13^{a}	1.31 ± 0.16^{abc}	69.0 ± 7.4^{abc}
Farm 3	Kitui west	1.21 ± 0.13^{a}	1.45 ± 0.26^{ab}	55.2 ± 3.7^{d}
Farm 4	Kitui west	0.73 ± 0.10^{a}	0.85 ± 0.09^{c}	61.3 ± 3.8^{abcd}
Farm 5	Kitui west	0.87 ± 0.12^{a}	1.21 ± 0.12^{abc}	58.5 ± 5.6^{cd}
Farm 6	Machang'a	0.98 ± 0.10^{a}	1.69 ± 0.14^{a}	71.9 ± 5.5^{ab}
Farm 7	Machang'a	1.32 ± 0.18^{a}	1.47 ± 0.14^{ab}	73.1 ± 5.2^{a}
Farm 8	Machang'a	0.75 ± 0.06^{a}	0.84 ± 0.12^{c}	60.8 ± 4.5^{bcd}
Farm 9	Machang'a	0.88 ± 0.20^{a}	0.96 ± 0.09^{bc}	63.5 ± 4.4^{abcd}
Farm 10	Machang'a	0.87 ± 0.21^{a}	1.12 ± 0.12^{bc}	55.0 ± 5.3^{d}
Maize genotype				
M1		0.71 ± 0.05^{b}	0.95 ± 0.07^{b}	58.1 ± 2.9^{b}
M2		0.96 ± 0.11^{ab}	1.27 ± 0.13^{a}	65.3 ± 3.1^{ab}
M3		1.05 ± 0.10^{ab}	1.28 ± 0.09^{a}	68.9 ± 3.2^{a}
M4		1.09 ± 0.13^{a}	1.15 ± 0.09^{ab}	57.8 ± 4.0^{b}
P values				
Farm		0.0417	<0.0001	0.0136
Maize genotype		0.0286	0.0128	0.0084
Farm × Maize genotype		0.0981	0.0007	<0.0001

[*]Values with dissimilar letters along the columns are significantly different at $P < 0.05$ according to Tukey's HSD test. M1 and M2, open pollinated maize varieties; M3, landrace; M4, hybrid maize variety

shoot dry weight differed significantly ($p < 0.0001$) across farms and between maize genotypes. The highest shoot dry weights were recorded in Farm 6 (1.69 g plant^{-1}) from Machang'a and maize genotype M3 (1.28 g plant^{-1}). Furthermore, there was significant interaction ($p = 0.0007$) between farm and maize genotype in determining maize shoot dry weight (Table 4).

Percentage AMF root colonization varied significantly ($p = 0.0136$) between farms (Table 4). Farm 7 had the highest AMF colonization (73.1%), while Farms 10 and 3 recorded the lowest AMF colonization (Table 4). In addition, % AMF colonization differed significantly ($p = 0.0084$) between maize genotypes (Table 4). Interestingly, the landrace (M3) showed higher root AMF colonization than the other three maize genotypes. These results correspond with studies by Chu et al. (2013) and Njeru et al. (2013), who reported significant variation in percentage mycorrhizal colonization among different maize genotypes. Moreover, since the interaction between maize genotypes and indigenous AMF has not been a concern for many modern maize breeders (Chu et al. 2013), there is a need to screen for mycorrhizal responsiveness during maize breeding programs, as this could affect the overall crop productivity.

Relationship Between Soil Characteristics, Maize AMF Root Colonization and Growth Parameters

Based on RDA, there was a positive correlation between AMF colonization potential and soil pH, calcium ions, soil organic matter, nitrogen and shoot dry weight (Fig. 3). As expected, there was a negative correlation between available phosphates in the soil and maize AMF colonization, since high available P is known to suppress AMF colonization potential. Nevertheless, available phosphates and potassium ions correlated positively with root dry weights of the plants. According to Jordan-meille and Pellerin (2018), phosphorus and potassium are important for plant energetics. Potassium is necessary to plants due to its ability to activate enzymes necessary for the metabolic processes, and its reduction has been reported to reduce shoot and root dry weight accumulation in maize plants (Jordan-meille and Pellerin 2018).

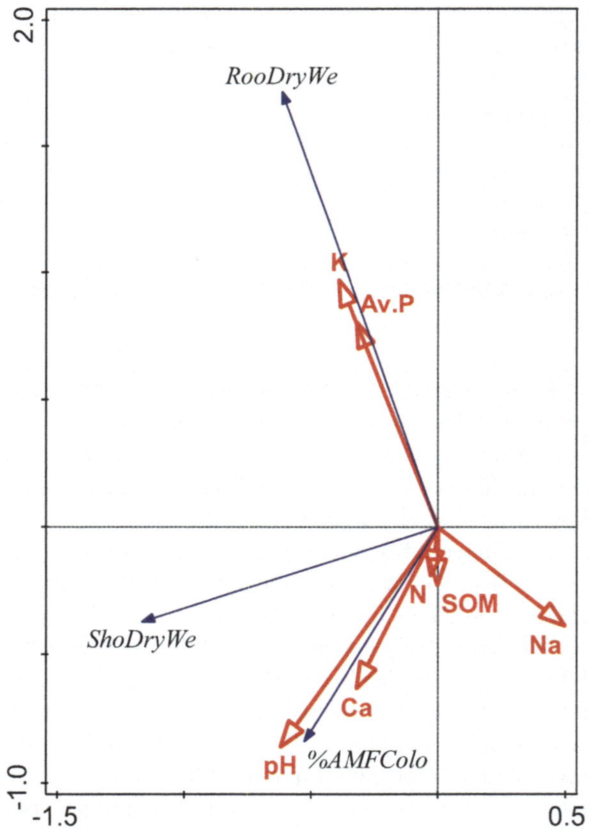

Fig. 3 *Redundancy analysis* (RDA) relating the variation in percentage AMF colonization, shoot dry weight, root dry weight and the soil properties in each farm

4 Conclusions

This study established that cowpea nodulation is dependent on the cowpea genotype and soil characteristics. The landrace cowpea variety performed poorly in nodule colonization when compared to the open pollinated varieties (OPVs). Moreover, it was established that percentage AMF colonization is dependent on the soil type and the maize genotype. Interestingly, the landrace maize crop performed well in AMF colonization, compared to the OPVs and hybrid. On the other hand, cowpea nodulation correlated positively with available phosphates and negatively with total nitrogen in the soil. This differed from AMF colonization, which correlated positively with soil organic matter, pH and nitrogen but negatively with available phosphates. Therefore, to enhance efficient delivery of essential agroecosystem services by rhizobia and AMF, future experiments should elucidate the greater role of soil characteristics and genotype under field conditions.

Acknowledgements This work was funded by *The World Academy of Science* (TWAS) and Kenyatta University. The authors wish to thank the farmers at Machang'a and Kitui West for their contributions and assistance during soil sampling.

References

Agoyi EE, Odong TL, Tumuhairwe JB, Chigeza G, Diers BW, Tukamuhabwa P (2017) Genotype by environment effects on promiscuous nodulation in soybean (*Glycine max* L. Merrill). Agric Food Secur 6(1):29. https://doi.org/10.1186/s40066-017-0107-7

Antunes PM, Goss MJ (2005) Communication in the tripartite symbiosis formed by arbuscular mycorrhizal fungi, rhizobia and legume plants: a review. In: Zobel R, Wright S (eds), Roots and soil management: interactions between roots and the soil. Agron Monogr 48. ASA, CSSA and SSSA, Madison, WI, pp 199–222

Aquino S, Scabora MH, Antonio J, Maria S, Maltoni KL, Maria A, Cassiolato R (2015) Mycorrhizal colonization and diversity and corn genotype yield in soils of the Cerrado region, Brazil. Colonização e diversidade micorrízica e produtividade de genótipos de milho, em solo de Cerrado:4107–4118. https://doi.org/10.5433/1679-0359.2015v36n6Supl2p4107

Argaw A (2013) Evaluation of symbiotic effectiveness and size of resident *Rhizobium leguminosarum* var. viciae nodulating lentil (*Lens culinaris* medic) in some Ethiopian soils. Arch Agron Soil Sci 59(7):929–945. https://doi.org/10.1080/03650340.2012.690144

Argaw A, Muleta D (2018) Effect of genotypes-Rhizobium-environment interaction on nodulation and productivity of common bean (*Phaseolus vulgaris* L.) in Eastern Ethiopia. Environ Syst Res 6(1):14. https://doi.org/10.1186/s40068-017-0091-8

Bargaz A, Faghire M, Abdi N, Farissi M, Sifi B, Drevon J, Ghoulam C (2012) Low soil phosphorus availability increases acid phosphatases activities and affects P partitioning in nodules, seeds and rhizosphere of *Phaseolus vulgaris*. Agriculture 2:139–153. https://doi.org/10.3390/agriculture2020139

Battini F, Grønlund M, Agnolucci M, Giovannetti M, Jakobsen I (2017) Facilitation of phosphorus uptake in maize plants by mycorrhizosphere bacteria article. Sci Rep 7(1):1–11. https://doi.org/10.1038/s41598-017-04959-0

Berta G, Copetta A, Gamalero E, Bona E, Cesaro P, Scarafoni A, D'Agostino G (2013) Maize development and grain quality are differentially affected by mycorrhizal fungi and a growth promoting pseudomonad in the field. Mycorrhiza 24(3):161–170. https://doi.org/10.1007/s00572-013-0523-x

Boon E, Zimmerman E, Lang BF, Hijri M, Cozzolino V, Di Meo V, Chen H (2013) Impact of arbuscular mycorrhizal fungi applications on maize production and soil phosphorus availability. PLoS One 129(3):40–44. https://doi.org/10.1016/j.gexplo.2013.02.006.

Castellane L, Maria A, Bueno M (2015) Characterization of exopolysaccharides produced by rhizobia species. Rev Bras Ciênc Solo 39(6):1566–1575. https://doi.org/10.1590/01000683rbcs20150084

Cerozi S, Fitzsimmons K (2016) The effect of pH on phosphorus availability and speciation in an aquaponics nutrient solution. Bioresour Technol 219:778–781. https://doi.org/10.1016/j.biortech.2016.08.079

Chen M, Graedel TE (2016) A half-century of global phosphorus flows, stocks, production, consumption, recycling, and environmental impacts. Glob Environ Chang 36:139–152. https://doi.org/10.1016/j.gloenvcha.2015.12.005

Chu Q, Wang X, Yang Y, Chen F, Zhang F, Feng G (2013) Mycorrhizal responsiveness of maize (*Zea mays* L.) genotypes as related to releasing date and available P content in soil. Mycorrhiza 23(6):497–505. https://doi.org/10.1007/s00572-013-0492-0

Clair SBS, Lynch JP (2010) The opening of Pandora's Box: climate change impacts on soil fertility and crop nutrition in developing countries. Plant Soil 335:101–115. https://doi.org/10.1007/s11104-010-0328-z

Cozzolino V, Di Meo V, Piccolo A (2013) Impact of arbuscular mycorrhizal fungi applications on maize production and soil phosphorus availability. J Geochem Explor 129:40–44. https://doi.org/10.1016/j.gexplo.2013.02.006

Erbas BC (2017) In the presence of climate change, the use of fertilizers and the effect of income on agricultural emissions. Sustainability 9(11):1–17. https://doi.org/10.3390/su9111989.

Garg N, Geetanjali (2007) Symbiotic nitrogen fixation in legume nodules: process and signaling. A review. Agron Sustain Dev 27(1):59–68. https://doi.org/10.1051/agro.2006030

Gianinazzi S, Gollotte A, Binet MN, van Tuinen D, Redecker D, Wipf D (2010) Agroecology: the key role of arbuscular mycorrhizas in ecosystem services. Mycorrhiza 20(8):519–530. https://doi.org/10.1007/s00572-010-0333-3

Giovannetti M, Mosse B (1980) An evaluation of techniques for measuring arbuscular mycorrhiza infection in roots. New Phytol 84:489–500

Gomes EA, Oliveira CA, Lana UGP, Noda RW, Marriel IE, De Souza FA, Lagoas S (2015) Arbuscular mycorrhizal fungal communities in the roots of maize lines contrasting for Al tolerance grown in limed and non-limed Brazilian Oxisoil. J Microbiol Biotechnol 25(7):978–987

Jida M, Assefa F (2011) Phenotypic and plant growth promoting characteristics of *Rhizobium leguminosarum* bv. *viciae* from lentil growing areas of Ethiopia. Afr J Biotechnol 5(24):4133–4142. https://doi.org/10.5897/AJMR11.400.

Jordan-meille AL, Pellerin S (2018) Shoot and root growth of hydroponic maize (*Zea mays* L.) as influenced by K deficiency. Plant Soil 304(1):157–168. https://www.jstor.org/stable/42951816

Leghari SJ, Wahocho NA, Laghari GM, Hafeez Laghari A, Mustafa Bhabhan G, Hussain Talpur K, Bhutto TA, Wahocho SA, Lashari AA (2016) Role of nitrogen for plant growth and development: a review. Adv Environ Biol 10(9):2019–2219

Lindstrom K, Murwira M, Willems A, Altier N (2010) The biodiversity of beneficial microbe-host mutualism: the case of rhizobia. Res Microbiol 161:453–463. https://doi.org/10.1016/j.resmic.2010.05.005

Liu W, Zhang Y, Jiang S, Deng Y, Christie P, Murray PJ, Zhang J (2016) Arbuscular mycorrhizal fungi in soil and roots respond differently to phosphorus inputs in an intensively managed calcareous agricultural soil. Sci Rep 6(1):24902. https://doi.org/10.1038/srep24902

Motaroki EM, Njeru EM, Koskey G, Maingi J (2018) Rhizobial inoculation methods affect the nodulation and plant growth traits of host plant genotypes: a case study of Common bean Phaseolus vulgaris L. germplasms cultivated by smallholder farmers in Eastern Kenya. *Advances in Agricultural.* Science 6(3):77–94

Mothapo NV, Grossman JM, Sooksa-nguan T, Maul J, Bräuer SL, Shi W (2013) Cropping history affects nodulation and symbiotic efficiency of distinct hairy vetch (*Vicia villosa* Roth.) genotypes with resident soil rhizobia. Biol Fertil Soils 49(7):871–879. https://doi.org/10.1007/s00374-013-0781-y

Mwendwa P, Giliba RA (2012) Climate change impacts and adaptation strategies in Kenya. Chin J Popul Resour and Environ 10(4):22–29. https://doi.org/10.1080/10042857.2012.10685104

Njeru EM (2018) Exploiting diversity to promote arbuscular mycorrhizal symbiosis and crop productivity in organic farming systems. AIMS Agric Food 3(3):280–294

Njeru E, Avio L, Sbrana C, Turrini A, Bocci G, Bàrberi P, Giovannetti M (2013) First evidence for a major cover crop effect on arbuscular mycorrhizal fungi and organic maize growth. Agron Sustain Dev 34(4):841–848. https://doi.org/10.1007/s13593-013-0197-y

Okalebo RJ, Gathua KW, Woomer PL (2002) Laboratory methods of soil and plant analysis: a working manual, 2nd edn. TSBF-CIAT and Sacred Africa, Nairobi, Kenya

Oliveira JP, Galli-Terasawa LV, Enke CG, Cordeiro VK, Armstrong LCT, Hungria M (2010) Genetic diversity of rhizobia in a Brazilian oxisol nodulating Mesoamerican and Andean genotypes of common bean (*Phaseolus vulgaris* L.). World J Microbiol Biotechnol 27(3):643–650

Ondieki DK, Nyaboga EN, Wagacha JM, Mwaura FB (2017) Morphological and genetic diversity of rhizobia nodulating cowpea (*Vigna unguiculata* L.) from agricultural soils of lower Eastern Kenya. Int J Microbiol 2017(8684921):9. https://doi.org/10.1155/2017/8684921

Onono PA, Wawire NWH, Ombuki C (2013) The response of maize production in Kenya to economic incentives. Int J Dev Sustain 2(2):530–543

Oruru MB, Njeru EM (2016) Upscaling arbuscular mycorrhizal symbiosis and related agroecosystems services in smallholder farming systems. Biomed Res Int 2016:1–12. https://doi.org/10.1155/2016/4376240

Oruru MB, Njeru EM, Pasquet R, Runo S (2018) Response of a wild-type and modern cowpea cultivars to arbuscular mycorrhizal fungi inoculation in sterilized and non-sterilized soil. J Plant Nutr 48(1):90–101. https://doi.org/10.1080/01904167.2017.1381728

Otieno PE, Muthomi JW, Chemining'wa GN, Nderitu JH (2009) Effect of rhizobia inoculation, farmyard manure, nitrogen fertilizer on nodulation and yield of food grain legumes. J Biol Sci 9:326–332

Philippot L, Raaijmakers JM, Lemanceau P, van der Putten WH (2013) Going back to the roots: the microbial ecology of the rhizosphere. Nat Rev Microbiol 11(11):789–799. https://doi.org/10.1038/nrmicro3109

Razaq M, Zhang P, Shen H, Salahuddin (2017) Influence of nitrogen and phosphorous on the growth and root morphology of Acer mono. PLOS ONE *12*(2):e0171321. https://doi.org/10.1371/journal.pone.0171321

Rondon MA, Lehmann J, Ramírez J, Hurtado M (2006) Biological nitrogen fixation by common beans (*Phaseolus vulgaris* L.) increases with bio-char additions. Biol Fertil Soils 43(6):699–708. https://doi.org/10.1007/s00374-006-0152-z

Slattery J, Pearce P (2002) Development of elite inoculant *Rhizobium* strains in South Eastern Australia. In: Herridge D (ed) Inoculants and Nitrogen Fixation of Legumes in Vietnam, ACIAR Proceedings, pp 86–94

Smith SE, Read DJ (2008) Mycorrhiza Symbioses. Elsevier, London

Ulzen J, Abaidoo RC, Mensah NE, Masso C, AbdelGadir AH (2016) *Bradyrhizobium* inoculants enhance grain yields of soybean and cowpea in Northern Ghana. Front Plant Sci 7:1770. https://doi.org/10.3389/fpls.2016.01770

van Tuinen D, Bouffaud M, Bernaud E, Colombet A, Tuinen D, Van Wipf D, Redecker D (2016) Regional-scale analysis of arbuscular mycorrhizal fungi: the case of Burgundy vineyards, International Molecular Mycorrhiza Meeting, Sept 2015, Cambridge, UK

Wambugu PW, Mathenge PW, Auma EO, van Rheenen HA, Wambugu PW, Mathenge PW, van Rheenen HA (2012) Constraints to on-farm maize (*Zea mays* L.) seed production in Western Kenya: plant growth and yield. ISRN Agron 2012:1–7. https://doi.org/10.5402/2012/153412

Yang Y, Liang Y, Han X, Chiu T, Ghosh A, Chen H, Tang M (2016) The roles of arbuscular mycorrhizal fungi (AMF) in phytoremediation and tree-herb interactions in Pb contaminated soil. Sci Rep 6(February):1–14. https://doi.org/10.1038/srep20469

Availability, Access and Use of Weather and Climate Information by Smallholder Farmers in the Kilombero River Catchment, Tanzania

E. P. Ruth, J. J. Kashaigili, and A. E. Majule

Abstract Climate information is vital in national development planning for the management of development opportunities and risks, as well as mitigation and adaptation. Nevertheless, little is known about the obstacles and opportunities for decision-making in Tanzania, particularly for smallholder farmers. This study sought to: (i) understand the demand for, availability and usefulness of weather and climate information and (ii) assess diffusion methods – how information reaches farmers and whether information is used for decision-making. Qualitative and quantitative approaches were used to collect data in the two study villages. Quantitative data were analysed statistically and qualitative data by content analysis. Findings revealed that rainfall, temperature and sunshine were the most commonly accessed weather information. The main use of information was for preparedness and the timing of crop planting, as revealed by about 95% and 70% of respondents, respectively. Barriers to the use of climate information included lack of localised weather and climate information, generalisation of the information and the content being too technical (ca. 60% of respondents). Socio-economic characteristics (level of education, age and gender) appeared to affect access to and use of information. This study recommends the provision of more timely and accurate information on climate, the training of farmers on using this information and improved cooperation with stakeholders such as researchers, meteorologists and extension officers on climate adaptation strategies.

E. P. Ruth (✉) · J. J. Kashaigili
Department of Forest Resources Assessment and Management, Sokoine University of Agriculture, Morogoro, Tanzania

A. E. Majule
Institute of Resources Assessment, University of Dar es Salaam, Dar es Salaam, Tanzania

B. R. Singh et al. (eds.), *Climate Impacts on Agricultural and Natural Resource Sustainability in Africa*, https://doi.org/10.1007/978-3-030-37537-9_28

Abbreviations

CIS	Climate information service
FGD	Focus group discussion
GDP	Gross domestic product
IITA	International Institute of Tropical Agriculture
IPCC	Intergovernmental Panel on Climate Change
IUFRO	International Union of Forest Research Organizations
KRC	Kilombero River Catchment
NGO's	Non-governmental organizations
SAGCOT	Southern Agricultural Growth Corridor of Tanzania
SPSS	Statistical Package for Social Sciences
TBC	Tanzania Broadcasting Corporation
TMA	Tanzania Meteorological Agency
UMFULA	Uncertainty Reduction in Models for Understanding Development Applications (*UMFULA means* "river" in Zulu)
URT	United Republic of Tanzania

1 Introduction

Agriculture is the basis of the economy in Africa and is among the major economic activities that support more than 75% of the world population (AGRA 2017; Kiers et al. 2008). Recently, climate variability and change is emerging as a major threat to agricultural development in Africa, constituting one of the biggest challenges of the twenty-first century, and it is expected to continue throughout the century and beyond if it is not mitigated effectively (Badar and Farha 2012).

Increasing temperatures, and other changes that lead to either droughts or floods, are already affecting many aspects of natural resources and human societies, particularly livelihoods (Parry et al. 2004). According to the Intergovernmental Panel on Climate Change (IPCC 2018), 1.5–2.0 °C of global warming is expected by 2100, accompanied by variations in rainfall. The increasingly unpredictable and unreliable nature of weather systems on the continent has placed an extra burden on food security and rural livelihoods (IPCC 2018). While agriculture is projected to be affected by the impacts of climate variability and change, it is expected to produce about 50% more to feed approximately 9 billion people in 2050 (World Bank 2015). It has been established that Africa has approximately 33 million smallholder farms, representing 80% of all farms in the region. In Africa, climate variability and change is likely to exacerbate existing problems for farmers and create new risks if it is not mitigated (IITA 2006). While the climate information service (CIS) addresses the timely provision of tailored, climate-related information that can be used to reduce losses and enhance profits, knowledge is still scarce regarding the identification of suitable approaches to effectively engage people with its use. In most cases where CIS has been introduced, the disconnection between users and producers of the CIS has undercut large-scale uptake (Singh et al. 2017). There is therefore a need to analyse the existing level of uptake of CSI and its products, and how the information delivered is being used, and to identify users' demands in terms of type and timeframe.

Africa may be more vulnerable to the effects of climate change and variability due to the following reasons: widespread poverty; limited access to information on climate variability and change; dependence on the natural environment and agriculture for the majority of people; complex governance and institutional systems; limited access to capital including markets, infrastructure, technology and ecosystem degradation; and complex disasters and conflicts (Lemos et al. 2012). The majority of African farmers are smallholders, with two-thirds of all farms below 2 ha and 90% of farms below 10 ha. Agriculture accounts for about one-third of gross domestic product (GDP) and three-quarters of employment in sub-Saharan Africa (IUFRO 2010).

In Tanzania, agriculture remains the largest sector in the Tanzanian economy, accounting for about half of the country's GDP and employing more than 75% of the labour force (URT 2011). It plays a critical role in the livelihoods of smallholder farmers in the country, as they rely on agriculture for their daily subsistence and for income generation (URT 2011). Agriculture contributes over 60% of export earnings in Tanzania, and over 70% of the population lives in rural areas where agriculture and related non-farm activities are their main occupation (URT 2004). While the contribution of the agriculture sector is immense, the sector has been seriously affected by the vagaries of climate variability and change. It is now evident that climate variability and change pose a significant threat to food security in Tanzania through reductions of agricultural production due to increased temperatures, drought, floods, reduced water availability and changes in rainfall patterns (Pamela 2009). The performance of the agriculture sector, which has historically been the backbone of Tanzania's economy, is projected to decline as a result of negative effects of on-going global climate change (Pamela 2009). The introduction of weather forecasting systems for risk management in agricultural practices is relevant as they provide timely weather information, helping producers to take appropriate decisions (Lemos et al. 2012). In this regard, adapting to climate variability and change and the provision of timely climate information to manage climate risks are crucial at all levels (global, national, regional, local) and economic sectors – particularly agriculture, which is sensitive to climate conditions. The effective use of weather and climate information to manage climate risks and assist in the preparation of adaptive actions is thus urgently needed (Troccoli et al. 2008; Hewitson et al. 2014). Existing information on climate change and variability can help to inform decision-making at various levels, by providing a deeper understanding of the impacts and risks, as well as supporting actions to be taken to reduce those risks (Troccoli et al. 2008). While the seasonal variability of climate is a major source of production risks, significant benefits have arisen from the use of climate information in decision-making, especially in agriculture (Fraisse et al. 2006). While climate information can be a powerful tool to help farmers and other rural communities adapt to climate risks (Patt et al. 2005), its widespread use faces many practical challenges.

This paper presents an investigation of the availability, access to and use of weather and climate information by smallholder famers in the Kilombero River Catchment in Morogoro, Tanzania. The aim of the study is to: (i) understand the demand for, availability and usefulness of weather and climate information; (ii) assess diffusion methods, if and how information reaches farmers, and if not why; and (iii) assess if weather and climate information is used for decision-making.

2 Materials and Methods

2.1 Description of the Study Area

The study was conducted in Kilombero River Catchment (KRC), located in the southern part of Tanzania and forming one of the four principal sub-basins of Rufiji River Basin (Fig. 1). It has an area of about 39,990 km^2 (ERB 2006). The catchment is oriented from SW to NE and situated between longitudes 34°33″E and 37°20″E and between latitudes 7°39″S and 10°01″S (Hughes 1992). The study was carried out near Kilombero Sugar Company Limited and in two nearby villages (Msolwa Ujamaa and Sanje). KRC was selected for a number of reasons: First, the area has

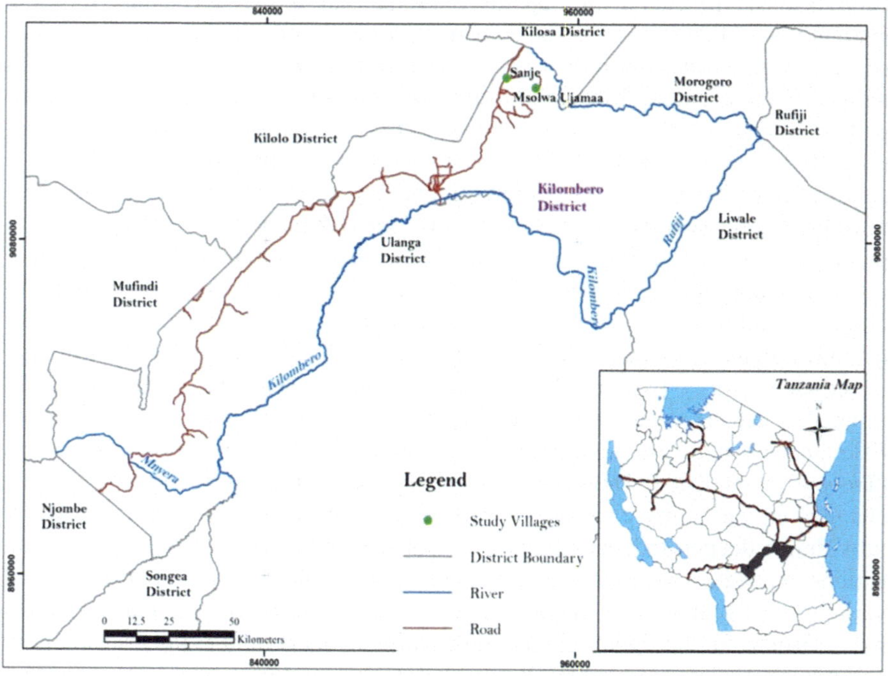

Fig. 1 Map of Tanzania showing the location of the study. (Source: Cartographic Unit, University of Dar es Salaam)

semi-arid climate and it is thus one of the most productive agricultural areas in Tanzania, with high production of sugarcane, paddy maize and bananas (ERB 2006; SAGCOT 2012). Second, KRC is now targeted for large-scale agricultural development initiatives under SAGCOT (Smalley et al. 2014). Third, the business model at Kilombero Sugar Company Limited has been identified by the government as the model for development of commercial agriculture elsewhere (ERB 2006; SAGCOT 2012). Furthermore, the two selected villages have a large number of out-growers of sugarcane and paddy. Approximately 32,960 ha are irrigated.

2.2 *Research Design*

This study made use of the mixed methods research approach (both qualitative and quantitative approaches) to ascertain how weather and climatic information are being used in decision-making processes in the agricultural sector. It has been argued that a mixed methods design answers research questions that could not be answered in other ways (Tashakkori and Teddlie 2003). Bryman (2008) remarks that, "mixed methods research may provide a better understanding of a phenomenon than if just one method was used".

2.3 *Sample and Sample Size*

The sampling frame for this study was the population of farmers. The sample for the study was selected randomly with a proportion of 5% whereby each person had an equal chance of being included in the sample, after Boyd et al. (1981). Msolwa Ujamaa village has a total of 1200 households, while Sanje has a total number of 1000 households, providing a sample size of 60 and 50 from the respective villages.

2.4 *Data Collection*

Both secondary and primary data were collected. The former were obtained from various sources such as Internet, libraries, Tanzania Meteorological Agency (TMA), published research papers, books, proceedings, workshop reports and various policies, strategies and acts, whereas the latter were collected through focus group discussions (FGDs), household survey, key informant interviews and field observation. The key informant interviews and FGDs were conducted to gather more information on how climate information was used in decision-making processes, the kind of climatic information available and information on the effectiveness, barriers and opportunities of climate information use. The FGD were conducted under the guidance of a moderator through a checklist of questions prepared according to research objectives. A total of 15 participants came from each village with members representing each sub-village, providing a total of 30 people for both villages. The groups involved village community members and village government leaders, with a balance in gender, wealth and age. The FGD method is effective since it is highly flexible, socially oriented and time saving, gives room to generate new ideas and provides insights on the current position of the topic at hand (Berg and Lune 2012).

2.5 Data Analysis and Presentation of Results

Information obtained from different sources was analysed using descriptive and inferential statistical tests. Quantitative data were descriptively computed and presented in the form of percentages, graphs, tables, bar charts, pie charts and line graphs, with the aid of Statistical Package for Social Sciences (SPSS) and Microsoft Excel. Prior to quantitative analysis, the household survey questionnaire was coded. The coding was generated after compiling a list of all responses to ensure that a single code is applied to each response without repetition, for open-ended questions. A coded template was designed in SPSS software and completed survey questionnaires were transferred into the software for data analysis. Data cleaning was performed to ensure that data values are complete and accurate. Descriptive data analysis, cross-tabulation and chi-square statistical analysis were conducted to draw inferences on the collected data.

Qualitative data were transcribed and organised into discussion topics. Content analysis of the transcribed data was carried out using Excel. In Excel, data were first sorted into themes and patterns were generated across themes to show relationships across key issues, such as education level, age and gender. Differences in views on climate services and information based on farmers' education level, age, gender and income were summarised before analysis and then compared by using the chi-square test (Cronk 2008). The chi-square test of independence was also used to find out whether two or more attributes were independent ($p > 0.05$).

3 Results

3.1 Socio-Economic Characteristics of the Respondents

Age of Respondents

The age of respondents in the two study villages of Sanje and Msolwa Ujamaa showed that 26% of respondents in Sanje and 32% of respondents in Msolwa Ujamaa were aged between 18 and 35 years, 46% in Sanje and 36% in Msolwa Ujamaa were aged between 36 and 60, while 28% in Sanje and 32% in Msolwa Ujamaa were aged above 60 years. These findings indicate that the number of respondents aged below 35 years was low. The mean age of respondents between 36 and 60 was larger than other age categories because this is referred to as the "working age" population, which provides more support to families. The results are in line with a study by Okwu and Loorka (2011), who found that the age group between 35 and 60 years was the most prevailing among farmers due to their responsibility for feeding and taking care of their families. This suggests that the active group was engaged in farming as their major economic activity. The mean age of respondents who were 60 and above was 30%. This is because this age group was targeted to obtain information on longer-term climate change and to include indigenous or traditional knowledge used, such as traditional climatic information.

Gender of Respondents

The results on the gender of respondents in the two study villages indicated that 54% were males and 46% were females in Sanje village, while in Msolwa Ujamaa, males comprised 60% and females 40%. This shows that both genders were involved in farming activities, as both depend on agriculture as the main provider for their familial needs. According to (Scott et al. 2005), males and females participate equally in agricultural production. These results reflect a degree of gender equality in the area. According to the World Bank (2015), rural women provide 50% of the agricultural labour force in Tanzania.

Respondents' Education Level

The findings on the education level of respondents showed that 22% in Sanje village had informal education compared to 28% in Msolwa Ujamaa. The biggest proportion of farmers, 50% in Sanje and 44% in Msolwa Ujamaa, had primary education, while those who had secondary and post-secondary education accounted for 28% in both villages. Results conform to the findings of Churi et al. (2012), who reported that the majority of farmers in rural areas have not attained a higher education level. In a discussion with the District Extension Officer, it was noted that low education among farming communities affects their ability to understand and adopt knowledge, including technologies on climate variability and change. Most farmers are not confident to try new innovations in their farms. A study by Voh (2002) reports the presence of a significant positive relationship between formal education and the adoption of new agricultural technologies. The study also noted that farmers with limited education cannot easily use new technologies for accessing and employing agricultural information, such as new improved seeds and cultivation systems. All such situations result in limited ability to access timely and relevant information for farming activities.

3.2 *Available Climate Services to Smallholder Farmers and Their Accessibility and Use*

Available Weather Services and Sources

The types of weather information available in the study area, to which people had access, were: rainfall forecasts (100% of respondents in Sanje and Msolwa Ujamaa), sunshine (72% of respondents in Sanje and 76% of respondents in Msolwa Ujamaa) and temperature (92% of respondents in Sanje and 94% of respondents in Msolwa Ujamaa). This indicates that farmers are interested in rainfall and temperature weather data mostly because it is easily available (Fig. 2). The range of this data starts from daily, weekly and monthly but they have no access to the climate information which covers a long time even though they are interested to access long-time climate information.

Timely access to accurate climate information and knowledge plays a major role in farmers' adaptation to climate change and variability, as well as in reducing vulnerability (Ngigi 2009). However, despite findings indicating that farmers have increased their annual yields as a result of access to climate information, particularly rainfall data, Tumbo et al. (2010) found that 90% of Tanzanian farmers depend solely on rain-fed agriculture. Timely and highly focused climate and weather forecasts to end users are therefore of great significance. Kandji et al. (2006) pointed out that appropriate seasonal forecast information helps governments and local people cope with climate change as well as climate variability. Availability and access of climate information make farmers aware of the next season. It also enhances their adaptive capacity through choice of what and when to cultivate including proper use of water. Such measures help in mitigating some of the adverse effects of climate change and variability.

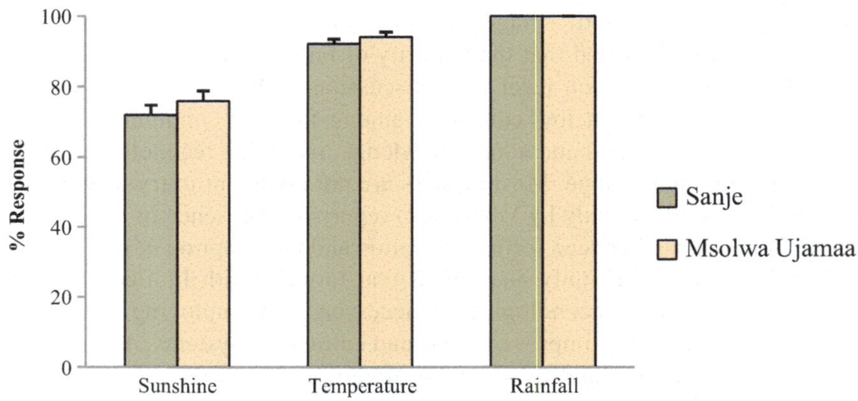

Fig. 2 Percentage of respondents on types of weather information available per village

Sources of Weather and Climate Information and Means of Access

In both study villages, it was disclosed that weather and climate information are mainly received from the Tanzania Meteorological Agency (TMA), as attested by 86% respondents in Sanje and 80% respondents in Msolwa Ujamaa (Fig. 3). The Illovo weather station was another source of climate information in both villages, whereby all (100%) respondents in Sanje and all (100%) respondents in Msolwa Ujamaa attested that they received information from this source through the extension officers. The last source of weather information in the study area was traditional sources, whereby 90% of respondents in Sanje and 84% of respondents in Msolwa Ujamaa used traditional signs and indicators to determine if there would be rain, drought or any other climatic events. Traditional signs or indicators, such as flowering of specific trees, particular animal sounds from, for example, certain frog species (among several others), are mostly practiced by older people in the area. A similar study to this one, conducted on farmers in semi-arid areas of Kenya (World Bank/CIAT 2015), found that local farmers used mostly radio, television and interpersonal sources, such as extension officers, village leaders and indigenous (traditional) leaders, as channels for accessing climate information.

Results further showed that climate information in both villages was accessed through television, such as the Tanzania Broadcasting Corporation (TBC), with 74% respondents in Sanje and 66% respondents in Msolwa Ujamaa obtaining information through television broadcasts and radio broadcasts such as Radio Ulanga and Radio Abood. Radio broadcast was revealed to be a cheap and easy option for many people. Another way of accessing climate information in the study area was through Illovo workers who live in the respective villages, whereby all (100%) respondents in Sanje and all (100%) respondents in Msolwa Ujamaa said that they were receiving information through people working at Illovo Sugar Company, who receive daily weather forecasts from the company in relation to their work.

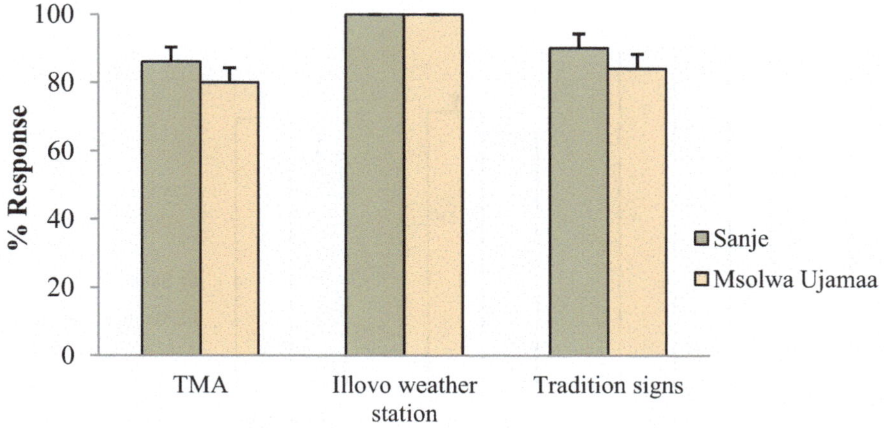

Fig. 3 Sources of weather and climate information per village by percentage

Use of Illovo climatic data was the most common way of receiving information because the villages are in close proximity to Illovo Sugar Company and most farmers grow sugarcane to sell at Illovo. Illovo is dependent on sugarcane from nearby villages, giving the company an incentive to ensure that villagers receive relevant climate information. Farmers also received climate information from extension officers, with 76% respondents in Sanje and 66% respondents in Msolwa Ujamaa receiving information from them.

Other methods of receiving climate information included the use of the traditional signs. Traditional signs were mostly used by older farmers, who were also found to have limited awareness about technical information services such as mobile phone, radio and television, as they had limited access to information via such communication channels compared to the younger members of their communities. These findings are in line with previous studies (Abubakar et al. 2009; Manyozo 2009), which found that access to agricultural information in most developing countries was provided through radio, television and mobile phone. The high usage of radio to access agricultural information in the area was due to a high number of radio stations broadcasting to the study area, low cost and appropriate airing time. Such information assisted in planning and decision-making regarding what and when to cultivate, to deal with any shortages of rainfall or other situations caused by climate variability.

Use of Weather Information in Agriculture

Findings from the study revealed that climate information had various uses in the studied villages. Ninety-six percent of respondents in Sanje and 96% of respondents in Msolwa Ujamaa used climate information for deciding whether or not to cultivate or prepare farms, seeds and fertilizers (Fig. 4). Another use of weather and climate information include timing, whereby 72% of respondents in Sanje and 76% of

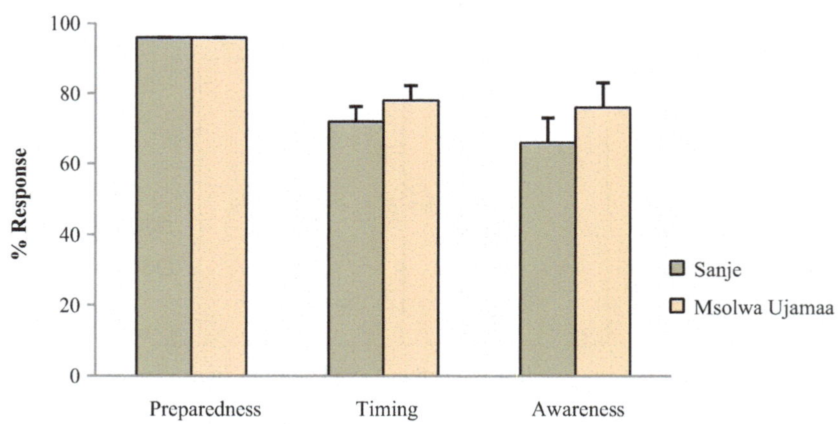

Fig. 4 Uses of weather and climate information per village by percentage

respondents in Msolwa Ujamaa used climate information to plan when to start cultivation, plant seeds, apply fertilisers and plan the exact time to sow, particularly regarding sugarcane. For sugarcane farmers, forecast data is critical due to the fact that the harvesting of sugarcane involves burning, which cannot be conducted when it is raining. Other uses included awareness, whereby 66% of respondents in Sanje and 76% of respondents in Msolwa Ujamaa used weather and climate information to gain awareness of any climate events likely to happen, such as floods or drought, and prepare accordingly in terms of both timing and action. Such preparation may include making decisions on the type of farming to employ, choosing the right seeds to grow (for example, early maturing or drought-resistant varieties) and employing the use of contours, ridges and furrows, mulching, and intercropping. These measures are to some extent dependent on weather and climate forecasts, for both the short and long term. Previous studies have also identified that farmers are more likely to use climate data for the timing of planting, cultivation and harvest (Rodriguez 2008; Vincent et al. 2017). Choosing suitable locations to farm, varying planting dates, diversifying crops and varieties, looking for alternative sources of income and choosing sustainable farming techniques are among the uses and advantages of weather and climate information that are relevant to smallholder farmers (Ambani and Percy 2014).

3.3 Respondents' Distribution by Age and Means of Access to Weather and Climate Information

Table 1 presents test statistics on selected socio-economic characteristics (age, gender and education level) in relation to means of access to weather and climate information. An association was found between the age of respondents and the use of radio in receiving climate information ($X^2 = 12.667$; $p = 0.001$). During the survey, it was observed that majority farmers aged between 36 and 60 years preferred using radio in accessing information compared to other information channels, due to their familiarity with radio, having used it for many years. These results also showed an association between the age of respondents and the use of traditional signs as a way of receiving weather and climate information.

3.4 Respondents' Gender and Means of Access to Weather and Climate Information

Findings in Table 1 indicate that males access weather information through radio and television more than females. Chi-square test results indicate that there is an association between the gender of farmers and access to weather and climate information through radio and television ($X^2 = 16.188$, $p = 0.000$ and $X^2 = 5.983$, $p = 0.014$).

Table 1 Socio-economic characteristics of farmers by percent in relation to means of accessing weather information

	Different means of receiving weather information			
Socio-economic characteristics	Radio	Television	Extension officers	Traditional signs
Gender				
Male	52 (59.8)	42 (60.0)	36 (50.7)	37 (50.7)
Female	35 (40.2)	28 (40.0)	35 (49.3)	36 (49.30)
Chi-square estimates 1	0.000[a]	0.014[a]	0.685	0.665
Age (years)				
18–35	15 (19.5)	10 (14.7)	9 (12.7)	5 (6.8)
36–60	42 (54.5)	45 (66.2)	36 (50.7)	38 (52.1)
Above 60	20 (26.0)	13 (19.1)	26 (36.6)	30 (41.1)
Chi-square estimates 2	0.001[a]	0.000[a]	0.071	0.000[a]
Education				
Informal	15 (18.7)	8 (11.4)	21 (29.6)	21 (28.8)
Primary	45 (56.3)	47 (67.1)	32 (45.1)	34 (46.6)
Secondary and above	20 (25.0)	15 (21.4)	18 (25.4)	18 (24.7)
Chi-square estimates 3	0.000[a]	0.000[a]	0.086	0.184

Source: Survey data (2017)
Numbers in brackets are percentages of respondents using a particular means for receiving weather information
[a]Significant at 5%

The study found that more men were likely to watch television programmes than female farmers because of differences in the roles they perform; discussions with farmers revealed that it was tricky for women to watch television, especially in the evening, due to their family roles. In most cases, weather/climate information broadcasting is aired after news programmes, a time when women are often involved in family matters. At such times, women often fetch water for domestic use, take care of children and prepare food, all of which are termed as female-related chores in the study areas. Results also showed that there was no association between the gender of respondents and access of climate information through traditional methods. In addition, both men and women obtained equal access to climate information through extension officers, with gender having no influence on this type of information access.

3.5 Respondents' Education Level and Means of Accessing Weather and Climate Information

The results listed in Table 1 further indicate that formal education is important in information access. Chi-square estimates indicate an association between farmers' education level and access to climate information through television and radio ($X^2 = 14.82, p = 0.000$; $X^2 = 27.730, p = 0.000$, respectively). These findings suggest that the ability to access information using television and radio differed between highly educated people and those with limited education. Limited education sometimes leads to low ability to access the information, as supported with findings by Berg and Lune (2012), who reported that literate farmers effectively use different ways to access climate information. Moreover, access to information using extension officers and local knowledge (traditional signs) was found to be independent of education level ($X^2 = 4.902$; $p = 0.086$ and $X^2 = 3.388$; $p = 0.184$). This implies that use of extension officers and traditional knowledge in Kilombero might not be affected by lack of education among smallholder farmers.

3.6 Barriers to Using Weather and Climate Information for Decision-Making of Smallholder Farmers

Although there are many advantages of using weather and climate information in crop production, such as effective agricultural planning that may increase production, findings from the study area revealed that smallholder farmers in both villages faced difficulties in using climate information due to various factors. The first major issue was that information was too general to understand, with 66% of respondents in Sanje and 66% of respondents in Msolwa Ujamaa reporting that they had difficulties in understanding where/when exactly the forecasted events would occur. For example, many said that forecasts by TMA cover the whole Kilombero area, which is huge. Therefore, pinpointing where a predicted weather event would occur is difficult. Another challenge is that climate information was too technical to comprehend, with 60% and 58% respondents in Sanje and Msolwa Ujamaa, respectively, disclosing that climate information obtained was sometimes very difficult to understand due to low education and difficult (technical) language. For example, some people said that a weather report might predict rain in the "southern highlands zone", but a typical person might not understand where exactly that zone is located. Another challenge was the inaccuracy of information received, with 80% of respondents in Sanje and 74% of respondents in Msolwa Ujamaa reporting that TMA weather forecasts proved mostly unreliable. One respondent stated that, "*they can predict that there will be enough rainfall and therefore we should prepare ourselves on farming and we can sow our seeds in the farms and wait for the rainfall. But it will never rain at the time they predicted, hence the seeds die*".

Timely availability of weather and climate information is crucial to farmers' success. Farmers need to be provided with information that is comprehensible, at the right time, to enable them to apply that information to their farming activities. This is in line with the findings of Kandji et al. (2006) in the context of the importance of climate information services to address the growing demand for food in a changing climate. Other studies verify that timely availability and access to information are vital for effective managerial functions such as planning, organising, leading and control, in the context of agriculture (Daron 2015; Daron et al. 2015).

3.7 Factors for Enhancing Access and Use of Weather and Climate Information

During the field visit in both villages, it was found that not all farmers were using weather and climate information in decision-making processes, due to so many challenges including a lack of education. In order to improve the use of climate and weather information in planning for agricultural production, the following are proposed:

- NGOs should provide education so that they can be more aware of the importance of using climate information in decision-making processes, especially in agriculture, as suggested by all (100%) respondents in Sanje and all (100%) respondents in Msolwa Ujamaa.
- The government should increase the density of weather stations down to the village level, so that villagers and farmers can access timely climate information, relevant at the local level.
- Weather and climate information should be more clearly specified in terms of area.
- Weather and climate information should be provided in simple language, as suggested by 88% of respondents in Sanje and 92% of respondents in Msolwa Ujamaa and also supported by Daron et al. (2015).
- An increased amount of related equipment may be necessary, as some respondents reported a lack of such equipment in some weather stations.
- Increase communication with scientists specialised in climate, so that farmers can access expert help in the use of weather and climate information.

4 Conclusions

This study investigated the availability, access and use of weather and climate information by smallholder farmers in the Kilombero River Catchment. Specifically, the study: (i) identified the available climate services and their accessibility to smallholder farmers and stakeholders in the study area, (ii) assessed the barriers to the use of weather and climate information in decision-making by smallholder farmers and (iii) investigated the factors for enhancing access to and use of information by smallholder farmers.

Regarding the available climate services, their accessibility and use, the study revealed that weather and rainfall information was available for rainfall, temperature and sunshine. Such information is accessed through radio, television, traditional signs and extension officers. Available climate information is mainly used in planning, preparedness and awareness of climate events such as floods. Respondents also used the information for informing their decisions on the timing of agricultural activities, such as when to start cultivation, when to apply fertilisers and when to harvest. Availability, access and use of weather and climate information, as well as climate services, were reported to be variable across institutional levels. At district and local levels, it is particularly important to ensure that efforts are made to improve access and availability to address inequality in the systematic provision of weather and climate information. However, improving availability, awareness and access to weather and climate information, including services, will not be enough in itself; these findings show that small-scale farmers in both villages encounter difficulties in using weather and climate information, due to various barriers. The first major one is that the information provided is too generic, making it difficult for users to make decisions due to a lack of detail. Another barrier is the technical nature of the provided weather and climate information, with respondents reporting that information received is difficult to understand due to their low levels of education and also due to the language used in the information being too complex and not simple enough for everyone to understand. A further barrier is the low reliability of information, with most respondents claiming that the weather forecasts and climate information received was mostly unreliable.

The findings on factors for enhancing access to and use of weather and climate information by smallholder farmers revealed that the issue of reliability of weather and climate information is vital for user satisfaction. It is well known that establishing credibility and trust in such information systems takes a long time. It is therefore important to ensure that users are not disappointed, to ensure that they continue to use and benefit from them. In general, respondents recognised that there was always uncertainty involved when using weather and climate information, whether scientific or traditional. There may also be differences in the credibility of information. In this regard, the TMA may assess forecast accuracy using scientific measures, but not be aware if such measures are relevant to users at the local level. Until this is addressed, scientists may continue to believe that they are producing reliable information, while local users mistrust it.

It important to note that the study was based on a very particular case that does not represent all of Tanzania. The geographic area of the study was small and a site for commercial agriculture, with more access to climate information due to the presence of Illovo Sugar Company. These findings are thus limited to the context of the study. Despite this limitation, the study provides key considerations for working towards enhancing farmers' access to and use of weather and climate information for improved agricultural production.

5 Recommendations

Based on the above findings and conclusions, the following recommendations are made:

Provide education/training on climate services and information. Training should be provided to smallholder farmers on the interpretation and use of weather and climate information, for its use decision-making regarding their agricultural activities.

Weather information should be communicated in easily understandable language and for area-specific forecasts. There is a need to develop user-friendly formats for communicating climate services and information, to enable the general public to and interpret weather and climate information without expert assistance. More spatially detailed forecasts will increase the relevance and credibility of the information and also trust among the users of such information.

Create and strengthen opportunities to discuss various approaches to validating weather and climate information. It is suggested that climate change programme partners and Tanzania Metrological Agencies should create opportunities and forums to engage in open discussions about various ways of assessing credibility of weather and climate information and how it can shape the use of climate services.

Capacity building for actors at the district and local level, to tailor climate information and advisories to its users. There is a need to ensure that the advice provided within climate services is appropriate through district- and local-level actors with in-depth relevant knowledge. It is recommended that climate change programme partners and TMA explore participatory approaches to training to enhance the capacity of district and local actors to interpret weather and climate information. Institutional and technical capacity should be built by concentrating on mechanisms that improve better interfaces with users; decentralising climate services to be closer to user needs could be considered.

A comparison between male and female respondents indicates that males have more means of accessing weather and climate information compared to female farmers. There is therefore a need for innovative solutions towards enhancing access and use of weather and climate information to female farmers. In this context, a revision in the timing of broadcasts of radio and TV programmes is advised, including the use of other communication channels such as mobile telephones.

Acknowledgements This work was carried out under the Future Climate for Africa UMFULA Project, with the financial support from the UK Natural Environment Research Council (NERC), grant ref. No: NE/M020037/1, and the UK Government's Department for International Development (DfID).

References

Abubakar BZ, Ango AK, Buhari U (2009) The roles of mass media in disseminating agricultural information to farmers in Birnin Kebbi Local Government area of Kebbi State: a case study of State Fadama II Development Project. J Agric Ext 13(2):42–54

AGRA (2017) Africa Agriculture Status Report 2017: the business of smallholder agriculture in Sub-Saharan Africa. Nairobi, Kenya: Alliance for a Green Revolution in Africa (AGRA)

Ambani M, Percy F (2014) Facing uncertainty: the value of climate information for adaptation, risk reduction and resilience in Africa. Adaptation Learning Programme (ALP) for Africa, CARE International, Nairobi, Africa

Badar AI, Farha NG (2012) Climate change: the biggest challenge in 21st century. In: Proceedings of the 1st International Technology, Education and Environment Conference. Human Resource

Management Academic Research Society, India, African Society for Scientific Research (ASSR), pp 497–508

Berg BL, Lune H (2012) Qualitative research methods for the social sciences, 8th edn. Pearson, Boston, MA, p 432

Boyd HK, Westfall R, Stasch SF (1981) Marketing research: text and cases. Homewood, Richard D, Irwin, Inc, p 180

Bryman A (2008) Social research methods, 3rd edn. Oxford University Press, Oxford/New York, p 748

Churi AJ, Mlozi MRS, Tumbo SD, Casmir R (2012) Understanding farmers information communication strategies for managing climate risks in rural semi-arid areas, Tanzania. Int J Inf Commun Technol Res 11:842

Cronk BC (2008) How to use SPSS: a step-by-step guide to analysis and interpretation, 5th edn. Pyrczak Publishing, Glendale, CA, p 90

Daron JD (2015) Challenges in using a robust decision making approach to guide climate change adaptation in South Africa. Clim Chang 132(3):459–473

Daron JD, Lorenz S, Wolski P, Blamey RC, Jack C (2015) Interpreting climate data visualisations to inform adaptation decisions. Clim Risk Manag 10:17–26

ERB (2006) Economic Research Bureau: a study to establishing mechanism for payments for water environmental Services for the Rufiji River Basin in Tanzania. Report to Tanzania MNRT: Forest and Beekeeping Division, 67pp

Fraisse CW, Paz JO, Hoogenboom G, Hatch A, Garcia YG, Guerra LC, Jones JW (2006) Peanut irrigation management using climate-based information. St. Joseph, ASAE, 177pp

Hewitson BC, Coauthors, Regional context (2014) Climate change 2014: impacts, adaptation, and vulnerability. In: Barros VR et al (eds) Part B: regional aspects: contribution of working group II to the fifth assessment report of the intergovernmental panel of climate change. Cambridge University Press, New York, pp 1133–1197

Hughes JA (1992) Data collection in context. Longman 1992, Indiana University, London/New York, 115pp

International Institute of Tropical Agriculture, IITA (2006) International research for development organization. Annual Report 2006, Croydon, UK. Available online: http://newint.iita.org/wp-content/uploads/2016/04/Annual-Report-2006-full-version.pdf. Accessed on 6 Oct 2017

International Union of Forest Research Organizations, IUFRO (2010) Does communal forest intervention management enhance forest benefits of smallholder farmers? Valdivia, Chile. Available online: https://earthzine.org/2013/01/08/iufro-the-international-union-of-forest-research-organizations/. Accessed on 12 Nov 2017

IPCC (2018) Summary for Policymakers. In: Masson-Delmotte V, Zhai P, Pörtner HO, Roberts D, Skea J, Shukla PR, Pirani A, Moufouma-Okia W, Péan C, Pidcock R, Connors S, Matthews JBR, Chen Y, Zhou X, Gomis MI, Lonnoy E, Maycock T, Tignor M, Waterfield T (eds) Global warming of 1.5°C – an IPCC Special Report on the impacts of global warming of 1.5°C above pre-industrial levels and related global greenhouse gas emission pathways, in the context of strengthening the global response to the threat of climate change, sustainable development, and efforts to eradicate poverty. World Meteorological Organization, Geneva, Switzerland. Available online: https://report.ipcc.ch/sr15/pdf/sr15_spm_final.pdf. Accessed on 13 Feb 2019

Kandji JW, Hansen JW, Royce FS, Messina CD (2006) Potential benefits of climate forecasting to agriculture. Agriculture, Ecosystems and Environment, New York/Melbourne, pp 169–184

Kiers ET, Leakey RRB, Izac AM, Heinemann JA, Rosenthal E, Nathan D, Jiggins J (2008) Agriculture at a crossroads. Science 320:320–321

Lemos MC, Kirchhoff CJ, Ramprasad V (2012) Narrowing the climate information usability gap. Nat Clim Chang 2:789

Manyozo L (2009) Mobilizing rural and community radio in Africa. Ecquid Novi Afr Journal Stud 30(1):1–23

Ngigi NS (2009) Climate change adaptation strategies. Water resources management options for smallholder farming systems in sub-Saharan Africa. Columbia University, New York, pp 3–10

Okwu LW, Loorka NS (2011) Productivity of farmers at various ages. North Cent J Agric Econ 6:81–87

Pamela WL (2009) Climate change impact in agriculture sector in Tanzania and its mitigation measures. Tanzania Meteorological Agency, Department of Climatology and Climate Change, Dar es Salaam, Tanzania

Parry ML, Rosenzweig C, Iglesias A, Livermore M, Fischer G (2004) Effects of climate change on global food production under SRES emissions and socio-economic scenarios. Glob Environ Chang 14:53–67

Patt A, Suarez P, Gwata C (2005) Effects of seasonal climate forecasts and participatory workshops among subsistence farmers in Zimbabwe. PNAS 102(2005):12623–12628

Rodriguez PL (2008) Framework for effective communication over mobile networks. Int J Mob Commun 2(1):97–102

SAGCOT (2012) SAGCOT Investment Partnership Program: opportunities for investors in the sugar sector. SAGCOT Centre Limited, Dar es Salaam, Tanzania. Available online: http://www.sagcot.com/uploads/media/Tanzania_SAGCOT_Sugar_Investor_Presentation.pdf. Accessed on 12 Nov 2017

Scott M, Swortzel KA, Taylor WN (2005) The relationships between selected demographic factors and the level of job satisfaction of extension agents. J South Agric Educ Res 55(1):102–115

Singh C, Daron J, Bazaz A, Ziervogel G, Spear D, Krishnaswamy J, Zaroug M, Kituyi E (2017) The utility of weather and climate information for adaptation decision-making: current uses and future prospects in Africa and India. Climate Dev. Available online: https://www.tandfonline.com/doi/full/10.1080/17565529.2017.1318744. Accessed on 12 Nov 2017

Smalley R, Sulle E, Malale L (2014) The role of the state and foreign capital in agricultural commercialization: the case of sugarcane out growers in Kilombero District, Tanzania. FAC Working Paper 106. Brighton, UK: Future Agricultures Consortium. Available online: http://www.plaas.org.za/sites/default/files/publicationspdf/FAC_Working_Paper_106.pdf. Accessed on 6 Oct 2017

Tashakkori A, Teddlie C (eds) (2003) Handbook of mixed methods in social and behavioural research. Sage, Thousand Oaks, CA, pp 315–345

Troccoli A, Harrison M, Anderson D, Mason J (2008) Seasonal climate, forecasting and managing risk, vol. 82. NATO science series on earth and environmental sciences. Springer, New York, pp 89–90

Tumbo SD, Kahimba FC, Mbilinyi BP, Rwehumbiza FB, Mahoo HF, Mbungu WB, Enfors EE (2010) Impact of projected climate change on agricultural production in semi-arid areas of Tanzania: a case of same district. Sokoine University of Agriculture, Tanzania, pp 453–463

United Republic of Tanzania, URT (2011) Agricultural Sector Development Programme (ASDP). Government Programme Document, Dar es Salaam. Available on: http://www.tdb.go.tz/uploads/documents/sw/1488801422-ASDP%20FINAL%2025%2005%2006%20(2).pdf. Accessed on 9 Oct 2017

URT (2004) National Environmental Action Plan (NEAP) 2013–2018. Vice President's Office, Dar es Salaam, Tanzania

Vincent K, Dougill AJ, Dixon JL, Stringer LC, Cull T (2017) Identifying climate services needs for national planning: insights from Malawi. Clim Pol 17(2):189–202

Voh JP (2002) Information sources and awareness of selected recommended farm practices: a case study of Kaduna State, Nigeria. Afr J Agric Sci 8(1,2):87

World Bank (2015) Working for a world free of poverty. Washington, DC. Available online: www.worldbank.org/en/publication/global-monitoring-report/poverty-forecasts-2015. Accessed on 6 Oct 2017

World Bank, CIAT (2015) Climate-smart agriculture in Kenya. CSA country profiles for Africa, Asia, and Latin America and the Caribbean series. The World Bank Group, Washington DC

Gender Differentiation in the Adoption of Climate Smart Agriculture Technologies and Level of Adaptive Capacity to Climate Change in Malawi

Tasokwa Kakota Chibowa, Gry Synnevag, Beston Maonga, and Michael Mainje

Abstract The gender gap in agriculture and its consequences for household food security and livelihoods are well documented. In Malawi, women perform between 50% and 70% of all agricultural tasks and produce almost 70% of the food. This study was conducted to assess gender differences in the adoption of climate smart agriculture (CSA) technologies and level of adaptive capacity of smallholder farmers in Malawi. The study used mixed methods in both data collection and analysis. Data were collected through 8 focus group discussions and individual household interviews with 229 male- and female-headed households from Dowa, Nkhotakota and Phalombe districts. A four-step adaptive capacity ladder to climate change impacts was developed. The results showed that gender gaps still exist in adoption and adaptive capacities. Over 70% of the male farmers adopted the climate smart agriculture technologies, while less than 30% of the female farmers adopted the technologies. The main constraints leading to low adoption among female farmers were high input demand and cost of inputs, labour requirement, lack of credit opportunities and income. Female farmers adopted technologies that required less input and were labour saving. In general, female farmers had lower adaptive capacity than male farmers. Farmers who had low adaptive capacity adopted few technologies. There is an urgent need for cost-effective and labour-saving technologies, policies and plans that enhance adaptive capacity of farmers, particularly female farmers.

T. K. Chibowa (✉) · B. Maonga
Lilongwe University of Agriculture and Natural Resources (LUANAR), Lilongwe, Malawi

G. Synnevag
Noragric, University of Life Sciences, Ås, Norway

M. Mainje
Development Fund of Norway, Lilongwe, Malawi

B. R. Singh et al. (eds.), *Climate Impacts on Agricultural and Natural Resource Sustainability in Africa*, https://doi.org/10.1007/978-3-030-37537-9_29

1 Introduction

1.1 Adoption of Climate Smart Agriculture Technologies in Malawi

Climate change effects threaten to reduce maize production, which is a staple food crop in most parts of Malawi, and threaten already vulnerable rural livelihoods (Msowoya et al. 2016). The increased focus on the negative impact of climate change and the associated need for adaptation of agronomic technologies and practices are spelt out in research and development efforts to promote climate smart agriculture (CSA) technologies in Malawi (Government of Malawi (GOM) 2012, 2016). The support to smallholders places strong emphasis on CSA practices as tools for increasing yields, while building resilience towards impact of climate change and at the same time reducing carbon emissions (FAO 2017). Lipper et al. (2014) state that CSA is 'an approach for transforming agricultural systems, primarily aimed at supporting food security under the new realities of climate change.' The most common CSA practice in Malawi is conservation agriculture (CA) based on minimal soil mechanical disturbance, permanent organic soil cover by crop residues and/or cover crops and diversified crop rotations with legumes (Ngwira et al. 2014). A range of other climate resilient technologies are also promoted, in association or alone, by non-governmental organisations (NGOs), Ministry of Agriculture through its extension system, international organisations and multi- and bilateral donor agencies. Adoption of CSA technologies by smallholder farmers is, however, constrained by limited access to inputs, agricultural resources and information (Lipper et al. 2014). It is argued that smallholder farmers in sub-Saharan Africa possess inadequate resource base required for proper adoption of CSA technologies (Pingali 2012). Lack of credit and market facilities also results in limited adaptive capacity to adopt agricultural technologies. Although CSA is promoted as a useful approach for transforming livelihoods and for reducing climate change effects, its success is dependent on the adaptive capacity of smallholder farmers. Increasing the adaptive capacity of farmers to climate change impacts is therefore necessary to improve productivity and household food security.

It is widely acknowledged that climate change impacts are gendered. The gender-differentiated impacts of climate change are directly linked to gender-differentiated vulnerabilities, coping and adaptation capacities and strategies (WEDO 2010; Goh 2012; Kakota et al. 2011; IPCC 2014). A case study in semi-arid areas of Tanzania showed that increased climate variability and change contributed to transformation of traditional gender roles, causing smallholder women farmers to become the main producers of staple food and in some cases, for cash crop production (Synnevåg et al. 2014). Although female farmers constitute the main producers of staple food, gender differences exist in both the level of adaptive capacity and adoption of technologies. Gender-differentiated adaptation strategies and adaptive capacity to climate change influence the adoption of the new CSA technologies.

1.2 Gender Differences in Adoption and Adaptive Capacities to Climate Change Impacts

In Malawi, women contribute 70% of the country's labour force and produce 70% of the household food. Malawian women also perform between 50% and 70% of all agricultural tasks (World Bank 2014; MGDS 111 2017–2022; GOM 2013). However, women continue to have poor access to and control over the means of agricultural production, including agricultural inputs, improved technologies, extension services, credit and land (Goh 2012; Banda et al. 2011), and the adoption of CSA technologies is still low, especially among female farmers. Most female smallholder farmers have restricted access to and ownership of agricultural inputs in sub-Saharan Africa (Peterman et al. 2014). This affects agricultural production and productivity. Findings from Malawi revealed that male-managed plots produce on average 25% more per hectare than female-managed plots (World Bank 2014). Inadequate capacity is also caused by small land holding size, low access to extension services, education, information and low participation in local institutions (Chirwa 2005; Williams 2008; Mlamba 2010; Kakota et al. 2013, Kakota et al. 2017). Triple roles played by women as well as lack of enabling environment are the other factors influencing their ability to cope with climate change and to enhance resilience and adaptive capacity. Culturally assigned roles within the household expose them to more challenges and barriers when it comes to access and control of natural, financial and social resources (Kakota et al. 2013; Lipper et al. 2014; Murray et al. 2016). Women and men therefore have different capabilities to implement adaptation strategies that ensure sustainable household food security. Understanding of the local context is necessary to improve women's adaptive capacity and adoption of CSA technologies (Murray et al. 2016). A study by Asfaw et al. (2017) showed that land tenure insecurity is one of the causes that exacerbates women smallholder farmers' vulnerability to weather shocks in patrilineal districts of Malawi. The same study also indicates that in matrilineal districts where women have more property rights, the impact of weather shocks on women is significantly lower (Asfaw et al. 2017). The higher labour demand in using agricultural technologies can also be a constraint for women in adopting CSA technologies. The use of zero or minimum tillage in conservation agriculture (CA), for instance, can increase weeds if the appropriate amount of herbicides are not used, thereby increasing women's labour load. Policy and institutional gaps also influence female farmers' capability to adapt to climate change.

It is known that women and men are not only victims of climate change, but also effective agents of change in relation to both mitigation and adaptation, and women are key agents of climate change adaptation and mitigation (WEDO 2007). However, there are few studies in Malawi on gender differentiation in the level of adaptive capacity and adoption of agricultural technologies. To fill the gap, this study was conducted to analyse gender and socio-economic issues affecting adoption of technologies and level of adaptive capacity. The study further investigates the current user status of CSA technologies by gender and gendered constraints that affect adoption of technologies by women and men farmers in three districts of Malawi where CSA technologies are promoted.

2 Methodology

2.1 Study Area

The study was conducted in the three districts of Phalombe, Nkhotakota and Dowa. Two of the districts, Phalombe and Nkhotakota, were identified based on vulnerability assessment by the Climate Adaptation for Rural Livelihoods and Agriculture (CARLA). The third district, Dowa, was identified from the districts where Development Fund of Norway is implementing a sustainable agriculture project using the lead farmer model. The choice of the districts was also based on the cultural and marriage beliefs that may have an influence in the access to resources, decision-making processes and adoption of CSA technologies.

Phalombe district is one of the districts that is vulnerable to climate change effects, especially floods, prolonged dry spells and stormy winds. The study was conducted in two extension planning areas (EPAs), Naminjiwa and Kasongo. The main staple crops grown are maize, rice and sweet potatoes. In addition, cash crops like tobacco, groundnuts, soybeans, sunflower, sesame and pepper are also grown. The communities follow the matrilineal type of marriage where women are given preferences in the ownership of assets such as land, and they remain in the villages while men migrate to other places to marry. In their culture, a woman has control over children and in the case of divorce, the children stay with their mother. Men in matrilineal districts have ownership rights to agricultural land they cultivate but may not possess user rights to land. It is difficult to be conclusive on whether the marriage system is entirely empowering women in controlling all agricultural inputs and outputs, as men in most rural Africa have a significant influence on land ownership, irrespective of the type of marriage system that was being practised within the community (Whitehead and Tsikata 2003).

Nkhotakota district is vulnerable to climate change effects such as dry spells and strong winds. Being located along the Lake Malawi lakeshore, the communities also depend on fishing for their livelihood. The study was conducted in Linga EPA. The main staple crops grown here are rice, cassava and maize. Groundnuts and rice are grown as cash crops. The communities are composed of Christians and Muslims, and they follow a patrilineal system of marriage where men are given preference in ownership of resources. Polygamy was also common among the Muslims in the area. Most of the participants in the female focus group were Muslims, because Islam is the most popular religion in Choto and surrounding villages where the study was conducted. The marriage system and religion may have an influence on the uptake of technologies and access to technologies between men and women.

Dowa is a patrilineal society and the dominating religion is Christianity. The study was conducted in Chibvala EPA, where projects on sustainable agriculture by Development Fund of Norway were being implemented. The EPA is accessible by most of the non-governmental organisations (NGOs) because of its proximity to Lilongwe, the capital city of Malawi. This being the case, Chibvala is one of the EPAs where trials of new technologies and practices in agriculture are implemented.

The main source of livelihood is subsistence agriculture, but some men migrate to the city in search of employment and business. Of the three districts, Dowa has the highest percentages of households growing groundnuts followed by sorghum. Intercropping cassava with maize is also common.

2.2 *Study Design and Data Collection Methods*

The first three stages of sampling involved selection of districts, extension planning areas and villages. The districts and EPAs were purposively sampled based on their vulnerability to climate change, marriage systems and implementation of the CSA technologies. Finally, using systematic and proportional sampling, households for interviews were selected from a list of villages that was obtained from each EPA in order to have proportionate representation of subsamples across villages. The study used mixed methods in both data collection and data analysis. The main data collection methods used were household survey, key informants interviews (KII) and focus group discussions (FGDs). A qualitative approach was used to investigate the major motives behind men and women's adoption and disadoption of CSA technologies and to prioritise men and women's CSA technology preferences. Gendered constraints that affect women and men's CSA adoption were also identified through FGDs. The FGDs were conducted with randomly selected groups of smallholder farmers ranging between 8 and 15 members per group, drawn from the sampled EPAs, in order to maintain consistence in the data flow. The discussions were conducted with separate groups of female and male farmers and two FGDs were conducted in each EPA. A total of eight FGDs were conducted, four for females only and four for males only. FGDs were intended to generate qualitative data necessary to substantiate quantitative data collected through semi-structured questionnaires. During the focus group discussion with female and male farmers, each group was requested to construct the ladder of adaptive capacity based on their communities. Members were also asked to draw the community vulnerability line (CVL) and choose a step where most female-headed households and male-headed households belong. This activity helped to establish the level of adaptive capacity among female and male farmers and factors that contributed to the movement up and down the ladder of adaptive capacity.

The household survey was conducted with 229 randomly sampled respondents from male- and female-headed households. The sample size was different in each district, but the total number per each district exceeded 30, which is a recommended minimum sample size. There were 60 respondents in Phalombe, 107 in Nkhotakota and 62 in Dowa. The female respondents were sampled from both female- and male-headed households, and in total, 60% of the respondents were females. The respondents included both male and female farmers from male-headed households and female farmers from female-headed households to allow comparison of factors for each group. A ladder of adaptive capacity with five steps was presented in the questionnaire where step 5 represented households with high capacity to adapt to climate change, while step 1 represented households with no capacity to adapt to climate change. Respondents were asked to position themselves on the ladder of adaptive capacity during the interviews.

A triangulation of data analysis methods was used to analyse the qualitative and quantitative data. Qualitative data were coded and analysed through content and thematic analysis using both preset and emergent codes/themes. Qualitative data analysis was conducted to explain the adoption and adaptive capacity trends between male and female farmers, while quantitative data analysis provided the magnitude on the adoption trends and adaptive capacity between men and women. A ladder of adaptive capacity was used to assess the level of adaptive capacity for male and female farmers.

3 Results and Discussion

3.1 Adoption of Technologies

The participants were requested to report the CSA technologies that they were implementing in their respective areas. The CSA technologies were ranked according to the preference in the adoption between male and female farmers, and the reasons for disadoption were also provided. The results were analysed according to sex of household head as well as sex of the respondent. This section provides results on CSA technologies, based on information provided by male and female farmers from both male- and female-headed households, priorities in adoption and constraints for adoption of technologies.

Adoption of Climate Smart Agriculture Technologies by Sex of the Household Head

The results from the household interviews showed that fewer CSA technologies are adopted by female-headed households than by their male counterparts. The main technologies that were adopted by most of the male-headed households were rainwater harvesting, conservation agriculture and pit planting; however, the female-headed households did not adopt any of these three technologies (Fig. 1). The discussions revealed that most of the CSA technologies are expensive and labour intensive, and most of the female-household heads do not adopt them.

The discussions with male and female farmers revealed that there are preferences on the types of technologies and crops by the male and female farmers. It was observed that conservation agriculture is labour intensive at the beginning of the season because of the scarcity of stalks for mulching, but it has many benefits, especially during dry spells. For example, moisture content in the soil is high and weeds are minimal, so instead of weeding, farmers just uproot the few weeds with hands. The labour and time that is saved from this farming is invested in other farming activities, income-generating activities and household chores. However, CA is also challenging because of pests and diseases that attack crops if herbicides are not applied and most rural female farm-

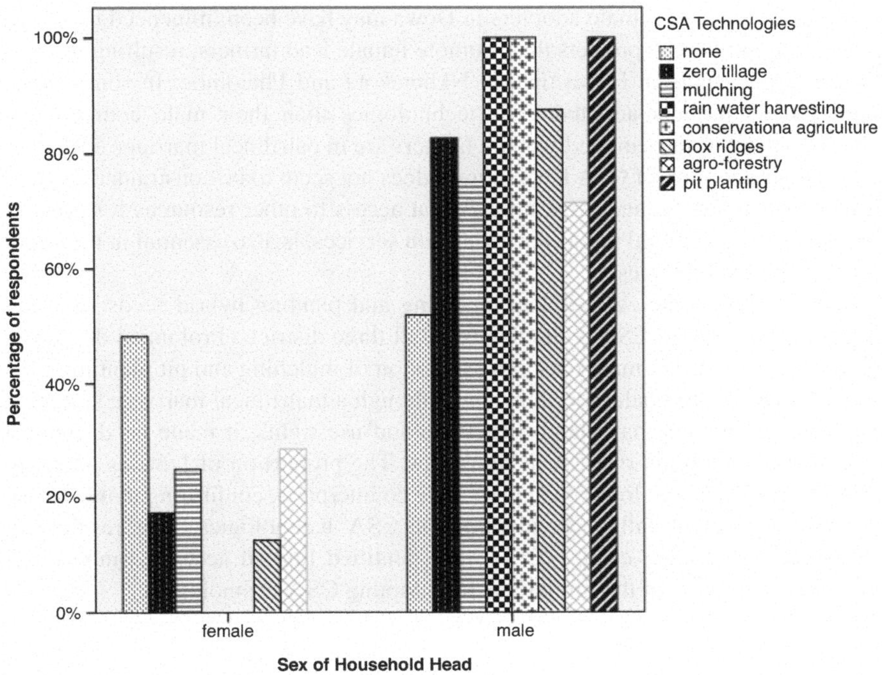

Fig. 1 CSA technologies adopted by sex of household head

ers do not afford the herbicides to complement the CA technology. Consequently, very few adopt the technology. The results are consistent with other studies that found that women continue to have low access to and control over the means of agricultural production, including agricultural inputs, improved technologies, extension services, credit and land (Peterman et al. 2014; Goh 2012; Banda et al. 2011).

The discussions also revealed that male and female farmers' preferences with regard to the types of crops and technologies to adopt are influenced by the climatic conditions of the area. Both female and male farmers in Phalombe preferred irrigation farming because it allows them to plant rice twice a year. When rains fail or fall within a short period of time, there is a serious shortage of water for irrigation because most rivers dry up. When there is poor rainfall distribution or low precipitation, crops are likely to fail because of reduced water access, as was the case during the survey in 2018.

Adoption of Climate Smart Agriculture Technologies Among Male and Female Smallholder Farmers

The household survey revealed that among all participants in this study, 80% men and 58% women, were aware of the CSA technologies. The rate of adoption of CSA technologies was lower among female farmers, from both male- and female-headed households, than male farmers. However, the percentage of female farmers adopting CSA technologies was higher in Dowa than in the other *districts* (Table 1). The higher percentage of female adopters in Dowa may have been influenced by a high concentration of NGO projects that promote female lead farmers, resulting in more female lead farmers in Dowa than in Nkhotakota and Phalombe. In some cases, more female farmers adopted CSA technologies than their male counterparts. Although most of the sampled women farmers are in patrilineal marriage system in Dowa, their adoption of CSA technologies does not seem to be constrained by their limited user rights to land. This implies that access to other resources for production, such as agricultural inputs and extension services, is also essential in the adoption of CSA technologies.

Table 1 shows that zero tillage, mulching and planting hybrid seeds were the most commonly used CSA technologies in all three districts. Prolonged dry spells were mentioned as the main reason for adoption of mulching and pit planting.

The results from Phalombe show that although a matrilineal marriage system is practised and women have land ownership and use rights, it made no difference regarding the adoption of CSA technologies. The proportion of females adopting CSA technologies was lower than their male counterparts, confirming that land use rights alone cannot influence adoption of CSA technologies. The results are consistent with Lipper et al. (2014) who identified limited access to inputs and information as some of the limitations for adopting CSA technologies.

Preferences in Adoption of Climate Smart Agriculture Technologies

Focus group discussions showed that the reasons for preferences for CSA technologies among male farmers were profitability, marketability and high yield while for female farmers the reasons were low cost of inputs, low labour requirement, food and high income. Although the decisions on which technologies to adopt are done jointly between men and women in male-headed households, men usually are the main decision-makers because they have resources for production that women may not have. This is different for female-headed households where women are the main

Table 1 Female (W) and male (M) users of CSA technologies in the three study areas

	Phalombe (60)		Nkhotakhota (107)		Dowa (62)	
CSA technologies in use	W %	M %	W%	M%	W %	M %
Mulching	49	88	41	65	72	94
Planting hybrid	44	65	37	92	67	65
Zero tillage	41	82	30	50	70	50
Crop diversification	36	70	35	10	34	50
Crop rotation	36	59	25	45	68	55
Agroforestry	29	45	18	25	60	32
Pit planting	16	25	12	100	73	69
Box ridges	18	59	22	26	58	38
Treadle irrigation	10	26	16	48	60	28
Rainwater harvesting	3	12	1	3	21	5

decision-makers; however, decisions to adopt the new technologies are still constrained by the availability of resources for production. In a matrilineal society, where women own land, men have control over the technologies to adopt because they mainly have access to agricultural inputs and extension services. This implies that as long as women lack means for agricultural production and productivity, men will continue to control the decisions to adopt the technologies.

Prioritised agricultural technologies among male and female farmers in the three districts are shown in Table 2. The prioritised preferences of technologies by women are primarily guided by their ability to afford inputs. The most preferred technologies are mulching, pit planting, agroforestry and manure use. These technologies demand relatively less amount of resources compared to the least prioritised ones such as crop rotation, planting hybrid seeds and irrigation. Both women and men smallholder farmers prioritised mulching as their first preference, not only because it enhances soil fertility and prevents soil erosion, but also because it demands less input. Those farmers who adopted mulching indicated that they sometimes get maize stalks from the fields of non-adopters. The prioritised order of technologies generally indicates that the technology preferences of women start from those demanding fewer inputs to those demanding higher inputs, whereas men prioritised even those demanding relatively higher inputs. Male farmers felt that pit planting is too labour demanding. The indicated labour constraint by men farmers is mainly due to their interest to devote more time to other off-farm income-generating activities (such as *Ganyu*) to fulfil their immediate needs.

Except for the use of mulching, which is preferred by women as a first priority, women judged other CA components as less advantageous. According to their description, the use of CA technologies, such as zero tillage, bring certain risk fac-

Table 2 Prioritised technologies among female and male smallholders in the three districts

	Phalombe (n = 60)		Nkhotakhota (n = 107)		Dowa (n = 62)	
	Women	Men	Women	Men	Women	Men
1st	Mulching	Mulching	Mulching	Pit planting	Manure use	Crop rotation
2nd	Pit planting	Manure use	Pit planting	Planting hybrid	Zero tillage	Mulching
3rd	Box ridges	Pit planting	Box ridges	Mulching	Mulching	Agroforestry
4th	Agroforestry	Zero tillage	Planting hybrid	Zero tillage	Pit planting	Pit planting
5th	Sasakawa	Crop rotation	Sasakawa	Crop rotation	Crop diversification	Sasakawa
6th	Crop diversification	Agroforestry	Treadle irrigation	Box ridges	Crop rotation	Treadle irrigation
7th	Crop rotation	Pit planting	Agroforestry	Irrigation	Sasakawa	Crop diversification
8th	Planting hybrid	Box ridges	Crop diversification	Agroforestry	Irrigation	Planting hybrid
9th	Irrigation	Irrigation	Rainwater harvesting	Rainwater harvesting	Agroforestry	Zero tillage

tors compared to conventional agriculture. Firstly, applying zero tillage may not pay off with regard to short-term outcome for their immediate household food demand. CSA technologies are often promoted to enhance maize production, and they reduce the ground for growing cassava on the same ridges, which is considered as an important crop for increasing household food security during dry seasons. If CA fails, women would lose the opportunity of growing other crops on their small piece of land. Secondly, applying zero tillage generally requires more inputs compared to other CSA technologies, and without application of herbicides, it causes more weeds that increase women's labour. Also male farmers indicated input scarcity as the main problem. They also associated conservation agriculture technologies with the free distribution of inputs from NGOs. Withdrawal of NGOs from distributing inputs demotivates most farmers, and they tend to discontinue the use of CA practices.

Constraints to Adoption of Climate Smart Agriculture Technologies by Male and Female Farmers

Table 3 shows the five main constraints to the adoption of CSA technologies identified by women and men in the three districts. The focus group discussions showed that all women and men across districts invariably indicate that lack of inputs and income are primary constraints that affect adoption of CSA technologies. Lack of inputs such as fertiliser, hybrid seeds, ox-carts to carry residues and treadle pumps for irrigation are some of the common constraints for both men and women smallholders. Most women listed lack of access to drought-resistant varieties and fertiliser as major barriers. Limited access to income, information, training, extension services and restricted access to water are the major gendered constraints. Access to land was not identified as a big barrier for women in all study areas. Although the land is available for women, scarcity of inputs, information and training are major challenges that constrain women's adoption of CSA technologies more than their access to land.

Limited credit opportunities or microloans from village savings loans (VSLs) are other major constraints to the adoption of technologies among male and female farmers. Women in polygamous marriages in Nkhotakota listed lack of labour support from husbands and women's inability to join women's local development groups due to the number of co-wives, as the first wives often join the membership, as one of the constraints. The interviews in Phalombe showed that although the matrilineal marriage system seems to provide a better decision-making ability to women over land, it does not bring significant difference in the adoption of CSA technologies between women in patrilineal and matrilineal societies. The constraints for adoption of CSA technologies apply to women in both societies.

Several programmes provide communities with food aid such as maize and beans, and if they have a bad harvest, they know that they will still get food. This has created a dependency syndrome and restricts farmers' motivation to use diversified farming methods.

Table 3 Top five constraints to the adoption of CSA technologies

	Phalombe		Nkhotakhota		Dowa	
	Women	Men	Women	Men	Women	Men
1.	Lack of inputs	Absence of income source	Lack of inputs	Lack of inputs	Lack of inputs	Lack of inputs
2.	Absence of income source	Lack of inputs	Absence of income source	Lack of water	Absence of income source	Limited access to credit
3.	Limited access to credit	Restricted access to training	Restricted access to training	Labour constraints	Limited access to credit	Absence of income source
4.	Restricted access to information	Limited access to credit	Limited access to land	Restricted access to training	Restricted access to training	Restricted access to information
5.	Restricted access to training	Limited access to land	Restricted access to information	Absence of Institutional support	Limited access to land	Limited access to education

3.2 Level of Adaptive Capacity

The ladder of adaptive capacity that was developed through focus group discussions had four steps, while the ladder that was preconstructed in the questionnaire for individual interviews had five steps. The focus group discussions provided a general overview of the level of adaptive capacity for women and men in their respective communities. The description on each step of the ladder was based on knowledge and experience of the community members. This section provides the socio-economic characteristics on each step of the ladder and also presents the level of adaptive capacity between men and women and also between female-headed and male-headed households.

Level of Adaptive Capacity and Socio-Economic Characteristics

The discussions with male and female groups in the three districts came up with four steps on the ladder of adaptive capacity, representing the four different groups of people that are found in the community, according to their capacity to adapt to climate change impacts and effects. The steps were named using the local names from step 1, representing households with no capacity to adapt up to step 4, representing a household with highest capacity to adapt. The discussions also provided characteristics for each group on the steps of the ladder that relates to their level of capacity (Fig. 2). The criteria to place households on each step of the ladder depended on the socio-economic characteristics that included: the type and quantity of livestock, type of house and housing materials, crop yield, farm equipment, ownership of resources such as land, means of transport, access to farm inputs and ability to take care of children in terms of dressing and send-

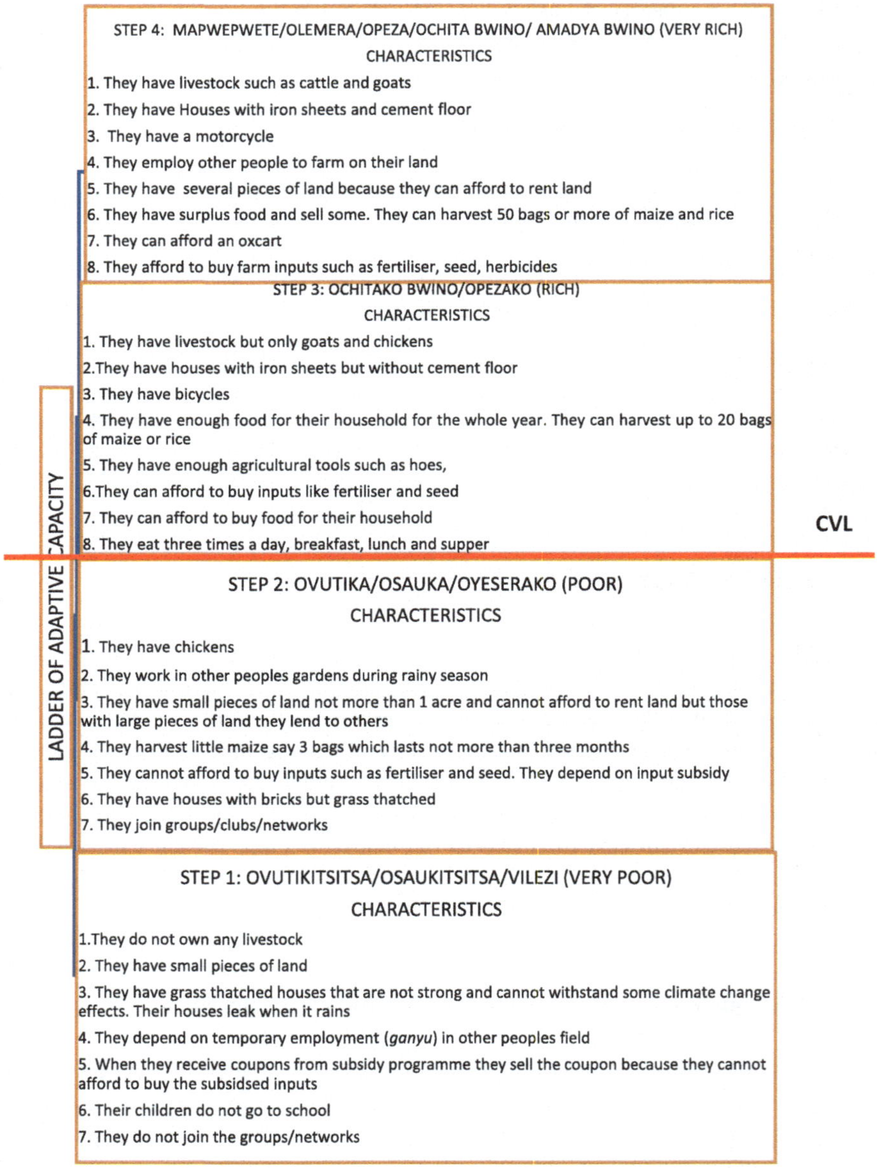

Fig. 2 Ladder of adaptive capacity and characteristics on each step

ing them to school. The ladder of adaptive capacity was developed based on the characteristics of the households in the study areas. It was noted that there was consistency among the female and male groups across the districts in constructing the ladder of adaptive capacity, and they all came up with four steps on the ladder.

The step 1 on the ladder of adaptive capacity was characterised by households that have small sizes of land, which cannot afford to buy inputs and mainly depend on input subsidy and are mainly dependent on piece work from other people's fields. These households do not own livestock and rarely harvest from their fields. They were categorised as households with the lowest capacity to adapt to climate change. On the other hand, step 4 of the ladder of adaptive capacity was characterised by households with reasonable sizes of land and can also afford to rent extra land from the community. The households on this step can afford to buy inputs like fertilisers and hybrid seeds and can hire labour from the community, especially from members on step 1. Households on step 4 have livestock such as cattle and goats in addition to chickens, and they mainly sell their agricultural produce in urban markets because they can afford transport. They were categorised as households with the highest capacity to adapt. The household characteristics kept on improving as you move up the ladder. The members of the discussion groups unanimously agreed that people who are considered to be vulnerable are those on steps 1 and 2; hence, the community vulnerability line was drawn on step 2.

Level of Adaptive Capacity by Sex of Respondents

The respondents were divided into three categories: women from female-headed households, women from male-headed households and men from male-headed households; however, the results were categorised into females and males, implying that women were categorised together whether from male- or female-headed households.

The results from the individual interviews showed that most female respondents were on step 2, implying that they have little capacity to adapt to climate change, while most of the male respondents were on steps 1, 2 and 3, implying that some have little or no capacity to adapt, while others have capacity to adapt to some of the climate change impacts and effects (Fig. 3). The reasons why most female farmers were on steps 1 and 2 were: low access to resources such as land, low access to information and technologies due to high illiteracy levels and limited access to extension services.

However, when asked about their adaptive capacity in 2013, female respondents indicated that they were on step 1 (22%) and step 2 (19%), while an equal number of male respondents were on steps 1 and 2 (Fig. 4). These findings imply that most male and female respondents have jumped up on the ladder from step 1 in 2013 to either step 2 or 3 in 2018. In general, both female and male farmers have some improved capacity to adapt to climate change than they were in 2013.

Although some households had moved up the ladder, the results showed that most of the female respondents were still on steps 1 and 2 in 2018, implying that they were below the community vulnerability line. This was different from male respondents who had moved upward to step 3 in 2018. The increase in the level of adaptive capacity for females was attributed to increased access to village saving and loan groups that provide soft loans to buy agricultural inputs. But the males

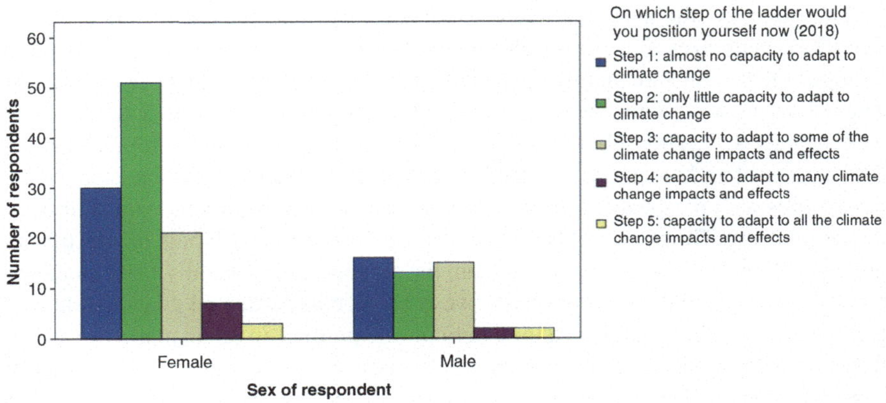

Fig. 3 Level of adaptive capacity in 2018 by sex of respondents

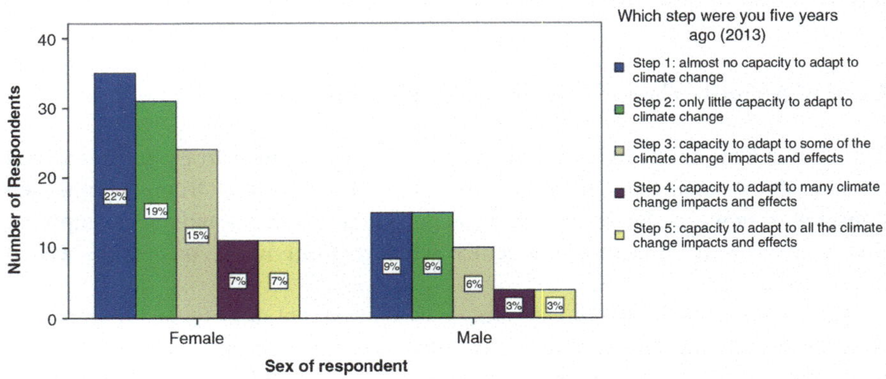

Fig. 4 Level of adaptive capacity in 2013 by sex of the respondents

attributed their movement to the access to other sources of livelihoods such as businesses. Men from Phalombe and Dowa also migrate to either urban areas or Mozambique (in the case of Phalombe) to look for informal employment, and this in turn improves the adaptive capacity of their households.

Level of Adaptive Capacity by Sex of the Head of Household

The level of adaptive capacity was also analysed according to the sex of the head of household, and the results showed that most of the male- and female-headed households were on step 2 (Fig. 5). However, there was still a reasonable percentage of male-headed households on step 1. The results show that the majority of the households have little or no capacity to adapt to climate change.

However, it was noted that, in 2013, there were more male-headed households on step 1 (24%) than on steps 2 and 3 (Fig. 6). The trend was similar for female-headed

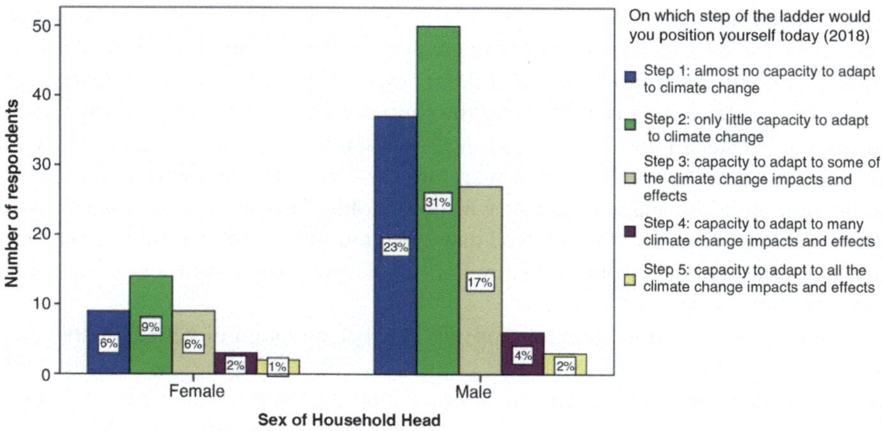

Fig. 5 Level of adaptive capacity in 2018 by sex of the household head

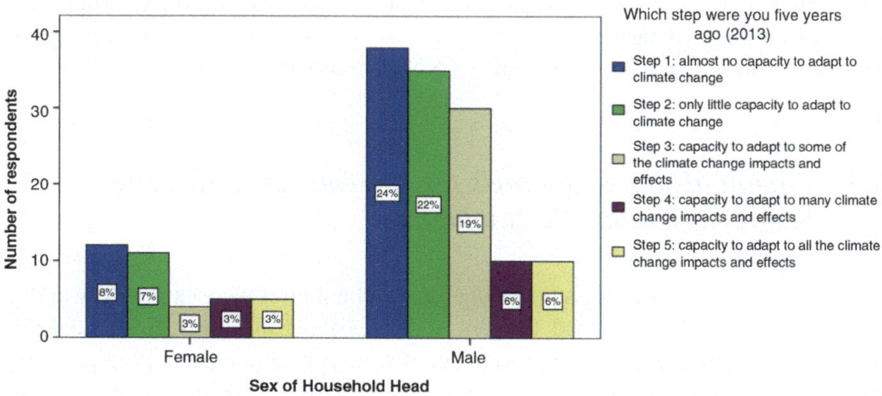

Fig. 6 Level of adaptive capacity in 2013 by sex of the household head

households. Some households had moved from step 1 to step 2 in 2018. The findings imply that some households that had no capacity to adapt in 2013 had acquired some capacity to adapt in 2018, implying that they had moved up the ladder of adaptive capacity. The results are consistent with adaptive capacity for male and female respondents that revealed that both had moved up the ladder. The reasons for increased adaptive capacity ranged from climate to socio-economic and technical factors including favourable rainfall, access to extension services, good farming practices and access to farm inputs. All these factors had a positive effect on crop production, but rainfall patterns and good farming practices were highlighted as major factors.

Some of the reasons why female-headed households were found down on the steps of the ladder are: poor distribution of rainfall that has rendered agriculture unprofitable, and most women depend on farming while men look for alternative sources of livelihood. The sizes of land are now smaller because land has been distributed among many children, and children, especially girls, are getting married at a younger age because it is difficult for the parents to provide for their children, they mostly fall on steps 1 and 2. The trend showed that there were more households with no capacity to adapt in 2013 than there are now in 2018, implying that there is improvement in the adaptive capacity of households. This result is consistent with the individual interviews that showed that most households have moved up the ladder. Both the individual questionnaire and focus group discussion questions used the same time period of 5 years.

The results showed a direct relationship between adoption of CSA technologies and level of adaptive capacity. This is consistent with the findings by Pingali 2012 who noted that low adaptive capacity affects adoption of new technologies. A farmer with low adaptive capacity will be unlikely to adopt new agricultural technologies. In general, female farmers had a low level of adaptive capacity, and this also affected adoption of CSA technologies. As such, interventions that implement new technologies should assess and address gaps in the adaptive capacity of farmers. Improving adaptive capacity of farmers, especially female farmers, is key to increasing the adoption of new technologies including CSA technologies.

3.3 Communication Channels for Disseminating Climate Smart Agriculture Technologies

The common channels for disseminating CSA technologies were extension workers (government and NGOs), lead farmers, radios, village meetings, field days and demonstration plots. But, the main channel of communication was through government extension workers and lead farmers (Fig. 7). However, there were vacant positions for extension workers in all the EPAs visited, and some sections within the EPA had no extension worker. Indeed, almost half of the positions of the extension workers at Linga EPA were filled with field assistants, who were students from Natural Resources College. In scenarios like these, which are common in most of the EPAs, lead farmers continue to play a significant role in disseminating technologies and good agricultural farming practices to the other farmers. Their proximity to farmers, in addition to their knowledge and experience of the local area, enhances trust and confidence among farmers.

In general, the results showed that female respondents had low access to common channels of communication such as radio and cell phone. Only 33% of the respondents in female-headed households owned a cell phone as compared to 59% of the respondents in male-headed households who owned one. Similarly, more male-headed households (70%) owned a radio than their female-headed household counterparts in which only 24% owned a radio. This implies that fewer females are able to access messages sent through cell phones and radio. The results showed that more respondents in Nkhotakota (39%) owned a radio than in Phalombe (35%).

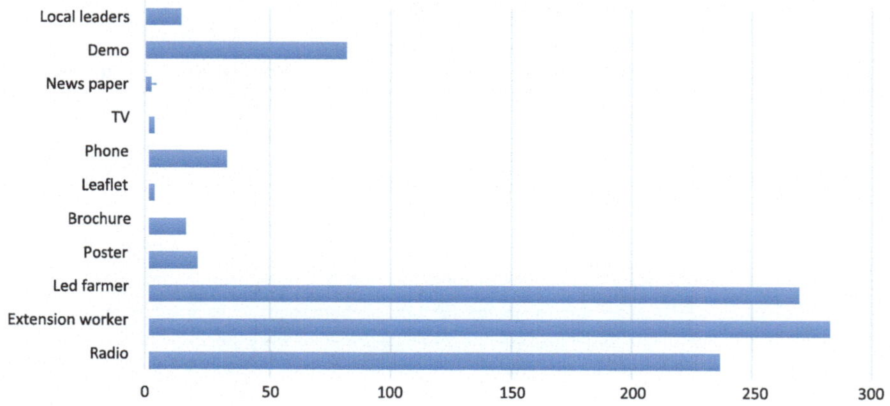

Fig. 7 Channels for disseminating information

However, 65% of the respondents in Phalombe did not own a cell phone, unlike 44% in Nkhotakota. The results also showed that 31% of the female respondents and 17% of the male respondents cannot read and write Chichewa, while 65% of the female respondents and 56% of the male respondents cannot read and write English. As such, if the information is presented in a written form and in English, very few female farmers will be able to access it. Phalombe had the highest percentage of respondents who can read and write Chichewa (65%), while Nkhotakota had 55%. The results are consistent with the baseline findings that showed that illiteracy was higher in Nkhotakota than Phalombe and Dowa districts. These findings provide an indicator of the channels that can be used to disseminate CSA technologies in each of the districts. It is also important to note that all the lead farmers can read and write Chichewa and most of them have cell phones and radios. Hence, they can also be used to bridge the communication gap.

The farmers also expressed their preference with regard to the sex of the extension workers or lead farmers that they were comfortable to work with. The majority of the farmers in Phalombe (56.8%) preferred the male extension workers/lead farmers, while more farmers in Nkhotakota (39.7%) and Dowa (46.1%) preferred the female extension worker/lead farmers (Table 4). Most of the female farmers indicated preference for female extension worker/lead farmer because of the social norms associated with the relationships and interaction between males and females. The interaction between a female farmer and a male extension worker/lead farmer may be misunderstood by the community as a love affair. However, there were some farmers who were comfortable with both female and male extension workers/lead farmers. The main reason for the choice of male extension workers was that they have more technical knowledge and skills on most of the agricultural practices.

This finding has serious implications for extension services in Malawi, and it underscores the need to balance up both female and male farmers in rural EPAs. However, most of the agricultural extension workers in Malawi are men. The few female extension workers prefer not to work and stay in rural areas, and as a result,

Table 4 Preferred sex of extension worker/lead farmer

District	Preferred sex of extension worker/lead farmer	Percentage of respondents
Phalombe	Male	56.8
	Female	10.8
	Both	32.4
Nkhotakota	Male	32.4
	Female	39.7
	Both	27.9
Dowa	Male	25.5
	Female	46.1
	Both	28.5

they are concentrated in EPAs that are close to urban areas. This makes it very challenging for female farmers to access extension services. In addition, the means of transport for extension workers are bicycles and motorcycles, and culturally, women in rural areas do not ordinarily ride motorcycles. But even with a motorcycle, it is still very challenging to access farmers in very remote areas because of the poor road network and infrastructure. While male extension workers may overcome some of these challenges, female extension workers run away from such challenges, and this further limits the options of female farmers in accessing extension services. This is why lead farmers play a significant role in providing extension services in remote areas. Unfortunately, they are few female lead farmers, and most of them have inadequate capacity to train farmers on new technologies. Providing training to lead farmers is key to improving extension services.

4 Conclusion and Recommendation

Female farmers have low adaptive capacity that translates into low adoption of technologies. Male farmers adopt technologies that are profitable, marketable and have short-term benefits, while female farmers adopt technologies that require less input and are cost effective and labour saving. Although lack of input was listed as the major constraint for adoption of technologies for both male and female farmers, there were additional challenges that are more pronounced among female farmers. These include: limited access to land, high illiteracy levels and low access to facilities that help in accessing information such as radio and cell phone. The results on adoption of technologies and level of adaptive capacity were similar in patrilineal and matrilineal societies. This implies that although women have land ownership rights through customary laws, land user rights and decision-making on technologies and inputs are influenced by gender and adaptive capacity. Overall, the level of adaptive capacity plays an important role in the adoption of technologies. Households that have a high capacity to adapt are likely to adopt many technologies, regardless of whether they are female-headed or male-headed households.

Since female and male farmers have different preferences on the sex of extension worker/lead farmer, imposing an extension worker/lead farmer of the opposite sex may lead to low or dis-adoption of technologies. Overall, female farmers adopted fewer technologies and had lower adaptive capacity than their male counterparts. This implies that interventions on new technologies should include gender analysis on the level of adaptive capacity. Based on the findings, it is recommended that:

- Government should encourage scientists, in consultation with farmers, to develop technologies that are cost-effective and labour saving.
- The agricultural extension system should design a programme that particularly addresses female extension agents and lead farmers to reach out to rural women smallholder farmers.
- The government and non-governmental organisations should increase access to loans and credit on agricultural inputs, especially for female farmers.
- The department of extension services should increase training of lead farmers to ensure that they acquire up-to-date knowledge and skills required in the implementation of CSA technologies.
- Researchers should conduct a gender analysis on levels of adaptive capacity before introducing interventions on agricultural practices to identify gaps and constraints.

This is very convincing and well analysed research. The authors should note however that the technical presentation of figure 2 on page 15 needs to be straightened out. Note that I did some editing within the figure itself, but the technology of it does not allow for changes to be tracked. Step two of the adaptive ladder does not show, so I did not do any editing in it.
While the writing itself is very good, there quite a number of typos, which made it necessary for me to move slowly so as not to miss any.

References

Asfaw S et al (2017) Gender, weather shocks and welfare: evidence from Malawi. J Dev Stud 52(2):271–291

Banda KM, Njikho KV, Malera G, Mauluka G, Mazinga MK, Ndhlovu S (2011, March) Women's access to land and household bargaining power: a comparative action research project in partrilineal and matrilineal societies in Malawi. Women's Legal Resource Centre (WOLREC)

Chirwa E (2005) Macroeconomic policy choices for growth and poverty reduction: access to land growth and poverty reduction in Malawi. Policy brief. Chancellor College. University of Malawi

FAO (2017) Climate-Smart-Agriculture Retrieved from http://www.fao.org/climate-smart-agriculture/en/ 11-05-2017

Goh AHX (2012) A literature review of the gender-differentiated impacts of climate change on women's and men's assets and well-being in developing countries. CAPRi Working Paper No. 106. Washington, D.C.: International Food Policy Research Institute. September, 2012

Government of Malawi (GOM) (2016) The National resilience plan: breaking the cycle of food insecurity in Malawi. Department of Disaster and management affairs

Government of Malawi (GOM) (2013) Malawi labour force survey 2013. National Statistics Office

Government of Malawi (GOM) (2012) National climate change policy. Ministry of Environment and Climate Change Management

Intergovernmental Panel on Climate Change (IPCC) (2014) Impacts, adaptation and vulnerability. In: Parry ML, Canziani OF, Palutikof JP, van der Linden PJ, Hanson CE (eds) Contribution of Working Group II to the fourth assessment report of the intergovernmental panel on climate change. Cambridge University Press, Cambridge, UK/New York

Kakota T, Maonga B, Synnevag G, Chonde C, Mainje M (2017) Harmonization of extension messages on climate smart agriculture in Malawi: do we speak with one voice, and to whom? J Agric Ext Rural Dev 9(11):255–261

Kakota T, Nyariki D, Mkwambisi D, Kogi-Makau W (2013) Determinants of household vulnerability to food insecurity: a case study of semi-arid districts in Malawi. J Int Dev 25:73–84. https://doi.org/10.1002/jid.2958

Kakota T, Nyariki DM, Mkwambisi D, Kogi-Makau W (2011) Gender vulnerability to climate variability and household food insecurity. Clim Dev 3(4):298–309

Lipper L et al (2014) Climate smart agriculture for food security. Nat Clim Chang 4:1068–1072

Mlamba (2010) Factors affecting adoption of conservation agriculture. MSc. Thesis. University of College Dublin

Murray U et al (2016) Smallholder farmers and climate smart agriculture: technology and labour productivity constraints amongst women smallholders in Malawi. Gend Technol Dev 20(2):117–148

Msowoya et al (2016) Climate change impacts on maize production in the Warm heart of Africa. Water Resour Manag 30(14):5299–5312

Ngwira A et al (2014) Adoption and extent of conservation agriculture among small holder farmers in Malawi. J Soil Water Conserv 69(2):107–119

Peterman A, Behrman JA, Quisumbing AR (2014) A review of empirical evidence on gender differences in non-land agricultural inputs, technology, and services in developing countries. In: Agnes QR, Ruth M-D, Terri LR, Andre C, Behrman JA, Amber P (eds) Gender in agriculture: closing the knowledge gap. Springer, New York/London, pp 145–186

Pingali PL (2012) Green revolution: impacts, limits, and the path ahead. Proc Natl Acad Sci 109(31):12302–12308

Synnevåg G, Kabote SJ, Nombo CI, Mamiro D, Mattee AZ (2014). Smallholder adaptation to climate change in semiarid areas of Tanzania:Experiences from Iramba and Meatu districts in Sustainable intensification to advance food security and enhance climate resilience in Africa. Springer International Publishing, Switzerland. pp 467–485

Women Environment and Development Organisation (WEDO) (2007). Changing the climate: why women's perspectives matter? Available on-line in http://www.wedo.org/files/climatechange-orng2ebook.pdf [01/09/2011]

WEDO (2010) Gender differentiated impacts of climate change in the economic community of West African States (ECOWAS) Sub-region: evidence and implications. WEDO

Whitehead A, Tsikata D (2003) Policy discourses on women's land rights in Sub Saharan Africa. The implications of the return to the customary. J Agrar Chang 3:67–112

Williams J (2008) Adoption of conservation agriculture in Malawi. MSc Thesis. Duke University

World Bank (2014) Levelling the field: improving opportunities for women farmers in Africa. World Bank and ONE Campaign, Washington DC

Smallholder Farming in Mara and Iringa Regions, Tanzania: Current Practices, Constraints and Opportunities

J. J. Kashaigili

Abstract Access to irrigation water is vital for increased productivity of smallholder farmers, especially in arid and semi-arid areas where rainfall is erratic. The majority of the people depend on groundwater from shallow wells for household water supply and farming, using simple technologies of water lifting such as buckets and handpumps. The water lifting technologies and household financial capability are said to have influence on water access and consequently the smallholder farming, the extent of which has not been well documented. This chapter presents research findings from a case study in Tanzania where primary data were collected using a semi-structured questionnaire which was administered to 135 respondents in Mara and Iringa regions, and key informant interviews were conducted using a checklist. Secondary data from various sources were used to supplement the primary data. The collected data were quantitatively analysed using a computer program to generate descriptive statistics, and qualitative data were analysed using content analysis. The financial returns analysis was conducted by considering the total number of families under each water control, the total area cultivated, the total investment and costs of inputs. Findings indicate that cultivated area varied considerably with the technologies used to lift water and the household financial capability. Manual pumps comprising treadle pumps, concrete pedal pumps (PePs) and buckets were found to be used by many people. Few people (16.29%) were found using motor pumps, and this could be attributed to the high capital investment. The crops grown were found differing across water control and were largely determined by the type of water sources and means of water access. The manual pump irrigator under both surface and groundwater and the motor pump irrigation using surface water were found to be comparatively profitable. The profitability is attributed to the ease of access to irrigation water at all times. Constraints pertaining to farmers' reliance on rain-fed farming pose production risks following unreliability and uncertainties of water supply. Other constraints include: land availability and tenure, inadequate working capital, poor market and marketing of produce. Despite the constraints,

J. J. Kashaigili (✉)
Department of Forest Resources Assessment and Management, Sokoine University of Agriculture, Morogoro, Tanzania

B. R. Singh et al. (eds.), *Climate Impacts on Agricultural and Natural Resource Sustainability in Africa*, https://doi.org/10.1007/978-3-030-37537-9_30

opportunities exist that could be considered for enhancing farmers' capabilities, including microcredit schemes for loans to enable farmers to buy pumps and agricultural inputs, existence of market opportunities because of expanding urban centres, rapid economic growth and the accompanying demand for more diversified products, mainly fruits and vegetables and existence of local manufacturer of manual pumps. Conclusively, improving access to irrigation water is instrumental for minimizing agricultural production risks and enhancing income of smallholder farmers.

Abbreviations and Acronyms

ASDP	Agriculture Sector Development Programme
ASDS	Agricultural Sector Development Strategy
BRN	Big Results Now
CAADP	Comprehensive Africa Agriculture Development Programme
DADPs	District Agricultural Development Plans
FAO	Food and Agriculture Organization of the United Nations
GDP	Gross Domestic Product
NGO	Non-Governmental Organization
NIMP	National Irrigation Master Plan
NIP	National Irrigation Policy
NPES	National Poverty Eradication Strategy
NSGRP	National Strategy for Growth and Reduction Poverty
PRSP	Poverty Reduction Strategy Paper
RDS	Rural Development Strategy
SAGCOT	Southern Agricultural Growth Corridor of Tanzania
TAFSIP	Tanzania Agriculture and Food Security Investment Plan
TDV	Tanzania Development Vision
URT	United Republic of Tanzania
WC	Water control

1 Introduction

1.1 Description of the Study Area

The Tanzanian economy largely depends on agriculture, and the contribution of the agricultural sector to the economic growth and the development of Tanzanians has continued to increase. In 2015, the agricultural sector contributed 29% of the GDP, compared to 28.8% in 2014 (Doloitte 2016). Currently, the sector alone provides employment to 65.5% of Tanzanians; in favourable seasons, it covers more than 100% of the domestic food needs. The 2012 Population and Housing Census showed that Tanzania had 44.9 million people in 2012, with an average intercensal growth rate of 2.7% per annum; in 2016, it was estimated to be 50.1 million people (URT 2016a). The human population projections show that Tanzania is expected to reach 63.5 million in 2025 (URT 2006). Rapid population growth impacts the water resources coverage, and because of that, the per capita water use has been decreasing over time. According to the key note presentation by the Permanent Secretary of the Ministry of Water, which was delivered at the 1st *Maji* Week Annual Scientific Conference in Dodoma (18–19 March 2019), Tanzania had 7862 m^3 per capita water use per year in 1962; however, due to population growth, increase in water demand to meet requirements for various socio-economic activities, pollution and climate change, this amount had decreased to 2600 m^3 in 2015; it is projected to further decrease to 1985 m^3, 1605 m^3 and 434 m^3 for the years 2025, 2035 and 2050, respectively. The decrease in water availability is expected to result in adverse effects on agriculture.

According to the National Irrigation Master Plan (NIMP) (URT 2018), Tanzania has a total of about 7.1 million ha of high and medium potential land (2.3 and 4.8 million ha, respectively) suitable for irrigation, supported by rivers, lakes, wetlands and aquifers (URT 2016b). Of the 2.3 million ha classified as high potential, only 461,326 ha had improved irrigation infrastructure in 2015, accounting for only 1.6% of the total land with irrigation potential (URT 2016b). Agriculture provides work for 14.7 million people, or 79% of the total economically active population, and 54% of agricultural workers are female (URT 2010b). Small-scale subsistence farmers comprise more than 90% of the farming population, with medium- and large-scale farmers accounting for the rest. Almost 80% of Tanzanians depend directly or indirectly on agriculture for their livelihood. The majority of these are smallholder farmers; their productivity is remarkably low and they are highly vulnerable to food insecurity. Unreliable rainfall, both within the season and from 1 year to another, is a constraint to the development of rain-fed agriculture in Tanzania in most parts of the country.

The main feature of the smallholder farming in Tanzania is characterized by the plot sizes, which are in many cases less than 1 ha. A study by Kashaigili (2007) revealed plots of between 0.42 and 0.66 acres (0.17–0.27 ha) for individuals under participatory farming groups and between 2.63 (1.06 ha) and 3.0 acres (1.21 ha) for private farmers. The majority of farmers rely on hand hoes as their main cultivating tools, although a small percentage use tractors and draft animals. The main food crops grown are maize, okra, cucumber, tomatoes, sorghum, millet, paddy, wheat, sweet potato, cassava, pulses and bananas. The yields of most crops are generally low, primarily due to the dependence on rain-fed agriculture that is highly variable. In very rare cases, groundwater has been used for irrigation as there has been very little attention to it nationwide. In many areas, especially in arid and semi-arid areas, groundwater development has always been for domestic water supply.

1.2 Plans for Irrigation Development in Tanzania

The government of Tanzania is determined to enhance agricultural production. This can be seen from the various strategies and policies formulated. For example, in 1994, the National Irrigation Development Plan (NIDP) was prepared to enhance agricultural production and ensure food security. The NIDP proposed 147 irrigation schemes to help in the alleviation of the many agricultural production constraints. Since then, several policies and strategies have been formulated, which include the Tanzania Development Vision 2025, National Poverty Eradication Strategy (NPES), Poverty Reduction Strategy Paper (PRSP), Rural Development Strategy (RDS), The National Strategy for Growth and Reduction Poverty (NSGRP), The National Irrigation Master Plan (NIMP), Agriculture Sector Development Programme (ASDP), The Southern Agricultural Growth Corridor of Tanzania (SAGCOT), The Comprehensive Africa Agriculture Development Programme (CAADP) and The National Irrigation Policy of 2010. The National Irrigation Master Plan (NIMP) of 2002 (revised in 2018) was launched to achieve the Agricultural Sector Development Strategy (ASDS) objective of sustainable irrigation development for increased agricultural productivity and profitability. The NIMP addresses development of irrigation schemes, with emphasis on smallholder farmers. The ASDP was launched in 2006 with the main goal to achieve a sustained agricultural growth rate of 5% per annum through transformation from subsistence to commercial agriculture. The transformation was to be private-sector led through public–private partnerships and implementation of the District Agricultural Development Plans (DADPs).

Tanzania has recently launched the Agricultural Development Plan Second Sector Phase Two (ASDP II). The 5-year plan is one of the key instruments that the government uses to meet Tanzania Development Vision (TDV) 2025 and will be implemented until 2023. The objective of the ASDP II is to transform the agricultural sector (crops, livestock and fisheries) towards higher productivity, commercialization level and smallholder farmer income for improved livelihood, 'food security and nutrition' (URT 2016a, b). ASDP II targets are to be achieved by 2024–2025: (i) inclusive and sustainable agricultural growth of 6% per annum; (ii) reduced rural poverty (reduce the percentage of the rural population living below the poverty line from 33.3% in 2011–2012 to 24% in 2025); and (iii) enhanced "food security and nutrition" (reduce the percentage of rural households (HHs) living below food poverty line from 11.3% in 2011–2012 to 5% in 2025). According to the plan, the government of Tanzania would finance about 38% of the programme, development partners would provide 57% (37% on budget) and the beneficiaries/farmers would finance the remaining 5%. The ASDS-2 reflects the changes in the overall economic environment and the policies and programmes that emerged over the years. ASDS-2 sets a new direction for the development of the sector, integrates the Comprehensive Africa Agriculture Development Programme (CAADP) objectives and reflects most of the vision and principles enunciated in the Tanzania Agriculture and Food Security Investment Plan (TAFSIP). The key priorities for ASDS-2 include: (i) the role of science and technology (research, extension and fertilizer use by small-scale commercial farmers); (ii) further priorities such as irrigation, finance, mechanization, agro-processing and access to markets; and also (iii) strong articulation with other sector initiatives, such as Big Results Now (BRN) and the Southern Agricultural Growth Corridor of Tanzania (SAGCOT).

The National Irrigation Policy of 2010 provides a baseline for a focused development of the irrigation sector in Tanzania. The Policy covers the activities and interventions required for the sector to effectively contribute towards enhancement of production and productivity in the agriculture sector (URT 2010a, b). The NIP (section 2.4), among other areas, highlights the need for the development and management of irrigation schemes and the need to maximize use of available water resources.

1.3 Micro-irrigation Technologies and Water Access by Smallholder Farmers

Improving access to water resources is among the prerequisites for enhancing smallholder farming in the face of the existing challenges. Of importance to increasing the agricultural productivity of smallholder farmers is access to affordable and efficient irrigation technologies. Previous studies (Kashaigili 2007; SWMRG 2005) revealed the increasing challenges facing the agriculture sector, which include: rapid population growth, decreasing availability of land and competition for scarce water resources. Owing to decreasing investments and declining performance of many large-scale irrigation schemes, interest has been developing in recent years in seeking ways to improve the productivity and livelihoods of the world's small-scale farmers (Frausto 1999). Low-cost micro-irrigation technologies such as concrete pedal pump, bush pumps, rope and washer pumps, rower pumps, treadle pumps, pitcher pot systems, drag-hose sprinklers, hydraulic ram pumps, micro-irrigation systems, windmills, water harvesting techniques and a host of other technologies with mixed success have been developed. These offer a breakthrough in the efforts to use technology to increase smallholder irrigation potential. Nevertheless, despite the mushrooming of such technologies, the low adoption rate largely contributed by limited capital, markets and marketing skills, and lack of extension services (Kashaigili 2007) has been a major challenge.

While that has been noted, studies by McCartney et al. (2007) and Kashaigili (2007) indicate that smallholders are capable of managing irrigation systems efficiently, provided they have access to affordable technologies that are easy to operate, maintain and repair. It is important to note that small-scale systems and technologies are attractive because they put the operation, maintenance and management of systems directly in the hands of the individual farmers, thus eliminating any need for centralized control or management. Smallholder farmers are thus empowered because they are able to apply water when and where they need it

(Frausto 1999). In many cases, the capital costs are lower and local labour and skills are employed. Accordingly, smallholders can be more productive with their yields and more efficient in water use than larger irrigation schemes (Ruotsi 1999; McCartney et al. 2007). This has also been accentuated by Shah (2009) who notes that there are several positive microeconomic benefits of pump irrigation over gravity flow irrigation in India, and most are related to the fact that water can be better controlled. This study examines the type of water control technologies that are mostly preferred by smallholder farmers and which ones maximize returns to investment.

2 Materials and Methods

2.1 Description of the Study Area and Approaches

The study was carried out in two regions of Tanzania mainland, namely Iringa and Mara (Fig. 1). Iringa region is located between latitudes 6°55′ and 10°30′ south of the Equator and between longitudes 33°45′ and 36°55′ east of Greenwich. The region's total surface area is 58,936 km^2 at an altitude varying between 900 m asl. and 3000 m asl. It is made up of six districts, namely Iringa Rural, Kilolo, Makete, Mufindi, Njombe, Ludewa and Iringa Urban, with Iringa Rural district being the largest. The human population was estimated at 941,238 in 2012 (URT 2016a) and 948,882 in 2016, with an intercensal growth rate of 1.1% between 2012 and 2016. The region experiences one rainy season, mainly from November through May each year. The off season/dry season farming activities take place in the valley-bottom wetlands (locally known as *vinyungu*), utilizing residual moisture in the dry season. In some areas, the wetlands are drained to allow dry season farming. The main horticultural crops grown in the dry season include tomatoes, beans, okra, spinach and maize (green), while the main field crop grown in wet season is maize.

Mara region is in the northern part of Tanzania. The region is located between latitude 1 and 2 degrees south of the Equator and between longitudes 33°10′ and 35°15′ east of Greenwich. It has an area of 30,150 km^2 of which 7750 is covered by Lake Victoria water and 7000 km^2 by Serengeti National Park. The area available for human settlement and agricultural production is 14,799 km^2. Only about 3000 km^2 of this area is used for crop cultivation. On average, the altitude of the region varies between 1100 m asl. and 1800 m asl. The five administrative districts in the region include Musoma Urban, Musoma Rural, Serengeti, Bunda and Tarime. According to the 2012 national census, the population was estimated at 1,743,830; in 2016, it was estimated to be 1,924,230 with an annual intercensal growth rate of about 2.5% between 2012 and 2016. A large part of the region receives rainfall less than 900 mm per year, with few areas along the Lake Victoria and in the highlands receiving more than 1500 mm per year. The major crops grown include maize, cassava and sorghum.

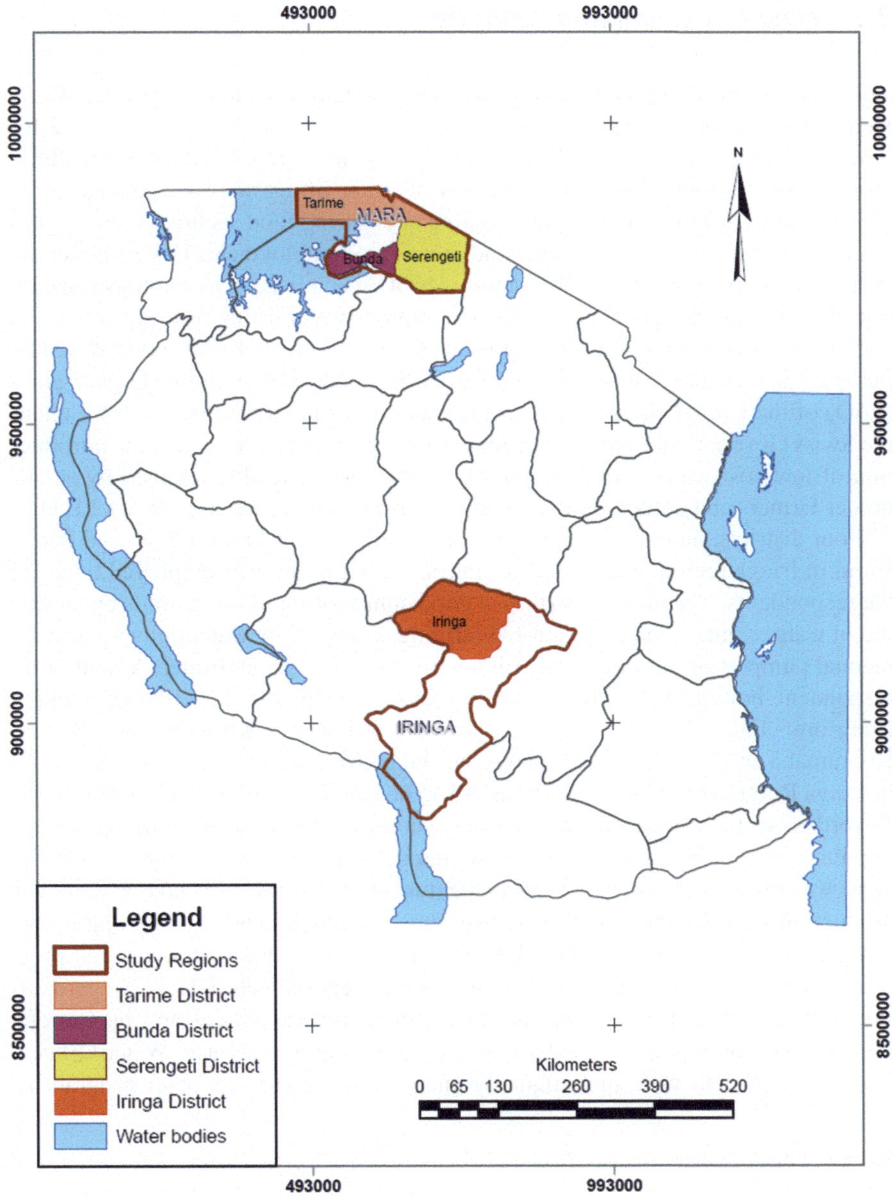

Fig. 1 Map showing location of the study areas

2.2 Data Collection and Analysis

This study adopted purposive sampling, and two regions, namely Iringa and Mara, were selected, considering their history in surface water and groundwater use for domestic use and small-scale farming. The two regions are known for horticultural crops production and different techniques of water lifting were introduced in the regions under different programmes aiming at ensuring food security, income generation and diversification of crops and creation of employment. For example, the Water for the Third World (W3W), a Swiss NGO specialized in technology transfer, introduced a concrete pedal pump (PeP) to improve water lifting from shallow wells and surface water bodies at a lifting head of about 7 m; it also introduced treadle pumps. The technology was widely spread by the Food and Agriculture Organization (FAO) of the United Nations Special Programme for Food Security (SPFS), under the Swiss Government funding. The main focus of the programme was the introduction of low-cost water control technologies that can be readily adopted by smallholder farmers and contribute to a rapid increase in crop production (FAO 2011).

Four districts, namely Tarime, Serengeti and Bunda in Mara region, and Iringa Rural in Iringa region, were selected. Purposive sampling was employed to select the respondents. The aim was to have a good composition of households under different water controls, namely rain-fed farming, gravity flow water delivery, private manual pump or motor pump irrigation with surface or groundwater. A total of 67 respondent households were purposely selected from the 12 villages, namely Bong'ore, Ilala Simba, Kiagata, Kidamali, Kilabera, Kirabera 'A', Sazira, Kitaramaka and Serengeti in Mara region, and 68 from Magubike, Nyatwari, Nzihi in Iringa Rural district, making a total of 135 households, structured as per Table 1. Nevertheless, there were two groundwater motor pump irrigators, thus necessitating altering the criteria to include more manual irrigators. A structured questionnaire was used to capture the local contextual specifics for the study, namely the production systems, costs, returns, constraints, challenges and existing opportunities. Secondary data were gathered through review of sectoral, strategic and projects documents, published and unpublished reports/materials in regard to smallholder agriculture in Tanzania. In addition, several offices and institutions were visited, including the local government authorities and Basin Water Offices. The collected data were quantitatively analysed using the computer program to

Table 1 Targeted sample size per water control (WC)

Water control	Total households
WC1: Rain-fed farming	30
WC2: Gravity/canal flow	30
WC3: Private manual pump irrigation from surface (WC3a) or groundwater (WC3b)	25
WC4a: Private motor pump irrigation with groundwater	25
WC4b: Private motor Pump irrigation from surface water	25
Total	135

generate descriptive statistics, and qualitative data were analysed using content analysis. The financial returns analysis was conducted by considering the total number of households under each water control, the total area cultivated, the total investment and costs of inputs.

3 Research Findings

3.1 Characteristics of Respondents

The respondents interviewed consisted of 73% males and 27% females (Table 2). All the respondents were adults, with the majority falling within the 36–50 (42.2%) age category, followed by 18–35 (37.8%). Therefore, majority of the respondents were within the active working age. As regards education, most of the respondents had completed primary school education (69.6%) while only a few (6.7%) had gone to secondary education. Also, 12.6% of the respondents had not received any formal education while 11.1% were primary school dropouts. Therefore, the majority of the people in the study villages had attained primary school education, and this is not surprising, as in Tanzania, primary education is regarded as a basic right for every citizen.

3.2 Household Assets

The results on household assets (Table 3) indicate differences in asset ownership between the water controls. Overall, manual irrigators using manual pumps (e.g. treadle, concrete pedal pumps and buckets) were found to own more assets. The results show that majority of people using canal have their dwelling houses made of cement and bricks, as compared to other water controls. Farmers who own motor pumps but pumping from surface water have on average more land to cultivate (acreage), as compared to other water controls followed by rain-fed farming. In comparison, motor pump irrigators are able to operate larger farming units than manual pump irrigators because it is less labour intensive, considering the same farming strategies. Overall, one would have vividly expected to see more assets ownership for motor pump irrigators than any other water control. Surprisingly, that has not been the case following the increased price of fuel. The survey period (2009–2010 season) coincided with the scarcity of fuel supply. Demand for fuel increased during the survey period, while the supply went low, leading to shooting up of fuel price. This was also echoed by farmers during the interviews that the hiking fuel price was among the limiting factors on the use of motor pumps.

Table 2 Characteristics of respondents with respect to water control

Variable	Rain-fed (WC1)	Canal (WC2)	Surface Water (SW) manual (WC3a)	Ground Water (GW) manual (WC3b)	Surface Water (SW) motor pump (WC4a)	Ground Water (GW) motor pump (WC4b)
Total respondents per WC	18	47	19	29	20	2
Number of male respondents	13	38	15	22	19	2
Number of female respondents	5	9	4	7	1	0
% of sample households with women as head of the household	3.7	6.7	3.0	5.2	0.7	0.0
Average age of the head of the household	48.6	39.7	40.3	40.1	42.0	55.5
Average years of education of the head of the household	5	6	6	6	6	7
Average number of full-time farm workers/ family	2.1	1.9	2.2	1.8	2.0	2.5
No. of family members above 16 years of age	46	101	54	73	58	8
No. of family members who are in trade or have a regular job	6	2	4	8	7	0
No. of family members who are engaged in farming or farm labour or non-farm casual work	27	63	41	50	39	6

Table 3 Respondents' household assets

Variables/asset type	Rain-fed (WC1)	Canal (WC2)	SW manual (WC3a)	GW manual (WC3b)	SW motor pump (WC4a)	GW motor Pump (WC4b)
Sample size	18	47	19	29	20	2
Number of respondents who have homes with thatched roof	3	11	9	12	3	2
Number of respondents who have homes made with cement and bricks	13	31	12	15	17	2
Number of respondents with formal title to land	11	29	6	10	13	0
Average land held in hectares	8.1	5.4	7.4	6.1	9.1	4.5
Average number of small, large and work animals owned	31	17	15	15	21	10
Number of respondents with boreholes for irrigation	0	3	5	1	4	1
Number of respondents with dug wells for irrigation	9	18	4	22	1	1
Number of respondents owning a treadle pump or bucket drum kit	1	4	11	17	0	1
Number of respondents owning a motor pump	0	0	0	0	20	2
Number of respondents owning flexible rubber pipe for water conveyance	1	2	8	17	20	2
Number of respondents owning bicycle	10	23	14	18	16	2
Number of respondents owning pesticide spray pump	10	24	11	18	16	1
Number of respondents owning mobile phone	10	28	16	25	19	1
Number of respondents owning motorcycle	1	2	1	2	0	0
Number of respondents owning motor car	0	1	0	1	0	0
Number of respondents owning colour TV	1	4	3	1	2	0
Number of respondents owning transistor radio	5	16	13	23	18	2

3.3 *Mode of Water Access for Crop Cultivation*

Table 4 presents the respondents' views on water access for irrigation. The results indicate that lifting using pumps was more preferred (48.1%) followed by gravity flow (26.7%) and lastly rain-fed farming. Gravity flow form of irrigation was found

Table 4 Distribution of respondents on irrigation profile in the study area

Mode of water access	Mara		Iringa		Total	
	n	%	*n*	%	*n*	%
Rainfall	7	5.1	27	20	34	25.2
Lift	60	44.5	5	3.7	65	48.1
Gravity flow	0	0	36	26.7	36	26.7
Total					135	100.0

$N = 135$

to be much used in Iringa (26.7%) and was not used at all in Mara. This is largely attributed to the nature of farming practices that normally take place in the valley bottom (*vinyungu*) in Iringa. In such areas, water from river streams is diverted to *vinyungu* fields by earth canals. Several wells are dug along the earth channels for collection and storage of water to irrigate nearby *vinyungu* fields using plastic buckets. It is also evident that the majority of respondents use manual irrigation methods because they are affordable and easy to operate. The methods such as treadle pumps and concrete pedal pumps are low-cost technologies which are affordable to many people and do not require specialized knowledge to operate (Kashaigili 2007).

3.4 Sources of Irrigation Water

Several sources of water for irrigation were reported by respondents as summarized in Table 5. It is apparent that people depend on canal water for irrigation as revealed by 30% of the respondents, followed by rain-fed agriculture (25.2%) and own well (22%). Comparing the two study sites, the majority of respondents in Mara and Iringa regions depend on canal water as their main source of irrigation water. It is important to note that the choice of water source among others is dictated by the water availability in the area and the available water controls. In areas with limited surface water (rivers and streams), ground water was found to be a major source.

3.5 Crops and Acreage on Various Water Control Classes

Table 6 presents the three high-priority crops under each water control, together with their corresponding cultivated areas during the 2009–2010 period in the study area. Comparing the water controls, the area cultivated under manual irrigation was much higher as compared to others. Overall, maize was found to be the most important crop under rain-fed farming and likewise for gravity flow irrigators. On the other hand, tomatoes and cabbage were ranked the first under motor pump irrigation using groundwater and surface water, respectively; the same applied to

Table 5 Source of irrigation water in the study area

Water sources	Mara		Iringa		Total	
	n	%	*n*	%	*n*	%
Own well	17	12.6	12	8.9	29	21.5
Own borehole	3	2.2	0	0	3	2.2
Community borehole	3	2.2	0	0	3	2.2
Small reservoir	3	2.2	0	0	3	2.2
Irrigation canal	18	13.3	22	16.3	40	29.6
River/stream	16	11.9	7	5.2	23	17
Rain-fed	7	5.2	27	20	34	25.2

Table 6 Crops preferred by adopters of different irrigation technologies in the study area

Variable/crop type	Rain-fed (WC1)	Canal (WC2)	SW manual (WC3a)	GW manual (WC3b)	SW motor pump (WC4a)	GW motor pump (WC4b)
Sample size (no. of farmers)	18	47	19	29	20	2
Name of the most important crop under largest area for the category	Tomatoes	Tomatoes	Tomatoes	Tomatoes	Watermelon	Tomatoes
Total acres under the crop for the WC	8.5	25.5	13.88	8.63	8.75	0.25
Average number of watering given to this crop/family cultivating this crop	–	26	64	42	23	60
Name of the second most important crop under largest area for the category	Maize	Maize	Maize	Maize	Maize	Cucumber
Total acres under the crop for the water control	4	10.25	4	5.5	6	0.5
Average number of watering given/family cultivating this crop	–	19	20	15	17	35
Name of the third most important crop under largest area for the category	Rice	Beans	Maize	Onions	Cabbage	Chinese
Total acres under the crop for the WC	3	5.75	3	2.5	3	0.25
Average number of watering given/family cultivating this crop	–	19	0	58	24	50

cucumber and cabbage for manual irrigation using groundwater and surface water, respectively. Cabbage is the most dominant crop in terms of acreage, and it is common with people who use manual pump for irrigation with surface water. The lifting devices commonly in use are buckets and manually operated pumps such as treadle pumps and concrete pedal pumps.

It is apparent that the types of crops grown differ across water controls. Farmers in rain-fed agriculture cultivate seasonal crops such as maize, beans, cassava, sorghum and cotton; green maize and tomato, on the other hand, are dominant crops with gravity flow irrigators as the market demand of green maize and tomatoes justifies the extra investment costs. Horticultural crops such as cabbage, tomatoes and onions are the dominant ones in motor and manual pump irrigators. This is so because horticultural crops have a good return on investment. For this reason, watermelon and cabbage are observed to be more dominant in motor pump irrigation. Overall, the results show that there is limited use of groundwater and gravity flow for irrigation in both study sites, Mara and Iringa. This can be attributed to the fact that the required capital investment for making wells or drilling boreholes is high. Kashaigili (2010) noted that there are limited uses of groundwater for irrigation in Tanzania due to several factors, including the fact that the majority of people in the community have inadequate information and understanding of groundwater resources. This has also been echoed by Mato (2002), Sadiki (2009) and Kongola (2008), who recommend further research before groundwater can be put to large-scale use for irrigation or other uses.

3.6 Financial Returns to Irrigation Under Different Water Control Technologies

Table 7 presents the financial returns under different water control technologies in the study area. The total income per household derived from the sale of crops under rain-fed farming, WC1, is US$1,011.2 while for canal irrigation, WC2, is US$1,447.1; under manual pump irrigation from surface water, WC3a, it is US$970.3, under manual pump irrigation from groundwater, WC3b, it is US$703.6, under motor pump irrigation from surface water, WC4a, it is US$1,609.9 and under motor pump irrigation from groundwater, WC4b, it is US$436.9. The results show that the total income from the sale of crops under motor pump irrigation from surface water is higher than that for other water control technologies. Nevertheless, the value added per family under surface and groundwater manual irrigation is comparatively higher, unlike that for other categories.

The study has revealed the existence of local markets for the agricultural crops for both small and large scale. In Mara, horticultural crops have all-round market in the village, especially during market days in the villages. The vegetable and fruits are sold to hotels and lodges in Serengeti district. The owners of hotels and lodges purchase the produce in bulk and pay farmers promptly. This has a direct benefit to farmers. Vegetables and fruits from Iringa are sold in the local markets in Iringa, but most of the supplies of vegetables and fruits (e.g. tomatoes, onions, cabbages, pep-

Table 7 Financial returns to irrigation under different water control technologies in the study area

Variable	Rain-fed (WC1)	Canal (WC2)	SW manual (WC3a)	GW manual (WC3b)	SW motor pump (WC4a)	GW motor pump (WC4b)
Sample size (number of farmer families)	18	47	19	29	20	2
Total investment made in pumps, pipes and water source development by all farmers in the Water Control Class	–	1101.6	1771.5	2418.1	9764.5	689.3
Total income (US$) from the sale of crops/families in the Water Control Class	1011.2	1447.1	970.3	703.6	1609.9	436.9
Total cash (US$) expenditure on farm inputs including fertilizer, pesticide, seeds, fuel and hired labour/families in the Water Control Class	132.0	127.2	126.0	107.5	148.0	46.1
Value-added (US$)/family in Water Control Class	879	1320	1524	1606	1462	391
Total area of farmland under cultivation (acres)	102.50	193.75	78.00	105.25	103.75	4.75
Value added/acre (US$)	154	320	371	442	282	165
Value added/irrigation investment (%)	–	120	86	66	15	57
Value added as % of working capital	–	1037	1210	1494	988	847
Value added per family worker (US$)	32.6	20.9	37.2	32.1	37.5	65.1

pers, watermelon) are sold in Dar es Salaam and other neighbouring towns such as Morogoro and Dodoma come from Iringa. This is facilitated by the presence of a good road network from Iringa to other regions where the products are sold. Irrigation enables farmers to have several crop cycles in a year, since they are confident of the markets all the time. Most of these crops are common for WC1 to WC4. Crops like maize, especially dry maize, sorghum and cassava, which are common for WC1, are mostly seasonal. Farmers depend on rainfall and do not invest in water lifting devices and pumps because of the weather and market uncertainty, unlike those in the other water control technologies. The farmers in WC3 and WC4 normally have all-round market exposure, compared to those in WC1.

4 Discussion of the Findings

The study has examined the several technologies that farmers use to irrigate their crops, ranging from simple technologies such as buckets and manual pumps (e.g. treadle, concrete) to motorized pumps. It has revealed that the disparities in water access, and hence income, are largely due to the financial capability of the individual household, along with the nature of the water sources. In areas where groundwater table is closer to the surface, people opt to use shallow wells, and the majority use buckets for delivering water to crops, and this is in line with the observation by Kashaigili (2007). While this has been noted, there is a greater need for improving access to water for irrigation by smallholder farmers for enhanced crop production and income. As such, understanding the problems and constraints to the current farming practices, and the existing opportunities for improvement, is imperative for enhanced agricultural production.

4.1 Problems and Constraints to the Current Smallholder Farming Practices

There are some problems and constraints in the farming practices. Table 8 summarizes the key challenges facing smallholder farming under different water controls.

(i) *Land tenure, access rights and land management*

The uncertainties regarding land tenure and the inadequate access to land have been a critical challenge to smallholder farming in Tanzania. As Salami et al. (2010) pointed out the constraints related to the tenure system, such as insecurity of land tenure, unequal access to land, lack of a mechanism to transfer rights and consolidate plots, have resulted in underdeveloped agriculture, high landlessness, food insecurity and degraded natural resources. This is in conformity with the findings of the study, where the majority of farmers were found working by renting plots from others. Under such arrangement, the majority of the smallholder farmers become voiceless; their rights are eroded, since the terms are prescribed by the land owner (Kadigi et al. 2017). The key

Table 8 Key problems and challenges in the study area

Water Control	Cross-cutting problems and challenges (for all WC)	Other problems and challenges
WC1: Rain-fed farming WC2: Gravity flow WC3: Private manual pump irrigation from surface or groundwater WC4a: Private motor Pump irrigation from surface water WC4b: Private motor pump irrigation with groundwater	Inadequate working capital Unreliable and insufficient water availability Poor market and marketing of produce Land availability and tenure	Dependency on family labour Fake fungicides and fertilizers High cost of inputs Lack of agricultural inputs Poor means of transport of crops Poor education Limited extension services Pests and diseases Climate change

concern is that it is hard to plan any long-term investment as most land owners are unpredictable and keep on changing farming conditions. It came out very clearly during interviews that in the event of good yield, landlords tend to inflate the leasing costs, and sometimes stop the renter from farming, pretending that they want to cultivate in the same area for the next season. Therefore, in the absence of secure tenure and access rights, investing in smallholder irrigation farming is quite problematic.

(ii) *Inadequate working capital*

Inadequate working capital is among the cross-cutting constraints. Almost all respondents under different water control were equally affected by capital. Therefore, the majority of smallholder farmers are incapable of expanding their farming investment as their savings are too limited, with very marginal surplus that does not warrant reinvesting. According to Salami et al. (2010), smallholder farmers depend on savings from their low incomes, which limits opportunities for expansion. It was further noted that over half of total rural household income in Tanzania comes from farming, 46.6% from non-farm employment (wages and self-employment) and less than 4% from remittances. Because of the lack of collateral and/or credit history, most farmers are bypassed not only by commercial and national development banks but also by formal microcredit institutions.

(iii) *Insufficient water availability*

Almost all respondents cited insufficient water availability as a critical constraint. The majority reported that surface water sources are no longer reliable, as a result of the increasing competing demands for surface water and the uncertainties resulting from perturbations of climate variability and change. It came out clear from interviews that most surface water sources have been overstressed, and almost all available run-of-the-river has already been allocated. Also, along this is the limited capital to enable investment in water supply solutions such as damming and construction of boreholes for groundwater mining. Therefore, ensuring reliable water source through water supply solutions is imperative for the enhancement of smallholder farming.

(iv) *Poor market and marketing of produce*

The study revealed existence of local markets for the agricultural crops for both small and large scale, mainly for horticultural crops, including vegetable and fruits that are sold to hotels and lodges. This was more evident in Mara region than in Iringa. Nevertheless, not all smallholder farmers have access to these markets, and the majority lack marketing skills. As was accentuated by Salami et al. (2010), improved access to input and output markets is a key precondition for the transformation of the agricultural sector from subsistence to commercial production. Smallholder farmers must be able to benefit more from efficient markets and local-level value addition and be more exposed to competition. Adequate storage facilities constitute another constraint to both marketing and food security. This was more evident in Iringa region, where tomato production under manual and pump irrigation is high, but owing to lack of rural electrification, there are no storage facilities and most of it rots away.

Kamara et al. (2002) reveal that large quantities of agricultural commodities produced by farmers rot away unmarketed, while the smallholder farmers do not have the technology for timely consumption. Equally notable is the inability of most smallholder farmers to get linked to the supermarket chains as they cannot meet the high quality and safety demands as well as delivery schedules that international value chains require, preventing them to compete in such markets (Kamara et al. 2002). In addition, inadequate and poor conditions of the market facilities, lack of reliable electricity and transportation systems, including road and rail, are serious impediments to the smallholder farmers.

4.2 Opportunities for Enhancing Smallholder Farming

Tanzania has a potential for attaining sustainable agricultural development and improving smallholders farming and income generation. Such opportunities include:

- *Existence of land-related policies and laws that govern access and use rights.* Such laws include Land Act of 1999 (URT 1999a); Village Land Act of 1999 (URT 1999b); The National Irrigation Policy of 2010 (URT 2010a); National Water Policy of 2002 (URT 2002); and The Water Resources Management Act, Supplement No. 11 of 2009 (URT 2009) to mention a few. Farmers should be made aware of the existing policies and laws governing land access and rights. More specifically, they should be made aware of their land tenure rights. Land tenure is an important part of social, political and economic structures. It is multidimensional, bringing into play social, technical, economic, institutional, legal and political aspects that are often ignored but must be taken into account. Land tenure relationships are well-defined and are enforceable in a formal court of law or through customary structures in a community.
- *Existence of microcredit schemes* (e.g. Savings and Credit Cooperative Organization Society [SACCOS]). Smallholder farmers should be made aware of the possibility of accessing loans to enable them to buy pumps and agricultural inputs. This could be in a form of hire purchase programme, which is used by many people who are willing to buy the pump but cannot raise the required amount at once. Having such an arrangement in place would make it possible for many people in the rural area to own a pump. And if there could be a loan facility where one would pay with the income from the farm after the harvest, this would improve their methods of farming, crop quality and increase acreage.
- *Availability of reliable rainfall in many parts of the country.* Implementation of rainwater harvesting as an alternative water supply solution for agricultural production could ensure availability of water for smallholder farmers and reduce on the dependence on surface water. Several of these rainwater harvesting technologies have been tested and are available for implementation in smallholder farming area. Also, effective use of groundwater, rather than the present overdependence on water surface water sources, needs to be considered.

- *Existence of extension services.* More support and institutional strengthening is fundamental for ensuring extension services to all the smallholder farmers.
- *Existence of market opportunities.* Farmers should be helped to market their crops by building their capacities, also value adding should be considered as a way forward towards improving farmers' income and poverty eradication. Crops like tomatoes (perishable crop) fail to fetch good market when the supply becomes higher than the demand, causing farmers to sell at very low prices. Furthermore, smallholder agriculture is projected to be economically sustainable in the future because of expanding urban centres, rapid economic growth in most of the case study countries and the accompanying demand for more diversified products, mainly fruits and vegetables.
- *Investment Opportunities.* The potential of agriculture and smallholder farming can be illustrated by the enhanced income generation in several Tanzanian agricultural export sub-sectors.
- *Existence of local manufacturer of manual pumps.* Examples of these include the local NGOs like Water for the Third World (W3W), which is engaged in manufacturing of Concrete Pedal Pump. This pump has been widely disseminated and is relatively affordable by smallholder farmers. The cost of the pump, including accessories, amounts to about US$150. There also exist other manual pumps such as treadle pump.

5 Conclusions

The study has examined current practices of smallholder farmers, including the challenges and available opportunities towards realization of sustainable smallholder farming. It has been revealed that the area cultivated varies considerably, depending on the technologies used to lift water, as well as the financial capability. Few people were found using motor pumps, and this could be attributable to the substantial amount of capital required to purchase them, as well as the skills to operate them. Manual pumps comprising of treadle pumps, concrete pedal pumps and buckets were found to be used by a good number of people, because of affordability and ease of operation. The crops grown were found to differ significantly across 'water control classes' and largely determined by the type of water sources and means of water access. The manual pump irrigator, under both surface and groundwater, and the motor pump irrigation using surface water were found to be comparatively profitable. The high profitability could be due to the ability of pump irrigators to take control of water and risk mitigation resulting from water scarcity. Reliance on rain-fed farming poses considerable production risks, following unreliability and uncertainties of water supply. Alongside the other constraints such as land availability and tenure, inadequate working capital and poor marketing of produce have been identified. The available opportunities that are presented in this research can go a long way in helping to overcome the constraints.

Acknowledgements This work was carried out as part of the research project led by the International Water Management Institute (IWMI) on 'Groundwater in Sub-Saharan Africa: Implications for Food Security'. Project Ref.4500010162.

References

Doloitte (2016) Tanzania Economic Outlook 2016. The story behind the numbers. Available online: https://www2.deloitte.com/content/dam/Deloitte/tz/Documents/tax/Economic%20Outlook%202016%20TZ.pdf. Accessed on 02/06/2019

FAO (2011) Review of water control technologies in the FAO programmes for food security – experiences and best practices in the introduction of water control technologies under the FAO assisted food security programmes with a focus on Sub-Saharan Africa. Available online: http://www.fao.org/3/i2176e/i2176e00.pdf. Accessed on 02/06/2019

Frausto K (1999) Developing irrigation options for small farmers, contributing paper, prepared for thematic review IV.2: Assessment of Irrigation Options Institute Development Enterprises, USA. Available online: http://www.dams.org/. Accessed on 12/05/2010

Kadigi RMJ, Kashaigili JJ, Sirima A, Kamau F, Sikira A, Mbungu W (2017) Land fragmentation, agricultural productivity and implications for agricultural investments in the Southern Agricultural Growth Corridor of Tanzania (SAGCOT) region, Tanzania. J Dev Agric Econ 9(2):26–36. https://doi.org/10.5897/JDAE2016.0797

Kamara AB, van Koppen B, Magingxa L (2002) Economic viability of small-scale irrigation systems in the context of state withdrawal: the Arabie scheme in the Northern Province of South Africa. Phys Chem Earth 27:815–823

Kashaigili JJ (2007) Survey on adoption and technical performance evaluation of the Swiss Concrete Pedal Pump (CPP). Report to the Food and Agriculture Organisations of the United Nations (FAO), 89 pp

Kashaigili JJ (2010) Assessment of groundwater availability and its current and potential use and impacts in Tanzania (Project Ref.4500010162). Report submitted to International Water Management Institute (IWMI), 67 pp

Kongola LRE (2008) Groundwater management in Sub-Saharan Africa under climate variability with a rural focus. Paper presented at the international conference on groundwater and climate in Africa, Speke Conference Centre, Munyonyo, Kampala, Uganda, 24th–28th June 2008

Mato RRAM (2002) Groundwater pollution in Urban Dar es Salaam, Tanzania: assessing vulnerability and protection priorities. PhD Thesis, Eindhoven University of Technology

McCartney MP, Lankford BA, Mahoo H (2007) Agricultural water management in a water stressed catchment: lessons from the RIPARWIN project. Colombo, Sri Lanka: International Water Management Institute (IWMI Research Report 116), 52 pp

Ruotsi J (1999) Financing irrigation development and private sector initiative with special reference to Sub-Saharan Africa. Rome: FAO E-mail conference 15 March to 23 April, 1999

Sadiki H (2009) Groundwater management in Africa: the case of Pangani Basin, Tanzania. 5th World Water Forum – Istanbul, Turkey 20th March 2009

Salami A, Kamara AB, Brixiova Z (2010) Smallholder agriculture in East Africa: trends, constraints and opportunities. Working Papers Series No. 105. African Development Bank, Tunis, Tunisia

Shah T (2009) Taming the anarchy: groundwater governance in South Asia. Resources for the Future/IWMI, Washington, DC/Colombo. Chapter 4

SWMRG (Soil Water Management Research Group) (2005) Experiences with micro agricultural water management technologies: Tanzania. Report submitted to International Water Management Institute (IWMI)

URT (United Republic of Tanzania) (1999a) National Land Act. Printed by the Government Printer Dar es Salaam-Tanzania, 507 pp
URT (United Republic of Tanzania) (1999b) The Village Land Act. Printed by the Government Printer Dar es Salaam-Tanzania, 285 pp
URT (United Republic of Tanzania) (2002) National Water Policy (NAWAPO). Ministry of Water and Irrigation, Dar Es Salaam, Tanzania, 88 pp
URT (United Republic of Tanzania) (2006) National Population Policy. Ministry of Planning, Economy and Empowerment, Dar es Salaam, 42 pp
URT (United Republic of Tanzania) (2009) Water Resource Management Act No. 11 of 2009 (WRMA). Ministry of Water, Government Printers, Dar es Salaam
URT (United Republic of Tanzania) (2010a) National Irrigation Policy. Ministry of Agriculture Food Security and Cooperative, Dar es Salaam
URT (United Republic of Tanzania) (2010b) The Economic Survey 2009. Ministry of Finance and Economic Affairs, Dar Es Salaam – Tanzania. Printed By Kiuta Dar Es Salaam, 258 pp
URT (United Republic of Tanzania) (2016a) 2016 Tanzania in figures. The National Bureau of Statistics. Available online: https://www.nbs.go.tz/nbs/takwimu/references/Tanzania_in_Figures_2016.pdf. Accessed on 02/06/2019
URT (United Republic of Tanzania) (2016b) Agricultural Sector Development Programme Phase Two (ASDP II). Available online: http://www.tzdpg.or.tz/fileadmin/documents/external/national_development_frameworks/ASDP2_Final_Document_20_May._2016__after_edit__1_.pdf. Accessed on 02/06/2019
URT (United Republic of Tanzania) (2018) The project on the revision of National Irrigation Master Plan in the United Republic of Tanzania, final report volume-I, main report. National Irrigation Commission, Ministry of Water and Irrigation, prepared by Japan International Cooperation Agency (JICA), Nippon Koei Co., Ltd. and International Development Center of Japan Inc. Available online: http://open_jicareport.jica.go.jp/pdf/12309308_01.pdf. Accessed on 09/06/2019

Impact of Farm Input Subsidies Vis-à-Vis Climate-Smart Technologies on Maize Productivity: A Tale of Smallholder Farmers in Malawi

Samson Pilanazo Katengeza

Abstract Farm input subsidies in Malawi are historically a strategic agriculture policy tool, particularly for enhancing maize production for national and household food security. This chapter presents a review of the impact of access to inorganic fertilizer through the Farm Input Subsidy Program (FISP) on maize productivity, and compares it with the impact of climate-smart agriculture (CSA) technologies, specifically the integration of inorganic and organic fertilizers. Results show a modest impact of FISP on maize productivity. Maize–fertilizer response rates, reported as nitrogen use efficiency (NUE) among FISP beneficiaries, ranges from 3 to 14 kg of maize per 1 kg of nitrogen (N) (kg/kgN) fertilizer used, which is below the expected agronomic average of 15 kg/kgN. Conversely, the NUE is 17–36 kg/kgN on experimental plots with integration of inorganic and organic fertilizer. These CSA technologies ensure efficient and optimal nutrient uptakek and drought resilience. This suggests that the impact of FISP can be enhanced if application of subsidized inorganic fertilizer is integrated with CSA technologies. FISP implementation strategy should therefore consider abandoning the current farmer-based targeting system and subsidize the soil by targeting adopters of CSA technologies. This approach has potential to provide the Government of Malawi with a sustainable exit strategy from FISP.

1 Introduction

Sustainable maize production and productivity dominates agriculture policy priorities in many developing countries, especially in sub-Saharan Africa (SSA) where a majority of the rural population depends on maize for food security. There is a knock-on effect that starts from poor and low maize production and productivity to limited access to quality and nutritious food (Hawkes and Ruel 2008). In Malawi, a

S. P. Katengeza (✉)
Lilongwe University of Agriculture and Natural Resources, Lilongwe, Malawi
e-mail: skatengeza@luanar.ac.mw

B. R. Singh et al. (eds.), *Climate Impacts on Agricultural and Natural Resource Sustainability in Africa*, https://doi.org/10.1007/978-3-030-37537-9_31

country that is heavily dependent on rain-fed agriculture, low levels of soil fertility and frequent dry spells are two major causes of low maize production and productivity (Ngwira et al. 2013a; Weber et al. 2012). Increasing nitrogen (N) application, maintenance and use efficiency, and drought resilience are important avenues to achieve sustainable increase in maize harvests (Snapp et al. 2014). The Malawi National Agriculture Policy (NAP), therefore, puts emphasis on timely and equitable access to high-quality productive inputs such as inorganic fertilizer and promotion of climate-smart agriculture (CSA) technologies (Government of Malawi 2016b).

Among key policy instruments is the Farm Input Subsidy Program (FISP) that enhances access to and use of inorganic fertilizer (Dorward et al. 2008). The program started in the 2005–2006 cropping season and officially subsidizes two 50-kg bags of inorganic fertilizer for maize production (Chirwa and Dorward 2013). The expectation was that the program would break the low-input/low-output poverty trap among smallholder farmers, kick-start growth, and raise agricultural incomes and food security (Ricker-Gilbert and Jayne 2017). Early evidence showed that the country achieved 53% national food surplus in 2006–2007 from a food deficit of 43% in 2004–2005. Average maize productivity jumped from 1.05 Mg per hectare (ha) in 2003–2004 and 0.76 Mg/ha in 2004–2005 to 1.59 Mg/ha in 2005–2006 and 2.04 Mg/ha in 2006–2007 (Denning et al. 2009).

This chapter reviews empirical evidence on impacts of subsidized inorganic fertilizer and related CSA technologies on maize productivity. Specifically, the chapter reviews: (a) impact of subsidized inorganic fertilizer on maize productivity; (b) marginal impact of subsidized inorganic fertilizer on maize productivity; and (c) impact of integrating inorganic and organic fertilizer on maize productivity. Inorganic and organic fertilizer technologies are potentially complements and are vital for sustainable maize production (Munthali 2007), but they could also be substitutes and compete for resources (Holden and Lunduka 2012).

2 Conceptual Approach

Figure 1 conceptualizes the symbiotic relationship between inorganic and organic CSA technologies, specifically focusing on inorganic fertilizer accessed through FISP and adoption of organic fertilizer through investment in agricultural research and extension services. Path 1 shows that government investment in FISP would reduce costs for inorganic fertilizer, thereby increasing input profitability for smallholder farmers (Lunduka et al. 2013). This would then encourage participating farmers to invest in other productivity-enhancement inputs such as CSA practices (Karamba and Winters 2015; Lunduka et al. 2013). Investment in CSA technologies would increase soil response to subsidized inorganic fertilizer as complements and improve input use efficiency (Karamba and Winters 2015; Ricker-Gilbert and Jayne 2017; Snapp et al. 2014). This cyclic and mutual relationship would result in parallel increase in use intensity of organic and inorganic fertilizer and other CSA technologies, thereby increasing maize production and productivity.

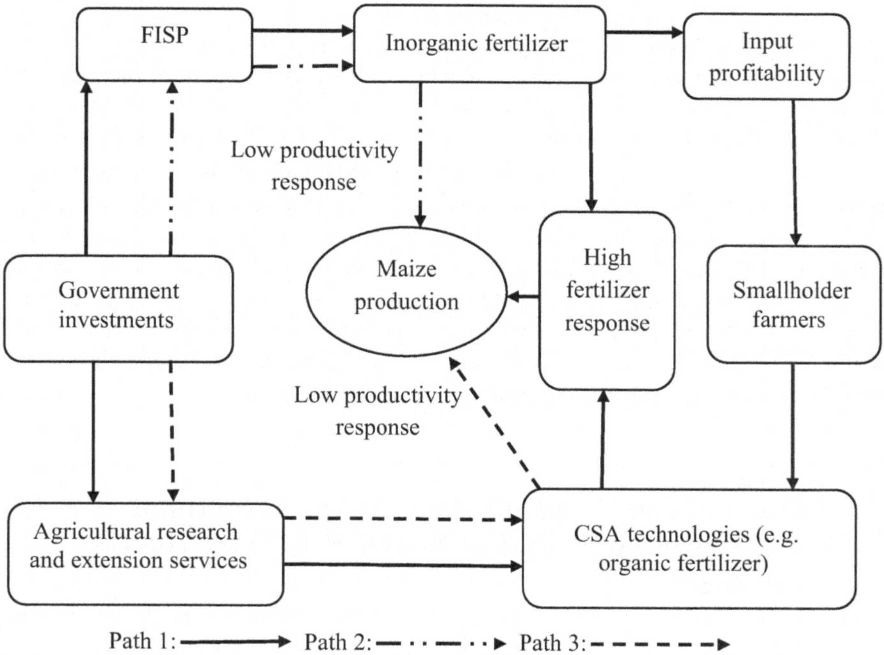

Fig. 1 Symbiotic relationship between inorganic fertilizer and climate-smart agriculture technologies

However, investments in large-scale input subsidies tend to have high costs that may outweigh the long-term program benefits. To fund such programs, governments are likely to relocate resources from other agricultural programs (Carr 2014; Lunduka et al. 2013). For instance, as shown in path 2, the government may scale down (or fail to increase) investments in agricultural research and extension services to fund input subsidies, and this may affect the adoption of CSA technologies. Although FISP may increase profitability of inputs as reported by Lunduka et al. (2013), farmers' response is lagged, and without extension services, propensity to adopt CSA technologies is low. Consequently, soil responsiveness to inorganic fertilizer will be low and long-term returns to input subsidies will be insignificant (Ricker-Gilbert and Jayne 2017). On the other hand, high investment in programs that promote adoption of CSA practices with minimal investment in inorganic fertilizer is not a panacea (Carr 2014) as shown in path 3. The soils in Malawi are too depleted of vital nutrients (Carr 1997; Drechsel et al. 2001) and require supplementary input of inorganic fertilizer (Snapp et al. 1998, 2014).

From this discussion, it is clear that investment in FISP affects investment in other programs that, in turn, affect the adoption of CSA technologies and vice versa. It was noted by Holden (2018) that subsidies tend to lead to complex substitutions, suggesting crowding in/out investments in other programs. Given limited resources, there is a need for wise investment in FISP without suffocating other programs.

Conditioning access to input subsidies on adoption of soil fertility management and climate-smart technologies (Snapp et al. 2014) is potentially a remedy to solve the problem of resource substitution and enhance the positive impact of the program. This chapter discusses the impact of subsidized inorganic fertilizer and integration of inorganic and organic fertilizer on maize productivity. The chapter further discusses suggestions on enhancing the impact of FISP by targeting the application of subsidized inorganic fertilizer on soils where CSA technologies have been used.

This review is necessary for Malawi as FISP remains a priority policy tool for the government (Snapp et al. 2014), at least in the foreseeable future. The results are also of policy relevance beyond Malawi given the global popularity of the current Malawi FISP. Some countries in SSA have used (Messina et al. 2017) or may use the Malawi FISP model to introduce their own farm input subsidies, and the conclusions of this chapter may be applicable in their settings.

3 Background and Implementation of the Farm Input Subsidy Program Vis-à-Vis Climate-Smart Agriculture Technologies

Farm input subsidies are government programs through which farmers receive farm inputs at below-market prices (Jayne et al. 2018). Smallholder farmers, particularly in SSA, face the low-input/low-output poverty trap exacerbated by market failures (Dorward et al. 2008). Input subsidies break this trap by reducing input cost and overcome market failures. The first set of input subsidies were introduced in the late 1960s in most African countries including Malawi (Chirwa and Dorward 2013; Wiggins and Brooks 2010). Such subsidies were implemented as *across-the-board* price subsidies and were universal, accessed by all producers.

In Malawi, the government funded a universal fertilizer subsidy in addition to providing cheap credit for smallholders through the state-run Agricultural Development and Marketing Corporation (ADMARC) and Smallholder Agricultural Credit Administration. Such policy instruments were responsible for the national food self-sufficiency in non-drought years, until early 1990s when the economic policy regimes changed and the country adopted economic liberalization, and farm input subsidies were removed. Agricultural markets were liberalized, the smallholder agricultural credit system collapsed, and the local currency was sharply devalued (Chinsinga and Poulton 2014). Farm inputs eventually became unaffordable to most smallholder farmers, and this resulted in dismal maize production. Exacerbated by an influx of Mozambican refugees and regular droughts, the country shifted from national food self-sufficiency to food deficit and reliance on food imports (Chinsinga 2008; Chinsinga and Poulton 2014).

In response, the government introduced the Starter Pack Scheme (SPS) in 1998–1999, which was heavily supported by the UK's Department for International Development (DFID) (Chirwa 2010). The SPS consisted of free inputs containing

0.1 ha worth of fertilizer and improved maize and legume seeds and covered all rural farming families, estimated at 2.86 million. The aim of the program was to improve food security and substitute maize imports (Crawford et al. 2006). The scheme was repeated in 1999–2000, but for the purpose of sustainability and exit strategy, it was scaled down to a Targeted Input Program (TIP) from 2000 to 2001 covering only half of the rural farming families. In 2001–2002, the country faced a severe food crisis affecting 3.2 million people and partly blamed it on the downscaling of TIP. The program was then implemented as an extended TIP (ETIP) from 2002 to 2003, reaching 2.8 million farmers (Pauw and Thurlow 2014). The ETIP phased out in 2004–2005 following the withdrawal of DFID's financial support (Chinsinga 2008).

In 2004–2005, about 4 million Malawians were affected by a severe hunger crisis as a result of poor maize harvests (Denning et al. 2009). The government's response was the reintroduction of fertilizer input subsidies, named Agricultural Input Subsidy Program (AISP), which was later renamed Farm Input Subsidy Program (FISP). The aim was to promote access to and use of inorganic fertilizer primarily for maize production in order to achieve sustainable increase in agricultural productivity and food security (Dorward et al. 2008). The program was to target resource-constrained but productive smallholders. It was intended to provide fertilizer, not as a safety net, but to encourage people who have land and human resource to make effective use of them: something they would otherwise not afford at commercial prices (Chinsinga 2008).

The administration of the program is via coupon vouchers given to targeted beneficiaries (Dorward and Chirwa 2011). Farmers are officially entitled to two coupons of 50-kg bags of inorganic fertilizer, although practically beneficiaries receive less than the intended 100 kg (Arndt et al. 2015). Other inputs have also been included in some years such as tobacco fertilizer, improved maize, legume, and cotton seeds, storage pesticides for maize, and cotton chemicals. In the first year of implementation in 2005–2006, approximately 166,000 Mg of inorganic fertilizer was subsidized (Fig. 2). The highest subsidized inorganic fertilizer subsidy was reported in 2007–2008 of about 216,000 Mg, but the quantity has been decreasing since then.

Farm input subsidies have effectively increased access to and use of inorganic fertilizer (Carr 2014; Chibwana et al. 2014; Karamba and Winters 2015; Lunduka et al. 2014), thereby enhancing nitrogen uptake. The annual fertilizer application was estimated at 26 kg/ha, approximately 8.6 kg nitrogen (N) in the early 1990s (Drechsel et al. 2001; Stoorvogel et al. 1993) after the removal of input subsides. Crawford et al. (2006) reported just over 25 kg/ha between 1996 and 2002, while the World Bank (2004) estimated a nutrient uptake of 15.9 kg/ha. Following the introduction of FISP in 2005–2006, Sheahan and Barrett (2017) reported an increase in annual inorganic fertilizer nitrogen consumption to 23.1–53.1 kg/ha, with a total Nitrogen, Phosphorous and Potassium (NPK) consumption of 31.9–56.3 kg/ha.

In addition to FISP efforts, several interventions are also underway in the country to promote the use of CSA technologies among smallholder farmers. The government through the Agricultural Sector-Wide Approach (2011–2015) program (ASWAp)

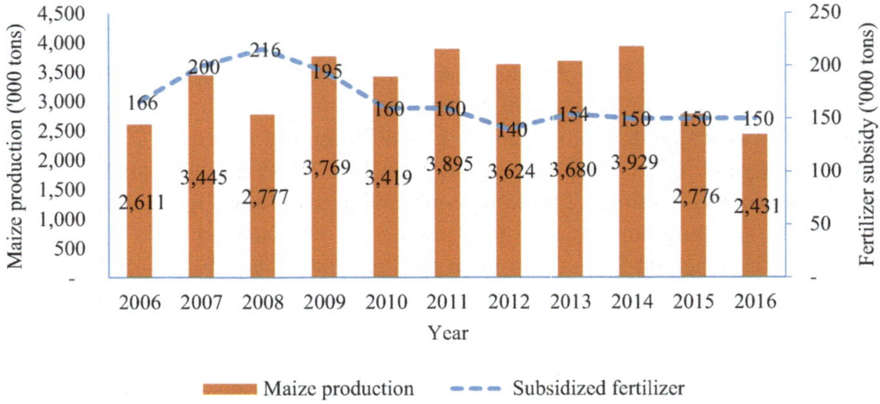

Fig. 2 Maize production and fertilizer subsidy over time in Malawi. (*Source*: Malawi's Ministry of Agriculture, Irrigation and Water Development)

prioritized soil and water conservation technologies such as contour and box ridges, organic manure, minimum tillage, and agroforestry (Government of Malawi 2011). These CSA technologies were aimed at building soil fertility, preventing soil erosion, and conserving rainwater. The ASWAp also placed emphasis on the Green Belt Initiative to establish rainwater harvesting systems to increase the level of irrigation farming. Other CSA technologies the government is promoting are conservation agriculture (CA), drought-tolerant crops, and precision agriculture (Government of Malawi 2015; Kaczan et al. 2013). The CSA technologies are also emphasized in the National Agriculture Policy, and they include: CA, agroforestry, improved seeds, irrigation, organic and inorganic fertilizer, and other integrated soil fertility management (ISFM) technologies (Government of Malawi 2016b). The focus of this chapter is ISFM, particularly the integration of inorganic and organic fertilizer, as a CSA technology. The chapter has used this narrow definition of CSA in order to have a focus and make a critical comparative review of the impacts of inorganic fertilizer accessed through FISP and the impact of CSA technologies on maize productivity.

4 Maize Productivity

4.1 Impacts of Subsidized Inorganic Fertilizer

Government estimates show a remarkable increase in maize production over the period of FISP implementation (Chirwa and Dorward 2013; Government of Malawi 2016a; Wiggins and Brooks 2010). In 2005–2006, which is the first year of FISP implementation, maize production more than doubled the 2004–2005 harvests of 1.26 million Mg to 2.61 million Mg. In the second year (2006–2007), official estimates were 3.44 million Mg, of which 300,000–600,000 Mg was an incremental production due to the subsidy program (Arndt et al. 2015; Dorward and Chirwa 2011). Overall maize yield increased by 2.6 times between 2005 and 2007 because of subsidies (Sanchez 2015). Denning et al. (2009) further reported that Malawi moved from 43% national food deficit in 2005 to 53% food surplus in 2007. The positive impact continued in later years as Dorward and Chirwa (2011) reported an increase in FISP-related maize production from 406,348 Mg in 2006 to 968,900 Mg in 2008–2009. Production estimates for years 2009–2010 to 2015–2016 (Fig. 2) were well above 3 million Mg, except for the drought years of 2014–2015 and 2015–2016.

At the household level, some studies support the macro-level statistics of positive impact of subsidized inorganic fertilizer on maize production and yields, while some authors contradict it. Holden and Lunduka (2010) used three waves of panel data from six districts in Central and Southern Malawi and reported a positive and significant impact of FISP on maize production and productivity. Using 6 years of panel data, Ricker-Gilbert and Jayne (2011) found that an additional kilogram of subsidized fertilizer in the current year and past three seasons increases maize yield by 1.82 and 3.16 kg, respectively. With reference to the standard package of FISP of 100 kg fertilizer, these figures suggest an incremental maize production due to FISP of 182 and 316 kg. The positive impact of past receipt of subsidized fertilizer could be because subsidized fertilizer gives farmers a learning experience on the importance of inorganic fertilizer, so that they may continue using the input in subsequent years. Another possible reason is that FISP generates enough income that enables farmers to invest in other yield-enhancing technologies (Lunduka et al. 2013).

In another study, Ricker-Gilbert and Jayne (2012) found that an additional kilogram of subsidized fertilizer contributes 2.43 kg of maize for an average farmer, and 2.61 and 0.75 kg for households at the 75% and 10% percentile of maize production, respectively. This suggests that better-off households benefit more from FISP as opposed to poor households. The possible explanation is that fertilizer response is very low on plots for poor households than for rich households. This result, however, raises questions on the program's ability to generate substantial gains in maize production among the poor. This further threatens the capacity of FISP to reduce the productivity and poverty gap between the poor and the rich. Another possible reason is that poor households are less likely (than better- off households) to receive subsidy coupons due to targeting errors and corruption in the program (Holden and Lunduka 2013).

Evaluating the 2012–2013 FISP, Dorward et al. (2013) found that access to a full subsidy pack (two 50-kg bags) gives an incremental benefit of at least 500 kg of maize, while a 50-kg fertilizer access increases maize production by 200–400 kg. In Kasungu and Machinga districts, Chibwana et al. (2014) reported a maize yield increase of 477 kg/ha by seed and fertilizer subsidy beneficiaries, while it was 249 kg/ha for fertilizer subsidy beneficiaries only. Using data from the Third Integrated Household Survey (IHS3), Karamba and Winters (2015) reported a modest increase in gross output value of 17%/ha due to FISP among male and female farmers. The results, however, failed to establish the significant impact of FISP on reducing the gender productivity gap between female and male farmers. The modest impact is possibly due to displacement of commercial inputs, input diversion, and receipt of less than the required amount.

Using the computable general equilibrium (CGE) model calibrated to empirical evidence from household-level evaluations, Arndt et al. (2015) reported the Malawi FISP as pro-poor. The findings, however, contradicted Jayne and Rashid (2013) who, using partial equilibrium assessments, reported FISP as low potential and deeply grounded in political interests. Nonetheless, Arndt et al. (2015) observed that FISP generates double dividends through higher and drought-resilient crop yields. Using the integrated household survey data for 2010 and 2013, Sibande et al. (2017) reported that access to FISP increases maize productivity, thereby leaving farmers with a surplus for sale. However, the maize supply quantities are too small to affect maize prices. Again, such quantities may not generate enough income to allow participating farmers self-finance input purchase in subsequent years, casting doubts on the program's long-term impact and sustainability.

Ricker-Gilbert and Jayne (2017) used four waves of panel data to estimate enduring effects of fertilizer subsidies on maize production and reported a modest positive impact of fertilizer subsidies in a given year. One-kilogram increase in subsidized fertilizer increases maize yield by 1.00–1.46 kg in year of access after accounting for contemporaneous crowding out of commercial fertilizer, representing 100–150 kg at standard FISP package. However, unlike their previous results of 2011, the 2017 findings by Ricker-Gilbert and Jayne did not find evidence of the impact of lagged access to fertilizer subsidies on maize yield. This suggests that the Malawi fertilizer subsidy program has limited long-term impacts on maize production.

Messina et al. (2017) used a novel approach to assess yield response to fertilizer through remote sensing to identify the spatiotemporal performance of agricultural fields and reported a national decline in the annual maize productivity trend. The authors (ibid) suggest that the positive evidence reported by some authors could be due to data error on maize production estimates and maize area cultivated. Farm-level data often suffer from respondent measurement errors (Jayne and Rashid 2013). An evaluation of the 2016–2017 subsidy by the Centre for Development Management (2017) also shows that production and productivity has stagnated in recent years. While there was a substantial increase in maize productivity from 1.5 Mg/ha to 2.4 Mg/ha in the early years of the program, from 2013 to 2014 productivity has declined to 1.9 Mg/ha, suggesting low maize yield–fertilizer response rate and subsequent reduction in the returns from FISP investments.

4.2 Marginal Maize Productivity Impact of Subsidized Inorganic Fertilizer

Finding positive impact of FISP on beneficiaries vis-à-vis non-beneficiaries is not enough as the success of the program depends on marginal return with respect to what is agronomically expected (Pauw and Thurlow 2014). This is related to nitrogen use efficiency (NUE) (Vanlauwe et al. 2001a) defined in this chapter as *amount of additional grain harvested per kilogram of nitrogen applied to the grain crop* (Snapp et al. 2014: P1). Based on agronomic evidence, NUE is 14–50 kg/kgN on on-station and on-farm trial plots, but it is —two to three times less on farmer-managed plots (Snapp et al. 2014). On farmer-managed plots, Snapp et al. (2014) reported NUE of 17.7 kg/kgN, Pauw and Thurlow (2014) found an average of 16.8 kg/kgN, while Carr (2014) reported 15.2 kg/kgN. Variety disaggregation shows 10–12 kg/kgN for local maize, 15 kg/kgN for open pollinated varieties, and 18–20 kg/kgN for hybrid maize varieties (Dorward et al. 2008).

The Malawi FISP contains 50 kg of NPK (with 23% nitrogen, i.e., 11.5 kgN) and 50 kg urea (with 46% nitrogen, i.e., 23 kgN) giving an average of 34.5 kgN (Pauw and Thurlow 2014). Using the average benchmark of 16.8 kg/kgN, a standard fertilizer subsidy package is expected to increase maize yield by 580 kg. However, evidence from household surveys shows less maize–fertilizer response among FISP beneficiaries, as shown in Table 1. Ricker-Gilbert and Jayne (2011, 2012) and Jayne and Rashid (2013) reported an overall low NUE of 5.5–9.6 kg/kgN for beneficiaries of subsidized inorganic fertilizer, with the poorest households reporting 2.4 kg/kgN. Dorward et al. (2013) found a relatively higher NUE of 15.2 kg/kgN, while Snapp et al. (2014), Pauw and Thurlow (2014), Chibwana et al. (2014), and Arndt et al. (2015) found maize–fertilizer response of 7–14.5 kg/kgN. Ricker-Gilbert and Jayne (2017) reported the smallest NUE of 3–4.4 kg/kgN. These findings suggest huge inefficient use of subsidized inorganic fertilizer. Jayne and Rashid (2013) indicated that with these inefficiencies, the incremental impact of subsidized inorganic fertilizer is insignificant and unprofitable.

In view of these inefficiencies, one would be tempted to dismiss FISP as an economic failure (Pauw and Thurlow 2014). It may therefore make no economic sense to continue investing in a program whose marginal benefit is negligible. However, history tells otherwise. Apart from drought, removal of fertilizer subsidies has been one major reason for recurrent food deficits in the country. In early to mid-1990s, food deficits were rampant after the removal of fertilizer subsidies (Chinsinga 2008;

Table 1 Nitrogen use efficiency among the Farm Input Subsidy Program beneficiaries in Malawi

Authors	Year	Nitrogen use efficiency (kg/kgN)
Ricker-Gilbert and Jayne	2011	5.5 for current year and 9.6 for past three seasons
Ricker-Gilbert and Jayne	2012	7.4 for average farmers, 7.9 for rich farmers, and 2.3 for poor farmers
Dorward et al.	2013	15.2
Jayne and Rashid	2013	8.1
Snapp et al.	2014	7.0–14.0
Pauw and Thurlow	2014	9.0–12.0
Chibwana et al.	2014	7.5–14.5
Arndt et al.	2015	11.8
Messina et al.	2017	10.0
Ricker-Gilbert and Jayne	2017	3.0–4.4

Chinsinga and Poulton 2014). In the early 2000s, the scaling down of TIP contributed to food deficits (Wiggins and Brooks 2010). Food deficits in the early 2010s have also been partly blamed on the downscaling of FISP (Pauw and Thurlow 2014). There could be several reasons why maize–fertilizer response or nitrogen use efficiency is low for beneficiaries of FISP such that addressing those bottlenecks would be necessary for the success of the program. Jayne and Rashid (2013), Snapp et al. (2014), Messina et al. (2017), and Ricker-Gilbert and Jayne (2017) show that poor soil fertility, loss of soil organic matter, water stress, and poor agronomic practices are among key bottlenecks. Furthermore, poor timing of input delivery, input diversion, recipient of less than the required amount, and beneficiary targeting errors have also limited efficient use of farm input subsidies (Centre for Development Management 2017; Holden and Lunduka 2013; Lunduka et al. 2014; Lunduka et al. 2013).

Poor Soil Fertility

The soils are highly degraded in Malawi due to nutrient mining, resulting in the depletion of nitrogen and other essential nutrients. Simply increasing nutrient application through subsidized inorganic fertilizer without first addressing soil degradation issues does not solve soil fertility problems (Ngwira et al. 2013b; Tchale and Sauer 2007). The current implementation strategy of FISP where soil conditions are grossly ignored will keep on decreasing its marginal return. Although it is expected that the program would have residual effect in nutrient buildup, especially phosphorous, this expectation is unlikely to materialize because of the heavy nutrient mining (Branca et al. 2011). Snapp et al. (2014) therefore recommended targeting inorganic fertilizer application on plots with sufficient soil quality or encouraging farmers to first adopt soil fertility management technologies. The discussion in Sect. 4.3 provides evidence that maize–fertilizer response is high on good-quality soils. Mueller et al. (2012) further indicated that maize would increase by 50% if soil nutrient deficiencies are addressed. Thus, if FISP targets the soil, the impact is likely to be consistently higher than the current targeting of the farmers who, in most cases, cultivate degraded soils.

Loss of Soil Organic Matter

Related to nutrient depletion is the reduction of soil organic matter (SOM), and hence soil organic carbon (SOC) (Chilimba et al. 2005; Snapp et al. 2014). Soils in Malawi are highly depleted of SOM (though not well quantified for specific sites) to the extent that they cannot effectively support crop productivity (Messina et al. 2017). The responsiveness of maize yield to inorganic fertilizer on farmer-managed plots has decreased by half of what is expected and is only 20% of the agronomic average because of the decline in SOM (Snapp et al. 2014). Unfortunately, sole reliance on subsidized inorganic fertilizer does not solve the problem of SOM (Mafongoya et al. 2006; Ngwira et al. 2013b). Inorganic fertilizer has high nutrient content with rapid release into the soil, but the retention rate is low when soils are deprived of organic matter, and its use efficiency is reduced. The loss of SOM in maize-based farming systems with continuous cropping therefore needs urgent attention if interventions such as FISP are to be effective. Building SOM requires application of organic amendments.

Water Stress

Maize production in many countries in SSA, including Malawi, is rain-fed in a single rainy season that is characterized by frequent dry spells resulting in water stress (Denning et al. 2009). This significantly affects maize–fertilizer response. Evidence has shown that when a drought occurs, even fertilizer subsidy does not save millions of Malawians from the hunger crisis that results from the poor harvests. For example, in 2001–2002, there was a hunger crisis despite the implementation of TIP. Although some critics blamed the scaling down of TIP for the hunger (Wiggins and Brooks 2010), the associated drought significantly contributed to the crisis. In 2004–2005 despite the ETIP, about 4 million Malawians were affected by hunger, mainly because of drought (Chabvunguma and Munthali 2008; Chinsinga 2008). In 2014–2015 and 2015–2016, maize production was respectively estimated at 2.8 million Mg and 2.4 million Mg down from 3.9 million Mg in 2013–2014 despite 150,000 Mg of inorganic fertilizer being subsidized in all the three consecutive years (Fig. 2). The difference in maize production was mainly due to regular dry spells in 2014–2015 and 2015–2016 cropping seasons. The Famine Early Warning Systems Network Malawi (2015) also reported that maize harvests in 2014–2015 were 25–30% lower than the previous five-year average, and a deficit of 500,000 Mg was recorded mainly due to late-season droughts.

This means that access to inorganic fertilizer through FISP has insignificant impact on maize production and productivity in the presence of drought, because the majority of the beneficiaries lack alternative technologies to sustain production (Giller et al. 1997 in Chilimba et al., 2005). Farmers also tend to reduce the use of productivity-enhancing inputs such as inorganic fertilizer in response to past adverse conditions (Sesmero et al. 2017). Controlling for water stress is fundamental to achieving higher fertilizer response rate. While irrigation is rarely used (Barrett et al. 2017) due to high investment and maintenance costs, integrating inorganic and organic fertilizer would enhance water and inorganic fertilizer use efficiency, thereby increasing the impact of subsidized inorganic fertilizer. Addressing both nutrient and water deficiencies can increase maize yield by 75% (Mueller et al. 2012).

4.3 Impact of Integrating Inorganic and Organic Fertilizers

Having identified key factors associated with poor maize–fertilizer response, studies have recommended integration of inorganic and organic fertilizer as a remedy (e.g., Chilimba et al. 2005; Mafongoya et al. 2006; Tchale and Sauer 2007).

Integrating inorganic and organic fertilizer can generate multiple benefits for the production of maize, and has the potential to increase NUE (Holden 2018). The chapter discusses evidence of the impact of integrating inorganic and organic fertilizer on maize productivity using on-farm and household survey studies. On-farm studies have compared maize yield on control plots where neither organic nor inorganic fertilizer was used with plots where only inorganic or organic fertilizer was used and plots where organic and inorganic fertilizers were combined.

Compared with control plots, Vanlauwe et al. (2001b) reported that maize yield was 750, 1000, and 2000 kg/ha higher in plots with inorganic fertilizer, organic fertilizer and combined inorganic and organic fertilizer, respectively. These findings were consistent with Sakala et al. (2003) who reported highest maize productivity impact on plots where green manure was integrated with inorganic fertilizer, compared with plots with only green manure or with continuous cropping. Fandika et al. (2007) reported similar findings where optimum maize yield was found in fields where inorganic nitrogen fertilizer (urea) was mixed with either compost manure or farmyard manure, unlike fields with sole application of inorganic nitrogen fertilizer or manure. A review by Munthali (2007) showed that sole application of inorganic fertilizer and organic fertilizer yielded 6.5 Mg/ha and 6.9 Mg/ha, respectively. On the other hand, integration of organic and inorganic fertilizer yielded a maximum of 7.5 Mg/ha.

Using household survey data, Tchale and Sauer (2007) found that use of ISFM where organic and inorganic fertilizer were combined had higher levels of technical efficiency (91%) than when inorganic fertilizer was used in isolation (79%). There was a loss of yield of 143 kg/ha in plots with chemical fertilizer only due to technical inefficiency, while there was only 58 kg/ha yield loss for households that used ISFM. In a related on-farm study, Vanlauwe et al. (2011) reported nitrogen agronomic use efficiency of maize between 17 and 26 kg/kgN depending on maize variety. When fertilizer was mixed with organic manure or compost, the result of NUE was 36 kg/kgN.

A meta-analysis by Chivenge et al. (2011) showed that application of organic resources, inorganic nitrogen fertilizer, and integration of organic and inorganic fertilizer increased maize yield by 60%, 84%, and 114%, respectively. The combined effect was due to extra nitrogen available from inorganic fertilizer with organic resources inducing nitrogen use efficiency and water retention (Vanlauwe et al. 2001a). In contrast, an on-farm study by Ngwira et al. (2013b) showed that maize yield was higher with sole fertilizer application followed by compost plus fertilizer and compost only relative to no fertilizer no compost. However, the impact was highest in areas with a long history of compost use, good rainfall, and good soils. This is because good soils mean that all essential nutrients are available, including soil organic matter that increases nitrogen use efficiency. With poor soils and low rainfall, the soils require compost manure to increase nitrogen use efficiency, thereby increasing maize yield. In a related study, Kaczan et al. (2013) observed that application of crop residues increased maize yield response to inorganic fertilizer by 86–216%.

5 Synthesis of the Findings and Way Forward

The discussion in Sect. 4 provides evidence that the impact of farm input subsidies on maize productivity is modest. While aggregate national maize production appears to have improved in non-drought years due to FISP implementation, marginal maize productivity impact of FISP remains below the agronomic average. Furthermore, the long-term impact of the program appears uncertain. The main reason is that the soils are highly degraded of vital nutrients and organic matter. The situation is worsened by regular weather shocks, especially droughts and prolonged mid-season dry spells, poor timing of subsidized input delivery, beneficiaries receiving less than the required amount of inorganic fertilizer, and targeting errors of the beneficiaries. The combined effect of poor soil fertility, low soil organic matter, and water stress due to droughts caused low maize–fertilizer response. The nutrient and water retention capacity are low, resulting in poor maize productivity impact. Increasing nitrogen application through input subsidies without first addressing soil condition does not solve the problem of poor maize productivity in such instances because the soils are not responsive.

On the other hand, maize productivity is high and stable with enduring effects on experimental plots where inorganic and organic fertilizer are integrated. These technologies enhance nutrient uptake, nutrient maintenance, and drought resilience of the maize crop. Integration of inorganic and organic fertilizer has been achieved on experimental plots, but achieving the same on farmer-managed plots is still a challenge. There are several implementation hiccups that need urgent attention through policy and integration process.

5.1 Policy Synchronization

The National Agriculture Policy (NAP) (Government of Malawi 2016b) recognizes the importance of integrating organic and inorganic fertilizer through ISFM as a CSA technology. However, integration on farmer-managed plots remains low because policy instruments that enhance access to and use of CSA technologies are not well synchronized. FISP as a key policy instrument has successfully increased access to and use of inorganic fertilizer (Karamba and Winters 2015; Lunduka et al. 2014), but its implementation is not well connected to the promotion of adoption of CSA technologies. This affects the integration process of organic and inorganic fertilizer. On the other hand, the government promotes CSA through investment in agricultural research and agricultural extension services that are often deprived of adequate funding. The two agriculture policy strategies—FISP and agricultural research and extension services—appear to compete for the meager government resources, thereby affecting operations of one or both. The huge budget requirement for FISP (Branca et al. 2011), for example, makes it difficult to increase budget allocation to agricultural research and extension (Carr 2014). Although recent reforms have resulted in reduction in FISP budget (Centre for Development Management 2017), the savings have not been reallocated to other agricultural programs such as extension services because the government has pressing needs in sectors other than agriculture. Making adoption of CSA technologies a prerequisite to accessing FISP would harmonize the operations of FISP and agricultural research and extension, thereby enhancing integration.

Another challenge to the integration of organic and inorganic fertilizer is that the problem of declining soil fertility receives less attention than the problem of low fertilizer use (Jayne and Rashid 2013). There has been great emphasis on low fertilizer use as a reason for low maize production and hence implementation of FISP, but little emphasis has been put on declining soil fertility. Policy discussions have been dominated by the problem of market failures of inputs and credit but how soil fertility is declining due to high population densities, continuous cropping, and mono-cropping is given little attention. The issues of declining soil fertility and low fertilizer use need equal policy attention.

Furthermore, while subsidized inorganic fertilizer has immediate production benefits, technologies that address declining soil fertility concerns such as organic fertilizer have lagged but long-term benefits (Snapp et al. 1998; Thierfelder et al. 2017). Unfortunately, many smallholder farmers are impatient as they give more weight to immediate gains over future benefits. As a result, they tend to dis-adopt technologies with delayed benefits after one or two seasons. This inconsistent use affects the impact and threatens potential integration of inorganic and organic technologies. Overcoming the risk of delayed production benefits can improve by consistent use of these technologies, thereby achieving the long-term integration process of organic and inorganic technologies. Following the emphasis in the NAP on the need to increase efficient and timely access to organic and inorganic fertilizer (Government of Malawi 2016b), policy tools should reflect the same to allow farmers to adopt integration.

One option as proposed by Snapp et al. (2014) is that FISP should be redesigned in such a way that it is conditioned on the adoption of soil fertility-enhancing and climate-smart technologies. The authors propose *conditional universal subsidy* where farmers will only access subsidized inorganic fertilizer upon the adoption of such complementary technologies. Conditioning access to subsidized inorganic fertilizer on adoption of organic fertilizer could efficiently increase the adoption of both. Agricultural extension officers would concentrate on core services of training and advising farmers on appropriate agronomic activities such as use of organic fertilizer and identify FISP beneficiaries from adopters of such CSA technologies. While these technologies have delayed production benefits, the potential access to input subsidies after adoption could be an incentive for the farmers to use the technologies consistently. The result will be high nutrient buildup, an increase in soil organic matter, as well as high water and nutrient retention, thereby increasing maize–fertilizer response. Potentially, this can provide the government with a sustainable exit strategy from FISP without compromising on sustainable maize production and productivity.

5.2 Integration Process

Sustainable and effective integration would require development of appropriate methods of integration. Certainly different organic sources for organic fertilizer contain different quantities of nutrients (Chilimba et al. 2005; Ngwira et al. 2013b). The appropriate combination rates of inorganic fertilizer with organic inputs such as compost manure, farmyard manure, green manure, agroforestry prunings, or other organic sources are not known. Smallholder farmers do not have the required technical know-how to effectively and efficiently adopt the integration of inorganic and organic fertilizer. Snapp et al. (2014) therefore recommended investments in agronomy to develop site-specific recommended rates for integration. This should be built on the site-specific inorganic fertilizer application rates developed by the Government of Malawi (2012). Providing farmers with proper knowledge of preparation and integration is paramount to having a good quality of integrated inorganic and organic inputs.

The next phase is to invest in agricultural extension services to train farmers in organic fertilizer preparation (Ngwira et al. 2013b). In addition, farmers should be made aware of potential short-term, medium-term, and long-term benefits of these technologies including the integration of organic and inorganic fertilizer (Munthali (2007). Such vital information should accompany the implementation of FISP to allow farmers to adopt relevant agronomic technologies besides adopting inorganic fertilizer (Munthali 2007; Snapp et al. 2014). Agricultural extension services should ensure that farmers appropriately follow agronomic recommendations on applying the integration of organic and inorganic fertilizer.

The implementation strategy of FISP should abandon the current farmer-based targeting system and provide the subsidies to all farmers who adopt CSA. That is, targeting the soil and not the farmer. This targeting will ensure that farmers adopt CSA first before accessing subsidized inorganic fertilizer. This will allow targeting of inorganic fertilizer application on soils that are of good quality to achieve high nitrogen and water use efficiency and subsequent increase in maize production and productivity. Due to budget constraints, the program should maintain the target of 0.4 ha of land for maize production. Thus, identification of FISP beneficiaries should follow those farmers that have appropriately adopted CSA technologies for a minimum of 0.4 ha of land. Agricultural extension officers should closely follow up with beneficiaries to allow proper exit from FISP.

6 Conclusion

The Farm Input Subsidy Program remains the strategic agriculture policy tool for Malawi in the near future. There is no doubt that the program has increased access to and use of inorganic fertilizer with modest impact on maize productivity. The current implementation strategy where soil fertility issues are grossly ignored and beneficiaries continue cultivating on degraded soils reduces enduring effects of the program and locks farmers into low maize response rates to fertilizer use. The incremental impact of the program is low, and the situation is worsened because of frequent dry spells. Addressing soil fertility and drought-resilience issues through adoption of integration of inorganic and organic fertilizer is potentially a magic bullet to increase maize–fertilizer response and achieve sustainable maize productivity gains from FISP. Nonetheless, achieving integration on farmer-managed plots is unlikely because policy instruments promoting access to and use of inorganic and organic fertilizer are not firmly harmonized. The chapter recommends synchronization of policy tools by conditioning access to FISP on adoption of CSA technologies such as organic fertilizer. Strategically, the harmonization should involve investment in agronomic research and agricultural extension services and transformation of FISP implementation strategy. This approach can provide the Government of Malawi with an opportunity for a sustainable exit strategy from FISP. Although the evidence in this chapter can be applicable in other countries in SSA, a comprehensive review comparing findings from many countries could be more appropriate for more evidence.

Acknowledgements I would like to acknowledge the valuable comments and suggestions from Prof. Stein Holden and Ms. Sarah Tione of the Norwegian University of Life Sciences and Mr. Osborne Tsoka from Malawi's Ministry of Agriculture, Irrigation and Water Development. I also sincerely thank the anonymous referee for the valuable comments on an earlier draft. Last but not least, I acknowledge the support from the Norwegian Programme for Capacity Development in Higher Education and Research for Development (NORHED)-Capacity Building for Climate Smart Natural Resource Management and Policy (CLISNARP) project for supporting the production of this chapter.

References

Arndt C, Pauw K, Thurlow J (2015) The economy-wide impacts and risks of Malawi's farm input subsidy program. Am J Agric Econ 98:962–980

Barrett CB, Christiaensen L, Sheahan M, Shimeles A (2017) On the structural transformation of rural Africa. J Afr Econ 26:i11–i35

Branca G, McCarthy N, Lipper L, Jolejole MC (2011) Climate-smart agriculture: a synthesis of empirical evidence of food security and mitigation benefits from improved cropland management. In: *Mitigation of Climate Change in Agriculture*. Food and Agriculture Organisation, Rome

Carr S (1997) A green revolution frustrated: lessons from the Malawi experience. Afr Crop Sci J 5:93–98

Carr S (2014) The challenge of Africa s nitrogen drought: some indicators from the Malawian experience. International Food Policy Research Institute, Washington, DC

Centre for Development Management (2017) Evaluation of the 2016/2017 Farm Input Subsidy Programme (FISP). Ministry of Agriculture, Irrigation and Water Development, Lilongwe

Chabvunguma S, Munthali G (2008) Determination of maize planting dates using some meteorological factors-Case Study Chitipa. Proceedings of the 8th National Research Council of Malawi Conference, 11–13 2008

Chibwana C, Shively G, Fisher M, Jumbe C, Masters W (2014) Measuring the impacts of Malawi's farm input subsidy programme. Afr J Agric Resour Econ 9:132–147

Chilimba A, Shano B, Chigowo M, Komwa M (2005) Quality assessment of compost manure produced by smallholder farmers in Malawi. Lilongwe

Chinsinga B (2008) Reclaiming policy space: lessons from Malawi's 2005/2006 fertilizer subsidy programme future agricultures. Institute of Development Studies, Brighton

Chinsinga B, Poulton C (2014) Beyond technocratic debates: the significance and transience of political incentives in the Malawi farm input subsidy programme (FISP). Dev Policy Rev 32:s123

Chirwa T (2010) Program evaluation of agricultural input subsidies in Malawi using treatment effects: methods and practicability based on propensity scores. MPRA Paper No. 21236. Accessed September 2011

Chirwa E, Dorward A (2013) Agricultural input subsidies: the recent Malawi experience. Oxford University Press, Oxford

Chivenge P, Vanlauwe B, Six J (2011) Does the combined application of organic and mineral nutrient sources influence maize productivity? A meta-analysis. Plant Soil 342:1–30

Crawford EW, Jayne TS, Kelly VA (2006) Alternative approaches for promoting fertilizer use in Africa. Agriculture & Rural Development Department, World Bank, Washington, DC

Denning G, Kabambe P, Sanchez P, Malik A, Flor R, Harawa R, Nkhoma P, Zamba C, Banda C, Magombo C (2009) Input subsidies to improve smallholder maize productivity in Malawi: toward an African Green Revolution. PLoS Biol 7:0002–0010

Dorward A, Chirwa E (2011) The Malawi agricultural input subsidy programme: 2005/06 to 2008/09. Int J Agric Sustain 9:232–247

Dorward A, Chirwa E, Kelly VA, Jayne TS, Slater R, Boughton D (2008) Evaluation of the 2006/7 agricultural input subsidy programme, Malawi. Final Report. Ministry of Agriculture and Food Security, Lilongwe

Dorward A, Chirwa E, Matita M, Mhango W, Mvula P, Taylor EJ, Thorne K (2013) Evaluation of the 2012/13 farm input subsidy programme. Ministry of Agriculture and Food Security, Malawi

Drechsel P, Gyiele L, Kunze D, Cofie O (2001) Population density, soil nutrient depletion, and economic growth in sub-Saharan Africa. Ecol Econ 38:251–258

Famine Early Warning Systems Network Malawi (2015) Malawi Food Security Outlook July to December 2015. Famine Early Warning Systems Network Malawi

Fandika I, Kadyampakeni D, Bottomani C, Kakhiwa H (2007) Comparative response of varied irrigated maize to organic and inorganic fertilizer application. Phys Chem Earth Parts A/B/C 32:1107–1116

Giller KE, Cadisch G, Ehaliotis C, Adams E, Sakala WD, Mafongoya PL (1997) Building soil nitrogen capital in Africa. In: Buresh RJ, Sanchez PA, Calhoun F (eds) *Replenishing soil fertility in Africa*. SSSA Spec. Publ. 51. SSSA and ASA, Madison

Government of Malawi (2011) Malawi agricultural sector wide approach (ASWAp) 2011–2015. In: Ministry of Agriculture and Food Security (ed). Ministry of Agriculture and Food Security, Lilongwe

Government of Malawi (2012) Guide to agricultural production and natural resources management. Ministry of Agriculture and Food Security, Lilongwe

Government of Malawi 2015. Intended nationally determined contribution (INDC). Lilongwe

Government of Malawi (2016a) Annual Economic Report 2016. Ministry of Finance, Economic Planning and Development, Department of Economic Planning and Development, Lilongwe

Government of Malawi (2016b) National Agriculture Policy. Ministry of Agriculture, Irrigation and Water Development, Lilongwe

Hawkes C, Ruel MT (2008) From agriculture to nutrition: Pathways, synergies and outcomes. In: *Agricultural and Rural Development Notes*. The World Bank, Washington, DC. 20433

Holden ST (2018) Fertilizer and sustainable intensification in Africa. Centre for Land Tenure Studies, Norwegian University of Life Sciences, Ås

Holden S, Lunduka R (2010) Impacts of the fertilizer subsidy programme in Malawi: targeting, household perceptions and preferences. Norwegian University of Life Sciences, Ås
Holden S, Lunduka R (2012) Do fertilizer subsidies crowd out organic manures? The case of Malawi. Agric Econ 43:303–314
Holden ST, Lunduka RW (2013) Who benefit from malawi's targeted farm input subsidy program? Forum for Development Studies. Taylor & Francis, pp 1–25
Jayne T, Rashid S (2013) Input subsidy programs in sub-Saharan Africa: a synthesis of recent evidence. Agric Econ 44:547–562
Jayne TS, Sitko NJ, Mason NM, Skole D (2018) Input subsidy programs and climate smart agriculture: current realities and future potential. In: Climate Smart Agriculture. Springer, Cham
Kaczan D, Arslan A, Lipper L (2013) Climate-smart agriculture? A review of current practice of agroforestry and conservation agriculture in Malawi and Zambia. Food and Agriculture Organization of the United Nations, Rome
Karamba RW, Winters PC (2015) Gender and agricultural productivity: implications of the Farm Input Subsidy Program in Malawi. Agric Econ 46:357–374
Lunduka R, Ricker-Gilbert J, Fisher M (2013) What are the farm-level impacts of Malawi's farm input subsidy program? A critical review. Agric Econ 44:563–579
Lunduka R, Ricker-Gilbert J, Shively G, Jayne T (2014) Redefining the goals and objectives of the Farm Input Subsidy Program (FISP) in Malawi. *Policy Brief I*. Michigan State University, Department of Agricultural, Food, and Resource Economics, East Lansing
Mafongoya P, Bationo A, Kihara J, Waswa BS (2006) Appropriate technologies to replenish soil fertility in southern Africa. Nutr Cycl Agroecosyst 76:137–151
Messina JP, Peter BG, Snapp SS (2017) Re-evaluating the Malawian farm input subsidy programme. Nature Plants 3:17013
Mueller ND, Gerber JS, Johnston M, Ray DK, Ramankutty N, Foley JA (2012) Closing yield gaps through nutrient and water management. Nature 490:254–257
Munthali MW (2007) Integrated soil fertility management technologies: a counteract to existing milestone in obtaining achievable economical crop yields in cultivated lands of poor smallholder farmers in Malawi. In: Bationo A, Waswa B, Kihara J, Kimetu J (eds) Advances in integrated soil fertility management in sub-Saharan Africa: challenges and opportunities. Springer Netherlands, Dordrecht
Ngwira A, Thierfelder C, Eash N, Lambert DM (2013a) Risk and maize-based cropping systems for smallholder Malawi farmers using conservation agriculture technologies. Exp Agric 49:483–503
Ngwira AR, Nyirenda M, Taylor D (2013b) Toward sustainable agriculture: an evaluation of compost and inorganic fertilizer on soil nutrient status and productivity of three maize varieties across multiple sites in Malawi. Agroecol Sustain Food Syst 37:859–881
Pauw K, Thurlow J (2014) Malawi's farm input subsidy program: where do we go from here? International Food Policy Research Institute (IFPRI), Washington, DC
Ricker-Gilbert J, Jayne T (2011) What are the enduring effects of fertilizer subsidy programs on recipient farm households? Evidence from Malawi. Department of Agricultural, Food and Resource Economics Michigan State University, East Lansing
Ricker-Gilbert J, Jayne TS (2012) Do fertilizer subsidies boost staple crop production and reduce poverty across the distribution of smallholders in Africa? Quantile regression results from Malawi. International Association of Agricultural Economists (IAAE) Triennial Conference, 2012 Foz Do Iguaçu, Brazil, 18–24 Aug 2012
Ricker-Gilbert J, Jayne TS (2017) Estimating the enduring effects of fertiliser subsidies on commercial fertiliser demand and maize production: panel data evidence from Malawi. J Agric Econ 68:70–97
Sakala W, Kumwenda JD, Saka A (2003) The potential of green manures to increase soil fertility and maize yields in Malawi. Biol Agric Hortic 21:121–130
Sanchez PA (2015) En route to plentiful food production in Africa. Nature Plants 1:14014

Sesmero J, Ricker-Gilbert J, Cook A (2017) How do African farm households respond to changes in current and past weather patterns? A structural panel data analysis from Malawi. Am J Agric Econ 100:115–144

Sheahan M, Barrett CB (2017) Ten striking facts about agricultural input use in sub-Saharan Africa. Food Policy 67:12–25

Sibande L, Bailey A, Davidova S (2017) The impact of farm input subsidies on maize marketing in Malawi. Food Policy 69:190–206

Snapp S, Mafongoya P, Waddington S (1998) Organic matter technologies for integrated nutrient management in smallholder cropping systems of southern Africa. Agric Ecosyst Environ 71:185–200

Snapp S, Jayne TS, Mhango W, Benson T, Ricker-Gilbert J (2014) Maize yield response to nitrogen in Malawi's smallholder production systems. Malawi Strategy Support Program. International Food Policy Institute, Washington, DC

Stoorvogel J, Smaling EA, Janssen B (1993) Calculating soil nutrient balances in Africa at different scales. Nutr Cycl Agroecosyst 35:227–235

Tchale H, Sauer J (2007) The efficiency of maize farming in Malawi. A bootstrapped translog frontier. Cahiers d'Economie et de Sociologie Rurales 82:33–56

Thierfelder C, Chivenge P, Mupangwa W, Rosenstock TS, Lamanna C, Eyre JX (2017) How climate-smart is conservation agriculture (CA)?–its potential to deliver on adaptation, mitigation and productivity on smallholder farms in southern Africa. Food Sec 9:537–560

Vanlauwe B, Aihou K, Aman S, Iwuafor EN, Tossah B, Diels J, Sanginga N, Lyasse O, Merckx R, Deckers J (2001a) Maize yield as affected by organic inputs and urea in the West African moist savanna. Agron J 93:1191–1199

Vanlauwe B, Wendt J, Diels J (2001b) Combined application of organic matter and fertilizer. Sustain Soil Fertil West Africa 58:247–279

Vanlauwe B, Kihara J, Chivenge P, Pypers P, Coe R, Six J (2011) Agronomic use efficiency of N fertilizer in maize-based systems in sub-Saharan Africa within the context of integrated soil fertility management. Plant Soil 339:35–50

Weber VS, Melchinger AE, Magorokosho C, Makumbi D, Bänziger M, Atlin GN (2012) Efficiency of managed-stress screening of elite maize hybrids under drought and low nitrogen for yield under rainfed conditions in Southern Africa. Crop Sci 52:1011–1020

Wiggins S, Brooks J (2010) The use of input subsidies in developing countries. Global Forum on Agriculture, 29–30 2010

World Bank (2004) Malawi country economic memorandum: policies for accelerating growth. World Bank, Washington, DC

Digital Storytelling as an Agricultural Extension Communication Tool in Smallholder Farming and Fishing Communities in Malawi

Neil Gordon Davey and Michael Kirby Moulton

Abstract Digital storytelling can overcome many of the challenges faced when communicating agricultural extension messages. It can represent a focused farmer-to-farmer communication, document the impacts of projects, and offer data for analysis. Our method focuses on beneficiaries' perspectives and how new agricultural technologies impact their lives. Through workshops, 11 final-year agricultural extension communication students attending Lilongwe University of Agriculture and Natural Resources were trained on how to collaboratively gather images with beneficiaries and how to use these images to drive a personal narrative. Flexibility allowed for data to be gathered efficiently. An image/discussion dynamic led to information that would hopefully reflect beneficiaries' perspectives on specific projects. Images were first captioned to create photo-stories. Subsequently, recorded narratives were added to create what became known as "image-based digital stories." Fourteen image-based digital stories were produced between November 2017 and March 2018 covering two agricultural projects: "Reducing carbon footprint through breeding and feeding based technologies for improved productivity" in Dedza district and "Sustainable environment and enterprise development for climate change adaptation in fisheries" in Nkhotakota district. The stories reflect what beneficiaries perceive as the most significant drivers and impacts of the respective projects, for example, the financial and health benefits of improved productivity in dairy farming; school fees can be paid and children's health improves through consuming milk. Participants became empowered as part of the research and dissemination process; the methodology enabled us to reach and interact with the farmers and fishers of Dedza and Nkhotakota in a meaningful way. Moreover, students and faculty became motivated to further use and develop the methodology. The resulting data in the form of (a) photo-stories and (b) image-based digital stories can be used

N. G. Davey (✉)
Department of International Environment & Development Studies, Norwegian University of Life Sciences, Ås, Norway
e-mail: neildavey@clanphoto.com

M. K. Moulton
Learning Centre, Norwegian University of Life Sciences, Ås, Norway

B. R. Singh et al. (eds.), *Climate Impacts on Agricultural and Natural Resource Sustainability in Africa*, https://doi.org/10.1007/978-3-030-37537-9_32

to validate extension messages and assist with future policymaking. The methodology could also aid future project design as a useful tool within monitoring and evaluation and project management.

1 Introduction: Digital Storytelling in Agricultural Extension Communication

Digital storytelling (DS) is an innovative tool that can communicate agricultural extension messages and help address the commonly occurring gap between knowledge and action in research and education. Within appropriate contexts, DS can form the core of baseline studies, document the impacts of projects, and supplement projects' qualitative data. We utilised DS to highlight the impact, through the eyes of the beneficiaries, of two selected projects in Malawi: "Reducing carbon footprint through breeding and feeding based technologies for improved productivity" (REDCAP) at Dzaonewekha Milk Bulking Group in Dedza district – approximately an hour's drive south-east of the capital, Lilongwe; and "Sustainable environment and enterprise development for climate change adaptation in fisheries" (SEED-Fish) in Vinthenga village on the shores of Lake Malawi in Nkhotakota district – approximately a 3-hour drive north-east of Lilongwe. The two projects are part of the wider Capacity Building for Managing Climate Change in Malawi (CABMACC) programme which is funded by the Royal Kingdom of Norway.

Our method was developed to shift focus from the science of a project to the reality of a project for the beneficiary. It should be able to effectively communicate agricultural extension messages in a farmer-to-farmer capacity, overcome some of the challenges associated with such a task in rural Malawi, and provide a record of a project's impact for all stakeholders. Furthermore, from a capacity-building perspective, it combines classroom learning and field practice to highlight how DS as a potentially powerful new tool in agricultural communication can help develop a context-specific knowledge base. The resulting stories showcase small-scale farmers' and fishers' perspectives of the two projects and reveal how new agricultural technologies impact their lives. This chapter presents an overview of the context, methodology and results of DS as it was developed and used as part of agricultural extension communication and capacity building.

2 Image-Based Digital Storytelling: A Hybrid Methodology

Traditional storytelling as a presentation form can help increase access to information, generate, and share knowledge, especially in communities with lower education levels (Mkwizu 2016). Digital stories add value by being both auditive and visual. The narratives and images that are the foundation of digital stories stimulate our senses (Tønnessen 2012). This can evoke an emotional response that conventional approaches to agricultural extension communication often lack. In turn, these emotions can help shape our understanding of the world around us (Ribeiro 2017). Furthermore, knowledge comes from the process of creating, interpreting, and reflecting upon the stories and the underlying meanings they hold. This is true for both storyteller and audience (Jamissen and Moulton 2017); often what we see and hear can offer meaning instead of, or in addition to, conventional measurements.

The participatory and collaborative process of DS can strengthen relations between project partners, in this case, Norwegian University of Life Sciences (NMBU), students and faculty of Lilongwe University of Agriculture & Natural Resources (LUANAR), and subsequently, the extension services and project beneficiaries – the storytellers. The experience helps build trust and rapport between partners. Furthermore, it highlights how relatively simple technologies can be used to effectively educate and communicate in today's digital world. However, this can still prove challenging, for example, in areas that lack electricity. The methodology therefore needed to be flexible and adaptable, in the context of our project goals but within our means and resources. Nevertheless, as Ribeiro (2017) claims, even using the simplest of modern technologies can help increase digital literacy, knowledge, and information.

The method also needed to encourage dialogue and discussion with participants. This moves away from structured interviews, surveys, or questionnaires and digs deeper into the meanings attributed to the phenomena being explored. Shor and Friere (1987) claimed that dialogue evokes reflection, and the ensuing narratives express subjective realities. It was our task to plant a seed and allow the beneficiaries to become storytellers. It was also our task to present the stories from the storyteller's perspective (Alterio and McDrury 2003). Therefore, they could both share their experiences with their peers and provide new or previously overlooked insight into the subject matter, that is, project-specific goals. Alterio and McDrury (2003, p. 157) claim that insights gained through listening to and discussing stories can help "discover new ways to understand past actions or find appropriate professional responses to situations".

Photography has long been a useful tool in research situations, in particular social anthropology. Photographs can be used in different ways throughout the research process. They can raise questions and help develop insight into the subject matter (Collier and Collier 1986); they can be illustrative and informative; and they provide a lasting visual record that can be revisited and scrutinised, connecting the viewer to the data (Szto et al. 2005). Moreover, inserting photographs into a research situation can help inspire dialogue and discussion, induce responses from participants, and encourage reflexivity and participant subjectivity (Davey 2018). Through collaborative interpretation, they help attribute meaning to phenomena, offering questions about the nature of reality from different perspectives (Pink 2013). Thus, we anticipated that an image/discussion dynamic could elicit both the narratives we sought for the stories as well as the dialogue, reflection and understanding among peers and other target groups that we hoped would result from the finished stories.

Using photographs allows researchers and participants to interact differently than when using conventional discursive research methods. According to Clark (2012), when photographs are used to elicit narratives, participants become empowered as part of the research process rather than being objects of the research. Moreover, participants are actively encouraged to become researchers instead of research subjects. Furthermore, combining photographs with captions, narratives, or conversation excerpts adds an experiential and human element – something which scientific research often lacks. Using photographs can be an effective and efficient method to gather information when research presents constraints such as a lack of resources or time. Thus, our method uses theoretical strands from documentary photography, photovoice methodology, and DS, and we refer to it as image-based digital storytelling.

Photovoice was developed as a participatory action research (PAR) methodology. Developed by Caroline Wang and Mary Ann Burris it was originally known as photo novella (Wang and Burris 1994). Its theoretical underpinnings stem from Feminism, Freierian Education for Critical Consciousness, and participatory documentary photography (Latz 2017). Essentially, its premise is that participants express themselves through the photographs they take. As a PAR method, it empowers participants and, therefore, according to White et al. (1991), it avoids treating participants passively. Participants often caption their photographs or provide a narrative which can lead to digital stories. In contrast to conventional DS which often begins with a narrative and brings in the visual element later, Photovoice is image driven.

We wanted the method to stay within the PAR philosophy in so much as participants were as actively involved as possible. Nevertheless, we recognised potential challenges at the early stages of developing the methodology and which are in line with the limitations often encountered when undertaking this type of research. Within academia, these often include "demands on the researcher, externally defined and mandated projects, […], lack of resources, and lack of time" (Latz 2017, p 29). In the villages where we were to apply our methodology, language and lack of education and electricity can be overriding issues. Thus, due to contextual resource constraints, we opted to elicit a narrative from participants by using the image/discussion dynamic in the hope that it would overcome these potential obstacles.

The methodology involved active cooperation of 11 final-year Agriculture Development Communication (ADC) students at LUANAR. The students were trained on how to collaboratively gather images with project beneficiaries and how to use these images to drive a personal narrative. The hybrid method retained many of the important characteristics of both Photovoice and DS but was tailored to fit our context.

3 Image-Based Digital Storytelling: Workshops and Field Visits

The training and storytelling took place during two separate visits to Malawi – the first of which was in November 2017. This involved 3 days of classroom workshops with the 11 students and some faculty members, along with one field visit to

Dzaonewekha Milk Bulking Group in Dedza District. Our intention was to pair each student with a farmer and through a discussion identify what was important for the farmer within the context of the project, along with choosing appropriate pictures to take. The first 2 days of workshops included an introduction to the method, introducing visual literacy, getting to grips with the cameras, editing, and using text and captioning. After coming to grips with the cameras on day 1, students were assigned exercises including: "Self-portraits Without a Face" (take a photograph of something that represents or shows who they are) and "Treasure Hunt" (take five photographs in 30 min, each one representing a high angle, a low angle, a detail, a response to words and a feeling). Students were encouraged to be creative, think outside the box, and interpret as they wished. The photographs from "Treasure Hunt," along with the techniques and interpretations, were reviewed and discussed on day 2 to prepare students for their coming task in the field.

Day 2 also involved an introduction to editing. For example, students were divided into three groups with each given a pack of photographs. The groups then identified themes and connections between the photographs, arranging them in order to tell a story. This encouraged metaphorical thinking and looking beyond what the photograph was of, but rather what it was about or what it represented. This led to a captioning exercise, where students had to match captions, or short narratives, with photographs to help explain the context and meaning of a particular image, why it was taken and what it represents. This also provided an example of what we hoped for after the field visit – captioned photo-essays that acted as the raw material for image-based digital stories.

Prior to the field visit, we raised an open discussion with the team of students asking how they thought they could build trust with participants. They raised several good ethically driven points that could help gain narratives that reflected the true perspectives of the participants. For example, since the students were to cooperate with farmers in a rural setting, the female students decided to wear traditional wrapped skirts and no make-up. It was also pointed out that they should not brandish "fancy" labelled bottles of water. Furthermore, we all considered it important to respect the community and its customs and remain polite at all times. As final year communication students, the group was already familiar with the ethics that should take precedence during social research. Visual research often brings its own ethical challenges. Nevertheless, as Hitchcock, in conversation with Collier and Collier (1986) stated, respect is always paramount and especially so when working in indigenous or small communities. Moreover, Diener and Crandal's four areas of ethics which concern harm to participants, informed consent, privacy, and trust, are just as applicable to visual research as they are to conventional methods (Davey 2018).

Day 3 involved a field trip to Dzaonewekha Milk Bulking Group and dairy farmers in Dedza District who are part of the REDCAP project. One student was elected to introduce the group to the community. After formal introductions involving village elders and the project steering committee, each student was paired with a farmer. They then travelled to the respective farms or residences to collaborate with the farmers. For 2–3 hours they discussed with the farmers what to take pictures of and what the pictures' content meant to them in the context of their life, their livelihood, and their role in the project. Some students recorded farmers' narratives on their phones or used the video setting on the digital cameras we provided to record sound. Others did not have this opportunity, so they simply took hand-written notes to accompany the photographs. At the end of the field visit, students were happy with what they had gathered in a relatively short space of time. Farmers were happy to actively cooperate, but we respected that their time was valuable and we did not want to overstay our welcome.

The group spent the final day of the first week in the classroom at LUANAR's Bunda campus. The group worked individually, selecting a series of photographs from the previous day. Individual photographs were captioned accordingly with students working either from their recordings or from their notes. Captions were both in the native Chichewa language and translated to English. Photographs were printed using a Canon Selphy portable printer, accompanying captions were typed up and printed, and the 11 resulting photo-essays were displayed, discussed and refined. This provided a valuable learning experience for the students and the photo-essays provided the basis for the subsequent image-based digital stories.

We visited the second project site in February 2018. Three of the students accompanied us to the SEED-Fish project in Vinthenga village, Nkhotakota District on the shore of Lake Malawi. This field visit followed a similar format to the previous one, with the students being paired with fishers and fish processors. One student collaborated with a husband and wife, one with a local fisherman, and the other with a group of fish processors. However, after the initial pairing and photography, photographs were printed using the mobile printer before students returned to their respective partners to record the narratives. Not only did the printed photographs then truly act as an elicitation tool, but they were also left with the participants as a record of the day's activities.

4 Results: Stories from Dedza and Nkhotakota

From Dedza, there are specific themes that are common throughout the photo-stories. Firstly, they all include a portrait of the beneficiary with an accompanying caption (see Fig. 1 for an example). Prominent themes include family (Fig. 2), livestock (Fig. 3), financial benefits from dairy farming (Fig. 4), crops (Fig. 5), and agricultural technologies (Figs. 6 and 7). Other themes include the health benefits of milk (Fig. 8).

The photographs from Nkhotakota contain similar themes. However, one notable difference is that participants highlight some of the major challenges involved in their project. This includes the damage to and subsequent lack of maintenance of the solar dryer tent (Fig. 9).

Once all the raw material was gathered and the photo stories finalised, students and a facilitator collaboratively produced the digital stories. Several of them recorded third-person narratives to accompany their photo-essays. Others worked to structure the digital stories using their previously recorded first-person narratives.

Fig. 1 Samuel Kambuku is from Khombe village, Dedza. He has a wife, and three children who are at primary school. Mr Kambuku joined Dzaonewekha milk bulking group in 2007. (Photograph: Mary Mwanza)

Fig. 2 Mr Kambuku has also managed to pay school fees for his son who has finished secondary education. (Photograph: Marcelline Kaunda)

Fig. 3 To curb the low milk yield, Mr Liweta supplements the cows' diets by feeding them green forage, legume straws, and maize stalks. (Photograph: Lay-Vee Chiteya)

Fig. 4 Mrs Kennedy's milk sales to Dzaonewekha milk bulking group have helped her to buy a bicycle, build a house, and pay school fees for her children. She has also bought pigs, goats, and a chicken, along with three acres of farmland. (Photograph: Jannepher Godfrey)

Fig. 5 The project also encouraged me to plant multipurpose trees like Leucaena, Grillicidia, and Tephrosia vogelii to help maintain the soil as well as to provide protein to the cows. This has helped increase milk production and provide more revenue. (Photograph: Linda Mtambo)

Fig. 6 Training on dairy mash has enabled me to now milk eighteen litres per day. (Photograph: Linda Mtambo)

Fig. 7 Mr Liweta also uses maize stalks to make manure. He combines them with dung in a pit next to the kraal. Once the manure is ready, he takes the manure to the maize field in an ox-cart. (Photograph: Lay-Vee Chiteya)

Fig. 8 In my household we consume a lot of milk. My mother is 84 years old and is still strong and healthy. She helps with household chores like cooking, and cleans her room herself. (Photograph: Maureen Makoko)

Fig. 9 We were happy to receive this solar dryer as we knew it would help our business succeed. However, we now face challenges. After 2 months, the dryer started to tear at the sides, allowing in water and flies. (Photograph: Ethel Chavula)

All the narratives were recorded in Chichewa, but each photo had an accompanying English caption. The narratives were added and the digital stories were produced using Adobe Premiere Pro. However, professional editing software is not required to successfully produce digital stories and adequate free versions are widely available. The digital stories were finished in Norway but shared on Vimeo at several stages so that students could provide feedback and have input into the digital stories' finishing touches. WhatsApp was an effective way of communicating with the group during this process.

The result is a total of 14 image-based digital stories. Ten stories have a third-person narrative and four have a first-person narrative, of which one story has eight voices. Eight of the 14 stories include women and 5 are told solely by female farmers. Stories range from 1.37 min to 5.45 min in length but are typically between 2 and 5 min. The stories from Nkhotakota are typically longer and use more photographs than those from Dedza. Between 4 and 10 photographs are used per story. The stories reveal some significant drivers and impacts of the two projects from the perspectives of the Dedza dairy farmers and the Nkhotakota fishers and fish processors. The stories, we believe, are a good example of the product we hoped to create. Both in their digital and more tangible form of printed and captioned photographs, the stories can hopefully be utilised in many ways by various stakeholders.

5 Image-Based Digital Stories: Value, Validity, Capacity Building, Ownership and Empowerment

By using images as a foundation for the stories, the story makers created a lasting visual record of the two projects. Furthermore, when accompanied by the audio narratives or captions, the audience will not only forever be connected to the data contained within the photographs, for which Szto et al. (2005) claim visual research is useful, but the viewer will also gain insight into how the participants interpret that data in accordance with how the projects impact their lives. In their 1939 book, *An American Exodus*, photographer Dorothea Lange in collaboration with her husband, Paul S. Taylor printed excerpts from participants' conversations that took place at the time they took their photographs depicting rural poverty during the depression era (Newhall 1982). This avoided the need for a third-person narrative and added an experiential element to the photographs – a technique that documentary photographers have since commonly used. Moreover, using captions can help provide viewers with the "desired ethnographic meanings" if used correctly, according to Pink (2013, p 171). Ultimately, this technique results in rich data which the viewer can relate to, learn from, or analyse in the context of the respective research questions in the case of research projects. How the audience uses the data contained within the images and narratives depends on their requirements and the depth of their analysis. This chapter is neither meant to be an analysis of the stories' content, nor is it our place to do so; instead, we aim to reflect upon the value of the product and the methodology within the context in which it was used. Nevertheless, it is clear to us that the stories hold valuable information for all stakeholders – from farmers and their peers to extension services, policymakers and future project designers.

Certain elements of the stories make them useful in a variety of contexts. For instance, the experiential element provided by the beneficiaries has value in peer-to-peer farmer communication. In the examples above, Fig. 1 introduces the farmer and his story, and the portrait of him in front of his home and the accompanying narrative about his family is a life situation to which many farmers from that area can presumably relate. This can help to instantly engage an audience of his peers if they recognise something – either the farmer or similarities with themselves – within the picture and narrative. Figure 5 tells of what the farmer has learned from being part of the project, further sharing her newly acquired knowledge of which trees to plant to improve soil quality and increase dairy cows' protein intake. Other stories – the images and captions – share equally valuable information such as on new foraging systems or how to make mash feed. There are also many instances of clearly visible or audible emotions within the images and narratives – something which is not always evident from conventional research methods. These include but are not limited to pride, humour, and frustration. Overall, the data contained within the stories is rich and plentiful. In respect to the main motivations, drivers, and impacts of being part of either of the two projects, participants did not directly mention reducing carbon footprint or adapting to climate change, although the farming techniques that were introduced, such as new ways of making cattle feed for the dry season, helps them to adapt in addition to improving their livelihoods. Moreover, the common focus throughout all stories was on improved food security and improved livelihoods, specifically how increased cash income and purchases have benefitted not only their own lives but also those of their families.

At the end of the day, the stories are anecdotal. Indeed, one researcher asked how we can validate the information they contain. However, do we need to? The methodological approach as we used it relies on participants' subjective interpretations. Harper (2002) emphasises the phenomenological nature of such an approach. Moreover, visual ethnographer Sarah Pink (2013) fully embraces a sensory approach to elicitation but nevertheless questions the truthfulness of knowledge produced in this manner. Participants used the platform to talk openly and share their stories, and thus their own subjective realities. The 14 stories provide insight into the projects as the beneficiaries see them and the meanings they attribute to their experiences. The truths are theirs.

The image-based digital storytelling project was largely practice-led. What the students learned through the classroom workshop was used immediately in the field. This would not have been possible without the cooperation of all those involved, that is, the institutions, students, beneficiaries and extension services, and highlights that active cooperation among stakeholders within an innovative and engaging project can efficiently yield tangible results such as the image-based digital stories.

The LUANAR students' motivation to take part in the project was high. Those who became involved did not know what to expect at first but soon became engaged with the methodology and were eager to use it in the field both in this project and in their future studies (Songola 2017). As final year ADC students, they already had previous experience with photography through a photojournalism course. Nevertheless, neither students nor faculty members had experienced a hybrid method such as this one to collect and disseminate information to farmers and other stakeholders. LUANAR faculty members recognised the method's value for relaying information to those who needed it the most. One Malawian colleague, who is involved with agricultural extension communication education, enthusiastically remarked that he had never encountered such an innovative and effective methodology and instantly saw its potential. The stories also stimulated discussion with potential future partners such as farm radio on how they could utilise the stories in agricultural extension broadcasts, further enhancing the cooperative nature of the methodology. Furthermore, the method illustrated the potential of using relatively simple digital tools within agricultural extension communication.

The students' background in communication undoubtedly contributed to what we perceived as being a successful project. Through short training, most students became confident in applying their newly acquired skills and knowledge in the field. Furthermore, previous experience in areas such as ethics helped; we discussed issues including how to gain the farmers' trust and how to best bring forth their stories while adhering to ethical guidelines and without influencing the story content. Students were able to adapt, think, and act quickly, in order to gather data efficiently since time was often of concern to the participants. Throughout the process, the students actively shared ideas and thoughts, and this helped create a rich learning experience relevant to their studies.

As facilitators, we took somewhat of a back seat in the field, allowing for the students take the lead in what they were trained to do. We believe this further enhanced acquiring the "true" perspectives of the beneficiaries and avoided potentially predetermined narratives which can often be the case in this type of research situation (Latz 2017). Essentially, the practice-led methodology, which evolved through cooperation among stakeholders, helped address the gap between knowing and doing that Pfeffer and Sutton (2000) emphasise is prevalent within organisations. In the context of the REDCAP and SEED-Fish projects, respectively, we hope the image-based digital stories can help put the acquired knowledge into appropriate actions. From research to dissemination, the image-based digital stories appear to have added value in terms of efficacy, data quality, and capacity building over what could have been achieved through using more conventional or traditional methods.

One of the most important elements of both DS and our hybrid methodology is sharing the stories. This helps create ownership and thus empowerment among participants. In our case, this meant returning to the communities to share and discuss the stories. This also highlighted the flexibility of the method; it would have been logistically difficult to show the finished digital stories in Nkhotakota and Dedza, but the photographs could still be utilised.

For example, in Nkhotakota where three stories were made, each story's images were hung on a clothesline for the community to view (Fig. 10). The three ADC students who accompanied us to Nkhotakota and who had previously collaborated with some of the fishers and fish processors to make the stories then facilitated a discussion session with the community. The students asked the storytellers to share his or her story and experiences through the images. The individual interpretations of the images fuelled a dialogue among participants, who included project beneficiaries, extension agents, and academics. The images were clearly something participants could identify and connect with, and the discussions were fruitful.

Another empowering component of the methodology was listening to the stories. Although the stories could not be shown to the communities in their video format due to logistical and resource constraints, the recorded narratives were played from a flash drive through a portable battery or solar powered radio – technology that is simple, relatively inexpensive, and becoming widely available. After storytellers had shared their stories through the photographs, participants listened to the original recorded narratives whilst continuing to view the images. Reflection is a key element of DS according to Jamissen and Moulton (2017), and this awarded the audience the opportunity to reflect upon what they were both seeing and hearing – to interpret, create meaning and therefore knowledge – in a similar way to how they would with a video format. Furthermore, to hear their own voice emanating from the radio, or that of a third person narrating their story, was empowering for not only the storyteller but also for the audience – the stories were being heard.

Although the three ADC students facilitated the group discussions well, the images played their role as we hoped. Likewise, this is true for the entire methodol-

Fig. 10 Sharing the stories at Nkhotakota. (Photograph: Neil Gordon Davey)

ogy, from eliciting narratives, empowering participants (Clark 2012; Latz 2017), connecting the audience to the data (Szto et al. 2005), to triggering dialogue and discussion. Participants did not hold back during the community viewing and sharing of the stories. Instead, they used it as a platform to talk openly. Moreover, we did not experience any significant gender imbalances throughout the process. However, it is unclear if the methodology helped address any gender imbalance that otherwise could have been prevalent. Nevertheless, women were prominent and active throughout, and in total, females outnumbered males who were involved in the process, both as storytellers and as story makers. Thus, during the discussions over the photographs at Nkhotakota, women were equally, if not more, active than men.

6 Where Do We Go from Here?

We believe the stories play an important role in agricultural extension communication and capacity-building strategies. We hope they can bring forward the contextual and subjective perspectives of the project beneficiaries to policymakers. Furthermore, the method included participants to an extent that is often overlooked and for which many development projects of a similar nature are criticised. This created a sense of ownership and pride and the willingness, we believe, of beneficiaries to share their stories led to richer and deeper understanding of their specific situations. Presentation packs consisting of the photographs and recorded narratives were left with the communities so that participants' experiences can not only be shared among their closest peers but also further afield with the help of the extension services. This is just one use for the many elements of the stories that hopefully can be used in different combinations appropriate to specific contexts and situations.

There is great potential for all stakeholders to learn through the sharing of the stories. Moreover, the stories are just the beginning. Trialling image-based digital storytelling in this context opened up questions such as: How can we best integrate the methodology into agricultural communication education? What type of structures can we put into place to integrate and upscale the use of digital stories by the extension services and by farmers? Can we collaborate with other initiatives such as farm radio to bring the stories to communities? Can we use digital stories as a project management tool so that farmers' voices always influence project aims and impacts? Thus, further exploration can fully exploit the potential and flexibility of this digital storytelling methodology.

References

Alterio M, McDrury J (2003) Learning through storytelling in higher education: using reflection and experience to improve learning. Routledge, London

Clark A (2012) Visual ethics in a contemporary landscape. In: Pink S (ed) Advances in visual methodology. Sage, London, pp 17–35

Collier J, Collier M (1986) Visual anthropology: photography as a research method. UNM Press, Albuquerque

Davey NG (2018) Transitions to sustainability: a photographic exploration into the motivations, impacts, and expectations of community renewable energy production on the Isle of Eigg. MSc thesis, Norwegian University of Life Sciences

Harper D (2002) Talking about pictures: a case for photo elicitation. Vis Stud 17(1):13–26

Jamissen G, Moulton M (2017) "Now I See": digital storytelling for mediating interprofessional collaboration. In: Jamissen G, Hardy P, Nordkvelle Y, Pleasants H (eds) Digital storytelling in higher education, Palgrave. Macmillan, London, pp 243–259

Latz AO (2017) Photovoice research in education and beyond: a practical guide from theory to exhibition. Taylor & Francis, New York

Mkwizu Y (2016) The role of storytelling presentation technique in enabling access and sharing of climate change knowledge in Tanzania. University of Dar es salaam

Newhall B (1982) The history of photography. The Museum of Modern Art, New York

Pfeffer J, Sutton RI (2000) The knowing-doing gap: how smart companies turn knowledge into action. Harvard business press, Boston

Pink S (2013) Doing visual ethnography. Sage, Los Angeles

Ribeiro SP (2017) Digital storytelling: learning to be in higher education. In: Jamissen G, Hardy P, Nordkvelle Y, Pleasants H (eds) Digital storytelling in higher education. Palgrave Macmillan, London, pp 207–203

Shor L, Friere P (1987) A pedagogy for liberation: dialogues on transforming education. Bergen and Garvey/Greenwood Press, Westport

Songola I (2017) LUANAR ADC students trained in digital story telling. http://www.bunda.luanar.mw/luanar/news_detail.php?id=381#.XAbnH12yVWE.email. Accessed 07 Mar 2019

Szto P, Furman R, Langer C (2005) Poetry and photography: an exploration into expressive/creative qualitative research. Qual Soc Work 4(2):135–156

Tønnessen ES (2012) Digitale fortellinger som multimodal tekst. In: Haug KH, Jamissen G, Ohlmann C (eds) Digitalt fortalte historier. Refleksjoner for læring. Cappelen Damm Akademiske, Oslo, ch 4

Wang C, Burris MA (1994) Empowerment through photo novella: portraits of participation. Health Educ Q 21(2):171–186

White WF, Greenwood DJ, Lazes P (1991) Participatory action research: through practice to science in social science research. In: Whyte WF (ed) Participatory action research. Sage, Newsbury Park, pp 19–56

Assessing the Role of Storytelling Presentation in Knowledge Transfer from Climate Change Projects in Tanzania: The Case of the EPINAV Programme

Y. B. Mkwizu, N. G. Davey, L. O. Eik, R. Kangalawe, and L. R. Kurwijila

Abstract The *Enhancing Pro-Poor Innovations in Natural Resources and Agricultural Value Chains* (EPINAV) programme was implemented in 2010–2016 by Sokoine University of Agriculture (SUA) to enhance productivity, livelihood security, and human capacity to utilise pro-poor and climate change adaptive innovations in agricultural and natural resources value chains. This chapter examines the contribution of storytelling in sharing knowledge from the EPINAV programme. The assessment of knowledge-sharing and access is conducted using the knowledge conductivity framework. The framework indicates that storytelling presentation has a greater likelihood of enabling knowledge movement between and among project stakeholders. We examined knowledge conductivity among the targeted stakeholders who were involved in the implementation of the programme, namely, implementation team, researchers, farmers, policymakers, external reviewers, and funding agency. Issues and interests of each group are identified based on their mandates and areas of work relevant to the programme. From specific category interests, we established their role in the project story. This chapter examines the contribution of storytelling in empowering the active engagement of stakeholders to increase the achievement of project results. It concludes by proposing increased use of a storytelling presentation framework in the management and communication of knowledge from EPINAV to similar programmes.

Y. B. Mkwizu (✉)
Dar es Salaam Nature Study (DarNature), Dar es Salaam, Tanzania

N. G. Davey · L. O. Eik
Norwegian University of Life Sciences, Department of International Environment and Development Studies, Noragric, As, Norway

R. Kangalawe
University of Dar Salaam, Dar es Salaam, Tanzania

L. R. Kurwijila
Sokoine University of Agriculture, Morogoro, Tanzania

B. R. Singh et al. (eds.), *Climate Impacts on Agricultural and Natural Resource Sustainability in Africa*, https://doi.org/10.1007/978-3-030-37537-9_33

1 Introduction

There is an increasing need to improve sharing and access to project information among stakeholders as a means to ensure engagement and buy-in in climate change and agriculture initiatives. In Tanzania, implementation teams are expected to share results and lessons from the field that will enable utilisation and application of the knowledge generated, and upscaling of innovations. Communication strategies to achieve these goals include the National Environmental Education and Communication Strategy (NEMC 2004); MKUKUTA I&II communication strategy (URT 2005; URT 2010); Information, Education; the Communication Strategy for Eastern Arc Mountains (TFCG 2008); communication strategy for International Union for Conservation of Nature (IUCN) project on Climate Change and Development (Gordon-Maclean et al. 2009); and the National Climate Change Communication Strategy (URT 2012).

In order for research on forests and climate change to be useful, the results should be communicated to potential users and should enable stakeholders to access reliable and updated information, including local communities (FAO 2012). The Africa Adaptation Programme for Tanzania, for example, includes a knowledge management system (UNDP 2010). Through consultations, the processes involved identification of the type of information and media of communication preferred by intended users. Researchers at The Centre for Research on Environmental Decisions (CRED) studying communication in climate change initiatives point out the importance of knowing the social identities and affiliations of the target audience as essential for successful participation (CRED 2009).

A study to examine the role of storytelling revealed that politicians, researchers, district officers, and communities preferred the storytelling presentation technique when reporting on climate change initiatives in Tanzania. Co-production of project information and real characters were also preferred (Mkwizu 2016). Interactively and collaboratively, storytelling may enable the integration of cultural and livelihood aspects (Denning 2005).

Storytelling presents the opportunity for co-production and sharing of information at all stages of a programme, while at the same time attracting the audience through entertainment and message delivery. Therefore, the application of storytelling may enhance meaningful engagement of stakeholders, thus improving communication and better realisation of project goals.

2 Conceptual Framework and Storytelling in Projects

Information sharing includes more than physical components. In this chapter, as proposed by Burnett et al. (2008), we use three elements of information access to the material: physical access is the written documents, intellectual access involves understanding that information, and social access is the cultural environment that influences the decision of access to information. Project stakeholders have access to report if they have developed sufficient capacity to convert availability to accessibility at all three levels. Hence, the production and sharing of project information in climate change initiatives should use approaches and techniques that will enable the appropriate information to be accessed by the relevant people at the right time. In Fig. 1, we propose a framework to enhance the accessibility of information to stakeholders during the EPINAV programme.

Communication in climate change initiatives may involve concepts that involve specific categories of stakeholders, farmers in particular. Stories can help us to both present and understand complex concepts (BSI 2003; Scholtz 2003; Snowden 1999; Sole and Wilson 2002). According to Denning (2005), storytelling enables communication that is quick, natural, clear, truthful, collaborative, persuasive, accurate, intuitive, entertaining, moving, and interactive, as well as more easily remembered (Wilkins 1984; Tversky and Kahneman 1973). Methods such as digital storytelling further enhance communication by being both auditive and visual, stimulating our senses and provoking reflection on the meanings that the stories hold (Davey and Moulton 2019). The process also helps build capacity while transferring ideas, meanings, and knowledge among stakeholders through gathering and sharing the stories.

Climate change initiatives make good storytelling topics because of the possibility to create a set of events in a way that touches the interests of stakeholders (Sikes and Gale 2006). We may translate the project's objectives to a set of challenges faced by the existing community, which require active participation from the stakeholders (Pretty 1995). By providing a platform upon which to express themselves, storytelling can be used to uncover the issues and realities of stakeholders and to understand their needs (Merla 2009).

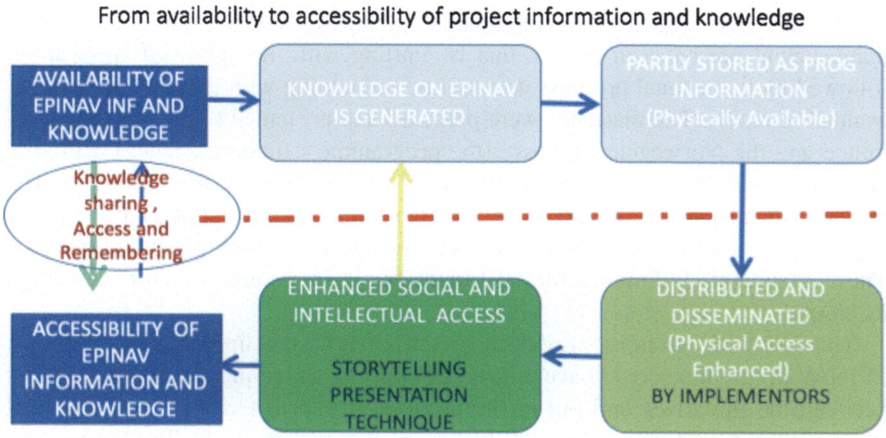

Fig. 1 Conceptual framework showing the availability and accessibility of EPINAV information

3 EPINAV Programme and Its Associated Storytelling

3.1 The EPINAV Programme

The EPINAV programme started in December 2011 through the signing of the agreement between the Government of the United Republic of Tanzania and the Kingdom of Norway. It was implemented for 6 years and was a continuation of the Norwegian Government's assistance to Sokoine University of Agriculture (SUA), which started in 1973. The programme involved a collaboration between SUA and the Norwegian University of Life Sciences (NMBU). Previous projects and programmes aimed at improving the teaching and learning environment at SUA, improving the livelihoods of smallholder farmers, and the development and dissemination of appropriate technologies to the end beneficiaries. They included the Future Opportunities and Challenges in Agricultural Learning (FOCAL) programme (2003–2006) and the Programme for Agricultural and Natural Resources Transformation for Improved Livelihoods (PANTIL) (2006–2010).

EPINAV involved three main components: Research and strategic interventions (RSI); capacity building and institutional collaboration (CBIC); and planning, monitoring, and evaluation (PME). The RSI component had two subcomponents: The first of these was research based and focused on promoting the more widespread use of best agricultural and natural resources management practices; the second subcomponent was concentrated in strategic intervention and was designed to demonstrate the best practices and technologies generated from SUA research. The CBIC component focused on capacity building within human resources, the improvement of selected infrastructures, gender equality, and other cross-cutting issues. The PME mandate was to assess and carry out baseline studies for all project areas and conduct periodic monitoring and evaluation of project activities as well as impact assessment.

The EPINAV programme's communication strategy utilised traditional project management presentation modes, that is, starting with the approved application, followed by five annual progress documents, and ending with the final project and evaluation report. The materials were prepared and presented to the steering committee and the Norwegian Embassy (the programme's funding agency). EPINAV also monitored its interventions and prepared various reports for internal use. The end review and audit reports provided an independent assessment of the programme regarding the use of programme funds and the approved work plans and budgets. These documents followed a formal format and used a conventional presentation technique.

The following sections provide a review of how the implementation of the EPINAV programme could have utilised storytelling techniques to strengthen links between the initiatives laid out in the proposal, the progress reporting, the final report, and the evaluation. An introduction of the project storytelling framework (PSF) could have increased the level of engagement of key stakeholders from the project coordination team. Merla (2009) points out that storytelling is useful in strengthening the relationship between stakeholders and the project team by making them more personable and entertaining. Davey and Moulton (2019) echo this sentiment and add that the collaborative and participatory nature of (digital) storytelling helps build trust and rapport between project partners. Project stories may include telling about a sequence of events and also integrate aspects of conflict, such as unsuccessful actions, broken promises, or wrongful interpretations (Munk-Madsen and Andersen 2006; Ryan 1991). Before reviewing how the different stages of the EPINAV could have benefited from PSF, let us look at the programme itself in more detail.

3.2 Planning Process and the Programme Document

As noted earlier, the EPINAV programme was a continuation of the Norwegian Government's assistance to the Sokoine University of Agriculture. To understand gaps in the planning, let us look at a few points from the programme reviews. The end review mentioned a design gap related to missing linkages between the implemented EPINAV projects. Ideally, these projects should have been nested under the programme, each showing a link to the overarching programme goal. While acknowledging the successes of individual projects and interventions, the review findings also highlighted the complexities around the many projects constituting the EPINAV programme.

At the initial stage, the project team needed to develop and communicate the project goal. This included fostering a shared understanding of the goal hierarchy, and how the inputs fed into the activities could be connected to the outputs and outcomes to bring about the intended impact (Merla 2009). For this purpose, storytelling could have helped by relating information from these projects as individual narratives to a common link within the hierarchy. This initial stage could have enabled the production of a base story.

The programme document was the final product of the planning, verified with the signing of the agreement. For the EPINAV story, this would have been the same as the base story. Project teams did consult with stakeholders but mostly for informing about the planning and design. Teams were not required to document various views from the beneficiaries regarding duration, budget, and geographical areas. In preparation, negotiation took place between the actors. Through storytelling, each 'side' could better understand the needs of the other, thereby creating a long-lasting relationship during implementation (Merla 2009).

Introducing PSF would suggest a story setting. When the programme proposal was finally approved and the agreement signed, the head of the implementation team (the subject) took responsibility for the programme goal (the object) being implemented within 4 years, using an agreed budget, reaching mentioned beneficiaries, and in identified geographical areas. In the process, the subject had to work with stakeholders with different functions and roles. This stage of the planning process represented a suitable platform for an EPINAV story setting, followed by stakeholders' characterisation as pointed out by Akysheva (2017), who discussed the potential of storytelling in project implementation.

The EPINAV implementation team took on the task of achieving the programme goal to enhance productivity, livelihood security, and human capacity to utilise pro-poor and climate change adaptive innovations in agricultural and natural resources value chains. The protagonist and antagonist aspects would then be introduced, with stakeholders presenting opposing views about the realisation of the goal and each collecting evidence to support their case. For example, a stakeholder may suggest that a goal is unattainable, and the leader would then be obliged to prove this wrong. The EPINAV story would eventually be built up as an equivalent to midterm and end reviews, progress, and final reports as well as audit reports.

3.3 *Stakeholders and Characterisation*

Characterisation enables a storyteller to create roles that define the sphere of actions. Munk-Madsen and Andersen (2006) provide an essential grouping of characters in projects. They suggest a subject who is striving for the object. This subject is equivalent to representatives of the entity entrusted to oversee implementation of the plan until the realisation of the project goal. This character is, therefore, a protagonist. The receiver is grouped on the same 'side' as the subject because both need the results, even though only the latter is looking for ways to achieve it. This protagonist is equivalent to the implementing entity and the end beneficiaries.

The characterisation of stakeholders helps the implementation team and the funder to understand players and their roles. The planning and design stage benefits from imagination of the way that these characters will likely behave once the project has started and created a platform for them to join the project story.

From the programme document, six categories of stakeholders could be characters in the EPINAV story. They include implementation team, researchers, farmers, policymakers, external reviewers, and the funding agency. Mkwizu (2016) divided them into categories of (a) protagonist and associates (PAs) and (b) antagonist and associates (AAs). The PA characters would support the implementation team leader (the coordinator) who is the protagonist and who also represents the main character. There would be three subcategories: The coordination team, the funder, and the beneficiaries under the PAs listed in the left column of Table 1. The antagonist and associate (AAs) characters would present different and more opposing views about the programme activities. In addition to expressing opinions, they would perform actions that promote the programme plan. They would also act as silent monitors of processes and operations. Their list includes characters shown in the right column of Table 1.

Funder

The funder supported the project after reviews of draft proposals and an appraisal report that showed how the programme would attain the intended goal. The donor supported the project under the condition that it would fulfil the expected results. An

Table 1 Stakeholders' characterisation in supporting and opposing categories

Protagonist and associates (PAs)	Antagonist and associates (AAs)
PCT – Protagonists from the coordination team	ACT – Antagonists from the coordination team (when presenting internal risks)
PF – Protagonists from funders (after programme approval)	AF – Antagonists from funders (before programme approval)
PFB – Protagonists from small farmers as beneficiaries (those supporting the initiative)	AFB – Antagonists from small farmers as beneficiaries (those sceptical about the initiative)
PRB – Protagonists from researchers as beneficiaries	ARB – Antagonists from researchers as beneficiaries
PLB – Protagonists from local authorities as beneficiaries	ALB – Antagonists from local authorities as beneficiaries
PPB – Protagonists from policymakers as beneficiaries	APB – Antagonist policymakers as beneficiaries
PCB – Protagonists from private companies as beneficiaries	ACB – Antagonists from private companies as beneficiaries
AARA – Protagonists from appraisal, reviews, and audits (if their findings are positive, they are neutral before discoveries)	AARA – Antagonists from appraisal, reviews, and audits (if their conclusions are negative, they are neutral before discoveries)

Below, we provide a review of the characterisation of the EPINAV stakeholders

appraisal was commissioned by the donor to verify whether the proposed framework would lead to the intended results. Through PSF, the EPINAV story would have the donor as part of the PAs after having decided to support the programme. In the beginning, however, when SUA first presented the request for funding, the donor would have played the AAs role. It is for this reason that an appraisal was commissioned by the financier to verify whether the proposed framework would lead to the intended results. However, the appraisal report presented concerns from the beneficiaries. The PSF would have recipients offering their views and concerns themselves, as part of the PAs or AAs. The setting would open the opportunity for adding realities about the programme to the funder. More important is the role of the EPINAV story in creating the connection between the donor, the implementation team, and the recipients in realising the agreed results (Merla 2009; Mkwizu 2016).

Main Beneficiaries

EPINAV's primary beneficiaries were farmers. Other beneficiaries included researchers, postgraduate students, local authorities, policymakers, and private companies. On the one hand, recipients present the supporting and opposing views offered by the programme implementation team (PIT). They had a reason to support the intentions of the PIT, thereby ensuring that the project with affiliated benefits materialises. On the other hand, based on their reserved assumptions and the risks surrounding the connection between proposed activities and expected outcomes and impacts, they will also be inclined to challenge, thereby checking the realism of the results framework.

The EPINAV base story would then end by having the sceptical group joining the AAs and remaining active to track their doubts during implementation. This would enhance active participation, and as Merla (2009) argues, storytelling enables customisation of the process and real engagement. Stories will depict conflicting views between the differing opinions on the initiative, as well as between the stakeholders and the implementers of the programme.

Researchers

As characters, researchers would work together with farmers to undertake the planned research activities. During the implementation period, they could reveal weaknesses overlooked at the design stage. Researchers who decided to join the AAs at the beginning could select research questions that challenged the proposed results framework at the base story. That was to be an example of conflict from the research perspective, existing internally in the team and eventually revealed between one researcher and another in a story (Munk-Madsen and Andersen 2006). The antagonising research role could have changed based on the findings from the field or may contribute objectively in giving both PAs and AAs additional information. Characterising researchers in PAs and AAs brings an essential attribute to other characters regarding the importance of owning different views and presenting concerns about the programme framework. As pointed by Merla (2009), differing opinions communicated through stories can, in turn, help to create a common objective and understanding.

Equally important are researchers who select research questions to hypothesise the success of programme interventions in delivering the results. Their participation as part of the PAs would add to a need to increase the documentation of success stories from the field, especially using digital storytelling.

Appraisal, Review, and Audit Teams

The appraisal, review, evaluation, and audit groups were neutral. Their mandate was to review and assess the implementation of planned activities and the realisation of expected results. Findings from their work could, therefore, be used by the supporting groups, thus joining the PAs, or the opposing group, thus joining AAs. As a result, the EPINAV story would help to uncover the issues and realities of the beneficiaries and to understand their concerns, thereby increasing the chances of successful project outcomes (Merla 2009).

Programme Implementation Team

This team, led by the coordinator, provided secretariat functions for the project. Ideally, all members of this team would be part of the PAs. However, some members may possess AA elements, either intentionally or unintentionally, thereby failing to perform their duties. Setting a story scenario with a possibility for members of this group to fall under AA characters would increase the likelihood of capturing different views. As a result, the team would be more aware of the realities that unfolded from the project at its various stages, thereby increasing the prospects of achieving project goals.

3.4 Results Framework and Building a Conflict

Based on a standard log frame with indicators prepared at the design stage, the EPINAV programme used the outcomes and impacts of its projects to report on results. This results framework is what the EPINAV story would apply to stage a 'building conflict'. The AA opinions and views would differ with the PIT, based on what antagonists from small farmer beneficiaries (AFB) consider to be lacking about the proposed activities versus the benefits they will get. These assumptions present a platform for building a conflict-setting scenario, along with the expectation that the activities and outputs would be connected to the findings and effects, and deliver the expected results to the beneficiaries.

It is in the interest of the beneficiaries falling under AAs to challenge the PIT as part of PAs so that activities are conducted in a manner that will create a connection to the results that benefit them. As a result, the AAs would have pressured the PIT to observe the criteria of effectiveness, efficiency, and sustainability when implementing project activities. The role of the EPINAV story in using AAs to challenge PAs to see these criteria would be useful in the reviews and final reporting. Both quantitative and qualitative baseline information would be viewed differently by PAs and AAs. The EPINAV story would employ digital storytelling of prospective beneficiaries as a supporting baseline to be updated during progress reporting. The targets could also provide a common focus for characters when measuring progress. The EPINAV story would suggest a common approach and a means to assure the beneficiaries under AAs and PAs that they will reap the benefits 'promised' by the programme.

Setting up a 'building conflict' from a results framework relates to Munk-Madsen and Andersen's (2006) suggestion of applying characterisation to create actors who define the sphere of actions in a project. The PIT was striving for the programme goal, standing as a protagonist, while other members fall under either the supporting or the opposing category.

The low-level indicators used to measure progress in the implementation of activities and delivered outputs were well-utilised in the progress reporting, reviews, and evaluation, as well as in the final reporting by the PIT. The amount of training provided at the community level is an example of an indicator used and training reports as a means of verification. The percentage increase in crop productivity and income levels were used as high-level indicators to measure the outcomes and impact, while monitoring reports were means of verification. In order to ensure the authenticity of confirmation, digital storytelling would be used in the building of the EPINAV story.

3.5 *Resolution of the Base Story*

The ending of EPINAV design and planning is manifested in the approval for funding by the Norwegian Embassy. The conclusion was also displayed in the approval by the Tanzanian authority, the Ministry of Finance and Planning. Based on PSF proposed by Mkwizu (2016), support of the EPINAV proposal would have revealed a 'winning side' of the PAs. Mostly, the AAs views and opinions that doubted and challenged the PAs were not successful. But the AAs would have promised to prove the PAs wrong and to provide evidence for it in the next 'episode', that is, at a relevant time of the project. Thus, the programme would have remained a vital process, presenting the opportunity for future approvals, where both PAs and AAs offer their cases. AAs, with farmers as the primary beneficiaries, would have concluded at this stage that they would carefully follow how activities implemented under strategic interventions were to lead to their increased productivity and enhanced income. The next verification would have happened during progress reporting and midterm review.

4 Implementation of Activities, Midterm Review, and Progress Reporting

4.1 *Storytelling During Implementation*

The EPINAV programme communicated its implemented actions and results through regular progress reports. The coordination team prepared the documents under the leadership of the coordinator, showing achievements during each reporting period. Considering that EPINAV was comprised of several projects, progress reports informed on the implemented activities and achievements of a combination of individual projects. The reporting was required to list all activities conducted and to show how those activities were linked to higher-level results, and how they contributed towards attaining the programme goal.

The EPINAV story provides an alternative narrative to support reporting and to verify the information presented on implemented activities. One major challenge that the coordination team faces is the danger of assuming that reports from individual projects were accurate. Otherwise, the programme story helps the PIT in reporting to the Steering Committee and the Embassy and in receiving reports from the field.

The EPINAV programme had a substantial geographical coverage and broad stakeholders' involvement, which made the monitoring of fieldwork challenging, both for the coordination team and for the funder. A monitoring team was constituted and assigned the role of undertaking field-based verification on an annual basis. Reports from these missions indicated a need for informing the coordination team on what was happening in the area. The monitoring team reports provided an alternative narrative to the reports from the coordinators of research and strategic intervention projects. However, because the monitoring teams were appointed by the coordination team and reported to them, the alternative narrative they provided may not have been sufficient to reveal all the essential realities surrounding the success or failure of the project, for example, at its design. This is explained by the differences in the findings from monitoring team reports and review reports, where the programme was considered to be too ambitious and 'spread out' at the design, a weakness revealed by the review reports, and not by the monitoring teams.

4.2 Storytelling from Progress Reporting and Midterm Review

Progress reports applied a results framework to show results achieved from planned work. Because research and strategic intervention projects had just begun, fewer results were reported at the end of the second year when the midterm review was conducted. Hence, progress reports during the 4th and 5th years of the programme, the latter falling under the extension period, were the documents that revealed the contribution of implemented activities and recorded outputs to the results.

The EPINAV story presents the potential of using progress reporting to build on the base story, which was set following programme approval. Progress reports would be equivalent to mid-stories. On the point of departure between PAs and AAs following programme approval, the first progress reports, together with the midterm review, were to be subjected to challenges from the AFB because they would not have realised the results (outcomes and impacts) by that time. These beneficiaries would update their digital stories, which they set as a baseline, and use them to confirm that no changes occurred from their side. The first mid-stories would, therefore, have given power to AAs to put pressure on the PAs, mainly the programme coordinator, to increase efficiency in resource use, both time and monetary. Progress reports in the last 2 years would have more results to show and would be supported by the beneficiaries from both AAs and PAs, verified by their updated digital stories. This level of participation from stakeholders is vital for the success of an initiative that targets farmers (Pretty 1995).

According to the EPINAV agreement, if the midterm review and audit reports indicated poor performance, the programme could be discontinued. In this case, neither PAs nor AAs would have won. However, even if there were no conclusive findings, verification from beneficiaries who joined both AAs and PAs would have provided added information to validate the findings. Equally important is the final programme and audit reports. Results and conclusions of the two stories were the essential judgement of the programme failure or success when supported directly by verification from beneficiaries. Essentially, the storytelling from midterm reviews and progress reporting could have helped in portraying the final winning of either PAs or AAs. More importantly, there was a possibility to reveal and clarify issues that might have emerged after the programme had ended. This possibility is an essential reminder to the PAs to ensure sustainability, as well as to take mitigation measures to manage identified and new risks during implementation.

The project storytelling framework (PSF) suggests that investment in the design of the EPINAV base story, which touched the interests of partners, would facilitate the updating of mid- and end stories. The stakeholders themselves would update stakeholders' stories that constituted a project story, and the coordination team's task would mainly be to merge them to produce a combined and coordinated project progress report. The story facilitator's central role would be to guide the process of collecting views from stakeholders to ensure proper documentation of activities. Storytelling would allow the participation of beneficiaries and ensure increased ownership and sustainability of the initiatives (Merla 2009; Mkwizu 2016).

5 Finalisation: Final Reports and Evaluation

Completion of phased programmes presents the opportunity for the direct application of lessons learnt from the ending phase to the new one. When reviewing the finalisation of EPINAV, it would have been natural to include a review of its predecessor, PANTIL, to establish the extent to which the lessons learnt were utilised. This would have been achieved by introducing a storytelling presentation featuring PA and AA characters from PANTIL to EPINAV.

Consider a scenario in which the end review of PANTIL is used to inform EPINAV's programme document. In a situation where the goal remained the same from the coordination team and funder's perspective, it would be the approaches that would be changed. By the end of PANTIL, the PAs would have won, and most of AAs would have joined the supporting camp. There would be a few characters that remained in the AAs, and new ones would enter, for example, from a research group, who would challenge the proposed approach.

The absence of PSF in the planning of EPINAV meant that the beneficiaries who joined the AAs were missing, and this reduced the possibilities of revealing if there were specific cases where the new phase failed to apply the lessons learnt. Even when the coordination team considered and agreed to utilise experiences and recommendations from the ending stage, there was still a possibility of not utilising the previous findings in their entirety. As such, the PA team would have been strengthened at PANTIL's finalisation and the initialisation of EPINAV, because some AAs, as well as peripheral beneficiaries or outsiders, may be motivated to become part of the PAs in the new EPINAV.

The above scenario explains some of the reasons why the finalisation of EPINAV may not have fully utilised previous programme lessons in the design of the next or follow-up phases. Phased projects and programmes such as PANTIL and EPINAV have similarities with episodic TV shows. Project members were mainly PAs with no real presence of AAs, despite their distinct existence. Communication of project information was dominated by PAs reporting to the funder, from the design to the finalisation, portraying a one-sided narrative. However, there were other narratives of the programme by unnamed AAs, as well as elements of AAs hidden within PAs. In phased projects and programmes, the other 'tales' would bring different realities regarding the programme's success or failure and consequently lead to 'opposing stories' in the design of the follow-up programme. The opposing stories at the design of a follow-up plan can affect some members of the PAs by introducing elements of the AAs, thus contributing to balanced discussions with different views rather than mainly supporting functions.

The project storytelling framework should be used to prevent hidden narratives by AAs. The AAs are allowed to surface in the transformation from the end phase to a new one by allowing the realities, as perceived by the AAs, to be reflected in the design. In this way, the PAs are faced with different views, which require more serious consideration in applying lessons learnt, and this would ensure that winning is attained by competing with AAs. The story provides a means to bring differences in modalities such as obligations, knowledge, intentions, and desires (Ryan 1991) and to have them addressed in the following phase. This is similar to what Davey and Moulton (2019) set-out to achieve by using image-based digital storytelling as part of agricultural extension communication in small-scale farming and fishing communities in Malawi to communicate and reflect upon the perspectives of the programme beneficiaries.

The application of PSF in phased projects and programmes presents a vital scenario that can be applied in the upscaling and adoption of piloted initiatives. When tested actions are to be introduced in a different area, considerations should be made for its customisation into the social, cultural, and economic needs of the new environment. Although the PAs in the new area may easily take the roles from those in the pilot area, the implementing team needs to capture the realities of the AAs, because facts about the needs, ownership, and useful approaches for the success of the initiative will be revealed by the AAs rather than the PAs.

6 The Framework of Using Storytelling in Similar Actions

The steps in introducing the project storytelling framework in project management are presented in Table 2. From the project storytelling framework (PSF), this chapter proposes two ways in which the EPINAV story could be developed. One could be through co-production between the story facilitator and the stakeholders and the other by the story facilitator doing it alone. The first approach is a real story and would require the integration of the EPINAV story at the design stage. The story would then be set and constituted with real and actual characters. This approach is suitable for projects with development as the primary emphasis, where beneficiaries appear more strongly as essential characters. Pre-planning arrangement would be needed to agree on modalities and resources. The EPINAV story was not

Table 2 Steps to introduce the project storytelling framework in project management

<table>
<tr><th>Conventional</th><th>Project storytelling framework</th></tr>
<tr><td>Programme document</td><td>Base story</td></tr>
<tr><td>Presentation of stakeholders</td><td>Characterisation: Two categories, protagonist and associates (PAs) and antagonist and associates (AAs), are proposed with appraisal, review, and audit teams as neutral players</td></tr>
<tr><td>Results framework (RF)</td><td>Setting up a 'building conflict'.</td></tr>
<tr><td>Conclusion
Usually missing in programme document. Reflected in the RF</td><td>Resolution. End of base story. PAs win and the project is approved. AAs promise to prove PAs wrong.
Evidence follows in the next episode – the first progress report.</td></tr>
<tr><td colspan="2">A storytelling presentation of the approved project document is prepared. Characters who are real stakeholders are engaged in the preparation of the story. They will be involved again to validate progress reports, midterm reviews, and final reports. Hence, they will own the results and increase the likelihood of applying and sustaining the results for more significant lasting impacts</td></tr>
<tr><td>Progress reports and midterm reviews</td><td>Mid-stories: Build on the base story</td></tr>
<tr><td colspan="2">If reports are approved, PAs win at the margin similar to the quality and reflection of the RF. If the reports are not accepted, the AAs win</td></tr>
<tr><td>Final report and evaluation</td><td>End story (the final resolution)</td></tr>
<tr><td colspan="2">If the final reports and reviews are positive and are approved, the PAs win. After the end story, the AAs tell PAs not to celebrate the winning. Within the coming days, some unknown realities, good or bad, related to the project may surface. They may prevent future support. AAs also promise a close watch on the sustainability of results.</td></tr>
</table>

included in the design. Therefore, the second approach of preparing the imaginary EPINAV story has been used to portray what the programme could have achieved. The fictitious programme story can then be used to guide the preparation of a related story of a similar plan, either as an imaginary story or an actual story. Both approaches apply PSF as a tool in programme management. However, the resources needed would also be different, with the fictitious story utilising fewer resources than the real one.

The imaginary story has the additional advantage of allowing opposite views to be presented by relating them to a character rather than a real-life individual. This opens more possibilities to capture opinions from more targeted beneficiaries, who might have been reluctant to participate if they were to appear as the source of information, especially from the opposing side. Data collection methods could then be used to verify the accurate representation of the views of stakeholders.

The foundation of PSF should be made at the design stage. Although the framework can be applied at implementation and finalisation, setting it at these stages misses the capturing of AAs issues, concerns, and assumptions at the beginning, which would serve as a baseline and reference. Hence, for mid- and end stories to deliver the intended project management contribution, a base story as an output of the design or plan is proposed. With imaginary stories, the facilitator would use the end of previous programme reports, the programme document of the newly introduced programme and the appraisal report to form a base story. Therefore, for an initiative that has moved a few steps in its planning or is even approved, a PSF can be introduced similar to the review proposed above. Once a base story is successfully prepared, the building of a story can be accomplished by the facilitator as an additional input to the project coordination team's efforts.

Storytelling is an efficient means of gathering massive amounts of data in a short time. The PSF has the potential to enable the project to shorten the time needed to implement a scheme, where the additional time used on the production of the base story is compensated for in the progress and final reporting. This includes pressure from the AAs to the PAs to observe the effectiveness and efficiency in attaining results. This is also made possible by setting a platform for stakeholders to feed in their stories. Individual and categorised stories are combined, leading to the conclusion of the story. The overall work plan in the EPINAV programme's project documents, for example, would become a tale of each category's stories, which would then be updated based on the progress made.

7 Conclusion and Recommendations

This chapter reviewed the contribution of the Project Storytelling Framework (PSF) and its application in similar future initiatives using the completed EPINAV programme as a case example. The review shows that PSF has the potential to enable the project to improve efficiency and effectiveness in delivering results and enhancing sustainability. Storytelling also helps stakeholders to present their cases, which then opens up further possibilities for revealing relevant information that may not otherwise have been detected using conventional methods. It is further suggested that introducing the EPINAV story would provide a platform for the implementing institutions to demonstrate its contribution of facilitating farmers to improve their livelihoods through innovations in agricultural value chains. Presentations from the implementing institutions could have been backed by digital stories recorded at the beginning as a baseline and later updated to provide factual verification from beneficiaries. Similar initiatives should, therefore, consider applying PSF to their design, implementation, and finalisation. Regarding resources for the facilitation of PSF application, the same treatment as allocation for audits and reviews should be adopted. This means that funds for the PSF facilitator should be budgeted in the project with clear evidence of its contribution to the enhanced engagement and ownership, as well as efficiency, effectiveness, sustainability, and impact. To reduce the influence of the coordination team, who are generally expected to deliver results to the beneficiaries, such resources should be managed by the funder. A frame agreement with result-based payment approach should be applied to an entity or individual subjected to the delivery of the base story, mid-stories, and end story.

References

Akysheva U (2017) Storytelling practices in project management exploratory study in new business process implementation in information and communication technology projects. Master Thesis, Umeå School of Business and Economics, Umeå University

British Standards Institute – BSI (2003) Managing culture and knowledge: guide to good practice. British Standards Institute, London

Burnett G, Jaeger PT, Thompson KM (2008) Normative behaviour and information: the social aspects of information access. Libr Inf Sci Res 30(1):56–66

Centre for Research on Environmental Decisions – CRED (2009) The psychology of climate change communication: a guide for scientists, journalists, educators, political aides, and the interested public. New York. Available at: http://guide.cred.columbia.edu/pdfs/CREDguide_full-res.pdf. Accessed on 30 June 2015

Denning S (2005) The leader's guide to storytelling. Jossey-Bass, San Francisco

FAO (2012) Building resilience for adaptation to climate change in the agriculture sector. In: Proceedings of a Joint FAO/OECD Workshop 23–24 April 2012. Food and Agriculture Organization of the United Nations, Rome

Gordon-Maclean A, Harrison PJ, Shemdoe R, Wielandt K (2009) Developing a communications strategy for IUCN's climate change and development project. A Guidance Booklet for Practitioners. IUCN Eastern and Southern Africa Regional Office, Nairobi, Tanzania

Merla E (2009) Storytelling is for kids and project managers. Paper presented at PMI® Global Congress 2009 Asia Pacific, Kuala Lumpur, Malaysia. Project Management Institute, Newtown Square. Available at: https://www.pmi.org/learning/library/storytelling-project-managers-six-core-elements-6877

Mkwizu Y (2016) The role of storytelling presentation technique in enabling access and sharing of climate change knowledge in Tanzania. PhD Thesis, University of Dar es Salaam

Munk-Madsen and Andersen (2006) Storytelling in projects: transforming project plans into stories. A paper presented in the Euram 2006 Conference, Oslo, 17–20 May 2006. Available at: http://www.euram2006.no

National Environment Management Council – NEMC (2004) National environmental education and communication strategy 2005–2009. NEMC, Dar es Salaam

Pretty J (1995) Regenerating agriculture. Politics and practice for sustainability and self reliance. Earthscan, London, p 320

Ryan M-L (1991) Possible worlds, artificial intelligence and narrative theory. Indiana University Press, Bloomington

Scholtz H (2003) Why business needs a story. People Dyn 2003(6):8–11

Sikes P, Gale K (2006) Narrative approaches educational research: research in education. Available at: http://www.edu.plymouth.ac.uk/resined/narrative/narrativehome.htm

Snowden DJ (1999) A framework for creating a sustainable programme. In: Rock S (ed) Knowledge management: a real business guide. Caspian Publishing, London, pp 7–17

Sole D, Wilson D (2002) Storytelling in Organizations: the power and traps of using stories to share knowledge in organizations. In: The knowledge management advantage. Available at: www.kmadvantage.com/docs/km_articles/Storytelling_in_orgnaization.pdf

Tanzania Forest Conservation Group – TFCG (2008) Information, education and communication strategy for the eastern arc mountains. Tanzania Forest Conservation Group, Dar es Salaam

Tversky A, Kahneman D (1973) Availability: a heuristic for judging frequency and probability. Cogn Psychol 5:207–232

UNDP (2010) The Africa Adaptation Programme (AAP). Project document – Tanzania. Available at: www.undp-aap.org

United Republic of Tanzania – URT (2005) National strategy for growth and reduction of poverty I. Ministry of Finance and Planning, Dar es Salaam

United Republic of Tanzania – URT (2010) National strategy for growth and reduction of poverty II. Ministry of Finance and Planning, Dar es Salaam

United Republic of Tanzania – URT (2012) National climate change communication strategy (NCCCS). Vice President's Office, Division of Environment, Dar es Salaam

Wilkins AL (1984) The creation of company cultures: the role of stories and human resource systems. Hum Resour Manag 23(1):41–60

Part VII
Conclusion

Knowledge Gaps and Research Priorities

Bal Ram Singh, Andy Safalaoh, Nyambilila A. Amuri, Lars Olav Eik, Bishal K. Sitaula, and Rattan Lal

Abstract The major challenge related to sustainable management of natural resources, agricultural and livestock production, and the assessment of innovative technologies and policies is to identify solutions for these problems. Soil degradation, a serious problem in sub-Saharan Africa, is affected by climate change through emission of greenhouse gases (GHGs). Thus, restoration and sustainable management of soil to minimize risks of soil degradation are important to adaptation and mitigation of climate change and to advancing sustainable development goals of the United Nations. Livestock is a source of animal proteins for humans on the one hand and a source of nutrients and energy (biogas) on the other. However, livestock is vulnerable to many climate change-induced disasters such as prolonged droughts and floods, resulting in loss of animals and feed resources. In order to fully benefit from livestock, the targeted interventions needed include adoption of feed preservation technologies, including hay and silage for use during lean periods, controlled planning and management of communal grazing areas, rehabilitation of degraded communal grazing areas, and rangelands as a way of improving pasture availability. Research and development priorities and emerging issues include conservation agriculture (CA), use of legume-based cropping systems, integrated nutrient management (INM), climate-resilient livestock and feed systems, value addition policies, and adoption of innovative technologies.

B. R. Singh (✉)
Faculty of Environmental Sciences and Natural Resource Management,
Norwegian University of Life Sciences, Ås, Norway
e-mail: balram.singh@nmbu.no

A. Safalaoh
Animal Science Department, Lilongwe University of Agriculture and Natural Resources (LUANAR), Lilongwe, Malawi

N. A. Amuri
Department of Soil Science and Geological Sciences, Sokoine University of Agriculture, Morogoro, Tanzania

L. O. Eik · B. K. Sitaula
Department of International Environment and Development Studies, Norwegian University of Life Sciences, Ås, Norway

R. Lal
School of Environment and Natural Resources, Carbon Management and Sequestration Center, The Ohio State University, Columbus, OH, USA

B. R. Singh et al. (eds.), *Climate Impacts on Agricultural and Natural Resource Sustainability in Africa*, https://doi.org/10.1007/978-3-030-37537-9_34

1 Introduction

Countries in sub-Saharan Africa (SSA) primarily depend on agriculture for economic growth and livelihoods. However, a majority of the households, especially rural smallholder farmers, are continually food insecure and have limited livelihood opportunities in the wake of climate change. This is caused by many factors including political and institutional challenges as well as unsustainable practices in agriculture, agroecology, and overall natural resource management.Thus, the need for implementation of knowledge-based sustainable agro-ecological and natural resource management systems, and of the best management practices (BMPs), cannot be overemphasized.

The conference, on the proceedings of which this book is based, was organized in Malawi to present and discuss the major challenges related to sustainable management of natural resources, agricultural and livestock production, and assess the importance of innovative technologies and policies in order to identify solutions for these problems. The major issues covered in the book include: (i) conservation agriculture, carbon sequestration, and soil and water management, (ii) sustainable crop/livestock/aquaculture/fish production, (iii) policy and institutions for sustainable agriculture (CA) and natural resource management, (iv) value addition options for smallholder's market access and integration, and (v) up-scaling innovative technologies on smallholder farms. These issues are presented in the face of changing climate in SSA countries.

Therefore, the main aim of this chapter is to provide a brief synthesis of the topics covered in the book and to identify some emerging development issues and research priorities.

2 Rationale and Objectives of the Conference

The research project on the Capacity Building for Managing Climate Change in Malawi (CABMACC) was implemented through the Lilongwe University of Agriculture and Natural Resources (LUANAR) in collaboration with the Norwegian University of Life Sciences (NMBU) during 2014–2018. The overall objective of the program was to improve livelihoods and food security through innovative responses and enhanced capacity of adaptation to climate change. The CABMACC was aimed at strengthening the teaching, training, research, technology development, and outreach for climate change adaptation and mitigation planning at LUANAR. To document and synthesize the results obtained during this program and to identify and share the major findings, which could result in identification of recommendations and policy-driven documents, an international conference on "Sustainable Agricultural and Natural Resource Management under Changing Climate in Sub Saharan Africa" was organized in Lilongwe, Malawi in October 2018. The objectives of the conference were to:

(i) present the key findings from research projects undertaken in the CABMACC Program and other research projects conducted by LUANAR and its partners,
(ii) provide an opportunity for researchers from universities and other institutions in SSA region including those under the NORHED program to share their research findings in several areas including conservation agriculture, sustainable soil management, soil and crop health, integrated crop/livestock/aquaculture/agroforestry systems on smallholder farms, market access, climate change and food and nutrition security,
(iii) explore possibilities for increased collaboration between African institutions (south-south collaboration), and
(iv) produce quality scientific publications in a book form by a reputed publisher.

3 Soil Resources and Degradation

Soil degradation, decline in quality and functionality of the soil due to natural and anthropogenic factors, is a major global issue of the twenty-first century. Soil degradation, a biophysical process but driven by the human dimensions (Lal 2008), is also likely to increase with the increase in climate change. Indeed, soil degradation is affected by the climate change. By exacerbating emission of greenhouse gases (GHGs), soil degradation creates a positive feedback to climate change. Thus, restoration and sustainable management of soil to minimize risks of soil degradation is important to adaptation and mitigation of climate change and advancing sustainable development goals of the United Nations (Lal et al. 2018b). It is estimated that a third of Earth's soil is severely degraded due to agricultural and other anthropogenic activities, and the fertile topsoil is being lost at a high rate (Watts 2018). There are several processes of soil degradation (Fig. 1). The physical process involves loss of soil structure, resulting in crusting, compaction, hard setting, runoff, and erosion by water and wind. Accelerated soil erosion, affecting 1.1 Bha by water erosion and 0.55 Bha by wind erosion (Oldeman 1994), is the most widespread problem. Human activities related to land-use change are the primary cause of accelerated soil erosion, and the annual loss of topsoil may be much higher than 35.9 Gt/yr. (Borrelli et al. 2017). Some have argued that degradation is so serious that only 60 years of farming will be left if the soil degradation continues at the same rate (Arsenault 2014). Global soil erosion has a strong impact on global carbon cycling (Lal 2003), with the attendant impact on changes in the hydrological and elemental cycling. The process of soil chemical degradation encompasses acidification, salinization, nutrient depletion, elemental imbalance, and reduction in cation/anion exchange capacity of the soil. Salinization, increase in concentration of soluble salts in the root zone, affects a large tract of land in arid and semi-arid regions. Secondary salinization, already affecting 20% of irrigated lands and likely to increase as climate change increases, is a major global issue. Severe adverse

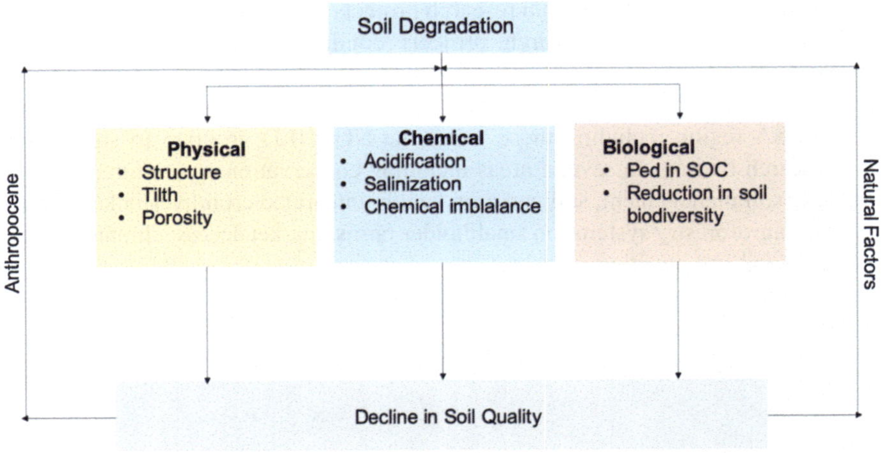

Fig. 1 Types of soil degradation (SOC = soil organic carbon). (Redrawn and adapted from Lal et al. 1989, Lal 2020)

effects on crop yields can be alleviated by plant growth-promoting bacteria (Shrivastava and Kumar 2015). Soil biological degradation involves decline in activity and species diversity of soil biota, especially those of microorganisms, which play critical roles in cycling of nutrients, decomposition of wastes, and denaturing of pollutants (Sims 1990). Severe biological degradation can lead to drastic reduction/depletion of soil organic carbon (SOC) concentration, with strong impact on the global C cycle (Lal 2004). Biological degradation is exacerbated by a combination of environmental factors (e.g., soil temperature, moisture regimes, acidity, and salinity) and anthropogenic activities (Mishra and Dhar 2004).

Aggravation of soil degradation by physical, chemical, and biological processes (Fig. 1) leads to ecological degradation. The latter includes disruption in cycling of water and elements and decoupling of the coupled cycling of carbon with that of water, nitrogen, phosphorus, and sulfur. The anthropogenic global warming (Lal 2010) can aggravate such a decoupling with severe implications to provisioning of essential ecosystem services or creation of disservices. Therefore, sustainable management of soil resources is essential to reversing the degradation trends and strengthening of the coupled cycling. Enhancing SOC concentration to above the threshold level of ~2% in the root zone is critical to restoration of degraded soils and ecosystems. Restoration of the depleted SOC stock of 135 Pg, since the dawn of settled agriculture, can have a strong drawdown of atmospheric CO_2 and generate numerous co-benefits (Lal et al. 2018a, b; Lal 2018). A widespread adoption of a system-based conservation agriculture (Lal 2015), and other systems that create a positive soil/ecosystem C budget, can restore soil health and reverse the degradation trend. The strategy is to reconcile the need for meeting the growing demands of increasing and progressively affluent world population with the necessity of restoring degraded soils and ecosystems and improving the environment.

4 Climate Smart Agriculture

Climate change has a great impact on agriculture; thus, considering the role of agriculture in food and fiber security and as source of livelihood for majority of population, FAO saw the need for recognition of agriculture in the climate change agenda. Climate smart agriculture (CSA) consists of multiple approaches toward sustainable agriculture to attain food and nutritional security, in the face of climate change. The Concept of CSA hinges on three pillars: agricultural productivity, adaptation, and reduction of greenhouse gases (FAO 2013; Lipper et al. 2014). To achieve CSA, there must be improved agricultural productivity (crop, livestock and fisheries) without degrading the environment and ensuring adaptation to climate change through sustaining livelihood. The concept also includes enabling farming communities to be resilient to climate shocks and contributing to mitigation of climate change through reducing emission of greenhouse gases (GHGs) per kg of food or fiber produced, build up C sinks, or participate in C trade. CSA places agriculture in the middle of the climate change agenda and calls for committed effort to make agriculture more adaptive and resilient to climate change so that it can sustain productivity and feed the growing population. In sub-Sahara Africa, CSA is of high importance because the agricultural productivity has been low, and the population increase is high. Although adoption of modern agricultural technology can increase productivity, the effects of climate change can compromise the benefit of modern agricultural technologies, leading to even lower adoption of these technologies among farmers.

One of the most noteworthy impacts of climate change is increased events of extreme weather variability, which significantly affects productivity and food security, even when other good agricultural practices are employed. The increased weather extremes (floods and droughts and extended dry spell, shifting of weather pattern – onset and end of rainy season) have further challenged the efforts to increase productivity in holders – farming systems in SSA. Climate change has also increased risks in agriculture, making adoption of technologies to improve productivity an even more unattractive investment option. Furthermore, the quick and easy coping strategies to low productivity due to decreased fertility, degradation due to erosion, and extreme weather have resulted in shifting cultivation and/or expansion of agricultural land to increase production. This has led to increased deforestation and encroachment on wetlands and other protected areas. Therefore, a holistic approach through climate smart agriculture is needed to do agriculture more smartly to ensure, sustain, and enhance productivity, without causing natural resources degradation or contributing to further greenhouse gas emissions.

4.1 Climate Smart Agriculture to Enhance Productivity

CSA technologies and practices must ensure improved productivity, even in the presence of climate change. There are various proven technologies that increase productivity with minimum negative effects on the environment such as appropriate use of fertilizers; soil and moisture conservation; use of improved high-yielding crops, pasture, and animal varieties/breeds; and resistance/tolerance to pest and diseases. Girvetz et al. (2019) showed that climate change impact led to decreased precipitation for the most part of Africa, except for few parts in South Africa and East Africa where precipitation increased. The evaporation stress has also increased due to increase in temperature, which outweighs the benefit of increased total precipitation in those parts of South and East Africa, resulting in drier soils than there would be without climate change (Girvetz et al. 2019). Therefore, for each of the sustainable agriculture intensification technologies and practices, one must ensure that they give the same good results even in the face of the above-mentioned climate change impact. Such a guarantee is possible with use of smartly chosen conservation agriculture (CA) to ensure infiltration and retention of rainwater in the soil profile, without creating water logging conditions for vulnerable crops. Alternatively, plans, mechanisms, and practices for supplemental irrigation must be available when needed. Soil fertility management must also go hand in hand with sustainable pest management to ensure adequate productivity. The choice of crop and varieties to be grown must be based on the season weather forecast, while other soil fertility, crop and pest management choices will depend on season and soil conditions. The decision on either microdosing or full site-specific fertilizer use will depend on intra-season amount and distribution of rainfall. This is because, when rainfall is low and/or poorly distributed, microdose can produce the same yield as full dose due to limited moisture to ensure all nutrients supplied by full dose fertilizer rate are absorbed. Lack of reliable weather/climate information and forecast negatively affects farmers' decision on investing in agricultural input use. Given low soil fertility, supplementing the deficient nutrients, and use of other amendments to optimize physical and chemical soil properties, it is necessary to find out how best climate/weather information can be factored in the planning of agricultural production among smallholder farmers.

Irrigation for Vertical Intensification and Supplemental Irrigation

Irrigation has been suggested to be a quick fix to unreliable and low rainfall. Although irrigation has been proven to increase productivity at smallholder farming system, sustaining irrigation is still a major problem due to climate change impacts (Lefore et al. 2019). The unsustainability of irrigation is due to excessive use of diminishing water resources, mostly groundwater and surface river water (Lefore et al. 2019), increased salinity in irrigated systems, and increased evaporation due to increased temperature. Lack of or weak institutional set-up to manage water resources is a challenge to accessing irrigation water. Studies that shed light on best institutional arrangements and multi-sector water resource management to ensure CSA are still scanty.

Research and practice for supplemental irrigation to correct moisture deficit for rain-fed agriculture in smallholder farming systems is almost not practiced in SSA. Rainwater harvesting (RWH) and management has the potential to sustain rain-fed agriculture in SSA in the face of weather variability due to climate change (Biazin et al. 2012). Rainwater harvesting for domestic use has been practiced to some extent, but for supplemental irrigation it is limited to horticultural crops. Supplemental irrigation for field crops at farm and landscape levels is limited. Studies that integrate climate, weather information, crop requirement, land morphology, rainwater harvesting infrastructure like reservoir type, capacity, and associated cost and benefit are limited. In situ rainwater harvesting and moisture conservation have been studied (*zai*, *ngoro*, *chololo* pits, tied ridges, sunken beds) (Hatibu and Mahoo 1999) and well reviewed by Biazin et al. (2012) and Pachpute et al. (2009). In most cases, the low success of the in situ rainwater harvesting is related to extended dry spell and short rain seasons, as well as the need to balance adequate moisture availability for crop without adverse effect due to extended water logging conditions. Scaling up and out of rainwater harvesting, either micro- or macro-catchment, is still a problem due to mismatch of technical knowledge, selection of appropriate technology, socio-economic factors, and acceptability by local community (Biazin et al. 2012; Pachpute et al. 2009). As a result, despite the potential of RWH, it is estimated that only 5% of rainwater can be harvested (Pachpute et al. 2009). Comprehensive studies on rainwater harvesting and management at farm and landscape scale are needed to enable wide adoption of this technology as a strategy for achieving CSA in SSA.

Site-Specific Fertilizer Recommendations and Landscape Soil Management

For CSA to be achieved, response to fertilizers (organic or inorganic) applied in low fertility soils must be guaranteed, even when rainfall is lower, evaporation is high, and dry spell frequencies increase. Heisse and Morimoto (2019) reported that farmers' imperfect information due to lack of reliable weather forecast affects farmers' decision to apply fertilizer and makes them likely not to apply fertilizer in a year in which rainfall seems erratic. Reliable information about climate requires investment on weather database and data management, to enable understanding of the past changes, monitor and predict the future climate scenarios for preparedness, and early warning (Girvetz et al. 2019). Investment in studies on agrometeorology with automated weather data are limited in SSA. For the CSA to be achieved, research and practice that integrate all good agronomic practices packaged for specific landscape, and availability of reliable weather information relevant for agriculture, are needed.

CSA studies advocate that CSA must be site specific, based on specific requirements and capability of specific area (FAO 2013). Characterization of soils to deduce soil quality and required nutrients is needed for CSA, to provide a risk factor that will enhance farmers' decision on the type and quantity of fertilizer to apply. This requires systematic research on agronomy, decision-making, and soil information. Studies to assess and integrate the rainfall forecast, soil-water conservation, and fertilizer to apply are highly needed to achieve CSA.

Sustainable Pest Management Practices and Improved Crop Varieties and Breeds

Climate change and weather variability is also linked to increased pest and diseases. CSA must also ensure that pest management is timely and done in an ecological manner to ensure low environmental and food and fiber contamination. Where necessary, judicious use of pesticides is recommended at appropriate rates and observing safety period to ensure that there are no pesticide residues in food and fiber. To attain CSA and prevent crop losses due to pests at farm level, studies on surveillances and forecast of pest are needed. Selection of the right crop varieties and livestock breeds resilient to drought, rising temperatures, pest, and diseases is crucial to attain CSA (Scherr et al. 2012). While the improved crop varieties and livestock breeds require specific studies, studies that integrate the choice of the varieties and livestock breeds based on the environmental condition and socio-economic analysis are also needed.

4.2 Climate Smart Agriculture and Greenhouse Gas Emissions

The GHG emissions from agriculture include N_2O, CO_2, and CH_4 for submerged crops, especially rice. Emission of N_2O from agriculture is related to transformation of soil N and SOM, which tend to increase due to use of N-containing fertilizers, soil physical manipulations which favor SOM decomposition, or burning of crop residues to release CO_2 to the atmosphere. The release of N_2O, CO_2, and CH_4 from agricultural soils is a natural phenomenon mediated by soil microorganism; CSA requires that these emissions must be accompanied by high productivity, and that the emissions for each kilogram of food and fiber produced are kept low (Lipper et al. 2014). Despite these recommendations, the practice to achieve low emissions in CSA is hampered by limited studies that measure emissions versus productivity in the tropical conditions. Bwana (2019) measured N_2O emissions in 30 kg N/ha and 60 kg N/ha in organic and conventional cotton production practices. He reported low cumulative yield scale emissions of 0.09 g N_2O-N/kg to 1.65 g N_2O-N/kg seed cotton in organic and 0.06 g N_2O-N/kg to 1.48 g N_2O-N/kg seed cotton in conventional practices associated with low cotton yields of 0.38 Mg/ha to 1.42 Mg/ha for organic and 0.43 Mg/ha to 1.56 Mg/ha for conventional practices. Thus, to ensure no increase in emissions, the same N input rate can be used in conjunction with soil moisture conservation and practices to increase SOC to boost yield and to reduce yield emissions. Rosenstock et al.'s (2019) analysis showed that most studies addressing CSA examined one pillar of CSA, mostly productivity, while about 32% study two pillars (mostly productivity and adaptation) and less than 1% study all three of CSA, making mitigation and emissions of CSA least studied. Thus, estimation of GHG emissions and mitigation is essential to optimize the three pillars of CSA.

Agricultural practices with low GHG emissions are needed. There has been a dilemma on best practices that will increase productivity with minimum GHG emissions. It is urged that increased use of agricultural inputs (i.e., agricultural intensification) will simultaneously increase GHG emissions. The GHG emissions reported to occur from most SSA areas are mainly estimates from models or simulated conditions in Europe and USA (Rosenstock et al. 2019). Few studies have estimated GHG emissions from fields in SSA. It is reported that N rates lower than 150 kg N/ha have lower emissions of nitrous oxides (Palm et al. 2017).

4.3 Climate Smart Agriculture and Weather Forecast

The most important climate change impact in SSA agriculture is increased inter-annual and inter-season variability in climate, including increased severity of heat stress, drought, and floods (Burian et al. 2019; Trenberth et al. 2014). To enhance agriculture adaptation to such climate change impact and enhance resilience to climate change, studies on adaptive weather forecast for agricultural intensification are critical. Integration of weather forecast to address inter-annual and inter-season variability in climate is one of the areas that has been least studied (Thornton et al. 2014). Studies to enhance prediction of rainfall incidences are important for decisions on use of inputs based on weather forecast. Time of planting must be updated based on onset of rainfall in a particular season and in a particular year. The quantities of fertilizer and type of seed variety to be used can be properly guided by the updated climate information for agriculture.

The analysis by Heisse and Morimoto (2019) showed some concern that use of mineral fertilizers in countries like Tanzania might not bring sustainable development due to fossil fuel and energy consumption in manufacturing of fertilizers, environmental degradation due to N runoff, and transaction costs in distribution and an uneven wealthy distribution due to subsidies. Amuri (2015) argued that environmental degradation due to depletion of soil fertility leads to the need to open up new farms to increase yield by expanding farmland, which in turn leads to increased vulnerability to wind and water erosion, and low productivity. Reports show that from 2000 to 2005, deforestation in Africa was 4 million ha, estimated to be 55% of global forest loss (Lupala et al. 2014). Thus, more research to integrate environmental externalities, weather variability, and prediction and agronomic performance of technologies is needed. The agronomic technology should ensure efficient utilization of inputs (fertilizer, pesticides) in the face of a given weather condition and potential for environmental degradation (both due to excessive or deficiencies of essential nutrients).

Heisse and Morimoto (2019) also highlighted the potential negative impact of policies that support subsidies and distribution of mineral fertilizers as increasingly risky due to dependency on imports of fertilizers and hence burden to balance of trade. This should also be analyzed against the risk of increased nutrient mining degradation, reduction of food and fiber yields, increased risk of food insecurity, and increased food importation that might exacerbate poverty. Thus, a smart agricultural research that will develop an adoptive technology to the agro-ecological conditions of SSA is needed.

4.4 Climate Smart Agriculture and Adaptation to Climate Change Impact

CSA requires that agricultural technologies must be efficient and effective, and the environmental impact and socioeconomic consequences of adopting the technology should be analyzed at both macro and micro levels. CSA must ensure that the livelihood of farming community is sustained even in the presence of climate change. This will include availability of reliable market, good prices for agricultural produce, to enable farmers to have sufficient income, and they should have savings to sustain their lives in case of climate shock and absence of own crop and livestock production. Investment in building resilience through farmers' institution capacity (cooperatives), storage, agroprocessing, and value addition is necessary.

4.5 Mapping Areas for Agricultural Potential Areas

Precise use of nutrients based on the soil nutrient status and physical and chemical properties requires that soil information be generated and be mapped at farm scale. These maps will not only serve in planning but also in management of fields by farmers, extension services, and local government and national levels. Spatial information will be a handy tool to guide decision-making on how to focus areas for targeting intensification such as fertilizer type and quantities to apply to achieve sustainable agricultural intensification (Palm et al. 2017).

Successful CSA implementation requires careful planning at landscape level, taking into account land morphology and associated variability of soils and moisture. Lack of sufficient information at landscape scale climate impact planning has been identified as the one of the drawbacks toward landscape planning to attain CSA success (Hunter and Crespo 2019). For example, Tanzania has an agro-ecological map that shows the agro-ecological zone (AEZ) at a very coarse scale and has not been updated. The fertilizer recommendations available are outdated and are based on the same AEZ maps (Mowo et al. 1993). There is insufficient information on fertilizer trials with integrated climate/weather, and soil testing information is lacking, hindering the attempts to update fertilizer recommendations.

5 Livestock Production and Mitigation Potential for Climate Change

As pointed earlier in this book, livestock (including poultry) is widely recognized as an important source of nourishment and as a livelihood asset for the poor worldwide. Livestock is a source of animal proteins including milk, meat, eggs, and blood (contributing to food and nutrition security and dietary diversity). It also provides manure and is important for cultural and social uses such as dowry and rituals. In addition, it is a moving bank, default insurance for diversifying risk and source of income and draught power. However, livestock is both a culprit and contributor of greenhouse gas emissions, global warming, and subsequently climate change (Gerber et al. 2013; Thornton et al. 2011). It is estimated that livestock contributes about 18% of global anthropogenic greenhouse-gas (GHG) emissions, including methane from animals (25%), carbon dioxide from land use and its changes (32%), and nitrous oxide from manure and slurry (31%) (Steinfeld et al. 2006). Gerber et al. (2013) reported that livestock contributes about 7.1 Gt CO_2-eq per annum, representing 14.5% of human-induced GHG emissions, a majority of which is contributed by beef and cattle milk production.

Livestock is also a victim of climate change-related effects in many ways. Due to their poor resource base and the wider pervasive poverty that characterize the context in which they are embedded, the poor are particularly vulnerable to many climate change-induced disasters such as prolonged droughts and floods. It is therefore imperative that resilient interventions be put in place to mitigate any related negative consequences that may arise with respect to smallholder agriculture including livestock. Negative effects of climate change include but not limited to the following:

Loss of animals and infrastructure While drought and dry spells are more common than floods, recent evidence shows that floods can be equally as devastating. For example, the Malawi Post-floods assessment report (Government of Malawi 2015) showed that an estimated 195,032 livestock were either reported as lost or damaged, with the total economic losses estimated at MK5849 million (~US$12.3 million). Increases in incidences of livestock parasites and diseases were also reported, resulting in further losses.

Loss of feed resource base Negative effects of climate change on livestock include variability in availability, quality, distribution, composition, and cost of fodder or herbage and concentrates. Floods and droughts reduce access to feed and fresh water, in addition to negatively affecting quantity and quality of feed. Floods sweep away trees and shrubs and grass (pasture), which can sometimes be covered in mud, making the fodder unavailable to animals. The situation is aggravated during droughts and continued dry spells, where all the pasture and shrubs are easily destroyed by uncontrolled bushfires. Shortages of feed force animals to walk long distances in search of food and water as most water points completely dry up. Unfortunately, the movement of animals in search of water and food further aggravates the conditions of the animals, as it depletes the already low energy body reserves. Degradation of pastures in periods of drought may lead to complete loss of productive pasture species and desperate consumption of toxic plants by animals and indirectly leads to soil degradation. Due to lack of access to fresh water and feed, animals easily succumb to disease (eating contaminated soil, toxic plants, drinking dirty water, operating in muddy environments), leading to poor body condition scores which negatively affect productivity through reduced growth and reproduction.

Discomfort and Distress During droughts, temperatures are always high, leading to discomfort and distress for the animals. Due to poor body conditions, scores of animals (caused by lack of nutritious and adequate feed), animals become susceptible to many diseases, further jeopardizing their chances of survival.

In order to fully benefit from livestock, there is need to develop and adopt system-wide interventions. For drought-declared districts or zones where farmers are too impecunious and resource-poor to cope with the effects of prolonged drought and severe livestock-threatening floods, governments should consider formulating policies and establishing institutions that proactively promote and protect livestock as an important productive asset. It should also recognize livestock production as a crucial livelihood strategy, especially for the vulnerable poor. Policy options include deliberate provision of extension packages for management of livestock under drought and floods. These options emphasize interventions of adoption of feed preservation technologies, such as hay and silage for use during lean periods, controlled planning and management of communal grazing areas, as well as rehabilitation of degraded communal grazing areas and rangelands as way of improving pasture availability. In addition, government needs to provide emergency veterinary support and services and establish livestock survival feeding programs through supply and distribution of emergency feed, including fodder and feed supplements. The alternative is to move animals to other areas of the country where feed is sufficient during prolonged droughts. In order to reduce competition with humans, the option of using of unconventional drought tolerant sorghum, millets and quinoa should be considered. There is also the option of using other emerging sources such as insects as animal protein for feeding monogastric animals. In primarily grass or rangeland-based systems, use of multipurpose fodder trees and adoption of new technologies such as hydroponically produced fodder is crucial.

In addition, deliberate action is required to ensure provision of water by establishing emergency livestock watering points and provision of animal shelters to reduce effects of heat stress. Communities should also take advantage of floods by setting up mechanisms such as dam construction for capturing floodwaters for use by livestock during periods of drought.

Because severe droughts or floods can lead to loss of animals, livestock insurance can be considered as one of the best strategies for farmers to cope with such climatic hazards. Alternatively, governments can also help farmers to rebuild their livestock assets through implementation of livestock relief restocking schemes targeting the vulnerable and affected poor smallholder livestock keepers.

Farmers should also be willing to make drastic and pragmatic decisions and to adopt management strategies such as culling or destocking through selling of animals. This is particularly important in cases where farmers are unable to provide the required feed and are therefore unable to maintain all the animals in reasonable condition. Where chances of survival are limited, ethical killing of animals should be considered. These options can work better with active and working legislation on animal welfare. From the foregoing, it is evident that government should pursue mitigation trajectories that integrate both proactive and reactive climate change adaptation and mitigation approaches in a bid to reduce vulnerability of poor smallholder livestock producers. For proper planning, governments should establish drought forecasting and early warning systems, including routine monitoring of livestock conditions.

In order to mitigate against emission of greenhouse gases, several options that address adoption constraints, costs, and numerous trade-offs should be considered (Herrero et al. 2016). First, there is need to consider dietary change by reducing demand for livestock products, which could reduce pressures on land and natural resources in developing countries, which in turn helps to reduce GHG emissions (Thornton et al. 2011). Low amounts of animal source foods, in combination with a diversity of plant-based foods, has been linked to healthy diets (Willett et al. 2019). There is also a need to adopt rapid implementation of agricultural mitigation options to reduce greenhouse gas emissions, adoption of land management practices that shift agriculture from a carbon source to sink, and a fundamental shift in production priorities. For livestock, these options include improved formulation and efficiency of ruminant diets combined with livestock numbers to help mitigate emission of methane. Reduction of numbers of livestock entails switching to use of animals that are more productive. Switching from ruminants to monogastric animals such as poultry and pigs could also reduce methane emissions. The foregoing suggest that mitigation of GHG emissions requires interventions that reduce demand hence consumption of livestock products, sustainably intensify livestock production, promote carbon sequestration in rangelands and reduce emissions from manures. However, moderating consumption of livestock products, avoidance of negative impacts on livelihoods, economic activities and the environment, and increasing affordability and adoption of mitigation practices require further research and investments (Herrero et al. 2016).

6 Transformative and Transgressive Approaches

Education promotes and enhances awareness of personal transformation that is relevant for various arenas of development including agriculture. Our current framing of the agriculture and climate-related problem is not adequate because such framing does not address awareness of actual growers on the ground. The farming communities are separated practically, cognitively, and emotionally from the framing and their understanding of climate change and associated impacts on agriculture. To counter this, the concept of personal transformation is increasingly useful while advocating individual and systemic change relevant for agriculture and climate change.

7 Research and Development Priorities and Emerging Issues

Climate change is one of the key factors impinging on the achievement of sustainable development goals (SDGs), especially SDGs 1, 2, 3, and 6. It is, therefore, important to ensure that climate change-related issues are addressed in a holistic and systematic way, including social, economic, and environmental issues and the political will. Some of the interventions include selecting animal species and crop varieties that are suitable for the weather conditions; diversifying crops and animals, sustainable soil and water conservation, and natural resource management including climate smart agriculture, and afforestation and natural tree regeneration. Emerging issues such as gender and empowerment and involvement of the youth are also among the key factors. A need for embracing transformative and transgressive approaches for resilience to climate change was also emphasized. Characteristics of transformation include radical change that requires new knowledge, innovation, identification, and implementation of alternatives to institutionalized positions and radical changes in the decision-making processes.

The following key issues and research priorities were identified from the sessions and papers presented at the conference:

(i) CA is beneficial and should be promoted. It provides numerous benefits including increased soil depth, concentration of carbon in the soil, and increased soil aggregation and water retention. However, its efficacy must be validated under site-specific soil and climatic conditions as the process may not necessarily be applicable under all conditions.
(ii) Legume-based cropping systems (e.g., pigeon pea with cassava) and integrated soil fertility management can increase nitrogen fixation, enhance nitrogen use efficiency, and improve yield and protein content of the subsequent crop of maize grown on Alfisols.
(iii) To obtain wide adoption of CSA and policies to successfully achieve CSA, there is a need of more multidisciplinary studies involving multi-sectors to address knowledge and practice gaps.
(iv) Landscape approach is recommended for attaining CSA in SSA, with proper use of land capability for agriculture but also taking advantage of land morphology, site-specific fertilizer recommendations, soil management, rainwater harvesting, and supplemental irrigation.
(v) Attaining CSA requires comprehensive planning and joint effort at national level through research evidence policies, implementation strategies, and responsibilities at small-scale farming community.
(vi) Both smallholder and commercial livestock farmers need alternative, sustainable, and climate smart feeds and feeding systems. Therefore, there is a need for interventions, which include feed preservation, and technologies for production of high quality feed and fodders including hydroponically produced fodder.
(vii) Use of locally adapted crops including indigenous vegetables which can contribute to sustainable food and nutrition security, source of income, general livelihoods, and resilience of smallholder farmers in rural areas when supported with interventions such as sustainable production and distribution and marketing of indigenes vegetable seeds.
(viii) Livestock, apart from being sources of animal-based food (e.g., meat and milk), can also be used as a source of energy through production of manure-

based biogas. However, interventions by government and non-governmental organizations (NGO) are needed in the form of technical and financial support for building such biogas plants.

(ix) Strategies of mitigating the global warming (e.g., proper feeding systems, stocking rates, and choice of animal breeds) must also be implemented.

(x) Payments for carbon credits or ecosystem services as an incentive to farmers to adopt climate-smart practices should be encouraged. This initiative can lead to farmers 'triple win' outcomes for climate adaptation, food security, and livelihoods of participating farmers via the provision of carbon payments.

(xi) Technologies such as biochar can be added to soil to enhance its agricultural and environmental value since it is stable in the soil, acts as a good liming agent, and increases soil moisture when combined with CA.

(xii) Value addition, access to markets, and farmer-centric business models are key to sustainable production systems and building resilience of crop and livestock smallholder farmers.

(xiii) Adoption of innovative technologies for smallholder farmers (e.g., CA, integrated nutrient management, integrated dairy and cropping systems) can enhance and sustain farm-productivity and income.

(xiv) There is also a potential of adopting integrated aquaculture for improving food security and ensuring diversification of diets, as the systems generate both animal proteins and vegetables.

(xv) There is a need for up-scaling and disseminating the proven technologies/ innovations to smallholders and seeking clearance from the agricultural technology clearing committees of the respective countries.

(xvi) Development of evidence-based policy briefs for consideration by relevant policy institutions is necessary for the application of technologies by smallholder farmers. To enable farmers to take full responsibilities to manage their lands, empowerment of farmers through associations and extension services is unavoidable, while ensuring profitability of farming enterprises through reliable markets.

(xvii) Conducting research on adoption of climate smart agriculture technologies and systems must be given a high priority to build resilience of smallholder farmer communities.

References

Amuri NA (2015) Enhancing Resilience of Food Production Systems under Changing Climate and Soil Degradation in Semi-arid and Highlands of Tanzania. In: R. Lal et al. (eds.), Sustainable Intensification to Advance Food Security and Enhance Climate Resilience in Africa. Springer International Publishing Switzerland, pp 385–405. https://doi.org/10.1007/978-3-319-09360-4

Arsenault C (2014) Only 60 years of farming left if soil degradation continues. Sci Am (https://www.scientificamerican.com/article/only-60-year-of-farming-left-if-soil-degradation-continues)

Biazin B, Sterk G, Temesgen M, Abdulkedir A, Stroosnijder L (2012) Rainwater harvesting and management in rained agricultural systems in sub-Saharan Africa – A review. Phys Chem Earth 47–48:139–151

Borrelli P, Robinson DA, Fleischer LR, Lugato E, Ballabio C, Alewell C, Meusburger K, Modugno S, Schutt B, Ferro V, Bagarello V, Van Oost K, Montanarella L, Panagos P (2017) An assessment of the global impact of 21st century land use change on soil erosion. Nat Commun 8(1):2013

Burian A, Karaya R, Wernerssond JEV, Egberthf M, Lokorwah B, Nyberg G (2019) A community-based evaluation of population growth and agropastoralist resilience in Sub-Saharan drylands. Environmental Science & Policy 92: 323–330.

Bwana TN (2019) Assessment of environmental and agronomic performance of smallholder organic and conventional cotton production systems in Meatu, Tanzania. PhD Thesis, Sokoine University of Agriculture

FAO (2013) Climate-smart agriculture source book. FAO, Rome

Gerber PJ, Steinfeld H, Henderson B, Mottet A, Opio C, Dijkman J, Falcucci A, Tempio G (2013) Tackling climate change through livestock – a global assessment of emissions and mitigation opportunities. Food and Agriculture Organization of the United Nations (FAO), Rome

Girvetz E, Ramirez-Villegas J, Claessens L, Lamanna C, Navarro-Racines C, Nowak A, Thornton P, Rosenstock TS (2019) Future climate projections in Africa: where are we headed. In: Rosenstock TS, Nowak A, Girvetz E (eds) The climate-smart agriculture papers-investigating the business of a productive, resilient and low emission future. Springer Nature Switzerland AG. https://doi.org/10.1007/978-3-319-92798-5

Government of Malawi (2015) Malawi 2015 floods post disaster needs assessment report. Available at: https://www.gfdrr.org/sites/default/files/publication/pda-2015-malawi.pdf-report. Accessed 4 June 2019

Hatibu N, Mahoo H (1999) Rainwater harvesting technologies for agricultural production: a case for Dodoma, Tanzania. Sokoine University of agriculture, Department of Agricultural Engineering and Land Planning.

Heisse C, Morimoto R (2019) Climate change, chemical fertilisers, and sustainable development – panel evidence from Tanzanian Maize farmers. SOAS Department of Economics Working Paper No. 217. SOAS University of London, London

Herrero M, Henderson B, Havlík P, Thornton PK, Conant RT, Smith P, Wirsenius S, Hristov AN, Gerber P, Gill M, Butterbach-Bahl K (2016) Greenhouse gas mitigation potentials in the livestock sector. Nat Clim Chang 6(5):452

Hunter R, Crespo O (2019) Large scale crop suitability assessment under future climate using the Ecocrop model: the case of six provinces in Angola's Planalto region. In: Rosenstock TS, Nowak A, Girvetz E (eds) The climate-smart agriculture papers-investigating the business of a productive, resilient and low emission future. Springer Nature Switzerland AG. https://doi.org/10.1007/978-3-319-92798-5

Lal R (2003) Soil erosion and the global carbon budget. Environ Int 29:437–450

Lal R (2004) Soil carbon sequestration impacts on global climate change and food security. Science 304:1623–1627

Lal R (2008) Laws of sustainable soil management. Agron Sustain Dev 29:7–9

Lal R (2010) Managing soils and ecosystems for mitigating anthropogenic carbon emissions and advancing global food security. Bioscience 60(9):708–721

Lal R (2015) A system approach to conservation agriculture. J Soil Water Conserv 70:82A–88A

Lal R (2018) Digging deeper: a holistic perspective of factors affecting SOC sequestration. Glob Chang Biol 24(8):3285–3301. https://doi.org/10.1111/gcb.14054

Lal R (2020) Managing soils for reducing the dependence on chemicals and import of resources into agro-ecosystems. In: Lal R (ed) "Soil and fertilizer". Advances in soil science. Taylor and Francis, Boca Raton. (In Press)

Lal R, Hall GF, Miller FP (1989) Soil degradation: l. Basic processes. Land Degrad Rehabil 1:51–69

Lal R, Smith P, Jungkunst HF, Mitsch W et al (2018a) The carbon sequestration potential of terrestrial ecosystems. J Soil Water Conserv 73:145 A–152 A

Lal R, Horn R, Kosaki T (eds) (2018b) Soil and sustainable development goals. Catena Schweizerbart, Stuttgart, 196 pp
Lefore N, Giordano M, Ringler C, Barron J (2019) Sustainable and equitable growth in farmer-led irrigation in sub-Saharan Africa: what will it take? Water Altern 12:156–168
Lipper L, Thornton PK, Campbell B, Baedeker T, Braimoh A, Bwalya M, Caron P, Cattaneo A, Garrity D, Henry K, Hottle R, Jackson L, Jarvis A, Kossam F, Mann W, McCarthy N, Meybeck A, Neufeldt H, Remington T, Sen PT, Sessa R, Shula R, Tibu A, Torquebiau EF (2014) Climate smart agriculture for food security. Nat Clim Chang 4:1068–1072
Lupala ZJ, Lusambo LP, Ngaga YM (2014) Management, growth, and carbon storage in Miombo woodlands of Tanzania. Int J For Res. Vol.2014 Article Id.629317 p11.
Mishra,U & Dhar D.W. 2004. Biodiversity and biological degradation of soil.Resonance 9:26-33.
Mowo JG, Floor J, Kaihura FBS, Magoggo JP (1993) Review of fertilizer recommendations in Tanzania: Part 2, Revised fertilizer recommendations for Tanzania. Soil Fertility Report No.6. National Soil Service, ARI–Mlingano, Tanga, Tanzania
Oldeman R (1994) Global extent of soil degradation. ISRIC Bi-Annual Report 1991–1992. ISRIC, Wageningen, Holland, pp 19–36
Pachpute JS, Tumbo SD, Sally H, Mul ML (2009) Sustainability of rainwater harvesting systems in rural catchment of Sub-Saharan Africa. Water Resour Manag 23:2815–2839
Palm C, Neill C, Lefebvre P, Tully K (2017) Targeting sustainable intensification of maize-based agriculture in East Africa. Trop Conserv Sci 10:1–4
Rosenstock TS, Lamanna C, Namoi N, Arslan A, Richards M (2019) What is the evidence base for climate-smart agriculture in East and Southern Africa? A systematic map. In: Rosenstock TS et al (eds) The climate-smart agriculture papers. https://doi.org/10.1007/978-3-319-92798-5_12
Scherr SJ, Shames S, Friedman R (2012) From climate-smart agriculture to climate-smart landscapes. Agric Food Secur 1:12
Shrivastava P, Kumar R (2015) Soil salinity: a serious environmental issue and plant growth-promoting bacteria as one of the tools for its alleviation. Saudi J Soil Biol 22:123–131
Sims GK (1990) Biological degradation of soil. In: Lal R, Stewart BA (eds) "Soil Degradation". Advances in soil science. Springer, New York, pp 289–330
Steinfeld H, Gerber P, Wassenaar TD, Castel V, Rosales M, Rosales M, de Haan C (2006) Livestock's long shadow: environmental issues and options. Food & Agriculture Organization (FAO), Rome
Thornton PK, Herrero MT, Ericksen PJ (2011) Livestock and climate change. Issue Brief. November 2011. International Livestock Research Institute, Nairobi, Kenya. Available at: Accessed 7 July 2019
Thornton PK, Ericksen PJ, Herrero M, Challinor AJ (2014) Climate variability and vulnerability to climate change: a review. Glob Change Biol 20:3313–3328
Trenberth KE, Dai A, van der Schrier G, Jones PD, Barichivich J, Briffa KR, Sheffield J (2014) Global warming and changes in drought. Nature Climate Change, 4:17–22.
Watts J (2018) Third of earth's soil is acutely degraded due to agriculture. https://www.theguardian.com/environment/2017/sep/12/third-of-earths-soil-acutely-degraded-due-to-agriculture-study
Willett W, Rockström J, Loken B, Springmann M, Lang T, Vermeulen S, Garnett T, Tilman D, DeClerck F, Wood A, Jonell M (2019) Food in the Anthropocene: the EAT–Lancet Commission on healthy diets from sustainable food systems. Lancet 393(10170):447–492

Index

B. R. Singh et al. (eds.), *Climate Impacts on Agricultural and Natural Resource Sustainability in Africa*, https://doi.org/10.1007/978-3-030-37537-9

D

T

U

The manufacturer's authorised representative in the EU is Springer Nature Customer Service Centre GmbH, Europaplatz 3, 69115 Heidelberg, Germany. If you have any concerns regarding our products, please contact ProductSafety@springernature.com

Printed and bound by CPI Group (UK) Ltd, Croydon, CR0 4YY
30/09/2026
02225556-0018